Potential Theory and Dynamics on the Berkovich Projective Line

Mathematical
Surveys
and
Monographs

Volume 159

Potential Theory and Dynamics on the Berkovich Projective Line

Matthew Baker
Robert Rumely

American Mathematical Society
Providence, Rhode Island

EDITORIAL COMMITTEE

Jerry L. Bona Michael G. Eastwood
Ralph L. Cohen, Chair J. T. Stafford
Benjamin Sudakov

2000 Mathematics Subject Classification. Primary 14G20;
Secondary 14G22, 14G40, 37F10, 37F99, 31C05, 31C15, 31C45.

For additional information and updates on this book, visit
www.ams.org/bookpages/surv-159

Library of Congress Cataloging-in-Publication Data

Baker, Matthew, 1973–
  Potential theory and dynamics on the Berkovich projective line / Matthew Baker, Robert Rumely.
    p. cm. — (Mathematical surveys and monographs ; v. 159)
  Includes bibliographical references and index.
  ISBN 978-0-8218-4924-8 (alk. paper)
  1. Potential theory (Mathematics) 2. Topological spaces. 3. Topological dynamics. I. Rumely, Robert S. II. Title.

QA404.7.B35 2010
515′.96—dc22
                                                                       2009036372

Copying and reprinting. Individual readers of this publication, and nonprofit libraries acting for them, are permitted to make fair use of the material, such as to copy a chapter for use in teaching or research. Permission is granted to quote brief passages from this publication in reviews, provided the customary acknowledgment of the source is given.

Republication, systematic copying, or multiple reproduction of any material in this publication is permitted only under license from the American Mathematical Society. Requests for such permission should be addressed to the Acquisitions Department, American Mathematical Society, 201 Charles Street, Providence, Rhode Island 02904-2294 USA. Requests can also be made by e-mail to reprint-permission@ams.org.

© 2010 by the American Mathematical Society. All rights reserved.
The American Mathematical Society retains all rights
except those granted to the United States Government.
Printed in the United States of America.

∞ The paper used in this book is acid-free and falls within the guidelines
established to ensure permanence and durability.
Visit the AMS home page at http://www.ams.org/

10 9 8 7 6 5 4 3 2 1   15 14 13 12 11 10

Contents

Preface	ix
History	x
Related works	xi
Acknowledgments	xii
Differences from the preliminary version	xiii
Introduction	xv
Notation	xxix
Chapter 1. The Berkovich unit disc	**1**
1.1. Definition of $\mathcal{D}(0,1)$	1
1.2. Berkovich's classification of points in $\mathcal{D}(0,1)$	2
1.3. The topology on $\mathcal{D}(0,1)$	7
1.4. The tree structure on $\mathcal{D}(0,1)$	9
1.5. Metrizability	17
1.6. Notes and further references	18
Chapter 2. The Berkovich projective line	**19**
2.1. The Berkovich affine line $\mathbb{A}^1_{\text{Berk}}$	19
2.2. The Berkovich "Proj" construction	23
2.3. The action of a rational map φ on $\mathbb{P}^1_{\text{Berk}}$	30
2.4. Points of $\mathbb{P}^1_{\text{Berk}}$ revisited	35
2.5. The tree structure on $\mathbb{H}_{\text{Berk}}$ and $\mathbb{P}^1_{\text{Berk}}$	38
2.6. Discs, annuli, and simple domains	40
2.7. The strong topology	42
2.8. Notes and further references	47
Chapter 3. Metrized graphs	**49**
3.1. Definitions	49
3.2. The space $\text{CPA}(\Gamma)$	50
3.3. The potential kernel $j_z(x,y)$	52
3.4. The Zhang space $\text{Zh}(\Gamma)$	54

3.5.	The space BDV(Γ)	56
3.6.	The Laplacian on a metrized graph	61
3.7.	Properties of the Laplacian on BDV(Γ)	68

Chapter 4. The Hsia kernel — 73

4.1.	Definition of the Hsia kernel	73
4.2.	The extension of $j_z(x,y)$ to $\mathbb{P}^1_{\text{Berk}}$	76
4.3.	The spherical distance and the spherical kernel	79
4.4.	The generalized Hsia kernel	81
4.5.	Notes and further references	85

Chapter 5. The Laplacian on the Berkovich projective line — 87

5.1.	Continuous functions	87
5.2.	Measures on $\mathbb{P}^1_{\text{Berk}}$	91
5.3.	Coherent systems of measures	95
5.4.	The Laplacian on a subdomain of $\mathbb{P}^1_{\text{Berk}}$	97
5.5.	Properties of the Laplacian	101
5.6.	The Dirichlet pairing	104
5.7.	Favre–Rivera-Letelier smoothing	113
5.8.	The Laplacians of Favre, Jonsson, and Rivera-Letelier, and of Thuillier	116
5.9.	Notes and further references	119

Chapter 6. Capacity theory — 121

6.1.	Logarithmic capacities	121
6.2.	The equilibrium distribution	123
6.3.	Potential functions attached to probability measures	128
6.4.	The transfinite diameter and the Chebyshev constant	136
6.5.	The Fekete-Szegö theorem	141
6.6.	Notes and further references	144

Chapter 7. Harmonic functions — 145

7.1.	Harmonic functions	145
7.2.	The Maximum Principle	150
7.3.	The Poisson formula	155
7.4.	Uniform convergence	161
7.5.	Harnack's principle	162
7.6.	Green's functions	163
7.7.	Pullbacks	174
7.8.	The multi-center Fekete-Szegö theorem	177

7.9.	A Bilu-type equidistribution theorem	184
7.10.	Notes and further references	191

Chapter 8. Subharmonic functions 193
8.1.	Subharmonic and strongly subharmonic functions	193
8.2.	Domination subharmonicity	199
8.3.	Stability properties	207
8.4.	The Domination Theorem	213
8.5.	The Riesz Decomposition Theorem	214
8.6.	The topological short exact sequence	219
8.7.	Convergence of Laplacians	227
8.8.	Hartogs's lemma	230
8.9.	Smoothing	234
8.10.	The Energy Minimization Principle	240
8.11.	Notes and further references	248

Chapter 9. Multiplicities 249
9.1.	An analytic construction of multiplicities	249
9.2.	Images of segments and finite graphs	271
9.3.	Images of discs and annuli	278
9.4.	The pushforward and pullback measures	285
9.5.	The pullback formula for subharmonic functions	287
9.6.	Notes and further references	290

Chapter 10. Applications to the dynamics of rational maps 291
10.1.	Construction of the canonical measure	293
10.2.	The Arakelov-Green's function $g_{\mu_\varphi}(x,y)$	299
10.3.	Adelic equidistribution of dynamically small points	306
10.4.	Equidistribution of preimages	318
10.5.	The Berkovich Fatou and Julia sets	328
10.6.	Equicontinuity	333
10.7.	Fixed point theorems and their applications	340
10.8.	Dynamics of polynomial maps	354
10.9.	Rational dynamics over $\mathbb{C}_p$	357
10.10.	Examples	370
10.11.	Notes and further references	374

Appendix A. Some results from analysis and topology 377
A.1.	Convex functions	377
A.2.	Upper and lower semicontinuous functions	378

A.3.	Nets	379
A.4.	Measure-theoretic terminology	381
A.5.	Radon measures	381
A.6.	Baire measures	382
A.7.	The Portmanteau theorem	383
A.8.	The one-point compactification	384
A.9.	Uniform spaces	385
A.10.	Newton polygons	387

Appendix B. $\mathbb{R}$-trees and Gromov hyperbolicity — 393

B.1.	Definitions	393
B.2.	An equivalent definition of $\mathbb{R}$-tree	394
B.3.	Geodesic triangles	395
B.4.	The Gromov product	397
B.5.	$\mathbb{R}$-trees and partial orders	401
B.6.	The weak and strong topologies	402

Appendix C. A Brief overview of Berkovich's theory — 405

C.1.	Motivation	405
C.2.	Seminorms and norms	406
C.3.	The spectrum of a normed ring	406
C.4.	Affinoid algebras and affinoid spaces	409
C.5.	Global k-analytic spaces	412
C.6.	Properties of k-analytic spaces	414

Bibliography — 417

Index — 423

Preface

This book is a revised and expanded version of the authors' manuscript "Analysis and Dynamics on the Berkovich Projective Line" ([**91**], July 2004). Its purpose is to develop the foundations of potential theory and rational dynamics on the Berkovich projective line.

The theory developed here has applications in arithmetic geometry, arithmetic intersection theory, and arithmetic dynamics. In an effort to create a reference which is as useful as possible, we work over an arbitrary complete and algebraically closed non-Archimedean field. We also state our global applications over an arbitrary product formula field whenever possible. Recent work has shown that such generality is essential, even when addressing classical problems over $\mathbb{C}$. As examples, we note the first author's proof of a Northcott-type finiteness theorem for the dynamical height attached to a nonisotrivial rational function of degree at least 2 over a function field [**5**] and his joint work with Laura DeMarco [**6**] on finiteness results for preperiodic points of complex dynamical systems.

We first give a detailed description of the topological structure of the Berkovich projective line. We then introduce the *Hsia kernel*, the fundamental kernel for potential theory (closely related to the Gromov kernel of [**47**]). Next we define a Laplacian operator on $\mathbb{P}^1_{\text{Berk}}$ and construct theories of capacities, harmonic functions, and subharmonic functions, all strikingly similar to their classical counterparts over $\mathbb{C}$. We develop a theory of multiplicities for rational maps and give applications to non-Archimedean dynamics, including the construction of a canonical invariant probability measure on $\mathbb{P}^1_{\text{Berk}}$ analogous to the well-known measure on $\mathbb{P}^1(\mathbb{C})$ constructed by Lyubich and by Freire, Lopes, and Mañé. Finally, we investigate Berkovich space analogues of the classical Fatou-Julia theory for rational iteration over $\mathbb{C}$.

In §7.8, we give an updated treatment (in the special case of $\mathbb{P}^1$) of the Fekete and Fekete-Szegö theorems from [**88**], replacing the somewhat esoteric notion of "algebraic capacitability" with the simple notion of compactness. In §7.9, working over an arbitrary product formula field, we prove a generalization of Bilu's equidistribution theorem [**24**] for algebraic points which are 'small' with respect to the height function attached to a compact Berkovich adelic set. In §10.3, again working over a product formula field, we prove an adelic equidistribution theorem for algebraic points which are 'small' with respect to the dynamical height attached to a rational function of degree at least 2, extending results in [**9**], [**35**], and [**47**].

A more detailed overview of the results in this book can be found in the first author's lecture notes from the 2007 Arizona Winter School [4], and in the Introduction below.

History

This book began as a set of lecture notes from a seminar on the Berkovich projective line held at the University of Georgia during the spring of 2004. The purpose of the seminar was to develop the tools needed to prove an adelic equidistribution theorem for small points with respect to the dynamical height attached to a rational function of degree $d \geq 2$ defined over a number field (Theorem 10.24). Establishing such a theorem had been one of the main goals in our 2002 NSF proposal DMS-0300784.

In [8], the first author and Liang-Chung Hsia had proved an adelic equidistribution theorem for points of $\mathbb{P}^1(\overline{\mathbb{Q}})$ having small dynamical height with respect to the iteration of a polynomial map. Two basic problems remained after that work. First, there was the issue of generalizing the main results of [8] to *rational functions*, rather than just polynomials. It occurs frequently in complex dynamics and potential theory that one needs heavier machinery to deal with rational maps than with polynomials. Second, because the filled Julia set in $\mathbb{P}^1(\mathbb{C}_p)$ of a polynomial over $\mathbb{C}_p$ is often non-compact, the authors of [8] were unable to formulate their result as a true "equidistribution" theorem. Instead, they introduced a somewhat artificial notion of "pseudo-equidistribution" and showed that when the filled Julia set is compact, then pseudo-equidistribution coincides with equidistribution.

The second author, upon learning of the results in [8], suggested that Berkovich's theory might allow those results to be formulated more cleanly. Several years earlier, in [36], he had proposed that Berkovich spaces would be a natural setting for non-Archimedean potential theory.

We thus set out to generalize the results of [8] to a true equidistribution theorem on $\mathbb{P}^1_{\text{Berk}}$, valid for arbitrary rational maps. An important step in this plan was to establish the existence of a canonical invariant measure on $\mathbb{P}^1_{\text{Berk}}$ attached to a rational function of degree at least 2 defined over $\mathbb{C}_p$, having properties analogous to those of the canonical measure in complex dynamics (see [54, 72]). It was clear that even defining the canonical measure would require significant foundational work.

At roughly the same time, Antoine Chambert-Loir posted a paper to the arXiv preprint server proving (among other things) non-Archimedean Berkovich space analogues of Bilu's equidistribution theorem and the Szpiro-Ullmo-Zhang equidistribution theorem for abelian varieties with good reduction. In the summer of 2003, the first author met with Chambert-Loir in Paris and learned that Chambert-Loir's student Amaury Thuillier had recently defined a Laplacian operator on Berkovich curves. Not knowing exactly what Thuillier had proved, nor when his results might be publicly available, we undertook to develop a measure-valued Laplacian and a theory of subharmonic functions on $\mathbb{P}^1_{\text{Berk}}$ ourselves, with a view toward applying them in a dynamical setting. The previous year, we had studied Laplacians and their spectral theory on metrized graphs, and that work made it plausible that a Laplacian operator could be constructed on $\mathbb{P}^1_{\text{Berk}}$ by taking an inverse limit of graph Laplacians.

The project succeeded, and we presented our equidistribution theorem at the conference on Arithmetical Dynamical Systems held at CUNY in May 2004. To our surprise, Chambert-Loir, Thuillier, and Pascal Autissier had proved the same theorem using an approach based on Arakelov theory. At the same conference, Rob Benedetto pointed us to the work of Juan Rivera-Letelier, who had independently rediscovered the Berkovich projective line and used it to carry out a deep study of non-Archimedean dynamics. Soon after, we learned that Charles Favre and Rivera-Letelier had independently proved the equidistribution theorem as well.

The realization that three different groups of researchers had been working on similar ideas slowed our plans to develop the theory further. However, over time it became evident that each of the approaches had merit: for example, our proof brought out connections with arithmetic capacities; the proof of Chambert-Loir, Thuillier, and Autissier was later generalized to higher dimensions; and Favre and Rivera-Letelier's proof yielded explicit quantitative error bounds. Ultimately, we, at least, have benefitted greatly from the others' perspectives.

Thus, while this book began as a research monograph, we now view it mainly as an expository work whose goal is to give a systematic presentation of foundational results in potential theory and dynamics on $\mathbb{P}^1_{\text{Berk}}$. Although the approach to potential theory given here is our own, it has overlaps with the theory developed by Thuillier for curves of arbitrary genus. Many of the results in the final two chapters on the dynamics of rational functions were originally discovered by Rivera-Letelier, though some of our proofs are new.

Related works

Amaury Thuillier, in his doctoral thesis [**94**], established the foundations of potential theory for Berkovich curves of arbitrary genus. Thuillier constructs a Laplacian operator and theories of harmonic and subharmonic functions and gives applications of his work to Arakelov intersection theory. Thuillier's work has great generality and scope, but it is written in a sophisticated language and assumes a considerable amount of machinery. Because this book is written in a more elementary language and deals only with $\mathbb{P}^1$, it may be a more accessible introduction to the subject for some readers.

Juan Rivera-Letelier, in his doctoral thesis [**81**] and subsequent papers [**82, 83, 84, 80**], has carried out a profound study of the dynamics of rational maps on the Berkovich projective line (though his papers are written in a rather different terminology). Section 10.9 contains an exposition of Rivera-Letelier's work.

Using Rivera-Letelier's ideas, we have simplified and generalized our discussion of multiplicities in Chapter 9 and have greatly extended our original results on the dynamics of rational maps in Chapter 10. It should be noted that Rivera-Letelier's proofs are written with $\mathbb{C}_p$ as the ground field. One of the goals of this book is to establish a reference for parts of his theory which hold over an arbitrary complete and algebraically closed non-Archimedean field.

Charles Favre and Mattias Jonsson [45] have developed a Laplacian operator, and parts of potential theory, in the general context of $\mathbb{R}$-trees. Their definition of the Laplacian, while ultimately yielding the same operator on $\mathbb{P}^1_{\text{Berk}}$, has a rather different flavor from ours. As noted above, Chambert-Loir [35] and Favre and Rivera-Letelier [46, 47] have given independent proofs of the adelic dynamical equidistribution theorem, as well as constructions of the canonical measure on $\mathbb{P}^1_{\text{Berk}}$ attached to a rational function. Recently Favre and Rivera-Letelier [48] have investigated ergodic theory for rational maps on $\mathbb{P}^1_{\text{Berk}}$. In Section 10.4, we prove a special case of their theorem on the convergence of pullback measures to the canonical measure, which we use as the basis for our development of Fatou-Julia theory.

Acknowledgments

We would like to thank all the people who assisted us in the course of this project, in particular our wives for their patience and understanding.

We thank Robert Varley and Mattias Jonsson for useful suggestions and Sheldon Axler, Antoine Chambert-Loir, Robert Coleman, Xander Faber, William Noorduin, Daeshik Park, Clay Petsche, Juan Rivera-Letelier, Joe Silverman, and Steve Winburn for proofreading parts of the manuscript. Aaron Abrams and Brian Conrad gave useful suggestions on Appendices B and C, respectively. Charles Favre suggested several improvements in Chapter 8. The idea for the Hsia kernel as the fundamental kernel for potential theory on the Berkovich line was inspired by a manuscript of Liang-Chung Hsia [62]. We thank Rob Benedetto for directing us to the work of Juan Rivera-Letelier. We also thank Rivera-Letelier for many stimulating conversations about dynamics of rational maps and the Berkovich projective line. Finally, we thank the anonymous referees who read through this work and made a number of valuable suggestions and everyone at the AMS who assisted with the production of this book, including Marcia Almeida, Barbara Beeton, and Ina Mette. We especially thank Arlene O'Sean for her careful and thorough editing.

We are grateful to the National Science Foundation for its support of this project, primarily in the form of the research grant DMS-0300784, but also through research grants DMS-0600027 and DMS-0601037, under which the bulk of the writing was carried out. Any opinions, findings, conclusions, or recommendations expressed in this work are those of the authors and do not necessarily reflect the views of the National Science Foundation.

Differences from the preliminary version

There are several differences between the present manuscript and the preliminary version [**91**] posted to the arXiv preprint server in July 2004. For one thing, we have corrected a number of errors in the earlier version.

In addition, we have revised all of the statements and proofs so that they hold over an arbitrary complete, algebraically closed field K endowed with a nontrivial non-Archimedean absolute value, rather than just over the field $\mathbb{C}_p$. The main difference is that the Berkovich projective line over $\mathbb{C}_p$ is *metrizable* and has *countable* branching at every point, whereas in general the Berkovich projective line over K is nonmetrizable and has uncountable branching. Replacing $\mathbb{C}_p$ by K throughout required a significant reworking of many of our original proofs, since [**91**] relies in several places on arguments valid for metric spaces but not for an arbitrary compact Hausdorff space. Consequently, the present book makes more demands on the reader in terms of topological prerequisites; for example we now make use of nets rather than sequences in several places.

In some sense, this works against the concrete and "elementary" exposition that we have striven for. However, the changes seem desirable for at least two reasons. First, some proofs become more natural once the crutch of metrizability is removed. Second, and perhaps more importantly, the theory for more general fields is needed for many applications. The first author's paper [**5**] is one example of this: it contains a Northcott-type theorem for dynamical canonical heights over a general field k endowed with a product formula; the theorem is proved by working locally at each place v on the Berkovich projective line over $\mathbb{C}_v$ (the smallest complete and algebraically closed field containing k and possessing an absolute value extending the given one $|\ |_v$ on k). As another example, Kontsevich and Soibelman [**68**] have recently used Berkovich's theory over fields such as the completion of an algebraic closure of $\mathbb{C}((T))$ to study homological mirror symmetry. We mention also the work of Favre and Jonsson [**45**] on the valuative tree, as well as the related work of Jan Kiwi [**66**], both of which have applications to complex dynamics.

Here is a summary of the main differences between this work and [**91**]:

- We have added a detailed Introduction summarizing the work.
- We have added several appendices in order to make the presentation more self-contained.
- We have added a symbol table and an index and updated the bibliography.
- We give a different construction of $\mathbb{P}^1_{\text{Berk}}$ (analogous to the "Proj" construction in algebraic geometry) which makes it easier to understand the action of a rational function.
- We have changed some of our notation and terminology to be compatible with that of the authors mentioned above.

- We have added sections on the Dirichlet pairing and Favre–Rivera-Letelier smoothing.
- We compare our Laplacian with those of Favre, Jonsson, and Rivera-Letelier, and of Thuillier.
- We have included a discussion of $\mathbb{R}$-trees, and in particular of $\mathbb{P}^1_{\text{Berk}}$ as a "profinite $\mathbb{R}$-tree".
- We have expanded our discussion of the Poisson formula on $\mathbb{P}^1_{\text{Berk}}$.
- We have added a section on Thuillier's short exact sequence describing subharmonic functions in terms of harmonic functions and positive σ-finite measures.
- We have added a section on Hartogs's lemma, a key ingredient in the work of Favre and Rivera-Letelier.
- We state and prove Berkovich space versions of Bilu's equidistribution theorem, the dynamical equidistribution theorem for small points, and the arithmetic Fekete-Szegö theorem for $\mathbb{P}^1$.
- We have simplified and expanded the discussion in Chapter 9 on analytic multiplicities.
- We have greatly expanded the material on dynamics of rational maps, incorporating the work of Rivera-Letelier and the joint work of Favre and Rivera-Letelier, and including a section on examples.

Introduction

This book has several goals. The first goal is to develop the foundations of potential theory on $\mathbb{P}^1_{\text{Berk}}$, including the definition of a measure-valued Laplacian operator, capacity theory, and a theory of harmonic and subharmonic functions. A second goal is to give applications of potential theory on $\mathbb{P}^1_{\text{Berk}}$, especially to the dynamics of rational maps defined over an arbitrary complete and algebraically closed non-Archimedean field K. A third goal is to provide the reader with a concrete introduction to Berkovich's theory of analytic spaces by focusing on the special case of the Berkovich projective line.

We now outline the contents of the book.

The Berkovich affine and projective lines. Let K be an algebraically closed field which is complete with respect to a nontrivial non-Archimedean absolute value. The topology on K induced by the given absolute value is Hausdorff, but it is also totally disconnected and not locally compact. This makes it difficult to define a good notion of an analytic function on K. Tate dealt with this problem by developing the subject now known as *rigid analysis*, in which one works with a certain Grothendieck topology on K. This leads to a satisfactory theory of analytic functions, but since the underlying topological space is unchanged, difficulties remain for other applications. For example, using only the topology on K, there is no evident way to define a Laplacian operator analogous to the classical Laplacian on $\mathbb{C}$ or to work sensibly with probability measures on K.

However, these difficulties, and many more, can be resolved in a very satisfactory way using Berkovich's theory. The Berkovich affine line $\mathbb{A}^1_{\text{Berk}}$ over K is a locally compact, Hausdorff, and path-connected topological space which contains K (with the topology induced by the given absolute value) as a dense subspace. One obtains the Berkovich projective line $\mathbb{P}^1_{\text{Berk}}$ by adjoining to $\mathbb{A}^1_{\text{Berk}}$ in a suitable manner a point at infinity; the resulting space $\mathbb{P}^1_{\text{Berk}}$ is a compact, Hausdorff, path-connected topological space which contains $\mathbb{P}^1(K)$ (with its natural topology) as a dense subspace. In fact, $\mathbb{A}^1_{\text{Berk}}$ and $\mathbb{P}^1_{\text{Berk}}$ are more than just path-connected: they are *uniquely* path-connected, in the sense that any two distinct points can be joined by a unique arc. The unique path-connectedness is closely related to the fact that $\mathbb{A}^1_{\text{Berk}}$ and $\mathbb{P}^1_{\text{Berk}}$ are endowed with a natural tree structure. (More specifically, they are $\mathbb{R}$-*trees*, as defined in §1.4.) The tree structure on $\mathbb{A}^1_{\text{Berk}}$ (resp. $\mathbb{P}^1_{\text{Berk}}$) can be used to define a *Laplacian operator* in terms of the classical Laplacian on a finite graph. This in turn leads to a theory of harmonic and subharmonic functions which closely parallels the classical theory over $\mathbb{C}$.

The definition of $\mathbb{A}^1_{\text{Berk}}$ is quite simple and makes sense with K replaced by an arbitrary field k endowed with a (possibly Archimedean or even trivial) absolute value. As a set, $\mathbb{A}^1_{\text{Berk},k}$ consists of all multiplicative seminorms on the polynomial ring $k[T]$ which extend the usual absolute value on k. (A *multiplicative seminorm* on a ring A is a function $[\ ]_x : A \to \mathbb{R}_{\geq 0}$ satisfying $[0]_x = 0, [1]_x = 1, [fg]_x = [f]_x \cdot [g]_x$, and $[f+g]_x \leq [f]_x + [g]_x$ for all $f, g \in A$.) By an aesthetically desirable abuse of notation, we will identify seminorms $[\ ]_x$ with points $x \in \mathbb{A}^1_{\text{Berk},k}$, and we will usually omit explicit reference to the field k, writing simply $\mathbb{A}^1_{\text{Berk}}$. The topology on $\mathbb{A}^1_{\text{Berk},k}$ is the weakest one for which $x \mapsto [f]_x$ is continuous for every $f \in k[T]$.

To motivate this definition, we observe that in the classical setting, every multiplicative seminorm on $\mathbb{C}[T]$ which extends the usual absolute value on $\mathbb{C}$ is of the form $f \mapsto |f(z)|$ for some $z \in \mathbb{C}$. (This can be deduced from the well-known Gelfand-Mazur theorem from functional analysis.) It is then easy to see that $\mathbb{A}^1_{\text{Berk},\mathbb{C}}$ is homeomorphic to $\mathbb{C}$ itself and also to the Gelfand spectrum (i.e., the space of all maximal ideals) of $\mathbb{C}[T]$.

In the non-Archimedean world, K can once again be identified with the Gelfand space of maximal ideals in $K[T]$, but now there are many more multiplicative seminorms on $K[T]$ than just the ones given by evaluation at a point of K. The prototypical example arises by fixing a closed disc $D(a,r) = \{z \in K : |z - a| \leq r\}$ in K and defining $[\ ]_{D(a,r)}$ by

$$[f]_{D(a,r)} = \sup_{z \in D(a,r)} |f(z)|.$$

It is an elementary consequence of Gauss's lemma that $[\ ]_{D(a,r)}$ is *multiplicative*, and the other axioms for a seminorm are trivially satisfied. Thus each disc $D(a,r)$ gives rise to a point of $\mathbb{A}^1_{\text{Berk}}$. Note that this includes discs for which $r \notin |K^\times|$, i.e., "irrational discs" for which the set $\{z \in K : |z - a| = r\}$ is empty. We may consider the point a as a "degenerate" disc of radius zero. (If $r > 0$, then $[\ ]_{D(a,r)}$ is not only a seminorm, but a norm.) It is not hard to see that distinct discs $D(a,r)$ with $r \geq 0$ give rise to distinct multiplicative seminorms on $K[T]$, and therefore the set of all such discs embeds naturally into $\mathbb{A}^1_{\text{Berk}}$.

Suppose $x, x' \in \mathbb{A}^1_{\text{Berk}}$ are distinct points corresponding to the (possibly degenerate) discs $D(a,r), D(a',r')$, respectively. The unique path in $\mathbb{A}^1_{\text{Berk}}$ between x and x' has a very intuitive description. If $D(a,r) \subset D(a',r')$, it consists of all points of $\mathbb{A}^1_{\text{Berk}}$ corresponding to discs containing $D(a,r)$ and contained in $D(a',r')$. The set of all such "intermediate discs" is totally ordered by containment, and if $a = a'$, it is just $\{D(a,t) : r \leq t \leq r'\}$. If $D(a,r)$ and $D(a',r')$ are disjoint, the unique path between x and x' consists of all points of $\mathbb{A}^1_{\text{Berk}}$ corresponding to discs which are either of the form $D(a,t)$ with $r \leq t \leq |a - a'|$ or of the form $D(a',t)$ with $r' \leq t \leq |a - a'|$.

The disc $D(a, |a - a'|) = D(a', |a - a'|)$ is the smallest one containing both $D(a, r)$ and $D(a', r')$, and if $x \vee x'$ denotes the point of $\mathbb{A}^1_{\text{Berk}}$ corresponding to $D(a, |a - a'|)$, then the path from x to x' is just the path from x to $x \vee x'$ followed by the path from $x \vee x'$ to x'.

In particular, if a, a' are distinct points of K, one can visualize the path in $\mathbb{A}^1_{\text{Berk}}$ from a to a' as follows: increase the "radius" of the degenerate disc $D(a, 0)$ until a disc $D(a, r)$ is reached which also contains a'. This disc can also be written as $D(a', s)$ with $s = |a - a'|$. Now decrease s until the radius reaches zero. This "connects" the totally disconnected space K by adding points corresponding to closed discs in K. In order to obtain a *compact* space, however, it is necessary in general to add even more points, for K may not be *spherically complete* (this happens, e.g., when $K = \mathbb{C}_p$): there may be decreasing sequences of closed discs with empty intersection. Intuitively, we need to add in points corresponding to such sequences in order to obtain a space which has a chance of being compact. More precisely, returning to the definition of $\mathbb{A}^1_{\text{Berk}}$ in terms of multiplicative seminorms, if $\{D(a_i, r_i)\}$ is any decreasing nested sequence of closed discs, then the map

$$f \mapsto \lim_{i \to \infty} [f]_{D(a_i, r_i)}$$

defines a multiplicative seminorm on $K[T]$ extending the usual absolute value on K. One can show that two sequences of discs with empty intersection define the same seminorm if and only if the sequences are *cofinal*. This yields a large number of additional points of $\mathbb{A}^1_{\text{Berk}}$. According to *Berkovich's classification theorem*, we have now described all the points of $\mathbb{A}^1_{\text{Berk}}$: each point $x \in \mathbb{A}^1_{\text{Berk}}$ corresponds to a decreasing nested sequence $\{D(a_i, r_i)\}$ of closed discs, and we can categorize the points of $\mathbb{A}^1_{\text{Berk}}$ into four types according to the nature of $D = \bigcap D(a_i, r_i)$:

(I) D is a point of K.
(II) D is a closed disc with radius belonging to $|K^\times|$.
(III) D is a closed disc with radius *not* belonging to $|K^\times|$.
(IV) $D = \emptyset$.

As a set, $\mathbb{P}^1_{\text{Berk}}$ can be obtained from $\mathbb{A}^1_{\text{Berk}}$ by adding a type I point denoted ∞. The topology on $\mathbb{P}^1_{\text{Berk}}$ is that of the one-point compactification.

Following Rivera-Letelier, we write $\mathbb{H}_{\text{Berk}}$ for the subset of $\mathbb{P}^1_{\text{Berk}}$ consisting of all points of type II, III, or IV (Berkovich "hyperbolic space"). Note that $\mathbb{H}_{\text{Berk}}$ consists of precisely the points in $\mathbb{P}^1_{\text{Berk}}$ for which $[\ ]_x$ is a norm. We also write $\mathbb{H}^{\mathbb{Q}}_{\text{Berk}}$ for the set of type II points and $\mathbb{H}^{\mathbb{R}}_{\text{Berk}}$ for the set of points of type II or III.

The description of points of $\mathbb{A}^1_{\text{Berk}}$ in terms of closed discs is very useful, because it allows one to visualize quite concretely the abstract space of multiplicative seminorms which we started with. It also allows us to understand in a more concrete way the natural partial order on $\mathbb{A}^1_{\text{Berk}}$ in which $x \leq y$ if and only if $[f]_x \leq [f]_y$ for all $f \in K[T]$. In terms of discs, if x, y are points of type I, II, or III, one can show that $x \leq y$ if and only if the disc

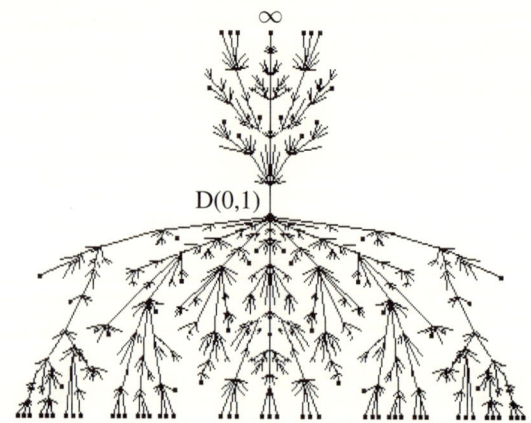

FIGURE 0.1. The Berkovich Projective Line

corresponding to x is contained in the disc corresponding to y. (We leave it as an exercise to the reader to extend this description of the partial order to points of type IV.) For any pair of points $x, y \in \mathbb{A}^1_{\text{Berk}}$, there is a unique least upper bound $x \vee y \in \mathbb{A}^1_{\text{Berk}}$ with respect to this partial order. We can extend the partial order to $\mathbb{P}^1_{\text{Berk}}$ by declaring that $x \leq \infty$ for all $x \in \mathbb{A}^1_{\text{Berk}}$.

Writing
$$[x, x'] \;=\; \{z \in \mathbb{P}^1_{\text{Berk}} \,:\, x \leq z \leq x'\} \cup \{z \in \mathbb{P}^1_{\text{Berk}} \,:\, x' \leq z \leq x\} \,,$$
it is easy to see that the unique path between $x, y \in \mathbb{P}^1_{\text{Berk}}$ is just
$$[x, x \vee y] \cup [x \vee y, y] \,.$$

There is a canonical metric ρ on $\mathbb{H}_{\text{Berk}}$ which is of great importance for potential theory. To define it, we first define a function diam : $\mathbb{A}^1_{\text{Berk}} \to \mathbb{R}_{\geq 0}$ by setting $\text{diam}(x) = \lim r_i$ if x corresponds to the nested sequence $\{D(a_i, r_i)\}$. This is easily checked to be well-defined, independent of the choice of nested sequence. If $x \in \mathbb{H}^{\mathbb{R}}_{\text{Berk}}$, then $\text{diam}(x)$ is just the diameter (= radius) of the corresponding closed disc. Because K is complete, if x is of type IV, then $\text{diam}(x) > 0$. Thus $\text{diam}(x) = 0$ for $x \in \mathbb{A}^1_{\text{Berk}}$ of type I, and $\text{diam}(x) > 0$ for $x \in \mathbb{H}_{\text{Berk}}$.

If $x, y \in \mathbb{H}_{\text{Berk}}$ with $x \leq y$, we define
$$\rho(x, y) \;=\; \log_v \frac{\text{diam}(y)}{\text{diam}(x)} \,,$$
where $\log_v$ denotes the logarithm to the base q_v, with $q_v > 1$ a fixed real number chosen so that $-\log_v |\cdot|$ is a prescribed normalized valuation on K.

More generally, for $x, y \in \mathbb{H}_{\text{Berk}}$ arbitrary, we define the *path distance metric* $\rho(x, y)$ by
$$\rho(x, y) \;=\; \rho(x, x \vee y) + \rho(y, x \vee y) \,.$$
It is not hard to verify that ρ defines a metric on $\mathbb{H}_{\text{Berk}}$. One can extend ρ to a singular function on $\mathbb{P}^1_{\text{Berk}}$ by declaring that if $x \in \mathbb{P}^1(K)$ and $y \in \mathbb{P}^1_{\text{Berk}}$,

we have $\rho(x,y) = \infty$ if $x \neq y$ and 0 if $x = y$. However, we usually only consider ρ as being defined on $\mathbb{H}_{\text{Berk}}$.

It is important to note that the topology on $\mathbb{H}_{\text{Berk}}$ defined by the metric ρ is *not* the subspace topology induced from the Berkovich (or Gelfand) topology on $\mathbb{P}^1_{\text{Berk}} \supset \mathbb{H}_{\text{Berk}}$; it is strictly finer than the subspace topology.

The group $\text{PGL}(2, K)$ of *Möbius transformations* acts continuously on $\mathbb{P}^1_{\text{Berk}}$ in a natural way compatible with the usual action on $\mathbb{P}^1(K)$, and this action preserves $\mathbb{H}_{\text{Berk}}, \mathbb{H}^{\mathbb{Q}}_{\text{Berk}}$, and $\mathbb{H}^{\mathbb{R}}_{\text{Berk}}$. Using the definition of $\mathbb{P}^1_{\text{Berk}}$ in terms of multiplicative seminorms (and extending each $[\ ]_x$ to a seminorm on its local ring in the quotient field $K(T)$), we have $[f]_{M(x)} = [f \circ M]_x$ for each $M \in \text{PGL}(2, K)$. The action of $\text{PGL}(2, K)$ on $\mathbb{P}^1_{\text{Berk}}$ can also be described concretely in terms of Berkovich's classification theorem, using the fact that each $M \in \text{PGL}(2, K)$ takes closed discs to closed discs. An important observation is that $\text{PGL}(2, K)$ acts *isometrically* on $\mathbb{H}_{\text{Berk}}$, i.e.,

$$\rho(M(x), M(y)) = \rho(x, y)$$

for all $x, y \in \mathbb{H}_{\text{Berk}}$ and all $M \in \text{PGL}(2, K)$. This shows that the path distance metric ρ is "coordinate-free".

The diameter function diam can also be used to extend the usual distance function $|x - y|$ on K to $\mathbb{A}^1_{\text{Berk}}$. We call this extension the *Hsia kernel* and denote it by $\delta(x, y)_\infty$. Formally, for $x, y \in \mathbb{A}^1_{\text{Berk}}$ we have

$$\delta(x, y)_\infty = \text{diam}(x \vee y) \ .$$

It is easy to see that if $x, y \in K$, then $\delta(x, y)_\infty = |x - y|$. More generally, one has the formula

$$\delta(x, y)_\infty = \limsup_{(x_0, y_0) \to (x, y)} |x_0 - y_0| \ ,$$

where $(x_0, y_0) \in K \times K$ and the convergence implicit in the lim sup is with respect to the product topology on $\mathbb{P}^1_{\text{Berk}} \times \mathbb{P}^1_{\text{Berk}}$. The Hsia kernel satisfies all of the axioms for an ultrametric with one exception: we have $\delta(x, x)_\infty > 0$ for $x \in \mathbb{H}_{\text{Berk}}$.

The function $-\log_v \delta(x, y)_\infty$, which generalizes the usual potential theory kernel $-\log_v |x - y|$, leads to a theory of capacities on $\mathbb{P}^1_{\text{Berk}}$ which generalizes that of [88] and which has many features in common with classical capacity theory over $\mathbb{C}$.

There is also a *generalized Hsia kernel* $\delta(x, y)_\zeta$ with respect to an arbitrary point $\zeta \in \mathbb{P}^1_{\text{Berk}}$; we refer the reader to §4.4 for details.

We now come to an important description of $\mathbb{P}^1_{\text{Berk}}$ as a *profinite $\mathbb{R}$-tree*. An $\mathbb{R}$-*tree* is a metric space (T, d) such that for each distinct pair of points $x, y \in T$, there is a unique arc in T from x to y, and this arc is a geodesic. (See Appendix B for a more detailed discussion of $\mathbb{R}$-trees.) A *branch point* is a point $x \in T$ for which $T \backslash \{x\}$ has either one or more than two connected components. A *finite $\mathbb{R}$-tree* is an $\mathbb{R}$-tree which is compact and has only finitely many branch points. Intuitively, a finite $\mathbb{R}$-tree is just a finite tree in the usual graph-theoretic sense, but where the edges are thought of as line segments having definite lengths. Finally, a *profinite $\mathbb{R}$-tree* is an inverse limit of finite $\mathbb{R}$-trees.

Let us consider how these definitions play out for $\mathbb{P}^1_{\text{Berk}}$. If $S \subset \mathbb{P}^1_{\text{Berk}}$, define the *convex hull* of S to be the smallest path-connected subset of $\mathbb{P}^1_{\text{Berk}}$ containing S. (This is the same as the set of all paths between points of S.) By abuse of terminology, a *finite subgraph* of $\mathbb{P}^1_{\text{Berk}}$ will mean the convex hull of a finite subset $S \subset \mathbb{H}^{\mathbb{R}}_{\text{Berk}}$. Every finite subgraph Γ, when endowed with the induced path distance metric ρ, is a finite $\mathbb{R}$-tree, and the collection of all finite subgraphs of $\mathbb{P}^1_{\text{Berk}}$ is a directed set under inclusion. Moreover, if $\Gamma \leq \Gamma'$, then by a basic property of $\mathbb{R}$-trees, there is a continuous *retraction map* $r_{\Gamma',\Gamma} : \Gamma' \to \Gamma$. In §1.4, we will show that $\mathbb{P}^1_{\text{Berk}}$ is homeomorphic to the inverse limit $\varprojlim \Gamma$ over all finite subgraphs $\Gamma \subset \mathbb{P}^1_{\text{Berk}}$. (Intuitively, this is just a topological formulation of Berkovich's classification theorem.) This description of $\mathbb{P}^1_{\text{Berk}}$ as a profinite $\mathbb{R}$-tree provides a convenient way to visualize the topology on $\mathbb{P}^1_{\text{Berk}}$: two points are "close" if they retract to the same point of a "large" finite subgraph. For each Γ, we let $r_{\mathbb{P}^1_{\text{Berk}},\Gamma}$ be the natural retraction map from $\mathbb{P}^1_{\text{Berk}}$ to Γ coming from the universal property of the inverse limit.

A fundamental system of open neighborhoods for the topology on $\mathbb{P}^1_{\text{Berk}}$ is given by the *open affinoid subsets*, which are the sets of the form $r^{-1}_{\mathbb{P}^1_{\text{Berk}},\Gamma}(V)$ for Γ a finite subgraph of $\mathbb{H}^{\mathbb{R}}_{\text{Berk}}$ and V an open subset of Γ. We will refer to a connected open affinoid subset of $\mathbb{P}^1_{\text{Berk}}$ as a *simple domain*. Simple domains can be completely characterized as the connected open subsets of $\mathbb{P}^1_{\text{Berk}}$ having a finite (nonzero) number of boundary points, all of which are contained in $\mathbb{H}^{\mathbb{R}}_{\text{Berk}}$. If U is an open subset of $\mathbb{P}^1_{\text{Berk}}$, a *simple subdomain* of U is defined to be a simple domain whose closure is contained in U.

Laplacians. The profinite $\mathbb{R}$-tree structure on $\mathbb{P}^1_{\text{Berk}}$ leads directly to the construction of a Laplacian operator. On a finite subgraph Γ of $\mathbb{P}^1_{\text{Berk}}$ (or, more generally, on any 'metrized graph'; see Chapter 3 for details), there is a natural Laplacian operator Δ_Γ generalizing the well-known combinatorial Laplacian on a weighted graph. If $f : \Gamma \to \mathbb{R}$ is continuous, and $\mathcal{C}^2$ except at a finite number of points, then there is a unique Borel measure $\Delta_\Gamma(f)$ of total mass zero on Γ such that

$$(0.1) \qquad \int_\Gamma \psi \, \Delta_\Gamma(f) = \int_\Gamma f'(x)\psi'(x) \, dx$$

for all continuous, piecewise affine functions ψ on Γ. The measure $\Delta_\Gamma(f)$ has a discrete part and a continuous part. At each $P \in \Gamma$ which is either a branch point of Γ or a point where $f(x)$ fails to be $\mathcal{C}^2$, $\Delta_\Gamma(f)$ has a point mass equal to the negative of the sum of the directional derivatives of $f(x)$ on the edges emanating from P. On the intervening edges, it is given by $-f''(x)dx$. (See Chapter 3 for details.)

We define $\mathrm{BDV}(\Gamma)$ to be the space of all continuous real-valued functions f on Γ for which the distribution defined by

$$\psi \mapsto \int_\Gamma f\, \Delta_\Gamma(\psi)\ , \tag{0.2}$$

for all ψ as above, is represented by a bounded signed Borel measure $\Delta_\Gamma(f)$. A simple integration by parts argument shows that this measure coincides with the one defined by (0.1) when f is sufficiently smooth. The name "BDV" is an abbreviation for "Bounded Differential Variation". We call the measure $\Delta_\Gamma(f)$ the *Laplacian* of f on Γ.

The Laplacian satisfies an important compatibility property with respect to the partial order on the set of finite subgraphs of $\mathbb{P}^1_{\mathrm{Berk}}$ given by containment: if $\Gamma \leq \Gamma'$ and $f \in \mathrm{BDV}(\Gamma')$, then

$$\Delta_\Gamma(f|_\Gamma) \;=\; \left(r_{\Gamma',\Gamma}\right)_* \Delta_{\Gamma'}(f)\ . \tag{0.3}$$

We define $\mathrm{BDV}(\mathbb{P}^1_{\mathrm{Berk}})$ to be the collection of all functions $f : \mathbb{P}^1_{\mathrm{Berk}} \to \mathbb{R} \cup \{\pm\infty\}$ such that:

- $f|_\Gamma \in \mathrm{BDV}(\Gamma)$ for each finite subgraph Γ.
- The measures $|\Delta_\Gamma(f)|$ have uniformly bounded total mass.

Note that belonging to $\mathrm{BDV}(\mathbb{P}^1_{\mathrm{Berk}})$ imposes no condition on the values of f at points of $\mathbb{P}^1(K)$.

Using the compatibility property (0.3), one shows that if $f \in \mathrm{BDV}(\mathbb{P}^1_{\mathrm{Berk}})$, then the collection of measures $\{\Delta_\Gamma\}$ "cohere" to give a unique Borel measure $\Delta(f)$ of total mass zero on the inverse limit space $\mathbb{P}^1_{\mathrm{Berk}}$ satisfying

$$\left(r_{\mathbb{P}^1_{\mathrm{Berk}},\Gamma}\right)_* \Delta(f) \;=\; \Delta_\Gamma(f)$$

for all finite subgraphs Γ of $\mathbb{P}^1_{\mathrm{Berk}}$. We call $\Delta(f)$ the *Laplacian* of f on $\mathbb{P}^1_{\mathrm{Berk}}$.

Similarly, if U is a domain (i.e., a nonempty connected open subset) in $\mathbb{P}^1_{\mathrm{Berk}}$, one defines a class $\mathrm{BDV}(U)$ of functions $f : U \to \mathbb{R} \cup \{\pm\infty\}$ for which the Laplacian $\Delta_{\overline{U}}(f)$ is a bounded Borel measure of total mass zero supported on the closure of U. The measure $\Delta_{\overline{U}}(f)$ has the property that

$$\left(r_{\overline{U},\Gamma}\right)_* \Delta(f) \;=\; \Delta_\Gamma(f)$$

for all finite subgraphs Γ of $\mathbb{P}^1_{\mathrm{Berk}}$ contained in U.

As a concrete example, fix $y \in \mathbb{A}^1_{\mathrm{Berk}}$ and let $f : \mathbb{P}^1_{\mathrm{Berk}} \to \mathbb{R} \cup \{\pm\infty\}$ be defined by $f(\infty) = -\infty$ and

$$f(x) \;=\; -\log_v \delta(x,y)_\infty$$

for $x \in \mathbb{A}^1_{\mathrm{Berk}}$. Then $f \in \mathrm{BDV}(\mathbb{P}^1_{\mathrm{Berk}})$, and

$$\Delta f \;=\; \delta_y - \delta_\infty \tag{0.4}$$

is a discrete measure on $\mathbb{P}^1_{\mathrm{Berk}}$ supported on $\{y,\infty\}$. Intuitively, the explanation for the formula (0.4) is as follows. The function f is locally constant away from the path $\Lambda = [y,\infty]$ from y to ∞; more precisely, we have $f(x) = f(r_{\mathbb{P}^1_{\mathrm{Berk}},\Lambda}(x))$. Moreover, the restriction of f to Λ is linear (with respect to the distance function ρ) with slope -1. For every suitable test function ψ, we therefore have the "heuristic" calculation

$$\int_{\mathbb{P}^1_{\text{Berk}}} \psi \Delta f = \int_\Lambda f'(x) \psi'(x) \, dx = -\int_y^\infty \psi'(x) \, dx = \psi(y) - \psi(\infty) \; .$$

(To make this calculation rigorous, one needs to exhaust $\Lambda = [y, \infty]$ by an increasing sequence of line segments $\Gamma \subset \mathbb{H}^{\mathbb{R}}_{\text{Berk}}$ and then observe that the corresponding measures $\Delta_\Gamma f$ converge weakly to $\delta_y - \delta_\infty$.)

Equation (0.4) shows that $-\log_v \delta(x, y)_\infty$, like its classical counterpart $-\log|x-y|$ over $\mathbb{C}$, is a fundamental solution (in the sense of distributions) to the Laplace equation. This "explains" why $-\log_v \delta(x, y)_\infty$ is the correct kernel for doing potential theory.

More generally, let $\varphi \in K(T)$ be a nonzero rational function with zeros and poles given by the divisor $\text{div}(\varphi)$ on $\mathbb{P}^1(K)$. The usual action of φ on $\mathbb{P}^1(K)$ extends naturally to an action of φ on $\mathbb{P}^1_{\text{Berk}}$, and there is a continuous function $-\log_v[\varphi]_x : \mathbb{P}^1_{\text{Berk}} \to \mathbb{R} \cup \{\pm \infty\}$ extending the usual map $x \mapsto -\log_v |\varphi(x)|$ on $\mathbb{P}^1(K)$. One derives from (0.4) the following version of the *Poincaré-Lelong formula*:

$$\Delta_{\mathbb{P}^1_{\text{Berk}}} \left(-\log_v[\varphi]_x \right) = \delta_{\text{div}(\varphi)} \; .$$

Capacities. Fix $\zeta \in \mathbb{P}^1_{\text{Berk}}$, and let E be a compact subset of $\mathbb{P}^1_{\text{Berk}} \backslash \{\zeta\}$. (For concreteness, the reader may wish to imagine that $\zeta = \infty$.) By analogy with the classical theory over $\mathbb{C}$ and also with the non-Archimedean theory developed in [**88**], one can define the *logarithmic capacity* of E with respect to ζ. This is done as follows.

Given a probability measure ν on $\mathbb{P}^1_{\text{Berk}}$ with support contained in E, we define the energy integral

$$I_\zeta(\nu) = \iint_{E \times E} -\log_v \delta(x, y)_\zeta \, d\nu(x) d\nu(y) \; .$$

Letting ν vary over the collection $\mathbb{P}(E)$ of all probability measures supported on E, one defines the *Robin constant*

$$V_\zeta(E) = \inf_{\nu \in \mathbb{P}(E)} I_\zeta(\nu) \; .$$

The *logarithmic capacity* of E relative to ζ is then defined to be

$$\gamma_\zeta(E) = q_v^{-V_\zeta(E)} \; .$$

For an arbitrary set H, the logarithmic capacity $\gamma_\zeta(H)$ is defined by

$$\gamma_\zeta(H) = \sup_{\text{compact } E \subset H} \gamma_\zeta(E) \; .$$

A countably supported probability measure must have point masses, and $\delta(x,x)_\zeta = 0$ for $x \in \mathbb{P}^1(K) \backslash \{\zeta\}$; thus $V_\zeta(E) = +\infty$ when $E \subset \mathbb{P}^1(K)$ is countable, so every countable subset of $\mathbb{P}^1(K)$ has capacity zero. On the other hand, for a "nonclassical" point $x \in \mathbb{H}_{\text{Berk}}$ we have $V_\zeta(\{x\}) < +\infty$, since $\delta(x,x)_\zeta > 0$, and therefore $\gamma_\zeta(\{x\}) > 0$. In particular, a singleton set can have positive capacity, a phenomenon which has no classical analogue. More generally, if $E \cap \mathbb{H}_{\text{Berk}} \neq \emptyset$, then $\gamma_\zeta(E) > 0$.

As a more elaborate example, if $K = \mathbb{C}_p$ and $E = \mathbb{Z}_p \subset \mathbb{A}^1(\mathbb{C}_p) \subset \mathbb{A}^1_{\text{Berk},\mathbb{C}_p}$, then $\gamma_\infty(E) = p^{-1/(p-1)}$. Since $\delta(x,y)_\infty = |x-y|$ for $x, y \in K$, this follows from the same computation as in [**88**, Example 4.1.24].

For fixed E, the property of E having capacity 0 relative to ζ is independent of the point $\zeta \notin E$.

If E is compact and $\gamma_\zeta(E) > 0$, we show that there is a *unique* probability measure $\mu_{E,\zeta}$ on E, called the *equilibrium measure* of E with respect to ζ, which minimizes energy (i.e., for which $I_\zeta(\mu_{E,\zeta}) = V_\zeta(E)$). As in the classical case, $\mu_{E,\zeta}$ is always supported on the boundary of E.

Closely linked to the theory of capacities is the theory of *potential functions*. For each probability measure ν supported on $\mathbb{P}^1_{\text{Berk}} \backslash \{\zeta\}$, one defines the potential function $u_\nu(z,\zeta)$ by

$$u_\nu(z,\zeta) = \int -\log_v \delta(z,w)_\zeta \, d\nu(w) \ .$$

As in classical potential theory, potential functions need not be continuous, but they do share several of the distinguishing features of continuous functions. For example, $u_\nu(z,\zeta)$ is lower semicontinuous, and it is continuous at each $z \notin \text{supp}(\nu)$. Potential functions on $\mathbb{P}^1_{\text{Berk}}$ satisfy the following analogues of Maria's theorem and Frostman's theorem from complex potential theory:

THEOREM (Maria). *If $u_\nu(z,\zeta) \leq M$ on $\text{supp}(\nu)$, then $u_\nu(z,\zeta) \leq M$ for all $z \in \mathbb{P}^1_{\text{Berk}} \backslash \{\zeta\}$.*

THEOREM (Frostman). *If a compact set E has positive capacity, then the equilibrium potential $u_E(z,\zeta)$ satisfies $u_E(z,\zeta) \leq V_\zeta(E)$ for all $z \in \mathbb{P}^1_{\text{Berk}} \backslash \{\zeta\}$, and $u_E(z,\zeta) = V_\zeta(E)$ for all $z \in E$ outside a set of capacity zero.*

As in capacity theory over $\mathbb{C}$, one can also define the *transfinite diameter* and the *Chebyshev constant* of E, and they both turn out to be equal to the logarithmic capacity of E. (In fact, we define three different variants of the Chebyshev constant and prove that they are all equal.)

As an arithmetic application of the theory of capacities on $\mathbb{P}^1_{\text{Berk}}$, we formulate generalizations to $\mathbb{P}^1_{\text{Berk}}$ of the Fekete and Fekete-Szegö theorems from [**88**]. The proofs are easy, since they go by reducing the general case to the special case of RL-domains, which was already treated in [**88**]. Nonetheless, the results are aesthetically pleasing because in their statement, the simple notion of compactness replaces the awkward concept of "algebraic capacitability". The possibility for such a reformulation is directly related to the fact that $\mathbb{P}^1_{\text{Berk}}$ is compact, while $\mathbb{P}^1(K)$ is not.

Harmonic functions. If U is a domain in $\mathbb{P}^1_{\text{Berk}}$, a real-valued function $f : U \to \mathbb{R}$ is called *strongly harmonic* on U if it is continuous, belongs to $\text{BDV}(U)$, and if $\Delta_{\overline{U}}(f)$ is supported on ∂U. The function f is *harmonic* on U if every point $x \in U$ has a connected open neighborhood on which f is strongly harmonic.

Harmonic functions on domains $U \subseteq \mathbb{P}^1_{\text{Berk}}$ satisfy many properties analogous to their classical counterparts over $\mathbb{C}$. For example, a harmonic function which attains its maximum or minimum value on U must be constant. There is also an analogue of the Poisson formula: if f is a harmonic function on an open affinoid U, then f extends uniquely to the boundary ∂U, and the values of f on U can be computed explicitly in terms of $f|_{\partial U}$. A version of Harnack's principle holds as well: the limit of a monotonically increasing sequence of nonnegative harmonic functions on U is either harmonic or identically $+\infty$. Even better than the classical case (where a hypothesis of uniform convergence is required), a *pointwise* limit of harmonic functions is automatically harmonic. As is the case over $\mathbb{C}$, harmonicity is preserved under pullbacks by meromorphic functions.

Fix $\zeta \in \mathbb{P}^1_{\text{Berk}}$, and let E be a compact subset of $\mathbb{P}^1_{\text{Berk}} \backslash \{\zeta\}$. We define the *Green's function of E relative to ζ* to be

$$G(z, \zeta; E) = V_\zeta(E) - u_E(z, \zeta)$$

for all $z \in \mathbb{P}^1_{\text{Berk}}$. We show that the Green's function is everywhere nonnegative and that it is strictly positive on the connected component U_ζ of $\mathbb{P}^1_{\text{Berk}} \backslash E$ containing ζ. Also, $G(z, \zeta; E)$ is finite on $\mathbb{P}^1_{\text{Berk}} \backslash \{\zeta\}$, with a logarithmic singularity at ζ, and it is harmonic on $U_\zeta \backslash \{\zeta\}$. Additionally, $G(z, \zeta; E)$ is identically zero on the complement of U_ζ outside a set of capacity zero. The Laplacian of $G(z, \zeta; E)$ on $\mathbb{P}^1_{\text{Berk}}$ is equal to $\delta_\zeta - \mu_{E,\zeta}$. As in the classical case, the Green's function is symmetric as a function of z and ζ: we have

$$G(z_1, z_2; E) = G(z_2, z_1; E)$$

for all $z_1, z_2 \notin E$. In a satisfying improvement over the theory for $\mathbb{P}^1(\mathbb{C}_p)$ in [88], the role of $G(z, \zeta; E)$ as a reproducing kernel for the Berkovich space Laplacian becomes evident.

As an arithmetic application of the theory of Green's functions and capacities on $\mathbb{P}^1_{\text{Berk}}$, we prove a Berkovich space generalization of Bilu's equidistribution theorem for a rather general class of adelic heights.

Subharmonic functions. We give two characterizations of what it means for a function on a domain $U \subseteq \mathbb{P}^1_{\text{Berk}}$ to be subharmonic. The first, which we take as the definition, is as follows. We say that a function $f : U \to \mathbb{R} \cup \{-\infty\}$ is *strongly subharmonic* if it is upper semicontinuous, satisfies a further technical semicontinuity hypothesis at points of $\mathbb{P}^1(K)$, and if the positive part of $\Delta_{\overline{U}}(f)$ is supported on ∂U. We say that f is *subharmonic* on U if every point of U has a connected open neighborhood on which f is strongly subharmonic. We also say that f is *superharmonic* on U if $-f$ is subharmonic on U. As an example, if ν is a probability measure on $\mathbb{P}^1_{\text{Berk}}$ and $\zeta \notin \text{supp}(\nu)$, the potential function $u_\nu(x, \zeta)$ is strongly superharmonic on $\mathbb{P}^1_{\text{Berk}} \backslash \{\zeta\}$ and is strongly subharmonic on $\mathbb{P}^1_{\text{Berk}} \backslash \text{supp}(\nu)$. A function f is harmonic on U if and only if it is both subharmonic and superharmonic on U.

As a second characterization of subharmonic functions, we say that $f : U \to \mathbb{R} \cup \{-\infty\}$ (not identically $-\infty$) is *domination subharmonic* on the domain U if it is upper semicontinuous and if for each simple subdomain V of U and each harmonic function h on V for which $f \leq h$ on ∂V, we have $f \leq h$ on V. A fundamental fact, proved in §8.2, is that f is subharmonic on U if and only if it is domination subharmonic on U.

Like harmonic functions, subharmonic functions satisfy the Maximum Principle: if U is a domain in $\mathbb{P}^1_{\text{Berk}}$ and f is a subharmonic function which attains its maximum value on U, then f is constant. In addition, subharmonic functions on domains in $\mathbb{P}^1_{\text{Berk}}$ are stable under many of the same operations (e.g., convex combinations, maximum, monotone convergence, uniform convergence) as their classical counterparts. There is also an analogue of the Riesz Decomposition Theorem, according to which a subharmonic function on a simple subdomain $V \subset U$ can be written as the difference of a harmonic function and a potential function. We also show that subharmonic functions can be well-approximated by continuous functions of a special form, which we call *smooth* functions.

In §8.10, we define the notion of an Arakelov-Green's function on $\mathbb{P}^1_{\text{Berk}}$ and establish an *energy minimization principle* used in the proof of the main result in [**9**]. We give two proofs of the Energy Minimization Principle, one using the theory of subharmonic functions and another using the Dirichlet pairing.

Multiplicities. If $\varphi \in K(T)$ is a nonconstant rational function, then as discussed above, the action of φ on $\mathbb{P}^1(K)$ extends naturally to an action of φ on $\mathbb{P}^1_{\text{Berk}}$. We use the theory of Laplacians to give an analytic construction of *multiplicities* for points in $\mathbb{P}^1_{\text{Berk}}$ which generalize the usual multiplicity of φ at a point $a \in \mathbb{P}^1(K)$ (i.e., the multiplicity of a as a preimage of $b = \varphi(a) \in \mathbb{P}^1(K)$). Using the theory of multiplicities, we show that the extended map $\varphi : \mathbb{P}^1_{\text{Berk}} \to \mathbb{P}^1_{\text{Berk}}$ is a surjective open mapping. We also obtain a purely topological interpretation of multiplicities, which shows that our multiplicities coincide with those defined by Rivera-Letelier. For each $a \in \mathbb{P}^1_{\text{Berk}}$, the multiplicity of φ at a is a positive integer, and if $\text{char}(K) = 0$, it is equal to 1 if and only if φ is locally injective at a. For each $b \in \mathbb{P}^1_{\text{Berk}}$, the sum of the multiplicities of φ over all preimages of b is equal to the degree of φ.

Using these multiplicities, we define the pushforward and pullback of a bounded Borel measure on $\mathbb{P}^1_{\text{Berk}}$ under φ. The pushforward and pullback measures satisfy the expected functoriality properties; for example, if f is subharmonic on U, then $f \circ \varphi$ is subharmonic on $\varphi^{-1}(U)$ and the Laplacian of $f \circ \varphi$ is the pullback under φ of the Laplacian of f.

Applications to the dynamics of rational maps. Though Berkovich introduced his theory of analytic spaces with rather different goals in mind, Berkovich spaces are well adapted to the study of non-Archimedean dynamics. The fact that the topological space $\mathbb{P}^1_{\text{Berk}}$ is both compact and connected means in practice that many of the difficulties encountered in "classical" non-Archimedean dynamics disappear when one defines the Fatou and Julia sets as subsets of $\mathbb{P}^1_{\text{Berk}}$. For example, the notion of a connected component is straightforward in the Berkovich setting, so one avoids the subtle issues involved in defining Fatou components in $\mathbb{P}^1(\mathbb{C}_p)$ (e.g., the D-components versus analytic components in Benedetto's paper [**14**], or the definition by Rivera-Letelier in [**83**]).

Suppose $\varphi \in K(T)$ is a rational function of degree $d \geq 2$. In §10.1, we construct a *canonical probability measure* μ_φ on $\mathbb{P}^1_{\text{Berk}}$ attached to φ, whose properties are analogous to the well-known measure on $\mathbb{P}^1(\mathbb{C})$ first defined by Lyubich and by Freire, Lopes, and Mañé. The measure μ_φ is φ-invariant (i.e., satisfies $\varphi_*(\mu_\varphi) = \mu_\varphi$) and also satisfies the functional equation $\varphi^*(\mu_\varphi) = d \cdot \mu_\varphi$.

In §10.2, we prove an explicit formula and functional equation for the Arakelov-Green's function $g_{\mu_\varphi}(x,y)$ associated to μ_φ. These results, along with the Energy Minimization Principle mentioned earlier, play a key role in applications of the theory to arithmetic dynamics over global fields (see [**5**] and [**9**]). In §10.3, we use these results to prove an adelic equidistribution theorem (Theorem 10.24) for the Galois conjugates of algebraic points of small dynamical height over a number field k.

We then discuss analogues for $\mathbb{P}^1_{\text{Berk}}$ of classical results in the Fatou-Julia theory of iteration of rational maps on $\mathbb{P}^1(\mathbb{C})$. In particular, we define the Berkovich Fatou and Julia sets of φ and prove that the Berkovich Julia set J_φ (like its complex counterpart, but unlike its counterpart in $\mathbb{P}^1(K)$) is always nonempty. We give a new proof of the Favre–Rivera-Letelier equidistribution theorem for iterated pullbacks of Dirac measures attached to nonexceptional points, and using this theorem, we show that the Berkovich Julia set shares many properties with its classical complex counterpart. For example, it is either connected or has uncountably many connected components, repelling periodic points are dense in it, and the "Transitivity Theorem" holds.

In $\mathbb{P}^1(K)$, the notion of equicontinuity leads to a good definition of the Fatou set. In $\mathbb{P}^1_{\text{Berk}}$, as was pointed out to us by Rivera-Letelier, this remains true when $K = \mathbb{C}_p$ but fails for general K. We explain the subtleties regarding equicontinuity in the Berkovich case and give Rivera-Letelier's proof that over $\mathbb{C}_p$ the Berkovich equicontinuity locus coincides with the Berkovich Fatou set. We also give an overview (mostly without proof) of some of Rivera-Letelier's fundamental results concerning rational dynamics over $\mathbb{C}_p$. While some of Rivera-Letelier's results hold for arbitrary K, others make special use of the fact that the residue field of $\mathbb{C}_p$ is a union of finite fields.

Appendices. In Appendix A, we review some facts from real analysis and point-set topology which are used throughout the text. Some of these (e.g., the Riesz Representation Theorem) are well known, while others (e.g., the Portmanteau theorem) are hard to find precise references for. We have provided self-contained proofs for the latter. We also include a detailed discussion of nets in topological spaces: since the space $\mathbb{P}^1_{\text{Berk},K}$ is not in general metrizable, sequences do not suffice when discussing notions such as continuity.

In Appendix B, we discuss $\mathbb{R}$-trees and their relation to Gromov's theory of hyperbolic spaces. This appendix serves two main purposes. On the one hand, it provides references for some basic definitions and facts about $\mathbb{R}$-trees which are used in the text. On the other hand, it provides some intuition for the general theory of $\mathbb{R}$-trees by exploring the fundamental role played by the Gromov product, which is closely related to our generalized Hsia kernel.

Appendix C gives a brief overview of some basic definitions and results from Berkovich's theory of non-Archimedean analytic spaces. This material is included in order to give the reader some perspective on the relationship between the special cases dealt with in this book (the Berkovich unit disc, affine line, and projective line) and the general setting of Berkovich's theory.

Notation

We set the following notation, which will be used throughout unless otherwise specified. Symbols are listed roughly in the order they are introduced in the book, except that related notations are grouped together.

$\mathbb{Z}$	the ring of integers.
$\mathbb{N}$	the set of natural numbers, $\{n \in \mathbb{Z} : n \geq 0\}$.
$\mathbb{Q}$	the field of rational numbers.
$\overline{\mathbb{Q}}$	a fixed algebraic closure of $\mathbb{Q}$.
$\mathbb{R}$	the field of real numbers.
$\mathbb{C}$	the field of complex numbers.
$\mathbb{Q}_p$	the field of p-adic numbers.
$\mathbb{Z}_p$	the ring of integers of $\mathbb{Q}_p$.
$\mathbb{C}_p$	the completion of a fixed algebraic closure of $\mathbb{Q}_p$ for some prime number p.
$\mathbb{F}_p$	the finite field with p elements.
$\overline{\mathbb{F}}_p$	a fixed algebraic closure of $\mathbb{F}_p$.
K	a complete, algebraically closed non-Archimedean field.
$K^\times$	the set of nonzero elements in K.
$\|\cdot\|$	the non-Archimedean absolute value on K.
$\|x,y\|$	the spherical distance on $\mathbb{P}^1(K)$ associated to $\|\cdot\|$, and also the spherical kernel, its canonical upper semicontinuous extension to $\mathbb{P}^1_{\text{Berk}}$ (see §4.3).
$\|(x,y)\|$	the norm $\max(\|x\|,\|y\|)$ of a point $(x,y) \in K^2$ (see §10.1).
q_v	a fixed real number greater than 1 associated to K, used to normalize $\|\cdot\|$ and $\text{ord}_v(\cdot)$.
$\log_v(t)$	shorthand for $\log_{q_v}(t)$.
$\text{ord}_v(\cdot)$	the normalized valuation $-\log_v(\|\cdot\|)$ associated to $\|\cdot\|$.
$\|K^\times\|$	the value group of K, that is, $\{\|\alpha\| : \alpha \in K^\times\}$.
$\mathcal{O}$	the valuation ring of K.
$\mathfrak{m}$	the maximal ideal of $\mathcal{O}$.
$\tilde{K}$	the residue field $\mathcal{O}/\mathfrak{m}$ of K.
$\tilde{g}(T)$	the reduction, in $\tilde{K}(T)$, of a function $g(T) \in \mathcal{O}(T)$.
$K[T]$	the ring of polynomials with coefficients in K.
$K(T)$	the field of rational functions with coefficients in K.
$K[[T]]$	the ring of formal power series with coefficients in K.
$K\langle T\rangle$	the Tate algebra of formal power series converging on the closed unit disc.

$\mathbb{A}^1$	the affine line over K.
$\mathbb{P}^1$	the projective line over K.
$\mathbb{A}^1_{\text{Berk}}$	the Berkovich affine line over K.
$\mathbb{P}^1_{\text{Berk}}$	the Berkovich projective line over K.
$\mathbb{H}_{\text{Berk}}$	the "hyperbolic space" $\mathbb{P}^1_{\text{Berk}} \backslash \mathbb{P}^1(K)$.
$\mathbb{H}^{\mathbb{Q}}_{\text{Berk}}$	the set of points of type II in $\mathbb{H}_{\text{Berk}}$ (corresponding to rational discs in K).
$\mathbb{H}^{\mathbb{R}}_{\text{Berk}}$	the set of points of type II or III in $\mathbb{H}_{\text{Berk}}$ (corresponding to either rational or irrational discs in K).
$\zeta_{a,r}$	the point of $\mathbb{A}^1_{\text{Berk}}$ corresponding to $D(a,r)$ under Berkovich's classification theorem.
ζ_{Gauss}	the 'Gauss point' $\zeta_{0,1}$, corresponding to $D(0,1)$.
$[\;]_x$	the seminorm associated to a point $x \in \mathbb{P}^1_{\text{Berk}}$.
$[x,y]$	the path (or arc) from x to y.
$x \vee_\zeta y$	the point where the paths $[x,\zeta]$, $[y,\zeta]$ first meet.
$x \vee_\infty y$	the point where the paths $[x,\infty]$, $[y,\infty]$ first meet.
$x \vee y$	shorthand for $x \vee_{\zeta_{\text{Gauss}}} y$.
T_a	the 'projectivized tangent space' at $a \in \mathbb{P}^1_{\text{Berk}}$, the set of equivalence classes of paths emanating from a which share a common initial segment (see §B.6 in Appendix B).
$\vec{v} \in T_a$	a tangent direction at a (see §B.6 in Appendix B).
$\delta(x,y)_\zeta$	the generalized Hsia kernel with respect to ζ (see §4.4).
$\text{diam}_\zeta(x)$	the number $\delta(x,x)_\zeta$.
$\text{diam}_\infty(x)$	the number $\delta(x,x)_\infty$, equal to $\lim_{i \to \infty} r_i$ for any nested sequence of discs $\{D(a_i, r_i)\}$ corresponding to $x \in \mathbb{A}^1_{\text{Berk}}$.
$\text{diam}(x)$	the number $\|x,x\| = \delta(x,x)_{\zeta_{\text{Gauss}}}$.
$\rho(x,y)$	the path distance metric on $\mathbb{H}_{\text{Berk}}$; see §2.7.
$\ell(Z)$	the total path length of a set $Z \subset \mathbb{H}_{\text{Berk}}$.
$j_\zeta(x,y)$	the fundamental potential kernel relative to the point ζ, given by $j_\zeta(x,y) = \rho(\zeta, x \vee_\zeta y)$.
$\overline{X}$	the closure of a set X in $\mathbb{P}^1_{\text{Berk}}$.
X^c	the complement $\mathbb{P}^1_{\text{Berk}} \backslash X$.
∂X	the boundary of a set X.
$X(K)$	the set of K-rational points in X, i.e., $X \cap \mathbb{P}^1(K)$.
$\text{cl}_{\mathbb{H}}(X)$	the closure of a set $X \subset \mathbb{H}_{\text{Berk}}$, in the strong topology.
$\partial_{\mathbb{H}}(X)$	the boundary of a set $X \subset \mathbb{H}_{\text{Berk}}$, in the strong topology.
$D(a,r)$	the closed disc $\{x \in K : |x-a| \leq r\}$ of radius $r \geq 0$ centered at a. If $r \in |K^\times|$, we call the disc *rational*; if $r \notin |K^\times|$, we call it *irrational*.
$D(a,r)^-$	the open disc $\{x \in K : |x-a| < r\}$ of radius r about a.
$\mathcal{D}(a,r)$	the closed Berkovich disc $\{x \in \mathbb{A}_{\text{Berk}} : [T-a]_x \leq r\}$ corresponding to the classical disc $D(a,r)$.
$\mathcal{D}(a,r)^-$	the open Berkovich disc $\{x \in \mathbb{A}_{\text{Berk}} : [T-a]_x < r\}$ corresponding to the classical disc $D(a,r)^-$.

$B(a,r)$	the closed ball $\{x \in \mathbb{P}^1(K) : \|x,a\| \leq r\}$ of radius r about a in $\mathbb{P}^1(K)$, relative to the spherical distance $\|x,y\|$.
$B(a,r)^-$	the open ball $\{x \in \mathbb{P}^1(K) : \|x,a\| < r\}$ of radius r about a in $\mathbb{P}^1(K)$, relative to the spherical distance.
$\mathcal{B}(a,r)_\zeta$	the closed ball $\{z \in \mathbb{P}^1_{\text{Berk}} : \delta(x,y)_\zeta \leq r\}$.
$\mathcal{B}(a,r)_\zeta^-$	the open ball $\{z \in \mathbb{P}^1_{\text{Berk}} : \delta(x,y)_\zeta < r\}$.
$\mathcal{B}(a,r)$	the closed ball $\{x \in \mathbb{P}^1_{\text{Berk}} : \|x,a\| \leq r\} = \mathcal{B}(a,r)_{\zeta_{\text{Gauss}}}$.
$\mathcal{B}(a,r)^-$	the open ball $\{x \in \mathbb{P}^1_{\text{Berk}} : \|x,a\| < r\} = \mathcal{B}(a,r)_{\zeta_{\text{Gauss}}}^-$.
$\mathcal{B}_a(\vec{v})^-$	the component of $\mathbb{P}^1_{\text{Berk}} \setminus \{a\}$ corresponding to $\vec{v} \in T_a$.
$\widehat{\mathcal{B}}(a,\delta)$	the set $\{z \in \mathbb{H}_{\text{Berk}} : \rho(a,z) \leq \delta\}$, a closed ball for the strong topology.
$\widehat{\mathcal{B}}(a,\delta)^-$	the set $\{z \in \mathbb{H}_{\text{Berk}} : \rho(a,z) < \delta\}$, an open ball for the strong topology.
$\widehat{\mathcal{B}}_X(a,\delta)^-$	for $X \subset \mathbb{H}_{\text{Berk}}$, the set $X \cap \widehat{\mathcal{B}}(a,\delta)^-$.
$X(\zeta,\delta)$	the set $\{z \in \mathbb{P}^1_{\text{Berk}} : \text{diam}_\zeta(x) \geq \delta\} = \widehat{\mathcal{B}}(\zeta, -\log_v(\delta))$.
Γ	a finite metrized graph.
$\text{CPA}(\Gamma)$	the space of continuous, piecewise affine functions on Γ.
$\text{Zh}(\Gamma)$	the Zhang space of Γ (see §3.4).
$\text{BDV}(\Gamma)$	the space of functions of 'bounded differential variation' on Γ (see §3.5).
$\langle f,g \rangle_{\Gamma,\text{Dir}}$	the Dirichlet pairing on Γ, for $f,g \in \text{BDV}(\Gamma)$.
$d_{\vec{v}}(f)$	the derivative of f in the tangent direction $\vec{v}$.
f'_+	the one-sided derivative of f in the positive direction along an oriented segment.
$f'_{\zeta,+}$	the derivative $d_{\vec{v}}(f)$ in the direction $\vec{v}$ towards ζ.
$\Delta_\Gamma(f)$	the Laplacian of $f \in \text{BDV}(\Gamma)$.
$\Delta(f)$	in Chapter 3, the Laplacian $\Delta_\Gamma(f)$; elsewhere, $\Delta_{\mathbb{P}^1_{\text{Berk}}}(f)$.
$r_{U,X}$	the retraction map from a domain U to a closed subset X, often written r_X.
$(r_{U,X})_*(\mu)$	the pushforward from U to X of a measure μ, under $r_{U,X}$.
$\mathcal{C}(U)$	the space of continuous functions on U.
$\mathcal{C}(\overline{U})$	the space of continuous functions on the closure $\overline{U}$.
$\text{CPA}(U)$	the space of functions of the form $f \circ r_{U,\Gamma}$, with $f \in \text{CPA}(\Gamma)$ for some finite graph $\Gamma \subset U$.
$\text{BDV}(U)$	the space of functions of 'bounded differential variation' on a domain U (see §5.4).
$\mathcal{C}(\overline{U}) \cap \text{BDV}(U)$	by abuse of notation, the space of functions $f \in \mathcal{C}(\overline{U})$ with $f|_U \in \text{BDV}(U)$ (see Definition 5.13).
$\mathcal{C}_c(U)$	the space of continuous functions vanishing outside a compact subset of U.
$\text{CPA}_c(U)$	the set of functions in $\text{CPA}(U)$ vanishing outside a compact subset of U.
$\text{BDV}_c(U)$	the set of functions in $\text{BDV}(U)$ vanishing outside a compact subset of U.

$\Delta_{\overline{U}}(f)$	the complete Laplacian of $f \in \mathrm{BDV}(U)$ (see §5.4).
$\Delta_U(f)$	the Laplacian $\Delta_{\overline{U}}(f)|_U$ of $f \in \mathrm{BDV}(U)$ (see §5.4).
$\Delta_{\partial U}(f)$	the boundary derivative $\Delta_{\overline{U}}(f)|_{\partial U}$ of $f \in \mathrm{BDV}(U)$.
$\langle f, g \rangle_{U,\mathrm{Dir}}$	the Dirichlet pairing on a domain U.
λ	the one-dimensional Hausdorff measure on $\mathbb{H}_{\mathrm{Berk}}$, which restricts to dx on each segment.
$\mathrm{supp}(\mu)$	the support of a measure μ.
$|\mu|$	the measure $\mu_1 + \mu_2$, if the Jordan decomposition of the measure μ is $\mu_1 - \mu_2$.
$I_\zeta(\mu)$	the 'energy integral' $\iint -\log_v(\delta(x,y)_\zeta)\,d\mu(x)d\mu(y)$ for μ.
$u_\mu(z,\zeta)$	the potential function $\int -\log_v(\delta(x,y)_\zeta)\,d\mu(y)$ for μ.
$V_\zeta(E)$	the Robin constant of a set E, relative to the point ζ.
$\gamma_\zeta(E)$	the logarithmic capacity of a set E, relative to ζ.
μ_E	the equilibrium distribution of a set E.
$u_E(z,\zeta)$	the potential function associated to μ_E.
$G(x,\zeta;E)$	the Green's function of a set E of positive capacity.
$d_\infty(E)_\zeta$	the transfinite diameter of a set E relative to ζ (see §6.4).
$\mathrm{CH}(E)_\zeta$	the Chebyshev constant of E relative to ζ (see §6.4).
$\mathrm{CH}^*(E)_\zeta$	the restricted Chebyshev constant of E relative to ζ.
$\mathrm{CH}^a(E)_\zeta$	the algebraic Chebyshev constant of E relative to ζ.
$\mathbb{E}$	an 'adelic set' $\prod_v E_v$ for a number field k (see §7.8).
$\mathfrak{X}$	a finite, Galois-stable set of points in $\mathbb{P}^1(\overline{k})$.
$\mathcal{P}^n$	the set of n-dimensional probability vectors.
$\Gamma(\mathbb{E},\mathfrak{X})$	the global Green's matrix of $\mathbb{E}$ relative to $\mathfrak{X}$.
$V(\mathbb{E},\mathfrak{X})$	the global Robin constant of $\mathbb{E}$ relative to $\mathfrak{X}$.
$\gamma(\mathbb{E},\mathfrak{X})$	the global capacity of $\mathbb{E}$ relative to $\mathfrak{X}$.
$\mathcal{H}(U)$	the space of harmonic functions on an open set U.
$\mathcal{SH}(U)$	the space of subharmonic functions on an open set U.
f^*	the upper semicontinuous regularization of a function f.
$\mathcal{M}^+(U)$	the space of positive, locally finite Borel measures on an open set U.
$\mathcal{M}_1^+(U)$	the space of Borel probability measures on U.
$g_\mu(x,y)$	the Arakelov-Green's function associated to a probability measure μ.
$\mathcal{AG}[\zeta]$	the space of Arakelov-Green's functions having a singularity at ζ.
$\varphi(T)$	a rational function in $K(T)$.
$\varphi^*(\mu)$	the pullback of a measure μ by the rational function φ.
$\varphi_*(\mu)$	the pushforward of a measure μ by φ.
$\varphi^{(n)}(T)$	the n-fold iterate $\varphi \circ \cdots \circ \varphi$.
$\deg(\varphi)$	the degree of the rational function $\varphi(T) \in K(T)$.
$m_\varphi(a)$	the multiplicity of $\varphi(T)$ at $a \in \mathbb{P}^1_{\mathrm{Berk}}$.
$m_\varphi(a,\vec{v})$	the multiplicity of $\varphi(T)$ at a in the tangent direction $\vec{v}$.
$r_\varphi(a,\vec{v})$	the rate of repulsion of $\varphi(T)$ at a in the direction $\vec{v}$.

$N_\beta(V)$	the number of solutions to $\varphi(z) = \beta$ in V, counting multiplicities.
$N^+_{\zeta,\beta}(V)$	the number $\max(0, N_\zeta(V) - N_\beta(V))$.
$\mathcal{A}_{a,c}$	the open Berkovich annulus with boundary points a, c.
$\mathrm{Mod}(\mathcal{A})$	the modulus of an annulus $\mathcal{A}$.
μ_φ	the 'canonical measure' associated to $\varphi(T)$.
$g_\varphi(x,y)$	another name for the Arakelov-Green's function $g_{\mu_\varphi}(x,y)$ (see §10.2).
$\hat{h}_{\varphi,v,(x)}$	the Call-Silverman local height function associated to $\varphi(T)$ and the point x.
H_F	the homogeneous dynamical height function associated to $F = (F_1, F_2)$, where $F_1, F_2 \in K[X,Y]$ are homogenous polynomials.
$\mathrm{Res}(F)$	the resultant of the homogeneous polynomials F_1, F_2.
$\mathrm{GO}(x)$	the 'grand orbit' of a point $x \in \mathbb{P}^1_{\mathrm{Berk}}$ under $\varphi(T) \in K(T)$.
E_φ	the 'exceptional set' of all points in $\mathbb{P}^1_{\mathrm{Berk}}$ having finite grand orbit under $\varphi(T)$.
F_φ	the Berkovich Fatou set of $\varphi(T)$.
J_φ	the Berkovich Julia set of $\varphi(T)$.
K_φ	the Berkovich filled Julia set of a polynomial $\varphi(T) \in K[T]$.
$\mathbb{H}_p$	the space $\mathbb{H}_{\mathrm{Berk}}$, when $K = \mathbb{C}_p$.
$\mathcal{A}_x(\varphi)$	the immediate basin of attraction of an attracting fixed point x for $\varphi(T) \in \mathbb{C}_p(T)$.
$\mathcal{E}(\varphi)$	the domain of quasi-periodicity of a function $\varphi(T) \in \mathbb{C}_p(T)$.

CHAPTER 1

The Berkovich unit disc

In this chapter, we recall a theorem of Berkovich which states that points of the Berkovich unit disc $\mathcal{D}(0,1)$ over K can be identified with equivalence classes of nested sequences of closed discs $\{D(a_i, r_i)\}_{i=1,2,\ldots}$ contained in the closed unit disc $D(0,1)$ of K. This leads to an explicit description of the Berkovich unit disc as an "infinitely branched tree"; more precisely, we show that $\mathcal{D}(0,1)$ is an inverse limit of finite $\mathbb{R}$-trees.

1.1. Definition of $\mathcal{D}(0,1)$

Let $\mathcal{A} = K\langle T \rangle$ be the ring of all formal power series with coefficients in K, converging on $D(0,1)$. That is, $\mathcal{A}$ is the ring of all power series $f(T) = \sum_{i=0}^{\infty} a_i T^i \in K[[T]]$ such that $\lim_{i \to \infty} |a_i| = 0$. Equipped with the *Gauss norm* $\|\ \|$ defined by $\|f\| = \max_i(|a_i|)$, $\mathcal{A}$ becomes a Banach algebra over K.

A *multiplicative seminorm* on $\mathcal{A}$ is a function $[\ ]_x : \mathcal{A} \to \mathbb{R}_{\geq 0}$ such that $[0]_x = 0$, $[1]_x = 1$, $[f \cdot g]_x = [f]_x \cdot [g]_x$, and $[f+g]_x \leq [f]_x + [g]_x$ for all $f, g \in \mathcal{A}$. It is a *norm* provided that $[f]_x = 0$ if and only if $f = 0$.

A multiplicative seminorm $[\ ]_x$ is called *bounded* if there is a constant C_x such that $[f]_x \leq C_x \|f\|$ for all $f \in \mathcal{A}$. It is well known (see [**49**, Proposition 5.2]) that boundedness is equivalent to continuity relative to the Banach norm topology on $\mathcal{A}$. The reason for writing the x in $[\ ]_x$ is that we will be considering the space of all bounded multiplicative seminorms on $\mathcal{A}$, and we will identify the seminorm $[\ ]_x$ with a point x in this space.

It can be deduced from the definition that a bounded multiplicative seminorm $[\ ]_x$ on $\mathcal{A}$ behaves just like a non-Archimedean absolute value, except that its kernel may be nontrivial. For example, $[\ ]_x$ satisfies the following properties:

LEMMA 1.1. *Let $[\ ]_x$ be a bounded multiplicative seminorm on $\mathcal{A}$. Then for all $f, g \in \mathcal{A}$,*

(A) $[f]_x \leq \|f\|$.
(B) $[c]_x = |c|$ *for all* $c \in K$.
(C) $[f+g]_x \leq \max([f]_x, [g]_x)$, *with equality if* $[f]_x \neq [g]_x$.

PROOF. (A) For each n, $([f]_x)^n = [f^n]_x \leq C_x \|f^n\| = C_x \|f\|^n$, so $[f]_x \leq C_x^{1/n} \|f\|$, and letting $n \to \infty$ gives the desired inequality.

(B) By the definition of the Gauss norm, $\|c\| = |c|$. If $c = 0$, then trivially $[c]_x = 0$; otherwise, $[c]_x \leq \|c\| = |c|$ and $[c^{-1}]_x \leq \|c^{-1}\| = |c^{-1}|$, while multiplicativity gives $[c]_x \cdot [c^{-1}]_x = [c \cdot c^{-1}]_x = 1$. Combining these gives $[c]_x = |c|$.

(C) The binomial theorem shows that for each n,

$$\begin{aligned}([f+g]_x)^n &= [(f+g)^n]_x = [\sum_{k=0}^{n} \binom{n}{k} f^k g^{n-k}]_x \\ &\leq \sum_{k=0}^{n} |\binom{n}{k}| \cdot [f]_x^k [g]_x^{n-k} \leq \sum_{k=0}^{n} [f]_x^k [g]_x^{n-k} \\ &\leq (n+1) \cdot \max([f]_x, [g]_x)^n .\end{aligned}$$

Taking n^{th} roots and passing to a limit gives the desired inequality. If in addition $[f]_x < [g]_x$, then $[g]_x \leq \max([f+g]_x, [-f]_x)$ gives $[f+g]_x = [g]_x$. □

As a set, the Berkovich unit disc $\mathcal{D}(0,1)$ is defined to be the functional analytic *spectrum* of $\mathcal{A}$, i.e., the set of all bounded multiplicative seminorms $[\ ]_x$ on $K\langle T\rangle$. The set $\mathcal{D}(0,1)$ is clearly nonempty, since it contains the Gauss norm. By abuse of notation, we will often denote the seminorm $[\ ]_x \in \mathcal{D}(0,1)$ by just x. The topology on $\mathcal{D}(0,1)$ is taken to be the *Gelfand topology* (which we will usually refer to as the *Berkovich topology*): it is the weakest topology such that for all $f \in \mathcal{A}$ and all $\alpha \in \mathbb{R}$, the sets

$$\begin{aligned}U(f,\alpha) &= \{x \in \mathcal{D}(0,1) : [f]_x < \alpha\} , \\ V(f,\alpha) &= \{x \in \mathcal{D}(0,1) : [f]_x > \alpha\}\end{aligned}$$

are open. This topology makes $\mathcal{D}(0,1)$ into a compact Hausdorff space: see Theorem C.3. The space $\mathcal{D}(0,1)$ is connected, and in fact path-connected; this will emerge as a simple consequence of our description of it in §1.4 as a profinite $\mathbb{R}$-tree. See also [16, Corollary 3.2.3] for a generalization to higher dimensions.

1.2. Berkovich's classification of points in $\mathcal{D}(0,1)$

A useful observation is that each $x \in \mathcal{D}(0,1)$ is determined by its values on the linear polynomials $T - a$ for $a \in D(0,1)$. Indeed, fix $x \in \mathcal{D}(0,1)$. By the Weierstrass Preparation Theorem [26, Theorem 5.2.2/1], each $f \in K\langle T\rangle$ can be uniquely written as

$$f = c \cdot \prod_{j=1}^{m}(T - a_j) \cdot u(T) ,$$

where $c \in K$, $a_j \in D(0,1)$ for each j, and $u(T)$ is a unit power series, that is, $u(T) = 1 + \sum_{i=1}^{\infty} a_i T^i \in K\langle T\rangle$ with $|a_i| < 1$ for all $i \geq 1$ and $\lim_{i\to\infty} |a_i| = 0$. It is easy to see that $[u]_x = 1$. Indeed, $u(T)$ has a multiplicative inverse $u^{-1}(T)$ of the same form, and by the definition of the Gauss norm $\|u\| =$

$\|u^{-1}\| = 1$. Since $[u]_x \leq \|u\| = 1$, $[u^{-1}]_x \leq \|u^{-1}\| \leq 1$, and $[u]_x \cdot [u^{-1}]_x = [u \cdot u^{-1}]_x = 1$, we must have $[u]_x = 1$. It follows that

$$[f]_x = |c| \cdot \prod_{j=1}^{m} [T - a_j]_x .$$

Thus x is determined by the values of $[\ ]_x$ on the linear polynomials $T - a_j$.

Before proceeding further, we give some examples of elements of $\mathcal{D}(0,1)$. For each $a \in D(0,1)$ we have the *evaluation seminorm*

$$[f]_a = |f(a)| .$$

The boundedness of $[f]_a$ follows easily from the ultrametric inequality, and it is obvious that $[\ ]_a$ is multiplicative.

Also, for each subdisc $D(a,r) \subseteq D(0,1)$, we have the *supremum norm*

$$[f]_{D(a,r)} = \sup_{z \in D(a,r)} |f(z)| .$$

One of the miracles of the non-Archimedean universe is that this norm is *multiplicative*. This is a consequence of the Maximum Modulus Principle in non-Archimedean analysis (see [**26**, Propositions 5.1.4/2 and 5.1.4/3]), which tells us that if $f(T) = \sum_{i=0}^{\infty} a_i (T-a)^i \in K[[T]]$ converges on $D(a,r)$ (i.e., if $\lim |a_i| r^i = 0$), then

(1.1) $$[f]_{D(a,r)} = \sup |a_i| r^i .$$

The norm $[f] = \sup |a_i| r^i$ is easily verified to be multiplicative: just multiply out the corresponding power series and use the ultrametric inequality. More generally, for any decreasing sequence of discs $x = \{D(a_i, r_i)\}_{i \geq 1}$, one can consider the limit seminorm

$$[f]_x = \lim_{i \to \infty} [f]_{D(a_i, r_i)} .$$

Berkovich's classification theorem asserts that every point $x \in \mathcal{D}(0,1)$ arises in this way:

THEOREM 1.2 (Berkovich, [**16**], p.18). *Every* $x \in \mathcal{D}(0,1)$ *can be realized as*

(1.2) $$[f]_x = \lim_{i \to \infty} [f]_{D(a_i, r_i)}$$

for some sequence of nested discs $D(a_1, r_1) \supseteq D(a_2, r_2) \supseteq \cdots$. *If this sequence has a nonempty intersection, then either*

(A) *the intersection is a single point* a, *in which case* $[f]_x = |f(a)|$, *or*
(B) *the intersection is a closed disc* $D(a,r)$ *(where* r *may or may not belong to the value group of* K*), in which case* $[f]_x = [f]_{D(a,r)}$.

PROOF. Fix $x \in \mathcal{D}(0,1)$, and consider the family of (possibly degenerate) discs

$$\mathcal{F} = \{D(a, [T-a]_x) : a \in D(0,1)\} .$$

We claim that the family $\mathcal{F}$ is totally ordered by containment. Indeed, if $a, b \in D(0,1)$ and $[T - a]_x \geq [T - b]_x$, then

$$\begin{aligned}
|a - b| &= [a - b]_x = [(T - b) - (T - a)]_x \\
&\leq \max([T - a]_x, [T - b]_x) = [T - a]_x ,
\end{aligned} \tag{1.3}$$

with equality if $[T - a]_x > [T - b]_x$. In particular $b \in D(a, [T - a]_x)$, and

$$D(b, [T - b]_x) \subseteq D(a, [T - a]_x) .$$

Put $r = \inf_{a \in D(0,1)} [T - a]_x$, and choose a sequence of points $a_i \in D(0,1)$ such that the numbers $r_i = [T - a_i]_x$ satisfy $\lim_{i \to \infty} r_i = r$.

We claim that for each polynomial $T - a$ with $a \in D(0,1)$, we have

$$[T - a]_x = \lim_{i \to \infty} [T - a]_{D(a_i, r_i)} . \tag{1.4}$$

Fix $a \in D(0, 1)$. By the definition of r, we have $[T - a]_x \geq r$.

If $[T - a]_x = r$, then for each a_i we have $r_i = [T - a_i]_x \geq |a_i - a|$ (by (1.3) applied to $a = a_i$ and $b = a$), so $a \in D(a_i, r_i)$. Hence

$$[T - a]_{D(a_i, r_i)} := \sup_{z \in D(a_i, r_i)} |z - a| = r_i .$$

Since $\lim_{i \to \infty} r_i = r$, (1.4) holds in this case. If $[T - a]_x > r$, then for each a_i with $[T - a]_x > [T - a_i]_x$ (which holds for all but finitely many a_i), we have $[T - a]_x = |a - a_i|$ by the strict case in (1.3), which means that $|a - a_i| > [T - a_i]_x = r_i$. Hence

$$[T - a]_{D(a_i, r_i)} := \sup_{z \in D(a_i, r_i)} |z - a| = |a - a_i| = [T - a]_x .$$

Thus the limit on the right side of (1.4) stabilizes at $[T - a]_x$, and (1.4) holds in this case as well. As noted previously, $[\ ]_x$ is determined by its values on the polynomials $T - a$, so for all $f \in K\langle T \rangle$,

$$[f]_x = \lim_{i \to \infty} [f]_{D(a_i, r_i)} . \tag{1.5}$$

Now suppose the family $\mathcal{F}$ has nonempty intersection, and let a be a point in that intersection. Formula (1.4) gives

$$[T - a]_x = \lim_{i \to \infty} [T - a]_{D(a_i, r_i)} \leq \lim_{i \to \infty} r_i = r ,$$

while the definition of r shows that $[T - a]_x \geq r$. Thus $[T - a]_x = r$. Hence the disc $D(a, r)$ (which may consist of a single point if $r = 0$) is a minimal element of $\mathcal{F}$. The arguments above show that formula (1.5) holds for *any* sequence of discs $D(a_i, r_i)$ such that $r_i = [T - a_i]_x$ satisfies $\lim r_i = r$. If we take $a_i = a$ for each i, then $r_i = [T - a]_x = r$, and (1.5) gives $[f]_x = [f]_{D(a,r)}$. If $r = 0$, it gives $[f]_x = |f(a)|$. $\square$

In the important case when $K = \mathbb{C}_p$ for some prime p, there *do* exist nested sequences of closed discs $\{D(a_i, r_i)\}$ with empty intersection. Such sequences necessarily satisfy $r = \lim r_i > 0$, since if $r = 0$, the completeness of K shows that the intersection is a point $a \in K$. To construct an example of such a sequence, fix $0 < r < 1$, and choose a sequence $\{r_i\}$ which decreases monotonically to r, with $r < r_i \leq 1$ for each i. The algebraic closure $\overline{\mathbb{Q}}$ of $\mathbb{Q}$ is countable and is dense in $\mathbb{C}_p$. Enumerate the elements of $\overline{\mathbb{Q}} \cap D(0,1)$ as $\{\alpha_j\}_{j \geq 1}$. Define a sequence of discs $D(a_i, r_i)$ as follows. Take $D(a_1, r_1) = D(\alpha_1, r_1)$. Suppose $D(a_i, r_i)$ has been constructed. Let j_i be the least index of an element with $\alpha_j \in D(a_i, r_i)$. Since $r_i > r_{i+1}$, $D(a_i, r_i) \backslash D(\alpha_{j_i}, r_{i+1})$ is nonempty; let a_{i+1} be any element of it. Then $\alpha_{j_i} \notin D(a_{i+1}, r_{i+1})$.

Clearly the sequence j_i increases to ∞; indeed one sees inductively that $j_i \geq i$. For each i, the construction has arranged that the α_j with $j \leq j_i$ do not belong to $D(a_{i+1}, r_{i+1})$. It follows that $\bigcap_{i=1}^{\infty} D(a_i, r_i)$ contains no elements of $\overline{\mathbb{Q}}$. If it were nonempty, it would be a disc $D(a, r)$. However, every such disc contains elements of $\overline{\mathbb{Q}}$. Hence $\bigcap_{i=1}^{\infty} D(a_i, r_i)$ must be empty.

This brings us to Berkovich's classification of elements of $\mathcal{D}(0,1)$:

Type I: Points corresponding to nested sequences $\{D(a_i, r_i)\}$ with $\lim r_i = 0$ are said to be of *type I*. As noted above, the completeness of K assures that the intersection of such a sequence is a point $a \in K$, and the corresponding seminorm is $[\ ]_a$. We will also call these points the "classical points".

Type II: Points corresponding to nested sequences $\{D(a_i, r_i)\}$ with nonempty intersection, for which $r = \lim r_i > 0$ belongs to the value group $|K^\times|$ of K, are said to be of *type II*; these correspond to a supremum norm $[\ ]_{D(a,r)}$. The corresponding discs $D(a, r)$ are called 'rational'.

Type III: Points corresponding to nested sequences $\{D(a_i, r_i)\}$ with nonempty intersection, but for which $r = \lim r_i > 0$ does not belong to the value group of K, are said to be of *type III*; the corresponding discs are called 'irrational'. These also correspond to a supremum norm $[\ ]_{D(a,r)}$; however the disc $D(a,r)$ is not an "affinoid domain" in the sense of classical rigid analysis. (Alternately, in Berkovich's terminology, $D(a,r)$ is not a *strict affinoid*.)

Type IV: Points corresponding to nested sequences $\{D(a_i, r_i)\}$ with empty intersection are said to be of *type IV*. As noted before, necessarily $\lim r_i > 0$.

A non-Archimedean field K is called *spherically complete* if every nested sequence of closed discs in K has nonempty intersection. Thus $\mathbb{P}^1_{\text{Berk},K}$ contains no points of type IV if and only if K is spherically complete. Finite extensions of $\mathbb{Q}_p$ are spherically complete, while as we have just seen, $\mathbb{C}_p$ is not spherically complete. It is easy to see that every spherically complete field is complete, and it is known (see [43, Chapter 7]) that every non-Archimedean field L is contained in an algebraically closed and spherically complete field whose absolute value extends the one on L.

We let $\zeta_{a,r}$ denote the point of $\mathcal{D}(0,1)$ of type II or III corresponding to the closed rational or irrational disc $D(a,r)$. Allowing degenerate discs (i.e., $r = 0$), we can extend this notation to points of type I: $\zeta_{a,0}$ will denote the point of $\mathcal{D}(0,1)$ corresponding to $a \in D(0,1)$.

Using the terminology of Chambert-Loir [35], we will call the distinguished point $\zeta_{0,1}$ in $\mathcal{D}(0,1)$, corresponding to the Gauss norm $\|f\| = [f]_{D(0,1)}$, the *Gauss point*. We will usually write ζ_{Gauss} instead of $\zeta_{0,1}$.

We will call two nested sequences of closed discs *equivalent* if they define the same point in $\mathcal{D}(0,1)$.

LEMMA 1.3. *Two nested sequences of closed discs* $\{D(a_i, r_i)\}, \{D(a'_j, r'_j)\}$ *are equivalent if and only if*

(A) *each has nonempty intersection, and their intersections are the same, or*

(B) *both have empty intersection, and each sequence is cofinal in the other. (To say that* $\{D(a'_j, r'_j)\}$ *is cofinal in* $\{D(a_i, r_i)\}$ *means that for each i, there is a j such that* $D(a'_j, r'_j) \subseteq D(a_i, r_i)$.)

PROOF. Since the limit (1.2) is a decreasing one, it is clear that two sequences $\{D(a_i, r_i)\}$ and $\{D(a'_j, r'_j)\}$ are equivalent if either condition (A) or (B) is satisfied.

These conditions are also necessary. Indeed, from Theorem 1.2 it is clear that two sequences with nonempty intersection are equivalent if and only if they have the same intersection. A sequence x with nonempty intersection $D(a, r)$ (possibly $r = 0$) cannot be equivalent to any sequence $y = \{D(a_i, r_i)\}$ with empty intersection, since for any i with $a \notin D(a_i, r_i)$ we have

$$[T - a_i]_x = [T - a_i]_{D(a,r)} = |a - a_i| > r_i ,$$

whereas

$$[T - a_i]_y \leq [T - a_i]_{D(a_i, r_i)} = r_i .$$

Finally, two sequences $x = \{D(a_i, r_i)\}$ and $y = \{D(a'_j, r'_j)\}$ with empty intersection which are not mutually cofinal cannot be equivalent. Suppose y is not cofinal in x. After removing some initial terms of x, we can assume that $D(a'_j, r'_j) \not\subseteq D(a_1, r_1)$ for any j. Since $\bigcap_j^\infty D(a'_j, r'_j) = \emptyset$, after deleting some initial terms of y, we can assume that $a_1 \notin D(a'_1, r'_1)$. As any two discs are either disjoint or one contains the other, it must be that $D(a_1, r_1)$ and $D(a'_1, r'_1)$ are disjoint. Since $D(a'_j, r'_j) \subseteq D(a'_1, r'_1)$ for all j, we have $[T - a_1]_{D(a'_j, r'_j)} = |a_1 - a'_1| > r_1$ for each j. Hence

$$[T - a_1]_x \leq [T - a_1]_{D(a_1, r_1)} = r_1 ,$$

while

$$[T - a_1]_y = \lim_{j \to \infty} [T - a_1]_{D(a'_j, r'_j)} = |a_1 - a'_1| > r_1 . \qquad \square$$

REMARK 1.4. More generally, every nested *family* (or *net*) of closed discs in $D(0,1)$ defines a point of $\mathcal{D}(0,1)$, and we call two families of nested discs equivalent if they define the same point in $\mathcal{D}(0,1)$. The proof of Theorem 1.2 shows that every nested family of closed discs is equivalent to a sequence of closed discs, so that equivalence classes of nested families and nested sequences coincide.

1.3. The topology on $\mathcal{D}(0,1)$

In this section we will give a geometric description of a basis for the open sets of $\mathcal{D}(0,1)$. By definition, the Berkovich topology on $\mathcal{D}(0,1)$ is generated by the sets $U(f,\alpha) = \{x \in \mathcal{D}(0,1) : [f]_x < \alpha\}$ and $V(f,\alpha) = \{x \in \mathcal{D}(0,1) : [f]_x > \alpha\}$, for $f \in K\langle T \rangle$ and $\alpha \in \mathbb{R}$. Since the value group of K is dense in $\mathbb{R}_{\geq 0}$, it suffices to consider α belonging to the value group of K. Also, by the Weierstrass Preparation Theorem and the fact that any unit power series $u(T)$ satisfies $[u]_x \equiv 1$, we can restrict to polynomials $f(T) \in K[T]$ with roots in $D(0,1)$.

Given a nonconstant polynomial $f(T) = c\prod_{i=1}^{n}(T-a_i)^{m_i} \in K[T]$ and $\alpha > 0$ belonging to the value group of K, there is a well-known description of the set $\{z \in K : |f(z)| \leq \alpha\}$ as a finite union of closed discs $\bigcup_{i=1}^{N} D(a_i, r_i)$ (cf. [**33**, Theorem 3.1.2, p. 180]). Here the centers can be taken to be roots of $f(T)$, and each r_i belongs to the value group of K. If desired, one can assume that the discs in the decomposition are pairwise disjoint. However, for us it will be more useful to allow redundancy and to assume that all the roots occur as centers, so that $N = n$. If a_i and a_j are roots with $a_j \in D(a_i, r_i)$, we require that $D(a_i, r_i) = D(a_j, r_j)$.

Taking the union over an increasing sequence of α, we can lift the requirement that α belongs to the value group of K. (Note that if α is not in the value group, then $\{z \in K : |f(z)| = \alpha\}$ is empty.) Thus, any $\alpha > 0$ determines a collection of numbers $r_i > 0$, which belong to the value group of K if α does, for which

$$\{z \in K : |f(z)| \leq \alpha\} = \bigcup_{i=1}^{n} D(a_i, r_i) \ .$$

Here we can assume as before that $a_1, \ldots, a_n$ are the roots of $f(z)$, and if $a_j \in D(a_i, r_i)$, then $r_j = r_i$. Using this and the factorization of $f(T)$, it is easy to see that

$$\{z \in K : |f(z)| < \alpha\} = \bigcup_{i=1}^{n} D(a_i, r_i)^- \ ,$$

where $D(a_i, r_i)^- = \{z \in K : |z - a_i| < r_i\}$.

For any closed disc $D(b,t) \subset D(a_i, r_i)^-$ with t in the value group of K, one sees readily that

$$\sup_{z \in D(b,t)} |f(z)| < \alpha \ .$$

Likewise, for any $D(b,t) \subset D(a_i, r_i) \setminus (\bigcup_{a_j \in D(a_i, r_i)} D(a_j, r_i)^-)$ one has $|f(z)| \equiv \alpha$ on $D(b,t)$, and for any $D(b,t)$ disjoint from $\bigcup_{i=1}^{n} D(a_i, r_i)$ one has $|f(z)| \equiv \beta$ on $D(b,t)$, for some $\beta > \alpha$.

Suppose $x \in \mathcal{D}(0,1)$ corresponds to a nested sequence of closed discs $\{D(b_j, t_j)\}$. Without loss of generality, we can assume that each t_j belongs to the value group of K. We will say that x is *associated to an open disc* $D(a,r)^-$ if there is some j such that $D(b_j, t_j) \subset D(a,r)^-$. We say that x

is *associated to a closed disc* $D(a, r)$ if there is some j such that $D(b_j, t_j) \subset D(a, r)$, or if $\bigcap_{j=1}^{\infty} D(b_j, t_j) = D(a, r)$. From the assertions in the previous paragraph, it follows that $[f]_x < \alpha$ if and only if x is associated to some $D(a_i, r_i)^-$. Likewise, $[f]_x > \alpha$ if and only if x is not associated to any of the $D(a_i, r_i)$.

LEMMA 1.5. *Let* $x \in \mathcal{D}(0, 1)$. *Then:*

(A) x *is associated to an open disc* $D(a, r)^-$ *if and only if* $[T - a]_x < r$.
(B) x *is associated to a closed disc* $D(a, r)$ *if and only if* $[T - a]_x \leq r$.

PROOF. Assume $x \in \mathcal{D}(0, 1)$ corresponds to a nested sequence $\{D(b_j, t_j)\}$ with each t_j in the value group of K. Here $[T - a]_{D(b_j, t_j)} = \max(t_j, |b_j - a|)$. Thus, if $[T-a]_x < r$, then there is some j for which $\max(t_j, |b_j - a|) < r$, and this implies that $D(b_j, t_j) \subset D(a, r)^-$. Conversely, if $D(b_j, t_j) \subset D(a, r)^-$, then $t_j < r$ and $|b_j - a| < r$, so $[T - a]_x \leq [T - a]_{D(b_j, t_j)} < r$. This proves (A).

For (B), first suppose x is associated to $D(a, r)$. If some $D(b_j, t_j) \subseteq D(a, r)$, then clearly $[T - a]_x \leq [T - a]_{D(b_j, t_j)} \leq r$. By Theorem 1.2, if $\bigcap_{j=1}^{\infty} D(b_j, t_j) = D(a, r)$, then $[T - a]_x = [T - a]_{D(a, r)} = r$. Conversely, suppose $[T - a]_x \leq r$. If $D(b_j, t_j) \subseteq D(a, r)$ for some j, then x is certainly associated to $D(a, r)$. Otherwise, for each j either $D(b_j, t_j)$ is disjoint from $D(a, r)$ or $D(b_j, t_j) \supset D(a, r)$. If some $D(b_{j_0}, t_{j_0})$ is disjoint from $D(a, r)$, then by the ultrametric inequality $|z - a| = |b_{j_0} - a| > r$ for all $z \in D(b_{j_0}, t_{j_0})$. Since the discs $D(b_j, t_j)$ are nested, this gives $[T - a]_{D(b_j, t_j)} = |b_{j_0} - a|$ for all $j \geq j_0$; hence $[T - a]_x = |b_{j_0} - a|$. This contradicts the fact that $[T - a]_x \leq r$, so it must be that each $D(b_j, t_j)$ contains $D(a, r)$. Thus, $\bigcap_{j=1}^{\infty} D(b_j, t_j) = D(a, t)$ for some $t \geq r$. By Theorem 1.2, $[T - a]_x = t$. Since we have assumed that $[T - a]_x \leq r$, this gives $t = r$, so x is associated to $D(a, r)$. $\square$

This leads us to define open and closed "Berkovich discs", as follows. For $a \in D(0, 1)$ and $r > 0$, write

$$\mathcal{D}(a, r)^- = \{x \in \mathcal{D}(0, 1) : [T - a]_x < r\},$$
$$\mathcal{D}(a, r) = \{x \in \mathcal{D}(0, 1) : [T - a]_x \leq r\}.$$

With this notation, our discussion above shows that

$$U(f, \alpha) = \bigcup_{i=1}^{N} \mathcal{D}(a_i, r_i)^-, \quad V(f, \alpha) = \mathcal{D}(0, 1) \setminus \bigcup_{i=1}^{N} \mathcal{D}(a_i, r_i).$$

Taking finite intersections of sets of the form $U(f, \alpha)$ and $V(f, \alpha)$ gives a basis for the open sets in the Berkovich topology. Since the value group $|K^\times|$ is dense in $\mathbb{R}$, it clearly suffices to take $\alpha \in |K^\times|$, in which case $r, r_i \in |K^\times|$ as well. Thus:

PROPOSITION 1.6. *A basis for the open sets of $\mathcal{D}(0,1)$ is given by the sets*

$$(1.6) \quad \mathcal{D}(a,r)^-, \quad \mathcal{D}(a,r)^- \setminus \bigcup_{i=1}^{N} \mathcal{D}(a_i,r_i), \quad \text{and} \quad \mathcal{D}(0,1) \setminus \bigcup_{i=1}^{N} \mathcal{D}(a_i,r_i),$$

where a and the a_i range over $D(0,1)$ and where r, r_i all belong to the value group $|K^\times|$.

The special subsets appearing in Proposition 1.6 are called *connected strict open affinoids* in $\mathcal{D}(0,1)$. If we lift the requirement that $r, r_i \in |K^\times|$, we obtain the more general class of *connected open affinoids* in $\mathcal{D}(0,1)$. We will also refer to these as *simple domains*. Equivalently, a connected open affinoid in $\mathcal{D}(0,1)$ is any finite intersection of open Berkovich discs. An *open affinoid* in $\mathcal{D}(0,1)$ is defined as a finite union of connected open affinoids.

We can define (connected, strict) *closed affinoids* in $\mathcal{D}(0,1)$ similarly, reversing the roles of the open and closed discs in (1.6). The connected strict closed affinoids form a fundamental system of compact neighborhoods for the Berkovich topology on $\mathcal{D}(0,1)$.

Note that an open affinoid V in $\mathcal{D}(0,1)$ is empty if and only if its intersection with $D(0,1)$ is empty. This follows from the fact that if $f \in K\langle T\rangle$ and $[f]_x < \alpha$ (resp. $[f]_x > \alpha$) for some $x \in \mathcal{D}(0,1)$, then $|f(z)| < \alpha$ (resp. $|f(z)| > \alpha$) for some $z \in D(0,1)$ by Berkovich's classification theorem (since x is a limit of sup norms on discs in $D(0,1)$). In particular, we conclude:

LEMMA 1.7. *$D(0,1)$ is dense in $\mathcal{D}(0,1)$.*

Similarly, for any $z \in D(0,1)$ and any $f \in K\langle T\rangle$, if x is a type II point of $\mathcal{D}(0,1)$ corresponding to a sufficiently small disc centered around z, then $|f(z)|$ and $[f]_x$ can be made as close as we please. Combined with Lemma 1.7, we therefore find:

LEMMA 1.8. *The set of type II points of $\mathcal{D}(0,1)$ is dense in $\mathcal{D}(0,1)$.*

1.4. The tree structure on $\mathcal{D}(0,1)$

In this section, we will use Berkovich's classification theorem to show that $\mathcal{D}(0,1)$ is homeomorphic to an $\mathbb{R}$-tree endowed with its weak topology and to an inverse limit of finite $\mathbb{R}$-trees. This "profinite $\mathbb{R}$-tree" structure is fundamental for all of the material that follows.

An $\mathbb{R}$-*tree* is a metric space (T,d) such that for any two points $x, y \in T$, there is a unique arc $[x,y]$ in T joining x to y, and this arc is a geodesic segment. A self-contained discussion of $\mathbb{R}$-trees is given in Appendix B. A point $a \in T$ is called *ordinary* if $T \setminus \{a\}$ has exactly two components; otherwise it is called a *branch point*. An $\mathbb{R}$-tree is called *finite* if it is compact and has finitely many branch points.

First, we define a partial order on $\mathcal{D}(0,1)$ as follows. For $x, y \in \mathcal{D}(0,1)$, define $x \preceq y$ iff $[f]_x \leq [f]_y$ for all $f \in K\langle T\rangle$. Clearly, the Gauss point ζ_{Gauss} is the unique maximal point with respect to this partial order: by Lemma 1.1, we have $x \preceq \zeta_{\text{Gauss}}$ for all $x \in \mathcal{D}(0,1)$.

It is useful to interpret this partial order in terms of Berkovich's classification of points of $\mathcal{D}(0,1)$. We begin with the following lemma (recall that $\zeta_{a,r}$ denotes the point of $\mathcal{D}(0,1)$ of type II or III corresponding to the closed disc $D(a,r)$):

LEMMA 1.9. *We have* $\zeta_{a,r} \preceq \zeta_{a',r'}$ *if and only if* $D(a,r) \subseteq D(a',r')$.

PROOF. Let $x = \zeta_{a,r}$ and $x' = \zeta_{a',r'}$. Recall that for $f \in K\langle T \rangle$, we have $[f]_x = \sup_{z \in D(a,r)} |f(z)|$ and $[f]_{x'} = \sup_{z \in D(a',r')} |f(z)|$. If $D(a,r) \subseteq D(a',r')$, it is therefore clear that $x \preceq x'$.

On the other hand, we see that

$$D(a,r) \subseteq D(a',r') \iff \sup_{z \in D(a,r)} |T - a'| \leq r' \iff [T-a']_x \leq r' .$$

So if $D(a,r) \not\subseteq D(a',r')$, then $[T-a']_x > r' = [T-a']_{x'}$, and thus we do not have $x \preceq x'$. □

Thus the partial order $\preceq$ corresponds to containment of discs on the points of types II and III. We can easily extend this observation to *all* points of $\mathcal{D}(0,1)$. For this, note that every point of $\mathcal{D}(0,1)$, except ζ_{Gauss}, can be represented by a *strictly decreasing* sequence of discs.

LEMMA 1.10. *Let* $x, y \in \mathcal{D}(0,1) \setminus \{\zeta_{\text{Gauss}}\}$ *be arbitrary, and choose* strictly decreasing *sequences of discs* $\{D(a_i, r_i)\}, \{D(a_i', r_i')\}$ *corresponding to* x *and* y, *respectively. Then* $x \preceq y$ *if and only if for each natural number* k, *there exist* $m, n \geq k$ *such that* $D(a_m, r_m) \subseteq D(a_n', r_n')$.

PROOF. Write $D_i = D(a_i, r_i)$ and $D_j' = D(a_j', r_j')$. Recall that for every $f \in K\langle T \rangle$, we have

$$[f]_x = \inf_i [f]_{D_i}$$

and similarly

$$[f]_y = \inf_j [f]_{D_j'} .$$

If for each k there exist $m, n \geq k$ such that $D_m \subseteq D_n'$, then for k sufficiently large and any $\varepsilon > 0$, we have

$$[f]_x \leq [f]_{D_m} \leq [f]_{D_n'} \leq [f]_y + \varepsilon .$$

As f and ε are arbitrary, it follows that $x \preceq y$.

Conversely, suppose that $x \preceq y$, and fix $k \geq 1$. Consider the function $f' = T - a_{k+1}'$. Since the sequence $\{D_j'\}$ is strictly decreasing, we have $[f']_{D_{k+1}'} = r_{k+1}' < r_k'$, so that $[f']_{D_{k+1}'} \leq r_k' - \varepsilon$ for some $\varepsilon > 0$. On the other hand, for m sufficiently large we have

$$[f']_{D_{k+1}'} \geq [f']_y \geq [f']_x \geq [f']_{D_m} - \varepsilon .$$

It follows that $[T - a_{k+1}']_{D_m} \leq r_k'$. Since $D_k' = D(a_k', r_k') = D(a_{k+1}', r_k')$, it follows that $D_m \subseteq D_k'$, and we may take $n = k$. □

It follows easily from Lemma 1.10 that if $x \in D(0,1)$ is a point of type I, then $y \preceq x$ if and only if $y = x$, and $x \preceq y$ if and only if $x \in D(a'_j, r'_j)$ for all j. Similarly, if $x \in \mathcal{D}(0,1)$ is a point of type IV, then $y \preceq x$ if and only if $y = x$. Thus we have:

COROLLARY 1.11. *Under the partial order $\preceq$, the Gauss point ζ_{Gauss} is the unique maximal point of $\mathcal{D}(0,1)$, and the points of type I and type IV are the minimal points.*

Let $(T, \leq)$ be a partially ordered set satisfying the following two axioms:
(P1) T has a unique maximal element ζ, called the *root* of T.
(P2) For each $x \in T$, the set $S_x = \{z \in T : z \geq x\}$ is totally ordered.

We say that T is a *parametrized rooted tree* if there is a function $\alpha : T \to \mathbb{R}_{\geq 0}$ with values in the nonnegative reals such that:
(P3) $\alpha(\zeta) = 0$.
(P4) α is *order-reversing*, in the sense that $x \leq y$ implies $\alpha(x) \geq \alpha(y)$.
(P5) The restriction of α to any full totally ordered subset of T gives a bijection onto a real interval. (A totally ordered subset S of T is called *full* if $x, y \in S$, $z \in T$, and $x \leq z \leq y$ implies $z \in S$.)

Let diam : $\mathcal{D}(0,1) \to \mathbb{R}_{\geq 0}$ be the function sending $x \in \mathcal{D}(0,1)$ to $\inf_{a \in D(0,1)}[T-a]_x$. If x corresponds to a nested sequence of closed discs $\{D(a_i, r_i)\}$ and $r = \lim_{i \to \infty} r_i$, then $\text{diam}(x) = r$. We call $\text{diam}(x)$ the *diameter* (or *radius*) of x. We claim that the partially ordered set $(\mathcal{D}(0,1), \preceq)$, equipped with the function $1 - \text{diam}(x)$ and the root ζ_{Gauss}, is a parametrized rooted tree. Indeed, axioms (P1), (P3), and (P4) are obvious, and axioms (P2) and (P5) follow easily from Lemma 1.10 and the fact that two discs in K are either disjoint or one contains the other.

An $\mathbb{R}$-tree (T, d) is called *rooted* if it has a distinguished point ζ. In §B.5 of Appendix B, it is shown that there is a one-to-one correspondence between parametrized rooted trees and rooted $\mathbb{R}$-trees. Using that correspondence, we see that (letting $x \vee y$ denote the least upper bound of x and y):

LEMMA 1.12. *The metric*
$$d(x, y) = 2\,\text{diam}(x \vee y) - \text{diam}(x) - \text{diam}(y)$$
makes $\mathcal{D}(0,1)$ into an $\mathbb{R}$-tree.

We call $d(x, y)$ the *small metric* on $\mathcal{D}(0,1)$. It is important to note that the Berkovich, or Gelfand, topology on $\mathcal{D}(0,1)$ is *not* the same as the topology induced by $d(x, y)$. In fact, as we now show, the Berkovich topology coincides with the *weak topology* on the $\mathbb{R}$-tree $(\mathcal{D}(0,1), d)$: the topology in which a subbasis for the open sets is given by the connected components of $\mathcal{D}(0,1)\backslash\{x\}$, as x varies over $\mathcal{D}(0,1)$. (See Appendix B for more discussion of the weak topology on an $\mathbb{R}$-tree.)

PROPOSITION 1.13. *$\mathcal{D}(0,1)$ with its Berkovich topology is homeomorphic to the $\mathbb{R}$-tree $(\mathcal{D}(0,1), d)$ with its weak topology.*

Before giving the proof, we make a few preliminary remarks.

Recall that in terms of the partial order on $\mathcal{D}(0,1)$, a closed Berkovich disc $\mathcal{D}(a,r)$ consists of all points $z \in \mathcal{D}(0,1)$ with $z \preceq \zeta_{a,r}$. The open Berkovich disc $\mathcal{D}(a,r)^-$ is the connected component of $\mathcal{D}(0,1)\backslash\zeta_{a,r}$ containing a, and the open Berkovich disc $\mathcal{D}(0,1)\backslash\mathcal{D}(a,r)$ is the connected component of $\mathcal{D}(0,1)\backslash\{\zeta_{a,r}\}$ containing ζ_{Gauss}. The point $\zeta_{a,r}$ is the unique boundary point of $\mathcal{D}(a,r)^-$ (resp. $\mathcal{D}(0,1)\backslash\mathcal{D}(a,r)$), as one sees easily from the fact that the complement of $\mathcal{D}(a,r)^-$ (resp. $\mathcal{D}(0,1)\backslash\mathcal{D}(a,r)$) is a union of open Berkovich discs.

In particular, *tangent vectors* at $\zeta_{a,r}$ in the $\mathbb{R}$-tree $(\mathcal{D}(0,1),d)$ can be identified with the open Berkovich discs having $\zeta_{a,r}$ as a boundary point. For $x \in \mathcal{D}(0,1)$, the space T_x of tangent vectors at x (or, more properly, tangent directions at x) is the set of equivalence classes of paths $[x,y]$ emanating from x, where y is any point of $\mathcal{D}(0,1)$ not equal to x. Here we declare that two paths $[x,y_1], [x,y_2]$ are *equivalent* if they share a common initial segment. (See Appendix B). There is a natural bijection between elements $\vec{v}$ of T_x and connected components of $\mathcal{D}(0,1)\backslash\{x\}$. We denote by $U(x;\vec{v})$ the connected component of $\mathcal{D}(0,1)\backslash\{x\}$ corresponding to $\vec{v} \in T_x$. As x and $\vec{v}$ vary, the open sets $U(x;\vec{v})$ generate the weak topology.

PROOF OF PROPOSITION 1.13. Let X denote $\mathcal{D}(0,1)$ equipped with the weak $\mathbb{R}$-tree topology. The natural map ι (which is the identity map on the underlying sets) from $\mathcal{D}(0,1)$ to X is continuous, since the above discussion shows that the inverse image of a set of the form $U(x;\vec{v})$ is an open Berkovich disc. Conversely every Berkovich open disc is of the form $U(x;\vec{v})$, which by Proposition 1.6 implies that the inverse of ι is also continuous. □

As a consequence of Proposition 1.13 and Corollary B.20, we obtain:

COROLLARY 1.14. *The space $\mathcal{D}(0,1)$ is uniquely path-connected.*

The different branches emanating from the Gauss point in $\mathcal{D}(0,1)$ (i.e., the tangent vectors $\vec{v} \in T_{\zeta_{\text{Gauss}}}$) are naturally in one-to-one correspondence with elements of the residue field $\tilde{K}$ of K, or equivalently with the open discs of radius 1 contained in $D(0,1)$. Each branch splits into infinitely many branches at each point of type II, and each new branch behaves in the same way. This incredible collection of splitting branches forms a sort of "witch's broom". However, the witch's broom has some structure: there is branching *only* at the points of type II, not those of type III; furthermore, the branches emanating from a type II point $\zeta_{a,r} \neq \zeta_{0,1}$ are in one-to-one correspondence with $\mathbb{P}^1(\tilde{K})$: there is one branch going "up" to the Gauss point, and the other branches correspond (in a noncanonical way) to elements of $\tilde{K}$, or equivalently with the open discs of radius r contained in $D(a,r)$. Some of the branches extend all the way to the bottom (terminating in points of type I), while others are "cauterized off" earlier and terminate at points of type IV, but every branch terminates either at a point of type I or type IV.

1.4. THE TREE STRUCTURE ON $\mathcal{D}(0,1)$

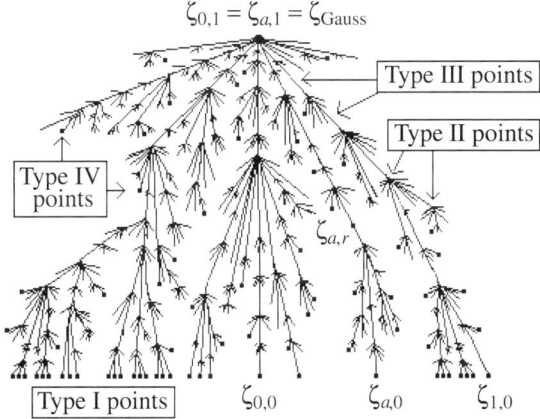

FIGURE 1.1. The Berkovich Unit Disc

There is another metric which is only defined on $\mathcal{D}(0,1)\backslash D(0,1)$, i.e., on the points of types II, III, and IV. It is given by the parametrization function $-\log_v(\text{diam})$ on $\mathcal{D}(0,1)\backslash D(0,1)$. The same verification as before shows that the pair $(\mathcal{D}(0,1)\backslash D(0,1), -\log_v(\text{diam}))$ is a parametrized rooted tree, and the corresponding metric ρ is called the *big metric* on the $\mathbb{R}$-tree $\mathcal{D}(0,1)\backslash D(0,1)$. In particular:

LEMMA 1.15. *The metric*

$$\rho(x,y) = 2\log_v(\text{diam}(x \vee y)) - \log_v(\text{diam}(x)) - \log_v(\text{diam}(y))$$

makes $\mathcal{D}(0,1)\backslash D(0,1)$ into an $\mathbb{R}$-tree.

Since we will often make use of the tree structures on $\mathcal{D}(0,1)$ and $\mathcal{D}(0,1)\backslash D(0,1)$, we will now try to make those structures more concrete.

Let $x \in \mathcal{D}(0,1)$ be a point of type I, II, or III in $\mathcal{D}(0,1)$, corresponding to a (possibly degenerate) disc $D(a,r) \subset D(0,1)$ of radius $r \geq 0$. For each $t \in [r,1]$, define $x(t) = \zeta_{a,t} \in \mathcal{D}(0,1)$. We define the *line of discs* $[x, \zeta_{\text{Gauss}}]$ to be the subset $\{x(t)\} \subset \mathcal{D}(0,1)$ corresponding to the collection of discs $\{D(a,t) : r \leq t \leq 1\}$ in $D(0,1)$.

LEMMA 1.16. *For $x \in \mathcal{D}(0,1)$ of type I, II, or III, the mapping $i_{a,r}(t) = x(t)$ is a continuous embedding of the real interval $[r,1]$ into $\mathcal{D}(0,1)$. In particular, the line of discs $[x, \zeta_{\text{Gauss}}] \subset \mathcal{D}(0,1)$ is homeomorphic to an interval.*

PROOF. Let $\|\ \|_{D(a,t)}$ be the multiplicative seminorm corresponding to $x(t)$. For any $f \in K\langle T \rangle$, it follows from (1.1) that the map from $[r,1]$ to $\mathbb{R}$ defined by $t \mapsto \|f\|_{D(a,t)}$ is continuous; by the definition of the Berkovich topology, this implies the continuity of $i_{a,r}$. Since $i_{a,r}$ is clearly injective, it suffices to prove that the inverse mapping is also continuous. This in turn follows from the fact that $i_{a,r}((\alpha,\beta)) = \{x \in \mathcal{D}(0,1) \mid \alpha < [T-a]_x < \beta\}$ for all open intervals $(\alpha,\beta) \subset [r,1]$. □

One can easily extend Lemma 1.16 to the situation where x is a point of type IV. If x corresponds to the nested sequence $D(a_i, r_i)$ of closed discs, with $\lim r_i = r$, define the line of discs $[x, \zeta_{\text{Gauss}}]$ to be x together with the set of all points $\zeta_{a',r'} \in \mathcal{D}(0,1)$ of type II or III with $D(a_i, r_i) \subseteq D(a', r')$ for some i. Using the fact that $r_i \to r$ and that two discs in $\mathcal{D}(0,1)$ are either disjoint or one contains the other, it follows easily that for each $r' > r$, there is a *unique* point $x(r') \in [x, \zeta_{\text{Gauss}}]$ corresponding to a disc of radius r'. We can thus define a mapping $i_x : [r, 1] \to [x, \zeta_{\text{Gauss}}]$ by sending r to x and r' to $x(r')$ for each $r < r' \leq 1$. Clearly $x(1) = \zeta_{\text{Gauss}}$. An argument similar to the proof of Lemma 1.16 shows:

LEMMA 1.17. *For $x \in \mathcal{D}(0,1)$ of type IV, the mapping $i_x : t \mapsto x(t)$ is a continuous embedding of the real interval $[r, 1]$ into $\mathcal{D}(0,1)$.*

This gives another way of seeing that the space $\mathcal{D}(0,1)$ is path-connected.

There are (at least) two ways to define a metric on the line of discs $[x, \zeta_{\text{Gauss}}]$. In the *small metric*, we declare that $[x, \zeta_{\text{Gauss}}]$ has length $1 - r$, and in the *big metric*, we declare that $[x, \zeta_{\text{Gauss}}]$ has length $\log_v(1/r)$. For the small metric the map $t \in [r, 1] \mapsto x(t)$ is an isometry, and for the big metric the map $s \in [0, \log_v(1/r)] \mapsto x(q_v^{-s})$ is an isometry. Even though it looks more cumbersome, the big metric turns out to be much more natural and useful than the small metric. (This will become clear in the next chapter.)

Now let $S = \{D(a_1, r_1), \ldots, D(a_n, r_n)\}$ be any finite set of (rational or irrational) discs of positive radius contained in $D(0,1)$. For simplicity, we assume that $D(0,1) \notin S$. To each disc $D(a_i, r_i)$, there is an associated point $\zeta_{a_i, r_i} \in \mathcal{D}(0,1)$, which is a point of type II or III.

Define the *graph of discs* $\Gamma_S \subset \mathcal{D}(0,1)$ to be the union of the associated lines of discs $[\zeta_{a_i, r_i}, \zeta_{\text{Gauss}}]$:

$$\Gamma_S = \bigcup_{i=1}^n [\zeta_{a_i, r_i}, \zeta_{\text{Gauss}}].$$

The subset Γ_S inherits both a small and big metric structure from the corresponding metrics on the lines of discs which comprise it. From Lemmas 1.12 and 1.15, it follows that Γ_S is an $\mathbb{R}$-tree with respect to both of these metrics.

An alternate description of Γ_S, which may make its structure clearer, is as follows. Define the *saturation* of S to be the set $\hat{S}$ obtained by adjoining to S all discs $D(a_i, |a_i - a_j|)$ with $D(a_i, r_i), D(a_j, r_j) \in S$ and also the disc $D(0, 1)$. Note that the radius of each disc in $\hat{S} \backslash S$ belongs to the value group of K. Then Γ_S is the finite $\mathbb{R}$-tree whose branch points correspond to the discs in $\hat{S}$ and which has an isometric path of length $|r_i - r_j|$ (in the small metric) between each pair of nodes $D(a_i, r_i), D(a_j, r_j) \in \hat{S}$ for which $D(a_i, r_i) \supseteq D(a_j, r_j)$ or $D(a_j, r_j) \supseteq D(a_i, r_i)$.

1.4. THE TREE STRUCTURE ON $\mathcal{D}(0,1)$

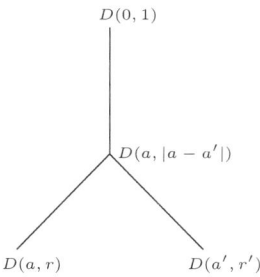

FIGURE 1.2. The tree Γ_S when $S = \{D(a,r), D(a',r')\}$ with $D(a,r)$ and $D(a',r')$ disjoint

It follows from this description that every branch point of Γ_S (not including the endpoints) is a type II point of $\mathcal{D}(0,1)$.

If S_1 and S_2 are any two finite sets of discs, then Γ_{S_1} and Γ_{S_2} are both $\mathbb{R}$-subtrees of $\Gamma_{S_1 \cup S_2}$. Moreover, the embedding of Γ_{S_i} in $\Gamma_{S_1 \cup S_2}$ is an isometry (with respect to either metric). Thus, the collection of graphs Γ_S is a directed set. Let

$$\Lambda = \varinjlim \Gamma_S = \bigcup_S \Gamma_S$$

be the union (or direct limit) of all of the $\mathbb{R}$-trees Γ_S; this is itself an $\mathbb{R}$-tree (equipped with two possible metrics). With respect to the small metric, Λ is a bounded metric space, but with respect to the big metric, Λ is unbounded.

By construction, each disc $D(a,r) \subset D(0,1)$ of positive radius corresponds to a unique point of Λ. In other words, as a set Λ just consists of the points of type II or III in $\mathcal{D}(0,1)$. To incorporate the points of types I and IV, we must enlarge Λ by adding "ends". Consider a strictly decreasing nested sequence of closed discs $x = \{D(a_i, r_i)\}$, and put $r = \lim r_i$. The union of the lines of discs $[\zeta_{a_i, r_i}, \zeta_{\text{Gauss}}]$ is a "half-open" line of discs, which we will write as $(\zeta_{a,r}, \zeta_{\text{Gauss}}]$. If $\bigcap_i D(a_i, r_i)$ is nonempty, then the intersection is either a disc $D(a,r)$ of positive radius, in which case $(\zeta_{a,r}, \zeta_{\text{Gauss}}]$ extends in a natural way to the closed path $[\zeta_{a,r}, \zeta_{\text{Gauss}}]$, or a point $a \in D(0,1)$, in which case the half-open path $(\zeta_{a,r}, \zeta_{\text{Gauss}}]$ extends to a closed path by adding on the corresponding type I point a as an endpoint. Similarly, if $\bigcap_i D(a_i, r_i) = \emptyset$, we must adjoin a new endpoint of type IV in order to close up $(\zeta_{a,r}, \zeta_{\text{Gauss}}]$. It is clear that cofinal sequences define the same half-open path, so the point closing up this path depends only on the corresponding seminorm $[\;]_x$ and not on the sequence defining it.

A cleaner way to say all of this is that points of $\mathcal{D}(0,1)$ are in one-to-one correspondence with points in the completion of Λ with respect to the small metric. Since the points of type I are infinitely far away with respect to the big metric, the completion of Λ with respect to the big metric coincides (on the level of points) with $\mathcal{D}(0,1) \backslash D(0,1)$.

REMARK 1.18. It is important to realize that $\mathcal{D}(0,1)$ is *not* homeomorphic to the completion of Λ with respect to the small metric, with topology given by the small metric. Rather, as shown above in Proposition 1.13, $\mathcal{D}(0,1)$ is the topological space whose underlying point set is the $\mathbb{R}$-tree given by the completion of Λ with respect to the small metric and whose topology is the *weak topology* on that $\mathbb{R}$-tree.

Still another way to understand the topology on $\mathcal{D}(0,1)$ is in terms of inverse limits. Let $\mathcal{F}_0$ be the collection of all finite $\mathbb{R}$-trees of the form Γ_S as above. We have already noted that $\mathcal{F}_0$ is a directed set under inclusion, and we write $\Gamma \leq \Gamma'$ if $\Gamma \subseteq \Gamma'$ as subsets of $\mathcal{D}(0,1)$. Thus whenever $\Gamma \leq \Gamma'$, there is an inclusion map
$$i_{\Gamma,\Gamma'} : \Gamma \to \Gamma' \ .$$
The set Λ is the direct limit of Γ_S over $\mathcal{F}_0$ with respect to the maps $i_{\Gamma,\Gamma'}$.

There is also a *retraction map* $r_{\Gamma',\Gamma} : \Gamma' \to \Gamma$ defined whenever $\Gamma \leq \Gamma'$. This is a general property of $\mathbb{R}$-trees: since there is a unique path between any two points of Γ', if $x \in \Gamma'$, we can define $r_{\Gamma',\Gamma}(x)$ to be, for any $y \in \Gamma$, the first point where the unique path in Γ' from x to y intersects Γ. This definition is independent of the choice of y, and one sees from the definition that $r_{\Gamma',\Gamma}(x) = x$ if and only if $x \in \Gamma$. (In particular, $r_{\Gamma',\Gamma}(x)$ is surjective.) Moreover, the retraction maps are compatible, in the sense that if $\Gamma \leq \Gamma' \leq \Gamma''$, then

(1.7) $$r_{\Gamma'',\Gamma} = r_{\Gamma',\Gamma} \circ r_{\Gamma'',\Gamma'} \ .$$

Note also that for each Γ, Γ', there is a Γ'' with $\Gamma \leq \Gamma''$, $\Gamma' \leq \Gamma''$.

By (1.7), the inverse limit
$$\widehat{\Lambda} = \varprojlim \Gamma \ ,$$
taken over all $\Gamma \in \mathcal{F}_0$, is well-defined. For each Γ, there is a natural retraction $r_{\widehat{\Lambda},\Gamma} : \widehat{\Lambda} \to \Gamma$. We endow $\widehat{\Lambda}$ with the *inverse limit topology*, in which the basic open sets are ones of the form $r_{\widehat{\Lambda},\Gamma}^{-1}(U)$, where $\Gamma \in \mathcal{F}_0$ is arbitrary and $U \subset \Gamma$ is any open set. Here each Γ is given its metric topology (which is the same for both the big and small metrics). It follows from Tychonoff's theorem that $\widehat{\Lambda}$ is a compact Hausdorff topological space. The space $\widehat{\Lambda}$ is an example of a *profinite $\mathbb{R}$-tree*, i.e., an inverse limit of finite $\mathbb{R}$-trees.

PROPOSITION 1.19. *The topological spaces $\mathcal{D}(0,1)$ (with the weak topology) and $\widehat{\Lambda}$ (with the inverse limit topology) are homeomorphic.*

PROOF. Since $\mathcal{D}(0,1)$ can be viewed as the completion of Λ with respect to the small metric (equipped with the weak topology), for each $\Gamma \in \mathcal{F}_0$ there is a retraction map $r_{\mathcal{D}(0,1),\Gamma} : \mathcal{D}(0,1) \to \Gamma$. This map is continuous, since one can easily verify that if $\Gamma \in \mathcal{F}_0$ and if $V \subset \Gamma$ is an open subset of Γ, then $r_{\mathcal{D}(0,1),\Gamma}^{-1}(V)$ is an open affinoid in $\mathcal{D}(0,1)$. By the universal property of inverse limits, there is a continuous map ψ from $\mathcal{D}(0,1)$ to $\widehat{\Lambda}$.

The map ψ is injective because elements of $\mathcal{F}_0$ "separate points": given two distinct points $x, y \in \mathcal{D}(0,1)$, there is a point $\zeta_{a,r}$ of type II such that either $x \preceq \zeta_{a,r}$ but $y \not\preceq \zeta_{a,r}$, or $x \not\preceq \zeta_{a,r}$ but $y \preceq \zeta_{a,r}$, and if $\Gamma = [\zeta_{a,r}, \zeta_{\text{Gauss}}]$, then
$$r_{\mathcal{D}(0,1),\Gamma}(x) \neq r_{\mathcal{D}(0,1),\Gamma}(y) \ .$$

The map ψ is also surjective, since by the definition of the indexing set $\mathcal{F}_0$, each point $y \in \widehat{\Lambda}$ corresponds to a nested family of closed discs: namely, for each $\Gamma \in \mathcal{F}_0$, the point $\zeta_\Gamma := r_{\widehat{\Lambda},\Gamma}(y)$ corresponds to a disc $D(a_\Gamma, r_\Gamma)$, and if $\Gamma_1 \subseteq \Gamma_2$, then $r_{\Gamma_2,\Gamma_1}(\zeta_{\Gamma_2}) = \zeta_{\Gamma_1}$, so $D(a_{\Gamma_2}, r_{\Gamma_2}) \subseteq D(a_{\Gamma_1}, r_{\Gamma_1})$. It is easy to check that if $x \in \mathcal{D}(0,1)$ corresponds to this nested family, then $\psi(x) = y$.

A continuous bijection between compact Hausdorff topological spaces is automatically a homeomorphism, so the desired result now follows. $\square$

In particular, we see that $\mathcal{D}(0,1)$ is homeomorphic to a profinite $\mathbb{R}$-tree, which we think of as rooted at the Gauss point.

1.5. Metrizability

If K has a countable dense subset K^0, then the basis in (1.6) has a countable subbasis, obtained by restricting to discs $\mathcal{D}(a,r)^-$ and $\mathcal{D}(a_i, r_i)$ with centers in $K^0 \cap D(0,1)$ and radii in $|(K^0)^\times| \cap (0,1]$. For instance, this holds when $K = \mathbb{C}_p$, with K^0 being the algebraic closure of $\mathbb{Q}$ in $\mathbb{C}_p$. However, it fails (for example) if either the residue field or the value group of K is uncountable.

COROLLARY 1.20. *If K has a countable dense subset, then $\mathcal{D}(0,1)$ is a metrizable space.*

PROOF. A compact Hausdorff space is "T3", that is, each point is closed, and for each point x and each closed set A with $x \notin A$, there are disjoint open neighborhoods U of x and V of A. Urysohn's metrization theorem [**65**, p. 125] says that any T3 space with a countable dense subset is metrizable. $\square$

As previously remarked, the metrics $d(x,y)$ and $\rho(x,y)$ do *not* define the Berkovich (i.e., Gelfand) topology. For example, the set $\{x \in \mathcal{D}(0,1) : d(x, \zeta_{\text{Gauss}}) < 1/2\}$ does not contain any point of type I, while every neighborhood of ζ_{Gauss} in the Berkovich topology contains infinitely many such points. This same example shows that $d(x,y)$ (resp. $\rho(x,y)$) is not even continuous for the Berkovich topology. (In fact, the topology defined by $d(x,y)$ (resp. $\rho(x,y)$) is strictly finer than the Berkovich topology.)

When K has a countable dense subset K^0, then by tracing through the proof of Urysohn's theorem, one can construct a metric defining the Berkovich topology as follows. First, define a 'separation kernel' for discs $D(a,r)$ and points $a' \in D(0,1)$ by

$$\Delta(D(a,r), a') = \sup_{z \in D(a,r)} (|z - a'|) = \max(r, |a - a'|) \ .$$

Extend it to pairs of discs $D(a,r)$, $D(a',r')$ by
$$\Delta(D(a,r), D(a',r')) = \sup_{\substack{z \in D(a,r) \\ w \in D(a',r')}} (|z-w|) = \max(r, r', |a-a'|)$$
and then to arbitrary points $x, y \in \mathcal{D}(0,1)$ by
$$\Delta(x, y) = \lim_{i \to \infty} \max(r_i, r_i', |a_i - a_i'|)$$
if x, y correspond to sequences of nested discs $\{D(a_i, r_i)\}$, $\{D(a_i', r_i')\}$.

Let $\{\alpha_j\}_{j \in \mathbb{N}}$ be an enumeration of $K^0 \cap D(0,1)$, and define a map φ from $\mathcal{D}(0,1)$ to the infinite-dimensional unit cube $[0,1]^{\mathbb{N}}$ by putting
$$\varphi(x) = (\Delta(x, \alpha_j))_{j \in \mathbb{N}} .$$
It can be checked that φ is a topological isomorphism from $\mathcal{D}(0,1)$ onto its image, equipped with the induced topology. Pulling back the metric on $[0,1]^{\mathbb{N}}$, one obtains a metric defining the Berkovich topology on $\mathcal{D}(0,1)$:
$$\delta(x,y) = \sum_{j=1}^{\infty} \frac{1}{2^j} |\Delta(x, \alpha_j) - \Delta(y, \alpha_j)| .$$
However, this formula seems nearly useless for understanding the Berkovich topology.

1.6. Notes and further references

Much of the material in this chapter is originally due to V. Berkovich and can be found in his book [16], though we have gone into more detail than Berkovich in describing the structure of $\mathcal{D}(0,1)$. It should be noted that Berkovich spaces of higher dimension cannot be described nearly as explicitly (see, however, [20, 21]).

The notational convention $\zeta_{a,r}$ was suggested by Joe Silverman. We have borrowed from Antoine Chambert-Loir the term "Gauss point" for the maximal point $\zeta_{\text{Gauss}} = \zeta_{0,1}$ of $\mathcal{D}(0,1)$.

Our notation $x \vee y$ for the least upper bound of $x, y \in \mathcal{D}(0,1)$ is borrowed from Favre and Rivera-Letelier ([46, 47]), except that they use $\wedge$ instead of $\vee$. Since the Gauss point is *maximal* with respect to the partial order $\preceq$ on $\mathcal{D}(0,1)$, our notation $x \vee y$ is compatible with the typical usage from the theory of partially ordered sets.

CHAPTER 2

The Berkovich projective line

In this chapter, we will define the Berkovich projective line $\mathbb{P}^1_{\text{Berk}}$ and give several constructions of it. Our primary definition is reminiscent of the "Proj" construction in algebraic geometry. Using the "Proj" definition, one sees easily that the action of a rational map φ on $\mathbb{P}^1(K)$ extends naturally to $\mathbb{P}^1_{\text{Berk}}$. Our main structural result is that $\mathbb{P}^1_{\text{Berk}}$, like $\mathcal{D}(0,1)$, is canonically homeomorphic to an inverse limit of finite $\mathbb{R}$-trees.

It should be noted that $\mathbb{P}^1_{\text{Berk}}$ depends on the ground field K, and when it is important to emphasize this, we will write $\mathbb{P}^1_{\text{Berk},K}$. However, usually we regard K as fixed and suppress it from the notation.

It should also be noted that Berkovich spaces, like schemes, are more than just topological spaces: for example, they are naturally endowed with a "sheaf of analytic functions". In our exposition of $\mathbb{P}^1_{\text{Berk}}$, we focus almost exclusively on the topological space, though some indications of the sheaf structure appear in §2.1 and §2.4 below. However, in considering curves of higher genus and spaces of higher dimension, it is essential to consider the sheaf of analytic functions as well. A brief introduction to general Berkovich spaces is given in Appendix C.

After studying $\mathbb{P}^1_{\text{Berk}}$, we introduce an important subset, the "Berkovich hyperbolic space" $\mathbb{H}_{\text{Berk}} = \mathbb{P}^1_{\text{Berk}} \backslash \mathbb{P}^1(K)$, and show that it is equipped with a natural metric $\rho(x, y)$ extending the "big metric" defined in the previous chapter. We prove that ρ is invariant under linear fractional transformations.

2.1. The Berkovich affine line $\mathbb{A}^1_{\text{Berk}}$

Before studying $\mathbb{P}^1_{\text{Berk}}$, we consider the Berkovich affine line $\mathbb{A}^1_{\text{Berk}}$.

As a topological space, $\mathbb{A}^1_{\text{Berk}}$ can be defined as follows. The underlying point set for $\mathbb{A}^1_{\text{Berk}}$ is the collection of all multiplicative seminorms $[\ ]_x$ on the polynomial ring $K[T]$ which extend the absolute value on K. The topology on $\mathbb{A}^1_{\text{Berk}}$ is the weakest one for which $x \mapsto [f]_x$ is continuous for all $f \in K[T]$.

In general, Berkovich spaces are not metrizable, so in studying them one must use *nets* rather than *sequences*. Recall that if X is a topological space, a *net* in X is a mapping $\alpha \mapsto x_\alpha$ from A to X, where A is an arbitrary directed set. A sequence is the same thing as a net indexed by $\mathbb{N}$. If $x \in X$, a net $\langle x_\alpha \rangle$ in X *converges* to x (written $x_\alpha \to x$) if for every neighborhood U of x, there exists $\alpha_0 \in A$ such that $x_\alpha \in U$ for all $\alpha \geq \alpha_0$. For a more detailed discussion of nets, see §A.3 of Appendix A.

It is easy to see that a net $\langle x_\alpha \rangle$ in $\mathbb{A}^1_{\text{Berk}}$ converges to x if and only if $[f]_{x_\alpha} \to [f]_x$ for all $f \in K[T]$, so the topology on $\mathbb{A}^1_{\text{Berk}}$ is the topology of *pointwise convergence* with respect to $K[T]$.

There is a natural continuous embedding of K into $\mathbb{A}^1_{\text{Berk}}$ which associates to a point $z \in K$ the evaluation seminorm $[f]_z = |f(z)|$.

Similarly to Lemma 1.1(B), we have:

LEMMA 2.1. *A multiplicative seminorm $[\ ]_x$ extends the absolute value on K if and only if its restriction to K is bounded (that is, there is a constant B such that $[c]_x \leq B|c|$ for each $c \in K$).*

PROOF. Suppose the restriction of $[\ ]_x$ to K is bounded. Since $[1]_x = |1| = 1$, we must have $B \geq 1$. If $c \in K$, then $([c]_x)^n = [c^n]_x \leq B|c^n| = B|c|^n$ for each $n \in \mathbb{N}$, so taking n^{th} roots and letting $n \to \infty$ gives $[c]_x \leq |c|$. If $c = 0$, this shows that $[c]_x = 0$. On the other hand, if $c \neq 0$, then both $[c]_x \leq |c|$ and $[c^{-1}]_x \leq |c^{-1}|$. Multiplicativity gives $[c]_x \cdot [c^{-1}]_x = [c \cdot c^{-1}]_x = 1$, from which it follows that $[c]_x = |c|$.

The other direction is trivial. $\square$

Berkovich's theory was inspired by rigid analysis. One of the key insights in rigid analysis is that one can algebraically define certain bounded sets by considering the maximal ideals in suitable commutative Banach rings. The polynomial ring $K[x]$ (equipped with its K-topology) is not a Banach ring, but there is a natural collection of Banach rings associated to it. (See Appendix C for basic properties of Banach rings.) For each real number $R > 0$, let

$$K\langle R^{-1}T \rangle = \{\sum_{k=0}^{\infty} c_k T^k \in K[[T]] : \lim_{k \to \infty} R^k |c_k| = 0\}$$

be the ring of formal power series with coefficients in K and having radius of convergence at least R. It is complete under the norm defined by $\|f\|_R = \max_{k \geq 0} R^k |c_k|$. When $R = 1$, $K\langle R^{-1}T \rangle$ is just the ring $K\langle T \rangle$ discussed in Chapter 1.

For an arbitrary complete normed ring $\mathcal{A}$, we denote by $\mathcal{M}(\mathcal{A})$ the space of all bounded multiplicative seminorms on $\mathcal{A}$, equipped with the weakest topology for which $x \mapsto [f]_x$ is continuous for all $f \in \mathcal{A}$. We define the "Berkovich disc of radius R about 0" by

$$\mathcal{D}(0, R) := \mathcal{M}(K\langle R^{-1}T \rangle).$$

When $R = 1$, $\mathcal{D}(0, 1)$ is precisely the Berkovich unit disc studied in Chapter 1. It is easy to see that for each R in the value group of $K^\times$, $\mathcal{D}(0, R)$ is homeomorphic to $\mathcal{D}(0, 1)$ as a topological space.

If $0 < r < R$, then $\mathcal{D}(0, r)$ can naturally be viewed as a subspace of $\mathcal{D}(0, R)$ via the map $\iota_{r,R}$ sending a bounded multiplicative seminorm $[\ ]_x$ on $K\langle r^{-1}T \rangle$ to the bounded multiplicative seminorm $[\ ]_{\iota_{r,R}(x)}$ on $K\langle R^{-1}T \rangle$

defined by
$$[f]_{\iota_{r,R}(x)} = [\pi(f)]_x,$$
where $\pi : K\langle R^{-1}T\rangle \to K\langle r^{-1}T\rangle$ is the natural K-algebra homomorphism. This map is easily checked to be continuous, with dense image, so $\iota_{r,R}$ is injective. It thus makes sense to consider the direct limit, or union, of the discs $\mathcal{D}(0,R)$. This union is homeomorphic to $\mathbb{A}^1_{\text{Berk}}$:

$$(2.1) \qquad \mathbb{A}^1_{\text{Berk}} \cong \bigcup_{R>0} \mathcal{D}(0,R).$$

We sketch a proof of (2.1) by specifying continuous maps in each direction which are inverse to one another. On the one hand, for each R there is an obvious map $\iota_R : \mathcal{D}(0,R) \to \mathbb{A}^1_{\text{Berk}}$ coming from the inclusion $K[T] \hookrightarrow K\langle R^{-1}T\rangle$, and if $r < R$, then $\iota_r = \iota_R \circ \iota_{r,R}$. We thus obtain a map $\iota : \varinjlim \mathcal{D}(0,R) \to \mathbb{A}^1_{\text{Berk}}$.

In the other direction, suppose $x \in \mathbb{A}^1_{\text{Berk}}$, and let $R = [T]_x$. Then one can define $\psi(x) \in \mathcal{D}(0,R)$ via

$$(2.2) \qquad [f]_{\psi(x)} = \lim_{N\to\infty} [\sum_{i=0}^{N} c_i T^i]_x$$

for $f = \sum_{i=0}^{\infty} c_i T^i \in K\langle R^{-1}T\rangle$. To see that this limit exists, note that for $M \leq N$, a telescoping series argument using the ultrametric inequality for $[\ ]_x$ gives

$$(2.3) \qquad |[\sum_{i=0}^{N} c_i T^i]_x - [\sum_{i=0}^{M} c_i T^i]_x| \leq \max_{M \leq i \leq N}\{|c_i|R^i\}.$$

Since $|c_i|R^i \to 0$, it follows that $\{[\sum_{i=0}^{N} c_i T^i]_x\}_{N\geq 0}$ is a Cauchy sequence and therefore that the limit in (2.2) exists. It is straightforward to check that the resulting map $\psi : \mathbb{A}^1_{\text{Berk}} \to \varinjlim \mathcal{D}(0,R)$ is continuous and that ι and ψ are inverses to one another. Therefore $\mathbb{A}^1_{\text{Berk}}$ is homeomorphic to $\bigcup_{R>0} \mathcal{D}(0,R)$ as claimed.

Berkovich's classification theorem admits the following extension to $\mathbb{A}^1_{\text{Berk}}$, with the same proof:

THEOREM 2.2. *Every $x \in \mathbb{A}^1_{\text{Berk}}$ can be realized as*

$$[f]_x = \lim_{i\to\infty} [f]_{D(a_i,r_i)}$$

for some sequence of nested closed discs $D(a_1,r_1) \supseteq D(a_2,r_2) \supseteq \cdots$ contained in K. If this sequence has a nonempty intersection, then either

(A) *the intersection is a single point a, in which case $[f]_x = |f(a)|$, or*

(B) *the intersection is a closed disc $D(a,r)$ of radius $r > 0$, in which case $[f]_x = [f]_{D(a,r)}$.*

Thus, it makes sense to speak of points of types I, II, III, and IV in $\mathbb{A}^1_{\text{Berk}}$, corresponding to the seminorms associated to classical points, 'rational' discs $D(a,r)$ with radii in $|K^\times|$, 'irrational' discs $D(a,r)$ with radii not in $|K^\times|$, and nested sequences $\{D(a_i, r_i)\}$ with empty intersection, respectively.

We will now give a more intrinsic characterization of the different types of points. For each $x \in \mathbb{A}^1_{\text{Berk}}$, define its *local ring* in $K(T)$ by

$$(2.4) \qquad \mathcal{R}_x = \{f = g/h \in K(T) : g, h \in K[T], [h]_x \neq 0\}.$$

There is a natural extension of $[\ ]_x$ to a multiplicative seminorm on $\mathcal{R}_x$, given by $[g/h]_x = [g]_x/[h]_x$ for g, h as in (2.4). Let $[\mathcal{R}_x^\times]_x$ be the value group of $[\ ]_x$. Put

$$\mathcal{O}_x = \{f \in \mathcal{R}_x : [f]_x \leq 1\}, \quad \mathfrak{m}_x = \{f \in \mathcal{R}_x : [f]_x < 1\},$$

and let $\tilde{k}_x = \mathcal{O}_x/\mathfrak{m}_x$ be the residue field. Recall that $|K^\times|$ denotes the value group of K and $\tilde{K}$ denotes the residue field of K. Recall also that we are assuming that K (and hence $\tilde{K}$) is algebraically closed.

The following proposition expands on some observations which are stated without proof in [**16**, p. 19].

PROPOSITION 2.3. *Given* $x \in \mathbb{A}^1_{\text{Berk}}$, *if* x *is of type* I, *then* $\mathcal{R}_x \subsetneq K(T)$ *and* $[\ ]_x$ *is a seminorm but not a norm. If* x *is of type* II, III, *or* IV, *then* $\mathcal{R}_x = K(T)$ *and* $[\ ]_x$ *is a norm. Indeed:*

(A) x *is of type* I *iff* $\mathcal{R}_x \subsetneq K(T)$, $[\mathcal{R}_x^\times]_x = |K^\times|$, *and* $\tilde{k}_x = \tilde{K}$.
(B) x *is of type* II *iff* $\mathcal{R}_x = K(T)$, $[\mathcal{R}_x^\times]_x = |K^\times|$, *and* $\tilde{k}_x \cong \tilde{K}(t)$ *where* t *is transcendental over* $\tilde{K}$.
(C) x *is of type* III *iff* $\mathcal{R}_x = K(T)$, $[\mathcal{R}_x^\times]_x \supsetneq |K^\times|$, *and* $\tilde{k}_x = \tilde{K}$.
(D) x *is of type* IV *iff* $\mathcal{R}_x = K(T)$, $[\mathcal{R}_x^\times]_x = |K^\times|$, *and* $\tilde{k}_x = \tilde{K}$.

PROOF. The possibilities for the triples $(\mathcal{R}_x, [\mathcal{R}_x^\times]_x, \tilde{k}_x)$ are mutually exclusive, so it suffices to prove all implications in the forward direction.

If x is of type I, the assertions are straightforward and are left to the reader.

Suppose x is of type II and corresponds to $D(a, r)$ with $r \in |K^\times|$. No nonzero polynomial can vanish identically on $D(a, r)$, so $\mathcal{R}_x = K(T)$. If $g \in K[T]$, the non-Archimedean maximum principle (see [**26**, Proposition 3, p. 197] shows that there is a point $p \in D(a, r)$ for which $[g]_x = |g(p)|$, from which it follows easily that $[\mathcal{R}_x^\times]_x = |K^\times|$. Let $c \in K^\times$ satisfy $|c| = r$, and let $t \in \tilde{k}_x$ be the reduction of $(T - a)/c$. It is easy to check that t is transcendental over $\tilde{K}$ and that $\tilde{k}_x = \tilde{K}(t)$.

Next suppose x is of type III and corresponds to $D(a, r)$ with $r \notin |K^\times|$. As before, no polynomial can vanish identically on $D(a, r)$, so $\mathcal{R}_x = K(T)$. Furthermore, $[T - a]_x = r$ so $[\mathcal{R}_x^\times]_x$ properly contains $|K^\times|$. To see that $\tilde{k}_x = \tilde{K}$, it suffices to note that each $f \in K(T)$ can be written as a quotient

$f = g/h$ for polynomials

$$g(T) = \sum_{i=0}^{m} b_i (T-a)^i, \quad h(T) = \sum_{j=0}^{n} c_j (T-a)^j, \quad b_i, c_j \in K,$$

and the fact that $r \notin |K^\times|$ means that the strict case of the ultrametric inequality applies to the terms in g, h. Thus there are indices i_0, j_0 for which $[g]_x = |b_{i_0}| r^{i_0}$, $[h]_x = |c_{j_0}| r^{j_0}$. If $[f]_x = 1$, we must have $i_0 = j_0$ and it follows that

$$f \equiv b_{i_0}/c_{i_0} \pmod{\mathfrak{m}_x} .$$

Finally, suppose x is of type IV and corresponds to a nested sequence of discs $\{D(a_i, r_i)\}$ with empty intersection. For each $0 \neq g \in K[T]$, there will be an N for which the ball $D(a_N, r_N)$ does not contain any of the zeros of g. The ultrametric inequality implies that $|g|$ is constant on $D(a_N, r_N)$, and then the fact that the $D(a_i, r_i)$ are nested means that $[g]_x = [g]_{D(a_N, r_N)} \in |K^\times|$. It follows that $\mathcal{R}_x = K(T)$ and that $[\mathcal{R}_x^\times]_x = |K^\times|$. To see that $\tilde{k}_x = \tilde{K}$, let $f \in K(T)$ satisfy $[f]_x = 1$, and write $f = g/h$ with $g, h \in K[T]$. As above, there will be an N for which $[g]_x = [g]_{D(a_N, r_N)}$ and $[h]_x = [h]_{D(a_N, r_N)}$; necessarily $[g]_x = [h]_x$. Via the theory of Newton polygons (see Corollary A.19 in Appendix A), the fact that g has no zeros in $D(a_N, r_N)$ implies that $|g(z)|$ is constant on $D(a_N, r_N)$ and that $|g(z) - g(a_N)| < |g(a_N)|$ for all $z \in D(a_N, r_N)$. Hence

$$[g - g(a_N)]_x \leq [g - g(a_N)]_{D(a_N, r_N)} < |g(a_N)| = [g]_{D(a_N, r_N)} = [g]_x .$$

Similarly $[h - h(a_N)]_x < [h]_x$, and therefore

$$f \equiv g(a_N)/h(a_N) \pmod{\mathfrak{m}_x} .$$

□

2.2. The Berkovich "Proj" construction

As a topological space, one can define the Berkovich projective line $\mathbb{P}^1_{\text{Berk}}$ to be the one-point compactification (cf. §A.8) of the locally compact Hausdorff space $\mathbb{A}^1_{\text{Berk}}$. The extra point is denoted, as usual, by the symbol ∞ and is regarded as a point of type I.

However, this definition is not very useful. For example, it does not make clear why or how a rational function $\varphi \in K(T)$ induces a map from $\mathbb{P}^1_{\text{Berk}}$ to itself, though in fact this does occur.

We therefore introduce an alternate construction of $\mathbb{P}^1_{\text{Berk}}$ which is reminiscent of the "Proj" construction in algebraic geometry (in the sense that both constructions produce $\mathbb{P}^1$ directly from its homogeneous coordinate ring $K[X, Y]$ without any gluing). We then discuss how $\mathbb{P}^1_{\text{Berk}}$, defined via the "Proj" construction, can be thought of either as $\mathbb{A}^1_{\text{Berk}}$ together with a point at infinity or as two copies of the Berkovich unit disc $\mathcal{D}(0, 1)$ glued together along the annulus $\mathcal{A}(1, 1) = \{x \in \mathcal{D}(0, 1) : [T]_x = 1\}$.

Let S denote the set of multiplicative seminorms on the two-variable polynomial ring $K[X, Y]$ which extend the absolute value on K and which are not identically zero on the maximal ideal (X, Y) of $K[X, Y]$. To emphasize that these are seminorms on the two-variable ring and not $K[T]$, we will write them as $[\![\ ]\!]$. We put an equivalence relation on S by declaring that $[\![\ ]\!]_1 \sim [\![\ ]\!]_2$ if and only if there exists a constant $C > 0$ such that for all $d \in \mathbb{N}$ and all homogeneous polynomials $G \in K[X, Y]$ of degree d, $[\![G]\!]_1 = C^d [\![G]\!]_2$.

As a set, we define $\mathbb{P}^1_{\text{Berk}}$ to be the collection of equivalence classes of elements of S.

Note that each $[\![\ ]\!] \in S$ is automatically non-Archimedean; this follows from the argument in Lemma 1.1(C). The condition that $[\![\ ]\!]$ is not identically zero on the maximal ideal (X, Y) is equivalent to requiring that $[\![X]\!]$ and $[\![Y]\!]$ are not both zero. We will call a seminorm $[\![\ ]\!]$ in S *normalized* if $\max([\![X]\!], [\![Y]\!]) = 1$. We claim there exist normalized seminorms in each equivalence class. Indeed, given $[\![\ ]\!] \in S$, suppose without loss of generality that $[\![Y]\!] \geq [\![X]\!]$, so in particular $[\![Y]\!] > 0$. First extend $[\![\ ]\!]$ to a seminorm on $K[X, Y, 1/Y]$ by multiplicativity, and then define $[\![\ ]\!]^*$ on $K[X, Y]$ by

$$[\![G(X, Y)]\!]^* = [\![G(X/Y, 1)]\!].$$

It is easy to check that $[\![\ ]\!]^*$ is a normalized seminorm on $K[X, Y]$, that it belongs to S, and that it is equivalent to $[\![\ ]\!]$.

By the definition of the equivalence relation, all the normalized seminorms in a class take the same values on homogeneous polynomials. Explicitly, if $z \in \mathbb{P}^1_{\text{Berk}}$ and $[\![\ ]\!]_z$ is any representative of the class z, then

$$[\![G]\!]^*_z = [\![G]\!]_z / \max([\![X]\!]_z, [\![Y]\!]_z)^d$$

for all homogeneous $G \in K[X, Y]$ of degree d, and the right side is independent of the representative chosen.

The topology on $\mathbb{P}^1_{\text{Berk}}$ is defined to be the weakest one such that $z \mapsto [\![G]\!]^*_z$ is continuous for each homogeneous $G \in K[X, Y]$.

We will now make explicit the realization of $\mathbb{P}^1_{\text{Berk}}$ as $\mathbb{A}^1_{\text{Berk}}$ together with a point at infinity. More precisely, we will see that $\mathbb{P}^1_{\text{Berk}}$ can be obtained by gluing together two copies of $\mathbb{A}^1_{\text{Berk}}$ along the open subset $\mathbb{A}^1_{\text{Berk}} \setminus \{0\}$ via a homeomorphism

$$\theta : \mathbb{A}^1_{\text{Berk}} \setminus \{0\} \to \mathbb{A}^1_{\text{Berk}} \setminus \{0\}.$$

The construction closely parallels the standard decomposition of the usual projective line $\mathbb{P}^1$ into the open charts $\mathbb{P}^1 \setminus \{\infty\}$ and $\mathbb{P}^1 \setminus \{0\}$, both of which are isomorphic to $\mathbb{A}^1$, with the transition from one chart to the other given by $t \mapsto 1/t$. Here ∞ denotes the point of $\mathbb{P}^1$ with homogeneous coordinates $(1 : 0)$. The Berkovich space version of this decomposition can be understood as follows.

Define the point ∞ in $\mathbb{P}^1_{\text{Berk}}$ to be the equivalence class of the seminorm $[\![\ ]\!]_\infty$ defined by $[\![G]\!]_\infty = |G(1, 0)|$. More generally, if $P \in \mathbb{P}^1(K)$ has homogeneous coordinates $(a : b)$, the equivalence class of the evaluation seminorm

$[\![G]\!]_P = |G(a,b)|$ is independent of the choice of homogeneous coordinates, and therefore $[\![\ ]\!]_P$ is a well-defined point of $\mathbb{P}^1_{\text{Berk}}$. This furnishes a natural embedding of $\mathbb{P}^1(K)$ into $\mathbb{P}^1_{\text{Berk}}$.

Up to equivalence, $[\![\ ]\!]_\infty$ is the unique element of S for which $[\![Y]\!] = 0$. Indeed, if $[\![Y]\!] = 0$, then by multiplicativity and the ultrametric inequality, if
$$G(X,Y) = a_0 X^d + a_1 X^{d-1} Y + \cdots + a_{d-1} XY^{d-1} + a_d Y^d$$
is homogeneous of degree d, we must have $[\![G(X,Y)]\!] = |a_0| \cdot [\![X]\!]^d$. Similarly, $[\![\ ]\!]_0 = [\![\ ]\!]_{(0:1)}$ is the unique element of S, up to equivalence, for which $[\![X]\!] = 0$.

Gluing two copies of $\mathbb{A}^1_{\text{Berk}}$. We will now show that $\mathbb{P}^1_{\text{Berk}} \backslash \{\infty\}$ and $\mathbb{P}^1_{\text{Berk}} \backslash \{0\}$ are both homeomorphic to $\mathbb{A}^1_{\text{Berk}}$. By symmetry, it suffices to consider the case of $\mathbb{P}^1_{\text{Berk}} \backslash \{\infty\}$. We define maps in each direction between $\mathbb{P}^1_{\text{Berk}} \backslash \{\infty\}$ and $\mathbb{A}^1_{\text{Berk}}$ as follows.

Given $z \in \mathbb{A}^1_{\text{Berk}}$, define a seminorm on $K[X, Y]$ by
$$[\![G]\!]_{\chi_\infty(z)} = [G(T, 1)]_z$$
for all $G \in K[X, Y]$. It is easy to see that $[\![\ ]\!]_{\chi_\infty(z)}$ is an element of S, with $[\![Y]\!]_{\chi_\infty(z)} = 1$. By abuse of notation, let $\chi_\infty : \mathbb{A}^1_{\text{Berk}} \to \mathbb{P}^1_{\text{Berk}} \backslash \{\infty\}$ be the map which takes z to the equivalence class of $[\![\ ]\!]_{\chi_\infty(z)}$.

Define $\psi_\infty : \mathbb{P}^1_{\text{Berk}} \backslash \{\infty\} \to \mathbb{A}^1_{\text{Berk}}$ by the formula
$$[g]_{\psi_\infty(z)} = [\![G]\!]_z / [\![Y]\!]_z^d$$
for each $g \in K[T]$, where $G \in K[X, Y]$ is homogeneous of degree d and satisfies $G(T, 1) = g(T)$ and $[\![\ ]\!]_z$ is any representative of the class z. It is immediate from the definition of equivalence and the fact that $[\![Y]\!]_z \neq 0$ for $z \in \mathbb{P}^1_{\text{Berk}} \backslash \{\infty\}$ that ψ_∞ is well-defined. It is straightforward to show that χ_∞ and ψ_∞ are inverses of one another. We will shortly see that they are both continuous, so ψ_∞ defines a homeomorphism
$$\mathbb{P}^1_{\text{Berk}} \backslash \{\infty\} \to \mathbb{A}^1_{\text{Berk}}\ .$$
Similarly, there is a homeomorphism
$$\psi_0 : \mathbb{P}^1_{\text{Berk}} \backslash \{0\} \to \mathbb{A}^1_{\text{Berk}}$$
defined for $h \in K[T]$ by
$$[h]_{\psi_0(z)} = [\![H]\!]_z / [\![X]\!]_z^d$$
for any homogeneous polynomial $H \in K[X, Y]$ satisfying $H(1, T) = h(T)$.

The continuity of χ_∞ and ψ_∞ can be verified as follows. If $\langle z_\alpha \rangle$ is a net in $\mathbb{A}^1_{\text{Berk}}$ converging to $z \in \mathbb{A}^1_{\text{Berk}}$, then for each $g \in K[T]$ we have $[g]_{z_\alpha} \to [g]_z$, so it follows that for each homogeneous $G \in K[X, Y]$,
$$\frac{[G(T,1)]_{z_\alpha}}{\max\{1, [T]_{z_\alpha}\}} \to \frac{[G(T,1)]_z}{\max\{1, [T]_z\}}\ .$$

This is equivalent to the assertion that $[\![G]\!]^*_{\chi_\infty(z_\alpha)} \to [\![G]\!]^*_{\chi_\infty(z)}$, and thus χ_∞ is continuous.

Similarly, if $z'_\alpha \to z'$ in $\mathbb{P}^1_{\text{Berk}} \backslash \{\infty\}$, then $[\![Y]\!]_{z'_\alpha} \to [\![Y]\!]_{z'} \neq 0$ and for each homogeneous $H \in K[X, Y]$, $[\![H]\!]_{z'_\alpha} \to [\![H]\!]_{z'}$, and therefore

$$\frac{[\![H]\!]_{z'_\alpha}}{[\![Y]\!]^d_{z'_\alpha}} \to \frac{[\![H]\!]_{z'}}{[\![Y]\!]^d_{z'}}.$$

This implies that ψ_∞ is continuous.

Unravelling the definitions, we see that there is a commutative diagram

$$\begin{array}{ccc} \mathbb{P}^1_{\text{Berk}} \backslash \{0, \infty\} & = & \mathbb{P}^1_{\text{Berk}} \backslash \{0, \infty\} \\ \psi_\infty \downarrow & & \downarrow \psi_0 \\ \mathbb{A}^1_{\text{Berk}} \backslash \{0\} & \xrightarrow{\theta} & \mathbb{A}^1_{\text{Berk}} \backslash \{0\} \end{array}$$

in which the involution θ of $\mathbb{A}^1_{\text{Berk}} \backslash \{0\}$ is defined for $g \in K[T]$ by

(2.5) $$[g]_{\theta(z)} = [\tilde{g}]_z / [T]^d_z,$$

where $d = \deg(g)$ and $\tilde{g} = T^d g(1/T)$ is the *reciprocal polynomial* of g.

In conclusion, one obtains $\mathbb{P}^1_{\text{Berk}}$ as a topological space by gluing two copies V_0, V_∞ of $\mathbb{A}^1_{\text{Berk}}$ together via the identification

$$z \in V_0 \backslash \{0\} \sim \theta(z) \in V_\infty \backslash \{0\}.$$

Identifying $\mathbb{P}^1_{\text{Berk}}$ with $\mathbb{A}^1_{\text{Berk}} \cup \{\infty\}$, we view the open and closed Berkovich discs

$$\begin{aligned} \mathcal{D}(a, r)^- &= \{x \in \mathbb{A}^1_{\text{Berk}} : [T - a]_x < r\}, \\ \mathcal{D}(a, r) &= \{x \in \mathbb{A}^1_{\text{Berk}} : [T - a]_x \leq r\} \end{aligned}$$

as subsets of $\mathbb{P}^1_{\text{Berk}}$. It is also useful to consider the sets $\mathbb{P}^1_{\text{Berk}} \backslash \mathcal{D}(a, r)$ and $\mathbb{P}^1_{\text{Berk}} \backslash \mathcal{D}(a, r)^-$ to be open (resp. closed) Berkovich discs.

When restricted to $\mathbb{P}^1(K) \backslash \{0, \infty\} \cong K^\times$, θ is just the map $t \mapsto 1/t$. Let $a \in K$ and $r > 0$. It is well known that if $D(a, r) \subset \mathbb{A}^1(K)$ is a disc not containing 0, then its image under $t \mapsto 1/t$ is $D(1/a, r/|a|^2)$. The same is true for the "Berkovich discs":

LEMMA 2.4. *Identify $\mathbb{P}^1_{\text{Berk}}$ with $\mathbb{A}^1_{\text{Berk}} \cup \{\infty\}$, and extend θ to an involution $\theta : \mathbb{P}^1_{\text{Berk}} \to \mathbb{P}^1_{\text{Berk}}$ which interchanges 0 and ∞. Then:*

(A) *For Berkovich discs $\mathcal{D}(a, r)^-, \mathcal{D}(a, r)$ not containing 0,*

$$\theta(\mathcal{D}(a, r)^-) = \mathcal{D}(1/a, r/|a|^2)^-, \quad \theta(\mathcal{D}(a, r)) = \mathcal{D}(1/a, r/|a|^2).$$

(B) *For the Berkovich discs $\mathcal{D}(0, r)^-, \mathcal{D}(0, r)$,*

$$\theta(\mathcal{D}(0, r)^-) = \mathbb{P}^1_{\text{Berk}} \backslash \mathcal{D}(0, 1/r), \quad \theta(\mathcal{D}(0, r)) = \mathbb{P}^1_{\text{Berk}} \backslash \mathcal{D}(0, 1/r)^-.$$

PROOF. We give the proof only for a disc $\mathcal{D}(a,r)^-$ not containing 0; the other cases are similar. Suppose $x \in \mathcal{D}(a,r)^-$. By assumption, $|a| \geq r > 0$. By the strict form of the ultrametric inequality, $[T]_x = \max([a]_x, [T-a]_x) = |a|$. By formula (2.5),

$$[T - \frac{1}{a}]_{\theta(x)} = \frac{[T \cdot (\frac{1}{T} - \frac{1}{a})]_x}{[T]_x} = \frac{[T-a]_x}{[a]_x [T]_x} < \frac{r}{|a|^2} .$$

It follows that $\theta(\mathcal{D}(a,r)^-) \subseteq \mathcal{D}(1/a, r/|a|^2)^-$. Similarly $\theta(\mathcal{D}(1/a, r/|a|^2)^-) \subseteq \mathcal{D}(a,r)^-$. Since θ is an involution, $\theta(\mathcal{D}(a,r)^-) = \mathcal{D}(1/a, r/|a|^2)^-$. $\square$

Seminorms on local rings in $K(T)$. Each $z \in \mathbb{P}^1_{\text{Berk}}$ corresponds to an equivalence class of seminorms on $K[X,Y]$, but by Proposition 2.3 and the isomorphism we have just shown, it also corresponds to a local ring $\mathcal{R}_z \subset K(T)$ and a unique seminorm $[\ ]_z$ on $\mathcal{R}_z$.

Given $z \in \mathbb{P}^1_{\text{Berk}}$, its local ring in $K(T)$ and seminorm $[\ ]_z$ can be recovered from any of the associated seminorms $[\![\ ]\!]_z$ on $K[X,Y]$ as follows. Identifying T with X/Y, each quotient of homogeneous polynomials $G(X,Y), H(X,Y) \in K[X,Y]$ of common degree d can be viewed as an element of $K(T)$. Then

$$\mathcal{R}_z = \{G/H \in K(T) : G, H \in K[X,Y] \text{ are homogeneous}, [\![H]\!]_z \neq 0\},$$

and for each $f = G/H \in \mathcal{R}_z$,

(2.6) $$[f]_z = [\![G]\!]_z / [\![H]\!]_z .$$

In particular, the point $\infty \in \mathbb{P}^1_{\text{Berk}}$ corresponds to the seminorm whose local ring is

$$\mathcal{R}_\infty = \{g/h \in K(T) : g, h \in K[T], h \neq 0, \deg(h) \geq \deg(g)\},$$

and $[f]_\infty = |f(\infty)|$ for $f \in \mathcal{R}_\infty$. Clearly $[\ ]_\infty$ is of type I in terms of the characterization in Proposition 2.3.

Gluing two copies of $\mathcal{D}(0,1)$. One can also construct $\mathbb{P}^1_{\text{Berk}}$ as a topological space by gluing together two copies of $\mathcal{D}(0,1)$ along the common annulus

$$\mathcal{A}(1,1) = \{x \in \mathcal{D}(0,1) : [T]_x = 1\} .$$

REMARK 2.5. In the terminology of Appendix C, we have

$$\mathcal{A}(1,1) = \mathcal{M}(K\langle T, T^{-1}\rangle) ,$$

where

$$K\langle T, T^{-1}\rangle = \{\sum_{i=-\infty}^{\infty} c_i T^i : \lim_{|i| \to \infty} |c_i| = 0\} .$$

In order to understand this description of $\mathbb{P}^1_{\text{Berk}}$, first note that $\mathbb{P}^1_{\text{Berk}} = W_0 \cup W_\infty$, where

$$W_0 = \{z \in \mathbb{P}^1_{\text{Berk}} : [X]_z \leq [Y]_z\},$$
$$W_\infty = \{z \in \mathbb{P}^1_{\text{Berk}} : [Y]_z \leq [X]_z\}.$$

There is a homeomorphism $\overline{\psi}_\infty : W_\infty \cong \mathcal{D}(0,1)$ defined by the property that for each $g = \sum_{i=0}^\infty c_i T^i \in K\langle T \rangle$,

$$[g]_{\overline{\psi}_\infty(z)} = \lim_{N \to \infty} [\sum_{i=0}^N c_i X^{N-i} Y^i]_z / [X]_z^N.$$

Here the limit on the right side exists, by a telescoping series argument similar to the one in (2.3). Similarly, one defines a homeomorphism $\overline{\psi}_0 : W_0 \cong \mathcal{D}(0,1)$, and the intersection $W_0 \cap W_\infty = \{z \in \mathbb{P}^1_{\text{Berk}} : [X]_z = [Y]_z\}$ is mapped by both ψ_∞ and ψ_0 to $\mathcal{A}(1,1)$. It follows that $\mathbb{P}^1_{\text{Berk}}$ can be obtained by gluing two copies W_0 and W_∞ of $\mathcal{D}(0,1)$ together along their common annulus $\mathcal{A}(1,1)$ via the gluing map $\overline{\theta} : \mathcal{A}(1,1) \to \mathcal{A}(1,1)$ defined for $g \in K\langle T, T^{-1}\rangle$ by

$$[g(T)]_{\overline{\theta}(z)} = [g(1/T)]_z.$$

Note that $\overline{\theta}$ is the restriction to $\mathcal{A}(1,1)$ of the involution $\theta : \mathbb{P}^1_{\text{Berk}} \to \mathbb{P}^1_{\text{Berk}}$ in Lemma 2.4.

We leave detailed verifications of all of these assertions to the reader.

The topology on $\mathbb{P}^1_{\text{Berk}}$. The isomorphisms above enable us to give a more concrete description of the topology on $\mathbb{P}^1_{\text{Berk}}$. Observe that a subset of $\mathbb{P}^1_{\text{Berk}}$ is open if and only if its intersections with both W_0 and W_∞ are open.

PROPOSITION 2.6. *$\mathbb{P}^1_{\text{Berk}}$ is a compact Hausdorff topological space.*

PROOF. By Theorem C.3, W_0 and W_∞ are compact and Hausdorff, and so is their intersection $\mathcal{A}(1,1)$. Hence $\mathbb{P}^1_{\text{Berk}}$, which is obtained by gluing W_0 and W_∞ along $\mathcal{A}(1,1)$, is compact and Hausdorff as well. □

We can explicitly describe the topology on $\mathbb{P}^1_{\text{Berk}}$ as follows:

PROPOSITION 2.7.
(A) *Identifying $\mathbb{P}^1_{\text{Berk}}$ with $\mathbb{A}^1_{\text{Berk}} \cup \{\infty\}$, a basis for the open sets of $\mathbb{P}^1_{\text{Berk}}$ is given by sets of the form*

(2.7) $$\mathcal{D}(a,r)^-, \quad \mathcal{D}(a,r)^- \backslash \bigcup_{i=1}^N \mathcal{D}(a_i, r_i), \quad \text{and} \quad \mathbb{P}^1_{\text{Berk}} \backslash \bigcup_{i=1}^N \mathcal{D}(a_i, r_i),$$

where the $a, a_i \in K$ and the $r, r_i > 0$. If desired, one can require that the r, r_i belong to $|K^\times|$.

(B) *Similarly, for each $x \in \mathbb{P}^1_{\text{Berk}}$, a basis for the closed neighborhoods of x is given by sets of the form*

$$(2.8) \qquad \mathcal{D}(a,r), \quad \mathcal{D}(a,r)\backslash \bigcup_{i=1}^{N} \mathcal{D}(a_i,r_i)^-, \quad \text{and} \quad \mathbb{P}^1_{\text{Berk}}\backslash \bigcup_{i=1}^{N} \mathcal{D}(a_i,r_i)^-$$

containing x, where the $a, a_i \in K$ and the $r, r_i > 0$. Again, one can require that the r, r_i belong to $|K^\times|$.

PROOF. To prove part (A), let B be the collection of sets of the form (2.7). By Lemma 2.4, B is stable under the involution θ. This remains true if we consider the subcollection $B_0 \subset B$ where r and the r_i are required to belong to $|K^\times|$. Proposition 1.6 gives a basis for open sets of $W_0 \cong \mathcal{D}(0,1)$. It is easy to check that for each set in B, its intersection with W_0 is empty or is one of the sets in Proposition 1.6 and conversely that each set in Proposition 1.6 arises as such an intersection. As B is stable under θ, the same is true for W_∞. Since $\mathbb{P}^1_{\text{Berk}} = W_0 \cup W_\infty$ and B is closed under finite intersections, the result follows.

The proof of part (B) is similar. $\square$

REMARK 2.8. When the r, r_i above belong to $|K^\times|$, the sets (2.7) (and finite unions of them) are called *strict open affinoids* in $\mathbb{P}^1_{\text{Berk}}$. Likewise the sets (2.8) (and finite unions of them) are called *strict closed affinoids* in $\mathbb{P}^1_{\text{Berk}}$. (See Appendix C for a more general discussion of affinoids.)

Decomposition into $\mathcal{D}(0,1)$ and $\mathcal{D}(0,1)^-$. There is one more description of $\mathbb{P}^1_{\text{Berk}}$ which will be useful to us. By what we have already said, the open set $\mathbb{P}^1_{\text{Berk}}\backslash\mathcal{D}(0,1)$ is homeomorphic to the open Berkovich unit disc

$$\mathcal{D}(0,1)^- \ = \ \{z \in \mathcal{D}(0,1) \ : \ [T]_z < 1\} \ .$$

As a set, we may therefore identify $\mathbb{P}^1_{\text{Berk}}$ with the disjoint union of a closed set $\mathcal{X}$ homeomorphic to $\mathcal{D}(0,1)$ and an open subset $\mathcal{Y}$ homeomorphic to $\mathcal{D}(0,1)^-$. Moreover, the closure of $\mathcal{Y}$ in $\mathbb{P}^1_{\text{Berk}}$ is $\overline{\mathcal{Y}} = \mathcal{Y} \cup \{\zeta_{\text{Gauss}}\}$. The topology on $\mathbb{P}^1_{\text{Berk}}$ is then determined by saying that a subset is open if and only if its intersections with $\mathcal{X}$ and $\overline{\mathcal{Y}}$ are both open.

This provides a useful way to visualize $\mathbb{P}^1_{\text{Berk}}$: one takes the previous description of $\mathcal{D}(0,1)$ and adds an "extra branch at infinity" corresponding to $\mathbb{P}^1_{\text{Berk}}\backslash\mathcal{D}(0,1)$.

The following lemma is clear from this description of $\mathbb{P}^1_{\text{Berk}}$, since we know the corresponding facts for $\mathcal{D}(0,1)$ (cf. Lemmas 1.7 and 1.8):

LEMMA 2.9. *Both $\mathbb{P}^1(K)$ and $\mathbb{P}^1_{\text{Berk}}\backslash\mathbb{P}^1(K)$ are dense in $\mathbb{P}^1_{\text{Berk}}$.*

In addition, it follows readily from this description and the corresponding fact for $\mathcal{D}(0,1)$ (cf. Corollary 1.14) that:

LEMMA 2.10. *The space $\mathbb{P}^1_{\text{Berk}}$ is uniquely path-connected. More precisely, given any two distinct points $x, y \in \mathbb{P}^1_{\text{Berk}}$, there is a unique arc $[x,y]$ in $\mathbb{P}^1_{\text{Berk}}$ from x to y, and if $x \in \mathcal{X}$ and $y \in \mathcal{Y}$, then $[x,y]$ contains ζ_{Gauss}.*

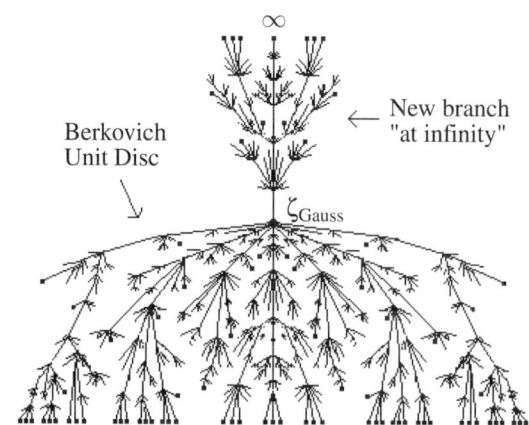

FIGURE 2.1. The Berkovich Projective Line

2.3. The action of a rational map φ on $\mathbb{P}^1_{\text{Berk}}$

Let $\varphi \in K(T)$ be a rational function of degree $d \geq 1$. In this section, we explain how the usual action of φ on $\mathbb{P}^1(K)$ extends to a continuous action $\varphi : \mathbb{P}^1_{\text{Berk}} \to \mathbb{P}^1_{\text{Berk}}$.

Choose a homogeneous lifting $F = (F_1, F_2)$ of φ, where $F_1, F_2 \in K[X, Y]$ are homogeneous of degree d and have no common zeros in K, with $\varphi = F_1/F_2$. (Recall that K is algebraically closed.) The condition that F_1 and F_2 have no common zeros is equivalent to requiring that the homogeneous resultant $\text{Res}(F) = \text{Res}(F_1, F_2)$ be nonzero. (For a definition of the resultant of two homogeneous polynomials, see [**71**, Chapter IX, §4].)

We define the action of φ on $\mathbb{P}^1_{\text{Berk}}$ by showing that F induces a map between seminorms on $K[X, Y]$ which respects S and which, upon passage to equivalence, is independent of the choice of the lifting.

For any seminorm $[\![\ ]\!]_z$ on $K[X, Y]$, define $[\![\ ]\!]_{F(z)}$ by

(2.9) $\qquad [\![G]\!]_{F(z)} := [\![G(F_1(X, Y), F_2(X, Y))]\!]_z$

for $G \in K[X, Y]$. It is easy to check that $[\![\ ]\!]_{F(z)}$ is a multiplicative seminorm on $K[X, Y]$ which extends the absolute value on K. We claim that if $z \in S$, then $F(z) \in S$. It suffices to show that $[\![X]\!]_{F(z)} = [\![F_1(X, Y)]\!]_z$ and $[\![Y]\!]_{F(z)} = [\![F_2(X, Y)]\!]_z$ cannot both be zero. This follows from the following well-known property of the homogeneous resultant (see [**71**, Chapter IX, §4], [**93**, §2.4], or [**98**, §5.8]):

LEMMA 2.11. *There are homogenous polynomials $G_1, G_2, H_1,$ and H_2 in $K[X, Y]$ of degree $d - 1$ such that*

$$G_1(X, Y)F_1(X, Y) + G_2(X, Y)F_2(X, Y) = \text{Res}(F)X^{2d-1},$$
$$H_1(X, Y)F_1(X, Y) + H_2(X, Y)F_2(X, Y) = \text{Res}(F)Y^{2d-1}.$$

By Lemma 2.11, the ultrametric inequality for $[\ ]_z$, and the fact that $|\mathrm{Res}(F)| \neq 0$, if $[\![F_1(X,Y)]\!]_z = [\![F_2(X,Y)]\!]_z = 0$, then $[\![X]\!]_z = [\![Y]\!]_z = 0$, a contradiction. Thus at least one of $[\![X]\!]_{F(z)}$ and $[\![Y]\!]_{F(z)}$ is nonzero, so $F(z) \in S$. It is easy to see that if $z, z' \in S$ and $z \sim z'$, then $F(z) \sim F(z')$. Likewise, if F' is another homogeneous lifting of φ, then $F(z) \sim F'(z)$. Thus, upon passage to equivalence, φ induces a map $\varphi : \mathbb{P}^1_{\mathrm{Berk}} \to \mathbb{P}^1_{\mathrm{Berk}}$.

The map which we have just defined can be described in terms of the local rings $\mathcal{R}_z \subset K(T)$ and seminorms $[\ ]_z$, as follows. Given $z \in \mathbb{P}^1_{\mathrm{Berk}}$, let $[\![\ ]\!]_{\tilde z}$ be a seminorm on $K[X,Y]$ which induces $[\ ]_z$. We claim that

(2.10) $$\mathcal{R}_{\varphi(z)} \ = \ \{f \in K(T) : f \circ \varphi \in \mathcal{R}_z\}$$

and that $[f]_{\varphi(z)} = [f \circ \varphi]_z$ for each $f \in \mathcal{R}_{\varphi(z)}$. Given such an f, write $f = G/H$ where $G, H \in K[X,Y]$ are homogeneous of common degree and $[\![H]\!]_{F(\tilde z)} = [\![H \circ F]\!]_{\tilde z} \neq 0$. Then $f \circ \varphi = (G \circ F)/(H \circ F)$ and

(2.11) $$[f \circ \varphi]_z \ = \ [\![G \circ F]\!]_{\tilde z}/[\![H \circ F]\!]_{\tilde z} \ = \ [\![G]\!]_{F(\tilde z)}/[\![H]\!]_{F(\tilde z)} \ = \ [f]_{\varphi(z)} \ .$$

Clearly (2.10) and (2.11) uniquely determine the action of φ on $\mathbb{P}^1_{\mathrm{Berk}}$. They also show this action extends the usual action of φ on $\mathbb{P}^1(K)$: if $z \in \mathbb{P}^1_{\mathrm{Berk}}$ corresponds to $a \in \mathbb{P}^1(K)$, then $[\ ]_{\varphi(z)} = [\ ]_{\varphi(a)}$.

The continuity of $\varphi : \mathbb{P}^1_{\mathrm{Berk}} \to \mathbb{P}^1_{\mathrm{Berk}}$ can be verified as follows: If $\langle z_\alpha \rangle$ is a net in $\mathbb{P}^1_{\mathrm{Berk}}$ converging to z, then for all homogeneous $G \in K[X,Y]$, we have $[\![G \circ F]\!]^*_{z_\alpha} \to [\![G \circ F]\!]^*_z$ and $[\![F_i]\!]^*_{z_\alpha} \to [\![F_i]\!]^*_z$, and thus

$$[\![G]\!]^*_{\varphi(z_\alpha)} \ = \ [\![G \circ F]\!]^*_{z_\alpha}/\max([\![F_1]\!]^*_{z_\alpha}, [\![F_2]\!]^*_{z_\alpha})$$

converges to

$$[\![G]\!]^*_{\varphi(z)} \ = \ [\![G \circ F]\!]^*_z/\max([\![F_1]\!]^*_z, [\![F_2]\!]^*_z)$$

as required.

In summary, we have:

DEFINITION 2.12. Let $\varphi \in K(T)$ be a nonconstant rational function. The action of φ on $\mathbb{P}^1_{\mathrm{Berk}}$ is the one induced by the action of a homogeneous lifting $F = (F_1, F_2)$ on seminorms of $K[X,Y]$, with $[\![G]\!]_{F(z)} = [\![G \circ F]\!]_z$. It is continuous and is characterized by the property that for each $z \in \mathbb{P}^1_{\mathrm{Berk}}$, we have $\mathcal{R}_{\varphi(z)} = \{f \in K(T) : f \circ \varphi \in \mathcal{R}_z\}$, and for each $f \in \mathcal{R}_{\varphi(z)}$

(2.12) $$[f]_{\varphi(z)} \ = \ [f \circ \varphi]_z \ .$$

We will see later (in Corollaries 9.9 and 9.10) that $\varphi : \mathbb{P}^1_{\mathrm{Berk}} \to \mathbb{P}^1_{\mathrm{Berk}}$ is an open surjective mapping and that every point $z \in \mathbb{P}^1_{\mathrm{Berk}}$ has at most d preimages under φ.

If $d = 1$, it is clear that φ has an algebraic inverse and thus induces an *automorphism* of $\mathbb{P}^1_{\mathrm{Berk}}$. Define $\mathrm{Aut}(\mathbb{P}^1_{\mathrm{Berk}})$ to be the group of automorphisms of $\mathbb{P}^1_{\mathrm{Berk}}$ coming from algebraic maps $\varphi(T) \in K(T)$.

COROLLARY 2.13. $\mathrm{Aut}(\mathbb{P}^1_{\mathrm{Berk}}) \cong \mathrm{PGL}_2(K)$, *the group of Möbius transformations (or linear fractional transformations) acting on* $\mathbb{P}^1(K)$. *In particular:*

(A) *Given triples of distinct points* (a_0, a_1, a_∞), $(\zeta_0, \zeta_1, \zeta_\infty)$ *in* $\mathbb{P}^1(K)$, *there is a unique* $\varphi \in \text{Aut}(\mathbb{P}^1_{\text{Berk}})$ *for which* $\varphi(a_0) = \zeta_0$, $\varphi(a_1) = \zeta_1$, *and* $\varphi(a_\infty) = \zeta_\infty$.

(B) *Given a triple* (a_0, A, a_∞) *where* $a_0, a_\infty \in \mathbb{P}^1(K)$ *are distinct points and A is a type* II *point on the arc* $[a_0, a_\infty]$ *and another triple* $(\zeta_0, Z, \zeta_\infty)$ *of the same kind, there is a* $\varphi \in \text{Aut}(\mathbb{P}^1_{\text{Berk}})$ *for which* $\varphi(a_0) = \zeta_0$, $\varphi(A) = Z$, *and* $\varphi(a_\infty) = \zeta_\infty$.

PROOF. By the discussion above, each linear fractional transformation $\varphi(T) = (aT+b)/(cT+d) \in K(T)$ induces an automorphism of $\mathbb{P}^1_{\text{Berk}}$.

Conversely, suppose that $\varphi(T) \in K(T)$ induces an automorphism of $\mathbb{P}^1_{\text{Berk}}$. In Proposition 2.15 below, we will see that $\varphi : \mathbb{P}^1_{\text{Berk}} \to \mathbb{P}^1_{\text{Berk}}$ takes $\mathbb{P}^1(K)$ to $\mathbb{P}^1(K)$ and $\mathbb{P}^1_{\text{Berk}} \backslash \mathbb{P}^1(K)$ to $\mathbb{P}^1_{\text{Berk}} \backslash \mathbb{P}^1(K)$. Hence $\varphi(T)$ induces an automorphism of $\mathbb{P}^1(K)$. It is well known that the group of algebraic automorphisms of $\mathbb{P}^1(K)$ is precisely the group of linear fractional transformations, which is isomorphic to $\text{PGL}_2(K)$.

Assertion (A) is a well-known fact.

For assertion (B), note that since each $\varphi \in \text{Aut}(\mathbb{P}^1_{\text{Berk}})$ is bijective and bicontinuous, it takes arcs to arcs. Thus, if distinct points $b_0, b_1, b_\infty \in \mathbb{P}^1(K)$ are given and if B is the uniquely determined type II point at the intersection of the three arcs $[b_0, b_1]$, $[b_0, b_\infty]$, and $[b_1, b_\infty]$, then $\varphi(B)$ is the uniquely determined type II point at the intersection of the arcs $[\varphi(b_0), \varphi(b_1)]$, $[\varphi(b_0), \varphi(b_\infty)]$, and $[\varphi(b_1), \varphi(b_\infty)]$. Given triples (a_0, A, a_∞), $(\zeta_0, Z, \zeta_\infty)$ as in the corollary, choose $a_1, \zeta_1 \in \mathbb{P}^1(K)$ in such a way that the arcs between the points a_i determine A and the arcs between the points ζ_i determine Z. If we choose $\varphi \in \text{Aut}(\mathbb{P}^1_{\text{Berk}})$ with $\varphi(a_0) = \zeta_0$, $\varphi(a_1) = \zeta_1$, and $\varphi(a_\infty) = \zeta_\infty$, then we also have $\varphi(A) = Z$. $\square$

REMARK 2.14. We will often view elements $\varphi \in \text{Aut}(\mathbb{P}^1_{\text{Berk}})$ as 'changes of coordinates'. Thus, Corollary 2.13 can be viewed as saying that there always exist changes of coordinates with the properties in (A) and (B).

However, coordinate changes of other types need not exist. For example, suppose (a_0, A, a_∞) and $(\zeta_0, Z, \zeta_\infty)$ are triples as in (B), except that A and Z are of type III. There may not be a $\varphi \in \text{Aut}(\mathbb{P}^1_{\text{Berk}})$ with the properties in (B). For instance, suppose $K = \mathbb{C}_p$ and $a_0 = \zeta_0 = 0$, $a_\infty = \zeta_\infty = \infty$. If $\varphi \in \text{Aut}(\mathbb{P}^1_{\text{Berk}})$ takes 0 to 0 and ∞ to ∞, then $\varphi(T) = cT$ for some $c \in \mathbb{C}_p^\times$. If $|c| = 1$, then $\varphi(T)$ fixes each point of the arc $[0, \infty]$. Since $|\mathbb{C}_p^\times|$ is countable, there are only countably many possible images $\varphi(A)$.

We now show that the action of φ on $\mathbb{P}^1_{\text{Berk}}$ preserves the "types" of points:

PROPOSITION 2.15. *If* $\varphi \in K(T)$ *is nonconstant, then* $\varphi : \mathbb{P}^1_{\text{Berk}} \to \mathbb{P}^1_{\text{Berk}}$ *takes points of each type* (I, II, III, IV) *to points of the same type. Thus* $\varphi(z)$ *has a given type if and only if z does.*

PROOF. We will use the characterization of points in Proposition 2.3.

2.3. THE ACTION OF A RATIONAL MAP φ ON $\mathbb{P}^1_{\text{Berk}}$

We first show that $z \in \mathbb{P}^1_{\text{Berk}}$ is of type I iff $\varphi(z)$ is of type I. This is to be expected, since the action of φ on $\mathbb{P}^1_{\text{Berk}}$ extends its usual action on $\mathbb{P}^1(K)$. Suppose z has type I and corresponds to the point $a \in \mathbb{P}^1(K)$. Take $0 \ne f \in K(T)$ which vanishes at $\varphi(a)$. Then $[f]_{\varphi(z)} = [f \circ \varphi]_z = |f(\varphi(a))| = 0$, so $[\ ]_{\varphi(z)}$ is a seminorm but not a norm, and $\varphi(z)$ has type I. Conversely, suppose $\varphi(z)$ has type I and corresponds to $b \in \mathbb{P}^1(K)$. Choose $0 \ne f \in K(T)$ with $f(b) = 0$. Then $[f \circ \varphi]_z = [f]_{\varphi(z)} = |f(b)| = 0$, so $[\ ]_z$ is a seminorm but not a norm, and z has type I.

If z has type II, III, or IV, then $\mathcal{R}_z = K(T)$ by Proposition 2.3. By what we have just shown, $\varphi(z) \in \mathbb{P}^1_{\text{Berk}} \backslash \mathbb{P}^1(K)$, and it follows that $\mathcal{R}_{\varphi(z)} = K(T)$ as well.

Suppose z is of type II. Since $[\mathcal{R}_z^\times]_z = |K^\times|$ by Proposition 2.3 and $[f]_{\varphi(z)} = [f \circ \varphi]_z$ for each $f \in K(T)$, it follows that $[\mathcal{R}_z^\times]_{\varphi(z)} = |K^\times|$. We claim that the residue field $\tilde{k}_{\varphi(z)}$ properly contains $\tilde{K}$. Under the classification in Theorem 2.2, z corresponds to a disc $D(a,r)$ with $r \in |K^\times|$. After replacing T by $(T-a)/b$ for an appropriate $b \in K^\times$, we can assume that $D(a,r) = D(0,1)$ and that $[\ ]_z$ is the Gauss norm. Given $f \in \mathcal{O}_z$, write $\tilde{f}$ for its image in the residue field $\tilde{k}_z = \mathcal{O}_z/\mathfrak{m}_z$. Then $\tilde{k}_z = \tilde{K}(t)$ where $t = \tilde{T}$ is transcendental over $\tilde{K}$. By Lemma 2.16 below, there are constants $c_1, c_2 \in K$ for which the function $f = c_1\varphi(T) - c_2$ belongs to $\mathcal{O}_z$ and has nonconstant reduction, that is, $\tilde{f} \in \tilde{k}_z \backslash \tilde{K}$. Since f is the pullback of $f_0 := c_1 T - c_2$ by $\varphi(T)$, the image of f_0 in the residue field $\tilde{k}_{\varphi(z)}$ must also be nonconstant. (Note that $\tilde{f}$ is nonconstant iff $[f-a]_z = 1$ for all $a \in K$ with $|a| \le 1$, which holds iff $[f_0 - a]_{\varphi(z)} = 1$ for all $a \in K$ with $|a| \le 1$.) Thus $\tilde{k}_{\varphi(z)} \supsetneq \tilde{K}$, and $\varphi(z)$ is of type II by Proposition 2.3.

Next suppose z is of type III and corresponds to a disc $D(a,r)$ with $r \notin |K^\times|$ under Theorem 2.2. By Proposition 2.3, $[\mathcal{R}_z^\times]_z \supsetneq |K^\times|$ and $\tilde{k}_z = \tilde{K}$. We first show that $\tilde{k}_{\varphi(z)} = \tilde{K}$. If $f \in K(T)$ and $[f]_{\varphi(z)} \le 1$, then $[f \circ \varphi]_z \le 1$. Since $\tilde{k}_z = \tilde{K}$, there is a $b \in K$ with $|b| \le 1$ for which $[(f \circ \varphi) - b]_z < 1$. But then $[f - b]_{\varphi(z)} < 1$, so $\tilde{f} = \tilde{b} \in \tilde{K} \subseteq \tilde{k}_{\varphi(z)}$. Since f was arbitrary, $\tilde{k}_{\varphi(z)} = \tilde{K}$.

We next show that $[\mathcal{R}_{\varphi(z)}^\times]_{\varphi(z)} \supsetneq |K^\times|$. If $[\varphi(T)]_z \notin |K^\times|$, then since $[T]_{\varphi(z)} = [\varphi(T)]_z$, we are done. If $[\varphi(T)]_z \in |K^\times|$, write $\varphi(T) = P(T)/Q(T)$ where $P, Q \in K[T]$ have no common factor, and expand

$$P(T) = a_0 + a_1(T-a) + \cdots + a_d(T-a)^d,$$
$$Q(T) = b_0 + b_1(T-a) + \cdots + b_d(T-a)^d.$$

Since z is of type III, $[T-a]_z = r \notin |K^\times|$ and the nonzero terms in $P(T)$ have distinct values under $[\ ]_z$. A similar assertion holds for $Q(T)$. Since $[P]_z/[Q]_z = [P/Q]_z \in |K^\times|$, the strict form of the ultrametric inequality shows there must be an index k for which $[P]_z = |a_k| \cdot r^k$, $[Q]_z = |b_k| \cdot r^k$. Now consider the function

$$f = b_k \varphi(T) - a_k = \frac{b_k P(T) - a_k Q(T)}{Q(T)}.$$

Since the numerator of f contains no $(T-a)^k$ term, it must be that $[b_k P(T) - a_k Q(T)]_z = |c| \cdot r^j$ for some $j \ne k$ and some $c \in K^\times$. Hence $[f]_z \notin |K^\times|$. Since f is the pullback of $f_0 := b_k T - a_k$ by $\varphi(T)$, we have $[f_0]_{\varphi(z)} = [f]_z$. Thus $[\mathcal{R}^\times_{\varphi(z)}]_{\varphi(z)} \supsetneq |K^\times|$, and by Proposition 2.3, $\varphi(z)$ is of type III.

If z is of type IV, then $[\mathcal{R}^\times_z]_z = |K^\times|$ and $\tilde{k}_z = \tilde{K}$. The arguments above, concerning $[\mathcal{R}^\times_z]_z$ for points of type II and $\tilde{k}_z$ for points of type III, show that $[\mathcal{R}^\times_{\varphi(z)}]_{\varphi(z)} = |K^\times|$ and $\tilde{k}_{\varphi(z)} = \tilde{K}$. Thus $\varphi(z)$ is of type IV. $\square$

LEMMA 2.16. *Let $\varphi(T) \in K(T)$ be nonconstant, and let $z = \zeta_{\text{Gauss}} \in \mathbb{P}^1_{\text{Berk}}$ be the Gauss point. Then there are constants $c_1, c_2 \in K$ for which $f(T) = c_1 \varphi(T) - c_2$ belongs to $\mathcal{O}_z$ and has nonconstant reduction in $\tilde{k}_z = \tilde{K}(\tilde{T})$.*

PROOF. After scaling $\varphi(T)$, we can assume that $[\varphi]_z = 1$. Write $\varphi(T) = P(T)/Q(T)$, where $P, Q \in K[T]$ are polynomials with no common factor, satisfying $[P]_z = [Q]_z = 1$. Thus
$$P(T) = a_0 + a_1 T + \cdots + a_d T^d, \quad Q(T) = b_0 + b_1 T + \cdots + b_d T^d$$
where $\max_i(|a_i|) = \max_i(|b_i|) = 1$. In particular, both $\tilde{P}(T)$ and $\tilde{Q}(T)$ are nonzero. If $\tilde{P}(T)$ is not a scalar multiple of $\tilde{Q}(T)$, then $\tilde{\varphi}(T) \in \tilde{k}_z$ is nonconstant, and we are done.

On the other hand, suppose $\tilde{P}(T) = \tilde{c} \cdot \tilde{Q}(T)$ for some $\tilde{c} \in \tilde{K}^\times$. We claim there is a number $c = c_0 \in K$ for which
$$\inf_{|c| \le 1} [P - cQ]_z = \inf_{|c| \le 1} (\max_{0 \le i \le d} |a_i - c b_i|)$$
is achieved. Assuming this, let $d_0 \in K^\times$ be such that $|d_0| = [P - c_0 Q]_z$, and put $P_0(T) = (P(T) - c_0 Q(T))/d_0$. Consider
$$f(T) = \frac{1}{d_0}(\varphi(T) - c_0) = \frac{P_0(T)}{Q(T)}.$$
Then $[P_0(T)]_z = [Q(T)]_z = 1$, and $\tilde{P}_0(T)$ is not a scalar multiple of $\tilde{Q}(T)$ by the minimality in the choice of c_0. Hence $\tilde{f}(T)$ is nonconstant in $\tilde{k}_z$.

To establish our claim, let $\mathcal{I} \subset \{0, 1, \ldots, d\}$ be the set of indices for which $b_i \ne 0$ and $|a_i| \le |b_i|$. Since $\tilde{P}(T) = \tilde{c} \cdot \tilde{Q}(T)$, $\mathcal{I}$ is nonempty. If $j \notin \mathcal{I}$, then $|a_j - c b_j| = |a_j|$ for all $|c| \le 1$. Hence it suffices to show that

(2.13) $$\inf_{|c| \le 1} (\max_{i \in \mathcal{I}} |a_i - c b_i|)$$

is achieved for some c_0. For each $k \in \mathcal{I}$, put
$$U_k = D(\frac{a_k}{b_k}, 1) \setminus \bigcup_{i \in \mathcal{I},\ i \ne k} D(\frac{a_k}{b_k}, \left|\frac{a_i}{b_i} - \frac{a_k}{b_k}\right|)^-.$$

Then $\bigcup_{k \in \mathcal{I}} U_k = D(0,1) = \{c \in K : |c| \le 1\}$, and since there are only finitely many U_k, the infimum in (2.13) is the infimum over $c \in U_k$, for some k.

Fix $k \in \mathcal{I}$. Then for each $i \in \mathcal{I}$ and each $c \in U_k$,

$$
\begin{aligned}
|a_i - cb_i| &= |b_i| \cdot |(\frac{a_i}{b_i} - \frac{a_k}{b_k}) - (c - \frac{a_k}{b_k})| \\
(2.14) \qquad &= |b_i| \cdot \max\left(|\frac{a_i}{b_i} - \frac{a_k}{b_k}|, |c - \frac{a_k}{b_k}|\right),
\end{aligned}
$$

and the right side of (2.14) is nondecreasing in $|c - a_k/b_k|$. Thus

$$\inf_{c \in U_k} (\max_{i \in \mathcal{I}} |a_i - cb_i|)$$

is achieved when $c = a_k/b_k$.

This shows that we can take $c_0 = a_k/b_k$ for some $k \in \mathcal{I}$. $\square$

As a complement to Lemma 2.16, we prove:

LEMMA 2.17. *Let $\varphi(T) \in K(T)$ be nonconstant. Then $\varphi(T)$ has a well-defined reduction $\tilde{\varphi}(T) \in \tilde{k}_{\zeta_{\text{Gauss}}} = \tilde{K}(\tilde{T})$ if and only if $\varphi(\zeta_{\text{Gauss}}) \in \mathcal{D}(0,1)$, and $\varphi(T)$ has nonconstant reduction if and only if $\varphi(\zeta_{\text{Gauss}}) = \zeta_{\text{Gauss}}$.*

PROOF. By definition, the reduction $\tilde{\varphi}(T)$ is defined exactly when $\varphi(T) \in \mathcal{O}_{\zeta_{\text{Gauss}}}$, i.e., when $[\varphi(T)]_{\zeta_{\text{Gauss}}} \leq 1$. This means precisely that $\varphi(\zeta_{\text{Gauss}}) \in \mathcal{D}(0,1)$, proving the first assertion.

For the second assertion, suppose first that φ has nonconstant reduction. In particular, we have $\varphi(\zeta_{\text{Gauss}}) \in \mathcal{D}(0,1)$. If $\varphi(\zeta_{\text{Gauss}}) \neq \zeta_{\text{Gauss}}$, then $\varphi(\zeta_{\text{Gauss}}) \in \mathcal{D}(a,1)^-$ for some $a \in D(0,1)$, and thus $\tilde{\varphi}(\tilde{T}) = \tilde{a}$ is constant, a contradiction. Thus $\varphi(\zeta_{\text{Gauss}}) = \zeta_{\text{Gauss}}$.

Conversely, if $\varphi(\zeta_{\text{Gauss}}) = \zeta_{\text{Gauss}}$, then $[\varphi(T)]_{\zeta_{\text{Gauss}}} = 1$ and thus we can write $\varphi(T) = P(T)/Q(T)$ with $[P(T)]_{\zeta_{\text{Gauss}}} = [Q(T)]_{\zeta_{\text{Gauss}}} = 1$. If $\tilde{\varphi}(\tilde{T})$ is constant, then $[\varphi(T) - a]_{\zeta_{\text{Gauss}}} < 1$ for some $a \in D(0,1)$, implying that $\varphi(\zeta_{\text{Gauss}}) \in \mathcal{D}(a,1)^-$, a contradiction. $\square$

2.4. Points of $\mathbb{P}^1_{\text{Berk}}$ revisited

In order to understand the action of a rational map on $\mathbb{P}^1_{\text{Berk}}$ in a more concrete way, let us begin by reinterpreting the points of $\mathbb{A}^1_{\text{Berk}}$. By Theorem 2.2, each $x \in \mathbb{A}^1_{\text{Berk}}$ corresponds to an equivalence class of nested discs $\{D(a_i, r_i)\}$ in K, and for each $f \in K[T]$ the corresponding seminorm $[f]_x$ is a limit of the sup norms $[f]_{D(a_i,r_i)}$. However, this association of $[f]_x$ with sup norms is misleading. A more accurate assertion is that $[f]_x$ is the *generic value* of $|f(z)|$ at x.

To understand this, suppose x is a point of type II, so that x corresponds to a disc $D(a,r)$ with r in the value group of $K^\times$. Theorem 2.2 asserts that for $f \in K[T]$,

$$[f]_x = \max_{z \in D(a,r)} |f(z)|.$$

If the zeros of $f(T)$ in $D(a,r)$ are $a_1, \ldots, a_m$, then by the Weierstrass Preparation Theorem $|f(z)|$ takes on its maximum value on $D(a,r)$ at each point of $D(a,r) \backslash \bigcup_{i=1}^m D(a_i,r)^-$. In other words, $[f]_x$ is the constant value which

$|f(z)|$ assumes 'almost everywhere' on $D(a,r)$. As observed in Proposition 2.3, the multiplicative seminorm $[\ ]_x$ extends in a unique way to the field of rational functions $K(T)$, with

$$[f/g]_x \;=\; \frac{[f]_x}{[g]_x} \;.$$

However, this extended seminorm is definitely *not* the supremum norm: if $(f/g)(z)$ has poles in $D(a,r)$, then $\sup_{z \in D(a,r)} |(f/g)(z)| = \infty$. Rather, if $f(z)$ has zeros $a_1, \ldots, a_m$ and $g(z)$ has zeros $b_1, \ldots, b_n$, then $[f/g]_x$ is the constant value which $|f(z)/g(z)|$ assumes everywhere on the *closed affinoid* $D(a,r) \backslash (\bigcup_{i=1}^m D(a_i, r)^-) \cup (\bigcup_{j=1}^n D(b_j, r)^-)$. This is best understood as the 'generic value' of $|(f/g)(z)|$ on $D(a,r)$.

For points x of type I, III, or IV, the notion of a generic value of $|f(z)|$ at x has to be understood in a slightly broader way. Let $x \in \mathbb{A}^1_{\text{Berk}}$ be arbitrary. Fix $f \in \mathcal{R}_x \subset K(T)$. By continuity, for each $\varepsilon > 0$ there is a neighborhood U of x such that for each $t \in U$, $|[f]_t - [f]_x| < \varepsilon$. In particular, for each type I point $z \in U$,

$$\bigl|\,|f(z)| - [f]_x \,\bigr| \;<\; \varepsilon\;.$$

By the description of the topology of $\mathbb{P}^1_{\text{Berk}}$ in Proposition 2.7, sets of the form $\mathcal{D}(a,r) \backslash \bigcup_{i=1}^m \mathcal{D}(a_i, r_i)^-$ are cofinal in the set of closed neighborhoods of x. Thus, $[f]_x$ is the unique number such that for each $\varepsilon > 0$, there is a closed affinoid neighborhood $\mathcal{D}(a,r) \backslash \bigcup_{i=1}^m \mathcal{D}(a_i, r_i)^-$ of x such that $|f(z)|$ is within ε of $[f]_x$ on that neighborhood. In this sense $[f]_x$ is the generic value of $|f(z)|$ at x.

The discussion above applies to the point $x = \infty$, if one replaces the affinoids $\mathcal{D}(a,r) \backslash \bigcup_{i=1}^m \mathcal{D}(a_i, r_i)^-$ with the affinoids $\mathbb{P}^1_{\text{Berk}} \backslash \mathcal{D}(0,r)^-$, as $r \to \infty$.

In rigid analysis, each closed affinoid $V = D(a,r) \backslash (\bigcup_{i=1}^m D(a_i, r_i)^-)$ with $r, r_1, \ldots, r_m$ in the value group of $K^\times$ corresponds to an *affinoid subdomain* of $\mathbb{A}^1$. More precisely, V is a *Laurent domain*, isomorphic to the set of maximal ideals $\text{Max}(\mathcal{A}_V)$ in the K-affinoid algebra

$$\mathcal{A}_V \;=\; K\langle T, r^{-1}T_1, r_1 S_1, \ldots, r_m S_m \rangle / \mathcal{I}_V \;,$$

where

$$\mathcal{I}_V \;=\; (T_1 - (T-a), (T-a_1)S_1 - 1, \ldots, (T-a_m)S_m - 1)$$

and for $r_1, \ldots, r_n \in |K^\times|$, we define $K\langle r_1^{-1}T_1, \ldots, r_n^{-1}T_n \rangle$ to be the Tate algebra

$$K\langle r_1^{-1}T_1, \ldots, r_n^{-1}T_n \rangle \;=\; \{f = \sum_\nu a_\nu T^\nu \,:\, |a_\nu| r^\nu \to 0 \text{ as } |\nu| \to \infty\}$$

of power series convergent on the polydisc $\{(z_1, \ldots, z_n) \in K^n : \max |z_i| \leq r_i\}$ (see Appendix C for further discussion). The relations generating $\mathcal{I}_V$ mean that $\text{Max}(\mathcal{A}_V)$ is isomorphic to the set of $x \in K$ for which $|x - a| \leq r$ and $|x - a_i| \geq r_i$ for $i = 1, \ldots, m$; this is precisely the closed affinoid V.

By elementary rigid analysis, $K\langle r_1^{-1}T_1, \ldots, r_n^{-1}T_n \rangle$ is a Banach algebra over K when equipped with the 'Gauss norm'
$$\|f\| = \max_\nu |a_\nu| r^\nu \;,$$
and $\mathcal{A}_V$ is a Banach algebra over K when equipped with the corresponding quotient norm. The discussion in §2.1 shows that the functional analytic spectrum $\mathcal{M}(\mathcal{A}_V)$ can be identified with the subset $D(a,r) \backslash \bigcup_{i=1}^m D(a_i,r_i)^-$ of $\mathbb{A}^1_{\text{Berk}}$: more precisely,

(2.15) $\mathcal{M}(\mathcal{A}_V) = \{x \in \mathbb{A}^1_{\text{Berk}} \;:\; [T-a]_x \leq r,\; [T-a_i]_x \geq r_1, i=1,\ldots,m\}$.

We have emphasized the interpretation of seminorms $[\;]_x$ as 'generic values' partly to explain why such an inclusion is reasonable. For affinoids V more complicated than a disc, one should not expect to identify seminorms $[\;]_x$ on $\mathcal{A}_V$ with sup norms on nested sequences of discs; however, one might hope to identify them with sup norms on nested sequences of closed affinoids contained in $D(a,r) \backslash \bigcup_{i=1}^m D(a_i,r_i)^-$. Such an identification can in fact be made, though we will not give the details. However, we do note that each closed affinoid $Y = D(b,R) \backslash \bigcup_{i=1}^n D(b_i,R)^-$ in which the deleted discs have the same radius as the outer disc determines a bounded multiplicative seminorm (actually a norm)

(2.16) $$[f]_Y = \sup_{z \in D(b,R) \backslash \bigcup_{i=1}^n D(b_i,R)^-} |f(z)|$$

on any ring $\mathcal{A}_V$ for which $V = D(a,r) \backslash \bigcup_{i=1}^m D(a_i,r_i)^-$ contains Y.

Another reason for emphasizing the interpretation of seminorms $[\;]_x$ as generic values is to be able to formulate a concrete description of the action of a rational function φ on $\mathbb{P}^1_{\text{Berk}}$ in terms of its action on $\mathbb{P}^1(K)$. One can deduce from the above discussion:

PROPOSITION 2.18. *Let $\varphi \in K(T)$ be a nonconstant rational function, and suppose $x \in \mathbb{A}^1_{\text{Berk}}$ is a point of type II, corresponding to a disc $D(a,r)$ in K under Berkovich's classification. Then $\varphi(x)$ (which is also of type II) corresponds to the disc $D(b,R)$ iff there exist $a_1,\ldots,a_m,b_1,\ldots,b_n \in K$ for which the closed affinoid $D(b,R) \backslash \bigcup_{i=1}^n D(b_i,R)^-$ is the image under φ of the closed affinoid $D(a,r) \backslash \bigcup_{i=1}^m D(a_i,r)^-$.*

PROOF. Put $X = D(a,r) \backslash \bigcup_{i=1}^m D(a_i,r)^-$, $Y = D(b,R) \backslash \bigcup_{i=1}^n D(b_i,R)^-$, and let $\mathcal{A}_X$, $\mathcal{A}_Y$ be the corresponding K-affinoid algebras. Let $[\;]_x$ be the bounded multiplicative seminorm on $\mathcal{A}_X$ corresponding to the sup norm on X via (2.16), and let $[\;]_y$ be the bounded multiplicative seminorm on $\mathcal{A}_Y$ corresponding to the sup norm on Y. Since $\varphi(X) = Y$, for each $f \in \mathcal{A}_Y$, we have $f \circ \varphi \in \mathcal{A}_X$, with $[f]_y = [f \circ \varphi]_x$. Viewing $[\;]_x$ and $[\;]_y$ as bounded multiplicative seminorms on $K[T]$ via (2.15), it follows that $\varphi(x) = y$. $\square$

In Rivera-Letelier's approach to $\mathbb{P}^1_{\text{Berk}}$, a fundamental result proved early in the theory is that for each $s < r$ sufficiently near r, the image under

φ of the open annulus $D(a,r)^-\backslash D(a,s) \subset \mathbb{P}^1(K)$ is another annulus [81, Proposition 3.1]. Rivera-Letelier's description of the action of $\varphi(T)$ on $\mathbb{P}^1_{\text{Berk}}$ is based on this.

We will not establish Rivera-Letelier's annulus theorem until much later (Lemma 9.33). However, if $x \in \mathbb{P}^1_{\text{Berk}}$ corresponds to $D(a,r)$, it is easy to see that knowing the action of φ on the collection of annuli $D(a,r)^-\backslash D(a,s)$ with $s \to r$ determines its action on x:

PROPOSITION 2.19. *Let $\varphi \in K(T)$ be a nonconstant rational function, and suppose $x \in \mathbb{A}^1_{\text{Berk}}$ is a point of type II or III which corresponds to a disc $D(a,r)$ under Berkovich's classification. Suppose also that for each $s < r$ sufficiently close to r, the image of the annulus $D(a,r)^-\backslash D(a,s)$ is an annulus $D(b,R)^-\backslash D(b,S)$ (resp. an annulus $D(b,S)^-\backslash D(b,R)$). Then $\varphi(x)$ corresponds to $D(b,R)$ under Berkovich's classification.*

Similarly, if $x \in \mathbb{P}^1_{\text{Berk}}$ is of type IV and corresponds to a sequence of nested discs $\{D(a_i,r_i)\}$ under Berkovich's classification and for sufficiently large i the image of $D(a_i,r_i)$ under φ is a disc $D(b_i,R_i)$, then $\varphi(x)$ corresponds to $\{D(b_i,R_i)\}$ under Berkovich's classification.

PROOF. First suppose x is of type II or III. Let $y = \varphi(x)$, and let y_1 be the point corresponding to $D(b,R)$. We will show that each neighborhood of y contains y_1 and hence that $y = y_1$.

Let U be a neighborhood of y. Since $\varphi : \mathbb{P}^1_{\text{Berk}} \to \mathbb{P}^1_{\text{Berk}}$ is continuous, $\varphi^{-1}(U)$ contains a neighborhood of x and hence contains an annulus $D(a,r)^-\backslash D(a,s)$ for some $s < r$. We can assume s is near enough to r that the hypotheses of the proposition hold. Hence U contains $D(b,R)^-\backslash D(b,S)$ (resp. $D(b,S)^-\backslash D(b,R)$). Since y_1 belongs to the closure of each of these annuli in $\mathbb{P}^1_{\text{Berk}}$, it follows that $y_1 \in U$.

Next suppose x is of type IV and that $\varphi(D(a_i,r_i)) = D(b_i,R_i)$ for each $i \geq N$. Since the $D(a_i,r_i)$ are nested, so are the $D(b_i,R_i)$ for $i \geq N$. Since $\bigcap_{i=N}^\infty D(a_i,r_i) = \emptyset$, we have $\bigcap_{i=N}^\infty D(b_i,R_i) = \emptyset$. Thus $\{D(b_i,R_i)\}_{i \geq N}$ corresponds to a point $y_2 \in \mathbb{A}^1_{\text{Berk}}$ of type IV. An argument like the one above shows that $\varphi(x) = y_2$. □

2.5. The tree structure on $\mathbb{H}_{\text{Berk}}$ and $\mathbb{P}^1_{\text{Berk}}$

Recall that there is a natural embedding of $\mathbb{P}^1(K)$ in $\mathbb{P}^1_{\text{Berk}}$ with dense image.

DEFINITION 2.20. The *Berkovich hyperbolic space* $\mathbb{H}_{\text{Berk}}$ is defined by
$$\mathbb{H}_{\text{Berk}} = \mathbb{P}^1_{\text{Berk}} \backslash \mathbb{P}^1(K) . \tag{2.17}$$

Thus $\mathbb{H}_{\text{Berk}}$ consists of the points of $\mathbb{P}^1_{\text{Berk}}$ of types II, III, and IV. We will write $\mathbb{H}^{\mathbb{Q}}_{\text{Berk}}$ for the set of all points of type II, and $\mathbb{H}^{\mathbb{R}}_{\text{Berk}}$ for the set of all points of type II or III.

Note that since ∞ is a type I point of $\mathbb{P}^1_{\text{Berk}}$, we can also define $\mathbb{H}_{\text{Berk}}$ as
$$\mathbb{H}_{\text{Berk}} = \mathbb{A}^1_{\text{Berk}} \backslash \mathbb{A}^1(K) .$$

Our next goal is to define the space of *finite subgraphs* of $\mathbb{P}^1_{\text{Berk}}$ and to prove that $\mathbb{P}^1_{\text{Berk}}$ can be identified with the inverse limit of all such subgraphs. For this, recall first that
$$\mathbb{P}^1_{\text{Berk}} = W_0 \cup W_\infty ,$$
where W_0, W_∞ are each homeomorphic to the closed Berkovich unit disc $\mathcal{D}(0,1)$ and $W_0 \cap W_\infty$ is homeomorphic to an annulus $\mathcal{A}(1,1)$ in each of the two pieces. Also, recall from Lemma 2.10 that $\mathbb{P}^1_{\text{Berk}}$ is uniquely path-connected, so that we may define the *convex hull* of a finite set of points $\zeta_1, \ldots, \zeta_n \in \mathbb{P}^1_{\text{Berk}}$ by either of the following equivalent definitions:

(A) the smallest connected subset of $\mathbb{P}^1_{\text{Berk}}$ containing $\zeta_1, \ldots, \zeta_n$,
(B) the union of all arcs $[\zeta_i, \zeta_j]$ connecting ζ_i to ζ_j, for $1 \leq i < j \leq n$.

By a *finite subgraph* of $\mathbb{P}^1_{\text{Berk}}$, we mean the convex hull of a finite set of points $\zeta_1, \ldots, \zeta_n$ in $\mathbb{H}_{\text{Berk}}$. As we will see in §2.7 below, there is a canonical metric on every finite subgraph Γ of $\mathbb{P}^1_{\text{Berk}}$ which makes Γ into a finite $\mathbb{R}$-tree in a natural way. However, for now, we just think of Γ as a compact Hausdorff topological space.

Write $\mathcal{F}$ for the collection of all finite subgraphs of $\mathbb{P}^1_{\text{Berk}}$. Similarly, write $\mathcal{F}^{\mathbb{Q}}$ (resp. $\mathcal{F}^{\mathbb{R}}$) for the subset of $\mathcal{F}$ consisting of graphs which are the convex hull of a finite set of points $\zeta_1, \ldots, \zeta_n \in \mathbb{H}^{\mathbb{Q}}_{\text{Berk}}$ (resp. $\mathbb{H}^{\mathbb{R}}_{\text{Berk}}$).

In §1.4, when we considered the structure of the space $\mathcal{D}(0,1)$ as an inverse limit of finite $\mathbb{R}$-trees, we only looked at finite subgraphs of $\mathcal{D}(0,1)$ which were rooted at the Gauss point. It is therefore convenient to define
$$\mathcal{F}_{\text{Gauss}} = \{\Gamma \in \mathcal{F} : \zeta_{\text{Gauss}} \in \Gamma\}$$
and to define $\mathcal{F}^{\mathbb{Q}}_{\text{Gauss}}$ and $\mathcal{F}^{\mathbb{R}}_{\text{Gauss}}$ accordingly. There is a natural bijection between $\mathcal{F}_{\text{Gauss}} \cap W_0$ (resp. $\mathcal{F}_{\text{Gauss}} \cap W_\infty$) and the set $\mathcal{F}_0$ defined in §1.4.

The set $\mathcal{F}$ is naturally a directed set under inclusion, i.e., $\Gamma \leq \Gamma'$ if and only if $\Gamma \subseteq \Gamma'$. Similarly, all of the other sets $\mathcal{F}^{\mathbb{Q}}, \mathcal{F}^{\mathbb{R}}, \mathcal{F}_{\text{Gauss}}, \mathcal{F}^{\mathbb{Q}}_{\text{Gauss}}, \mathcal{F}^{\mathbb{R}}_{\text{Gauss}}$ are directed under inclusion. Thus whenever $\Gamma \leq \Gamma'$, there is an inclusion
$$i_{\Gamma,\Gamma'} : \Gamma \to \Gamma' .$$

Just as in §1.4, there is also a *retraction map* $r_{\Gamma',\Gamma} : \Gamma' \to \Gamma$ defined whenever $\Gamma \leq \Gamma'$. These retraction maps are again compatible with the partial order on $\mathcal{F}$, in the sense that if $\Gamma \leq \Gamma' \leq \Gamma''$, then

(2.18) $$r_{\Gamma'',\Gamma} = r_{\Gamma',\Gamma} \circ r_{\Gamma'',\Gamma'} .$$

By (2.18), the inverse limit $\varprojlim_{\Gamma \in \mathcal{F}} \Gamma$ is well-defined, and by Tychonoff's theorem, this inverse limit is a compact Hausdorff topological space.

The main structural result concerning finite subgraphs of $\mathbb{P}^1_{\text{Berk}}$ is

THEOREM 2.21. *There is a canonical homeomorphism*
$$\mathbb{P}^1_{\text{Berk}} \cong \varprojlim_{\Gamma \in \mathcal{F}} \Gamma ,$$

i.e., $\mathbb{P}^1_{\text{Berk}}$ *can be identified with the inverse limit of all of its finite subgraphs. Moreover, the same result holds if the directed set $\mathcal{F}$ is replaced by any of $\mathcal{F}^{\mathbb{Q}}, \mathcal{F}^{\mathbb{R}}, \mathcal{F}_{\text{Gauss}}, \mathcal{F}^{\mathbb{Q}}_{\text{Gauss}}, \mathcal{F}^{\mathbb{R}}_{\text{Gauss}}$.*

PROOF. Since the directed sets $\mathcal{F}$ and $\mathcal{F}_{\text{Gauss}}$ are cofinal, it suffices to prove the first statement with $\mathcal{F}$ replaced by $\mathcal{F}_{\text{Gauss}}$. The result then follows from Proposition 1.19, using the identification $\mathbb{P}^1_{\text{Berk}} = W_0 \cup W_\infty$.

The last statement follows from the proof of Proposition 1.19, since the restriction to W_0 or W_∞ of each of $\mathcal{F}^{\mathbb{Q}}, \mathcal{F}^{\mathbb{R}}, \mathcal{F}_{\text{Gauss}}, \mathcal{F}^{\mathbb{Q}}_{\text{Gauss}}, \mathcal{F}^{\mathbb{R}}_{\text{Gauss}}$ will also separate points (by the argument given). □

Theorem 2.21 is conceptually quite important, since it allows us to visualize the complicated, infinitely branched tree $\mathbb{P}^1_{\text{Berk}}$ as a limit of *finite* $\mathbb{R}$-trees. This is the main way we will think of $\mathbb{P}^1_{\text{Berk}}$ in this book. For example, the existence of a Laplacian operator on $\mathbb{P}^1_{\text{Berk}}$ with nice properties will be deduced in a straightforward way from the existence of a natural Laplacian operator on a finite metrized graph, together with the fact that if $\Gamma_1 \leq \Gamma_2$, then the Laplacians on Γ_1 and Γ_2 are "compatible" in an appropriate sense.

As a complement to Theorem 2.21, we can also consider the *direct limit* $\varinjlim_{\Gamma \in \mathcal{F}} \Gamma$ with respect to the inclusion maps $i_{\Gamma, \Gamma'}$. In this case, the limit is isomorphic (as a set) to $\mathbb{H}_{\text{Berk}}$. We omit the straightforward proof of

THEOREM 2.22. *There are canonical bijections*

$$\mathbb{H}_{\text{Berk}} \cong \varinjlim_{\Gamma \in \mathcal{F}} \Gamma \cong \varinjlim_{\Gamma \in \mathcal{F}_{\text{Gauss}}} \Gamma,$$

$$\mathbb{H}^{\mathbb{R}}_{\text{Berk}} \cong \varinjlim_{\Gamma \in \mathcal{F}^{\mathbb{R}}} \Gamma \cong \varinjlim_{\Gamma \in \mathcal{F}^{\mathbb{R}}_{\text{Gauss}}} \Gamma, \quad \mathbb{H}^{\mathbb{Q}}_{\text{Berk}} \cong \varinjlim_{\Gamma \in \mathcal{F}^{\mathbb{Q}}} \Gamma \cong \varinjlim_{\Gamma \in \mathcal{F}^{\mathbb{Q}}_{\text{Gauss}}} \Gamma.$$

REMARK 2.23. Recall that a set $U \subset \mathbb{H}_{\text{Berk}}$ is open for the *direct limit topology* on $\mathbb{H}_{\text{Berk}} \cong \varinjlim_{\Gamma \in \mathcal{F}} \Gamma$ if and only if $U \cap \Gamma$ is open in Γ for each $\Gamma \in \mathcal{F}$. The direct limit topology is stronger than the subspace topology induced by the inclusion $\mathbb{H}_{\text{Berk}} \subseteq \mathbb{P}^1_{\text{Berk}}$ (cf. Remark 1.18). It is also stronger than the topology on $\mathbb{H}_{\text{Berk}}$ induced by the metric ρ introduced in §2.7. We will not use the direct limit topology in this book.

2.6. Discs, annuli, and simple domains

DEFINITION 2.24. A *domain* is a nonempty connected open set in $\mathbb{P}^1_{\text{Berk}}$. A *subdomain* of a domain U is a domain V with $\overline{V} \subset U$.

In this section we will introduce some important types of domains.

DEFINITION 2.25. An *open Berkovich disc* is a domain $U \subset \mathbb{P}^1_{\text{Berk}}$ with exactly one boundary point, which is of type II or III; that is, U is a connected component of $\mathbb{P}^1_{\text{Berk}} \backslash \{x\}$, for some $x \in \mathbb{H}^{\mathbb{R}}_{\text{Berk}}$. If x is of type II, U will be called a *strict open Berkovich disc*. A *generalized open Berkovich disc* is a connected component U of $\mathbb{P}^1_{\text{Berk}} \backslash \{x\}$, where $x \in \mathbb{P}^1_{\text{Berk}}$ is arbitrary.

It is easy to see that open Berkovich discs are precisely the sets of the form $\mathcal{D}(a,r)^-$ and $\mathbb{P}^1_{\text{Berk}}\backslash \mathcal{D}(a,r)$, and strict open Berkovich discs are those with $r \in |K^\times|$.

In Appendix B, we associate to each point $x \in \mathbb{P}^1_{\text{Berk}}$ a set T_x of *tangent vectors* or *tangent directions* (we use the terms interchangeably). By definition, a tangent vector $\vec{v} \in T_x$ is an equivalence class of paths originating at x, where two paths $[x,a]$ and $[x,b]$ are considered equivalent if they have a nontrivial common initial segment. The connected components of $\mathbb{P}^1_{\text{Berk}}\backslash\{x\}$ are in one-to-one correspondence with the tangent vectors $\vec{v} \in T_x$, and we write $\mathcal{B}_x(\vec{v})^-$ for the connected component of $\mathbb{P}^1_{\text{Berk}}\backslash\{x\}$ corresponding to $\vec{v}$:

$$(2.19) \qquad \mathcal{B}_x(\vec{v})^- \;=\; \{a \in \mathbb{P}^1_{\text{Berk}}\backslash\{x\} : [x,a] \in \vec{v}\}\ .$$

If x is of type II or III, then $\mathcal{B}_x(\vec{v})^-$ is an open Berkovich disc. If x is of type I or IV, there is only one tangent direction $\vec{v} \in T_x$, and in that case $\mathcal{B}_x(\vec{v})^- = \mathbb{P}^1_{\text{Berk}}\backslash\{x\}$ is a generalized open Berkovich disc.

Let Γ be a finite subgraph of $\mathbb{P}^1_{\text{Berk}}$, and let $r_\Gamma : \mathbb{P}^1_{\text{Berk}} \to \Gamma$ be the natural retraction map. If $V \subset \Gamma$ is an open set, then $U = r_\Gamma^{-1}(V)$ is an open subset of $\mathbb{P}^1_{\text{Berk}}$. Of particular interest is the case where $V = \Gamma^0 = \Gamma\backslash\partial\Gamma$ is the interior of Γ in its metric topology.

When Γ is a segment, $U = r_\Gamma^{-1}(\Gamma^0)$ is an annulus:

DEFINITION 2.26. An *open Berkovich annulus* is a domain $U \subset \mathbb{P}^1_{\text{Berk}}$ with exactly two boundary points x, y, which are of type II or III; that is, $U = r_\Gamma^{-1}((x,y))$, where $\Gamma = [x,y]$ is a segment in $\mathbb{H}^{\mathbb{R}}_{\text{Berk}}$. If x and y are both of type II, U will be called a *strict open Berkovich annulus*.

A *generalized open Berkovich annulus* is a domain with two arbitrary distinct boundary points in $\mathbb{P}^1_{\text{Berk}}$. Thus, a generalized open Berkovich annulus is a set of the form $U = r_\Gamma^{-1}((x,y))$ where $x,y \in \mathbb{P}^1_{\text{Berk}}$ are arbitrary distinct points and $\Gamma = [x,y]$.

For arbitrary finite subgraphs Γ, we identify three types of domains, depending on the nature of the endpoints of Γ:

DEFINITION 2.27. A *simple domain* is a domain $U \subset \mathbb{P}^1_{\text{Berk}}$ which is either an open Berkovich disc or has the form $U = r_\Gamma^{-1}(\Gamma^0)$ where Γ is a nontrivial finite subgraph of $\mathbb{H}^{\mathbb{R}}_{\text{Berk}}$; that is, the endpoints of Γ are of type II or III.

A *strict simple domain* is a domain U which is either a strict open Berkovich disc, or has the form $U = r_\Gamma^{-1}(\Gamma^0)$ where Γ is a nontrivial finite subgraph of $\mathbb{H}_{\text{Berk}}$ whose endpoints are all of type II.

A *finite-dendrite domain* is a domain U which is either a generalized open Berkovich disc with boundary point $x \in \mathbb{H}_{\text{Berk}}$ or has the form $U = r_\Gamma^{-1}(\Gamma^0)$ for an arbitrary nontrivial finite subgraph Γ of $\mathbb{H}_{\text{Berk}}$.

Clearly each strict simple domain is a simple domain, and each simple domain is a finite-dendrite domain. The reason for the terminology "finite-dendrite domain" will become clear in §7.1. It is easy to see that the simple domains (resp. strict simple domains, finite-dendrite domains) form a basis for Berkovich topology on $\mathbb{P}^1_{\text{Berk}}$.

Simple domains will be the open sets used most often in this book. The proof of the following lemma is straightforward, and we leave it to the reader as an exercise in unraveling the definitions.

LEMMA 2.28. *For a subset $U \subseteq \mathbb{P}^1_{\text{Berk}}$, the following are equivalent:*
- (A) *U is a simple domain.*
- (B) *$U = r_\Gamma^{-1}(V)$, where V is a proper connected open subset of a finite subgraph $\Gamma \subset \mathbb{H}^\mathbb{R}_{\text{Berk}}$.*
- (C) *U is a finite intersection of open Berkovich discs.*
- (D) *U is a connected open set whose boundary consists of a finite (but nonzero) number of points of type II or type III.*
- (E) *U is a connected open affinoid domain.*

Furthermore, U is a strict simple domain if and only if it is a connected strict open affinoid.

If $x \in \mathbb{P}^1(K)$ (i.e., x is of type I), then the open Berkovich discs containing x form a neighborhood base for x. If x is of type IV, corresponding to a nested sequence of discs $\{D(a_i, r_i)\}$, then the open Berkovich discs $\mathcal{D}(a_i, r_i)^-$ form a neighborhood base for x. However, if $x \in \mathbb{H}^\mathbb{R}_{\text{Berk}}$ is of type II or III, then open Berkovich discs do not suffice to form a neighborhood base: one must use more general simple domains. More precisely, if x is of type III (corresponding to the irrational disc $D(a, s)$), then the open Berkovich annuli of the form $\mathcal{D}(a, R)^- \backslash \mathcal{D}(a, r)$ with $r < s < R$ form a neighborhood base for x, while if x is of type II (corresponding to the rational disc $D(a, s)$) then the simple domains of the form $\mathcal{D}(a, R)^- \backslash (\mathcal{D}(a_1, r_1) \cup \cdots \cup \mathcal{D}(a_k, r_k))$ with $r_i < s < R$ for $i = 1, \ldots, k$ form a neighborhood base for x.

2.7. The strong topology

In this section we will introduce another topology on $\mathbb{P}^1_{\text{Berk}}$ and $\mathbb{H}_{\text{Berk}}$, called the *strong topology* or *path distance topology*, which is strictly finer than the Berkovich topology.

Recall that $\mathbb{P}^1_{\text{Berk}}$ can be identified with the disjoint union of $\mathcal{D}(0, 1)$ and the subset $\mathbb{P}^1_{\text{Berk}} \backslash \mathcal{D}(0, 1)$, which is homeomorphic to $\mathcal{D}(0, 1)^-$ via the map $\psi = \overline{\psi}_\infty$ (which on the level of type I points takes z to $1/z$). Using this identification, we can extend the "small" and "big" metrics $d(x, y)$ and $\rho(x, y)$ from the previous chapter to $\mathbb{P}^1_{\text{Berk}}$ and $\mathbb{H}_{\text{Berk}}$, respectively.

Following Favre and Rivera-Letelier, we will often call the Berkovich topology on $\mathbb{P}^1_{\text{Berk}}$ (and corresponding subspace topology on $\mathbb{H}_{\text{Berk}}$) the *weak topology*. We call the topology on $\mathbb{P}^1_{\text{Berk}}$ defined by the metric $d(x, y)$ (and the associated subspace topology on $\mathbb{H}_{\text{Berk}}$) the *strong topology*. It turns out that the metric $\rho(x, y)$ induces the strong topology on $\mathbb{H}_{\text{Berk}}$ as well: indeed, in the strong topology, each point of $\mathbb{H}_{\text{Berk}}$ has a neighborhood on which $d(x, y)$ and $\rho(x, y)$ are mutually bounded in terms of each other.

2.7. THE STRONG TOPOLOGY

In the strong topology, $\mathbb{P}^1_{\text{Berk}}$ is a complete metric space but is not even locally compact. By contrast, in the weak topology, it is compact but not in general metrizable.

We first define the "small metric" $d(x,y)$. Identify $\mathbb{P}^1_{\text{Berk}}$ with the disjoint union of $\mathcal{D}(0,1)$ and $\mathbb{P}^1_{\text{Berk}} \backslash \mathcal{D}(0,1) \cong \mathcal{D}(0,1)^-$ using the map ψ, as above. Recall that for $x, y \in \mathcal{D}(0,1)$, we have

$$\begin{aligned} d(x,y) &= (\text{diam}(x \vee y) - \text{diam}(x)) + (\text{diam}(x \vee y) - \text{diam}(y)) \\ &= 2\,\text{diam}(x \vee y) - \text{diam}(x) - \text{diam}(y)\,, \end{aligned}$$

which is the distance obtained by going from x to y along the geodesic path through their join $x \vee y$. In general, for $x, y \in \mathbb{P}^1_{\text{Berk}}$, put

$$(2.20) \qquad \text{diam}(x) = \begin{cases} \text{diam}(x) & \text{if } x \in \mathcal{D}(0,1)\,, \\ \text{diam}(\psi(x)) & \text{if } x \in \mathbb{P}^1_{\text{Berk}} \backslash \mathcal{D}(0,1) \end{cases}$$

and let $x \vee y$ be the point where the paths $[x, \zeta_{\text{Gauss}}]$ and $[y, \zeta_{\text{Gauss}}]$ first meet (which will be ζ_{Gauss} if x and y are in different components of $\mathbb{P}^1_{\text{Berk}} \backslash \{\zeta_{\text{Gauss}}\}$). For $x, y \in \mathbb{P}^1_{\text{Berk}}$, define

$$(2.21) \qquad d(x,y) = 2\,\text{diam}(x \vee y) - \text{diam}(x) - \text{diam}(y)\,.$$

Under the metric $d(x,y)$, each point of $\mathbb{P}^1_{\text{Berk}}$ has distance at most 1 from ζ_{Gauss}, and the maximal distance between two points is 2.

Alternately, $d(x,y)$ can be defined by viewing $\mathbb{P}^1_{\text{Berk}}$ as a parametrized rooted tree, with the root ζ_{Gauss} and the parametrization map $\alpha : \mathbb{P}^1_{\text{Berk}} \to \mathbb{R}_{\geq 0}$ given by

$$\alpha(z) = \begin{cases} 1 - \text{diam}(z) & \text{if } z \in \mathcal{D}(0,1)\,, \\ 1 - \text{diam}(\psi(z)) & z \notin \mathcal{D}(0,1)\,. \end{cases}$$

It is easy to verify that $\mathbb{P}^1_{\text{Berk}}$, equipped with this root and parametrization, satisfies axioms (P1)–(P5) in Appendix B for a parametrized rooted tree, and that the metric $d(x,y)$ we have just described is the associated distance function. In particular, $(\mathbb{P}^1_{\text{Berk}}, d)$ is itself an $\mathbb{R}$-tree.

The metric $d(x,y)$ is an extension of (twice) the classical spherical distance on $\mathbb{P}^1(K)$. If we identify $\mathbb{P}^1(K)$ with $K \cup \{\infty\}$ and put $0 = 1/\infty$, then for $x, y \in \mathbb{P}^1(K)$ the spherical distance is defined by

$$\|x,y\| = \begin{cases} |x - y| & \text{if } |x|, |y| \leq 1, \\ |1/x - 1/y| & \text{if } |x|, |y| > 1, \\ 1 & \text{otherwise.} \end{cases}$$

Given $x, y \in D(0,1)$, the smallest disc containing x and y is $D(x, |x-y|)$, so $\text{diam}(x \vee y) = |x - y| = \|x,y\|$. Furthermore, viewing x and y as degenerate discs of radius 0, we have $\text{diam}(x) = \text{diam}(y) = 0$. Thus $d(x,y) = 2\|x,y\|$. Analogous computations show this holds for all $x, y \in \mathbb{P}^1(K)$.

The metric $d(x,y)$ plays an important role in the work of Favre and Rivera-Letelier [46, 47, 48], because the action of a nonconstant rational function $\varphi(T) \in K(T)$ is Lipschitz continuous with respect to $d(x,y)$ (see

Proposition 9.37). However, for us it will be much more useful to consider the extension of the big metric to $\mathbb{H}_{\text{Berk}}$. The big metric is completely canonical (see Proposition 2.30 below), whereas the extension of the small metric to $\mathbb{P}^1_{\text{Berk}}$ depends on a choice of coordinates for $\mathbb{P}^1$. In addition, we will see later that functions of the form $\log|f|$, where $f \in K(T)$ is a nonzero rational function, are *piecewise linear* with respect to the big metric on $\mathbb{H}_{\text{Berk}}$.

For these reasons and others, from this point onward we will mainly consider the big metric, and rather than using the awkward terms "small metric" and "big metric", we will refer to the big metric $\rho(x,y)$ on $\mathbb{H}_{\text{Berk}}$ as the *path distance metric*. We now give a concrete description of this metric.

Recall from Theorem 2.22 that $\mathbb{H}_{\text{Berk}}$ can be identified with the union of all finite subgraphs $\Gamma \in \mathcal{F}$, and $\mathbb{H}^{\mathbb{R}}_{\text{Berk}}$ can be identified with the union of all finite subgraphs $\Gamma \in \mathcal{F}^{\mathbb{R}}$. For simplicity, we first describe the metric ρ on $\mathbb{H}^{\mathbb{R}}_{\text{Berk}}$, and then we will extend it to $\mathbb{H}_{\text{Berk}}$ by continuity.

Recall also that $\mathcal{D}(0,1) \cap \mathbb{H}_{\text{Berk}}$ was described in §1.4 as the union of all graphs of discs Γ_S, where $S = \{D(a_1, t_1), \ldots, D(a_n, t_n)\}$ runs over all finite sets of discs contained in $D(0,1)$. A similar construction applies for any $\mathcal{D}(0,r)$, taking

$$\Gamma_S = \bigcup_{i=1}^{n} [t_i, r]_{D(a_i, r_i)} ,$$

where $[t_i, r]_{D(a_i, t_i)} = \{D(a_i, t) : t_i \leq t \leq r\}$ is a 'line of discs'. Under this parametrization, the segment between $D(a, r_1)$ and $D(a, r_2)$ has length $|r_1 - r_2|$. However, as we have discussed, it is better to logarithmically reparametrize each line of discs $[r_1, r_2]_{D(a, r_1)}$, putting

$$\langle \log(r_1), \log(r_2) \rangle_{D(a, r_1)} = \{D(a, q_v^t) : \log_v(r_1) \leq t \leq \log_v(r_2)\}$$

and giving it length $|\log_v(r_2) - \log_v(r_1)|$. Carrying out this reparametrization for each $\mathcal{D}(0, r)$ and viewing $\mathbb{H}^{\mathbb{R}}_{\text{Berk}}$ as the union of $\mathbb{H}^{\mathbb{R}}_{\text{Berk}} \cap \mathcal{D}(0, r)$ over all $r > 0$, we obtain the path distance metric $\rho(x, y)$ on $\mathbb{H}^{\mathbb{R}}_{\text{Berk}}$. It follows easily from the definitions that ρ extends by continuity to all of $\mathbb{H}_{\text{Berk}}$ and that ρ is indeed a *metric* on $\mathbb{H}_{\text{Berk}}$. In particular, all points of $\mathbb{H}_{\text{Berk}}$ are at finite distance from each other.

Concretely, if $x, y \in \mathbb{H}^{\mathbb{R}}_{\text{Berk}}$ correspond to the discs $D(a, r)$ and $D(b, s)$, respectively, and if $D(a, R)$ is the smallest disc containing both $D(a, r)$ and $D(b, s)$, then $R = \max(r, s, |a - b|)$ and

$$\begin{aligned}\rho(x, y) &= (\log_v R - \log_v r) + (\log_v R - \log_v s) \\ &= 2\log_v R - \log_v r - \log_v s .\end{aligned}$$
(2.22)

This formula is easily extended to arbitrary points $x, y \in \mathbb{H}_{\text{Berk}}$: consider the paths from x to ∞ and y to ∞, and let $x \vee_\infty y$ be the point where they first meet. If x corresponds to a nested sequence of discs $\{D(a_i, r_i)\}$, put $\text{diam}_\infty(x) = \inf r_i$. Then
(2.23)
$$\rho(x, y) = 2\log_v(\text{diam}_\infty(x \vee_\infty y)) - \log_v(\text{diam}_\infty(x)) - \log_v(\text{diam}_\infty(y)) .$$

We can also extend $\rho(x,y)$ to a singular function on all of $\mathbb{P}^1_{\text{Berk}}$ by setting $\rho(x,\zeta) = \rho(\zeta,x) = \infty$ for each $x \in \mathbb{H}_{\text{Berk}}$ and each $\zeta \in \mathbb{P}^1(K)$, and $\rho(\zeta,\xi) = \infty$ for all distinct points $\zeta, \xi \in \mathbb{P}^1(K)$. We set $\rho(\zeta,\zeta) = 0$ for $\zeta \in \mathbb{P}^1(K)$. We will sometimes use this extension, although for the most part we think of ρ as defined only on $\mathbb{H}_{\text{Berk}}$.

As noted above, we call the topology on $\mathbb{H}_{\text{Berk}}$ defined by the path metric $\rho(x,y)$ (and the topology defined on $\mathbb{P}^1_{\text{Berk}}$ by $d(x,y)$) the *strong topology*. Using the definitions (2.21) and (2.23) and the fact that $\log_v(t)$ is uniformly continuous on each bounded interval $[1/C, C]$, it is easy to check that both $d(x,y)$ and $\rho(x,y)$ induce the strong topology on $\mathbb{H}_{\text{Berk}}$.

It is an important fact that $\mathbb{H}_{\text{Berk}}$ and $\mathbb{P}^1_{\text{Berk}}$ are complete in the strong topology:

PROPOSITION 2.29. *$\mathbb{H}_{\text{Berk}}$ is a complete metric space under $\rho(x,y)$. Likewise, $\mathbb{P}^1_{\text{Berk}}$ is a complete metric space under $d(x,y)$.*

PROOF. We first show that $\mathbb{H}_{\text{Berk}}$ is complete under $\rho(x,y)$. Let $\langle x_\alpha \rangle_{\alpha \in A}$ be a Cauchy net for $\rho(x,y)$. Thus, for each $\varepsilon > 0$, there is an N_ε such that $\rho(x_\alpha, x_\beta) < \varepsilon$ for all $\alpha, \beta > N_\varepsilon$. Since $\mathbb{P}^1_{\text{Berk}}$ is compact in the weak topology (Proposition 2.6), the net $\{x_\alpha\}_{\alpha \in A}$ has a limit point $\xi \in \mathbb{P}^1_{\text{Berk}}$ in the weak topology. We will show that ξ is unique and belongs to $\mathbb{H}_{\text{Berk}}$ and that $\{x_\alpha\}_{\alpha \in A}$ converges to ξ in the strong topology.

We first show that $\xi \in \mathbb{H}_{\text{Berk}}$. Suppose to the contrary that $\xi \in \mathbb{P}^1(K)$, and let N_1 be such that $\rho(x_\alpha, x_\beta) < 1$ for all $\alpha, \beta > N_1$. Fix $\eta > N_1$. Noting that the path from x_η to ξ has infinite length, let ζ be the point on that path with $\rho(\zeta, x_\eta) = 2$. We claim that the x_α for $\alpha > N_1$ all belong to the same component of $\mathbb{P}^1_{\text{Berk}} \setminus \{\zeta\}$ as x_η. If this were not so, some x_α would belong to a different component, and hence the path from x_α to x_η would contain ζ. Thus $\rho(x_\alpha, x_\eta) > 2$, which is impossible. Hence the component of $\mathbb{P}^1_{\text{Berk}}$ containing ξ is an open set which contains no x_α with $\alpha > N_1$, contradicting the fact that ξ is a limit point of $\{x_\alpha\}_{\alpha \in A}$ in the weak topology.

We now show that $\{x_\alpha\}_{\alpha \in A}$ converges to ξ in the strong topology. Put $\delta = \limsup_\alpha \rho(x_\alpha, \xi)$. Here δ is finite, since $\xi \in \mathbb{H}_{\text{Berk}}$. If $\delta > 0$, put $\varepsilon = \delta/3$ and let N_ε be such that $\rho(x_\alpha, x_\beta) < \varepsilon$ for $\alpha, \beta > N_\varepsilon$. Since $\delta > 0$, there is an $\eta > N_\varepsilon$ such that $\rho(x_\eta, \xi) > 2\delta/3 = 2\varepsilon$. Let ζ in the path from ξ to x_η be such that $\rho(\xi, \zeta) = \varepsilon$. By an argument similar to the one above, the component of $\mathbb{P}^1_{\text{Berk}} \setminus \{\zeta\}$ containing ξ cannot contain any x_α with $\alpha > N_\varepsilon$. This again contradicts the fact that ξ is a limit point of $\{x_\alpha\}_{\alpha \in A}$ in the weak topology. Hence, $\delta = 0$.

Finally, suppose ξ' were another limit point of $\{x_\alpha\}_{\alpha \in A}$. Since

$$\limsup_\alpha \rho(x_\alpha, \xi) = \limsup_\alpha \rho(x_\alpha, \xi') = 0 ,$$

the triangle inequality shows that $\rho(\xi, \xi') = 0$. Thus $\xi = \xi'$.

The proof that $\mathbb{P}^1_{\text{Berk}}$ is complete under $d(x,y)$ is similar. $\square$

The path distance metric ρ is canonical, in the following sense.

PROPOSITION 2.30. *The path distance metric $\rho(x,y)$ on $\mathbb{H}_{\text{Berk}}$ is independent of the choice of homogeneous coordinates on $\mathbb{P}^1$, in the sense that if $h(z) \in K(z)$ is a linear fractional transformation, then $\rho(h(x), h(y)) = \rho(x,y)$ for all $x, y \in \mathbb{H}_{\text{Berk}}$.*

PROOF. Let $h(z) = (az+b)/(cz+d) \in K(z)$ be a linear fractional transformation. It induces a continuous automorphism h of $\mathbb{P}^1_{\text{Berk}}$. Since points of type II are dense in each finite subgraph of $\mathbb{P}^1_{\text{Berk}}$, it suffices to show that $\rho(h(x), h(y)) = \rho(x, y)$ for all $x, y \in \mathbb{H}_{\text{Berk}}$ of type II. By formula (2.23), we may assume without loss of generality that x, y correspond to discs $D(p, r_1)$, $D(p, r_2)$ with $D(p, r_1) \subseteq D(p, r_2)$. Finally, since an arbitrary linear fractional transformation can be written as a composition of affine maps and inversions, it suffices to show that the path distance is preserved by these two types of maps.

First suppose $h(z) = az + b$ is affine. The image of $D(p,r)$ under $h(z)$ is $D(ap+b, |a|r)$. If $x, y \in \mathbb{P}^1_{\text{Berk}}$ correspond to $D(p, r_1) \subseteq D(p, r_2)$, it follows that
$$\begin{aligned}\rho(h(x), h(y)) &= |\log_v(|a|r_1) - \log_v(|a|r_2)| \\ &= |\log_v(r_1) - \log_v(r_2)| = \rho(x,y) .\end{aligned}$$

Next suppose $h(z) = 1/z$, and let $x, y \in \mathbb{P}^1_{\text{Berk}}$ correspond to discs $D(q, r_1) \subseteq D(q, r_2)$. Composing with an affine transformation, we may assume without loss of generality that $0 \notin D(q, r_2)$. Recall from Proposition 2.18 that if x corresponds to a disc $D(a,r)$ and $\varphi \in K(T)$, then $\varphi(x)$ corresponds to a disc $D(b, R)$ if and only if there is a closed affinoid of the form $D(b,R) \backslash \bigcup_{i=1}^m D(b_i, R)^-$ which is the image under φ of a closed affinoid of the form $D(a,r) \backslash \bigcup_{i=1}^n D(a_i, r)^-$ for an appropriate choice of $a_1, \ldots, a_n$. If $D(a,r)$ does not contain 0, then $h(D(a,r)) = D(1/a, r/|a|^2)$. Since we are assuming that neither of the discs corresponding to x and y contains 0 and since $D(1/q, r_1/|q|^2) \subseteq D(1/q, r_2/|q|^2)$, it follows that
$$\begin{aligned}\rho(h(x), h(y)) &= |\log_v(r_1/|q|^2) - \log_v(r_2/|q|^2)| \\ &= |\log_v(r_1) - \log_v(r_2)| = \rho(x,y) .\end{aligned}$$
□

A useful point of view, adopted by Rivera-Letelier in his papers, is that $\rho(x,y)$ can be interpreted in terms of the modulus of an annulus. Indeed, if $x, y \in \mathbb{H}^{\mathbb{R}}_{\text{Berk}}$ correspond to closed discs $D(a, r_2) \subseteq D(a, r_1)$ in K, then one naturally associates to x and y the open annulus $A^-_{xy} = D(a, r_1)^- \backslash D(a, r_2)$, and $\rho(x,y) = \log_v(r_1/r_2)$ is nothing other than the *modulus* of this annulus. More generally, if $A^- = D_1^- \backslash D_2$ with D_1^- an open disc in $\mathbb{P}^1(K)$ and D_2 a closed disc in $\mathbb{P}^1(K)$ contained in D_1^- (i.e., if A^- is an *open annulus in* $\mathbb{P}^1(K)$), then one can define the modulus of A^- to be the modulus of $h(A^-)$ for any linear fractional transformation h which takes A^- to an open annulus of the form $D(a, r_1)^- \backslash D(a, r_2) \subset K$. Proposition 2.30 is then equivalent to the assertion that the modulus of an open annulus in $\mathbb{P}^1(K)$ is well-defined (a known fact from non-Archimedean analysis which we have implicitly proved again above). Proposition 2.30 is thus the non-Archimedean analogue of the well-known fact from complex analysis that the modulus of an open annulus in $\mathbb{C}$ (or, more generally, in the Riemann sphere $\mathbb{P}^1(\mathbb{C})$) is invariant under Möbius transformations.

2.8. Notes and further references

The idea for defining $\mathbb{P}^1_{\text{Berk}}$ via a "Proj" construction is due to Berkovich and can be found in his paper [**18**].

The notation $\mathbb{H}_{\text{Berk}}$ for the "Berkovich hyperbolic space" $\mathbb{P}^1_{\text{Berk}} \backslash \mathbb{P}^1(K)$ is an adaptation of Rivera-Letelier's notation in [**82, 81**]. Also, the idea that one can profitably think of $\mathbb{H}_{\text{Berk}}$ as an analogue of the Poincaré disc, or of hyperbolic 3-space, is something which we learned during conversations with Rivera-Letelier. One should not confuse $\mathbb{H}_{\text{Berk}}$ with the "Drinfeld upper half-plane" (see, e.g., [**18**] or [**58**]), which is a rather different object despite superficial similarities.

The concrete description of the action of a rational map in terms of Berkovich's classification theorem given in Proposition 2.18 forms the foundation for Rivera-Letelier's study of dynamics on $\mathbb{P}^1_{\text{Berk}}$.

We learned of the connection between the path distance metric ρ and moduli of annuli from Rivera-Letelier.

CHAPTER 3

Metrized graphs

Our approach to potential theory on $\mathbb{P}^1_{\text{Berk}}$ is based on the existence of a Laplacian on its finite subgraphs, regarded as metrized graphs for the metric induced by the path distance $\rho(x, y)$.

In this chapter, we recall some basic facts about metrized graphs from [**10**]. For the reader's convenience, we have used (with minor modifications) certain ideas from [**7**], [**10**], and [**37**] to make our presentation of the theory as self-contained as possible.

3.1. Definitions

Intuitively, a metrized graph is a finite graph whose edges are thought of as line segments having a well-defined length. In particular, Γ is a one-dimensional manifold except at finitely many "branch points", where it looks locally like an n-pointed star for some positive integer n. The path-length function along each edge extends to a metric $\rho(x, y)$ on all of Γ, making it a compact metric space.

A rigorous definition of a metrized graph (following [**100**]) is as follows.

DEFINITION 3.1. A *metrized graph* Γ is a compact, connected metric space such that for each $p \in \Gamma$, there are an $r_p > 0$ and an integer $n_p \geq 1$ such that p has a neighborhood $V_p(r_p)$ isometric to the star-shaped set

$$S(n_p, r_p) = \{z \in \mathbb{C} : z = te^{k \cdot 2\pi i/n_p} \text{ for some } 0 \leq t < r_p \text{ and some } k \in \mathbb{Z}\},$$

equipped with the path metric.

A finite $\mathbb{R}$-tree, as defined in §1.4, is the same thing as a metrized graph possessing no nontrivial cycles. Although $\mathbb{R}$-trees suffice for the purpose of studying $\mathbb{P}^1_{\text{Berk}}$, the more general concept of a metrized graph is needed to study Berkovich curves of higher genus. Since the theory of Laplacians is not much more difficult to set up on an arbitrary metrized graph than on a finite $\mathbb{R}$-tree, we describe the theory in this more general context.

By a *vertex set* for Γ, we mean a finite set of points S such that $\Gamma \backslash S$ is a union of open intervals, each of which has two distinct endpoints in Γ. A vertex set necessarily contains all endpoints and branch points of Γ. If Γ has loops, it also contains at least one interior point from each loop.

There is a close connection between metrized graphs and finite weighted graphs. Given a metrized graph Γ, any choice of a vertex set S for Γ gives rise to a weighted graph G which one may call a *model* for Γ; the weight of each edge in G is the reciprocal of its length in Γ. A different choice of vertex set leads to an equivalent weighted graph. Conversely, every weighted graph G determines a metrized graph in the obvious way, so there is a one-to-one correspondence between metrized graphs and equivalence classes of finite weighted graphs (see [**10**] and [**7**] for further details).

Often, when given a metrized graph Γ, one chooses without explicit comment a vertex set S, together with distinguished parametrizations of the edges of the corresponding model G. The definition of the Laplacian given below is independent of these implicit choices.

By an *isometric path* (or simply a *path*) in Γ, we will mean an injective length-preserving continuous map from the real interval $[0, L]$ into Γ. We will say that an isometric path $\gamma : [0, L] \to \Gamma$ *emanates from* p and *terminates at* q if $\gamma(0) = p$ and $\gamma(L) = q$. We call two isometric paths emanating from p *equivalent* if they share a common initial segment. For each $p \in \Gamma$, we let $T_p(\Gamma)$ (the projectivized "tangent space" at p) denote the set of equivalence classes of isometric paths in Γ emanating from p. It is easy to see that $|T_p(\Gamma)| = n_p$, i.e., there is a bijection between elements of T_p and the "edges" of Γ emanating from p.

It is useful to associate to each element of $T_p(\Gamma)$ a formal "unit tangent vector" $\vec{v}$ and to write $p + t\vec{v}$ instead of $\gamma(t)$, where $\gamma : [0, L] \to \Gamma$ is a representative path. If $f : \Gamma \to \mathbb{R}$ is a function and $\vec{v}$ is a formal unit tangent vector at p, then we will define the "*(one-sided) derivative of f in the direction $\vec{v}$*" to be

$$d_{\vec{v}}f(p) = \lim_{t \to 0^+} \frac{f(p + t\vec{v}) - f(p)}{t} = \lim_{t \to 0^+} \frac{f(\gamma(t)) - f(p)}{t},$$

provided the limit exists as a finite number.

3.2. The space CPA(Γ)

Let CPA(Γ) be the space of continuous, piecewise affine real-valued functions on Γ. (By a *piecewise affine function* f, we mean that there is a vertex set S_f for Γ such that f is affine on each edge in $\Gamma \backslash S_f$ with respect to an arclength parametrization of that edge.) If $f \in \text{CPA}(\Gamma)$, then clearly the directional derivatives $d_{\vec{v}} f(p)$ are defined for all $p \in \Gamma$ and all $\vec{v} \in T_p(\Gamma)$.

Chinburg and Rumely [**37**] introduced a Laplacian operator on CPA(Γ). Their Laplacian is a map from CPA(Γ) to the space of discrete signed measures on Γ. We take our Laplacian to be the negative of theirs and put

$$(3.1) \qquad \Delta(f) = \sum_{p \in \Gamma} \left(- \sum_{\vec{v} \in T_p(\Gamma)} d_{\vec{v}} f(p) \right) \delta_p(x),$$

where δ_p is the Dirac unit measure at p. The operator Δ coincides, in a precise sense, with the usual combinatorial Laplacian on a finite weighted graph; see [**7**] for details. By abuse of notation, we will write $\Delta(f)(p)$ for $\Delta(f)(\{p\}) = -\sum_{\vec{v} \in T_p(\Gamma)} d_{\vec{v}} f(p)$.

3.2. THE SPACE CPA(Γ)

If $f, g \in \text{CPA}(\Gamma)$, let S be a vertex set for Γ such that f and g are both affine on each edge of $\Gamma \backslash S$. Fix an orientation for each edge e_i of $\Gamma \backslash S$, and let $\gamma_i : [0, L_i] \to e_i$ be the arclength parametrization. By abuse of notation, we view γ_i as an identification and write $f(x) = f(\gamma_i(x))$ and $g(x) = g(\gamma_i(x))$ on e_i. Let $f'(x), g'(x)$ be the derivatives of f, g on $\Gamma \backslash S$, relative to the given orientations of the edges, and let dx be the measure on Γ which restricts to Lebesgue measure on each edge.

Here are some elementary properties of Δ:

PROPOSITION 3.2. *Let $f, g \in \text{CPA}(\Gamma)$. Then:*
- (A) $\Delta(f) \equiv 0$ *if and only if* $f \equiv C$ *for some constant C.*
- (B) $\Delta(f) = \Delta(g)$ *if and only if* $f = g + C$ *for some constant C.*
- (C) *If f is nonconstant, then f achieves its maximum at a point p where $\Delta(f)(p) > 0$, and its minimum at a point q where $\Delta(f)(q) < 0$.*
- (D) $\int_\Gamma f \, d\Delta(g) = \int_\Gamma g \, d\Delta(f) = \int_\Gamma f'(x) g'(x) \, dx$.
- (E) *The total mass $\Delta(f)(\Gamma) = 0$.*

PROOF. For (A), if $f \equiv C$, clearly $\Delta(f) \equiv 0$. Conversely, suppose $\Delta(f) \equiv 0$. Put $M = \max_{x \in \Gamma} f(x)$, and let $G = \{x \in \Gamma : f(x) = M\}$. Since f is continuous, G is closed. To see that G is also open, note that if $p \in G$, then $d_{\vec{v}} f(p) \leq 0$ for each $\vec{v} \in T_p(\Gamma)$, since M is the maximum value of f. Since f is piecewise affine, if f were not locally constant at p, then we would have $d_{\vec{v}}(f)(p) < 0$ for some $\vec{v}$, but then $\Delta(f)(p) > 0$, contrary to our assumption. Thus, G is both open and closed, and since Γ is connected, $G = \Gamma$. Trivially (A) implies (B).

For (C), suppose f is nonconstant, and let M and G be as in (A). If $p \in \partial G$, then the argument above shows that $\Delta(f)(p) > 0$. Similarly, f achieves its minimum at a point where $\Delta(f)(q) < 0$.

For (D), let S be a vertex set for Γ such that f and g are both affine on each edge of $\Gamma \backslash S$. Given an edge e_i of $\Gamma \backslash S$, fix an orientation of it, and let $\gamma_i : [0, L_i] \to e_i$ be an arclength parametrization. By abuse of notation, write $f(x) = f(\gamma_i(x))$ and $g(x) = g(\gamma_i(x))$ on e_i. Integration by parts shows

$$\int_\Gamma f(x) \, d\Delta(g)(x) = \sum_i \int_{e_i} f'(x) g'(x) \, dx = \int_\Gamma f'(x) g'(x) \, dx \ .$$

By symmetry, $\int_\Gamma g(x) \, d\Delta(f)(x) = \int_\Gamma f'(x) g'(x) \, dx$ as well.

Part (E) follows from (D), taking $g(x) \equiv 1$. □

The quantity in Proposition 3.2(D) will be denoted $\langle f, g \rangle_{\Gamma, \text{Dir}}$ and is called the *Dirichlet pairing*:

$$(3.2) \quad \langle f, g \rangle_{\Gamma, \text{Dir}} = \int_\Gamma f \, d\Delta(g) = \int_\Gamma g \, d\Delta(f) = \int_\Gamma f'(x) g'(x) \, dx \ .$$

It is easy to see that $\langle f, g \rangle_{\Gamma, \text{Dir}} = \langle g, f \rangle_{\Gamma, \text{Dir}}$ and that $\langle f, f \rangle_{\Gamma, \text{Dir}} \geq 0$, with $\langle f, f \rangle_{\Gamma, \text{Dir}} = 0$ if and only if f is constant.

3.3. The potential kernel $j_z(x,y)$

There is a *potential kernel* $j_z(x,y)$ on Γ which (for fixed y,z) belongs to CPA(Γ) and "inverts the Laplacian". It is defined by requiring that for each fixed $y, z \in \Gamma$,

(3.3) $$\begin{cases} \Delta_x(j_z(x,y)) = \delta_y(x) - \delta_z(x), \\ j_z(z,y) = 0. \end{cases}$$

The uniqueness of $j_z(x,y)$ follows from Proposition 3.2(B). Its existence can be shown in general using electrical network theory (see [**37**]) or by linear algebra (see [**7**] or [**100**]).

However, when Γ is a finite $\mathbb{R}$-tree (the only case needed in this book), the existence of a function satisfying (3.3) is trivial: given $x, y, z \in \Gamma$, let w be the point where the path from x to z and the path from y to z first meet, and put $j_z(x,y) = \rho(z,w)$, the path length from z to w. Then along the path from z to y, $j_z(x,y) = \rho(z,x)$; on branches off that path, $j_z(x,y)$ is constant. One easily checks that the function thus defined satisfies (3.3).

Alternately, the potential kernel $j_z(x,y)$ on a finite $\mathbb{R}$-tree Γ coincides with the Gromov product $(x|y)_z$ (see Appendix B):

(3.4) $$j_z(x,y) = (x|y)_z.$$

The potential kernel has the following physical interpretation. View Γ as an electrical network with terminals at y and z, with the resistance of each edge given by its length. Then $j_z(x,y)$ is the voltage at x when current 1 enters at y and exits at z, with reference voltage 0 at z. By construction, $j_z(x,y)$ belongs to CPA(Γ), and its slope on any edge of Γ has absolute value at most 1.

PROPOSITION 3.3.

(A) *The function $j_z(x,y)$ is nonnegative, bounded, symmetric in x and y, and jointly continuous in x, y, z. If S is a vertex set for Γ, then for fixed y, z, the function $h(x) = j_z(x,y)$ is affine on edges of $\Gamma \backslash (S \cup \{y,z\})$, and for all p, q we have $|j_z(p,y) - j_z(q,y)| \leq \rho(p,q)$.*

(B) *If $f \in \mathrm{CPA}(\Gamma)$ satisfies $\Delta(f) = \sum_{i=1}^n c_i \delta_{p_i}$, then for each z there is a constant $C = C_z$ such that*

$$f(x) = \sum_{i=1}^n c_i j_z(x, p_i) + C.$$

(C) *For each $z, \zeta \in \Gamma$,*

(3.5) $$j_\zeta(x,y) = j_z(x,y) - j_z(x,\zeta) - j_z(y,\zeta) + j_z(\zeta,\zeta).$$

(D) *Fix $y, z \in \Gamma$, and let $h(x) = j_z(x,y)$. Then for each $x \in \Gamma$ and each $\vec{v} \in T_x(\Gamma)$, the directional derivative $d_{\vec{v}} h(x) = \partial_{x,\vec{v}} j_z(x,y)$ satisfies*

$$|d_{\vec{v}} h(x)| \leq 1.$$

(E) If x, z, and $\vec{v} \in T_x(\Gamma)$ are fixed, then the function $\ell(y) = \partial_{x,\vec{v}} j_z(x,y)$ is continuous on $\Gamma \setminus \{x\}$ and affine on edges of $\Gamma \setminus (S \cup \{x, z\})$, with at worst a jump discontinuity at x.

PROOF. (A) The nonnegativity of $j_z(x,y)$ follows from its defining properties (3.3) and Proposition 3.2(C).

To establish the symmetry of $j_z(x,y)$, fix a, b, z and apply Proposition 3.2(D) to $h(x) = j_z(x,b)$ and $k(x) = j_z(x,a)$. Then $\Delta(h) = \delta_b - \delta_z$ and $\Delta(k) = \delta_a - \delta_z$. Since $h(z) = k(z) = 0$,
$$j_z(a,b) = \int_\Gamma h\, d\Delta(k) = \int_\Gamma k\, d\Delta(h) = j_z(b,a)\,.$$

Clearly $j_z(x,y)$ is continuous in x for each fixed y, z, with $|j_z(p,y) - j_z(q,y)| \leq \rho(p,q)$ for all p, q, since the absolute value of the slope of $j_z(x,y)$ along each edge is at most 1. From this, one deduces that it is jointly continuous in x, y for each fixed z, and indeed that for given p, q

$$\begin{aligned}(3.6)\quad |j_z(x,y) - j_z(p,q)| &\leq |j_z(x,y) - j_z(x,q)| + |j_z(x,q) - j_z(p,q)| \\ &\leq \rho(y,q) + \rho(x,p)\end{aligned}$$

using the symmetry shown above. The fact that for fixed y, $h(x) = j_z(x,y)$ is affine on edges of $\Gamma \setminus (S \cup \{y, z\})$ follows from the fact that $h(x) \in \mathrm{CPA}(\Gamma)$ and $\Delta(h) = \delta_y - \delta_z$.

Finally, for any fixed z_0, we claim that
$$(3.7) \quad j_z(x,y) = j_{z_0}(x,y) - j_{z_0}(x,z) - j_{z_0}(z,y) + j_{z_0}(z,z)\,.$$
The joint continuity of $j_z(x,y)$ in x, y, and z follows from this and (3.6). Its boundedness follows from continuity and the compactness of Γ.

To prove (3.7), note that for fixed y, z
$$\Delta_x(j_z(x,y)) = \delta_y(x) - \delta_z(x) = \Delta_x(j_{z_0}(x,y) - j_{z_0}(x,z))\,,$$
so by Proposition 3.2(B) there is a constant $C_{y,z}$ such that $j_z(x,y) = j_{z_0}(x,y) - j_{z_0}(x,z) + C_{y,z}$. Taking $x = z_0$ shows $C_{y,z} = j_z(z_0, y)$. In other words, we have
$$(3.8) \quad j_z(x,y) = j_{z_0}(x,y) - j_{z_0}(x,z) + j_z(z_0,y)\,.$$
Similarly, applying Δ_y to $j_z(z_0, y)$ and $-j_{z_0}(z,y)$ (and using symmetry), one finds there is a constant C_z such that $j_z(z_0, y) = -j_{z_0}(z,y) + C_z$. Taking $y = z$ shows $C_z = j_{z_0}(z,z)$. Combining these gives (3.7). This also proves (C).

For part (B), first note that if $\Delta(f) = \sum_{i=1}^n c_i \delta_{p_i}$, then $\sum_{i=1}^n c_i = 0$ by Proposition 3.2(E). Using this, one sees that
$$\Delta\left(\sum_{i=1}^n c_i j_{z_0}(x, p_i)\right) = \sum_{i=1}^n c_i \delta_{p_i}(x)\,,$$
and the result follows from Proposition 3.2(B).

Part (D) follows easily from the estimate (3.6).

To prove part (E), note that
$$\ell(y) = \lim_{\tau \to 0^+} \frac{j_z(x+\tau\vec{v}, y) - j_z(x,y)}{\tau} \ .$$
We already know that $\ell(y)$ is defined and satisfies $|\ell(y)| \leq 1$, for each y. Assume τ is small enough that $x + \tau\vec{v}$ belongs to the edge emanating from x in the direction $\vec{v}$. Since $j_z(x,y)$ and $j_z(x+\tau\vec{v}, y)$ are continuous and affine on edges of $\Gamma \backslash (S \cup \{x, x+\tau\vec{v}, z\})$, the same holds for their difference quotient. Passing to the limit as $\tau \to 0^+$, we see that $\ell(y)$ is continuous and affine on edges of $\Gamma \backslash (S \cup \{x\})$. This, in turn, shows it has at worst a jump discontinuity when $y = x$. $\square$

3.4. The Zhang space Zh(Γ)

Recall that if $f \in \text{CPA}(\Gamma)$, then

(3.9) $$\Delta(f) = \sum_{p \in \Gamma} \Delta(f)(p)\, \delta_p \ ,$$

where

$$\Delta(f)(p) = - \sum_{\vec{v} \in T_p(\Gamma)} d_{\vec{v}} f(p) \ .$$

It is possible to define a Laplacian on larger classes of functions. For example, Zhang [**100**] defined a Laplacian on the space of continuous, piecewise $\mathcal{C}^2$ functions whose one-sided directional derivatives $d_{\vec{v}} f(p)$ exist for all $p \in \Gamma$. We will write Zh(Γ) for this space. Zhang's definition is

(3.10) $$\Delta_{\text{Zh}}(f) = -f''(x)\, dx + \sum_{p \in \Gamma} \Bigl(- \sum_{\vec{v} \in T_p(\Gamma)} d_{\vec{v}} f(p)\Bigr) \delta_p(x) \ ,$$

where $f''(x)$ is taken relative to the arclength parametrization on each segment in the complement of an appropriate vertex set X_f for Γ. In other words, $f''(x) = \frac{d^2}{dt^2} f(p + t\vec{v})$ for $x = p + t\vec{v} \in \Gamma \backslash X_f$.

Again, $\Delta_{\text{Zh}}(f)$ is a measure on Γ. The condition that the directional derivatives exist for all p is easily seen to be equivalent to requiring that $f'' \in L^1(\Gamma, dx)$. For a function $f \in \text{CPA}(\Gamma)$, $f'' = 0$ in the complement of a vertex set, so Zhang's Laplacian is compatible with the one defined by (3.9).

Zhang's Laplacian is a hybrid of the usual (positive definite) Laplacian $-f''$ on $\mathbb{R}$ and the combinatorial Laplacian on a weighted graph. It satisfies appropriate analogues of the properties given in Proposition 3.2. (See Proposition 3.14 for a more general statement.) The particular form of definition (3.10) is justified by the observation that $\int f\, d\Delta(g) = \int g\, d\Delta(f)$ for all $f, g \in \text{Zh}(\Gamma)$, a fact which one easily verifies using integration by parts.

It turns out that one can extend the Laplacian to a still broader class of functions. This observation is fundamental for the applications given in this book. To motivate the definition of the more general Laplacian operator, note that Zhang's Laplacian has the following property.

3.4. THE ZHANG SPACE Zh(Γ)

LEMMA 3.4 (Mass Formula). *Let $E \subset \Gamma$ be a set which is a finite union of points and closed intervals. Then for each $f \in \mathrm{Zh}(\Gamma)$,*

$$\Delta_{\mathrm{Zh}}(f)(E) = - \sum_{\substack{p \in \partial E}} \sum_{\substack{\vec{v} \in T_p(\Gamma) \\ \text{outward}}} d_{\vec{v}} f(p) \tag{3.11}$$

where ∂E is the boundary of E and a direction $\vec{v}$ at p is called "outward" if the edge it corresponds to leads away from E.

PROOF. Choose a vertex set S for Γ which includes the endpoints of all segments in E, all isolated points of E, and all points where $f''(x)$ is not defined. Put $S_E = S \cap E$; then $E \backslash S_E$ is a finite union of open intervals (a_i, b_i). By definition,

$$\Delta_{\mathrm{Zh}}(f)(E) = -\sum_i \int_{a_i}^{b_i} f''(x)\, dx + \sum_{p \in S_E} \left(- \sum_{\vec{v} \in T_p(\Gamma)} d_{\vec{v}} f(p) \right). \tag{3.12}$$

On the other hand, for each interval (a_i, b_i)

$$\int_{a_i}^{b_i} f''(x)\, dx = f'(b_i) - f'(a_i)$$

where $f'(a_i)$ and $f'(b_i)$ are one-sided derivatives at the endpoints. If a_i, b_i correspond to $p, q \in S_E$, then for the direction vectors $\vec{v}, \vec{w}$ at p, q leading into (a_i, b_i) we have $d_{\vec{v}}(f)(p) = f'(a_i)$, $d_{\vec{w}} f(q) = -f'(b_i)$. It follows that

$$-\int_{a_i}^{b_i} f''(x)\, dx = d_{\vec{w}} f(q) + d_{\vec{v}} f(p).$$

Thus, in (3.12) the sum over the integrals cancels all the directional derivatives for directions leading into E. What remains is the sum over directional derivatives in directions leading outward from E. □

The Mass Formula says that for a closed set E which is a finite union of segments and isolated points, $\Delta_{\mathrm{Zh}}(f)(E)$ is determined simply by knowing the directional derivatives of f at the points of ∂E. In particular, taking $E = \Gamma$, we see that $\Delta_{\mathrm{Zh}}(f)(\Gamma) = 0$, since $\partial \Gamma$ is empty. As the complement of an open interval U is a closed set E of the type we have been discussing, it follows that

$$\Delta_{\mathrm{Zh}}(f)(U) = -\Delta_{\mathrm{Zh}}(f)(E) = \sum_{\substack{p \in \partial U}} \sum_{\substack{\vec{v} \in T_p(\Gamma) \\ \text{inward}}} d_{\vec{v}} f(p). \tag{3.13}$$

Taking unions of open and closed sets, we see that for any $T \subset \Gamma$ which is a finite union of points and intervals (open, closed, or half-open) there is a formula for $\Delta_{\mathrm{Zh}}(f)(T)$ in terms of the directional derivatives $d_{\vec{v}}(f)$ at points of ∂T. (This formula is easy to write out, but messy; see (3.14) below.) The collection of such sets forms a Boolean algebra $\mathcal{A}(\Gamma)$.

Conversely, suppose $f : \Gamma \to \mathbb{R}$ is any function whose directional derivatives $d_{\vec{v}}f(p)$ exist for all p and $\vec{v}$. Taking Lemma 3.4 as the starting point, one could hope to *define* the Laplacian of f by (3.11) and (3.13). This does indeed give a finitely additive set function $\Delta(f)$ on the Boolean algebra $\mathcal{A}(\Gamma)$. However, $\Delta(f)$ does not necessarily extend to a measure; there are pathological examples (similar to Weierstrass's continuous but nowhere differentiable functions) which oscillate so much that $\Delta(f)$ is not countably additive.

An additional condition is necessary: we need to assume that f is of *bounded differential variation*; see (3.18) below for a precise definition. We will write BDV(Γ) for the space of functions of bounded differential variation. It follows from Lemma 3.7 below that CPA(Γ) $\subset$ Zh(Γ) $\subset$ BDV(Γ).

In Theorem 3.6 of the next section, we will show that if $f \in$ BDV(Γ), then the finitely additive function $\Delta(f)$, defined initially on $\mathcal{A}(\Gamma)$, extends to a finite signed Borel measure of total mass 0 on Γ. Conversely, we will see in Corollary 3.12 that if μ is a finite signed Borel measure on Γ with total mass 0, then there is an $f \in$ BDV(Γ), unique up to addition of an arbitrary real constant, for which $\Delta(f) = \mu$. In other words, there is a natural bijection between finite signed Borel measures of total mass zero on Γ and functions in BDV(Γ) modulo constant functions. This gives a precise meaning to the catchphrase "BDV(Γ) is the largest space of (continuous) functions f for which $\Delta(f)$ can be defined as a measure".

We will see in Proposition 3.14 that all the properties of the Laplacian proved in Proposition 3.2 for functions in CPA(Γ) remain valid for functions in BDV(Γ).

3.5. The space BDV(Γ)

For the convenience of the reader, we have reproduced portions of [10][1] in this section.

Let $\mathcal{D}(\Gamma)$ be the class of all functions on Γ whose one-sided derivatives exist everywhere, i.e.,

$$\mathcal{D}(\Gamma) = \{f : \Gamma \to \mathbb{R} : \ d_{\vec{v}}f(p) \text{ exists for each } p \in \Gamma \text{ and } \vec{v} \in T_p(\Gamma)\} \ .$$

It is easy to see that each $f \in \mathcal{D}(\Gamma)$ is continuous.

As in the previous section, let $\mathcal{A} = \mathcal{A}(\Gamma)$ be the Boolean algebra of subsets of Γ generated by the connected open sets. Each set $S \in \mathcal{A}$ is a finite disjoint union of sets isometric to open, half-open, or closed intervals (we consider isolated points to be degenerate closed intervals); conversely, all such sets belong to $\mathcal{A}$. Define the set $b(S)$ of *boundary points* of S to be the collection of points belonging to the closures of both S and $\Gamma \backslash S$. It is easy to see that each $S \in \mathcal{A}$ has only finitely many boundary points. Note

---

[1]Reprinted with permission of the Canadian Mathematical Society, original article by Matthew Baker and Robert Rumely, "Harmonic analysis on metrized graphs", Canadian Journal of Mathematics, Vol. 59, No. 2 (2007), pages 225–275.

that under this definition, if $\Gamma = [0,1]$ and $S = [0, \frac{1}{2}]$, for example, then the left endpoint 0 is not a boundary point of S.

For each $p \in b(S)$, let $\text{In}(p, S)$ be the set of 'inward-directed unit vectors at p': the set of all $\vec{v} \in T_p(\Gamma)$ for which $p + t\vec{v}$ belongs to S for all sufficiently small $t > 0$. Similarly, let $\text{Out}(p, S) = T_p(\Gamma) \backslash \text{In}(p, S)$ be the collection of 'outward-directed unit vectors at p'. If p is an isolated point of S, then $\text{In}(p, S) = \emptyset$ and $\text{Out}(p, S) = T_p(\Gamma)$.

If $f \in \mathcal{D}(\Gamma)$, then we can define a finitely additive set function m_f on $\mathcal{A}$ by requiring that for each $S \in \mathcal{A}$

$$(3.14) \quad m_f(S) = \sum_{\substack{p \in b(S) \\ p \notin S}} \sum_{\vec{v} \in \text{In}(p,S)} d_{\vec{v}} f(p) - \sum_{\substack{p \in b(S) \\ p \in S}} \sum_{\vec{v} \in \text{Out}(p,S)} d_{\vec{v}} f(p) \, .$$

Thus, for an open set $S \in \mathcal{A}$,

$$(3.15) \quad m_f(S) = \sum_{p \in b(S)} \sum_{\vec{v} \in \text{In}(p,S)} d_{\vec{v}} f(p) \, ,$$

for a closed set $S \in \mathcal{A}$,

$$(3.16) \quad m_f(S) = - \sum_{p \in b(S)} \sum_{\vec{v} \in \text{Out}(p,S)} d_{\vec{v}} f(p) \, ,$$

and for a set $S = \{p\}$ consisting of a single point,

$$(3.17) \quad m_f(\{p\}) = - \sum_{\vec{v} \in T_p(\Gamma)} d_{\vec{v}} f(p) \, .$$

The finite additivity is clear if one notes that each $S \in \mathcal{A}$ can be written as a finite disjoint union of open intervals and points and that for each boundary point p of S, the set $\text{In}(p, S)$ coincides with $\text{Out}(p, \Gamma \backslash S)$.

Note that
$$m_f(\emptyset) = m_f(\Gamma) = 0 \, ,$$
since by our definition, both the empty set and the entire graph Γ have no boundary points. It follows that for any $S \in \mathcal{A}$,
$$m_f(\Gamma \backslash S) = -m_f(S) \, .$$
If $f_1, f_2 \in \mathcal{D}(\Gamma)$ and $c_1, c_2 \in \mathbb{R}$, then it is easy to see that the set function corresponding to $c_1 f_1 + c_2 f_2$ is
$$m_{c_1 f_1 + c_2 f_2} = c_1 m_{f_1} + c_2 m_{f_2} \, .$$

We will say that a function $f \in \mathcal{D}(\Gamma)$ is of *bounded differential variation*, denoted $f \in \text{BDV}(\Gamma)$, if there is a constant $B > 0$ such that for any countable collection $\mathcal{F}$ of pairwise disjoint sets in $\mathcal{A}$,

$$(3.18) \quad \sum_{S_i \in \mathcal{F}} |m_f(S_i)| \leq B \, .$$

In particular, $\text{BDV}(\Gamma)$ is a linear subspace of $\mathcal{D}(\Gamma)$.

PROPOSITION 3.5. *Let Γ be a metrized graph.*

(A) *If $f \in \text{BDV}(\Gamma)$, then there are only countably many points $p_i \in \Gamma$ for which $m_f(p_i) \neq 0$, and $\sum_i |m_f(p_i)|$ converges.*

(B) *In the definition of $\text{BDV}(\Gamma)$ one can restrict to families of pairwise disjoint connected open sets or to connected closed sets. More precisely, a function $f \in \mathcal{D}(\Gamma)$ belongs to $\text{BDV}(\Gamma)$ if and only if*

   (1) *there is a constant B_1 such that for any countable family $\mathcal{F}$ of pairwise disjoint connected open sets $V_i \in \mathcal{A}$,*

$$\sum_{\substack{V_i \in \mathcal{F} \\ V_i \text{ connected, open}}} |m_f(V_i)| \leq B_1 \quad \text{or}$$

   (2) *there is a constant B_2 such that for any countable family $\mathcal{F}$ of pairwise disjoint connected closed sets $E_i \in \mathcal{A}$,*

$$\sum_{\substack{E_i \in \mathcal{F} \\ E_i \text{ connected, closed}}} |m_f(E_i)| \leq B_2 \ .$$

PROOF. For (A), note that for each positive integer n, there are at most $nB = B/(1/n)$ points $p \in \Gamma$ with $|m_f(\{p\})| \geq 1/n$, so there are at most countably many points with $m_f(\{p\}) \neq 0$. Moreover,

$$\sum_{\substack{p \in \Gamma \\ m_f(\{p\}) \neq 0}} |m_f(\{p\})| \leq B \ .$$

For (B), note that if $f \in \text{BDV}(\Gamma)$, then (1) and (2) hold trivially. Conversely, suppose first that (1) holds. We begin by showing that there are only countably many points $p \in \Gamma$ for which $m_f(p) \neq 0$ and that

$$\sum_{m_f(\{p\}) \neq 0} |m_f(\{p\})| < 2B_1 \ . \tag{3.19}$$

To see this, suppose that for some $0 < n \in \mathbb{Z}$, there were more than $2nB_1$ points p with $|m_f(\{p\})| \geq 1/n$. Then there would be either more than nB_1 points with $m_f(\{p\}) \geq 1/n$ or more than nB_1 points with $m_f(\{p\}) \leq -1/n$. Suppose for example that $p_1, \ldots, p_r$ satisfy $m_f(\{p_i\}) \leq -1/n$, where $r > nB_1$. Put $K = \{p_1, \ldots, p_r\}$, and let $V_1, \ldots, V_s$ be the connected components of $\Gamma \backslash K$. Then

$$\begin{aligned} B_1 &< -(m_f(\{p_1\}) + \cdots + m_f(\{p_r\})) = -m_f(K) \\ &= m_f(V_1 \cup \cdots \cup V_s) = m_f(V_1) + \cdots + m_f(V_s) \\ &\leq |m_f(V_1)| + \cdots + |m_f(V_r)| \leq B_1 \ . \end{aligned}$$

Hence there can be only countably many points p_i with $m_f(\{p_i\}) \neq 0$, and (3.19) holds.

Now let $\mathcal{F}$ be any countable collection of pairwise disjoint sets $S_i \in \mathcal{A}$. Each S_i can be decomposed as a finite disjoint union of connected open sets and sets consisting of isolated points, and if

$$S_i = V_{i1} \cup \cdots \cup V_{i,r_i} \cup \{p_{i1}\} \cup \cdots \cup \{p_{i,s_i}\}$$

is such a decomposition, then

$$\sum_{S_i \in \mathcal{F}} |m_f(S_i)| \leq \sum_{i,j} |m_f(V_{ij})| + \sum_{i,k} |m_f(\{p_{ik}\})| \leq 3B_1 \ .$$

It follows that $f \in \mathrm{BDV}(\Gamma)$.

Next suppose that (2) holds; we will show that (1) holds as well. Consider a countable collection $\mathcal{F}$ of pairwise disjoint connected open sets V_i. Decompose $\mathcal{F} = \mathcal{F}_+ \cup \mathcal{F}_-$, where $V_i \in \mathcal{F}_+$ if and only if $m_f(V_i) \geq 0$ and $V_i \in \mathcal{F}_-$ if and only if $m_f(V_i) < 0$. Relabel the sets so that $\mathcal{F}_+ = \{V_1, V_3, V_5, \ldots\}$. For each n, put $K_n = \Gamma \backslash (V_1 \cup V_3 \cup \cdots \cup V_{2n+1})$, and decompose K_n as a finite disjoint union of connected closed sets $E_1, \ldots, E_r$. Then

$$\sum_{i=0}^{n} m_f(V_{2i+1}) = m_f(\Gamma \backslash K_n) = |m_f(K_n)|$$

$$\leq \sum_{j=1}^{r} |m_f(E_j)| \leq B_2 \ .$$

Letting $n \to \infty$, we see that $\sum_{V_i \in \mathcal{F}_+} m_f(V_i) \leq B_2$.

Similarly $\sum_{V_i \in \mathcal{F}_-} |m_f(V_i)| \leq B_2$, so $\sum_{V_i \in \mathcal{F}} |m_f(V_i)| \leq 2B_2$, and we are done. $\square$

We now come to the main result of this section.

THEOREM 3.6. *If $f \in \mathrm{BDV}(\Gamma)$, then the finitely additive set function m_f extends to a finite, signed Borel measure m_f^* of total mass 0 on Γ.*

PROOF. We begin with a reduction. It suffices to show that the restriction of m_f to each edge e extends to a (finite, signed Borel) measure on e. Identifying e with its parametrizing interval, we can assume without loss of generality that $\Gamma = [a,b]$ is a closed interval and that $f : [a,b] \to \mathbb{R}$ is in $\mathrm{BDV}([a,b])$.

We next decompose m_f into an atomic and an atomless part.

Let $\{p_1, p_2, \ldots\}$ be the points in $[a,b]$ for which $m_f(\{p_i\}) \neq 0$. For brevity, write $c_i = m_f(\{p_i\})$; by hypothesis, $\sum_i |c_i|$ converges. Define $g : \mathbb{R} \to \mathbb{R}$ by

$$g(t) = -\frac{1}{2} \sum_i c_i |t - p_i| \ ,$$

and let $m_g^* = \sum_i c_i \delta_{p_i}$ be the Borel measure giving the point mass c_i to each p_i.

By a direct computation, one checks that for each $x \in \mathbb{R}$, both one-sided derivatives $g'_{\pm}(x)$ exist, with

$$g'_-(x) = \frac{1}{2}(\sum_{x \leq p_i} c_i - \sum_{x > p_i} c_i), \qquad g'_+(x) = \frac{1}{2}(\sum_{x < p_i} c_i - \sum_{x \geq p_i} c_i) \ .$$

For any closed subinterval $[c,d] \subset [a,b]$,

$$\begin{aligned} m_g([c,d]) &= g'_-(c) - g'_+(d) \\ &= \sum_{c \leq p_i \leq d} c_i = m_g^*([c,d]). \end{aligned}$$

In particular the set function m_g has precisely the same point masses as m_f. Similar computations apply to open and half-open intervals. Thus the measure m_g^* extends m_g.

Replacing $f(t)$ by $h(t) = f(t) - g(t)$, we are reduced to the situation where $h \in \mathrm{BDV}([a,b])$ has no point masses. This means that for each $p \in (a,b)$,

$$0 = m_h(\{p\}) = h'_-(p) - h'_+(p),$$

or in other words that $h'(p)$ exists. By hypothesis, both $h'_+(a)$ and $h'_-(b)$ exist, so $h(t)$ is differentiable on $[a,b]$. The fact that $h \in \mathrm{BDV}([a,b])$ means that $h'(t)$ is of bounded total variation.

We claim that $h'(t)$ is continuous on $[a,b]$. Suppose it were discontinuous at some point p. By [**87**, p. 109], the existence of $h'(t)$ for all t means that $h'(t)$ cannot have a jump discontinuity, so the discontinuity must be due to oscillation. Hence there would be an $\varepsilon > 0$ and a sequence of points $t_1, t_2, \ldots$, either monotonically increasing or monotonically decreasing to p, such that $|h'(t_i) - h'(t_{i+1})| > \varepsilon$ for each i. Assume for convenience that the t_i are monotonically increasing. Considering the intervals $[t_1,t_2], [t_3,t_4], \ldots$, we see that

$$\sum_{i=1}^{\infty} |m_h([t_{2i-1}, t_{2i}])| = \infty \ ,$$

contradicting the fact that $h \in \mathrm{BDV}([a,b])$ and proving our claim.

For $a \leq t \leq b$, let $T(t)$ be the cumulative total variation function of $h'(t)$: letting $\mathcal{Q}$ vary over all finite partitions $a = q_0 < q_1 < \cdots < q_m = t$ of $[a,t]$,

$$T(t) = \sup_{\mathcal{Q}} \sum_{j=1}^{m} |h'(q_j) - h'(q_{j-1})| \ .$$

Then $T : [a,b] \to \mathbb{R}$ is monotone increasing and is continuous since $h'(t)$ is continuous. By Proposition 12 on p. 301 of [**86**], there is a unique finite signed Borel measure ν on $[a,b]$ such that $\nu((c,d]) = T(d) - T(c)$ for each half-open interval $(c,d] \subset [a,b]$. Since $T(t)$ is continuous, ν has no point masses.

Next put

$$T_1(t) = T(t) - h'(t) \ .$$

Then $T_1(t)$ is also monotone increasing and continuous, so by the same proposition there is a bounded Borel measure ν_1 on $[a,b]$ such that for each half-open interval $(c,d] \subset [a,b]$,
$$\nu_1((c,d]) = T_1(d) - T_1(c) .$$

As before, ν_1 has no point masses. Now define the Borel measure $m_h^* = \nu_1 - \nu$. By construction, m_h^* has finite total mass. For any half-open interval $(c,d] \subset [a,b]$,
$$m_h^*((c,d]) = h'(c) - h'(d) = m_h((c,d]) .$$
Since the measures ν and ν_1 have no point masses,
$$m_h^*([c,d]) = m_h^*((c,d)) = m_h^*([c,d)) = m_h^*((c,d]) .$$
Thus m_h^* extends the finitely additive set function m_h.

Finally, adding the measures m_g^* extending m_g and m_h^* extending m_h, we obtain the desired measure $m_f^* = m_g^* + m_h^*$ extending m_f. Since $m_f(\Gamma) = 0$, we have $m_f^*(\Gamma) = 0$ as well. □

3.6. The Laplacian on a metrized graph

If $f \in \mathrm{BDV}(\Gamma)$, we define the Laplacian $\Delta(f)$ to be the measure given by Theorem 3.6:
$$\Delta(f) = m_f^* .$$

In (3.10) we defined a Laplacian $\Delta_{\mathrm{Zh}}(f)$ on the Zhang space $\mathrm{Zh}(\Gamma)$. We have:

LEMMA 3.7. $\mathrm{Zh}(\Gamma)$ is a subset of $\mathrm{BDV}(\Gamma)$, and for each $f \in \mathrm{Zh}(\Gamma)$,
$$\Delta(f) = \Delta_{\mathrm{Zh}}(f).$$

PROOF. Fix $f \in \mathrm{Zh}(\Gamma)$, and let X_f be a vertex set for Γ such that $f \in \mathcal{C}^2(\Gamma \backslash X_f)$. We will first show that the directional derivatives $d_{\vec{v}} f(p)$ exist for each $p \in X_f$ and each $\vec{v} \in T_p(\Gamma)$. Fix such a p and $\vec{v}$, and let t_0 be small enough that $p + t\vec{v} \in \Gamma \backslash X_f$ for all $t \in (0, t_0)$. By abuse of notation, we will write $f(t)$ for $f(p + t\vec{v})$. By hypothesis, $f \in \mathcal{C}^2((0, t_0))$. By the Mean Value Theorem, $d_{\vec{v}} f(p)$ exists if and only if $\lim_{t \to 0^+} f'(t)$ exists (in which case the two are equal). Given $\varepsilon > 0$, let $0 < \delta < t_0$ be small enough that $\int_{(0,\delta)} |f''(t)|\, dt < \varepsilon$; this is possible since $f'' \in L^1(\Gamma, dx)$. Then for all $t_1, t_2 \in (0, \delta)$,

$$(3.20) \qquad |f'(t_2) - f'(t_1)| \leq \int_{t_1}^{t_2} |f''(t)|\, dt < \varepsilon,$$

which proves that $\lim_{t \to 0^+} f'(t) = d_{\vec{v}} f(p)$ exists.

Inequality (3.20) also implies (in the notation of §3.5) that for every countable family $\mathcal{F}$ of pairwise disjoint connected closed sets $E_i \in \mathcal{A}$, we

have
$$\sum_{E_i \in \mathcal{F}} |m_f(E_i)| \le \sum_{p \in X_f} |m_f(\{p\})| + \int_\Gamma |f''(t)|\, dt \ < \ \infty,$$
so that $f \in \mathrm{BDV}(\Gamma)$ as desired.

We next show that $\Delta(f) = \Delta_{\mathrm{Zh}}(f)$. If $p \in X_f$, we have $\Delta(f)(\{p\}) = \Delta_{\mathrm{Zh}}(f)(\{p\})$. To show that $\Delta(f) = \Delta_{\mathrm{Zh}}(f)$, it suffices to show that for each open interval (c,d) contained in an edge of $\Gamma \backslash X_f$, $\Delta(f)((c,d)) = \Delta_{\mathrm{Zh}}(f)((c,d))$. For this, note that
$$\begin{aligned}\Delta(f)((c,d)) &= m_f((c,d)) = f'(c) - f'(d) \\ &= -\int_c^d f''(x)\, dx = \Delta_{\mathrm{Zh}}((c,d))\ .\end{aligned}$$
$\square$

Conversely, we have

PROPOSITION 3.8. *If $f \in \mathrm{BDV}(\Gamma)$ and $\Delta(f)$ has the form $\Delta(f) = g(x)dx + \sum_{p_i \in X} c_{p_i} \delta_{p_i}(x)$ for a piecewise continuous function $g \in L^1(\Gamma, dx)$ and a finite set X, then $f \in \mathrm{Zh}(\Gamma)$. Furthermore, let X_g be a vertex set for Γ containing X and the finitely many points where $g(x)$ is not continuous; put $c_{p_i} = 0$ for each $p_i \in X_g \backslash \{p_1, \ldots, p_m\}$. Then $f''(x) = -g(x)$ for each $x \in \Gamma \backslash X_g$, $f'(x)$ is continuous on the closure of each segment of $\Gamma \backslash X_g$ (interpreting $f'(x)$ as a one-sided derivative at each endpoint), and $\Delta(f)(p_i) = c_{p_i}$ for each $p_i \in X_g$.*

PROOF. Suppose $f \in \mathrm{BDV}(\Gamma)$ and $\Delta(f)$ has the given form. Consider an edge in $\Gamma \backslash X_g$, and identify it with an interval (a,b) by means of the distinguished parametrization. For each $x \in (a,b)$, we have $\Delta(f)(x) = 0$, so $f'(x)$ exists. If $h > 0$ is sufficiently small, then
$$f'(x+h) - f'(x) = -\Delta(f)([x, x+h]) = -\int_x^{x+h} g(t)\, dt\ ,$$
while if $h < 0$, then
$$f'(x+h) - f'(x) = \Delta(f)([x+h, x]) = \int_{x+h}^x g(t)\, dt = -\int_x^{x+h} g(t)\, dt\ .$$
Hence
$$f''(x) = \lim_{h \to 0} \frac{f'(x+h) - f'(x)}{h} = \lim_{h \to 0} \left(-\frac{1}{h} \int_x^{x+h} g(t)\, dt \right) = -g(x)\ .$$
The assertion that $\Delta(f)(p_i) = c_{p_i}$ for each $p_i \in X_g$ is clear. $\square$

COROLLARY 3.9. *If $f \in \mathrm{BDV}(\Gamma)$ and $\Delta(f) = \sum_{i=1}^k c_i \delta_{p_i}$ is a discrete measure, then $f \in \mathrm{CPA}(\Gamma)$.*

PROOF. Since $\Delta(f)$ is discrete, it follows by Proposition 3.8 that $f(x) \in \text{Zh}(\Gamma)$ and $\Delta(f) = \Delta_{\text{Zh}}(f)$. Fixing a vertex set X for Γ, we see that $f''(x) = 0$ on $\Gamma \backslash X$, so $f(x)$ is affine on each segment of $\Gamma \backslash X$, which means that $f \in \text{CPA}(\Gamma)$. $\square$

Our next goal is to show that every measure of total mass zero is the Laplacian of some function in $\text{BDV}(\Gamma)$. In order to do so, we will need a notion of convergence of measures which is stronger than weak convergence.

DEFINITION 3.10. If ν is a finite signed Borel measure on Γ, then a sequence of finite signed Borel measures $\{\nu_n\}_{n \geq 1}$ converges *moderately well* to ν if:

(A) There is a bound B such that each ν_n has total variation $|\nu_n|(\Gamma) \leq B$.
(B) For each segment $D \subset \Gamma$ (open, closed, half-open, or reduced to a point), we have $\lim_{n \to \infty} \nu_n(D) = \nu(D)$.

Note that since each set in $\mathcal{A}(\Gamma)$ is a finite disjoint union of segments, condition (B) implies that if $\{\nu_n\}$ converges moderately well to ν, then

$$(3.21) \qquad \lim_{n \to \infty} \nu_n(S) = \nu(S) \quad \text{for each } S \in \mathcal{A}(\Gamma).$$

In particular, $\lim_{n \to \infty} \nu_n(\Gamma) = \nu(\Gamma)$. It follows easily from (3.21) that if $\{\nu_n\}$ converges moderately well to ν, then it converges weakly to ν, and that ν has total variation $|\nu|(\Gamma) \leq B$.

For any ν, there exist sequences of discrete measures which converge moderately well to ν. To construct such a sequence, decompose ν as $\nu = \omega + \mu$ where ω is atomic and μ is atomless. Since there are only countably many points where ν has a point mass, we can write

$$\omega = \sum_{k=1}^{\infty} c_i \delta_{x_i}$$

where each $c_i \in \mathbb{R}$ and $\sum_i |c_i| \leq |\nu|(\Gamma)$. Without loss of generality, we can assume the points x_i are labeled in such a way that $|c_1| \geq |c_2| \geq \cdots$. For each n, put

$$\omega_n = \sum_{k=1}^{n} c_i \delta_{x_i}.$$

Likewise, for each n, partition Γ into a finite disjoint union of segments $e_{n,k}$, $k = 1, \ldots, m_n$ (which may be open, closed, or half-open), such that each $e_{n,k}$ has length at most $1/n$. For each k, choose a point $y_{n,k} \in e_{n,k}$ and put

$$\mu_n = \sum_{k=1}^{m_n} \mu(e_{n,k}) \delta_{y_{n,k}}.$$

Finally, put $\nu_n = \omega_n + \mu_n$ for each n, and let $B = |\nu|(\Gamma)$. It is not hard to check that $\{\nu_n\}$ has the desired properties.

The following proposition is the key technical result needed to see that every measure of total mass zero is the Laplacian of some function in BDV(Γ).

PROPOSITION 3.11. *Suppose ν is a finite signed Borel measure on Γ. Fix $z \in \Gamma$, and put $h(x) = \int_\Gamma j_z(x,y)\,d\nu(y)$. Let $M = |\nu|(\Gamma)$ be the total variation of ν. Then:*

(A) *$h \in \mathrm{BDV}(\Gamma)$ and $\Delta(h) = \nu - \nu(\Gamma)\delta_z$.*

(B) *For each $x \in \Gamma$ and each $\vec{v} \in T_x(\Gamma)$, $|d_{\vec{v}}h(x)| \leq M$.*

(C) *Let $\{\nu_n\}$ be any sequence of finite signed Borel measures which converges weakly to ν. For each n, put $h_n(x) = \int_\Gamma j_z(x,y)\,d\nu_n(y)$. Then $\{h_n\}$ converges pointwise to h on Γ, and if there is a bound B such that $|\nu_n|(\Gamma) \leq B$ for all n, the convergence is uniform.*

Furthermore, if $\{\nu_n\}$ converges moderately well to ν, then for each $x \in \Gamma$ and each $\vec{v} \in T_x(\Gamma)$,

$$\lim_{n \to \infty} d_{\vec{v}} h_n(x) = d_{\vec{v}} h(x). \tag{3.22}$$

We remark that (3.22) need not hold if $\{\nu_n\}$ merely converges weakly to ν. For example, take $\Gamma = [0,1]$, let $\nu = \delta_1 - \delta_0$, and let $\nu_n = \delta_1 - \delta_{1/n}$ for each n. Then $h(x) = x$, and $h_n(x) = \max(0, x - 1/n)$. Let $\vec{v}$ be the unique tangent vector at 0. Then $d_{\vec{v}} h(0) = 1$, while $d_{\vec{v}} h_n(0) = 0$ for each n.

PROOF OF PROPOSITION 3.11. Fix z, and put $h(x) = \int j_z(x,y)\,d\nu(y)$. We first show that $d_{\vec{v}} h(x)$ exists for each $x \in \Gamma$ and each $\vec{v} \in T_x(\Gamma)$, so that $h \in \mathcal{D}(\Gamma)$. Fix x and $\vec{v}$, and note that

$$d_{\vec{v}} h(x) = \lim_{\tau \to 0^+} \int_\Gamma \frac{j_z(x + \tau\vec{v}, y) - j_z(x,y)}{\tau}\,d\nu(y), \tag{3.23}$$

provided the limit exists.

Let S be a vertex set for Γ. We only consider τ small enough that $x + \tau\vec{v}$ lies on the edge of $\Gamma\backslash(S \cup \{x, z\})$ in the direction $\vec{v}$. Let $e_\tau = (x, x + \tau\vec{v})$ be the open segment from x to $x + \tau\vec{v}$ contained in that edge. By Proposition 3.3(A), the function $t \mapsto j_z(t, y)$ is continuous on Γ and has constant slope on edges of $\Gamma\backslash(S \cup \{y, z\})$, so if $y \notin e_\tau$, then $(j_z(x + \tau\vec{v}, y) - j_z(x, y))/\tau = \partial_{x,\vec{v}} j_z(x, y)$. This means that

$$\int_\Gamma \frac{j_z(x + \tau\vec{v}, y) - j_z(x, y)}{\tau}\,d\nu(y)$$
$$= \int_{\Gamma\backslash e_\tau} \partial_{x,\vec{v}} j_z(x,y)\,d\nu(y) + \int_{e_\tau} \frac{j_z(x + \tau\vec{v}, y) - j_z(x, y)}{\tau}\,d\nu(y). \tag{3.24}$$

If $y \in e_\tau$, Proposition 3.3(A) gives $|(j_z(x + \tau\vec{v}, y) - j_z(x, y))/\tau| \leq 1$, while if $y \notin e_\tau$, Proposition 3.3(D) gives $|\partial_{x,\vec{v}} j_z(x, y)| \leq 1$. Hence as $\tau \to 0^+$, the first integral in (3.24) converges to $\int_\Gamma \partial_{x,\vec{v}} j_z(x,y)\,d\nu(y)$, while the second is bounded by $|\nu|(e_\tau)$ and converges to 0. Thus the limit in (3.23) exists and

$$d_{\vec{v}} h(x) = \int_\Gamma \partial_{x,\vec{v}} j_z(x, y)\,d\nu(y). \tag{3.25}$$

Again using the fact that $|\partial_{x,\vec{v}} j_z(x,y)| \leq 1$ for each y, this immediately implies that $|d_{\vec{v}} h(x)| \leq |\nu|(\Gamma) = M$.

Next let $\{\nu_n\}$ be any sequence of measures which converges weakly to ν, and put $h_n(x) = \int_\Gamma j_z(x,y) \, d\nu_n(y)$ for each n. For each x, the kernel $F_x(y) = j_z(x,y)$ is continuous, so by the definition of weak convergence, $\{h_n\}$ converges pointwise to h. On the other hand, by Proposition 3.3(A) we have $|j_z(x_1, y) - j_z(x_2, y)| \leq \rho(x_1, x_2)$ for all $x_1, x_2 \in \Gamma$. If there is a bound B such that $|\nu_n|(\Gamma) \leq B$ for all n, we obtain

$$|h_n(x_1) - h_n(x_2)| \;=\; \int_\Gamma |j_z(x_1,y) - j_z(x_2,y)| \, d|\nu_n|(y) \;\leq\; B \rho(x_1,x_2) \;.$$

Hence the functions h_n are equicontinuous, and since Γ is compact, the convergence of $\{h_n\}$ to h is uniform.

Next we will show that if $\{\nu_n\}$ converges moderately well to ν, then for each $x \in \Gamma$ and each $\vec{v} \in T_x(\Gamma)$,

$$(3.26) \qquad \lim_{n \to \infty} d_{\vec{v}} h_n(x) \;=\; d_{\vec{v}} h(x) \;.$$

The proof uses (3.25) for h and for the h_n. The point is that even though $\partial_{x,\vec{v}} j_z(x,y)$ is not continuous and ν and the ν_n may have point masses, the conditions in the definition of converging moderately well imply that

$$(3.27) \qquad \lim_{n \to \infty} \int_\Gamma \partial_{x,\vec{v}} j_z(x,y) \, d\nu_n(y) \;=\; \int_\Gamma \partial_{x,\vec{v}} j_z(x,y) \, d\nu(y) \;,$$

which is equivalent to (3.26).

To prove (3.27), fix x, z, and $\vec{v}$, and put $D(y) = \partial_{x,\vec{v}} j_z(x,y)$. As noted above, Proposition 3.3(D) gives $|D(y)| \leq 1$ for all y. By Proposition 3.3(E), $D(y)$ is continuous on $\Gamma \backslash (S \cup \{x\})$ and affine on edges of $\Gamma \backslash (S \cup \{x,z\})$, with at worst a jump discontinuity when $y = x$. We will construct a sequence of step functions $\{D_m\}$ converging uniformly to D and use them to show (3.27).

For each $m \in \mathbb{N}$ and each $k \in \mathbb{Z}$, put

$$S_{m,k} \;=\; \{y \in \Gamma : \frac{k-1}{m} < D(y) \leq \frac{k}{m}\} \;.$$

The properties of $D(y)$ noted above imply that $S_{m,k}$ is a finite union of points and open, closed, and half-open segments, so $S_{m,k} \in \mathcal{A}(\Gamma)$. Let $\chi_{S_{m,k}}$ be the characteristic function of $S_{m,k}$ and put

$$D_m(y) \;=\; \sum_{k=-m}^{m} \frac{k}{m} \chi_{S_{m,k}}(y) \;.$$

By construction, $\{D_m\}_{m \geq 1}$ converges uniformly to D. Thus

$$(3.28) \qquad \lim_{m \to \infty} \int_\Gamma D_m(y) \, d\nu(y) \;=\; \int_\Gamma D(y) \, d\nu(y) \;.$$

Since $\{\nu_n\}$ converges moderately well to ν, there is a constant B such that $|\nu_n|(\Gamma) \leq B$ for all n. We first claim that for all m and n,

$$\text{(3.29)} \qquad \left| \int_\Gamma D_m(y) \, d\nu_n(y) - \int_\Gamma D(y) \, d\nu_n(y) \right| \leq \frac{B}{m}.$$

Indeed, $|D(y) - D_m(y)| \leq 1/m$ for all y, and the sets $S_{m,k}$ for a given m are pairwise disjoint, so

$$\left| \int_\Gamma D_m(y) \, d\nu_n(y) - \int_\Gamma D(y) \, d\nu_n(y) \right| \leq \sum_{k=-m}^m \left| \int_{S_{m,k}} (D_m(y) - D(y)) \, d\nu_n(y) \right|$$

$$\leq \sum_{k=-m}^m \frac{1}{m} \cdot |\nu_n|(S_{m,k}) \leq \frac{B}{m}.$$

We also claim that for each m, there is an $N = N(m)$ such that for all $n \geq N$,

$$\text{(3.30)} \qquad \left| \int_\Gamma D_m(y) \, d\nu_n(y) - \int_\Gamma D_m(y) \, d\nu(y) \right| \leq \frac{2B}{m}.$$

Indeed,

$$\left| \int_\Gamma D_m(y) \, d\nu_n(y) - \int_\Gamma D_m(y) \, d\nu(y) \right| \leq \sum_{k=-m}^m \left| \frac{k}{m} \right| \cdot |\nu_n(S_{m,k}) - \nu(S_{m,k})|.$$

Since $\{\nu_n\}$ converges moderately well to ν, by (3.21) there is an $N(m)$ such that for each $n \geq N(m)$ and each k, $|\nu_n(S_{m,k}) - \nu(S_{m,k})| \leq B/m^2$. For such n, we have (3.30).

Combining (3.29) and (3.30) gives that for each m and each $n \geq N(m)$,

$$\text{(3.31)} \qquad \left| \int_\Gamma D(y) \, d\nu_n(y) - \int_\Gamma D_m(y) \, d\nu(y) \right| \leq \frac{3B}{m}.$$

Fixing m and letting $n \to \infty$ in (3.31), we obtain

$$\left| \limsup_{n \to \infty} \int_\Gamma D(y) \, d\nu_n(y) - \liminf_{n \to \infty} \int_\Gamma D(y) \, d\nu_n(y) \right| \leq \frac{6B}{m}.$$

From this, letting $m \to \infty$, we see that $\lim_{n \to \infty} \int_\Gamma D(y) \, d\nu_n(y)$ exists.

Returning to (3.31), let $n \to \infty$ and then let $m \to \infty$. Using (3.31) and (3.28) gives

$$\lim_{n \to \infty} \int_\Gamma D(y) \, d\nu_n(y) = \lim_{m \to \infty} \int_\Gamma D_m(y) \, d\nu(y) = \int_\Gamma D(y) \, d\nu(y).$$

This yields (3.27) and (3.26), as desired.

To complete the proof of Proposition 3.11, it remains to show that h belongs to $\mathrm{BDV}(\Gamma)$ and $\Delta(h) = \nu - \nu(\Gamma)\delta_z$.

Let $\{\nu_n\}$ be a sequence of discrete measures converging moderately well to n. Such a sequence exists by the discussion following Definition 3.10.

Using (3.26) and (3.14), we see that for each $S \in \mathcal{A}(\Gamma)$, the finitely additive set functions m_{h_n} and m_h satisfy

$$\tag{3.32} \lim_{n \to \infty} m_{h_n}(S) = m_h(S) \ .$$

Since $h_n \in \mathrm{CPA}(\Gamma)$, we have $\Delta(h_n) = \nu_n - (\nu_n(\Gamma))\delta_z$, and so

$$m_{h_n}(S) = (\Delta h_n)(S) = \nu_n(S) - \nu_n(\Gamma)\delta_z(S) \ .$$

Passing to the limit as n approaches infinity gives $m_h(S) = \nu(S) - \nu(\Gamma)\delta_z(S)$.

For any countable collection $\mathcal{F}$ of pairwise disjoint sets $S_i \in \mathcal{A}(\Gamma)$ it follows that

$$\sum_{i=1}^{\infty} |m_h(S_i)| \leq 2|\nu|(\Gamma) \ ,$$

so $h \in \mathrm{BDV}(\Gamma)$. The measure $\Delta(h) = m_h^*$ attached to h is determined by its values on sets in $\mathcal{A}(\Gamma)$, so it must coincide with $\nu - \nu(\Gamma)\delta_z$. $\square$

As an immediate consequence of Proposition 3.11, we obtain:

COROLLARY 3.12. *If ν is a finite signed Borel measure of total mass zero on Γ, then there exists a function $h \in \mathrm{BDV}(\Gamma)$ such that $\Delta h = \nu$.*

By part (B) of Proposition 3.14 below, h is unique up to addition of an arbitrary real constant.

We can deduce another useful result from Proposition 3.11. For any finite signed Borel measure ν on Γ, put

$$j_\nu(x, y) = \int_\Gamma j_\zeta(x, y) \, d\nu(\zeta) \ .$$

Clearly $j_\nu(x, y)$ is symmetric and is jointly continuous in x and y. It also has the following useful property:

PROPOSITION 3.13. *For each $y \in \Gamma$, the function $F_y(x) = j_\nu(x, y)$ belongs to the space $\mathrm{BDV}(\Gamma)$ and satisfies*

$$\Delta_x(F_y) = \nu(\Gamma) \cdot \delta_y - \nu \ .$$

PROOF. From (3.5), we see that for any fixed $z \in \Gamma$, we have

$$F_y(x) = \int j_\zeta(x, y) \, d\nu(\zeta) = \nu(\Gamma) j_z(x, y) - \int j_z(x, \zeta) \, d\nu(\zeta) - C \ ,$$

where C is independent of x. Using Proposition 3.11, we find that

$$\Delta_x F_y(x) = \nu(\Gamma)(\delta_y - \delta_z) - (\nu - \nu(\Gamma)\delta_z) = \nu(\Gamma)\delta_y - \nu \ . \qquad \square$$

3.7. Properties of the Laplacian on BDV(Γ)

It is clear from the definitions that the Laplacian on BDV(Γ) is linear:
$$\Delta(\alpha f + \beta g) = \alpha \Delta(f) + \beta \Delta(g)$$
for all $f, g \in \text{BDV}(\Gamma)$ and $\alpha, \beta \in \mathbb{R}$.

In addition, the Laplacian on BDV(Γ) has the following properties, which extend those proved in Proposition 3.2 for functions in CPA(Γ).

PROPOSITION 3.14. *If $f, g \in \text{BDV}(\Gamma)$, then:*
- (A) $\Delta(f) \equiv 0$ *if and only if* $f \equiv C$ *for some constant C.*
- (B) $\Delta(f) = \Delta(g)$ *if and only if* $f = g + C$ *for some constant C.*
- (C) *If f is nonconstant, then f achieves its maximum at a point p in the support of $\Delta(f)^+$ and its minimum at a point q in the support of $\Delta(f)^-$.*
- (D) *For each $z \in \Gamma$,*

$$(3.33) \quad \int_\Gamma f \, d\Delta(g) = \int_\Gamma g \, d\Delta(f) = \iint_{\Gamma \times \Gamma} j_z(x, y) \, d\Delta(f)(x) \, d\Delta(g)(y) \ .$$

- (E) $\Delta(f)(\Gamma) = 0$.

PROOF. (A) Trivially $f = C$ implies $\Delta(f) = 0$. Conversely, if $\Delta(f) = 0$, then by Corollary 3.9, we have $f \in \text{CPA}(\Gamma)$. The result now follows from part (A) of Proposition 3.2.

(B) This follows from (A) by the linearity of Δ.

(C) Let $x_1 \in \Gamma$ be a point where the continuous function f achieves its maximum, and let Γ_1 be the connected component of $\{x \in \Gamma : f(x) = f(x_1)\}$ containing x_1. Since f is nonconstant, we have $\Gamma_1 \neq \Gamma$.

If x_0 is a boundary point of Γ_1, we claim that $x_0 \in \text{supp}(\Delta_\Gamma(f)^+)$. Suppose not. Letting $\rho(x, y)$ denote the metric on Γ, there is a neighborhood $\Gamma_{x_0}(\varepsilon) = \{x \in \Gamma : \rho(x, x_0) < \varepsilon\}$ with $\Delta_\Gamma(f)|_{\Gamma_{x_0}(\varepsilon)} \leq 0$. After shrinking ε if necessary, we can assume that $\Gamma_{x_0}(\varepsilon)$ is a star, a union of half-open segments $[x_0, x_0 + \varepsilon \vec{v}_i)$ where the $\vec{v}_i$ are the unit tangent vectors at x_0. Since $f(x_0)$ is the maximum value of $f(x)$ on Γ, necessarily $d_{\vec{v}_i} f(x_0) \leq 0$ for each i. If $d_{\vec{v}_i} f(x_0) < 0$ for some i, then

$$\Delta_\Gamma(f)(x_0) = -\sum_i d_{\vec{v}_i} f(x_0) > 0 \ ,$$

contradicting $x_0 \notin \text{supp}(\Delta_\Gamma(f)^+)$. Hence $d_{\vec{v}_i} f(x_0) = 0$ for each i.

Since x_0 is a boundary point of Γ_1, there is a point $y_0 \in \Gamma_{x_0}(\varepsilon)$ with $f(y_0) < f(x_0)$. Let i be such that $y_0 \in e_i = (x_0, x_0 + \varepsilon \vec{v}_i)$. As $\Delta_\Gamma(f)|_{\Gamma_{x_0}(\varepsilon)} \leq 0$, it follows that $\Delta_\Gamma(f)|_{e_i} \leq 0$. For each $y = x_0 + t\vec{v}_i \in e_i$, there are two unit tangent vectors at y: write $\vec{v}_+$ for the one that leads away from x_0 and $\vec{v}_-$ for the one that leads towards x_0.

By the definition of $\Delta_\Gamma(f)$, for each $y \in e_i$,

$$(3.34) \quad 0 \geq \Delta_\Gamma(f)(y) = -d_{\vec{v}_-} f(y) - d_{\vec{v}_+} f(y) \ ,$$

so $d_{\vec{v}_+}f(y) \geq -d_{\vec{v}_-}f(y)$. Similarly, for each open subsegment $(y_1, y_2) \subset e_i$,

(3.35) $\qquad 0 \geq \Delta_\Gamma(f)((y_1, y_2)) = d_{\vec{v}_+}f(y_1) + d_{\vec{v}_-}f(y_2)$,

so $-d_{\vec{v}_-}f(y_2) \geq d_{\vec{v}_+}f(y_1)$. By (3.34) and (3.35), together with Lemma A.1 of Appendix A, f is convex upward on e_i. Since $d_{\vec{v}_i}(f)(x_0) = 0$, $f(x_0 + t\vec{v}_i)$ is nondecreasing in t. Hence $f(y_0) \geq f(x_0)$, a contradiction. Thus $x_0 \in \text{supp}(\Delta_\Gamma(f)^+)$, as desired. The case of a minimum is similar.

(D) Set $\nu = \Delta(f)$, so that $\nu(\Gamma) = 0$ by part (E). For fixed y, z, we know from Proposition 3.11 and the symmetry of $j_z(x, y)$ that $h(y) = \int j_z(x, y) \, d\nu(x)$ belongs to $\text{BDV}(\Gamma)$ and $\Delta h = \nu$. It follows from (B) that $f = h + C$ for some constant C. By Fubini's theorem and the fact that $\Delta(g)(\Gamma) = 0$, we have

$$\int_\Gamma f \, d\Delta(g) = \int_\Gamma h \, d\Delta(g) = \iint_{\Gamma \times \Gamma} j_z(x, y) \, d\Delta(f)(x) \, d\Delta(g)(y).$$

(E) This has already been noted and follows immediately from the definition of the Laplacian on $\text{BDV}(\Gamma)$. $\qquad\square$

For future reference, we note the following fact, established in the course of the proof of Proposition 3.14(D):

COROLLARY 3.15. *Given $f \in \text{BDV}(\Gamma)$, put $\nu = \Delta(f)$. Fix $z \in \Gamma$, and let $h(x) = \int_\Gamma j_z(x, y) \, d\nu(y)$. Then there is a constant $C = C_z$ such that $f = h + C$.*

Using Proposition 3.14(D), we can extend the Dirichlet pairing to functions $f, g \in \text{BDV}(\Gamma)$, setting

(3.36) $\qquad \langle f, g \rangle_{\Gamma, \text{Dir}} = \int_\Gamma f \, d\Delta(g) = \int_\Gamma g \, d\Delta(f)$

$\qquad\qquad\qquad = \iint_{\Gamma \times \Gamma} j_z(x, y) \, d\Delta(f)(x) \, d\Delta(g)(y).$

For $f, g \in \text{CPA}(\Gamma)$, in Proposition 3.2(D) we had $\langle f, g \rangle_{\Gamma, \text{Dir}} = \int_\Gamma f'(x) g'(x) \, dx$ as well. Generalizing this formula requires some preparation.

Fix a vertex set S for Γ, and choose an orientation on each edge of $\Gamma \backslash S$. For each $x \in \Gamma \backslash S$, let $\vec{v}_+ \in T_x(\Gamma)$ be the tangent vector at x in the positive direction, and let $\vec{v}_- \in T_x(\Gamma)$ be the tangent vector in the negative direction. If $f \in \text{BDV}(\Gamma)$, let $Z \subset \Gamma$ be the countable set of points where $\Delta(f)$ has a point mass. Then the *derivative* f' is defined for $x \in \Gamma \backslash (S \cup Z)$, where

(3.37) $\qquad f'(x) = d_{\vec{v}_+}(f)(x) = -d_{\vec{v}_-}(f)(x)$.

Likewise, we define the *one-sided derivative* f'_+ for $x \in \Gamma \backslash S$ by

(3.38) $\qquad f'_+(x) = d_{\vec{v}_+}(f)(x)$.

We regard f' and f'_+ as functions on Γ, defined except on sets of dx-measure zero.

LEMMA 3.16. *Let Γ be a metrized graph. Fix a vertex set S and an orientation for each edge of $\Gamma\backslash S$. Take $f \in \mathrm{BDV}(\Gamma)$, and let $M = |\Delta(f)|(\Gamma)$. Then:*

(A) *For all x where $f'(x)$ is defined, we have $|f'(x)| \leq M$, and for all x where $f'_+(x)$ is defined, we have $|f'_+(x)| \leq M$.*
(B) *f' and f'_+ are Borel measurable, and for each $0 < p \leq \infty$, they belong to $L^p(\Gamma, dx)$.*

PROOF. Part (A) follows immediately from Corollary 3.15 and Proposition 3.11(B).

To prove (B), it suffices to establish the Borel measurability of f'_+, since f'_+ is defined on $\Gamma\backslash S$ and $f' = f'_+$ except on a countable set. The fact that f' and f'_+ belong to $L^p(\Gamma, dx)$ for $0 < p \leq \infty$ then follows from the bound in (A) and the fact that Γ has finite length.

Since Γ has finitely many edges, it suffices to consider the case where Γ is a segment $[0, L]$. Put $\mu = \Delta(f)$, and note that by formula (3.15), for each $x \in [0, L)$ we have $d_{\vec{v}_+}(f)(x) = \mu((x, L])$.

Let $\mu_1, \mu_2 \geq 0$ be the measures in the Jordan decomposition of μ, so that $\mu = \mu_1 - \mu_2$ and $\mu_1(\Gamma) = \mu_2(\Gamma) = M/2$. Let $\zeta = 0$ be the left endpoint of $\Gamma = [0, L]$, and put $\nu_1 = \mu_1 - (M/2)\delta_\zeta$, $\nu_2 = \mu_2 - (M/2)\delta_\zeta$. By Proposition 3.11, there are functions $h_1, h_2 \in \mathrm{BDV}(\Gamma)$ satisfying $\Delta(h_1) = \nu_1$, $\Delta(h_2) = \nu_2$. By Proposition 3.14(A), there is a constant C such that $f = h_1 - h_2 + C$.

By replacing f in turn by h_1, h_2, it suffices to prove the Borel measurability of f'_+ under the assumption that $\mu = \Delta(f)$ is positive on $(0, L]$. Under this hypothesis, for all $x < y$ in $[0, L)$,

$$d_{\vec{v}_+}(f)(x) = \mu((x, L]) \geq \mu((y, L]) = d_{\vec{v}_+}(f)(y).$$

Thus f'_+ is decreasing on $[0, L)$ and is Borel measurable. $\square$

PROPOSITION 3.17. *Let Γ be a metrized graph. For all $f, g \in \mathrm{BDV}(\Gamma)$, we have*

$$(3.39) \qquad \langle f, g \rangle_{\Gamma, \mathrm{Dir}} = \int_\Gamma f'(x) g'(x)\, dx = \int_\Gamma f'_+(x) g'_+(x)\, dx.$$

Consequently, for each $f \in \mathrm{BDV}(\Gamma)$:

(A) *$\langle f, f \rangle_{\Gamma, \mathrm{Dir}} \geq 0$, and $\langle f, f \rangle_{\Gamma, \mathrm{Dir}} = 0$ if and only if f is constant.*
(B) *$\langle f, g \rangle_{\Gamma, \mathrm{Dir}} = 0$ for all $g \in \mathrm{BDV}(\Gamma)$ if and only if f is constant.*

PROOF. We first prove (3.39). Since f' (resp. g') differs from f'_+ (resp. g'_+) on a set of dx-measure 0, the two integrals in (3.39) are equal, and it suffices to show that $\langle f, g \rangle_{\Gamma, \mathrm{Dir}} = \int_\Gamma f'_+(x) g'_+(x)\, dx$. This integral is finite since $f'_+, g'_+ \in L^2(\Gamma, dx)$.

Put $\nu = \Delta(f)$, $\mu = \Delta(g)$, and fix $z \in \Gamma$. After replacing f and g by $f - C_1$ and $g - C_2$ for suitable constants C_1, C_2, we can assume that

$$f(x) = \int_\Gamma j_z(x, y)\, d\nu(y), \qquad g(x) = \int_\Gamma j_z(x, y)\, d\mu(y).$$

Let $\{\nu_n\}$ be a sequence of discrete measures converging moderately well to ν and let $\{\mu_n\}$ be a sequence of discrete measures converging moderately well to μ. Such sequences exist by the discussion after Definition 3.10. For each n, put

$$f_n(x) = \int_\Gamma j_z(x,y)\, d\nu_n(y)\,, \quad g_n(x) = \int_\Gamma j_z(x,y)\, d\mu_n(y)\,.$$

Then $f_n, g_n \in \mathrm{CPA}(\Gamma)$.

By Proposition 3.11(A), $\{f_n\}$ converges uniformly to f on Γ, and $\{g_n\}$ converges uniformly to g on Γ. By Proposition 3.14(D), for each n we have

$$\begin{aligned}\langle f, g\rangle_{\Gamma,\mathrm{Dir}} - \langle f_n, g_n\rangle_{\Gamma,\mathrm{Dir}} &= \int_\Gamma f\, d\Delta(g) - \int_\Gamma f_n\, d\Delta(g_n) \\ &= \int_\Gamma (f - f_n)\, d\Delta(g) - \int_\Gamma (g - g_n)\, d\Delta(f_n)\,.\end{aligned}$$

As $n \to \infty$, $\int_\Gamma (f - f_n)\, d\Delta(g) \to 0$ since $f - f_n$ converges uniformly to 0. Likewise, $\int_\Gamma (g - g_n)\, d\Delta(f_n) \to 0$ since $g - g_n$ converges uniformly to 0 and the measures $\Delta(f_n) = \nu_n - \nu_n(\Gamma)\delta_z$ have uniformly bounded total variation. Thus,

$$(3.40) \qquad \lim_{n\to\infty} \langle f_n, g_n\rangle_{\Gamma,\mathrm{Dir}} = \langle f, g\rangle_{\Gamma,\mathrm{Dir}}\,.$$

On the other hand, by Proposition 3.11(C) the functions f'_+, g'_+, $(f_n)'_+$, and $(g_n)'_+$ are uniformly bounded on $\Gamma\setminus S$, the sequence $\{(f_n)'_+\}$ converges pointwise to f'_+ on $\Gamma\setminus S$, and the sequence $\{(g_n)'_+\}$ converges pointwise to g'_+ on $\Gamma\setminus S$. It follows that $(f_n)'_+(g_n)'_+$ converges pointwise to $f'_+ g'_+$ on $\Gamma\setminus S$. By the Dominated Convergence Theorem,

$$(3.41) \qquad \lim_{n\to\infty} \int_\Gamma (f_n)'_+(x)\,(g_n)'_+(x)\, dx = \int_\Gamma f'_+(x)\, g'_+(x)\, dx\,.$$

Finally, by Proposition 3.2(D) and the remarks at the beginning of the proof, for each n

$$(3.42) \quad \langle f_n, g_n\rangle_{\Gamma,\mathrm{Dir}} = \int_\Gamma (f_n)'_+(x)\,(g_n)'_+(x)\, dx = \int_\Gamma f'_n(x)\, g'_n(x)\, dx\,.$$

Combining (3.40), (3.41), and (3.42) gives (3.39).

It remains to establish assertions (A) and (B).

We first establish the forward direction in (B). Suppose $\langle f, g\rangle_{\Gamma,\mathrm{Dir}} = 0$ for all $g \in \mathrm{BDV}(\Gamma)$. If f were not constant, there would be points $a \neq b$ with $f(a) \neq f(b)$. Take $g(x) = j_a(x,b)$. Then $g \in \mathrm{CPA}(\Gamma) \subset \mathrm{BDV}(\Gamma)$, and $\Delta(g) = \delta_b - \delta_a$. Hence $\langle f, g\rangle_{\Gamma,\mathrm{Dir}} = \int f\, d\Delta(g) = f(b) - f(a) \neq 0$, a contradiction.

We next establish the forward direction in assertion (A). For each $f \in \mathrm{BDV}(\Gamma)$, formula (3.39) implies that $\langle f, f\rangle_{\Gamma,\mathrm{Dir}} \geq 0$. If $\langle f, f\rangle_{\Gamma,\mathrm{Dir}} = 0$, then

$\int (f'(x))^2 \, dx = 0$, so dx-almost everywhere $f'(x) = 0$. It follows that for each $g \in \mathrm{BDV}(\Gamma)$

$$\langle f, g \rangle_{\Gamma, \mathrm{Dir}} \ = \ \int_\Gamma f'(x) g'(x) \, dx \ = \ 0 \ .$$

As shown above, this means f is constant.

The reverse direction in both assertions is trivial, so we are done. $\square$

CHAPTER 4

The Hsia kernel

In this chapter, we introduce the Hsia kernel, the fundamental kernel for potential theory on $\mathbb{P}^1_{\text{Berk}}$. It is denoted $\delta(x,y)_\infty$ and is the natural extension to $\mathbb{A}^1_{\text{Berk}}$ of the distance function $|x-y|$ on $\mathbb{A}^1(K)$. We learned of it from an unpublished manuscript of Liang-Chung Hsia [**62**], where it was introduced as a function on a space of discs.

After discussing its properties, we relate it to other objects in potential theory: the potential kernel $j_z(x,y)$ on graphs, the spherical distance $\|x,y\|$ on $\mathbb{P}^1(K)$, the canonical distance $[x,y]_\infty$ from [**88**], and the Gromov product on an $\mathbb{R}$-tree (cf. §B.4).

We then introduce an extension of $\|x,y\|$ to $\mathbb{P}^1_{\text{Berk}}$ which we call the spherical kernel. Using it, for each $\zeta \in \mathbb{P}^1_{\text{Berk}}$, we define a generalized Hsia kernel $\delta(x,y)_\zeta$, which plays an analogous role for potential theory relative to ζ.

4.1. Definition of the Hsia kernel

It is an elementary fact about ultrametric spaces that the *radius* and *diameter* of a disc coincide.

If $x \in \mathbb{A}^1_{\text{Berk}}$ corresponds to a sequence of nested discs $\{D(a_i, r_i)\}$, we call $r = \lim_{i \to \infty} r_i$ the *diameter* (or *radius*) of x and write $r = \operatorname{diam}_\infty(x)$. (Note that the diameter depends on the choice of coordinates on $\mathbb{A}^1$.)

The *Hsia kernel* $\delta(x,y)_\infty$ is defined for $x, y \in \mathbb{A}^1_{\text{Berk}}$ as follows. Consider the paths $[x, \infty]$ and $[y, \infty]$, and as in §2.7, let $x \vee_\infty y$ be the point where they first meet. Then

$$(4.1) \qquad \delta(x,y)_\infty = \operatorname{diam}_\infty(x \vee_\infty y) \ .$$

More concretely, if x corresponds to a sequence of nested discs $\{D(a_i, r_i)\}$ and y to $\{D(b_i, s_i)\}$, then

$$(4.2) \qquad \delta(x,y)_\infty = \lim_{i \to \infty} \max(r_i, s_i, |a_i - b_i|) \ .$$

Indeed, if $R_i = \max(r_i, s_i, |a_i - b_i|)$, then $D(a_i, R_i) = D(b_i, R_i)$ is the smallest disc containing both $D(a_i, r_i)$ and $D(b_i, s_i)$, so $\operatorname{diam}_\infty(x \vee_\infty y) = \lim_{i \to \infty} R_i$.

Clearly the Hsia kernel is symmetric, and $\delta(x,x)_\infty = \operatorname{diam}_\infty(x)$ for each x. If $x, y \in \mathbb{A}^1(K)$ are points of type I, then $\delta(x,y)_\infty = |x-y|$. If $x, y \in \mathbb{A}^1_{\text{Berk}}$ are points of type I, II, or III, with x corresponding to $D(a,r)$

and y corresponding to $D(b,s)$, then
$$\delta(x,y)_\infty = \max(r,s,|a-b|) = \sup_{z\in D(a,r), w\in D(b,s)} |z-w|.$$

The *Berkovich open and closed discs* in $\mathbb{A}^1_{\text{Berk}}$, with center $a \in K$ and radius $r > 0$, are defined by

(4.3) $\qquad \mathcal{D}(a,r)^- = \{x \in \mathbb{A}^1_{\text{Berk}} : [T-a]_x < r\},$

(4.4) $\qquad \mathcal{D}(a,r) = \{x \in \mathbb{A}^1_{\text{Berk}} : [T-a]_x \leq r\}.$

The following result summarizes the main properties of $\delta(x,y)_\infty$:

PROPOSITION 4.1.

(A) *The Hsia kernel is nonnegative, symmetric, and continuous in each variable separately. As a function of two variables, it is upper semi-continuous. It is continuous off the diagonal and continuous at (x_0, x_0) for each point x_0 of type I, but it is discontinuous at (x_0, x_0) for each point of type II, III, or IV.*

(B) *The Hsia kernel is the unique extension of $|x-y|$ to $\mathbb{A}^1_{\text{Berk}}$ such that*

(4.5) $\qquad \delta(x,y)_\infty = \limsup_{\substack{(a,b)\to(x,y) \\ a,b\in \mathbb{A}^1(K)}} |a-b|$

for each $x,y \in \mathbb{A}^1_{\text{Berk}}$.

(C) *For all $x,y,z \in \mathbb{A}^1_{\text{Berk}}$*
$$\delta(x,y)_\infty \leq \max(\delta(x,z)_\infty, \delta(y,z)_\infty),$$
with equality if $\delta(x,z)_\infty \neq \delta(y,z)_\infty$.

(D) *For each $a \in \mathbb{A}^1_{\text{Berk}}$ and $0 \leq r \in \mathbb{R}$, the 'open ball' $\mathcal{B}(a,r)^-_\infty := \{x \in \mathbb{A}^1_{\text{Berk}} : \delta(x,a)_\infty < r\}$ is connected and open in the Berkovich space topology. It is empty if $r \leq \operatorname{diam}_\infty(a)$ and coincides with an open disc $\mathcal{D}(b,r)^-$ for some $b \in \mathbb{A}^1(K)$ if $r > \operatorname{diam}_\infty(a)$.*

For each $a \in \mathbb{A}^1_{\text{Berk}}$ and $0 \leq r \in \mathbb{R}$, the 'closed ball' $\mathcal{B}(a,r)_\infty := \{x \in \mathbb{A}^1_{\text{Berk}} : \delta(x,a)_\infty \leq r\}$ is connected and closed in the Berkovich space topology. It is empty if $r < \operatorname{diam}_\infty(a)$ and coincides with a closed disc $\mathcal{D}(b,r)$ for some $b \in \mathbb{A}^1(K)$ if $r > \operatorname{diam}_\infty(a)$ or if $r = \operatorname{diam}_\infty(a)$ and a is of type II or III. If $r = \operatorname{diam}_\infty(a)$ and a is of type I or IV, then $\mathcal{B}(a,r)_\infty = \{a\}$.

PROOF.

(A) Consider $\delta(x,y)_\infty$ as a function of two variables. We first show it is continuous off the diagonal.

Take $x_0, y_0 \in \mathbb{A}^1_{\text{Berk}}$ with $x_0 \neq y_0$. If there are disjoint open discs with $x_0 \in \mathcal{D}(a,r)^-$ and $y_0 \in \mathcal{D}(b,s)^-$, then $|a-b| > \max(r,s)$, so for each $x \in \mathcal{D}(a,r)^-$, $y \in \mathcal{D}(b,s)^-$,
$$\delta(x,y)_\infty = |a-b| = \delta(x_0, y_0)_\infty,$$
and $\delta(x,y)_\infty$ is continuous at (x_0, y_0).

Otherwise, either each open disc $\mathcal{D}(a,r)^-$ containing x_0 contains y_0 or each open disc $\mathcal{D}(b,s)^-$ containing y_0 contains x_0. By symmetry, assume the former holds. Then x_0 must be a point of type II or III corresponding to a disc $D(a,r)$ with $y_0 \in \mathcal{D}(a,r)$ (it cannot be of type IV, because every open disc containing x_0 also contains $y_0 \ne x_0$). After relabeling the center, we can assume that $y_0 \in \mathcal{D}(a,s)^-$ for some $s < r$. Take $0 < \varepsilon < r-s$. Then $U := \mathcal{D}(a,r+\varepsilon)^- \setminus \mathcal{D}(a,r-\varepsilon)$ and $V := \mathcal{D}(a,s)^-$ are disjoint open neighborhoods of x_0, y_0, respectively. Given $x \in U$, put $r_1 = \inf\{t : x \in \mathcal{D}(a,t)\}$. Then $r-\varepsilon < r_1 < r+\varepsilon$. Formula (4.1) shows that $\delta(x,y)_\infty = r_1$ for each $y \in V$, so
$$|\delta(x,y)_\infty - \delta(x_0,y_0)_\infty| = |r_1 - r| < \varepsilon .$$
Thus, $\delta(x,y)_\infty$ is continuous at (x_0, y_0) in this case as well.

Now consider points (x_0, x_0) on the diagonal. Suppose $\operatorname{diam}_\infty(x_0) = r$. For each $\varepsilon > 0$, there is an open disc $\mathcal{D}(a, r+\varepsilon)^-$ containing x_0, and (4.1) shows that $\delta(x,y)_\infty < r+\varepsilon$ for each $x,y \in \mathcal{D}(a, r+\varepsilon)^-$. If $r=0$, this implies that $\delta(x,y)_\infty$ is continuous at (x_0, x_0).

If $r > 0$, then $\delta(x,y)_\infty$ is not continuous at (x_0, x_0), since every neighborhood of (x_0, x_0) contains points (a,a) with a of type I, and for such points $\delta(a,a)_\infty = 0$, while $\delta(x_0, x_0)_\infty = r$. By the discussion above,
$$(4.6) \qquad \limsup_{(x,y) \to (x_0, x_0)} \delta(x,y)_\infty = r = \delta(x_0, x_0)_\infty .$$
Thus $\delta(x,y)_\infty$ is upper semicontinuous as a function of two variables.

Now fix x, and consider $\delta(x,y)_\infty$ as a function of y. By what has been shown above, $\delta(x,y)_\infty$ is continuous for $y \ne x$. For continuity at x, put $r = \operatorname{diam}_\infty(x)$ and consider a neighborhood $\mathcal{D}(a, r+\varepsilon)^-$ of x. For each $y \in \mathcal{D}(a, r+\varepsilon)^-$, the paths from x and y to ∞ meet at a point $w = x \vee_\infty y \in \mathcal{D}(a, r+\varepsilon)^-$, so
$$r = \operatorname{diam}_\infty(x) = \delta(x,x)_\infty \le \delta(x,y)_\infty = \operatorname{diam}_\infty(w) \le r+\varepsilon .$$
Thus $|\delta(x,y)_\infty - \delta(x,x)_\infty| < \varepsilon$.

(B) Note that each neighborhood $\mathcal{D}(a, r+\varepsilon)^-$ of x_0 contains points $a \ne b$ of type I with $r < |a-b| < r+\varepsilon$, so (4.6) remains true even if the lim sup is restricted to $x,y \in \mathbb{A}^1(K)$. The characterization of $\delta(x,y)_\infty$ follows from this, together with continuity off the diagonal.

(C) For discs $D(a,r)$, $D(b,s)$ it is easy to see that
$$\max(r, s, |a-b|) = \sup_{\substack{p \in D(a,r) \\ q \in D(b,s)}} (|p-q|) .$$
The ultrametric inequality for $\delta(x,y)_\infty$ follows from this, formula (4.2), and the classical ultrametric inequality.

(D) Given $a \in \mathbb{A}^1_{\text{Berk}}$ and $r \in \mathbb{R}$, consider the 'open ball' $\mathcal{B}(a,r)^-_\infty := \{x \in \mathbb{A}^1_{\text{Berk}} : \delta(x,a)_\infty < r\}$. If $r \le \operatorname{diam}_\infty(a)$, then $\mathcal{B}(a,r)^-_\infty$ is clearly empty. If $r > \operatorname{diam}_\infty(a)$, then $\mathcal{B}(a,r)^-_\infty$ contains a and is open by the upper semicontinuity of $\delta(x,a)_\infty$. Each nonempty open set contains points of type I. If $b \in \mathcal{D}(a,r)^- \cap \mathbb{A}^1(K)$, the ultrametric inequality for $\delta(x,y)_\infty$ shows that $\mathcal{B}(a,r)_\infty^- = \mathcal{D}(b,r)^-$. To see that $\mathcal{B}(a,r)_\infty^-$ is connected, recall that $\delta(a,x)_\infty = \mathrm{diam}_\infty(a \vee_\infty x)$. If $\mathrm{diam}_\infty(a \vee_\infty x) < r$, then $\mathrm{diam}_\infty(a \vee_\infty y) = \mathrm{diam}_\infty(a \vee_\infty x) < r$ for each y in the path $[a,x]$, so $[a,x] \subset \mathcal{B}(a,r)_\infty^-$.

Similarly, consider the 'closed ball' $\mathcal{B}(a,r)_\infty := \{x \in \mathbb{A}^1_{\mathrm{Berk}} : \delta(x,a)_\infty \le r\}$. If $r < \mathrm{diam}_\infty(a)$, then $\mathcal{B}(a,r)_\infty$ is empty. If $r = \mathrm{diam}_\infty(a)$ and a is of type I or IV, it is easy to see that $\mathcal{B}(a,r)_\infty = \{a\}$. If $r > \mathrm{diam}_\infty(a)$, then $\mathcal{B}(a,r)_\infty$ contains $\mathcal{B}(a,r)_\infty^-$ and hence contains points b of type I; this also holds if $r = \mathrm{diam}_\infty(a)$ and a is of type II or III. In either case the ultrametric inequality for $\delta(x,y)_\infty$ shows that $\mathcal{B}(a,r)_\infty = \mathcal{D}(b,r)$ and in particular that $\mathcal{B}(a,r)_\infty$ is closed. An argument similar to the one above shows that $\mathcal{B}(a,r)_\infty$ is connected. $\square$

The function-theoretic meaning of the Hsia kernel is as follows:

COROLLARY 4.2. *For each $x \in \mathbb{A}^1_{\mathrm{Berk}}$ and each $a \in K$, the function $T - a \in K[T]$ satisfies*
$$\delta(x,a)_\infty = [T-a]_x .$$

PROOF. By Proposition 4.1(B), if $x \in \mathbb{A}^1_{\mathrm{Berk}}$ corresponds to the nested sequence of discs $\{D(a_i, r_i)\}$, then
$$[T-a]_x = \lim_{i \to \infty} \|T-a\|_{D(a_i, r_i)} = \limsup_{\substack{b \to x \\ b \in K}} |b-a| = \delta(x,a)_\infty .$$
$\square$

4.2. The extension of $j_z(x,y)$ to $\mathbb{P}^1_{\mathrm{Berk}}$

The Hsia kernel $\delta(x,y)_\infty$ has a pole at $\zeta = \infty$. We now set out to define an analogous kernel $\delta(x,y)_\zeta$ for an arbitrary $\zeta \in \mathbb{P}^1_{\mathrm{Berk}}$.

To do this, we will use the theory of metrized graphs from Chapter 3. Recall from §2.5 and §2.7 that a *finite subgraph* Γ of $\mathbb{P}^1_{\mathrm{Berk}}$ is a finite $\mathbb{R}$-tree contained in $\mathbb{H}_{\mathrm{Berk}}$ which is equipped with the (logarithmic) path distance metric $\rho(x,y)$. Since there is a unique path between any two points of $\mathbb{P}^1_{\mathrm{Berk}}$, for every finite subgraph Γ of $\mathbb{P}^1_{\mathrm{Berk}}$ there is a natural retraction map $r_\Gamma : \mathbb{P}^1_{\mathrm{Berk}} \to \Gamma$ (Theorem 2.21).

For each Γ, the relation between the Hsia kernel and the potential kernel $j_z(x,y)$ on Γ is as follows. Fix a coordinate system on $\mathbb{P}^1$, so that $\mathbb{P}^1_{\mathrm{Berk}} = \mathbb{A}^1_{\mathrm{Berk}} \cup \{\infty\}$. Recall that ζ_{Gauss} denotes the "Gauss point", the point of $\mathbb{A}^1_{\mathrm{Berk}}$ corresponding to the sup norm over the closed unit disc $D(0,1)$.

PROPOSITION 4.3. *Let $z \in \mathbb{H}_{\mathrm{Berk}}$, and let Γ be any finite subgraph of $\mathbb{P}^1_{\mathrm{Berk}}$ containing z. Put $\infty_\Gamma = r_\Gamma(\infty)$. Then for all $x, y \in \Gamma$,*
(4.7)
$$-\log_v(\delta(x,y)_\infty) = j_z(x,y) - j_z(x, \infty_\Gamma) - j_z(y, \infty_\Gamma) - \log_v(\mathrm{diam}_\infty(z)) .$$
In particular, if $z = \zeta_{\mathrm{Gauss}}$, then $\mathrm{diam}_\infty(z) = 1$, so
(4.8) $\quad -\log_v(\delta(x,y)_\infty) = j_{\zeta_{\mathrm{Gauss}}}(x,y) - j_{\zeta_{\mathrm{Gauss}}}(x, \infty_\Gamma) - j_{\zeta_{\mathrm{Gauss}}}(y, \infty_\Gamma) .$

PROOF. Fix $y \in \Gamma$, and set $s = \text{diam}_\infty(y)$, $S = \text{diam}_\infty(\infty_\Gamma)$. The intersection of the path from y to ∞ with Γ is the finite subgraph $\Gamma_y = [y, \infty_\Gamma]$ of Γ, isometric to the real interval $[\log_v(s), \log_v(S)]$.

Consider the function $f_y(x) = -\log_v(\delta(x,y)_\infty)$ on Γ. The geometric description (4.1) of $\delta(x,y)_\infty$ shows that for each $x \in \Gamma$,
$$\delta(x,y)_\infty = \delta(r_{\Gamma_y}(x), y)_\infty .$$
Thus, $f_y(x)$ is constant on branches of Γ off Γ_y. For $x \in \Gamma_y$,
$$-\log_v(\delta(x,y)_\infty) = -\log_v(t) ,$$
where $t = \text{diam}_\infty(x)$. The arclength parameter along Γ_y is $\log_v(t)$ for $s \le t \le S$, so the restriction of $f_y(x)$ to Γ_y has constant slope -1. Hence $f_y(x) \in \text{CPA}(\Gamma)$, and
$$\Delta_x(f_y(x)) = \delta_y(x) - \delta_{\infty_\Gamma}(x) .$$
Also, by (3.3), we have
$$\Delta_x(j_z(x,y) - j_z(x, \infty_\Gamma)) = \delta_y(x) - \delta_{\infty_\Gamma}(x) .$$
It follows from Proposition 3.2(B) that there is a constant $C_z(y)$ such that for all $x \in \Gamma$,
$$(4.9) \quad -\log_v(\delta(x,y)_\infty) = f_y(x) = j_z(x,y) - j_z(x, \infty_\Gamma) + C_z(y) .$$

Fixing x and letting y vary, we see from (4.9) and the symmetry of $\delta(x,y)_\infty$ that the function $h(y) = C_z(y)$ belongs to $\text{CPA}(\Gamma)$. Let Δ_y be the Laplacian with respect to the variable y. Applying Δ_y to both sides of (4.9) gives
$$\delta_x(y) - \delta_{\infty_\Gamma}(y) = (\delta_x(y) - \delta_z(y)) - 0 + \Delta_y(h(y)) ,$$
so $\Delta_y(h(y)) = -(\delta_{\infty_\Gamma}(y) - \delta_z(y))$. By Proposition 3.2(B), there is a constant C_z such that $h(y) = -j_z(y, \infty_\Gamma) + C_z$. Hence
$$-\log_v(\delta(x,y)_\infty) = f_y(x) = j_z(x,y) - j_z(x, \infty_\Gamma) - j_z(y, \infty_\Gamma) + C_z .$$
Taking $x = y = z$, we see that $C_z = -\log_v(\delta(z,z)_\infty) = -\log_v(\text{diam}_\infty(z))$. $\square$

Now fix $z \in \mathbb{H}_{\text{Berk}}$, and let Γ vary over finite subgraphs of $\mathbb{P}^1_{\text{Berk}}$ containing z. Temporarily write $j_z(x,y)_\Gamma$ for the potential kernel on Γ. We will show that the functions $j_z(x,y)_\Gamma$ are compatible, allowing us to define a function $j_z(x,y)$ on $\mathbb{H}_{\text{Berk}} \times \mathbb{H}_{\text{Berk}}$. This in turn extends to a function on $\mathbb{P}^1_{\text{Berk}} \times \mathbb{P}^1_{\text{Berk}}$, valued in $\mathbb{R} \cup \{+\infty\}$, which is singular on (and only on) the diagonal $\text{Diag}(K)$ in $\mathbb{P}^1(K) \times \mathbb{P}^1(K)$.

Suppose $\Gamma_1 \subset \Gamma_2$. Since Γ_1 and Γ_2 are both trees, the potential kernel on Γ_2 extends the one on Γ_1. Indeed, given $x, y \in \Gamma_1$, the point $w = w_z(x,y)$ where the paths from x and y to z meet belongs to both Γ_1 and Γ_2, so
$$j_z(x,y)_{\Gamma_2} = \rho(z,w) = j_z(x,y)_{\Gamma_1} .$$

Thus the functions $j_z(x,y)_\Gamma$ cohere to give a well-defined function $j_z(x,y)$ on $\mathbb{H}_{\text{Berk}} \times \mathbb{H}_{\text{Berk}}$. In terms of the Gromov product on the $\mathbb{R}$-tree $(\mathbb{H}_{\text{Berk}}, \rho)$ (see §B.4), we have
$$j_z(x,y) = (x|y)_z.$$

We can extend $j_z(x,y)$ to all $x, y \in \mathbb{P}^1_{\text{Berk}}$ as follows. Given x, y, let Γ be any finite subgraph containing z and $w_z(x,y)$. Then

$$(4.10) \qquad j_z(x,y) = \begin{cases} j_z(r_\Gamma(x), r_\Gamma(y))_\Gamma & \text{if } (x,y) \notin \text{Diag}(K), \\ \infty & \text{if } (x,y) \in \text{Diag}(K). \end{cases}$$

Explicitly, for $x, y \in \mathbb{P}^1(K)$,

$$(4.11) \qquad j_z(x,y) = \begin{cases} \rho(z, w_z(x,y)) & \text{if } x \neq y, \\ \infty & \text{if } x = y. \end{cases}$$

By Proposition 4.3 and the continuity of $\delta(x,y)_\infty$ off the diagonal, for $x, y \in \mathbb{A}^1_{\text{Berk}}$ we can write

$$(4.12) \quad -\log_v(\delta(x,y)_\infty) = j_z(x,y) - j_z(x,\infty) - j_z(y,\infty) - \log_v(\text{diam}_\infty(z)).$$

In particular, if $z = \zeta_{\text{Gauss}}$ is the Gauss point, then

$$(4.13) \quad -\log_v(\delta(x,y)_\infty) = j_{\zeta_{\text{Gauss}}}(x,y) - j_{\zeta_{\text{Gauss}}}(x,\infty) - j_{\zeta_{\text{Gauss}}}(y,\infty).$$

Similarly, by Proposition 3.3(C), for each $z, \zeta \in \mathbb{H}_{\text{Berk}}$ we have

$$(4.14) \qquad j_\zeta(x,y) = j_z(x,y) - j_z(x,\zeta) - j_z(\zeta,y) + j_z(\zeta,\zeta).$$

If $\varphi \in K(T)$ is a rational function, we can view $-\log_v([\varphi]_x)$ as a singular function on $\mathbb{P}^1_{\text{Berk}}$, taking $-\log_v([\varphi]_x) = \infty$ if $[\varphi]_x = 0$ (that is, if x is a type I point corresponding to a zero of φ), and $-\log_v([\varphi]_x) = -\infty$ if $\varphi \notin \mathcal{R}_x$ (that is, if x is a type I point corresponding to a pole of φ). As a consequence of Proposition 4.3, we obtain:

COROLLARY 4.4. *Let $0 \neq \varphi \in K(T)$ be a rational function with divisor $\text{div}(\varphi) = \sum_{i=1}^m n_i(a_i)$, and fix $z \in \mathbb{H}_{\text{Berk}}$. Then:*

(A) *For all $x \in \mathbb{P}^1_{\text{Berk}}$,*

$$(4.15) \qquad -\log_v([\varphi]_x) = -\log_v([\varphi]_z) + \sum_{i=1}^m n_i j_z(x, a_i).$$

(B) *Let Γ be a finite subgraph of $\mathbb{P}^1_{\text{Berk}}$ containing z. Then for all $x \in \Gamma$,*

$$(4.16) \qquad -\log_v([\varphi]_x) = -\log_v([\varphi]_z) + \sum_{i=1}^m n_i j_z(x, r_\Gamma(a_i))_\Gamma.$$

PROOF. (A) There is a $B \in K$ such that $\varphi(T) = B \cdot \prod_{a_i \neq \infty}(T - a_i)^{n_i}$. By Corollary 4.2, for each $x \in \Gamma$

$$-\log_v([\varphi]_x) = -\log_v(|B|) + \sum_{a_i \neq \infty} -n_i \log_v(\delta(x, a_i)_\infty).$$

Inserting formula (4.12) and using $\sum_{i=1}^{m} n_i = 0$ gives

$$-\log_v([\varphi]_x) \;=\; \sum_{i=1}^{m} n_i \hat{j}_z(x, a_i) \;+\; C$$

for some constant C. Taking $x = z$ gives $C = -\log_v([\varphi]_z)$.

Part (B) follows from (4.15), using (4.10). □

For future reference, we note the following useful variant of (4.10):

PROPOSITION 4.5 (Retraction Formula). *Let Γ be a finite subgraph of $\mathbb{P}^1_{\text{Berk}}$, and suppose $x, z \in \Gamma$. Then for any $y \in \mathbb{P}^1_{\text{Berk}}$*

$$\hat{j}_z(x, y) \;=\; \hat{j}_z(x, r_\Gamma(y))_\Gamma \;.$$

PROOF. Since x and z belong to Γ, so does the path from x to z. Hence the point $w_z(x, y)$ where the paths from x and y to z first meet, which lies on the path from x to z, belongs to Γ as well. □

4.3. The spherical distance and the spherical kernel

Recall that $\mathcal{O}$ denotes the valuation ring of K. The spherical distance $\|x, y\|$ on $\mathbb{P}^1(K)$ is the unique $GL_2(\mathcal{O})$-invariant metric on $\mathbb{P}^1(K)$ such that $\|x, y\| = |x - y|$ for $x, y \in D(0, 1)$. If $x = (x_0 : x_1)$ and $y = (y_0 : y_1)$ in homogeneous coordinates, then

$$(4.17) \qquad \|x, y\| \;=\; \frac{|x_0 y_1 - x_1 y_0|}{\max(|x_0|, |x_1|) \max(|y_0|, |y_1|)} \;.$$

In affine coordinates, for $x, y \in \mathbb{A}^1(K)$ this becomes

$$(4.18) \qquad \|x, y\| \;=\; \frac{|x - y|}{\max(1, |x|) \max(1, |y|)} \;.$$

Thus if we identify $\mathbb{P}^1(K)$ with $\mathbb{A}^1(K) \cup \{\infty\}$ and put $1/\infty = 0$, then

$$(4.19) \qquad \|x, y\| \;=\; \begin{cases} |x - y| & \text{if } x, y \in D(0, 1), \\ |1/x - 1/y| & \text{if } x, y \in \mathbb{P}^1(K) \setminus D(0, 1), \\ 1 & \text{if exactly one of } x, y \in D(0, 1) \;. \end{cases}$$

It is well known, and easy to check using (4.19), that the spherical distance satisfies the ultrametric inequality. Clearly its values lie in the interval $[0, 1]$.

We will now extend $\|x, y\|$ to a function from $\mathbb{P}^1_{\text{Berk}} \times \mathbb{P}^1_{\text{Berk}}$ to $[0, 1]$, which we call the *spherical kernel*.

For $x, y \in \mathbb{P}^1(K)$, we claim that $-\log_v \|x, y\| = \hat{j}_{\zeta_{\text{Gauss}}}(x, y)$, i.e., that

$$(4.20) \qquad \|x, y\| \;=\; q_v^{-\hat{j}_{\zeta_{\text{Gauss}}}(x, y)} \;.$$

To see this, consider the various cases in (4.19). If $x, y \in D(0, 1)$, then the point $w = w_{\zeta_{\text{Gauss}}}(x, y)$ where the paths from x and y to ζ_{Gauss} first meet

corresponds to the ball $D(x,r) = D(y,r)$ with $r = |x-y|$. The path distance $\rho(\zeta_{\text{Gauss}}, w)$ is $\log_v(1/r)$, so $j_{\zeta_{\text{Gauss}}}(x,y) = -\log_v(r)$ and
$$q_v^{-j_{\zeta_{\text{Gauss}}}(x,y)} = |x-y| = \|x,y\|.$$
If $x, y \in \mathbb{P}^1(K) \backslash D(0,1)$, a similar argument applies, using the local parameter $1/T$ at ∞. Finally, if one of x, y belongs to $D(0,1)$ and the other to $\mathbb{P}^1(K) \backslash D(0,1)$, then the paths from x and y to ζ_{Gauss} meet at $w = \zeta_{\text{Gauss}}$, so $j_{\zeta_{\text{Gauss}}}(x,y) = 0$ and $q_v^{-j_{\zeta_{\text{Gauss}}}(x,y)} = 1 = \|x,y\|$.

This motivates our definition of the spherical kernel: for all $x, y \in \mathbb{P}^1_{\text{Berk}}$, we put

(4.21) $$\|x,y\| = q_v^{-j_{\zeta_{\text{Gauss}}}(x,y)}.$$

This extension is different from the function $d(x,y)$ discussed in §2.7. Note that for $x, y \in \mathcal{D}(0,1)$,
$$\|x,y\| = \delta(x,y)_\infty.$$

The spherical kernel has the following geometric interpretation. Consider the paths from x and y to ζ_{Gauss}, and let $w = w_{\zeta_{\text{Gauss}}}(x,y)$ be the point where they first meet. Then by (4.21),
$$\|x,y\| = q_v^{-\rho(\zeta_{\text{Gauss}}, w)}.$$
For each $x \in \mathbb{P}^1_{\text{Berk}}$, we write $\operatorname{diam}(x) = \|x,x\|$. Clearly $\operatorname{diam}(x) = 0$ if $x \in \mathbb{P}^1(K)$, while $\operatorname{diam}(x) > 0$ if $x \in \mathbb{H}_{\text{Berk}}$.

REMARK 4.6. Note that although $\|x,y\|$ is a metric on $\mathbb{P}^1(K)$, the spherical kernel is *not* a metric on $\mathbb{P}^1_{\text{Berk}}$ because $\|x,x\| > 0$ if x is not of type I. Nonetheless, for each $a \in \mathbb{P}^1(K)$, the balls
$$\mathcal{B}(a,r)^- = \{x \in \mathbb{P}^1_{\text{Berk}} : \|x,a\| < r\},$$
$$\mathcal{B}(a,r) = \{x \in \mathbb{P}^1_{\text{Berk}} : \|x,a\| \leq r\}$$
are indeed open (resp. closed) in the Berkovich topology. If $r > 1$, then $\mathcal{B}(a,r)^- = \mathbb{P}^1_{\text{Berk}}$. If $r \leq 1$ and $a \in D(0,1)$, then $\mathcal{B}(a,r)^- = \mathcal{D}(a,r)^-$, while if $a \notin D(0,1)$, then $\mathcal{B}(a,r)^- = \{x \in \mathbb{P}^1_{\text{Berk}} : [1/T - 1/a]_x < r\}$. Similar formulas hold for the closed balls.

PROPOSITION 4.7.

(A) *The spherical kernel $\|x,y\|$ is nonnegative, symmetric, continuous in each variable separately, and valued in $[0,1]$. As a function of two variables, it is upper semicontinuous. It is continuous off the diagonal and continuous at (x_0, x_0) for each point x_0 of type I, but it is discontinuous at (x_0, x_0) for each point of type II, III, or IV.*

(B) *For all $x, y \in \mathbb{P}^1_{\text{Berk}}$*

(4.22) $$\|x,y\| = \limsup_{\substack{(a,b) \to (x,y) \\ a,b \in \mathbb{P}^1(K)}} \|a,b\|.$$

(C) For all $x, y, z \in \mathbb{P}^1_{\text{Berk}}$
$$\|x, y\| \leq \max(\|x, z\|, \|y, z\|),$$
with equality if $\|x, z\| \neq \|y, z\|$.

(D) For each $a \in \mathbb{P}^1_{\text{Berk}}$ and $r \in \mathbb{R}$, the open ball $\mathcal{B}(a, r)^- := \{x \in \mathbb{P}^1_{\text{Berk}} : \|x, a\| < r\}$ is connected and open in the Berkovich topology. It is empty if $r \leq \operatorname{diam}(a)$ and coincides with an open ball $\mathcal{B}(b, r)^-$ for some $b \in \mathbb{P}^1(K)$ if $r > \operatorname{diam}(a)$.

Likewise, the closed ball $\mathcal{B}(a, r) := \{x \in \mathbb{P}^1_{\text{Berk}} : \|x, a\| \leq r\}$ is connected and closed in the Berkovich topology. It is empty if $r < \|a, a\|$ and coincides with $\mathcal{B}(b, r)$ for some $b \in \mathbb{P}^1(K)$ if $r > \operatorname{diam}(a)$ or if $r = \operatorname{diam}(a)$ and a is of type II or III. If $r = \operatorname{diam}(a)$ and a is of type I or IV, then $\mathcal{B}(a, r) = \{a\}$.

PROOF. Similar to Proposition 4.1. $\square$

The balls $\mathcal{B}(a, r)^-$ and $\mathcal{B}(a, r)$ will play an important role in this book.

The Hsia kernel and the spherical kernel can be obtained from each other using the following formulas:

PROPOSITION 4.8.
(A) For $x, y \in \mathbb{A}^1_{\text{Berk}}$, $\quad \delta(x, y)_\infty = \dfrac{\|x, y\|}{\|x, \infty\| \|y, \infty\|}$.

(B) For $x, y \in \mathbb{P}^1_{\text{Berk}}$,
$$\|x, y\| = \begin{cases} \dfrac{\delta(x, y)_\infty}{\delta(x, \zeta_{\text{Gauss}})_\infty \, \delta(y, \zeta_{\text{Gauss}})_\infty} & \text{if } x, y \neq \infty, \\ 1/\delta(x, \zeta_{\text{Gauss}})_\infty & \text{if } x \neq \infty, y = \infty, \\ 1/\delta(y, \zeta_{\text{Gauss}})_\infty & \text{if } x = \infty, y \neq \infty, \\ 0 & \text{if } x = y = \infty. \end{cases}$$

PROOF. Part (A) follows from (4.13). Part (B) follows formally from part (A), using the fact that $\|z, \zeta_{\text{Gauss}}\| = 1$ for each $z \in \mathbb{P}^1_{\text{Berk}}$. $\square$

4.4. The generalized Hsia kernel

Proposition 4.8 motivates our definition of a *generalized Hsia kernel* for an arbitrary $\zeta \in \mathbb{P}^1_{\text{Berk}}$. Fixing ζ, put

(4.23) $$\delta(x, y)_\zeta = \frac{\|x, y\|}{\|x, \zeta\| \|y, \zeta\|}$$

for all $x, y \in \mathbb{P}^1_{\text{Berk}} \setminus \{\zeta\}$.

If $\zeta \notin \mathbb{P}^1(K)$, the formula (4.23) makes sense for all $x, y \in \mathbb{P}^1_{\text{Berk}}$, since $\|x, \zeta\|, \|\zeta, y\| \geq \|\zeta, \zeta\| > 0$, giving $\delta(x, \zeta)_\zeta = \delta(\zeta, y)_\zeta = 1/\|\zeta, \zeta\|$. If $\zeta \in \mathbb{P}^1(K)$, we put $\delta(x, \zeta)_\zeta = \delta(\zeta, y)_\zeta = \infty$. In this way, we can regard $\delta(x, y)_\zeta$ as a function from $\mathbb{P}^1_{\text{Berk}} \times \mathbb{P}^1_{\text{Berk}} \to \mathbb{R} \cup \{\infty\}$.

When $\zeta = \infty$, Proposition 4.8 shows that definition (4.23) is consistent with our earlier definition of $\delta(x,y)_\infty$. When $\zeta = \zeta_{\text{Gauss}}$ is the Gauss point, we have $\|x, \zeta_{\text{Gauss}}\| = \|\zeta_{\text{Gauss}}, y\| = 1$, so that

$$(4.24) \qquad \delta(x,y)_{\zeta_{\text{Gauss}}} = \|x,y\| = q_v^{j_{\zeta_{\text{Gauss}}}(x,y)}$$

coincides with the spherical kernel.

For an arbitrary point $\zeta \in \mathbb{H}_{\text{Berk}}$, Proposition 3.3(C) shows that

$$(4.25) \qquad \delta(x,y)_\zeta = C_\zeta \cdot q_v^{-j_\zeta(x,y)}, \qquad \text{where } C_\zeta = q_v^{j_{\zeta_{\text{Gauss}}}(\zeta,\zeta)}.$$

Thus for any $\zeta \in \mathbb{H}_{\text{Berk}}$, $\delta(x,y)_\zeta$ can be thought of as a generalized spherical distance.

It is sometimes more natural to view the generalized Hsia kernel as only defined up to scaling. By the definition of the spherical kernel, we have

$$\delta(x,y)_\zeta = q_v^{-j_{\zeta_{\text{Gauss}}}(x,y) + j_{\zeta_{\text{Gauss}}}(x,\zeta) + j_{\zeta_{\text{Gauss}}}(y,\zeta)}.$$

However, for any $z \in \mathbb{H}_{\text{Berk}}$, formula (4.14) shows that there is a constant $C = C_{\zeta, z}$ such that

$$(4.26) \qquad C \cdot \delta(x,y)_\zeta = q_v^{-j_z(x,y) + j_z(x,\zeta) + j_z(y,\zeta)}.$$

Since the Gauss point is dependent on a choice of coordinates, the normalization of $\delta(x,y)_\zeta$ in (4.23) is noncanonical. Hence, for each $C > 0$, we can also regard $C \cdot \delta(x,y)_\zeta$ as a generalized Hsia kernel relative to the point ζ.

However, when we write $\delta(x,y)_\zeta$, we will mean the function defined by (4.23), unless explicitly noted otherwise.

For future reference, we rewrite (4.25) in the following form:

$$(4.27) \qquad -\log_v(\delta(x,y)_\zeta) = j_\zeta(x,y) - j_{\zeta_{\text{Gauss}}}(\zeta,\zeta),$$

valid for all $\zeta \in \mathbb{H}_{\text{Berk}}$ and $x, y \in \mathbb{P}^1_{\text{Berk}}$. In particular, for $x, y, \zeta \in \mathbb{H}_{\text{Berk}}$, we have the following formula relating the generalized Hsia kernel and the Gromov product (see §B.4) on the $\mathbb{R}$-tree $(\mathbb{H}_{\text{Berk}}, \rho)$:

$$(4.28) \qquad -\log_v(\delta(x,y)_\zeta) = (x|y)_\zeta - (\zeta|\zeta)_{\zeta_{\text{Gauss}}}.$$

It is useful also to note that for any distinct $\xi, \zeta \in \mathbb{P}^1_{\text{Berk}}$, it follows formally from (4.23) that with $C = C_{\xi, \zeta} = \|\xi, \zeta\|^{-2}$,

$$(4.29) \qquad \delta(x,y)_\xi = C \cdot \frac{\delta(x,y)_\zeta}{\delta(x,\xi)_\zeta \, \delta(y,\xi)_\zeta}$$

for all $x, y \in \mathbb{P}^1_{\text{Berk}} \backslash \{\zeta\}$. If $\zeta \in \mathbb{H}_{\text{Berk}}$, then $\delta(x,y)_\zeta$ is bounded, and (4.29) holds for all $x, y \in \mathbb{P}^1_{\text{Berk}}$.

For each $x \in \mathbb{P}^1_{\text{Berk}}$, put $\text{diam}_\zeta(x) = \delta(x,x)_\zeta$. The generalized Hsia kernel has the following geometric interpretation which extends (4.1). Given $x, y \in \mathbb{P}^1_{\text{Berk}}$, let $w = x \vee_\zeta y$ be the point where the paths from x and y to ζ first meet. Then

$$(4.30) \qquad \delta(x,y)_\zeta = \text{diam}_\zeta(w).$$

For $\zeta \in \mathbb{H}_{\text{Berk}}$, it follows from (4.25) and the definitions that

(4.31) $\qquad -\log_v \delta(x,y)_\zeta = \rho(w,\zeta) + \log_v(\|\zeta,\zeta\|)$

and in particular (setting $y = x$) that

(4.32) $\qquad \text{diam}_\zeta(x) = \dfrac{1}{\|\zeta,\zeta\|} \cdot q_v^{-\rho(x,\zeta)}$.

We also note that the path distance $\rho(x,y)$ can be expressed in terms of the function $\text{diam}_\zeta(x)$, by

(4.33) $\rho(x,y) = 2\log_v(\text{diam}_\zeta(x \vee_\zeta y)) - \log_v(\text{diam}_\zeta(x)) - \log_v(\text{diam}_\zeta(y))$.

This generalizes formula (2.23). Indeed, since $x \vee_\zeta y$ lies on the path from x to y,

$$\rho(x,y) = \rho(x, x \vee_\zeta y) + \rho(x \vee_\zeta y, y) ,$$

and (4.33) follows from (4.32) when $\zeta \in \mathbb{H}_{\text{Berk}}$. For arbitrary $\zeta \in \mathbb{P}^1_{\text{Berk}}$, it follows from a calculation using (2.23), (4.29), and (4.30).

REMARK 4.9. For $\zeta \in \mathbb{P}^1(K)$, the reader familiar with [**88**] will recognize (4.23) as the 'canonical distance' $[x,y]_\zeta$ for $x,y \in \mathbb{P}^1(K)$. Thus, the generalized Hsia kernel is a natural extension of the canonical distance to the Berkovich projective line.

The following result generalizes Propositions 4.1 and 4.7:

PROPOSITION 4.10.

(A) *For each $\zeta \in \mathbb{P}^1_{\text{Berk}}$, the generalized Hsia kernel $\delta(x,y)_\zeta$ is nonnegative, symmetric, and continuous in each variable separately. If $\zeta \in \mathbb{H}_{\text{Berk}}$, then $\delta(x,y)_\zeta$ is bounded and valued in the interval $[0, 1/\|\zeta,\zeta\|]$. For $\zeta \in \mathbb{P}^1(K)$ it is unbounded and extends the canonical distance $[x,y]_\zeta$ from [**88**]. We have $\delta(x,y)_\zeta = 0$ if and only if x and y are of type I and $x = y \neq \zeta$, and $\delta(x,y)_\zeta = +\infty$ if and only if ζ is of type I and $x = \zeta$ or $y = \zeta$.*

As a function of two variables, $\delta(x,y)_\zeta$ is upper semicontinuous. It is continuous off the diagonal and at (x_0, x_0) for each point x_0 of type I, but it is discontinuous at (x_0, x_0) for each point of type II, III, or IV.

(B) *For each $x, y \in \mathbb{P}^1_{\text{Berk}}$*

(4.34) $\qquad \delta(x,y)_\zeta = \limsup_{\substack{(a,b) \to (x,y) \\ a,b \in \mathbb{P}^1(K)}} \delta(a,b)_\zeta$.

(C) *For all $x, y, z \in \mathbb{P}^1_{\text{Berk}}$,*

$$\delta(x,y)_\zeta \leq \max(\delta(x,z)_\zeta, \delta(y,z)_\zeta) ,$$

with equality if $\delta(x,z)_\zeta \neq \delta(y,z)_\zeta$.

(D) *For each $a \in \mathbb{P}^1_{\text{Berk}}$ and $r > 0$, the 'open ball' $\mathcal{B}(a,r)^-_\zeta := \{x \in \mathbb{P}^1_{\text{Berk}} : \delta(x,a)_\zeta < r\}$ is connected and open in the Berkovich topology. It is empty if $r \leq \text{diam}_\zeta(a)$ and coincides with an open ball $\mathcal{B}(b,r)^-_\zeta$ for some $b \in \mathbb{P}^1(K)$ if $r > \text{diam}_\zeta(a)$.*

Likewise, the 'closed ball' $\mathcal{B}(a,r)_\zeta := \{x \in \mathbb{P}^1_{\text{Berk}} : \delta(x,a)_\zeta \leq r\}$ is connected and closed in the Berkovich topology. It is empty if $r < \text{diam}_\zeta(a)$ and coincides with $\mathcal{B}(b,r)_\zeta$ for some $b \in \mathbb{P}^1(K)$ if $r > \text{diam}_\zeta(a)$ or if $r = \text{diam}_\zeta(a)$ and a is of type II or III. If $r = \text{diam}_\zeta(a)$ and a is of type I or IV, then $\mathcal{B}(a,r)_\zeta = \{a\}$.

PROOF. Parts (A), (B), and (D) follow by arguments similar to those in the proof of Proposition 4.1. Part (C) follows from (4.30) or can be shown using Proposition 4.7(C) and a case-by-case analysis similar to the proof of [**88**, Theorem 2.5.1, p. 125]. □

REMARK 4.11. When $\zeta = \zeta_{\text{Gauss}}$, note that

$$(4.35) \qquad \mathcal{B}(a,r)^-_{\zeta_{\text{Gauss}}} = \mathcal{B}(a,r)^- \ , \quad \mathcal{B}(a,r)_{\zeta_{\text{Gauss}}} = \mathcal{B}(a,r) \ .$$

REMARK 4.12. Suppose that $\zeta \in \mathbb{H}_{\text{Berk}}$ and that $a \in \mathbb{H}_{\text{Berk}}$ with $a \neq \zeta$. For $r < \text{diam}_\zeta(a)$, the ball $\mathcal{B}(a,r)_\zeta$ is empty by Proposition 4.10(D). For $r = \text{diam}_\zeta(a)$, it is nonempty, and we claim that the complement of $\mathcal{B}(a,r)_\zeta$ is the unique connected component of $\mathbb{P}^1_{\text{Berk}}\backslash\{a\}$ containing ζ. Letting $\vec{v}_\zeta$ be the tangent direction at a in the direction of ζ, we will write $\mathcal{B}_a(\vec{v}_\zeta)$ for this component. Note that it is open and has a as its only boundary point.

Indeed, by (4.31) we have

$$-\log_v(\delta(a,x)_\zeta) \ = \ \rho(w_\zeta(a,x),\zeta) + \log_v(\|\zeta,\zeta\|)$$

for every $x \in \mathbb{P}^1_{\text{Berk}}$, from which it follows easily that $\delta(a,x)_\zeta > r = \delta(a,a)_\zeta$ if and only if $\rho(a,\zeta) > \rho(w_\zeta(a,x),\zeta)$. The latter condition holds if and only if $x \in \mathcal{B}_a(\vec{v}_\zeta)$.

REMARK 4.13. Now let $\zeta \in \mathbb{P}^1_{\text{Berk}}$ be arbitrary, and again let $a \in \mathbb{H}^{\mathbb{R}}_{\text{Berk}}$. If $r < \text{diam}_\zeta(a)$, then the balls $\mathcal{B}(a,r)_\zeta$ and $\mathcal{B}(a,r)^-_\zeta$ are empty. If $r > \text{diam}_\zeta(\zeta)$, then $\mathcal{B}(a,r)_\zeta = \mathcal{B}(a,r)^-_\zeta = \mathbb{P}^1_{\text{Berk}}$. If $r = \text{diam}_\zeta(\zeta)$, then $\mathcal{B}(a,r)_\zeta = \mathbb{P}^1_{\text{Berk}}$, while $\mathcal{B}(a,r)^-_\zeta$ is the connected component of $\mathbb{P}^1_{\text{Berk}}\backslash\{\zeta\}$ containing a.

Now suppose $\text{diam}_\zeta(a) \leq r < \text{diam}_\zeta(\zeta)$. Then the balls $\mathcal{B}(a,r)_\zeta$ have the following geometric interpretation. Consider the function $\text{diam}_\zeta(x)$ for x in the path $[a,\zeta]$ from a to ζ. By Proposition 4.10(A), $\text{diam}_\zeta(x)$ is continuous on that path, since $\text{diam}_\zeta(x) = \delta(x,x)_\zeta = \delta(x,a)_\zeta$. By (4.30), it is monotone increasing. Hence there is a unique $x = x_r \in [a,\zeta]$ with $\text{diam}_\zeta(x_r) = r$. The closed ball $\mathcal{B}(a,r)_\zeta$ is the set of all $z \in \mathbb{P}^1_{\text{Berk}}$ such that the path $[z,\zeta]$ from z to ζ contains x_r, while the open ball $\mathcal{B}(a,r)^-_\zeta$ is the connected component of $\mathbb{P}^1_{\text{Berk}}\backslash\{x_r\}$ containing a. The complement of $\mathcal{B}(a,r)_\zeta$ is the connected component of $\mathbb{P}^1_{\text{Berk}}\backslash\{x_r\}$ containing ζ.

Finally, we note that the generalized Hsia kernel can be used to decompose absolute values of rational functions on $\mathbb{P}^1_{\text{Berk}}$, just as the canonical distance does on $\mathbb{P}^1(K)$:

COROLLARY 4.14. *Let $0 \neq \varphi \in K(T)$ have divisor $\operatorname{div}(\varphi) = \sum_{i=1}^m n_i(a_i)$. For each $\zeta \in \mathbb{P}^1_{\text{Berk}}$ disjoint from the support of $\operatorname{div}(\varphi)$, there is a constant $C = C(\varphi, \zeta)$ such that for all $x \in \mathbb{P}^1_{\text{Berk}}$,*

$$(4.36) \qquad [\varphi]_x = C \cdot \prod_{i=1}^m \delta(x, a_i)_\zeta^{n_i} \ .$$

PROOF. The proof is similar to the proof of Corollary 4.4. $\square$

4.5. Notes and further references

Our definition of the Hsia kernel was inspired by the unpublished manuscript [**62**] of Hsia. The fact, discussed in Proposition 4.1, that the Hsia kernel is not continuous as a function of two variables was pointed out to us by Robert Varley.

The relation between the Hsia kernel and the Gromov product was shown to us by Juan Rivera-Letelier, who (along with Charles Favre) independently discovered its applications to potential theory on $\mathbb{P}^1_{\text{Berk}}$.

CHAPTER 5

The Laplacian on the Berkovich projective line

Recall that a *domain* U in $\mathbb{P}^1_{\text{Berk}}$ is a nonempty connected open subset of $\mathbb{P}^1_{\text{Berk}}$. Given a domain U, we will construct a Laplacian $\Delta_U(f)$ for functions in a suitable space $\text{BDV}(U)$. When $U = \mathbb{P}^1_{\text{Berk}}$, we will write $\Delta(f)$ for $\Delta_U(f)$.

Our approach to constructing the Laplacian on U is to use the Laplacian for metrized graphs, taking a limit of the Laplacians over finite subgraphs of the domain. For a function $f : U \to \mathbb{R}$ satisfying appropriate hypotheses, the Laplacian of the restriction of f to each finite subgraph $\Gamma \subset U$ is a measure, so it defines a linear functional on the space $\mathcal{C}(\Gamma)$. The graph Laplacians satisfy a compatibility condition called *coherence*, explained in §5.3 below. Passing to the limit, we obtain a linear functional on the continuous functions on the closure of the domain. By the Riesz Representation Theorem, this corresponds to a measure on $\overline{U}$; we define this measure to be the *complete Laplacian* $\Delta_{\overline{U}}(f)$, and we put $\Delta_U(f) = \Delta_{\overline{U}}(f)|_U$.

5.1. Continuous functions

Given a domain $U \subset \mathbb{P}^1_{\text{Berk}}$, let $\overline{U}$ be its closure. In this section, our goal is to understand the space $\mathcal{C}(\overline{U})$ of continuous functions $f : \overline{U} \to \mathbb{R}$.

Recall (Definition 2.27) that a domain $V \subset \mathbb{P}^1_{\text{Berk}}$ is called *simple* if ∂V is a nonempty finite set $\{x_1, \ldots, x_m\} \subset \mathbb{H}_{\text{Berk}}$, where each x_i is of type II or III.

When $\overline{U}$ is a closed Berkovich disc, there is an easy description of $\mathcal{C}(\overline{U})$, based on the Stone-Weierstrass theorem:

PROPOSITION 5.1. *Let $\mathcal{D}(z_0, R)$ be a closed Berkovich disc. Then linear combinations of functions of the form $x \to [F]_x$, for polynomials $F \in K[T]$, are dense in $\mathcal{C}(\mathcal{D}(z_0, R))$.*

PROOF. Without loss of generality, we can assume $z_0 = 0$. By the definition of the Berkovich topology, sets of the form $U_f(a, b) = \{x \in \mathcal{D}(0, R) : [f]_x \in (a, b)\}$ for $f \in K\langle R^{-1}T\rangle$ constitute a basis for the open sets of $\mathcal{D}(0, R)$. Since $\mathcal{D}(0, R)$ is compact, the Stone-Weierstrass theorem tells us that the algebra of functions generated by the $[f]_x$ is dense in $\mathcal{C}(\mathcal{D}(0, R))$. The multiplicativity of the seminorms $[\cdot]_x$ shows that each monomial $[f_1]_x^{k_1} \cdots [f_n]_x^{k_n}$ reduces to a single term $[f_1^{k_1} \cdots f_n^{k_n}]_x$. Finally, the Weierstrass Preparation Theorem shows that for each $f \in K\langle R^{-1}T\rangle$ there is a polynomial $F \in K[T]$ with $[f]_x = [F]_x$ for all $x \in \mathcal{D}(0, R)$. □

We now seek a description of $\mathcal{C}(\overline{U})$ for an arbitrary domain U. This will be achieved in Proposition 5.4 below.

Recall from §2.5 the notion of a *finite subgraph* Γ of $\mathbb{P}^1_{\text{Berk}}$, namely, the convex hull of finitely many points $\zeta_1, \ldots, \zeta_n \in \mathbb{H}_{\text{Berk}}$. Equivalently, a finite subgraph is a compact, connected subgraph of $\mathbb{H}_{\text{Berk}}$ with finitely many edges and vertices. Each such Γ becomes a metrized graph when equipped with the path distance metric $\rho(x,y)$ induced from $\mathbb{H}_{\text{Berk}}$.

The following lemma is an immediate consequence of Lemma 2.28, but since that argument was left to the reader, we provide a complete proof here.

LEMMA 5.2. *Let Γ be a finite subgraph of $\mathbb{P}^1_{\text{Berk}}$. Then:*
(A) *Γ is a closed subset of $\mathbb{P}^1_{\text{Berk}}$.*
(B) *The metric topology on Γ coincides with the relative (i.e., subspace) topology induced from $\mathbb{P}^1_{\text{Berk}}$.*

PROOF. Part (A) follows from part (B): since Γ is compact in the metric topology, it is compact in the relative topology, hence closed.

We will now prove part (B). Fix a system of coordinates on $\mathbb{P}^1_{\text{Berk}}$, and regard Γ as a subset of $\mathbb{A}^1_{\text{Berk}}$. Using the terminology of §1.4, we can view Γ as a union of 'lines of discs', partially ordered by the function $x \to \text{diam}_\infty(x)$:

$$\Gamma = \bigcup_{i=1}^m [r_i, R]_{x_i}$$

where the x_i are the finitely many points of locally minimal diameter in Γ, $r_i = \text{diam}_\infty(x_i)$ for each i, $R = \text{diam}_\infty(x)$ where x is the unique point of maximal diameter in Γ, and $[r_i, R]_{x_i}$ is the geodesic path from x_i to x in $\mathbb{P}^1_{\text{Berk}}$. (Similarly, write $(r_i, R)_{x_i}$, $[r_i, R)_{x_i}$, and $(r_i, R]_{x_i}$ for the corresponding open or half-open paths.)

Recall from Proposition 4.1(D) that for each $x \in \mathbb{A}^1_{\text{Berk}}$ and $r > 0$, the balls

$$\mathcal{B}(x,r)^-_\infty = \{z \in \mathbb{A}^1_{\text{Berk}} : \delta(z,x)_\infty < r\},$$
$$\mathcal{B}(x,r)_\infty = \{z \in \mathbb{A}^1_{\text{Berk}} : \delta(z,x)_\infty \leq r\}$$

are open (resp. closed) in the Berkovich topology.

We will first show that each $x \in \Gamma$ has a basis of neighborhoods in the metric topology which are open in the relative topology. There are several cases to consider.

(1) $x = x_i$ is an endpoint of Γ and is a point of minimal radius in its branch. In this case, a basis for the open neighborhoods of x in the metric topology on Γ is given by the half-open segments $[r_i, r_i + \varepsilon)_{x_i}$ for sufficiently small $\varepsilon > 0$. Such a segment is the intersection of the ball $\mathcal{B}(x_i, r_i + \varepsilon)^-_\infty = \{z \in \mathbb{A}^1_{\text{Berk}} : \delta(z, x_i)_\infty < r_i + \varepsilon\}$ with Γ.

(2) x is an interior point of an edge of Γ. Suppose $x \in [r_i, R]_{x_i}$, and put $r = \text{diam}_\infty(x)$. Then a basis of the open neighborhoods of x in the metric topology on Γ is given by the open segments $(r - \varepsilon, r + \varepsilon)_{x_i}$ for

sufficiently small $\varepsilon > 0$. Such a segment is the intersection of the open annulus $\mathcal{B}(x_i, r+\varepsilon)_\infty^- \backslash \mathcal{B}(x_i, r-\varepsilon)_\infty$ with Γ.

(3) x is a branch point of Γ but is not the point of maximal radius. In this case, after relabeling the branches if necessary, we can assume that $[r_1, R]_{x_1}, \ldots, [r_k, R]_{x_k}$ come together at x. Put $r = \operatorname{diam}_\infty(x)$. Then a basis of the open neighborhoods of x in the metric topology on Γ is given by star-shaped sets of the form $\bigcup_{i=1}^{k}(r-\varepsilon, r+\varepsilon)_{x_i}$ for sufficiently small $\varepsilon > 0$. Such a set is the intersection of the simple domain $\mathcal{B}(x_1, r+\varepsilon)_\infty^- \backslash (\bigcup_{i=1}^{k} \mathcal{B}(x_i, r-\varepsilon)_\infty)$ with Γ.

(4) x is the point of maximal radius R in Γ. Some of the branches of Γ may come together at points below x; after relabeling the branches if necessary, we can assume that $[r_1, R]_{x_1}, \ldots, [r_k, R]_{x_\ell}$ come together at x. Then a basis of the open neighborhoods of x in the metric topology on Γ is given by star-shaped sets of the form $\bigcup_{i=1}^{\ell}(R-\varepsilon, R]_{x_i}$ for sufficiently small $\varepsilon > 0$. Such a set is the intersection of $\mathbb{P}^1_{\text{Berk}} \backslash \left(\bigcup_{i=1}^{\ell} \mathcal{B}(x_i, R-\varepsilon)_\infty\right)$ with Γ.

Next we will show that each subset of Γ which is open in the relative topology is also open in the metric topology. It suffices to consider the intersection of Γ with a basic open set of the form $\mathcal{B}(a, s)_\infty^- \backslash (\bigcup_{j=1}^{k} \mathcal{B}(a_j, s_j)_\infty)$.

First assume $s \leq R$. Using the same notation as before, let T be the set of minimal points x_i belonging to $\mathcal{B}(a, s)_\infty^-$, and partition T into disjoint subsets $T_0, T_1, \ldots, T_k$ where T_0 is the set of $x_i \in \mathcal{B}(a, s)_\infty^- \backslash (\bigcup_{j=1}^{k} \mathcal{B}(a_j, s_j)_\infty)$ and for each $j = 1, \ldots, k$, T_j is the set of $x_i \in \mathcal{B}(a_j, s_j)_\infty$. Then the intersection of $\mathcal{B}(a, s)_\infty^- \backslash (\bigcup_{j=1}^{k} \mathcal{B}(a_j, s_j)_\infty)$ with Γ is

$$\left(\bigcup_{x_i \in T_0} [r_i, s)_{x_i}\right) \cup \bigcup_{j=1}^{k} \left(\bigcup_{x_i \in T_j} (s_i, s)_{x_i}\right)$$

which is open in Γ. If $s > R$, a similar argument applies with the $[r_i, s)_{x_i}$ (resp. $(s_i, s)_{x_i}$) replaced by $[r_i, R]_{x_i}$ (resp. $(s_i, R]_{x_i}$). $\square$

Now let $E \subset \mathbb{P}^1_{\text{Berk}}$ be an arbitrary nonempty connected closed set, equipped with the relative topology. Recall that a subset of $\mathbb{P}^1_{\text{Berk}}$ is connected if and only if it is path-connected and that there is a unique path between any two points of $\mathbb{P}^1_{\text{Berk}}$.

There is a *retraction map* $r_E : \mathbb{P}^1_{\text{Berk}} \to E$ defined as follows. Fix a point $p_0 \in E$. Given $x \in \mathbb{P}^1_{\text{Berk}}$, let $r_E(x)$ be the first point p in E on the path from x to p_0. Clearly $r_E(x) = x$ if $x \in E$. To see that $r_E(x)$ is independent of the choice of p_0 when $x \notin E$, suppose $p_0' \in E$ is another point and let p' be the first point in E on the path from x to p_0'. If $p' \neq p$, then since E is connected, there is a path from p' to p contained in E. There is another path from p' to p obtained by concatenating the path from p' to x with the path from x to p and eliminating backtracking. This second path lies outside of E, apart from its endpoints. This contradicts the fact that there is a unique path between any two points of $\mathbb{P}^1_{\text{Berk}}$, so it must be that $p' = p$.

LEMMA 5.3. *For each nonempty closed connected subset $E \subset \mathbb{P}^1_{\text{Berk}}$, the retraction map $r_E : \mathbb{P}^1_{\text{Berk}} \to E$ is continuous.*

PROOF. The connected open sets form a basis for the topology of $\mathbb{P}^1_{\text{Berk}}$. If $U \subset \mathbb{P}^1_{\text{Berk}}$ is a connected open set, then $U \cap E$ is also connected, since if $x, y \in U \cap E$, the unique path P from x to y is contained in U since U is connected, and it is contained in E since E is connected. Hence it is contained in $U \cap E$. It follows that the relative topology on E has a basis consisting of connected open sets.

Let $V \subset E$ be a connected open set, and let $U \subset \mathbb{P}^1_{\text{Berk}}$ be an open set with $V = U \cap E$. If U_0 is the connected component of U containing V, then U_0 is also open; so we can assume without loss of generality that U is connected. Let $\widetilde{U}$ be the union of all the connected open subsets $U \subset \mathbb{P}^1_{\text{Berk}}$ with $U \cap E = V$. Then $\widetilde{U}$ is itself connected and open, and $\widetilde{U} \cap E = V$, so it is the maximal set with these properties.

We claim that $r_E^{-1}(V) = \widetilde{U}$. First, we will show that $r_E^{-1}(V) \subset \widetilde{U}$. Suppose $x \in r_E^{-1}(V)$ and put $p = r_E(x) \in V$. Consider the path P from x to p. If $x \notin \widetilde{U}$, let $\overline{x}$ be the first point of P which lies in the closure of $\widetilde{U}$. Then $\overline{x} \notin \widetilde{U}$, since $\widetilde{U}$ is open; also, $\overline{x} \notin E$, since p is the only point of P in E and $p \in \widetilde{U}$. Since E is compact and $\mathbb{P}^1_{\text{Berk}}$ is Hausdorff, there is a neighborhood W of $\overline{x}$ which is disjoint from E. Without loss of generality, we can assume W is connected. Since $\overline{x}$ is in the closure of $\widetilde{U}$, $W \cap \widetilde{U}$ is nonempty. It follows that $W \cup \widetilde{U}$ is connected and open and that $(W \cup \widetilde{U}) \cap E = V$. By the maximality of $\widetilde{U}$ we must have $W \subset \widetilde{U}$. This contradicts the fact that $\overline{x} \notin \widetilde{U}$. Hence $x \in \widetilde{U}$.

Next, we will show that $\widetilde{U} \subset r_E^{-1}(V)$. Fix a point $p_0 \in V$, and let $x \in \widetilde{U}$ be arbitrary. Since $\widetilde{U}$ is connected, the unique path from x to p_0 must be entirely contained in $\widetilde{U}$. Hence, the point $p = r_E(x)$, which is the first point in E along that path, belongs to $\widetilde{U}$. But then $p \in \widetilde{U} \cap E = V$, so $x \in r_E^{-1}(V)$. □

If $E_1 \subset E_2 \subset \mathbb{P}^1_{\text{Berk}}$ are two nonempty connected closed subsets, the retraction map $r_{E_1} : \mathbb{P}^1_{\text{Berk}} \to E_1$ induces a retraction map $r_{E_2,E_1} : E_2 \to E_1$. Clearly
$$r_{E_1}(x) = r_{E_2,E_1}(r_{E_2}(x))$$
for all x. If E_1 and E_2 both have the relative topology, it follows from Lemma 5.3 that r_{E_2,E_1} is continuous.

PROPOSITION 5.4. *Let $U \subset \mathbb{P}^1_{\text{Berk}}$ be a domain.*

(A) *As Γ ranges over all finite subgraphs of U and as f ranges over $\mathcal{C}(\Gamma)$, the functions of the form $f \circ r_{\overline{U},\Gamma}(x)$ are dense in $\mathcal{C}(\overline{U})$.*

(B) *As Γ ranges over all finite subgraphs of U and as f ranges over $\text{CPA}(\Gamma)$, the functions of the form $f \circ r_{\overline{U},\Gamma}(x)$ are dense in $\mathcal{C}(\overline{U})$.*

PROOF. For part (A), we will apply the Stone-Weierstrass theorem. Functions of the form $f \circ r_{\overline{U},\Gamma}(x)$ separate points in $\overline{U}$: given distinct points $x, y \in \overline{U}$, take Γ to be a closed segment $[p, q]$ in the path from x to y, and let $f \in \mathcal{C}([p, q])$ be any function with $f(p) \neq f(q)$. Then $f \circ r_{\overline{U},\Gamma}(x) \neq f \circ r_{\overline{U},\Gamma}(y)$. Likewise, each constant function on $\overline{U}$ has the form $f \circ r_{\overline{U},\Gamma}(x)$, where $\Gamma \subset U$ is arbitrary and f is the corresponding constant function on Γ.

Take $F \in \mathcal{C}(\overline{U})$. By the Stone-Weierstrass theorem (see [**65**, p. 244]), since $\overline{U}$ is compact, the algebra of functions generated by the $f \circ r_\Gamma(x)$ is dense in $\mathcal{C}(\overline{U})$. Hence, for any $\varepsilon > 0$, there are a finite number of finite subgraphs $\Gamma_{ij} \subset U$, $i = 1, \ldots, m$, $j = 1, \ldots, n_i$, and functions $f_{ij} \in \mathcal{C}(\Gamma_{ij})$, for which

$$|F(x) - (\sum_{i=1}^{m} \prod_{j=1}^{n_i} f_{ij} \circ r_{\overline{U},\Gamma_{ij}}(x))| < \varepsilon$$

for all $x \in \overline{U}$. Let Γ be the smallest (connected) finite subgraph of $\mathbb{P}^1_{\text{Berk}}$ containing all the Γ_{ij}; then $\Gamma \subset U$. For each i, j, put $g_{ij} = f_{ij} \circ r_{\Gamma,\Gamma_{ij}}(x)$; then $g_{ij} \in \mathcal{C}(\Gamma)$ and $g_{ij} \circ r_{\overline{U},\Gamma}(x) = f_{ij} \circ r_{\overline{U},\Gamma_{ij}}(x)$ for all $x \in \overline{U}$. Put

$$g = \sum_{i=1}^{m}(\prod_{j=1}^{n_i} g_{ij}) \in \mathcal{C}(\Gamma) .$$

Then $|F(x) - g \circ r_{\overline{U},\Gamma}(x)| < \varepsilon$ for all $x \in \overline{U}$.

Part (B) follows from part (A), since CPA(Γ) is dense in $\mathcal{C}(\Gamma)$ under the sup norm, for each Γ. $\square$

5.2. Measures on $\mathbb{P}^1_{\text{Berk}}$

In this section, we will establish two fundamental properties of measures on $\mathbb{P}^1_{\text{Berk}}$. First, each finite signed Borel measure on $\mathbb{P}^1_{\text{Berk}}$, or on a finite subgraph $\Gamma \subset \mathbb{P}^1_{\text{Berk}}$, is actually a *Radon measure*. The concept of a Radon measure abstracts the properties needed to prove the Riesz Representation Theorem (see §A.5). Second, the support of any finite signed Borel measure on $\mathbb{P}^1_{\text{Berk}}$, equipped with its induced topology as a subspace of $\mathbb{P}^1_{\text{Berk}}$ with the weak topology, is actually a separable metric space. This fact is due to Favre and Jonsson [**45**, Lemma 7.15]. Finally, we introduce the one-dimensional Hausdorff measure λ on $\mathbb{P}^1_{\text{Berk}}$.

We begin by noting some facts about the σ-algebra of Borel sets.

PROPOSITION 5.5.
(A) *The σ-algebras of Borel sets on $\mathbb{P}^1_{\text{Berk}}$ relative to the strong and weak topologies coincide.*
(B) *The following sets are Borel measurable:*
    (1) *each open or closed ball, relative to the weak topology, or for either of the metrics $\rho(x, y)$ and $d(x, y)$ for the strong topology,*
    (2) *each finite graph Γ and each path $[a, b]$,*
    (3) *the sets $\mathbb{H}_{\text{Berk}}$ and $\mathbb{P}^1(K)$.*

PROOF. We first show (A). Since the strong topology on $\mathbb{P}^1_{\text{Berk}}$ is stronger than the weak topology, it suffices to show that each open or closed ball for the strong topology belongs to the Borel σ-algebra for the weak topology. Consider a closed ball
$$\widehat{\mathcal{D}}(x,r) = \{z \in \mathbb{P}^1_{\text{Berk}} : d(x,z) \leq r\} .$$
Each connected component of $\mathbb{P}^1_{\text{Berk}} \backslash \widehat{\mathcal{D}}(x,r)$ in the strong topology is an open ball in the weak topology (see Lemma B.18 of Appendix B for the connectedness in the weak topology; the openness follows from the fact that in the weak topology each component of the complement of a point is open). Thus $\widehat{\mathcal{D}}(x,r)$ is closed in the weak topology. Now consider an open ball
$$\widehat{\mathcal{D}}(x,r)^- = \{z \in \mathbb{P}^1_{\text{Berk}} : d(x,z) < r\} .$$
It is the union of the closed balls $\widehat{\mathcal{D}}(x,s)$ for rationals $0 < s < r$, so it belongs to the Borel σ-algebra for the weak topology as well.

We next show (B). Part (1) follows from the discussion above. Part (2) follows from the fact that each finite graph Γ and each path $[a,b]$ are closed in the weak topology: their complements are unions of open balls. For (3), note that
$$\mathbb{H}_{\text{Berk}} = \bigcup_{\substack{s \in \mathbb{Q} \\ 0 < s < 1}} \widehat{\mathcal{D}}(\zeta_{\text{Gauss}}, s)$$
and that $\mathbb{P}^1(K) = \mathbb{P}^1_{\text{Berk}} \backslash \mathbb{H}_{\text{Berk}}$. $\square$

We now recall the definition of a Radon measure.

Let X be a locally compact Hausdorff space, and let μ be a positive Borel measure on X. If E is a Borel subset of X, then μ is called *outer regular* on E if
$$\mu(E) = \inf\{\mu(U) : U \supseteq E, U \text{ open}\}$$
and *inner regular* on E if
$$\mu(E) = \sup\{\mu(E') : E' \subseteq E, E' \text{ compact}\} ;$$
μ is called *regular* if it is both inner and outer regular on all Borel sets.

A positive Borel measure μ on X is a *Radon measure* if it is finite on all compact sets, outer regular on all Borel sets, and inner regular on all open sets. More generally, a signed Borel measure is said to be a *signed Radon measure* iff its positive and negative parts are both Radon measures.

If X is *compact* (or more generally if X is a countable union of compact sets), then μ is Radon iff it is regular and finite on all compact sets.

LEMMA 5.6.
(A) *If U is a connected open subset of $\mathbb{P}^1_{\text{Berk}}$ and $\overline{U}$ is its closure, then every finite signed Borel measure on $\overline{U}$ is Radon.*
(B) *If Γ is a finite subgraph of $\mathbb{P}^1_{\text{Berk}}$, then every finite signed Borel measure on Γ is Radon.*

PROOF. (A) Recall that the *Baire σ-algebra* on a locally compact Hausdorff space X is the σ-algebra generated by the sets

$$\{x \in X \,:\, f(x) < \alpha\}\,, \quad \{x \in X \,:\, f(x) > \alpha\}\,,$$

for $\alpha \in \mathbb{R}$ and $f \in \mathcal{C}_c(X)$. Elements of this σ-algebra are called Baire sets. According to Proposition A.12 of Appendix A, if X is compact and the Baire σ-algebra coincides with the Borel σ-algebra, then every finite signed Borel measure on X is Radon.

Using the fact that $\mathbb{P}^1_{\text{Berk}}$ can be obtained by gluing together two copies of $\mathcal{D}(0,1)$, it suffices to prove that every open subset of $\mathcal{D}(0,1)$ is a Baire set. However, this is trivial since the Berkovich topology on $\mathcal{D}(0,1)$ is generated by the sets

$$\begin{aligned} U(f,\alpha) &= \{x \in \mathcal{D}(0,1) : [f]_x < \alpha\}\,, \\ V(f,\alpha) &= \{x \in \mathcal{D}(0,1) : [f]_x > \alpha\}\,, \end{aligned}$$

for $f \in K\langle T \rangle$ and $\alpha \in \mathbb{R}$, and the corresponding functions $[f]_x$ belong to $\mathcal{C}(\mathcal{D}(0,1))$. The result now follows from Proposition A.12.

(B) It is easy to see that each connected open subset of Γ is a Baire set. Hence the result again follows from Proposition A.12. $\square$

A topological space is called *separable* if it has a countable dense subset. When $K = \mathbb{C}_p$, for example, the space $\mathbb{P}^1_{\text{Berk}}$ is separable and metrizable. However, over more general fields K, the space $\mathbb{P}^1_{\text{Berk},K}$ is in general neither separable nor metrizable. Nonetheless, if ν is a finite (or σ-finite) signed Borel measure on $\mathbb{P}^1_{\text{Berk}}$, then $\text{supp}(\nu)$, equipped with the topology induced by the weak topology on $\mathbb{P}^1_{\text{Berk}}$, is always a separable metric space. As noted above, this fact is due to Favre and Jonsson [45, Lemma 7.15]. For applications, we state the result in a more general form:

LEMMA 5.7 (Favre–Jonsson). *Let $V \subset \mathbb{P}^1_{\text{Berk}}$ be a domain, and let $\{\nu_n\}_{n \geq 1}$ be a countable collection of finite signed Borel measures supported on $\overline{V}$. Then there is a connected compact set $X \subset \overline{V}$ which contains $\bigcup_{n=1}^\infty \text{supp}(\nu_n)$ and has the following properties:*

(A) *X is the closure of the union of a countable increasing collection of finite subgraphs $\Gamma_1 \subseteq \Gamma_2 \subseteq \cdots \subset V$.*
(B) *X, with the induced topology from the weak topology on $\mathbb{P}^1_{\text{Berk}}$, is a separable metric space.*

PROOF. We first prove assertion (A). Since an arbitrary countable union of finite subgraphs can be replaced by a countable increasing union, it suffices to deal with a single finite signed Borel measure ν. By considering the positive and negative parts in the Jordan decomposition of ν, we can further assume that ν is a positive measure with finite mass.

We will now show that if ν is a positive measure with finite mass, supported on $\overline{V}$, then supp(ν) is contained in the closure of a countable increasing collection of finite subgraphs $\Gamma_1 \subseteq \Gamma_2 \subseteq \cdots \subset V$. Fix a point $\zeta \in \mathbb{H}_{\text{Berk}} \cap V$.

First, note that for each $r > 0$, the closed ball of radius r about ζ relative to the path distance metric,
$$\hat{\mathcal{B}}(\zeta, r) = \{z \in \mathbb{H}_{\text{Berk}} : \rho(z, \zeta) \leq r\},$$
is closed in the Berkovich topology. Indeed, for each $w \in \partial \hat{\mathcal{B}}(\zeta, r)$, the set of all $z \in \mathbb{P}^1_{\text{Berk}}$ such that the geodesic path $[\zeta, z]$ contains w in its interior is an open ball in the Berkovich topology, and the complement of $\hat{\mathcal{B}}(\zeta, r)$ is the union of such balls. Clearly $\hat{\mathcal{B}}(\zeta, r)$ is connected.

For each r, let
$$\nu_r = (r_{\mathbb{P}^1_{\text{Berk}}, \hat{\mathcal{B}}(\zeta,r)})_*(\nu)$$
be the retraction of ν to $\hat{\mathcal{B}}(\zeta, r)$. Then ν_r is a positive Borel measure supported on $\overline{V} \cap \hat{\mathcal{B}}(\zeta, r)$, with the same total mass as ν. Since $\nu(\overline{V})$ is finite, there are only countably many points where ν_r has a point mass. Let Y_r be that set of points. Letting r vary over all positive rationals, put $Y = \bigcup_{0 < r \in \mathbb{Q}} Y_r$ and let $\widetilde{Y} = Y \backslash \partial V = Y \cap V$. Enumerate the countable set $\widetilde{Y}$ as $\{y_n\}_{n \geq 1}$.

Put $\Gamma_0 = \{\zeta\}$ and for each $n \geq 1$, inductively define $\Gamma_n = \Gamma_{n-1} \cup [\zeta, y_n]$. Then Γ_n is a finite subgraph contained in V, and $\bigcup_{n=1}^\infty \Gamma_n$ contains all the points y_n. Let $X \subseteq \overline{V}$ be the closure of $\bigcup_{n=1}^\infty \Gamma_n$. Trivially X is compact. Since each Γ_n is separable, X is separable.

We claim that supp(ν) $\subseteq X$. To see this, fix $z_0 \in$ supp(ν). Since $\zeta \in X$, we can assume that $z_0 \neq \zeta$. Consider the path $[\zeta, z_0]$. For each $0 < r \in \mathbb{Q}$ with $r < \rho(\zeta, z_0)$, let x_r be the unique point on $[\zeta, z_0]$ with $\rho(\zeta, x_r) = r$. We claim that x_r is one of the points y_n. Clearly $x_r \in V$, since x_r is an interior point of $[\zeta, z_0]$ and $z_0 \in \overline{V}$. By definition x_r is a boundary point of $\hat{\mathcal{B}}(\zeta, r)$. Let U be a neighborhood of z_0, disjoint from $\hat{\mathcal{B}}(\zeta, r)$. As $z_0 \in$ supp(ν), we have $\nu(U) > 0$. Since ν_r is the retraction of ν to $\hat{\mathcal{B}}(\zeta, r)$, ν_r has a point mass at x_r. Thus z_0 is in the closure of $\bigcup_{n=1}^\infty \Gamma_n$.

Assertion (B) follows from assertion (A). To see this, note that X is homeomorphic to the inverse limit $Y = \varprojlim \Gamma_n$, equipped with its natural topology as a subspace of the product $\prod \Gamma_n$ (this is analogous to Theorem 2.21). For each n, the retraction $r_{X,\Gamma_n} : X \to \Gamma_n$ is continuous and surjective, and the maps r_{X,Γ_n} are compatible with the retraction maps $r_{n,m} : \Gamma_n \to \Gamma_m$ for $n \geq m$. The maps r_{X,Γ_n} induce a continuous map $r : X \to Y$, which is injective since $\bigcup \Gamma_n$ is dense in X and is surjective since X is compact and contains $\bigcup \Gamma_n$. Since both X and Y are compact Hausdorff spaces, r is a homeomorphism. On the other hand, each Γ_n is a separable metric space (under the path distance metric), so $\prod \Gamma_n$ is a separable metric space, and Y is a separable metric space as well. $\square$

Although most Borel measures we will encounter on $\mathbb{P}^1_{\text{Berk}}$ are finite or σ-finite, there is one important measure which in general is not. This is the *one-dimensional Hausdorff measure* relative to the path distance on $\mathbb{H}_{\text{Berk}}$. Loosely, it is the measure whose restriction to each path $[a, b]$ is Lebesgue measure. Following Favre and Rivera-Letelier [**47**, §4.5], we will denote this measure by λ.

Rigorously, λ is defined as follows. Let S be any Borel subset of $\mathbb{P}^1_{\text{Berk}}$. If $\Gamma \subset \mathbb{P}^1_{\text{Berk}}$ is any finite subgraph, then $\Gamma \cap S$ is a Borel subset of Γ. The "length" of $\Gamma \cap S$ is $\ell(\Gamma \cap S) = \int_{\Gamma \cap S} 1 \, dx$, where dx is Lebesgue measure on Γ relative to the arc-length parametrization of the edges of Γ, as in Chapter 3. If $\Gamma_1 \subset \Gamma_2$, then $\ell(\Gamma_1 \cap S) \leq \ell(\Gamma_2 \cap S)$. Letting Γ run over all finite subgraphs of $\mathbb{P}^1_{\text{Berk}}$, we define

$$(5.1) \qquad \lambda(S) \;=\; \sup_{\Gamma} \ell(\Gamma \cap S) \ .$$

Since the collection of all finite subgraphs of $\mathbb{P}^1_{\text{Berk}}$ forms a directed set, λ is countably additive. Thus, it is a positive Borel measure on $\mathbb{P}^1_{\text{Berk}}$.

If Γ is any finite subgraph of $\mathbb{P}^1_{\text{Berk}}$, the restriction of λ to Γ is the measure dx on Γ; in particular $\lambda(\Gamma) = \ell(\Gamma)$. If $[a, b]$ is any path in $\mathbb{P}^1_{\text{Berk}}$, the restriction of λ to $[a, b]$ is Lebesgue measure on $[a, b]$ relative to the arc-length parametrization (for Borel subsets of $[a, b]$). However, $\lambda(\mathbb{P}^1(K)) = 0$, and if U is any open set in $\mathbb{P}^1_{\text{Berk}}$, then $\lambda(U) = \infty$.

PROPOSITION 5.8. *Let $U \subset \mathbb{P}^1_{\text{Berk}}$ be a domain, and let $F : U \to \mathbb{R} \cup \{\infty, -\infty\}$ be a Borel-measurable function, defined except on a set $S \subset U$ of λ-measure zero, for which the Lebesgue integral $\int_U F \, d\lambda$ exists (as a finite number or as $\pm \infty$). Letting Γ run over all finite subgraphs of U, we have*

$$(5.2) \qquad \int_U F \, d\lambda \;=\; \lim_{\overrightarrow{\Gamma}} \int_\Gamma F|_\Gamma \, dx \ .$$

PROOF. This follows by considering the definition of a Lebesgue integral. Let F^+ and F^- be the positive and negative parts of f. Since $\int_U F \, d\lambda = \int_U F^+ \, d\lambda - \int_U F^- \, d\lambda$ (where at least one of the integrals is finite), we can assume without loss that $F \geq 0$. In this situation $\int_U F \, d\lambda$ is an increasing limit of integrals of step functions. For each step function, (5.2) holds by the definition of λ and the fact that the collection of finite subgraphs of U is a directed set under containment. $\square$

5.3. Coherent systems of measures

Let $U \subset \mathbb{P}^1_{\text{Berk}}$ be a domain, and for each finite subgraph $\Gamma \subset U$, let μ_Γ be a finite signed Borel measure on Γ.

DEFINITION 5.9. A system of measures $\{\mu_\Gamma\}$ on the finite subgraphs of U is called *coherent* if:

(A) For each pair of finite subgraphs Γ_1, Γ_2 of U with $\Gamma_1 \subset \Gamma_2$, we have $(r_{\Gamma_2, \Gamma_1})_*(\mu_{\Gamma_2}) = \mu_{\Gamma_1}$. Equivalently, for each Borel subset $T \subset \Gamma_1$,
$$\mu_{\Gamma_2}(r_{\Gamma_2, \Gamma_1}^{-1}(T)) = \mu_{\Gamma_1}(T) .$$

(B) There is a constant B such that for each finite subgraph $\Gamma \subset U$,
$$|\mu_\Gamma|(\Gamma) \leq B .$$

The collection of finite subgraphs $\Gamma \subset U$ forms a directed set under containment: for any two graphs Γ_1, Γ_2, there is a unique minimal finite subgraph Γ_3 containing Γ_1 and Γ_2.

If μ is a finite signed Borel measure on $\overline{U}$ and $\mu_\Gamma = (r_{\overline{U}, \Gamma})_*(\mu)$ for each $\Gamma \subset U$, then $\{\mu_\Gamma\}$ is clearly a coherent system of measures on the finite subgraphs of U. In Proposition 5.10 below, we will show that every coherent system arises in this way: there is a one-to-one correspondence between finite signed Borel measures μ on $\overline{U}$ and coherent systems of finite signed Borel measures on finite subgraphs of U.

The proof uses the Riesz Representation Theorem. The crucial point is that each Borel measure on $\mathbb{P}^1_{\text{Berk}}$ is Radon (Lemma 5.6):

PROPOSITION 5.10. *If $\{\mu_\Gamma\}$ is a coherent system of measures in U, then the map*
$$\Lambda(F) = \varinjlim_\Gamma \int_\Gamma F(x) \, d\mu_\Gamma(x)$$
defines a bounded linear functional on $\mathcal{C}(\overline{U})$ and there is a unique Borel measure μ on $\overline{U}$ such that
$$\Lambda(F) = \int_{\overline{U}} F(x) \, d\mu(x)$$
for each $F \in \mathcal{C}(\overline{U})$. This measure is characterized by the fact that $(r_{\overline{U},\Gamma})_*(\mu) = \mu_\Gamma$ for each finite subgraph $\Gamma \subset U$.

In particular, if μ_0 is a finite signed Borel measure on $\overline{U}$ and we put $\mu_\Gamma = (r_{\overline{U},\Gamma})_*(\mu_0)$ for each finite subgraph $\Gamma \subset U$, then μ_0 is the measure associated to the coherent system $\{\mu_\Gamma\}$ by the construction above.

PROOF. Fix $\Gamma_0 \subset U$ and $f_0 \in \mathcal{C}(\Gamma_0)$. Put $F_0 = f_0 \circ r_{\overline{U}, \Gamma_0}(x)$.

Since $\{\mu_\Gamma\}$ is a coherent system of measures, for each finite subgraph $\Gamma \subset U$ containing Γ_0,
$$\int_\Gamma f_0(r_{\Gamma, \Gamma_0}(x)) \, d\mu_\Gamma(x) = \int_{\Gamma_0} f_0(x) \, d\mu_{\Gamma_0}(x) .$$

Then, since $F_0|_\Gamma = f_0 \circ r_{\Gamma, \Gamma_0}$ for each Γ containing Γ_0,
$$\Lambda(F_0) = \varinjlim_\Gamma \int_\Gamma F_0(x) \, d\mu_\Gamma(x) = \int_{\Gamma_0} f_0(x) \, d\mu_{\Gamma_0}(x)$$
exists.

By Proposition 5.4, functions of the form $F = f \circ r_{\overline{U},\Gamma}$ for $f \in \mathcal{C}(\Gamma)$ and $\Gamma \subset U$ are dense in $\mathcal{C}(\overline{U})$ under the sup norm, so Λ is defined on a dense subset of $\mathcal{C}(\overline{U})$.

On the other hand, since $|\mu_\Gamma|$ has total mass at most B for every $\Gamma \subset U$, for each $G \in \mathcal{C}(\overline{U})$ we have

$$\limsup_{\Gamma} \left| \int_\Gamma G(x) \, d\mu_\Gamma(x) \right| \leq B \cdot \|G\|_{\overline{U}} \,.$$

It follows that Λ extends to a bounded linear functional on $\mathcal{C}(\overline{U})$. The Riesz Representation Theorem (Theorem A.8) tells us that there is a unique Radon measure μ on $\overline{U}$ such that

$$\Lambda(F) = \int_{\overline{U}} F(x) \, d\mu(x)$$

for each $F \in \mathcal{C}(\overline{U})$.

We claim that $(r_{\overline{U},\Gamma})_* \mu = \mu_\Gamma$ for each finite subgraph $\Gamma \subset U$. Fixing Γ, this follows from the uniqueness portion of the Riesz Representation Theorem on Γ and the fact that each finite signed Borel measure on Γ is Radon, since for any $f \in \mathcal{C}(\Gamma)$, we have

$$\int_\Gamma f \, d(r_{\overline{U},\Gamma})_* \mu \;=\; \int_{\overline{U}} (f \circ r_{\overline{U},\Gamma}) \, d\mu \;=\; \int_\Gamma f \, d\mu_\Gamma \,.$$

The uniqueness in the Riesz Representation Theorem on $\overline{U}$ shows that this property characterizes μ among signed Radon measures. However, by Lemma 5.6 each finite Borel measure on $\overline{U}$ is Radon, so it also characterizes μ among finite signed Borel measures. $\square$

5.4. The Laplacian on a subdomain of $\mathbb{P}^1_{\text{Berk}}$

In this section we will construct a measure-valued Laplacian operator on a suitable class of functions $f : U \to \mathbb{R} \cup \{\pm\infty\}$, where U is a domain in $\mathbb{P}^1_{\text{Berk}}$. The Laplacian has two variants, $\Delta_{\overline{U}}(f)$ and $\Delta_U(f)$, as explained in Definitions 5.15 and 5.16 below.

If $\Gamma \subset U$ is a finite subgraph and $f|_\Gamma \in \text{BDV}(\Gamma)$, we will write $\Delta_\Gamma(f)$ for $\Delta(f|_\Gamma)$. For the definition of $\text{BDV}(\Gamma)$ and $\Delta(f|_\Gamma)$, see §3.5.

DEFINITION 5.11. Let $U \subset \mathbb{P}^1_{\text{Berk}}$ be a domain. We will say that a function $f : U \to \mathbb{R} \cup \{\pm\infty\}$ (possibly undefined at some points of $U \cap \mathbb{P}^1(K)$) is of *bounded differential variation* on U and write $f \in \text{BDV}(U)$ if

(A) $f|_\Gamma \in \text{BDV}(\Gamma)$ for each finite subgraph $\Gamma \subset U$ and
(B) there is a constant $B(f)$ such that for each finite subgraph $\Gamma \subset U$,

$$|\Delta_\Gamma(f)|(\Gamma) \;\leq\; B(f) \,.$$

REMARK 5.12. The definition of $\text{BDV}(U)$ only concerns the restrictions of functions $f \in \text{BDV}(U)$ to finite subgraphs $\Gamma \subset U$. Such subgraphs are contained in (and exhaust) $U \cap \mathbb{H}_{\text{Berk}}$, so requiring that $f \in \text{BDV}(U)$ means (in particular) that f is finite on $U \cap \mathbb{H}_{\text{Berk}}$ and continuous on each finite subgraph $\Gamma \subset U$. However, it imposes no conditions on the behavior of f on $U \cap \mathbb{P}^1(K)$. That behavior must be deduced from auxiliary hypotheses, such as continuity or upper semicontinuity.

DEFINITION 5.13. Let $U \subset \mathbb{P}^1_{\text{Berk}}$ be a domain. With a slight abuse of notation, we write $\mathcal{C}(\overline{U}) \cap \text{BDV}(U)$ for the space of functions $f \in \mathcal{C}(\overline{U})$ whose restrictions to U belong to $\text{BDV}(U)$.

PROPOSITION 5.14. *If $f \in \text{BDV}(U)$, the system of measures $\{\Delta_\Gamma(f)\}_{\Gamma \subset U}$ is coherent.*

PROOF. The boundedness condition in the definition of coherence is built into the definition of $\text{BDV}(U)$; it suffices to check the compatibility under pushforwards.

Let finite subgraphs $\Gamma_1 \subset \Gamma_2 \subset U$ be given. Since Γ_2 can be obtained by sequentially attaching a finite number of edges to Γ_1, it suffices to consider the case where $\Gamma_2 = \Gamma_1 \cup T$ and T is a segment attached to Γ_1 at a point p. By the definition of the Laplacian on a metrized graph, for any Borel set e contained in $T \backslash \{p\}$ we have $\Delta_{\Gamma_2}(f)(e) = \Delta_T(f)(e)$. For the point p, the set of tangent vectors at p in Γ_2 is the union of the corresponding sets in Γ_1 and T. It follows that $\Delta_{\Gamma_2}(f)(\{p\}) = \Delta_T(f)(\{p\}) + \Delta_{\Gamma_1}(f)(\{p\})$. Since $(r_{\Gamma_2,\Gamma_1})^{-1}(\{p\}) = T$ and $\Delta_T(f)(T) = 0$,

$$\Delta_{\Gamma_2}(f)(r_{\Gamma_2,\Gamma_1})^{-1}(\{p\}) = \Delta_T(f)(T) + \Delta_{\Gamma_1}(f)(\{p\}) = \Delta_{\Gamma_1}(f)(\{p\}) .$$

Trivially $\Delta_{\Gamma_2}(f)(e) = \Delta_{\Gamma_1}(f)(e)$ for any Borel set e contained in $\Gamma_1 \backslash \{p\}$. Hence $(r_{\Gamma_2,\Gamma_1})_*(\Delta_{\Gamma_2}(f)) = \Delta_{\Gamma_1}(f)$. □

DEFINITION 5.15. If $U \subset \mathbb{P}^1_{\text{Berk}}$ is a domain and $f \in \text{BDV}(U)$, the *complete Laplacian*

$$\Delta_{\overline{U}}(f)$$

is the finite signed Borel measure on $\overline{U}$ associated to the coherent system $\{\Delta_\Gamma(f)\}_{\Gamma \subset U}$ by Proposition 5.10, characterized by the property that $(r_{\overline{U},\Gamma})_*(\Delta_U(f)) = \Delta_\Gamma(f)$ for each finite subgraph $\Gamma \subset U$.

Note that when U is a proper subdomain of $\mathbb{P}^1_{\text{Berk}}$, the complete Laplacian $\Delta_{\overline{U}}(f)$ is supported on $\overline{U}$, not just U. It is useful to decompose the measure $\Delta_{\overline{U}}(f)$ into two parts, $\Delta_U(f)$ and $\Delta_{\partial U}(f)$:

DEFINITION 5.16. If $U \subset \mathbb{P}^1_{\text{Berk}}$ is a domain and $f \in \text{BDV}(U)$, the *Laplacian*

$$\Delta_U(f) := \Delta_{\overline{U}}(f)|_U$$

is the restriction of $\Delta_{\overline{U}}(f)$ to U. The *boundary derivative*

$$\Delta_{\partial U}(f) := \Delta_{\overline{U}}(f)|_{\partial U}$$

is the restriction of $\Delta_{\overline{U}}(f)$ to ∂U.

Thus, $\Delta_U(f)$ is the Borel measure on $\overline{U}$ with $\Delta_U(f)(S) = \Delta_{\overline{U}}(f)(S \cap U)$ for each Borel set $S \subseteq \overline{U}$. The measure $\Delta_{\partial U}(f)$ is defined similarly; by abuse of notation, it will also be regarded as a Borel measure on ∂U. Clearly

$$\Delta_{\overline{U}}(f) = \Delta_U(f) + \Delta_{\partial U}(f) .$$

As will be seen, $\Delta_U(f)$ is analogous to the classical Laplacian on a domain, while $\Delta_{\partial U}(f)$ is analogous to the classical outward normal derivative on the boundary. Throughout the book, the reader should be sensitive to the notational distinction between $\Delta_{\overline{U}}(f)$ and $\Delta_U(f)$.

Observe that when $U = \mathbb{P}^1_{\text{Berk}}$, the Laplacian and the complete Laplacian coincide, while the boundary derivative vanishes. When $U = \overline{U} = \mathbb{P}^1_{\text{Berk}}$, we will write $\Delta(f)$ for $\Delta_{\mathbb{P}^1_{\text{Berk}}}(f)$.

Here are some examples of functions in $\text{BDV}(\mathbb{P}^1_{\text{Berk}})$ and their Laplacians:

EXAMPLE 5.17. If $f(x) = C$ is constant, then $f \in \text{BDV}(\mathbb{P}^1_{\text{Berk}})$ and $\Delta(f) \equiv 0$. Indeed, $\Delta_\Gamma(f) = 0$ for each finite subgraph Γ, so the associated functional $\Lambda(F)$ is the 0 functional.

Conversely, if U is a domain, $f \in \text{BDV}(U)$, and $\Delta_{\overline{U}}(f) = 0$, then f is constant on $U \cap \mathbb{H}_{\text{Berk}}$: see Lemma 5.24 below.

EXAMPLE 5.18. If $\Gamma_0 \subset \mathbb{H}_{\text{Berk}}$ is a finite subgraph and $f_0 \in \text{BDV}(\Gamma_0)$, then the function $f = f_0 \circ r_{\Gamma_0}$ belongs to $\text{BDV}(\mathbb{P}^1_{\text{Berk}})$ and $\Delta(f) = \Delta_{\Gamma_0}(f_0)$.

Indeed, for each subgraph Γ containing Γ_0, $f|_\Gamma$ is constant on branches off Γ_0, so $\Delta_\Gamma(f) = \Delta_{\Gamma_0}(f) = \Delta_{\Gamma_0}(f_0)$. Taking the limit over all subgraphs Γ, we obtain the result.

EXAMPLE 5.19. If $f(x) = -\log_v(\delta(x,y)_\zeta)$, then $f \in \text{BDV}(\mathbb{P}^1_{\text{Berk}})$ and

$$\Delta(f) = \delta_y(x) - \delta_\zeta(x) .$$

To see this, let $\zeta_{\text{Gauss}} \in \mathbb{P}^1_{\text{Berk}}$ be the Gauss point. Given a finite subgraph Γ containing ζ_{Gauss}, put $\tilde{y} = r_\Gamma(y)$ and $\tilde{\zeta} = r_\Gamma(\zeta)$. For $x \in \Gamma$, it follows from the definition of the Hsia kernel and Proposition 4.5 that

$$\begin{aligned}
-\log_v(\delta(x,y)_\zeta) &= j_{\zeta_{\text{Gauss}}}(x,y) - j_{\zeta_{\text{Gauss}}}(x,\zeta) - j_{\zeta_{\text{Gauss}}}(y,\zeta) \\
&= j_{\zeta_{\text{Gauss}}}(x,\tilde{y}) - j_{\zeta_{\text{Gauss}}}(x,\tilde{\zeta}) - j_{\zeta_{\text{Gauss}}}(y,\zeta) .
\end{aligned}$$
(5.3)

Hence $\Delta_\Gamma(f) = \delta_{\tilde{y}} - \delta_{\tilde{\zeta}} = (r_\Gamma)_*(\delta_y - \delta_\zeta)$. The only measure on $\mathbb{P}^1_{\text{Berk}}$ with this property for all Γ is $\mu = \delta_y - \delta_\zeta$.

EXAMPLE 5.20 ("Poincaré-Lelong formula"). Take $0 \neq g \in K(T)$; suppose $\text{div}(g) = \sum_{i=1}^m n_i(a_i)$. Then $f(x) = -\log_v([g]_x) \in \text{BDV}(\mathbb{P}^1_{\text{Berk}})$ and

$$\Delta(-\log_v([g]_x)) = \sum_{i=1}^m n_i \delta_{a_i}(x) .$$

This follows from Example 5.19. Let $\zeta \in \mathbb{P}^1_{\text{Berk}}$ be arbitrary. By the decomposition formula (4.14) for the generalized Hsia kernel, there is a constant

C_ζ such that
$$[g]_x = C_\zeta \cdot \prod_{a_i \neq \zeta} \delta(x, a_i)_\zeta .$$
Taking logarithms, applying Example 5.19, and using $\sum_{i=1}^m n_i = 0$ gives the result.

EXAMPLE 5.21. Let ν be a signed Borel measure on $\mathbb{P}^1_{\text{Berk}}$. If $\zeta \in \mathbb{H}_{\text{Berk}}$ or $\zeta \notin \text{supp}(\nu)$, consider the *potential function*
$$(5.4) \qquad u_\nu(x, \zeta) = \int -\log_v(\delta(x, y)_\zeta)\, d\nu(y) .$$
If $\zeta \in \mathbb{P}^1(K) \cap \text{supp}(\nu)$, consider instead the function
$$(5.5) \qquad u_\nu(z, \zeta) = u_\nu(z, \zeta_{\text{Gauss}}) - \nu(\mathbb{P}^1_{\text{Berk}}) \log_v(\|z, \zeta\|) .$$
Then $u_\nu(x, \zeta) \in \text{BDV}(\mathbb{P}^1_{\text{Berk}})$ and
$$(5.6) \qquad \Delta(u_\nu(x, \zeta)) = \nu - \nu(\mathbb{P}^1_{\text{Berk}})\delta_\zeta(x) .$$

To see this for (5.4), let ζ_{Gauss} be the Gauss point, and let Γ be any finite subgraph containing ζ_{Gauss}. For $x \in \Gamma$, using the definition of $\delta(x, y)_\zeta$ and the restriction formula $j_{\zeta_{\text{Gauss}}}(x, y) = j_{\zeta_{\text{Gauss}}}(x, r_\Gamma(y))$, we have
$$\begin{aligned} u_\nu(x, \zeta) &= \int (j_{\zeta_{\text{Gauss}}}(x, y) - j_{\zeta_{\text{Gauss}}}(x, \zeta) - j_{\zeta_{\text{Gauss}}}(y, \zeta))\, d\nu(y) \\ &= \int j_{\zeta_{\text{Gauss}}}(x, r_\Gamma(y))\, d\nu(y) - \int j_{\zeta_{\text{Gauss}}}(x, r_\Gamma(\zeta))\, d\nu(y) - C_\zeta \\ (5.7) \quad &= \int_\Gamma j_{\zeta_{\text{Gauss}}}(x, t)\, d((r_\Gamma)_*(\nu))(t) - \nu(\mathbb{P}^1_{\text{Berk}}) j_{\zeta_{\text{Gauss}}}(x, r_\Gamma(\zeta)) - C_\zeta \end{aligned}$$
where C_ζ is a finite constant. (It is finite because $\zeta \in \mathbb{H}_{\text{Berk}}$ or $\zeta \notin \text{supp}(\nu)$, which implies that $j_{\zeta_{\text{Gauss}}}(y, \zeta)$ is bounded and continuous as a function of y on the compact set $\text{supp}(\nu)$.) In the case of (5.5), formula (5.7) holds with $C_\zeta = 0$ because $\delta(x, y)_{\zeta_{\text{Gauss}}} = \|x, y\|$ (see §4.4).

By Proposition 3.11, $u_\nu(x, \zeta)|_\Gamma \in \text{BDV}(\Gamma)$ and
$$\Delta_\Gamma(u_\nu(z, \zeta)) = (r_\Gamma)_*(\nu) - \nu(\mathbb{P}^1_{\text{Berk}})\delta_{r_\Gamma(\zeta)} = (r_\Gamma)_*(\nu - \nu(\mathbb{P}^1_{\text{Berk}})\delta_\zeta) .$$
It follows that $\Delta(u_\nu(x, \zeta)) = \nu - \nu(\mathbb{P}^1_{\text{Berk}})\delta_\zeta(x)$.

EXAMPLE 5.22. In particular, suppose ν is probability measure on $\mathbb{P}^1_{\text{Berk}}$. If $\zeta \in \mathbb{H}_{\text{Berk}}$ or $\zeta \notin \text{supp}(\nu)$, consider the potential function
$$(5.8) \qquad u_\nu(x, \zeta) = \int -\log_v(\delta(x, y)_\zeta)\, d\nu(y) .$$
If $\zeta \in \mathbb{P}^1(K) \cap \text{supp}(\nu)$, consider instead the function
$$(5.9) \qquad u_\nu(z, \zeta) = u_\nu(z, \zeta_{\text{Gauss}}) - \log_v(\|z, \zeta\|) .$$
Then $u_\nu(x, \zeta) \in \text{BDV}(\mathbb{P}^1_{\text{Berk}})$ and
$$(5.10) \qquad \Delta(u_\nu(x, \zeta)) = \nu - \delta_\zeta(x) .$$

Moreover, in case (5.8) there is a constant C_ζ such that for all $z \in \mathbb{P}^1_{\text{Berk}}$,

$$(5.11) \qquad u_\nu(z,\zeta) = u_\nu(z,\zeta_{\text{Gauss}}) - \log_v(\|z,\zeta\|) - C_\zeta \ .$$

The functions $u_\nu(z,\zeta)$ play an important role in potential theory and will be studied extensively in subsequent chapters. In Proposition 5.28 below, it will be shown that essentially "all" functions in $\text{BDV}(\mathbb{P}^1_{\text{Berk}})$, up to constants and functions on $\mathbb{P}^1(K)$, are potential functions.

REMARK 5.23. If ν is positive, then $u_\nu(x,\zeta)$ is well-defined for all $x \in \mathbb{P}^1_{\text{Berk}}$; however if ν is signed and $\nu = \nu_1 - \nu_2$ is its Jordan decomposition, then $u_\nu(x,\zeta)$ will be well-defined for all $x \in \mathbb{H}_{\text{Berk}}$ but it may be indeterminate at points of $\mathbb{P}^1(K)$ where both $u_{\nu_1}(x,\zeta)$ and $u_{\nu_2}(x,\zeta)$ are ∞.

5.5. Properties of the Laplacian

The Laplacian has a number of important properties. First, the complete Laplacian vanishes only for "constant functions":

LEMMA 5.24. *Let $U \subseteq \mathbb{P}^1_{\text{Berk}}$ be a domain, and let $f \in \text{BDV}(U)$. Then $\Delta_{\overline{U}}(f) = 0$ if and only if f is constant on $U \cap \mathbb{H}_{\text{Berk}}$.*

PROOF. If f is constant on $U \cap \mathbb{H}_{\text{Berk}}$, then $f|_\Gamma$ is constant for every finite subgraph $\Gamma \subset U$, so $\Delta_\Gamma(f) = 0$ for every Γ and hence $\Delta_{\overline{U}}(f) = 0$.

Conversely, if U is a domain, $f \in \text{BDV}(U)$, and $\Delta_{\overline{U}}(f) = 0$, then for each finite subgraph $\Gamma \subset U$ we have $\Delta_\Gamma(f) = (r_{\overline{U},\Gamma})_* \Delta_{\overline{U}}(f) = 0$, so that f is constant on Γ by Proposition 3.14(A). Since there is a finite subgraph $\Gamma \subset U$ containing any two given points $x, y \in U \cap \mathbb{H}_{\text{Berk}}$, f is constant on $U \cap \mathbb{H}_{\text{Berk}}$. $\square$

The complete Laplacian $\Delta_{\overline{U}}(f)$ has total mass 0:

PROPOSITION 5.25. *For each $f \in \text{BDV}(U)$, $\Delta_{\overline{U}}(f)(\overline{U}) = 0$. Hence the boundary derivative and the Laplacian have opposite total masses:*

$$\Delta_{\partial U}(\partial U) = -\Delta_U(U).$$

PROOF. Applying Proposition 3.14(E), for any finite subgraph $\Gamma \subset U$

$$\Delta_{\overline{U}}(f)(\overline{U}) = (r_{\overline{U},\Gamma})_*(\Delta_{\overline{U}}(f))(\Gamma) = \Delta_\Gamma(f)(\Gamma) = 0 \ .$$
$\square$

The Laplacian and the complete Laplacian are both compatible with restriction to subdomains:

PROPOSITION 5.26. *Suppose $U_1 \subset U_2 \subset \mathbb{P}^1_{\text{Berk}}$ are domains and $f \in \text{BDV}(U_2)$. Then $f|_{U_1} \in \text{BDV}(U_1)$ and*

$$\Delta_{U_1}(f) = \Delta_{U_2}(f)|_{U_1} \ , \quad \Delta_{\overline{U}_1}(f) = (r_{\overline{U}_2,\overline{U}_1})_*(\Delta_{\overline{U}_2}(f)) \ .$$

PROOF. If $f \in \mathrm{BDV}(U_2)$, then trivially $f \in \mathrm{BDV}(U_1)$. By the compatibility of the retraction maps, for each $\Gamma \subset U_1$ we have

$$(r_{\overline{U}_1,\Gamma})_* \left((r_{\overline{U}_2,\overline{U}_1})_*(\Delta_{\overline{U}_2}(f)) \right) \;=\; (r_{\overline{U}_2,\Gamma})_*(\Delta_{\overline{U}_2}(f)) \;=\; \Delta_\Gamma(f) \ .$$

Hence $\Delta_{\overline{U}_1}(f) = (r_{\overline{U}_2,\overline{U}_1})_*(\Delta_{\overline{U}_2}(f))$, by the characterization of $\Delta_{\overline{U}_1}(f)$. Since $r_{\overline{U}_2,\overline{U}_1}$ acts as the identity on U_1 and maps $\overline{U}_2\backslash U_1$ onto ∂U_1, we have $\Delta_{U_1}(f) = \Delta_{U_2}(f)|_{U_1}$. □

Similarly, the Laplacian and the complete Laplacian are compatible with finite covers:

PROPOSITION 5.27. *Let $U \subset \mathbb{P}^1_{\mathrm{Berk}}$ be a domain, and let $V_1,\ldots,V_r \subset U$ be subdomains such that $U = \bigcup_{i=1}^r V_i$. Then for any function f, we have $f \in \mathrm{BDV}(U)$ iff $f|_{V_i} \in \mathrm{BDV}(V_i)$ for $i = 1,\ldots,r$. Moreover, in the latter case, for each $i=1,\ldots,r$*

$$\Delta_U(f)|_{V_i} = \Delta_{V_i}(f) \quad \text{and} \quad (r_{\overline{U},\overline{V}_i})_*(\Delta_{\overline{U}}(f)) = \Delta_{\overline{V}_i}(f) \ .$$

PROOF. If $f \in \mathrm{BDV}(U)$, then for each i we have $f|_{V_i} \in \mathrm{BDV}(V_i)$ by Proposition 5.26.

Conversely, suppose $f \in \mathrm{BDV}(V_i)$ for each i. Let $\Gamma \subset U$ be a finite subgraph. There exist finite subgraphs $\Gamma_i \subset V_i$, $i = 1,\ldots,r$, such that $\Gamma = \bigcup_{i=1}^r \Gamma_i$ (the Γ_i may have overlaps). Since the V_i are open, we can choose the Γ_i in such a way that each branch point p of Γ belongs to the interior of each Γ_i which contains p. Since $f|_{\Gamma_i} \in \mathrm{BDV}(\Gamma_i)$ for each i, we have that $f|_\Gamma \in \mathrm{BDV}(\Gamma)$.

The Γ_i will in general have overlaps. However, we can find a finite partition of Γ into pairwise disjoint sets $W_1,\ldots,W_m$, where each W_i is a point or an open segment, such that for each W_j one has $W_j \subset \Gamma_i$ for some i. From this one sees easily that

$$|\Delta_\Gamma(f)|(\Gamma) \;\leq\; \sum_{i=1}^r |\Delta_{\Gamma_i}(f)|(\Gamma_i) \ .$$

For each i there is a constant C_i, depending only on V_i and f, such that $|\Delta_{\Gamma_i}(f)|(\Gamma_i) \leq C_i$. It follows that the $|\Delta_\Gamma(f)|(\Gamma)$ are uniformly bounded, so $f \in \mathrm{BDV}(U)$.

The assertions about restrictions of Laplacians follow from Proposition 5.26. □

In a certain sense, $\mathrm{BDV}(U)$ is the largest class of functions on U on which one can define a measure-valued Laplacian with finite mass (compare with Corollary 3.12):

PROPOSITION 5.28. *Let $U \subseteq \mathbb{P}^1_{\mathrm{Berk}}$ be a domain. Then there is a one-to-one correspondence between finite signed Borel measures ν of total mass zero on $\overline{U}$ and functions $f \in \mathrm{BDV}(U)|_{\mathbb{H}_{\mathrm{Berk}}}$ modulo constant functions.*

PROOF. Choose a base point $\zeta \in \mathbb{H}_{\text{Berk}}$. If ν is a measure of total mass zero on $\overline{U}$, let $f_\nu(x) = u_\nu(x, \zeta)$. By Example 5.21 and Proposition 5.26, $f_\nu \in \text{BDV}(U)$. Similarly, if $f \in \text{BDV}(U)$, let $\nu_f = \Delta_{\overline{U}}(f)$. By Proposition 5.25, ν_f has total mass zero on $\overline{U}$. Finally, using Example 5.21 and Proposition 5.26 again, we see that $\Delta_{\overline{U}}(f_\nu) = \nu$ and also that the potential function of $\Delta_{\overline{U}}(f)$ has the same complete Laplacian as f, so their restrictions to $\mathbb{H}_{\text{Berk}}$ agree up to a constant by Lemma 5.24. □

We note the following consequence of Proposition 5.28:

COROLLARY 5.29. *Let $U \subseteq \mathbb{P}^1_{\text{Berk}}$ be a domain, and let $f \in \text{BDV}(U)$. Then f can be extended to a function $\tilde{f} \in \text{BDV}(\mathbb{P}^1_{\text{Berk}})$ with $\Delta(\tilde{f}) = \Delta_{\overline{U}}(f)$.*

PROOF. By Example 5.21 and the proof of Proposition 5.28, if we fix a base point $\zeta \in \mathbb{H}_{\text{Berk}}$ and put $\nu = \Delta_{\overline{U}}(f)$, then the potential function $u = u_\nu(x, \zeta)$ belongs to $\text{BDV}(\mathbb{P}^1_{\text{Berk}})$ and $u = f + C$ on $U \cap \mathbb{H}_{\text{Berk}}$, for some constant C. Put $\tilde{f} = f$ on U, and put $\tilde{f} = u - C$ on $\mathbb{P}^1_{\text{Berk}} \backslash U$. □

There is a "Mass Formula" which computes the mass that the Laplacian assigns to a simple domain $V \subset U$ in terms of directional derivatives of f along inward-pointing tangent directions at points of ∂V. (Recall from Appendix B that we associate to each point $x \in \mathbb{P}^1_{\text{Berk}}$ a set T_x of *tangent directions*, and if $x \in \partial V$, there is a unique $\vec{v} \in T_x$ which "points inward".)

PROPOSITION 5.30. *Let U be a domain in $\mathbb{P}^1_{\text{Berk}}$, and let $V \subset U$ be a simple domain with $\overline{V} \subset U$. Let $x_1, \ldots, x_m$ be the boundary points of V, and let the corresponding inward-pointing tangent vectors be $\vec{v}_1, \ldots, \vec{v}_m$. Finally, let $f \in \text{BDV}(U)$. Then*

$$\Delta_{\overline{U}}(f)(V) = \Delta_U(f)(V) = \sum_{i=1}^m d_{\vec{v}_i} f(x_i) .$$

PROOF. Let $\Gamma \subset \mathbb{H}^{\mathbb{R}}_{\text{Berk}}$ be the convex hull of $\{x_1, \ldots, x_m\}$, and let $\Gamma^0 := \Gamma \backslash \{x_1, \ldots, x_m\}$. Then it is easy to see that $V = r^{-1}_{\overline{U}, \Gamma}(\Gamma^0)$. Using Definition 5.15, we compute that

$$\Delta_{\overline{U}}(f)(V) = \Delta_{\overline{U}}(f)(r^{-1}_{\overline{U},\Gamma}(\Gamma^0))$$
$$= (r_{\overline{U},\Gamma})_*(\Delta_{\overline{U}}(f))(\Gamma^0) = \Delta_\Gamma(f)(\Gamma^0) .$$

The result now follows from the Mass Formula for finite graphs (3.13). □

REMARK 5.31. The result in Proposition 5.30 applies to simple subdomains $V \subset U$ even without the hypothesis that $\overline{V} \subset U$, provided one interprets the terms $d_{\vec{v}_i} f(x_i)$ properly. By Corollary 5.29 there is a function $\tilde{f} \in \text{BDV}(\mathbb{P}^1_{\text{Berk}})$ extending f. If one puts $d_{\vec{v}_i} f(x_i) = d_{\vec{v}_i} \tilde{f}(x_i)$, then the formula in Proposition 5.30 holds. Alternately, if one knows a function $\widehat{f}$ on $U \cup \partial V$ which extends f and is continuous at each $x_i \in \partial V$, then one can take $d_{\vec{v}_i} f(x_i) = d_{\vec{v}_i} \widehat{f}(x_i)$ in Proposition 5.30.

We will use the following criterion for convergence of Laplacians several times in the sequel.

PROPOSITION 5.32. *Let U be a domain in $\mathbb{P}^1_{\text{Berk}}$, and let $h_\alpha, h \in \text{BDV}(U)$ for α in some directed set I. Suppose that $h_\alpha|_\Gamma \to h|_\Gamma$ uniformly for every finite subgraph $\Gamma \subset U$ and that the measures $\Delta_{\overline{U}} h_\alpha$ have uniformly bounded total mass. Then $\Delta_{\overline{U}} h_\alpha$ converges weakly to $\Delta_{\overline{U}} h$ on $\overline{U}$.*

PROOF. Write $r_\Gamma = r_{\overline{U},\Gamma}$, $\mu_\alpha = \Delta_{\overline{U}} h_\alpha$, and $\mu = \Delta_{\overline{U}} h$. Let M be a bound such that $|\mu_\alpha| \leq M$ for all α.

By Proposition 5.4(B), as Γ ranges over all finite subgraphs of U and G ranges over all functions in $\text{CPA}(\Gamma)$, the "smooth" functions of the form $g(x) = G \circ r_\Gamma(x)$ are dense in $\mathcal{C}(\overline{U})$.

Let $f \in \mathcal{C}(\overline{U})$ be arbitrary. Writing $\|\ \|$ for the sup-norm on $\mathcal{C}(\overline{U})$, for any smooth $g \in \mathcal{C}(\overline{U})$ we have

$$\left| \int f \, d\mu_\alpha - \int f \, d\mu \right| \leq \left| \int f \, d\mu_\alpha - \int g \, d\mu_\alpha \right|$$
$$+ \left| \int g \, d\mu_\alpha - \int g \, d\mu \right| + \left| \int g \, d\mu - \int f \, d\mu \right|$$
$$\leq (M + |\mu|(\overline{U})) \|f - g\| + \left| \int g \, d\mu_\alpha - \int g \, d\mu \right| .$$

Choosing g so that $\|f - g\|$ is sufficiently small, we see that it suffices to prove that

$$\int g \, d\mu_\alpha \to \int g \, d\mu$$

when $g = G \circ r_\Gamma$ is a smooth function. Since $\int_{\overline{U}} (G \circ r_\Gamma) \, d\mu_\alpha = \int_\Gamma G \, d(r_\Gamma)_* \mu_\alpha = \int_\Gamma G \, d\Delta_\Gamma h_\alpha$ and similarly $\int_{\overline{U}} (G \circ r_\Gamma) \, d\mu = \int_\Gamma G \, d\Delta_\Gamma h$, we are reduced to proving that

$$\int_\Gamma G \, d(\Delta_\Gamma h_\alpha) \to \int_\Gamma G \, d(\Delta_\Gamma h) .$$

But $\int_\Gamma G \, d(\Delta_\Gamma h_\alpha) = \int_\Gamma h_\alpha \, d\Delta_\Gamma G$ and $\int_\Gamma G \, d(\Delta_\Gamma h) = \int_\Gamma h \, d\Delta_\Gamma G$. The result now follows from our hypothesis that $h_\alpha \to h$ uniformly on Γ. □

5.6. The Dirichlet pairing

In classical harmonic analysis on $\mathbb{R}^2$, a prominent role is played by the *Dirichlet pairing* $\langle f, g \rangle_{U,\text{Dir}} = \int_U \frac{\partial f}{\partial x} \frac{\partial g}{\partial x} + \frac{\partial f}{\partial y} \frac{\partial g}{\partial y} \, dx \, dy$, which is defined and finite for functions f, g whose partials are square-integrable on a domain $U \subset \mathbb{R}^2$. In this section we study an analogue of the Dirichlet pairing for a domain $U \subset \mathbb{P}^1_{\text{Berk}}$.

In Section 3.7, we have studied the Dirichlet pairing on a metrized graph Γ, and formula (3.36) and Proposition 3.17 show that for $f, g \in \text{BDV}(\Gamma)$ and

$z \in \Gamma$ arbitrary,

$$\langle f, g \rangle_{\Gamma,\text{Dir}} = \int_\Gamma f \, d\Delta_\Gamma(g) = \int_\Gamma g \, d\Delta_\Gamma(f) \tag{5.12}$$

$$= \iint_{\Gamma \times \Gamma} j_z(x,y) \, d\Delta_\Gamma(f)(x) \, d\Delta_\Gamma(g)(y) \tag{5.13}$$

$$= \int_\Gamma f'(x) \, g'(x) \, dx \, , \tag{5.14}$$

where the derivatives f', g' are defined except at the countably many points where $\Delta_\Gamma(f)$ or $\Delta_\Gamma(g)$ has a point mass and their sign depends on a choice of orientation for the edges of Γ.

If $U \subset \mathbb{P}^1_{\text{Berk}}$ is a domain and if $f, g \in \text{BDV}(U)$, then for each finite subgraph $\Gamma \subset U$ the restrictions of f and g to Γ belong to $\text{BDV}(\Gamma)$. This motivates

DEFINITION 5.33. Let $U \subset \mathbb{P}^1_{\text{Berk}}$ be a domain. For functions $f, g \in \text{BDV}(U)$ we define the *Dirichlet pairing* by

$$\langle f, g \rangle_{U,\text{Dir}} = \varinjlim_{\Gamma \subset U} \langle f, g \rangle_{\Gamma,\text{Dir}} \, , \tag{5.15}$$

provided the limit exists (as an element of $\mathbb{R} \cup \{\pm\infty\}$). Here, the limit is taken over the net of all finite subgraphs in U.

Clearly the Dirichlet pairing is symmetric; it is bilinear provided all relevant terms are defined and finite. The following result describes its kernel; recall that we write $\mathcal{C}(\overline{U}) \cap \text{BDV}(U)$ for the space of functions $f \in \mathcal{C}(\overline{U})$ for which $f|_U \in \text{BDV}(U)$ (Definition 5.13).

PROPOSITION 5.34. *Let $f \in \text{BDV}(U)$. Then $\langle f, f \rangle_{U,\text{Dir}}$ is well-defined and belongs to $[0, \infty]$. Moreover, the following are equivalent:*

(A) *f is constant on $U \cap \mathbb{H}_{\text{Berk}}$.*
(B) *$\langle f, f \rangle_{U,\text{Dir}} = 0$.*
(C) *$\langle f, g \rangle_{U,\text{Dir}} = 0$ for all $g \in \text{BDV}(U)$.*
(D) *$\langle f, g \rangle_{U,\text{Dir}} = 0$ for all $g \in \mathcal{C}(\overline{U}) \cap \text{BDV}(U)$.*

PROOF. For each $f \in \text{BDV}(U)$, the self-pairing $\langle f, f \rangle_{U,\text{Dir}}$ is well-defined and belongs to $[0, \infty]$, since the pairings $\langle f, f \rangle_{\Gamma,\text{Dir}} = \int_\Gamma (f'(x))^2 \, dx$ are nonnegative and increasing with Γ. Trivially (A) implies (B), (C), and (D).

To show that (B) implies (A), note that if $\langle f, f \rangle_{U,\text{Dir}} = 0$, then since the integrals $\int_\Gamma (f'(x))^2 \, dx = \langle f, f \rangle_{\Gamma,\text{Dir}}$ are increasing with Γ, it follows that $\langle f, f \rangle_{\Gamma,\text{Dir}} = 0$ for each finite subgraph $\Gamma \subset U$. By Proposition 3.17, f is constant on each such Γ. This means f is constant on $U \cap \mathbb{H}_{\text{Berk}}$.

Trivially (C) implies (D). To show that (D) implies (A), suppose that $\langle f, g \rangle_{U,\text{Dir}} = 0$ for all $g \in \mathcal{C}(\overline{U}) \cap \text{BDV}(U)$, and consider functions of the form $g_{\xi, \zeta}(x) = -\log_v(\delta(z, \xi)_\zeta)$ with $\xi, \zeta \in U \cap \mathbb{H}_{\text{Berk}}$. Such functions are bounded and continuous by Proposition 4.10(A) and formula (4.27). They belong to $\text{BDV}(U)$ and satisfy $\Delta_{\overline{U}}(g_{\xi,\zeta}) = \delta_\xi - \delta_\zeta$, by Example 5.19 and Proposition

5.26. The finite graphs $\Gamma \subset U$ containing the path $[\xi,\zeta]$ are cofinal in the set of all finite subgraphs of U, and for any such Γ we have $\Delta_\Gamma(g_{\xi,\zeta}) = r_{\overline{U},\Gamma}(\Delta_{\overline{U}}(g_{\xi,\zeta})) = \delta_\xi - \delta_\zeta$. Formula (5.12) gives $\langle f, g_{\xi,\zeta}\rangle_{\Gamma,\mathrm{Dir}} = f(\xi) - f(\zeta)$. Passing to the limit, we obtain

$$0 = \langle f, g_{\xi,\zeta}\rangle_{U,\mathrm{Dir}} = f(\xi) - f(\zeta) \ .$$

Since ξ and ζ are arbitrary, f is constant on $U \cap \mathbb{H}_{\mathrm{Berk}}$. □

We now set out to give some global formulas for the Dirichlet pairing. We will obtain different expressions for functions satisfying different hypotheses.

The following result[1] shows that $\langle f, g\rangle_{U,\mathrm{Dir}}$ is defined and finite if either f or g belongs to $\mathcal{C}(\overline{U}) \cap \mathrm{BDV}(U)$:

PROPOSITION 5.35. *Let U be a domain. Suppose $f \in \mathcal{C}(\overline{U}) \cap \mathrm{BDV}(U)$. Then $\langle f, g\rangle_{U,\mathrm{Dir}}$ is defined and finite for each $g \in \mathrm{BDV}(U)$, and*

$$(5.16) \qquad \langle f, g\rangle_{U,\mathrm{Dir}} = \int_{\overline{U}} f\, d\Delta_{\overline{U}}(g) \ .$$

PROOF. Fix $g \in \mathrm{BDV}(U)$. Since $f \in \mathcal{C}(\overline{U})$, the integral $\int_{\overline{U}} f\, d\Delta_{\overline{U}}(g)$ exists and is finite, and

$$(5.17) \qquad \int_{\overline{U}} f\, d\Delta_{\overline{U}}(g) = \lim_{\overrightarrow{\Gamma \subset U}} \int_\Gamma f\, d\Delta_\Gamma(g)$$

by the definition of $\Delta_{\overline{U}}(g)$ (see Definition 5.15 and Proposition 5.10). On the other hand, for each finite subgraph $\Gamma \subset U$, formula (3.36) gives $\langle f, g\rangle_{\Gamma,\mathrm{Dir}} = \int_\Gamma f\, d\Delta_\Gamma(g)$. Hence (5.16) follows from the definition of $\langle f, g\rangle_{U,\mathrm{Dir}}$. □

If one fixes a point $\zeta \in \mathbb{P}^1_{\mathrm{Berk}}$, then for each $x \in (U \cap \mathbb{H}_{\mathrm{Berk}} \backslash \{\zeta\})$ there is a unique tangent vector $\vec{v}_{\zeta,+}$ in T_x pointing towards ζ. Favre and Rivera-Letelier [47] noticed that this means that for each function $f \in \mathrm{BDV}(U)$, there is a distinguished directional derivative

$$(5.18) \qquad f'_{\zeta,+}(x) = d_{\vec{v}_{\zeta,+}} f(x) \ .$$

The domain of $f'_{\zeta,+}$ is $(U \cap \mathbb{H}_{\mathrm{Berk}}) \backslash \{\zeta\}$. For each path $[a,\zeta]$, the restriction of $f'_{\zeta,+}$ to $[a,\zeta]$ is the one-sided derivative of $f|_{[a,\zeta]}$ in the direction of ζ.

Given a finite subgraph $\Gamma \subset U$, choose a vertex set T for Γ which contains $\zeta_\Gamma = r_\Gamma(\zeta)$, and give each edge of $\Gamma \backslash T$ the orientation in which the positive direction is the one leading towards ζ. With these orientations, $f'_{\zeta,+}|_\Gamma$ coincides with the one-sided derivative $(f|_\Gamma)'_+$ studied in §3.7. By Lemma 3.16, $f'_{\zeta,+}|_\Gamma$ is Borel measurable and belongs to $L^2(\Gamma, dx)$. Furthermore, by Proposition 3.17, for $f, g \in \mathrm{BDV}(U)$ the Dirichlet pairing $\langle f, g\rangle_{\Gamma,\mathrm{Dir}}$ is independent of the choice of orientation of the edges of Γ, and so

$$(5.19) \qquad \langle f, g\rangle_{\Gamma,\mathrm{Dir}} = \int_\Gamma f'_{\zeta,+}(x)\, g'_{\zeta,+}(x)\, dx$$

---

[1]Proposition 5.35 is similar to [46, Proposition 4.5] of Favre and Rivera-Letelier, but with different notation and slightly different hypotheses.

for any ζ. The independence also follows from Lemma 5.36(C) below.

Let λ be the one-dimensional Hausdorff measure on $\mathbb{P}^1_{\mathrm{Berk}}$ introduced in §5.2.

LEMMA 5.36. *Let U be a domain, and take $f \in \mathrm{BDV}(U)$.*

(A) *For each ζ, $f'_{\zeta,+}$ is Borel measurable, with domain $(U \cap \mathbb{H}_{\mathrm{Berk}}) \backslash \{\zeta\}$.*
(B) *$f'_{\zeta,+} \in L^2(U \cap \mathbb{H}_{\mathrm{Berk}}, d\lambda)$ if and only if $\langle f, f \rangle_{U,\mathrm{Dir}}$ is defined and finite.*
(C) *If $\zeta, \xi \in \mathbb{P}^1_{\mathrm{Berk}}$, then $f'_{\zeta,+}(x) = f'_{\xi,+}(x)$ for all $x \in \mathbb{H}_{\mathrm{Berk}} \backslash [\zeta, \xi]$, and there is a countable subset $S \subset [\zeta, \xi]$ such that $f'_{\xi,+}(x) = -f'_{\zeta,+}(x)$ for all $x \in [\zeta, \xi] \backslash S$.*

PROOF. We first establish (A). Fix ζ. We have already seen that $f'_{\zeta,+}$ is defined on $(U \cap \mathbb{H}_{\mathrm{Berk}}) \backslash \{\zeta\}$. By hypothesis, $\Delta_{\overline{U}}(f)$ is a signed Borel measure on $\overline{U}$ with finite total mass. Let X be the connected compact subset of $\overline{U}$ containing $\mathrm{supp}(\Delta_{\overline{U}}(f))$ given by Lemma 5.7. Then X is the closure of $\bigcup_{k=1}^{\infty} \Gamma_k$ for an increasing sequence of finite graphs $\Gamma_1 \subset \Gamma_2 \subset \cdots \subset U$. Fix a point $a \in \Gamma_1$. After adjoining the segment $[\zeta, a]$ to X and to each Γ_k, we can assume without loss of generality that $\zeta \in X$ and $\zeta \in \Gamma_k$ for each k.

Note that f is constant on branches off X, since X is connected and $\Delta_{\overline{U}}(f)$ is supported on X. Hence $f'_{\zeta,+}(x) = 0$ for each $x \in (U \cap \mathbb{H}_{\mathrm{Berk}}) \backslash X$. Thus, to show that $f'_{\zeta,+}$ is Borel measurable, it suffices to show that for each $t > 0$, the sets

$$\begin{aligned} Y_t &= \{z \in U \cap X \cap \mathbb{H}_{\mathrm{Berk}} : f'_{\zeta,+}(z) > t\}, \\ Z_t &= \{z \in U \cap X \cap \mathbb{H}_{\mathrm{Berk}} : f'_{\zeta,+}(z) < -t\} \end{aligned}$$

are Borel. By Lemma 3.16, the restriction of $f'_{\zeta,+}$ to each Γ_k is Borel measurable, so $Y_t \cap \Gamma_k$ and $Z_t \cap \Gamma_k$ are Borel. This implies

$$Y_t^0 := \bigcup_{k=1}^{\infty} (Y_t \cap \Gamma_k) \quad \text{and} \quad Z_t^0 := \bigcup_{k=1}^{\infty} (Z_t \cap \Gamma_k)$$

are Borel. However, if $y \in Y_t \backslash Y_t^0$, then y is a boundary point of X which belongs to $\mathbb{H}_{\mathrm{Berk}}$, and $f'_{\zeta,+}(y) > t > 0$. This means that $\Delta_{\overline{U}}(f)$ has a point mass at y, with $|\Delta_{\overline{U}}(f)(y)| > t$. There are only finitely many such points, so Y_t is Borel. Similarly Z_t is Borel.

To establish (B), consider the integral $\int_{U \cap \mathbb{H}_{\mathrm{Berk}}} (f'_{\zeta,+}(x))^2 \, d\lambda(x)$. Since $(f'_{\zeta,+}(x))^2$ is Borel measurable and $(f'_{\zeta,+}(x))^2 \geq 0$, the integral is well-defined as an element of $[0, \infty]$. By Proposition 5.8 and the fact that $\lambda(\mathbb{P}^1(K)) = 0$, we have

$$(5.20) \qquad \int_{U \cap \mathbb{H}_{\mathrm{Berk}}} (f'_{\zeta,+}(x))^2 \, d\lambda(x) = \lim_{\overrightarrow{\Gamma \subset U}} \int_{\Gamma} (f'_{\zeta,+}(x))^2 \, dx.$$

Hence (B) follows from (5.19) and the definition of $\langle f, f \rangle_{U,\mathrm{Dir}}$.

We now establish (C). Fix $\zeta, \xi \in \mathbb{P}^1_{\text{Berk}}$. If $x \in \mathbb{H}_{\text{Berk}} \backslash [\zeta, \xi]$, then the tangent vectors $\vec{v}_{\zeta,+}, \vec{v}_{\xi,+} \in T_x$ are equal, so $f'_{\zeta,+}(x) = f'_{\xi,+}(x)$. Give $[\zeta, \xi]$ its natural orientation from ζ towards ξ, and let S be the countable set of points in $[\zeta, \xi]$ where $r_{[\zeta,\xi]}(\Delta_{\overline{U}}(f))$ has a point mass. If $x \in [\zeta, \xi] \backslash S$, then the derivative $(f|_{[\zeta,\xi]})'(x)$ is defined, and $f'_{\xi,+}(x) = (f|_{[\zeta,\xi]})'(x) = -f'_{\zeta,+}(x)$. □

The following result establishes the analogy between the Dirichlet pairing on a domain $U \subset \mathbb{P}^1_{\text{Berk}}$ and the classical Dirichlet pairing:

PROPOSITION 5.37. *Let $U \subset \mathbb{P}^1_{\text{Berk}}$ be a domain. Fix $\zeta \in \mathbb{P}^1_{\text{Berk}}$. Then for all $f, g \in \text{BDV}(U)$ for which $\int_{U \cap \mathbb{H}_{\text{Berk}}} f'_{\zeta,+}(x) g'_{\zeta,+}(x) \, d\lambda(x)$ is defined, we have*

$$(5.21) \quad \langle f, g \rangle_{U, \text{Dir}} = \int_{U \cap \mathbb{H}_{\text{Berk}}} f'_{\zeta,+}(x) g'_{\zeta,+}(x) \, d\lambda(x) \, .$$

In particular, (5.21) applies to $\langle f, f \rangle_{U, \text{Dir}}$ for all f and to $\langle f, g \rangle_{U, \text{Dir}}$ if $\langle f, f \rangle_{U, \text{Dir}}$ and $\langle g, g \rangle_{U, \text{Dir}}$ are both finite.

PROOF. If $\int_{U \cap \mathbb{H}_{\text{Berk}}} f'_{\zeta,+}(x) g'_{\zeta,+}(x) \, d\lambda(x)$ takes a definite value in $\mathbb{R} \cup \{\pm\infty\}$, it follows from Proposition 5.8 and the fact that $\lambda(\mathbb{P}^1(K)) = 0$ that

$$(5.22) \quad \int_{U \cap \mathbb{H}_{\text{Berk}}} f'_{\zeta,+}(x) g'_{\zeta,+}(x) \, d\lambda(x) = \lim_{\substack{\longrightarrow \\ \Gamma \subset U}} \int_\Gamma f'_{\zeta,+}(x) g'_{\zeta,+}(x) \, dx \, .$$

Hence (5.21) follows from (5.19) and the definition of $\langle f, g \rangle_{U, \text{Dir}}$. The final assertion follows from Lemma 5.36. □

For $f \in \text{BDV}(U)$, Favre and Rivera-Letelier give the following measure-theoretic interpretation of $f'_{\zeta,+}$. Let $\preceq_\zeta$ be the partial order on $\overline{U} \backslash \{\zeta\}$ such that $y \preceq_\zeta x$ iff $x \in [y, \zeta]$. For each $x \in \overline{U} \backslash \{\zeta\}$, write $* \preceq_\zeta x$ for the set $\{y \in \overline{U} : y \preceq_\zeta x\}$. This is the set of points in $\overline{U}$ which are "at least as far from ζ" as x. Put $\nu_f = \Delta_{\overline{U}}(f)$. Then for each $x \in U \cap \mathbb{H}_{\text{Berk}} \backslash \{\zeta\}$,

$$(5.23) \quad f'_{\zeta,+}(x) = -\nu_f(* \preceq_\zeta x) \, .$$

To see this, fix a point $y \in (x, \zeta)$ and let $\Gamma \subset U$ be any finite graph containing $[x, y]$. Put $S_{\Gamma, x} = \{z \in \Gamma : z \preceq_\zeta x\}$. Then $\nu_f(* \preceq_\zeta x) = ((r_{\overline{U},\Gamma})_* \nu_f)(S_{\Gamma,x})$. The set $S_{\Gamma,x}$ is a finite union of closed segments in Γ. Its unique boundary point in Γ is x, and $\vec{v}_{\zeta,+} \in T_x(\Gamma)$ is the unique outward-directed tangent vector to $S_{\Gamma,x}$. By formula (3.16),

$$((r_{\overline{U},\Gamma})_* \nu_f)(S_{\Gamma,x}) = -d_{\vec{v}_{\zeta,+}}(f)(x) \, .$$

This yields (5.23). Thus, Proposition 5.37 can be reformulated as saying that

$$(5.24) \quad \langle f, g \rangle_{U, \text{Dir}} = \int_U \nu_f(* \preceq_\zeta x) \nu_g(* \preceq_\zeta x) \, d\lambda(x)$$

whenever the integral on the right is defined.

Let $V_{\text{Dir}}(U)$ be the set of functions $f|_{U \cap \mathbb{H}_{\text{Berk}}}$ for $f \in \text{BDV}(U)$ such that $\langle f, f \rangle_{U, \text{Dir}}$ is defined and finite; it is a vector space by Lemma 5.36(B).

5.6. THE DIRICHLET PAIRING

COROLLARY 5.38. *Let $U \subset \mathbb{P}^1_{\text{Berk}}$ be a domain. Then $\langle f, g \rangle_{U,\text{Dir}}$ induces an inner product on $V_{\text{Dir}}(U)/\{\text{constant functions}\}$.*

PROOF. This follows from Propositions 5.34 and 5.37. □

The following result, which follows by combining Propositions 5.35 and 5.37, shows that for functions $f, g \in \mathcal{C}(\overline{U}) \cap \text{BDV}(U)$, the complete Laplacian enjoys the same self-adjointness property as the Laplacian on $\text{BDV}(\Gamma)$ for a metrized graph Γ (cf. Proposition 3.14).

COROLLARY 5.39. *Let $U \subseteq \mathbb{P}^1_{\text{Berk}}$ be a domain. Fix $\zeta \in \mathbb{P}^1_{\text{Berk}}$, and suppose $f, g \in \mathcal{C}(\overline{U}) \cap \text{BDV}(U)$. Then*

$$\langle f, g \rangle_{U,\text{Dir}} = \int_{\overline{U}} f \, d\Delta_{\overline{U}}(g) = \int_{\overline{U}} g \, d\Delta_{\overline{U}}(f)$$
$$(5.25) \qquad = \int_{U \cap \mathbb{H}_{\text{Berk}}} f'_{\zeta,+}(x) \, g'_{\zeta,+}(x) \, d\lambda(x) \ .$$

There is one more expression for $\langle f, g \rangle_{U,\text{Dir}}$ we will need. To formulate it, we introduce a concept which plays an important role in later chapters.

DEFINITION 5.40. *A finite signed measure μ on $\mathbb{P}^1_{\text{Berk}}$ has continuous potentials if for each $\zeta \in \mathbb{H}_{\text{Berk}}$, the potential function*

$$u_\mu(z, \zeta) = \int -\log_v(\delta(z,w)_\zeta) \, d\mu(w)$$

is continuous (and hence bounded) on $\mathbb{P}^1_{\text{Berk}}$. The measure μ has bounded potentials if for each $\zeta \in \mathbb{H}_{\text{Berk}}$, $u_\mu(z, \zeta)$ is bounded on $\mathbb{P}^1_{\text{Berk}}$.

If $u_\mu(z, \xi)$ is continuous (resp. bounded) for even one $\xi \in \mathbb{H}_{\text{Berk}}$, then μ has continuous (resp. bounded) potentials. To see this, take $\zeta \in \mathbb{H}_{\text{Berk}}$ and recall that there is a constant $C = C_{\zeta,\xi}$ such that

$$\delta(z,w)_\zeta = C \cdot \delta(z,w)_\xi / (\delta(z,\zeta)_\xi \delta(w,\zeta)_\xi)$$

(see (4.29)). Inserting this into the definition of $u_\mu(z,\zeta)$ gives

$$u_\mu(z,\zeta) = \int -\log_v(\delta(z,w)_\zeta) \, d\mu(w)$$
$$= -\log_v(C) + u_\mu(z,\xi) + \mu(\mathbb{P}^1_{\text{Berk}}) \cdot \log(\delta(z,\zeta)_\xi) - u_\mu(\zeta,\xi) \ ,$$

which is continuous (resp. bounded) provided $\zeta \in \mathbb{H}_{\text{Berk}}$.

PROPOSITION 5.41. *Let $U \subset \mathbb{P}^1_{\text{Berk}}$ be a domain, and let $\zeta \in \mathbb{H}_{\text{Berk}}$. Let $f, g \in \text{BDV}(U)$, and put $\nu = \Delta_{\overline{U}}(f)$, $\mu = \Delta_{\overline{U}}(g)$. Suppose in the Jordan decompositions $\nu = \nu_1 - \nu_2$ and $\mu = \mu_1 - \mu_2$ that the measures ν_2 and μ_2 have bounded potentials. Then $\langle f, g \rangle_{U,\text{Dir}}$ is defined and belongs to $\mathbb{R} \cup \{\infty\}$, and*

$$(5.26) \qquad \langle f, g \rangle_{U,\text{Dir}} = \iint_{\overline{U} \times \overline{U}} -\log(\delta(z,w)_\zeta) \, d\nu(z) \, d\mu(w) \ .$$

PROOF. We first show that $\iint -\log_v(\delta(z,w)_\zeta)\, d\nu(z)\, d\mu(w)$ is defined. To see this, note that $-\log_v(\delta(z,w)_\zeta) = j_\zeta(z,w) - j_{\zeta_{\text{Gauss}}}(\zeta,\zeta)$ is bounded from below (see formula (4.27)) and is Borel measurable as a function of two variables since it is lower semicontinuous (see Proposition 4.10). By the definition of the Lebesgue integral,

$$\iint_{\overline{U}\times\overline{U}} -\log_v(\delta(z,w)_\zeta)\, d\nu(z)\, d\mu(w)$$
$$= \sum_{i,j=1}^{2} (-1)^{i+j} \iint_{\overline{U}\times\overline{U}} -\log_v(\delta(z,w)_\zeta)\, d\nu_i(z)\, d\mu_j(w).$$

Since $j_\zeta(z,w) \geq 0$ and the measures ν_i, μ_j have finite mass, Tonelli's theorem shows that $\iint -\log(\delta(z,w)_\zeta)\, d\nu_i(z)\, d\mu_j(w)$ exists for each pair (i,j), that $u_{\nu_i}(w,\zeta)$ and $u_{\mu_j}(z,\zeta)$ are Borel measurable, and that

$$\iint_{\overline{U}\times\overline{U}} -\log(\delta(z,w)_\zeta)\, d\nu_i(z)\, d\mu_j(w)$$
$$= \int_{\overline{U}} u_{\nu_i}(w,\zeta)\, d\mu_j(w) = \int_{\overline{U}} u_{\mu_j}(z,\zeta)\, d\nu_i(z).$$

Since ν_2 and μ_2 have bounded potentials, the three integrals for which $i=2$ or $j=2$ are finite. Hence $\iint -\log(\delta(z,w)_\zeta)\, d\nu(z)\, d\mu(w)$ is defined and belongs to $\mathbb{R} \cup \{\infty\}$.

We next show $\langle f, g\rangle_{U,\text{Dir}}$ exists and equals $\iint -\log(\delta(z,w)_\zeta)\, d\nu(z)\, d\mu(w)$. Since ν and μ have total mass 0, we can replace $-\log(\delta(z,w)_\zeta)$ by $j_\zeta(z,w)$ without changing the integral.

Since the graphs containing ζ are cofinal in the set of finite subgraphs $\Gamma \subset U$, by (5.13) and the definition of $\langle f, g\rangle_{U,\text{Dir}}$ we have

$$\langle f, g\rangle_{U,\text{Dir}} = \lim_{\zeta \in \Gamma \subset U} \iint_{\Gamma\times\Gamma} j_\zeta(x,y)\, d\Delta_\Gamma(f)(x)\, d\Delta_\Gamma(g)(y).$$

On the other hand, $\Delta_\Gamma(f) = r_{\Gamma*}(\Delta_{\overline{U}}(f)) = r_{\Gamma*}(\nu_1) - r_{\Gamma*}(\nu_2)$, and similarly $\Delta_\Gamma(g) = r_{\Gamma*}(\mu_1) - r_{\Gamma*}(\mu_2)$. Hence it suffices to show that for each pair (i,j),

$$\lim_{\zeta \in \Gamma \subset U} \iint_{\Gamma\times\Gamma} j_\zeta(z,w)\, d(r_{\Gamma*}\nu_i)(z)\, d(r_{\Gamma*}\mu_j)(z)$$
(5.27)
$$= \iint_{\overline{U}\times\overline{U}} j_\zeta(z,w)\, d\nu_i(z)\, d\mu_j(w).$$

Consider any Γ with $\zeta \in \Gamma \subset U$. By the definition of $r_{\Gamma*}(\nu_i)$ and $r_{\Gamma*}(\mu_j)$,

$$\iint_{\Gamma\times\Gamma} j_\zeta(z,w)\, d(r_{\Gamma*}\nu_i)(z)\, d(r_{\Gamma*}\mu_j)(w)$$
(5.28)
$$= \iint_{\overline{U}\times\overline{U}} j_\zeta(r_\Gamma(z), r_\Gamma(w))\, d\nu_i(z)\, d\mu_j(w).$$

5.6. THE DIRICHLET PAIRING

The function $j_\zeta(r_\Gamma(z), r_\Gamma(w))$ is constant on branches off Γ, while $j_\zeta(z,w)$ is nondecreasing on branches off Γ (see (4.31)), so

$$
\iint_{\overline{U} \times \overline{U}} j_\zeta(r_\Gamma(z), r_\Gamma(w))\, d\nu_i(z)\, d\mu_j(w)
$$
(5.29)
$$
\leq \iint_{\overline{U} \times \overline{U}} j_\zeta(z,w)\, d\nu_i(z)\, d\mu_j(w) .
$$

On the other hand, by Lemma 5.7 there is a compact connected set $X \subset \overline{U}$ which contains the supports of $\nu_1, \nu_2, \mu_1, \mu_2$ and is the closure of the union of an increasing sequence of finite subgraphs $\Gamma_1 \subset \Gamma_2 \subset \cdots \subset U$. Without loss of generality, we can assume that $\zeta \in \Gamma_n$ for each n. The functions $j_\zeta(r_{\Gamma_n}(z), r_{\Gamma_n}(w))$ increase monotonically to $j_\zeta(z,w)$ on $X \times X$, so by the Monotone Convergence Theorem

$$
\lim_{n \to \infty} \iint_{\overline{U} \times \overline{U}} j_\zeta(r_{\Gamma_n}(z), r_{\Gamma_n}(w))\, d\nu_i(z)\, d\mu_j(w)
$$
(5.30)
$$
= \iint_{\overline{U} \times \overline{U}} j_\zeta(z,w)\, d\nu_i(z)\, d\mu_j(w) .
$$

Combining (5.28), (5.29), and (5.30) gives (5.27). $\square$

The following result provides an important class of measures with continuous potentials.

PROPOSITION 5.42. *Fix $a \in \mathbb{H}_{\mathrm{Berk}}$ and $R > 0$. Let ν be a finite signed Borel measure supported on the ball $\widehat{\mathcal{B}}(a,R) = \{z \in \mathbb{H}_{\mathrm{Berk}} : \rho(a,z) \leq R\}$. Then ν has continuous potentials: for each $\zeta \in \mathbb{H}_{\mathrm{Berk}}$, the potential function $u_\nu(x,\zeta)$ is continuous on $\mathbb{P}^1_{\mathrm{Berk}}$.*

PROOF. We begin with some reductions. For all suitably large R', the ball $\widehat{\mathcal{B}}(a,R)$ is contained in $\widehat{\mathcal{B}}(\zeta_{\mathrm{Gauss}}, R')$, so after replacing $\widehat{\mathcal{B}}(a,R)$ with $\widehat{\mathcal{B}}(\zeta_{\mathrm{Gauss}}, R')$, we can assume that $a = \zeta_{\mathrm{Gauss}}$. By formula (5.11), there is a constant C_ζ such that for each $x \in \mathbb{P}^1_{\mathrm{Berk}}$,

$$
u_\nu(x,\zeta) = u_\nu(x,\zeta_{\mathrm{Gauss}}) - \log_v(\|x,\zeta\|) - C_\zeta .
$$

Since the function $h(x) = -\log_v(\|x,\zeta\|)$ is continuous and bounded for $\zeta \in \mathbb{H}_{\mathrm{Berk}}$, it suffices to prove the proposition when $\zeta = \zeta_{\mathrm{Gauss}}$.

Write $X = \widehat{\mathcal{B}}(\zeta_{\mathrm{Gauss}}, R)$. Clearly X is connected; since it is closed in the weak topology, it is compact. By hypothesis, $\mathrm{supp}(\nu) \subseteq X$. Recall that $\delta(x,y)_{\zeta_{\mathrm{Gauss}}} = \|x,y\|$ and $-\log_v(\|x,y\|) = j_{\zeta_{\mathrm{Gauss}}}(x,y)$. Thus

$$
u_\nu(x,\zeta_{\mathrm{Gauss}}) = \int j_{\zeta_{\mathrm{Gauss}}}(x,y)\, d\nu(y) .
$$

For all $x,y \in \mathbb{P}^1_{\mathrm{Berk}}$, if $w = w(x,y)$ is the point where the paths $[x,\zeta_{\mathrm{Gauss}}]$ and $[y,\zeta_{\mathrm{Gauss}}]$ first meet, then by (4.11),

(5.31)
$$
j_{\zeta_{\mathrm{Gauss}}}(x,y) = \rho(\zeta_{\mathrm{Gauss}}, w) .
$$

In particular, for all $x,y \in X$ we have $0 \leq j_{\zeta_{\mathrm{Gauss}}}(x,y) \leq \rho(\zeta_{\mathrm{Gauss}}, x) \leq R$.

Fix $x \in \mathbb{P}^1_{\text{Berk}}$. If $x_0 = r_{\mathbb{P}^1_{\text{Berk}}, X}(x)$, then for each $y \in X$, the point w where the path $[x, \zeta_{\text{Gauss}}]$ meets $[y, \zeta_{\text{Gauss}}]$ lies on $[x_0, \zeta_{\text{Gauss}}]$. Hence $j_{\zeta_{\text{Gauss}}}(x, y) = j_{\zeta_{\text{Gauss}}}(x_0, y)$, and in turn $u_\nu(x, \zeta_{\text{Gauss}}) = u_\nu(x_0, \zeta_{\text{Gauss}})$. Since the retraction $r_{\mathbb{P}^1_{\text{Berk}}, X}$ is continuous, to show that $u_\nu(x, \zeta_{\text{Gauss}})$ is continuous on $\mathbb{P}^1_{\text{Berk}}$, it suffices to show that it is continuous on X.

Write $\nu = \mu + \omega$, where μ is atomic and ω is atomless. It suffices to prove the theorem separately for μ and ω.

Since $|\nu|(\mathbb{P}^1_{\text{Berk}})$ is finite, ν has at most countably many point masses and we can write $\mu = \sum_{i=1}^\infty c_i \delta_{x_i}$, where $\sum_{i=1}^\infty |c_i| < \infty$, and some or all of the c_i may be 0. For each $n \geq 1$ put $\mu_n = \sum_{i=1}^n c_i \delta_{x_i}$. For each $\varepsilon > 0$, there is an N such that $\sum_{i=N}^\infty |c_i| < \varepsilon$. If $n \geq N$, then for each $x \in X$,
$$\left| u_\mu(x, \zeta_{\text{Gauss}}) - u_{\mu_n}(x, \zeta_{\text{Gauss}}) \right| \leq R \cdot \varepsilon .$$
This shows that the continuous functions $u_{\mu_n}(x, \zeta_{\text{Gauss}})$ converge uniformly to $u_\mu(x, \zeta_{\text{Gauss}})$, so $u_\mu(x, \zeta_{\text{Gauss}})$ is continuous.

Now consider the atomless part ω. By separately considering the positive and negative parts in the Jordan decomposition $\omega = \omega_1 - \omega_2$, we are reduced to the case where ω is positive. If $\omega = 0$, there is nothing to prove. Otherwise, fix $n \in \mathbb{N}$, and put $\varepsilon = 1/n$.

For each $p \in X$, there is a neighborhood V_p of p such that $\omega(V_p) < \varepsilon$. After shrinking V_p if necessary, we can assume that V_p is a simple domain. Since X is compact, it is covered by finitely many of the sets V_p, say $V_{p_1}, \ldots, V_{p_m}$. Let $\bigcup_{i=1}^m \partial V_{p_i} = \{z_1, \ldots, z_\ell\}$, and let Γ_n be the finite graph spanned by the points $\zeta_{\text{Gauss}}, p_1, \ldots, p_m$, and $z_1, \ldots, z_\ell$. Put

(5.32) $$\omega_n = (r_{\Gamma_n})_*(\omega) ,$$

noting that ω_n is supported on $X \cap \Gamma_n$. Note also that for each V_{p_i}, since Γ_n contains all the boundary points of V_{p_i} and $V_{p_i} \cap \Gamma_n$ is nonempty, we have

(5.33) $$V_{p_i} = r_{\Gamma_n}^{-1}(V_{p_i} \cap \Gamma_n) .$$

We claim that for each $x \in X$,

(5.34) $$\left| u_\omega(x, \zeta_{\text{Gauss}}) - u_{\omega_n}(x, \zeta_{\text{Gauss}}) \right| \leq 2R/n .$$

If this is granted, then $u_\omega(x, \zeta_{\text{Gauss}})$ is the uniform limit of the continuous functions $u_{\omega_n}(x, \zeta_{\text{Gauss}})$ and hence is itself continuous.

To show (5.34), note that for each $i = 1, \ldots, m$ we have

$$u_\omega(x, \zeta_{\text{Gauss}}) = \int_{V_{p_i}} j_{\zeta_{\text{Gauss}}}(x, y) \, d\omega(y) + \int_{X \setminus V_{p_i}} j_{\zeta_{\text{Gauss}}}(x, y) \, d\omega(y) ,$$

$$u_{\omega_n}(x, \zeta_{\text{Gauss}}) = \int_{\Gamma_n \cap V_{p_i}} j_{\zeta_{\text{Gauss}}}(x, y) \, d\omega_n(y) + \int_{\Gamma_n \setminus V_{p_i}} j_{\zeta_{\text{Gauss}}}(x, y) \, d\omega_n(y) .$$

The crucial fact needed for (5.34) is that for all $x \in V_{p_i}$ and all $y \in \mathbb{P}^1_{\text{Berk}} \setminus V_{p_i}$ we have

(5.35) $$j_{\zeta_{\text{Gauss}}}(x, y) = j_{\zeta_{\text{Gauss}}}(x, r_{\Gamma_n}(y)) .$$

Given (5.35), it follows that

$$\int_{X \setminus V_{p_i}} j_{\zeta_{\text{Gauss}}}(x,y)\, d\omega(y) = \int_{X \setminus V_{p_i}} j_{\zeta_{\text{Gauss}}}(x, r_{\Gamma_n}(y))\, d\omega(y)$$
$$= \int_{\Gamma_n \setminus V_{p_i}} j_{\zeta_{\text{Gauss}}}(x, t)\, d\omega_n(t) .$$

On the other hand, if $x \in V_{p_i} \cap X$, then trivially $0 \leq j_{\zeta_{\text{Gauss}}}(x,y) \leq R$ for all y. Since $\omega(V_{p_i}) = \omega_n(\Gamma_n \cap V_{p_i}) < \varepsilon = 1/n$, this gives

$$\left| u_\omega(x, \zeta_{\text{Gauss}}) - u_{\omega_n}(x, \zeta_{\text{Gauss}}) \right|$$
$$\leq \left| \int_{V_{p_i}} j_{\zeta_{\text{Gauss}}}(x, y)\, d\omega(y) - \int_{\Gamma_n \cap V_{p_i}} j_{\zeta_{\text{Gauss}}}(x, y)\, d\omega_n(y) \right| \leq 2R/n ,$$

which is (5.34).

To show (5.35), fix $x \in V_{p_i}$ and $y \notin V_{p_i}$. As above, let w be the point where the paths $[x, \zeta_{\text{Gauss}}]$ and $[y, \zeta_{\text{Gauss}}]$ first meet. We consider three cases, depending on the locations of ζ_{Gauss} and y relative to V_{p_i}.

First, suppose that $\zeta_{\text{Gauss}} \in V_{p_i}$. Since $[x, \zeta_{\text{Gauss}}] \subset V_{p_i}$, clearly $w \in V_{p_i}$. On the other hand, since $y \notin V_{p_i}$, it follows from (5.33) that $r_{\Gamma_n}(y) \notin V_{p_i}$. Since $[w, \zeta_{\text{Gauss}}]$ is a terminal segment of $[y, \zeta_{\text{Gauss}}]$, necessarily $r_{\Gamma_n}(y) \in [y, w]$, and hence w is the point where the paths $[x, \zeta_{\text{Gauss}}]$ and $[r_{\Gamma_n}(y), \zeta_{\text{Gauss}}]$ first meet. Formula (5.35) follows from this and (5.31).

Next, suppose that $\zeta_{\text{Gauss}} \notin V_{p_i}$ and that ζ_{Gauss} and y belong to different components of $\mathbb{P}^1_{\text{Berk}} \setminus V_{p_i}$. In this case, $[y, \zeta_{\text{Gauss}}] \cap V_{p_i}$ is an open segment (z_j, z_k), where z_j and z_k are points of ∂V_{p_i} and z_j is closer to y than z_k. This means that $w \in (z_j, \zeta_{\text{Gauss}}]$. However, by construction $[z_j, \zeta_{\text{Gauss}}]$ belongs to Γ_n, so $r_{\Gamma_n}(y)$, which is the point where the path $[y, \zeta_{\text{Gauss}}]$ first meets Γ_n, belongs to $[y, z_j]$. Thus w is the point where the paths $[x, \zeta_{\text{Gauss}}]$ and $[r_{\Gamma_n}(y), \zeta_{\text{Gauss}}]$ first meet, and (5.35) follows as before.

Lastly, suppose that $\zeta_{\text{Gauss}} \notin V_{p_i}$ and that ζ_{Gauss} and y belong to the same component of $\mathbb{P}^1_{\text{Berk}} \setminus V_{p_i}$. Call that component W, noting that it is closed and connected. By hypothesis, $[y, \zeta_{\text{Gauss}}] \subset W$. On the other hand, $[x, \zeta_{\text{Gauss}}] \cap W$ is a closed segment $[z_j, \zeta_{\text{Gauss}}]$, where z_j is a boundary point of V_{p_i}. This means that $w \in [z_j, \zeta_{\text{Gauss}}]$. Since $[w, \zeta_{\text{Gauss}}] \subseteq [z_j, \zeta_{\text{Gauss}}] \subset \Gamma_n$, and since $r_{\Gamma_n}(y)$ is the first point where the path $[y, \zeta_{\text{Gauss}}]$ meets Γ_n, it follows that $r_{\Gamma_n}(y) \in [y, w]$. Thus again w is the point where the paths $[x, \zeta_{\text{Gauss}}]$ and $[r_{\Gamma_n}(y), \zeta_{\text{Gauss}}]$ first meet, and the proof of (5.35) is complete. $\square$

5.7. Favre–Rivera-Letelier smoothing

Classically, smoothing of functions is carried out by convolution with a smoothing kernel. In the Berkovich setting, there seems to be no such convolution. However, in [**47**], Favre and Rivera-Letelier introduce a notion of smoothing based on retraction towards a central point, which is close to classical smoothing in spirit.

Given a number $0 < \delta < 1$, consider the set
$$X(\zeta_{\text{Gauss}}, \delta) = \{z \in \mathbb{P}^1_{\text{Berk}} : \text{diam}(z) \geq \delta\}.$$
It is easy to see that if $R = -\log_v(\delta)$, then $X(\zeta_{\text{Gauss}}, \delta)$ coincides with the ball $\widehat{\mathcal{B}}(\zeta_{\text{Gauss}}, R) = \{z \in \mathbb{P}^1_{\text{Berk}} : \rho(\zeta_{\text{Gauss}}, z) \leq R\}$ for the strong topology. More generally, given a point[2] $\zeta \in \mathbb{H}_{\text{Berk}}$, recall that $\delta(x,y)_\zeta = \|x,y\|/(\|x,\zeta\|\|y,\zeta\|)$ and $\text{diam}_\zeta(x) = \delta(x,x)_\zeta$. Take δ satisfying $0 < \delta < \text{diam}_\zeta(\zeta) = 1/\|\zeta,\zeta\|$, and put
$$X(\zeta, \delta) = \{z \in \mathbb{P}^1_{\text{Berk}} : \text{diam}_\zeta(z) \geq \delta\}.$$
Here, if $R = \log_v(\text{diam}_\zeta(\zeta)/\delta)$, then $X(\zeta, \delta)$ coincides with the ball $\widehat{\mathcal{B}}(\zeta, R) = \{z \in \mathbb{P}^1_{\text{Berk}} : \rho(\zeta, z) \leq R\}$. Each $X(\zeta, \delta)$ is connected and closed in both the strong and weak topologies; its complement is the union of the open balls $\mathcal{B}(a, \delta)_\zeta^- = \{z \in \mathbb{P}^1_{\text{Berk}} : \delta(a,z)_\zeta < \delta\}$, for $a \in \mathbb{P}^1_{\text{Berk}} \backslash X(\zeta, \delta)$.

Consider the retraction $r_{\zeta,\delta} = r_{\mathbb{P}^1_{\text{Berk}}, X(\zeta,\delta)} : \mathbb{P}^1_{\text{Berk}} \to X(\zeta, \delta)$, which has the property that if $\text{diam}_\zeta(z) \geq \delta$, then $r_{\zeta,\delta}(z) = z$, while if $\text{diam}_\zeta(z) < \delta$, then $r_{\zeta,\delta}(z)$ is the unique point t on the path $[z, \zeta]$ for which $\text{diam}_\zeta(t) = \delta$.

DEFINITION 5.43. Let $U \subset \mathbb{P}^1_{\text{Berk}}$ be a domain, and take $f \in \text{BDV}(U)$. Given $\zeta \in U \cap \mathbb{H}_{\text{Berk}}$ and $0 < \delta < \text{diam}_\zeta(\zeta)$, the *Favre–Rivera-Letelier smoothing* $f_{\zeta,\delta}$ of f is the function $f \circ r_{\zeta,\delta}|_U$.

THEOREM 5.44. *Let $U \subset \mathbb{P}^1_{\text{Berk}}$ be a domain, and let $f \in \text{BDV}(U)$. Fix $\zeta \in U \cap \mathbb{H}_{\text{Berk}}$, and let $0 < \delta < 1/\|\zeta,\zeta\|$. Then the Favre–Rivera-Letelier smoothing $f_{\zeta,\delta} = f \circ r_{\zeta,\delta}|_U$ has the following properties:*

(A) *$f_{\zeta,\delta}$ extends to a function in $\mathcal{C}(\overline{U}) \cap \text{BDV}(U)$; in particular $f_{\zeta,\delta}$ is continuous and bounded.*

(B) *$f_{\zeta,\delta}$ coincides with f on $U \cap X(\zeta, \delta)$ and is constant on components of $U \backslash X(\zeta, \delta)$. More precisely, for each $a \in U \backslash X(\zeta, \delta)$ the component containing a is $U \cap \mathcal{B}(a, \delta)_\zeta^-$, and on that component $f_{\zeta,\delta}(x) = f(r_{\zeta,\delta}(a))$.*

(C) *$\Delta_{\overline{U}}(f_{\zeta,\delta}) = (r_{\zeta,\delta})_*(\Delta_{\overline{U}}(f))$.*

PROOF. We first show that $f_{\zeta,\delta} \in \text{BDV}(U)$. We only sketch the argument, as the details are straightforward but tedious. Put $X = X(\zeta, \delta)$, and for each finite subgraph $\Gamma \subset U$, let $\Gamma_X = \Gamma \cap X$.

Fixing Γ, note that Γ_X is a finite subgraph of Γ since X is closed and connected and $\mathbb{P}^1_{\text{Berk}}$ is uniquely path-connected. Thus, Γ can be obtained from Γ_X by adjoining a finite number of trees $T_{p_1}, \ldots, T_{p_m}$ at points $p_1, \ldots, p_m$. The function $f_{\zeta,\delta}|_\Gamma = f|_{\Gamma_X} \circ r_{\Gamma,\Gamma_X}$ coincides with f on Γ_X and is constant on each tree T_{p_i}, with $f_{\zeta,\delta}(x) = f(p_i)$ for all $x \in T_{p_i}$. Using the fact that

---

[2] One can actually take ζ to be any point in $\mathbb{P}^1_{\text{Berk}}$. Favre and Rivera-Letelier use $\zeta = \infty$, but taking $\zeta \in \mathbb{H}_{\text{Berk}}$ gives a smoothing which is bounded as well as continuous.

5.7. FAVRE–RIVERA-LETELIER SMOOTHING

$f|_{\Gamma_X} \in \mathrm{BDV}(\Gamma_X)$, one sees that $f_{\zeta,\delta}|_\Gamma \in \mathrm{BDV}(\Gamma)$ and that for each Borel subset $S \subset \Gamma$,

$$
\begin{aligned}
\Delta_\Gamma(f_{\zeta,\delta})(S) &= \Delta_{\Gamma_X}(f)(S \cap \Gamma_X) = \Delta_\Gamma(f)(r_{\Gamma,\Gamma_X}^{-1}(S \cap \Gamma_X)) \\
&= (r_{\Gamma,\Gamma_X})_*(\Delta_\Gamma(f))(S \cap \Gamma_X) \ .
\end{aligned}
\tag{5.36}
$$

It follows from the first equality in (5.36) that

$$
|\Delta_\Gamma(f_{\zeta,\delta})|(\Gamma) = |\Delta_{\Gamma_X}(f)|(\Gamma_X) \ .
\tag{5.37}
$$

Since $f \in \mathrm{BDV}(U)$, as Γ ranges over all finite subgraphs of U, there is a uniform bound B_f for the total variations $|\Delta_{\Gamma_X}(f)|(\Gamma_X)$, so (5.37) shows that $f_{\zeta,\delta}|_U \in \mathrm{BDV}(\mathbb{P}^1_{\mathrm{Berk}})$.

We can now prove (C). Since $\Delta_\Gamma(f_{\zeta,\delta}) = (r_\Gamma)_*(\Delta(f_{\zeta,\delta}))$ and $\Delta_{\Gamma_X}(f) = (r_{\Gamma_X})_*(\Delta(f))$, it follows from (5.37) that for each $\Gamma \subset U$ and each $g \in \mathcal{C}(\overline{U})$,

$$
\int_\Gamma g(x)\, d(r_{\Gamma*}\Delta(f_{\zeta,\delta}))(x) = \int_{\Gamma_X} g(x)\, d(r_{\Gamma_X*}\Delta(f))(x) \ .
$$

Applying the Riesz Representation Theorem via Proposition 5.10 yields

$$
\Delta_{\overline{U}}(f_{\zeta,\delta}) = (r_{\mathbb{P}^1_{\mathrm{Berk}}, X(\zeta,\delta)})_*(\Delta_{\overline{U}}(f)) = (r_{\zeta,\delta})_*(\Delta_{\overline{U}}(f))
\tag{5.38}
$$

as desired.

We next prove (A). We already know $f_{\zeta,\delta} \in \mathrm{BDV}(U)$; we will show that $f_{\zeta,\delta}$ is actually the restriction to U of a function in $\mathcal{C}(\mathbb{P}^1_{\mathrm{Berk}})$. The compactness of $\overline{U}$ then shows that $f_{\zeta,\delta}$ is bounded. Put $\nu = (r_{\zeta,\delta})_*(\Delta_{\overline{U}}(f))$. By Proposition 5.42, the potential function $u_\nu(z, \zeta_{\mathrm{Gauss}})$ is continuous on $\mathbb{P}^1_{\mathrm{Berk}}$. As shown above, $\Delta_{\overline{U}}(f_{\zeta,\delta}) = \nu$, so by Lemma 5.24 there is a constant C such that $f_{\zeta,\delta}(x) = u_\nu(x, \zeta_{\mathrm{Gauss}}) + C$ for all $x \in U \cap \mathbb{H}_{\mathrm{Berk}}$. However, by construction $f_{\zeta,\delta}$ is constant on a neighborhood of each point $x \in U \cap \mathbb{P}^1(K)$. Since $u_\nu(z, \zeta_{\mathrm{Gauss}})$ is continuous, this shows that $f_{\zeta,\delta}(x) = u_\nu(x, \zeta_{\mathrm{Gauss}}) + C$ for all $x \in U$.

Finally, assertion (B) follows from the definition of $f_{\zeta,\delta}$. $\square$

In applications, one usually takes a sequence of numbers $\delta_n \to 0$ and considers the sequence of functions $f_n = f_{\zeta,\delta_n}$, which coincide with f on larger and larger sets as $n \to \infty$.

PROPOSITION 5.45. *Let $U \subset \mathbb{P}^1_{\mathrm{Berk}}$ be a domain. Fix $\zeta \in U \cap \mathbb{H}_{\mathrm{Berk}}$, and let $\{\delta_n\}_{n \geq 1}$ be a sequence with $\lim_{n \to \infty} \delta_n = 0$, such that $0 < \delta_n < 1/\|\zeta, \zeta\|$ for each n.*

(A) *If $f \in \mathrm{BDV}(U)$, then the sequence of functions $\{f_{\zeta,\delta_n}\}$ converges to f on $U \cap \mathbb{H}_{\mathrm{Berk}}$, and the sequence of measures $\Delta_{\overline{U}}(f_{\zeta,\delta_n})$ converges weakly to $\Delta_{\overline{U}}(f)$.*

(B) *If $f, g \in \mathrm{BDV}(U)$ and $\langle f, g \rangle_{U,\mathrm{Dir}}$ is defined, then $\langle f_{\zeta,\delta_n}, g_{\zeta,\delta_n} \rangle_{U,\mathrm{Dir}}$ is defined and finite for each n, and*

$$
\lim_{n \to \infty} \langle f_{\zeta,\delta_n}, g_{\zeta,\delta_n} \rangle_{U,\mathrm{Dir}} = \langle f, g \rangle_{U,\mathrm{Dir}} \ .
\tag{5.39}
$$

PROOF. We prove (A) by applying Proposition 5.32. For each n, put $f_n = f_{\zeta,\delta_n}$ and put $\nu_n = \Delta_{\overline{U}}(f_{\zeta,\delta_n}) = (r_{\zeta,\delta_n})_*(\Delta_{\overline{U}}(f))$. Clearly the functions f_n converge uniformly to f on each finite subgraph $\Gamma \subset U$; indeed, $f_n|_\Gamma = f|_\Gamma$ for all sufficiently large n. Since retraction preserves the mass of a positive measure, it cannot increase the total variation of a signed measure. Hence $|\nu_n|(\overline{U}) \leq |\nu|(\overline{U})$ for each n. Thus Proposition 5.32 shows that the ν_n converge weakly to $\Delta(f)$.

To prove (B), suppose $\langle f, g \rangle_{U,\mathrm{Dir}} = L \in \mathbb{R} \cup \{\pm\infty\}$. We will show (5.39) only when $L \in \mathbb{R}$; the cases when $L = \pm\infty$ are similar. Fix $\varepsilon > 0$. By the definition (5.15) of $\langle f, g \rangle_{U,\mathrm{Dir}}$, there is a finite graph $\Gamma_\varepsilon \subset U$ such that for any finite subgraph Γ with $\Gamma_\varepsilon \subset \Gamma \subset U$, we have

(5.40) $$|\langle f, g \rangle_{\Gamma,\mathrm{Dir}} - L| < \varepsilon.$$

Given $n \geq N$, put $X_n = X(\zeta, \delta_n)$. Let N be large enough that $\Gamma_\varepsilon \subset X_n$ for all $n \geq N$.

Suppose $n \geq N$. For each Γ with $\Gamma_\varepsilon \subset \Gamma \subset U$, the subgraph $\Gamma_n = \Gamma \cap X_n$ contains Γ_ε, and since f_{ζ,δ_n} and g_{ζ,δ_n} are constant on branches off X_n,

(5.41) $$\langle f_{\zeta,\delta_n}, g_{\zeta,\delta_n} \rangle_{\Gamma,\mathrm{Dir}} = \langle f, g \rangle_{\Gamma_n,\mathrm{Dir}}.$$

Hence $|\langle f_{\zeta,\delta_n}, g_{\zeta,\delta_n} \rangle_{\Gamma,\mathrm{Dir}} - L| < \varepsilon$. By Theorem 5.44, f_{ζ,δ_n} and g_{ζ,δ_n} extend to functions in $\mathcal{C}(\overline{U}) \cap \mathrm{BDV}(U)$, so by Proposition 5.35 the pairing $\langle f_{\zeta,\delta_n}, g_{\zeta,\delta_n} \rangle_{U,\mathrm{Dir}}$ is defined and finite. It follows from (5.40), (5.41), and the definition of the Dirichlet pairing (5.15) that $|\langle f_{\zeta,\delta_n}, g_{\zeta,\delta_n} \rangle_{U,\mathrm{Dir}} - L| \leq \varepsilon$.

Since $\varepsilon > 0$ is arbitrary, $\lim_{n \to \infty} \langle f_{\zeta,\delta_n}, g_{\zeta,\delta_n} \rangle_{U,\mathrm{Dir}} = \langle f, g \rangle_{\Gamma,\mathrm{Dir}}$. $\square$

5.8. The Laplacians of Favre, Jonsson, and Rivera-Letelier, and of Thuillier

In this section we will compare our Laplacian with those constructed by Favre, Jonsson, and Rivera-Letelier, and by Thuillier.

Favre and Rivera-Letelier [**47**, pp. 335–336], invoking work of Favre and Jonsson [**45**], construct a Laplacian on $\mathbb{P}^1_{\mathrm{Berk}}$ in the following way.

Let $\mathcal{M}$ be the vector space of all finite signed Borel measures on $\mathbb{P}^1_{\mathrm{Berk}}$. Fixing $\zeta \in \mathbb{H}_{\mathrm{Berk}}$, for each $\mu \in \mathcal{M}$ define the *potential* $\hat{g}_\mu : \mathbb{H}_{\mathrm{Berk}} \to \mathbb{R}$ by

(5.42) $$\hat{g}_\mu(x) = -\mu(\mathbb{P}^1_{\mathrm{Berk}}) - \int_{\mathbb{P}^1_{\mathrm{Berk}}} (x|y)_\zeta \, d\mu(y),$$

where $(x|y)_\zeta$ is the Gromov product [**47**, p. 356] (see §B.4). Let the *space of potentials* $\mathcal{P}$ be the set of all functions $\hat{g}_\mu$, as μ varies over $\mathcal{M}$. It is a real vector space, since the map $\mu \mapsto \hat{g}_\mu$ is linear. Note that the constant functions belong to $\mathcal{P}$; indeed, if δ_ζ is the Dirac probability measure at ζ, then $\mu = -C \cdot \delta_\zeta$ maps to C. Since each $\mu \in \mathcal{M}$ can be decomposed as $\mu = \mu(\mathbb{P}^1_{\mathrm{Berk}})\delta_\zeta + \mu_0$, where μ_0 has total mass 0, the space $\mathcal{P}$ is spanned by the constant functions and the potentials $\hat{g}_{\mu_0}$ for measures μ_0 of mass 0.

By [**45**, Theorem 7.50], the map $\mu \mapsto \hat{g}_\mu$ is a bijection from $\mathcal{M}$ to $\mathcal{P}$. Inverting this map, Favre and Rivera-Letelier define their Laplacian (which we denote Δ_{FJRL}) for functions $f \in \mathcal{P}$ by setting

$$\Delta_{\text{FJRL}}(f) = \mu - \mu(\mathbb{P}^1_{\text{Berk}})\delta_\zeta$$

if $f = \hat{g}_\mu \in \mathcal{P}$. The discussion above shows that $\Delta_{\text{FJRL}}(C) = 0$ for each constant function C; indeed, by the bijectivity, $\Delta_{\text{FJRL}}(\hat{g}) = 0$ if and only if $\hat{g}$ is constant. Furthermore, $\Delta_{\text{FJRL}}(\hat{g}_{\mu_0}) = \mu_0$ for each $\mu_0 \in \mathcal{M}$ of total mass 0. In [**47**, Proposition 4.1], it is shown that the space $\mathcal{P}$ and the Laplacian Δ_{FJRL} are independent of the choice of ζ.

We claim that when $U = \mathbb{P}^1_{\text{Berk}}$, the Laplacian Δ_{FJRL} is the negative of our Laplacian $\Delta = \Delta_{\mathbb{P}^1_{\text{Berk}}}$. Indeed, by (4.28), there is a constant B such that $-\log_v(\delta(x,y)_\zeta) = (x|y)_\zeta + B$. Hence for each $\mu_0 \in \mathcal{M}$ of total mass 0 and each $x \in \mathbb{H}_{\text{Berk}}$,

$$
\begin{aligned}
u_{\mu_0}(x,\zeta) &= \int_{\mathbb{P}^1_{\text{Berk}}} -\log_v(\delta(x,y)_\zeta)\,d\mu_0(y) \\
&= \int_{\mathbb{P}^1_{\text{Berk}}} (x|y)_\zeta\,d\mu_0(y) = -\hat{g}_{\mu_0}(x)\,.
\end{aligned}
$$
(5.43)

In particular, by Proposition 5.28 the space $\mathcal{P}$ coincides with the restriction of our space $\text{BDV}(\mathbb{P}^1_{\text{Berk}})$ to $\mathbb{H}_{\text{Berk}}$.

For each $f \in \text{BDV}(\mathbb{P}^1_{\text{Berk}})$, Lemma 5.24 shows that $\Delta(f) = 0$ if and only if the restriction of f to $\mathbb{H}_{\text{Berk}}$ is constant, while by (5.42), (5.43), and (5.6) in Example 5.21, for each μ_0 of mass 0,

$$\Delta_{\text{FJRL}}(-\hat{g}_{\mu_0}(x)) = \Delta(u_{\mu_0}(x,\zeta)) = \mu_0\,.$$

This shows that Δ and Δ_{FJRL} are negatives of each other, except for an inessential difference in their domains.

We next consider the distribution-valued Laplacian defined by Thuillier [**94**]. Our presentation is somewhat oversimplified, since Thuillier works in the context of Berkovich curves over an arbitrary (not necessarily algebraically closed) complete non-Archimedean field k and part of his definition involves functoriality under base change.

Given a domain U, let $A^0(U)$ be the space of all continuous functions $g: U \to \mathbb{R}$ with the following property: for each simple subdomain V of U, there are a finite subgraph $\Gamma_V \subset \overline{V} \cap \mathbb{H}^\mathbb{R}_{\text{Berk}}$ and a function $G_V \in \text{CPA}(\Gamma_V)$ such that for each $z \in \overline{V}$,

$$g(z) = G_V(r_{\overline{V},\Gamma_V}(z))\,.$$

Let $A^0_c(U) \subset A^0(U)$ be the subspace of functions $g \in A^0(U)$ which are compactly supported: there is a simple subdomain V of U such that $g(z) = G_V(r_{\overline{V},\Gamma_V}(z))$ for each $z \in \overline{V}$ and $g(z) = 0$ for all $z \in U \setminus \overline{V}$.

Let $A^1(U)$ be the space of locally finite discrete measures μ supported on $U \cap \mathbb{H}_{\text{Berk}}^{\mathbb{R}}$:

$$\mu = \sum_{i=1}^{\infty} a_i \delta_{x_i},$$

where each $x_i \in U \cap \mathbb{H}_{\text{Berk}}^{\mathbb{R}}$ and for each simple subdomain V of U, only finitely many x_i belong to V. Likewise, let $A_c^1(U) \subset A^1(U)$ be the space of finitely supported discrete measures.

Thuillier considers $A^0(U)$ (resp. $A^1(U)$) to be the space of smooth functions (resp. 1-forms) and $A_c^0(U)$ (resp. $A_c^1(U)$) to be the space of smooth, compactly supported functions (resp. 1-forms). For each $g \in A^0(U)$, he defines the Laplacian

$$(5.44) \qquad dd^c g = \sum_{x \in U \cap \mathbb{H}_{\text{Berk}}} \left(\sum_{\vec{v} \in T_x} d_{\vec{v}} g(x) \right) \delta_x.$$

Here, for each x there are at most finitely many $\vec{v} \in T_x$ with $d_{\vec{v}} g(x) \neq 0$, and there are at most countably many $x \in U$ for which some $d_{\vec{v}} g(x) \neq 0$, so $dd^c g$ is well-defined and belongs to $A^1(U)$. Furthermore, if $g \in A_c^0(U)$, then $dd^c g \in A_c^1(U)$.

If $g \in A_c^0(U)$, then clearly $g \in \text{BDV}(U)$, and by (3.1), in terms of our Laplacian

$$dd^c g = -\Delta_U(g).$$

Next, Thuillier defines the spaces of 'distributions' and '1-currents' to be the dual spaces $D^0(U) = A_c^1(U)^\vee$ and $D^1(U) = A_c^0(U)^\vee$. Since $A_c^1(U)$ is isomorphic to the real vector space whose basis is the set of Dirac measures δ_x for $x \in U \cap \mathbb{H}_{\text{Berk}}$, $D^0(U)$ can be identified with the space of all (continuous or not!) functions $G : (U \cap \mathbb{H}_{\text{Berk}}) \to \mathbb{R}$. For each $G \in D^0(U)$, he defines the Laplacian $dd^c G \in D^1(U)$ by

$$dd^c G(g) = G(dd^c g)$$

for all $g \in A_c^0(U)$. Of course, it remains to show that for 'interesting' functions G, the distribution $dd^c G$ can be identified with a measure or some other explicit linear functional. It follows from Corollary 5.39 that if $f \in \text{BDV}(U)$, then $dd^c f = -\Delta_U(f)$. Thuillier shows that if $f \in A^0(U)$ is identified with a distribution $F \in D^0(U)$ and if $dd^c f$ is the measure defined in (5.44), then for each $g \in A_c^0(U)$,

$$\int_U f \, dd^c g = \int_U g \, dd^c f.$$

This enables him to identify the distribution $dd^c F$ with the measure $dd^c f$. He thus has a commutative diagram

$$\begin{array}{ccc}
A_c^0(U) & \xrightarrow{dd^c} & A_c^1(U) \\
\downarrow & & \downarrow \\
A^0(U) & \xrightarrow{dd^c} & A^1(U) \\
\downarrow & & \downarrow \\
D^0(U) = A_c^1(U)^\vee & \xrightarrow{dd^c} & D^1(U) = A_c^0(U)^\vee .
\end{array}$$

Likewise, he defines a class of 'subharmonic functions' (closely related to the class $\mathcal{SH}(U)$ we will study in Chapter 8) and shows that for each $f \in \mathcal{SH}(U)$, $dd^c f$ is a positive Radon measure.

5.9. Notes and further references

Mattias Jonsson pointed out to us that in [91], we were using without proof the fact that every Borel measure on $\mathcal{D}(0,1)$ is Radon. This is now proved as Lemma 5.6.

As mentioned in the Preface, the idea of constructing a Berkovich space Laplacian operator as an inverse limit of Laplacians on finite metrized graphs is due independently to the second author and Amaury Thuillier.

CHAPTER 6

Capacity theory

Fix $\zeta \in \mathbb{P}^1_{\text{Berk}}$, and let $E \subset \mathbb{P}^1_{\text{Berk}}$ be a set not containing ζ. In the first four sections of this chapter, we develop the theory of the *logarithmic capacity* of E with respect to ζ (or, more correctly, with respect to a choice of a generalized Hsia kernel $\delta(x,y)_\zeta$; we fix the choice made in (4.23)). Most of the proof techniques are standard; the exposition is adapted from [88, §4.1]. As an application, in §6.5 we give Berkovich space versions of the Fekete and Fekete-Szegö theorems.

6.1. Logarithmic capacities

Recall that a probability measure is a nonnegative Borel measure of total mass 1. Given a probability measure ν with support contained in E, define the *energy integral*

$$I_\zeta(\nu) = \iint_{E \times E} -\log_v \delta(x,y)_\zeta \, d\nu(x) d\nu(y) \ .$$

Here, the integral is a Lebesgue integral. The kernel $-\log_v \delta(x,y)_\zeta$ is lower semicontinuous, and hence Borel measurable, since $\delta(x,y)_\zeta$ is upper semicontinuous.

Let ν vary over probability measures with support contained in E, and define the *Robin constant*

$$V_\zeta(E) = \inf_\nu I_\zeta(\nu) \ .$$

Define the *logarithmic capacity*

(6.1) $$\gamma_\zeta(E) = q_v^{-V_\zeta(E)} \ .$$

By its definition, the logarithmic capacity is monotonic in E: if $E_1 \subset E_2$, then $\gamma_\zeta(E_1) \leq \gamma_\zeta(E_2)$. It also follows from the definition that for each E,

(6.2) $$\gamma_\zeta(E) = \sup_{\substack{E' \subset E \\ E' \text{ compact}}} \gamma_\zeta(E') \ .$$

Indeed, the support of each probability measure ν is compact, so tautologically for each ν supported on E there is a compact set $E' \subset E$ with $q_v^{-I_\zeta(\nu)} \leq \gamma_\zeta(E') \leq \gamma_\zeta(E)$.

There is an important distinction between sets of capacity 0 and sets of positive capacity. If E contains a point $a \in \mathbb{H}_{\text{Berk}}$ (i.e., a point of type II, III, or IV), then $\gamma_\zeta(E) > 0$, since the point mass δ_a satisfies $I_\zeta(\delta_a) =$

$-\log_v(\operatorname{diam}_\zeta(a)) < \infty$. Thus, every set of capacity 0 is contained in $\mathbb{P}^1(K)$. The converse is not true: there are many sets in $\mathbb{P}^1(K)$ with positive capacity.

The following result shows that the property that a set has capacity 0 or positive capacity is independent of the base point ζ.

PROPOSITION 6.1. *Suppose $E \subset \mathbb{P}^1_{\text{Berk}}$. Then $\gamma_\zeta(E) = 0$ for some $\zeta \in \mathbb{P}^1_{\text{Berk}} \backslash E$ if and only if $\gamma_\xi(E) = 0$ for every $\xi \in \mathbb{P}^1_{\text{Berk}} \backslash E$.*

PROOF. By (6.2), it suffices to consider the case where E is compact. Take $\xi \notin E$. Since $\|x, \xi\|$ is continuous on E, there is a constant $C_\xi > 0$ such that
$$1/C_\xi \leq \|x, \xi\| \leq C_\xi$$
for all $x \in E$. Since
$$\delta(x,y)_\xi = \frac{\|x,y\|}{\|x,\xi\| \|y,\xi\|} \ , \quad \delta(x,y)_\zeta = \frac{\|x,y\|}{\|x,\zeta\| \|y,\zeta\|} \ ,$$
it follows that
$$\frac{1}{(C_\zeta C_\xi)^2} \delta(x,y)_\zeta \ \leq \ \delta(x,y)_\xi \ \leq \ (C_\zeta C_\xi)^2 \delta(x,y)_\zeta \ .$$

Hence for each probability measure ν on E, either $I_\zeta(\nu) = I_\xi(\nu) = \infty$ or $I_\zeta(\nu)$ and $I_\xi(\nu)$ are both finite. $\square$

Here are some examples of capacities.

EXAMPLE 6.2. If $E \subset \mathbb{P}^1(K)$ is a countable set and $\zeta \notin E$, then $\gamma_\zeta(E) = 0$. To see this, note first that if ν is a probability measure supported on E, then necessarily ν has point masses (if $\nu(\{x\}) = 0$ for each $x \in E$, then by countable additivity $\nu(E) = 0$, which contradicts $\nu(E) = 1$). If $p \in E$ is a point with $\nu(\{p\}) > 0$, then
$$\begin{aligned} I_\zeta(\nu) &= \iint_{E \times E} -\log_v \delta(x,y)_\zeta \, d\nu(x) d\nu(y) \\ &\geq \ -\log_v \delta(p,p)_\zeta \cdot \nu(\{p\})^2 \ = \ \infty \ , \end{aligned}$$
so $V_\zeta(E) = \infty$.

EXAMPLE 6.3. If $E = \{a\}$ where $a \in \mathbb{H}_{\text{Berk}}$ and $\zeta \neq a$, then
$$\gamma_\zeta(E) \ = \ \operatorname{diam}_\zeta(a) \ .$$

Indeed, the only probability measure supported on E is the point mass $\nu = \delta_a$, for which
$$\begin{aligned} I_\zeta(\nu) &= \iint_{E \times E} -\log_v \delta(x,y)_\zeta \, d\nu(x) d\nu(y) \\ &= \ -\log_v \delta(a,a)_\zeta \ = \ -\log_v \operatorname{diam}_\zeta(a) \ . \end{aligned}$$
Hence $V_\zeta(E) = -\log_v \operatorname{diam}_\zeta(a)$ and $\gamma_\zeta(a) = \operatorname{diam}_\zeta(a)$.

EXAMPLE 6.4. Suppose $K = \mathbb{C}_p$. If $\zeta = \infty$ and $E = \mathbb{Z}_p \subset \mathbb{A}^1(\mathbb{C}_p)$ and if we take $q_v = p$ (the natural arithmetic choice when $\mathbb{Q}_p$ is regarded as the base field), then $\gamma_\infty(E) = p^{-1/(p-1)}$. More generally, if $E = \mathcal{O}_v$ is the valuation ring in a finite extension $K_v/\mathbb{Q}_p$ with ramification index e and residue degree f and if we take $q_v = p^f$ (regarding K_v as the base field), then $\gamma_\infty(E) = (p^f)^{-1/e(p^f-1)}$.

For the proof, see [**88**, Example 4.1.24, p. 212]. The capacity is given by the same computation as in the classical case, since $\delta(x,y)_\infty = |x-y|$ for $x, y \in \mathbb{A}^1(\mathbb{C}_p)$.

6.2. The equilibrium distribution

If E is compact and if $\gamma_\zeta(E) > 0$, there is always a probability measure $\mu = \mu_\zeta$ on E for which $I_\zeta(\mu) = V_\zeta(E)$; it is called an *equilibrium distribution*, or an *equilibrium measure*, for E with respect to ζ. In Proposition 7.21, we will show it is unique. Anticipating this, we will speak of *the* equilibrium measure in major results below. We do so in order to state them in a form suitable for reference. A reader encountering the theory for the first time should note that the proofs are valid for *any* equilibrium measure.

To show the existence of the equilibrium measure, we will use the following lemma:

LEMMA 6.5. *Let E be a compact Hausdorff space, and suppose $\langle \nu_\alpha \rangle$ is a net of probability measures on E which converge weakly to a measure μ. Then $\langle \nu_\alpha \times \nu_\alpha \rangle$ converges weakly to $\mu \times \mu$ on $E \times E$.*

PROOF. We fill in the details of the argument given in [**59**, p. 283]. Since E is a compact Hausdorff space, the Stone-Weierstrass theorem asserts that linear combinations of functions of the form $f(x)g(y)$, with $f, g \in \mathcal{C}(E)$, are dense in $\mathcal{C}(E \times E)$ under the sup norm $\|\cdot\|_{E \times E}$. For such products,

$$\lim_\alpha \iint_{E \times E} f(x)g(y)\, d\nu_\alpha(x) d\nu_\alpha(y) = \int_E f(x)\, d\mu(x) \cdot \int_E g(y)\, d\mu(y)$$
$$= \iint_{E \times E} f(x)g(y)\, d\mu(x) d\mu(y),$$

so the same holds for their linear combinations.

Now let $F(x, y) \in \mathcal{C}(E \times E)$ be arbitrary. Given $\varepsilon > 0$, take $f(x,y) = \sum_i c_i f_i(x) g_i(y)$ so that $h(x,y) = F(x,y) - f(x,y)$ satisfies $\|h\|_{E \times E} < \varepsilon$. For any probability measure ν on E, clearly

$$\left| \iint_{E \times E} h(x,y)\, d\nu(x) d\nu(y) \right| \leq \|h\|_{E \times E}.$$

Let α_0 be large enough that

$$\left| \iint_{E \times E} f(x,y)\, d\nu_\alpha(x) d\nu_\alpha(y) - \iint_{E \times E} f(x,y)\, d\mu(x) d\mu(y) \right| < \varepsilon$$

for $\alpha \geq \alpha_0$. By a three-epsilon argument, for such α we have
$$\left| \iint_{E \times E} F(x,y) \, d\nu_\alpha(x) d\nu_\alpha(y) - \iint_{E \times E} F(x,y) \, d\mu(x) d\mu(y) \right| \leq 3\varepsilon \, .$$
Since ε was arbitrary, it follows that
$$\lim_\alpha \iint_{E \times E} F(x,y) \, d\nu_\alpha(x) d\nu_\alpha(y) = \iint_{E \times E} F(x,y) \, d\mu(x) d\mu(y) \, .$$
Thus $\langle \nu_\alpha \times \nu_\alpha \rangle$ converges weakly to $\mu \times \mu$. $\square$

We can now show the existence of an equilibrium measure.

PROPOSITION 6.6. *Let $E \subset \mathbb{P}^1_{\text{Berk}} \backslash \{\zeta\}$ be a compact set with positive capacity. Then there is a probability measure μ supported on E such that $I_\zeta(\mu) = V_\zeta(E)$.*

PROOF. Take a sequence of probability measures ν_n on E for which $\lim_{n \to \infty} I_\zeta(\nu_n) = V_\zeta(E)$. After passing to a subsequence, if necessary, we can assume that $\{\nu_n\}$ converges weakly to a measure μ. Clearly μ is a probability measure supported on E.

Since E is a compact subset of $\mathbb{P}^1_{\text{Berk}} \backslash \{\zeta\}$ and $-\log_v \delta(x,y)_\zeta$ is lower semicontinuous, there is a constant $0 \leq M \in \mathbb{R}$ such that $-\log_v \delta(x,y)_\zeta$ is bounded below on $E \times E$ by $-M$. By Proposition A.3 applied to the function $-\log_v \delta(x,y)_\zeta$ on $E \times E$, we see that
$$\iint -\log_v \delta(x,y)_\zeta \, d\mu(x) d\mu(y) = \sup_{\substack{g \in \mathcal{C}(E \times E) \\ -M \leq g \leq -\log_v \delta(x,y)_\zeta}} \iint g(x,y) \, d\mu(x) d\mu(y) \, .$$
Furthermore, for each $g \in \mathcal{C}(E \times E)$, it follows from Lemma 6.5 that
$$\iint_{E \times E} g(x,y) \, d\mu(x) d\mu(y) = \lim_{n \to \infty} \iint_{E \times E} g(x,y) \, d\nu_n(x) d\nu_n(y) \, .$$
If, moreover, we have $g(x,y) \leq -\log_v \delta(x,y)_\zeta$, then for each n
$$\begin{aligned}\iint_{E \times E} g(x,y) \, d\nu_n(x) d\nu_n(y) &\leq \iint_{E \times E} -\log_v \delta(x,y)_\zeta \, d\nu_n(x) d\nu_n(y) \\ &= I_\zeta(\nu_n) \, ,\end{aligned}$$
and thus for each such g, we have
$$\iint_{E \times E} g(x,y) \, d\mu(x) d\mu(y) \leq \lim_{n \to \infty} I_\zeta(\nu_n) = V_\zeta(E) \, .$$
Therefore
$$\begin{aligned}I_\zeta(\mu) &= \iint_{E \times E} -\log_v \delta(x,y)_\zeta \, d\mu(x) d\mu(y) \\ &= \sup_{\substack{g \in \mathcal{C}(E \times E) \\ -M \leq g \leq -\log_v \delta(x,y)_\zeta}} \iint_{E \times E} g(x,y) \, d\mu(x) d\mu(y) \leq V_\zeta(E) \, .\end{aligned}$$
The opposite inequality is trivial, so $I_\zeta(\mu) = V_\zeta(E)$. $\square$

REMARK 6.7. In the classical proof over $\mathbb{C}$, one considers the truncated logarithm $-\log^{(t)}(|x-y|) = \min(t, -\log|x-y|)$, which is continuous and approaches $-\log(|x-y|)$ pointwise as $t \to \infty$. Here $-\log_v^{(t)}(\delta(x,y)_\zeta)$ is lower semicontinuous but not continuous, so it was necessary to introduce the approximating functions $g(x,y)$.

Let U_ζ be the connected component of $\mathbb{P}^1_{\text{Berk}}\backslash E$ containing ζ. Write $\partial E_\zeta = \partial U_\zeta$ for the boundary of U_ζ, the part of ∂E in common with the closure $\overline{U}_\zeta$.

PROPOSITION 6.8. *Let $E \subset \mathbb{P}^1_{\text{Berk}}\backslash\{\zeta\}$ be compact with positive capacity. Then each equilibrium distribution μ_ζ is supported on ∂E_ζ.*

PROOF. Suppose μ_ζ is not supported on ∂E_ζ, and fix $x_0 \in \text{supp}(\mu_\zeta)\backslash \partial E_\zeta$. Let $r_\zeta : E \to \partial E_\zeta$ be the retraction map which takes each $x \in E$ to the last point in E on the path from x to ζ. Then r_ζ is easily seen to be continuous. Put $\mu_0 = (r_\zeta)_*(\mu_\zeta)$. We claim that $I_\zeta(\mu_0) < I_\zeta(\mu_\zeta)$.

For each $x, y \in E$, if $\overline{x} = r_\zeta(x)$, $\overline{y} = r_\zeta(y)$, then $\delta(x,y)_\zeta \le \delta(\overline{x},\overline{y})_\zeta$. This follows from the geometric interpretation of $\delta(x,y)_\zeta$: if w is the point where the paths from x to ζ and y to ζ first meet and $\overline{w}$ is the point where the paths from $\overline{x}$ to ζ and $\overline{y}$ to ζ first meet, then $\overline{w}$ lies on the path from w to ζ.

Now consider the point x_0, and put $x_1 = r_\zeta(x_0)$. Then $\text{diam}_\zeta(x_0) < \text{diam}_\zeta(x_1)$. Fix r with $\text{diam}_\zeta(x_0) < r < \text{diam}_\zeta(x_1)$, and put $U = \mathcal{B}(x_0, r)^-_\zeta$. For each $x \in U$, $r_\zeta(x) = x_1$. If $x, y \in U$, then
$$\delta(x,y)_\zeta < r < \text{diam}_\zeta(x_1) = \delta(x_1, x_1)_\zeta = \delta(r_\zeta(x), r_\zeta(y))_\zeta \ .$$
Since $x_0 \in \text{supp}(\mu_\zeta)$, necessarily $\mu_\zeta(U) > 0$. Hence
$$\begin{aligned} I_\zeta(\mu_\zeta) &= \int -\log_v \delta(x,y)_\zeta \, d\mu_\zeta(x) d\mu_\zeta(y) \\ &> \int -\log_v \delta(r_\zeta(x), r_\zeta(y))_\zeta \, d\mu_\zeta(x) d\mu_\zeta(y) \\ &= \int -\log_v \delta(z,w)_\zeta \, d\mu_0(z) d\mu_0(w) \; = \; I_\zeta(\mu_0) \ .\end{aligned}$$
This contradicts the minimality of $I_\zeta(\mu_\zeta)$, so μ_ζ is supported on ∂E_ζ. $\square$

The existence of equilibrium measures has the following important consequence.

COROLLARY 6.9. *Let $E \subset \mathbb{P}^1_{\text{Berk}}\backslash\{\zeta\}$ be compact. Then for any $\varepsilon > 0$ and any neighborhood U of E, there is a strict closed affinoid $W \subset U$, containing E in its interior, for which $\gamma_\zeta(W) \le \gamma_\zeta(E) + \varepsilon$.*

PROOF. After shrinking U if necessary, we can assume that $\zeta \notin U$. We first construct a closed neighborhood W' of E, contained in U and containing E in its interior, for which $\gamma_\zeta(W') \le \gamma_\zeta(E) + \varepsilon$.

Let $\{W_\alpha\}$ be any family of (not necessarily connected) compact neighborhoods of E whose intersection is precisely E. (Such a family always exists: for example, for each point $z \notin E$, there is a covering of E by open sets whose closure does not contain z, and by compactness this open cover has a finite subcover. For each such z, take W_z to be the corresponding finite union of closed sets.)

Let $W_0 \supset E$ be any fixed compact neighborhood of E contained in U, not containing ζ. After replacing each W_α with $W_0 \cap W_\alpha$, we can assume without loss of generality that each $W_\alpha \subseteq W_0 \subseteq U$. Then, by augmenting $\{W_\alpha\}$ with all finite intersections of the sets in it, we can consider $\{W_\alpha\}$ to be a net, with the W_α partially ordered by containment.

As each W_α has nonempty intersection with $\mathbb{H}_{\text{Berk}}$, we have $\gamma_\zeta(W_\alpha) > 0$ for all α. Therefore each W_α has an equilibrium measure μ_α for which $I_\zeta(\mu_\alpha) = V_\zeta(\mu_\alpha) < \infty$. (The measure μ_α is in fact unique, but we don't need this.)

Passing to a subnet, we may assume without loss of generality that the net $\langle \mu_\alpha \rangle$ converges weakly to some probability measure μ^*. As $\bigcap W_\alpha = E$, it follows easily that μ^* is supported on E. We claim that

$$(6.3) \qquad \lim_\alpha V_\zeta(W_\alpha) \geq I_\zeta(\mu^*) \ .$$

Given this claim, the proof of the existence of W' can be completed as follows. Since for all α we have

$$V_\zeta(W_\alpha) \leq V_\zeta(E) \leq I_\zeta(\mu^*) \ ,$$

it must be that

$$\lim_\alpha V_\zeta(W_\alpha) = I_\zeta(\mu^*) = V_\zeta(E) \ .$$

This implies that

$$\lim_\alpha \gamma_\zeta(W_\alpha) = \gamma_\zeta(E) \ ,$$

so we can take $W' = W_\alpha$ for an appropriate α.

We next justify (6.3), in several steps. First of all, since W_0 is a compact subset of $\mathbb{P}^1_{\text{Berk}} \backslash \{\zeta\}$, we know that $-\log_v \delta(x, y)_\zeta$ is bounded below on $W_0 \times W_0$ by some constant $-M$. By Proposition A.3, for any probability measure μ we have

$$I_\zeta(\mu) = \sup_{\substack{g \in \mathcal{C}(W_0 \times W_0) \\ -M \leq g \leq -\log_v \delta(x,y)_\zeta}} \iint_{W_0 \times W_0} g(x, y) \, d\mu(x) d\mu(y) \ .$$

By Lemma 6.5, for each $g \in \mathcal{C}(W_0 \times W_0)$

$$\iint_{W_0 \times W_0} g(x, y) \, d\mu^*(x) d\mu^*(y) = \lim_\alpha \iint_{W_0 \times W_0} g(x, y) \, d\mu_\alpha(x) d\mu_\alpha(y) \ .$$

When in addition $g(x, y) \leq -\log_v \delta(x, y)_\zeta$ on $W_0 \times W_0$, for each α we have

$$\iint_{W_0 \times W_0} g(x, y) \, d\mu_\alpha(x) d\mu_\alpha(y) \leq I_\zeta(\mu_\alpha) \ .$$

6.2. THE EQUILIBRIUM DISTRIBUTION

Thus for each such g,

$$\lim_\alpha V_\zeta(W_\alpha) = \lim_\alpha I_\zeta(\mu_\alpha) \geq \lim_\alpha \iint_{W_0 \times W_0} g(x,y)\, d\mu_\alpha(x) d\mu_\alpha(y)$$
$$= \iint_{E \times E} g(x,y)\, d\mu^*(x) d\mu^*(y) \ .$$

Taking the supremum over all such g, we obtain

$$\lim_\alpha V_\zeta(W_\alpha) \geq \sup_g \iint_{E \times E} g(x,y)\, d\mu^*(x) d\mu^*(y) = I_\zeta(\mu^*)$$

as claimed.

To construct the strict closed affinoid W in the proposition, let W' be as above. Proposition 2.7 shows that for each $x \in \mathbb{P}^1_{\text{Berk}}$, a basis for the closed neighborhoods of x is given by sets of the form

$$\mathcal{D}(a,r)\ ,\quad \mathcal{D}(a,r) \backslash \bigcup_{i=1}^N \mathcal{D}(a_i, r_i)^- \ ,\quad \text{and}\quad \mathbb{P}^1_{\text{Berk}} \backslash \bigcup_{i=1}^N \mathcal{D}(a_i, r_i)^- \ ,$$

where the $a, a_i \in K$ and the r, r_i belong to $|K^\times|$. Such sets (and their unions) are called strict closed affinoids. For each $x \in E$, let V_x be a neighborhood of x of the above form, contained in W'. Cover E with finitely many such sets and let W be their union. The monotonicity property of the capacity shows that $\gamma_\zeta(W) \leq \gamma_\zeta(W') \leq \gamma_\zeta(E) + \varepsilon$. By construction, W is a strict closed affinoid. $\square$

By a *nest* of sets $\langle E_\alpha \rangle_{\alpha \in A}$ we mean a net of sets such that $E_{\alpha_1} \supseteq E_{\alpha_2}$ whenever $\alpha_1 \leq \alpha_2$. As an immediate consequence of Corollary 6.9, we obtain the following fact:

COROLLARY 6.10. *Fix $\zeta \in \mathbb{P}^1_{\text{Berk}}$, and let $\langle E_\alpha \rangle_{\alpha \in A}$ be a nest of compact sets in $\mathbb{P}^1_{\text{Berk}} \backslash \{\zeta\}$. Put $E = \bigcap_{\alpha \in A} E_\alpha$. Then $\gamma_\zeta(E) = \lim_\alpha \gamma_\zeta(E_\alpha)$.*

PROOF. Since $\langle E_\alpha \rangle$ has the finite intersection property, E is compact and nonempty. Fix $\varepsilon > 0$ and let W be a strict closed affinoid with $\gamma_\zeta(E) \leq \gamma_\zeta(W) + \varepsilon$ whose interior W_0 contains E, as given by Corollary 6.9. We claim that there is an α_0 such that $E_\alpha \subset W_0$ for all $\alpha \geq \alpha_0$. If this were not so, then $\langle E_\alpha \backslash W_0 \rangle_{\alpha \in A}$ would be a nest of nonempty compact sets with the finite intersection property, whose intersection would contain a point $x \notin E$. This contradicts the fact that $E = \bigcap_{\alpha \in A} E_\alpha$. $\square$

The proof of Corollary 6.9 also shows:

COROLLARY 6.11. *Let $E \subset \mathbb{P}^1_{\text{Berk}} \backslash \{\zeta\}$ be a compact set with positive capacity, and let $\langle W_\alpha \rangle$ be a nest of compact neighborhoods of E whose intersection is E. Let μ_α be an equilibrium measure on W_α relative to ζ, and let $\mu_E = \mu_{E,\zeta}$ be an equilibrium measure on E. Assume that μ_E is the unique equilibrium measure on E relative to ζ. Then $\langle \mu_\alpha \rangle$ converges weakly to μ_E.*

PROOF. The proof of Corollary 6.9 shows that if μ^* is any weak limit of a convergent subnet of $\langle\mu_\alpha\rangle$, then μ^* is supported on E and
$$I_\zeta(\mu^*) \;=\; V_\zeta(E)\;.$$
By uniqueness, we must have $\mu^* = \mu_E$, i.e., every convergent subnet of $\langle\mu_\alpha\rangle$ converges weakly to μ_E. This implies that $\langle\mu_\alpha\rangle$ itself converges weakly to μ_E. □

Anticipating the uniqueness of the equilibrium measure (which will be proved in Proposition 7.21), in what follows we will speak of 'the' equilibrium measure μ_E, or equilibrium distribution, of a compact set E of positive capacity (relative to ζ). However, the arguments below would apply to any equilibrium measure.

6.3. Potential functions attached to probability measures

For each $\zeta \in \mathbb{P}^1_{\text{Berk}}$ and each probability measure ν on $\mathbb{P}^1_{\text{Berk}}$, the associated potential function is
$$u_\nu(z,\zeta) \;=\; \int -\log_v(\delta(z,w)_\zeta)\,d\nu(w)\;.$$

In Example 5.22 we saw that $u_\nu(z,\zeta) \in \text{BDV}(\mathbb{P}^1_{\text{Berk}})$ and $\Delta(u_\nu(z,\zeta)) = \nu - \delta_\zeta(z)$. In this section we will study properties of $u_\nu(z,\zeta)$ which do not depend on the Laplacian. We begin with properties which hold for all potential functions.

Recall (see §A.2) that if X is a topological space, then $f : X \to \mathbb{R} \cup \{-\infty, \infty\}$ is lower semicontinuous if for each $z \in X$, we have $f(z) \neq -\infty$ and

(6.4) $$\liminf_{t \to z} f(t) \;\geq\; f(z)\;.$$

PROPOSITION 6.12. *Let ν be a probability measure on $\mathbb{P}^1_{\text{Berk}}$, and suppose that either $\zeta \in \mathbb{H}_{\text{Berk}}$ or that $\zeta \in \mathbb{P}^1(K)$ and $\zeta \notin \text{supp}(\nu)$. Then as a function valued in $\mathbb{R} \cup \{-\infty, \infty\}$, $u_\nu(z,\zeta)$ is well-defined for each $z \in \mathbb{P}^1_{\text{Berk}}$. It is continuous on $\mathbb{P}^1_{\text{Berk}} \setminus \text{supp}(\nu)$ and achieves its minimum at $z = \zeta$. If $\zeta \notin \text{supp}(\nu)$, then*

(6.5) $$\lim_{z \to \zeta} (u_\nu(z,\zeta) - \log_v(\|z,\zeta\|)) \;=\; 0\;.$$

If $\zeta \in \mathbb{H}_{\text{Berk}}$, then $u_\nu(z,\zeta)$ is lower semicontinuous on $\mathbb{P}^1_{\text{Berk}}$ and is bounded below. If $\zeta \in \mathbb{P}^1(K)$, then $u_\nu(z,\zeta)$ is lower semicontinuous on $\mathbb{P}^1_{\text{Berk}} \setminus \{\zeta\}$ with $u_\nu(\zeta,\zeta) = -\infty$, and there is a neighborhood of ζ on which $u_\nu(z,\zeta) = \log_v(\|z,\zeta\|)$.

Furthermore, for each $z \in \mathbb{P}^1_{\text{Berk}}$ and each path $[y,z]$, we have

(6.6) $$\liminf_{t \to z} u_\nu(t,\zeta) \;=\; \liminf_{\substack{t \to z \\ t \in \mathbb{H}_{\text{Berk}}}} u_\nu(t,\zeta) \;=\; \lim_{\substack{t \to z \\ t \in [y,z]}} u_\nu(t,\zeta) \;=\; u_\nu(z,\zeta)\;.$$

6.3. POTENTIAL FUNCTIONS ATTACHED TO PROBABILITY MEASURES

REMARK 6.13. In the terminology of §A.2, (6.6) asserts in particular that $u_\nu(z,\zeta)$ is "strongly lower semicontinuous".

PROOF. We first show that under the given hypotheses, $u_\nu(z,\zeta)$ is well-defined. Write $E = \mathrm{supp}(\nu)$, and consider the decomposition

$$\delta(z,w)_\zeta \;=\; \frac{\|z,w\|}{\|z,\zeta\|\,\|w,\zeta\|}\;.$$

Inserting this into the definition of $u_\nu(z,\zeta)$, we see that

$$(6.7) \quad u_\nu(z,\zeta) \;=\; \int_E -\log_v \|z,w\|\,d\nu(w) + \int_E \log_v \|z,\zeta\|\,d\nu(w)$$
$$+ \int_E \log_v \|w,\zeta\|\,d\nu(w)\;.$$

Here the second integral is $\log_v \|z,\zeta\|$, since the integrand does not involve w. The third integral is a finite constant C_ζ, since if either $\zeta \in \mathbb{H}_{\mathrm{Berk}}$ or $\zeta \in \mathbb{P}^1(K)$ and $\zeta \notin E$, then $\log_v \|w,\zeta\|$ is a bounded, continuous function of $w \in E$. Thus

$$(6.8) \quad u_\nu(z,\zeta) \;=\; \int_E -\log_v \|z,w\|\,d\nu(w) + \log_v \|z,\zeta\| + C_\zeta\;.$$

Note that the first term belongs to $[0,\infty]$, and the second to $[-\infty,0]$. If $\zeta \in \mathbb{H}_{\mathrm{Berk}}$, the second term is finite for all z. If $\zeta \in \mathbb{P}^1(K)$ and $\zeta \notin E$, then for $z \notin E$ the first term is finite, while for $z \in E$ the second term is finite. Thus $u_\nu(z,\zeta)$ is well-defined in $[-\infty,\infty]$ for all z.

To show that $u_\nu(z,\zeta)$ is continuous on $\mathbb{P}^1_{\mathrm{Berk}} \setminus \mathrm{supp}(\nu)$, note that for each $z_0 \notin \mathrm{supp}(\nu)$, since $\|z,w\|$ is continuous and nonzero off the diagonal and z_0 has a closed neighborhood disjoint from $\mathrm{supp}(\nu)$, the integral in (6.8) is continuous and bounded at $z = z_0$. Furthermore $\log_v \|z,\zeta\|$ is continuous as a function valued in $[-\infty,0]$, since $\|x,y\|$ is continuous as a function of each variable separately. Hence $u_\nu(z,\zeta)$ is continuous off $\mathrm{supp}(\nu)$.

We next establish the lower semicontinuity. If $\zeta \in \mathbb{H}_{\mathrm{Berk}}$, put $W_0 = \mathbb{P}^1_{\mathrm{Berk}}$; if $\zeta \in \mathbb{P}^1(K)$, let W_0 be any compact neighborhood of $\mathrm{supp}(\nu)$ which does not contain ζ. Let M be a constant such that $-\log_v(\delta(x,y)_\zeta) \geq -M$ for all $x,y \in W_0 \times W_0$. By Proposition A.3, $u_\nu(z,\zeta)$ coincides with

$$\sup\left\{ \int_{W_0} g(z,w)\,d\nu(w) \,:\, g \in \mathcal{C}(W_0 \times W_0),\, -M \leq g \leq -\log_v \delta(z,w)_\zeta \right\}.$$

Since W_0 is compact, each function $\int g(z,w)\,d\nu(w)$ is continuous on W_0. By Lemma A.2, $u_\nu(z,\zeta)$ is lower semicontinuous on W_0. If $\zeta \in \mathbb{H}_{\mathrm{Berk}}$, it follows that $u_\nu(z,\zeta)$ is lower semicontinuous on $\mathbb{P}^1_{\mathrm{Berk}}$. If $\zeta \in \mathbb{P}^1(K)$, then for each $z \neq \zeta$, we can choose W_0 so as to contain z in its interior, so $u_\nu(z,\zeta)$ is lower semicontinuous on $\mathbb{P}^1_{\mathrm{Berk}} \setminus \{\zeta\}$.

If $\zeta \in \mathbb{H}_{\mathrm{Berk}}$, the representation of $-\log_v(\delta(x,y)_\zeta)$ in (4.31) shows that $u_\nu(z,\zeta)$ is bounded below and achieves its minimum at $z = \zeta$. For any fixed

$w \neq \zeta$, when $z \to \zeta$ we have $-\log_v(\|z,w\|/\|\zeta,w\|) \to 0$. If $\zeta \notin \operatorname{supp}(\nu)$, the convergence is uniform for $w \in \operatorname{supp}(\nu)$. Rewriting (6.7) as

$$(6.9) \qquad u_\nu(z,\zeta) = \int_E -\log_v(\|z,w\|/\|\zeta,w\|) \, d\nu(w) + \log_v \|z,\zeta\| \, ,$$

we obtain (6.5) in this case.

If $\zeta \in \mathbb{P}^1(K)$, then $\lim_{z \to \zeta} \|z,\zeta\| = \|\zeta,\zeta\| = 0$, and by hypothesis $\zeta \notin E$, so (6.7) shows that $\lim_{z \to \zeta} u_\nu(z,\zeta) = u_\nu(\zeta,\zeta) = -\infty$. Furthermore, $\|w,\zeta\|$ is continuous and nonzero for $w \in E$, so

$$r := \min_{w \in E} \|w,\zeta\| > 0 \, .$$

Put $U = \mathcal{B}(\zeta, r)^-$. Then for all $z \in U$ and $w \in E$, the ultrametric inequality shows that $\|z,w\| = \|w,\zeta\|$. By (6.7), $u_\nu(z,\zeta) = \log_v(\|z,\zeta\|)$ for all $z \in U$. In particular, (6.5) holds.

We now prove (6.6). Fixing z, we first show that for each path $[y,z]$,

$$(6.10) \qquad \lim_{\substack{t \to z \\ t \in [y,z]}} u_\nu(t,\zeta) = u_\nu(z,\zeta) \, .$$

If $z \in \mathbb{H}_{\mathrm{Berk}}$, this is trivial: $u_\nu(x,\zeta) \in \mathrm{BDV}(\mathbb{P}^1_{\mathrm{Berk}})$ by Example 5.22, so $u_\nu(x,\zeta)$ is continuous when restricted to any finite subgraph $\Gamma \subset \mathbb{P}^1_{\mathrm{Berk}}$. Suppose $z \in \mathbb{P}^1(K)$. Let $y \neq z$ be near enough to z that ζ_{Gauss} does not lie on the path $[y,z]$, and let t approach z along $[y,z]$. For each $w \in E$, $\|t,w\|$ decreases monotonically to $\|z,w\|$, and thus $-\log_v \|t,w\|$ increases monotonically to $-\log_v \|z,w\|$. By the Monotone Convergence Theorem, $\int_E -\log_v \|t,w\| \, d\nu(w)$ converges to $\int_E -\log_v \|z,w\| \, d\nu(w)$. Likewise, $\log_v \|t,\zeta\|$ converges to $\log_v \|z,\zeta\|$. At most one of $\int_E -\log_v \|z,w\| \, d\nu(w)$ and $\log_v \|z,\zeta\|$ is infinite, so it follows from (6.8) that

$$\lim_{\substack{t \to z \\ t \in [y,z]}} u_\nu(t,\zeta) = u_\nu(z,\zeta)$$

as desired.

Formula (6.6) follows formally from this. Write $f(x)$ for $u_\nu(x,\zeta)$. When $\zeta \in \mathbb{H}_{\mathrm{Berk}}$ or when $z \neq \zeta$, then $f(x)$ is lower semicontinuous at z. It follows from (6.10), trivial inequalities, and (6.4) that

$$f(z) = \liminf_{\substack{t \to z \\ t \in [y,z]}} f(t) \geq \liminf_{\substack{t \to z \\ t \in \mathbb{H}_{\mathrm{Berk}}}} f(t) \geq \liminf_{t \to z} f(t) \geq f(z) \, ,$$

so equality must hold throughout. These inequalities also hold when $\zeta \in \mathbb{P}^1(K)$ and $z = \zeta$, since in that case $f(\zeta) = -\infty$ and $f(t) = \log_v(\|t,\zeta\|)$ in a neighborhood of ζ. $\square$

REMARK 6.14. The potential function $u_\nu(z,\zeta)$ need not be continuous on $\operatorname{supp}(\nu)$. For an example, take $K = \mathbb{C}_p$, let $\mathcal{O}$ be the ring of integers of K, and let $\mathfrak{m}$ be its maximal ideal. Let $\{\alpha_n\}_{n \in \mathbb{N}}$ be a set of representatives for the residue field $\tilde{K} = \mathcal{O}/\mathfrak{m}$, and put $\nu = \sum 2^{-n} \delta_{\alpha_n}$. Then $\operatorname{supp}(\nu) =$

$\{\alpha_n\}_{n\in\mathbb{N}}\cup\{\zeta_{\text{Gauss}}\}$, since each neighborhood of ζ_{Gauss} contains infinitely many α_n, and there are no other points of $\mathbb{P}^1_{\text{Berk}}$ with that property. It is easy to see that $u_\nu(\zeta_{\text{Gauss}},\infty) = 0$, while $u_\nu(\alpha_n,\infty) = \infty$ for each α_n.

For another example, see [88, Example 4.1.29, p. 217]. That example describes a set $E \subset \mathbb{P}^1(K)$ with positive capacity, such that if μ^* is the equilibrium distribution of E relative to $\zeta = \infty$, then $u_{\mu^*}(z,\infty)$ is not continuous on E.

In the classical theory, the two main facts about potential functions are Maria's theorem and Frostman's theorem. We will now establish their analogues for the Berkovich line.

THEOREM 6.15 (Maria). *Let ν be a probability measure supported on $\mathbb{P}^1_{\text{Berk}}\backslash\{\zeta\}$. If there is a constant $M < \infty$ such that $u_\nu(z,\zeta) \le M$ on $\text{supp}(\nu)$, then $u_\nu(z,\zeta) \le M$ for all $z \in \mathbb{P}^1_{\text{Berk}}\backslash\{\zeta\}$.*

PROOF. Put $E = \text{supp}(\nu)$ and fix $z \in \mathbb{P}^1_{\text{Berk}}\backslash(E \cup \{\zeta\})$. Since $\delta(z,w)_\zeta$ is continuous as a function of w, there is a point $\overline{z} \in E$ such that $\delta(z,\overline{z})_\zeta \le \delta(z,w)_\zeta$ for all $w \in E$. By the ultrametric inequality, for each $w \in E$

$$\delta(\overline{z},w)_\zeta \le \max(\delta(z,\overline{z})_\zeta, \delta(z,w)_\zeta) = \delta(z,w)_\zeta .$$

Hence

$$u_\nu(z,\zeta) = \int -\log_v \delta(z,w)_\zeta \, d\nu(w) \le \int -\log_v \delta(\overline{z},w)_\zeta \, d\nu(w)$$
$$= u_\nu(\overline{z},\zeta) \le M . \qquad \square$$

The following lemma asserts that sets of capacity 0 are 'small' in a measure-theoretic sense.

LEMMA 6.16. *If $f \subset \mathbb{P}^1_{\text{Berk}}\backslash\{\zeta\}$ is a set of capacity 0, then $\nu(f) = 0$ for any probability measure ν supported on $\mathbb{P}^1_{\text{Berk}}\backslash\{\zeta\}$ such that $I_\zeta(\nu) < \infty$.*

PROOF. Recall that $\text{supp}(\nu)$ is compact. After scaling the Hsia kernel (if necessary), we can assume that $\delta(x,y)_\zeta \le 1$ for all $x,y \in \text{supp}(\nu)$. If $\nu(f) > 0$, then $\nu(e) > 0$ for some compact subset e of f, and $\eta := (1/\nu(e))\cdot\nu|_e$ is a probability measure on e. It follows that

$$I_\zeta(\eta) = \iint_{e\times e} -\log_v \delta(x,y)_\zeta \, d\eta(x)d\eta(y)$$
$$\le \frac{1}{\nu(e)^2}\iint -\log_v \delta(x,y)_\zeta \, d\nu(x)d\nu(y) < \infty ,$$

contradicting the fact that f (and hence e) has capacity zero. $\square$

COROLLARY 6.17. *Let $\{f_n\}_{n\ge 1}$ be a countable collection of Borel sets in $\mathbb{P}^1_{\text{Berk}}\backslash\{\zeta\}$ such that each f_n has capacity 0. Put $f = \bigcup_{n=1}^\infty f_n$. Then f has capacity 0.*

PROOF. If $\gamma_\zeta(f) > 0$, there is a probability measure ν supported on f such that $I_\zeta(\nu) < \infty$. Since each f_n is Borel measurable, so is f, and

$$\sum_{n=1}^{\infty} \nu(f_n) \geq \nu(f) = 1 \ .$$

Hence $\nu(f_n) > 0$ for some n, contradicting Lemma 6.16. □

Recall that an F_σ set is a countable union of compact sets.

Anticipating the uniqueness of the equilibrium distribution, as before, we define the *equilibrium potential* $u_E(z,\zeta)$ of a set E of positive capacity to be the potential function associated to the equilibrium measure $\mu_\zeta = \mu_{E,\zeta}$.

THEOREM 6.18 (Frostman). *Let $E \subset \mathbb{P}^1_{\text{Berk}} \backslash \{\zeta\}$ be a compact set of positive capacity, with equilibrium measure μ_ζ. Then the equilibrium potential $u_E(z,\zeta) = u_{\mu_\zeta}(z,\zeta)$ satisfies:*

(A) $u_E(z,\zeta) \leq V_\zeta(E)$ *for all $z \in \mathbb{P}^1_{\text{Berk}} \backslash \{\zeta\}$.*
(B) $u_E(z,\zeta) = V_\zeta(E)$ *for all $z \in E$, except possibly on an F_σ set $f \subset E$ of capacity 0.*
(C) $u_E(z,\zeta)$ *is continuous at each point z_0 where $u_E(z_0,\zeta) = V_\zeta(E)$.*

PROOF. First, using a quadraticity argument, we will show $u_E(z,\zeta) \geq V_\zeta(E)$ on E except on the (possibly empty) set f. Then, we will show that $u_E(z,\zeta) \leq V_\zeta(E)$ on the support of the equilibrium measure $\mu = \mu_\zeta$. Since $u_E(z,\zeta) = u_\mu(z,\zeta)$, it will follow from Maria's theorem that $u_E(z,\zeta) \leq V_\zeta(E)$ for all $z \neq \zeta$.

Put
$$\begin{aligned} f &= \{z \in E : u_E(z,\zeta) < V_\zeta(E)\} \ , \\ f_n &= \{z \in E : u_E(z,\zeta) \leq V_\zeta(E) - 1/n\} \ , \quad \text{for } n = 1,2,3,\ldots. \end{aligned}$$

Since $u_E(z,\zeta)$ is lower semicontinuous, each f_n is closed, hence compact, so f is an F_σ set. By Corollary 6.17, $\gamma_\zeta(f) = 0$ if and only if $\gamma_\zeta(f_n) = 0$ for each n.

Suppose $\gamma_\zeta(f_n) > 0$ for some n; then there is a probability measure σ supported on f_n such that $I_\zeta(\sigma) < \infty$. On the other hand, since

$$V_\zeta(E) = \iint_{E \times E} -\log_v \delta(z,w)_\zeta \, d\mu(w) d\mu(z) = \int_E u_E(z,\zeta) \, d\mu(z) \ ,$$

there is a point $q \in \text{supp}(\mu) \subseteq E$ with $u_E(q,\zeta) \geq V_\zeta(E)$. Since $u_E(z,\zeta)$ is lower semicontinuous, there is a neighborhood U of q on which $u_E(z,\zeta) > V_\zeta(E) - 1/(2n)$. After shrinking U if necessary, we can assume that its closure $\overline{U}$ is disjoint from f_n. Put $e_n = E \cap U$, so that $q \in e_n$. By the definition of $\text{supp}(\mu)$, it follows that $M := \mu(U) = \mu(e_n) > 0$. Define a measure σ_1 of total mass 0 on E by

$$\sigma_1 = \begin{cases} M \cdot \sigma & \text{on } f_n \ , \\ -\mu & \text{on } e_n \ , \\ 0 & \text{elsewhere} \ . \end{cases}$$

We claim that $I_\zeta(\sigma_1)$ is finite. Indeed

$$\begin{aligned}
I_\zeta(\sigma_1) &= M^2 \cdot \iint_{f_n \times f_n} -\log_v \delta(z,w)_\zeta \, d\sigma(z) d\sigma(w) \\
&\quad - 2M \cdot \iint_{f_n \times e_n} -\log_v \delta(z,w)_\zeta \, d\sigma(z) d\mu(w) \\
&\quad + \iint_{e_n \times e_n} -\log_v \delta(z,w)_\zeta \, d\mu(z) d\mu(w) \; .
\end{aligned}$$

The first integral is finite by hypothesis. The second is finite because e_n and $\overline{U}$ are disjoint, so $-\log_v \delta(z,w)_\zeta$ is bounded on $e_n \times f_n$. The third is finite because $I_\zeta(\mu)$ is finite.

For each $0 \le t \le 1$, the measure $\mu_t := \mu + t\sigma_1$ is a probability measure on E. By an expansion like the one above,

$$\begin{aligned}
I_\zeta(\mu_t) - I_\zeta(\mu) &= 2t \cdot \int_E u_E(z,\zeta) \, d\sigma_1(z) + t^2 \cdot I_\zeta(\sigma_1) \\
&\le 2t \cdot ((V_\zeta(E) - 1/n) - (V_\zeta(E) - 1/(2n))) \cdot M + t^2 \cdot I_\zeta(\sigma_1) \\
&= (-M/n) \cdot t + I_\zeta(\sigma_1) \cdot t^2 \; .
\end{aligned}$$

For sufficiently small $t > 0$, the right side is negative. This contradicts the fact that μ gives the minimum value of $I_\zeta(\nu)$ for all probability measures ν supported on E. It follows that $\gamma_\zeta(f_n) = 0$ and hence that $\gamma_\zeta(f) = 0$.

The next step involves showing that $u_E(z,\zeta) \le V_\zeta(E)$ for all $z \in \text{supp}(\mu)$. If $u_E(q,\zeta) > V_\zeta(E)$ for some $q \in \text{supp}(\mu)$, take $\varepsilon > 0$ such that $u_E(q,\zeta) > V_\zeta(E) + \varepsilon$. The lower semicontinuity of $u_E(z,\zeta)$ shows that there is a neighborhood U of q on which $u_E(z,\zeta) > V_\zeta(E) + \varepsilon$. Put $e = U \cap E$. Since $q \in \text{supp}(\mu)$, $T := \mu(e) > 0$. By Lemma 6.16, $\mu(f) = 0$. Therefore, since $u_E(z,\zeta) \ge V_\zeta(E)$ on $E \backslash f$,

$$\begin{aligned}
V_\zeta(E) &= \int_e u_E(z,\zeta) \, d\mu(z) + \int_{E \backslash e} u_E(z,\zeta) \, d\mu(z) \\
&\ge T \cdot (V_\zeta(E) + \varepsilon) + (1-T) \cdot V_\zeta(E) = V_\zeta(E) + T\varepsilon \; ,
\end{aligned}$$

which is obviously false.

Thus $u_E(z,\zeta) \le V_\zeta(E)$ on $\text{supp}(\mu)$, and Maria's theorem implies that $u_E(z,\zeta) \le V_\zeta(E)$ for all z.

The final assertion, that $u_E(z,\zeta)$ is continuous at each point z_0 where $u_E(z_0,\zeta) = V_\zeta(E)$, is now trivial. By lower semicontinuity,

$$\liminf_{z \to z_0} u_E(z,\zeta) \ge u_E(z_0,\zeta) = V_\zeta(E) \; .$$

On the other hand, since $u_E(z,\zeta) \le V_\zeta(E)$ for all z,

$$\limsup_{z \to z_0} u_E(z,\zeta) \le V_\zeta(E) = u_E(z_0,\zeta) \; . \qquad \square$$

COROLLARY 6.19. *For any compact set $E \subset \mathbb{P}^1_{\text{Berk}} \backslash \{\zeta\}$ of positive capacity and any probability measure ν supported on E,*

$$\inf_{z \in E} u_\nu(z, \zeta) \leq V_\zeta(E) \leq \sup_{z \in E} u_\nu(z, \zeta) .$$

PROOF. This follows immediately from the identity

$$\int_E u_\nu(z, \zeta) \, d\mu(z) = \iint_{E \times E} -\log_v \delta(z, w)_\zeta \, d\nu(w) d\mu(z)$$
$$= \int_E u_E(w, \zeta) d\nu(w) = V_\zeta(E) .$$

Here the second equality follows from Tonelli's theorem, and the third from Frostman's theorem and the fact that if f is the exceptional set on which $u_E(z, \zeta) < V_\zeta(E)$, then $\nu(f) = 0$ by Lemma 6.16. □

We will now show that adjoining or removing a set of capacity 0 from a given set F does not change its capacity. This is a consequence of the following quantitative bound:

PROPOSITION 6.20. *Let $\{F_m\}_{m \geq 1}$ be a countable collection of sets contained in $\mathbb{P}^1_{\text{Berk}} \backslash \{\zeta\}$, and put $F = \bigcup_{m=1}^\infty F_m$. Suppose there is an $R < \infty$ such that $\delta(x, y)_\zeta \leq R$ for all $x, y \in F$ (or, equivalently, that there is a ball $\mathcal{B}(a, R)_\zeta$ containing F). Then*

$$(6.11) \qquad \frac{1}{V_\zeta(F) + \log_v(R) + 1} \leq \sum_{m=1}^\infty \frac{1}{V_\zeta(F_m) + \log_v(R) + 1} .$$

PROOF. First suppose $R < 1$. In this case we will show that

$$(6.12) \qquad \frac{1}{V_\zeta(F)} \leq \sum_{m=1}^\infty \frac{1}{V_\zeta(F_m)} .$$

Note that our hypothesis implies that $V_\zeta(F) \geq -\log_v(R) > 0$.

If $\gamma_\zeta(F) = 0$, then $\gamma_\zeta(F_m) = 0$ for each m, so $V_\zeta(F) = V_\zeta(F_m) = \infty$ and (6.12) is trivial. Hence we can assume that $\gamma_\zeta(F) > 0$. Let $E \subset F$ be a compact set with $\gamma_\zeta(E) > 0$, and let μ be its equilibrium distribution. For each m, put $E_m = E \cap F_m$. Then

$$(6.13) \qquad \sum_{m=1}^\infty \mu(E_m) \geq \mu(E) = 1 .$$

We claim that for each m, $V_\zeta(E)/V_\zeta(E_m) \geq \mu(E_m)$. If $\mu(E_m) = 0$, there is nothing to prove, so suppose $\mu(E_m) > 0$. There are compact sets $e_{m,1} \subset e_{m,2} \subset \cdots \subset E_m$ such that $\lim_{i \to \infty} \mu(e_{m,i}) = \mu(E_m)$ and $\lim_{i \to \infty} V_\zeta(e_{m,i}) = V_\zeta(E_m)$. Without loss of generality, we can assume $\mu(e_{m,i}) > 0$ for each i. Put $\nu_{m,i} = (1/\mu(e_{m,i})) \cdot \mu|_{e_{m,i}}$. Applying Corollary 6.19, we see that

$$(6.14) \qquad \sup_{z \in e_{m,i}} u_{\nu_{m,i}}(z, \zeta) \geq V_\zeta(e_{m,i}) .$$

But also, since $-\log_v \delta(z,w)_\zeta \geq -\log_v(R) > 0$ on E, for each $z \in e_{m,i}$

$$\frac{1}{\mu(e_{m,i})} u_E(z,\zeta) = \frac{1}{\mu(e_{m,i})} \int_E -\log_v \delta(z,w)_\zeta \, d\mu(w)$$
(6.15)
$$\geq \int_{e_{m,i}} -\log_v \delta(z,w)_\zeta \, d\nu_{m,i}(w) = u_{\nu_{m,i}}(z,\zeta) .$$

Because $\mu(e_{m,i}) > 0$, Theorem 6.18 shows that there exist points $z \in e_{m,i}$ where $u_E(z,\zeta) = V_\zeta(E)$. Combining (6.14) and (6.15) gives

$$V_\zeta(E)/\mu(e_{m,i}) \geq V_\zeta(e_{m,i}) .$$

Transposing terms and letting $i \to \infty$ shows that $V_\zeta(E)/V_\zeta(E_m) \geq \mu(E_m)$.

By (6.13), we have

$$\frac{1}{V_\zeta(E)} \leq \sum_{m=1}^{\infty} \frac{1}{V_\zeta(E_m)} .$$

But E can be chosen so that $V_\zeta(E)$ is arbitrarily close to $V_\zeta(F)$ and also so that as many of the $V_\zeta(E_m)$ as we wish are arbitrarily close to $V_\zeta(F_m)$. Taking a limit over such E, we obtain (6.12).

The general case follows by scaling the Hsia kernel. Replace $\delta(x,y)_\zeta$ by $\delta'(x,y)_\zeta = 1/(q_v R) \cdot \delta(x,y)_\zeta$, so $\delta'(x,y)_\zeta \leq 1/q_v$ for all $x, y \in F$. For each $E \subset F$, this changes $V_\zeta(E)$ to $V'_\zeta(E) = V_\zeta(E) + \log_v(R) + 1$. Applying (6.12) to $V'_\zeta(F)$ and the $V'_\zeta(F_m)$ gives (6.11). □

COROLLARY 6.21. *Let $e \subset \mathbb{P}^1_{\text{Berk}} \backslash \{\zeta\}$ have capacity 0. Then for any $F \subset \mathbb{P}^1_{\text{Berk}} \backslash \{\zeta\}$,*

$$\gamma_\zeta(F \cup e) = \gamma_\zeta(F \backslash e) = \gamma_\zeta(F) .$$

PROOF. By Proposition 6.20, if F and e are contained in a ball $\mathcal{B}(a,R)_\zeta$ with $R < \infty$, then

$$\frac{1}{V_\zeta(F \cup e) + \log_v(R) + 1} \leq \frac{1}{V_\zeta(F) + \log_v(R) + 1} + \frac{1}{V_\zeta(e) + R + 1}$$
$$= \frac{1}{V_\zeta(F) + \log_v(R) + 1} ,$$

giving $V_\zeta(F) \leq V_\zeta(F \cup e)$. Since the opposite inequality is trivial, $V_\zeta(F) = V_\zeta(F \cup e)$. The general case follows from this, by replacing F and e by $F \cap \mathcal{B}(a,R)_\zeta$ and $e \cap \mathcal{B}(a,R)_\zeta$, fixing a center a, and letting $R \to \infty$. Each compact subset of $\mathbb{P}^1_{\text{Berk}} \backslash \{\zeta\}$ is contained in $\mathcal{B}(a,R)_\zeta$ for some R, so

$$\gamma_\zeta(F \cup e) = \lim_{R \to \infty} \gamma_\zeta((F \cup e) \cap \mathcal{B}(a,R)_\zeta)$$
$$= \lim_{R \to \infty} \gamma_\zeta(F \cap \mathcal{B}(a,R)_\zeta) = \gamma_\zeta(F) .$$

For the equality $\gamma_\zeta(F \backslash e) = \gamma_\zeta(F)$, apply what has been just shown to e and $F \backslash e$, noting that $(F \backslash e) \cup e = F \cup e$:

$$\gamma_\zeta(F \backslash e) = \gamma_\zeta(F \cup e) = \gamma_\zeta(F) .$$
□

6.4. The transfinite diameter and the Chebyshev constant

As in the classical theory, there are two other important capacitary functions: the *transfinite diameter* and the *Chebyshev constant*. The existence of these quantities, and the fact that for compact sets they are equal to the logarithmic capacity, will be established below.

Fix $\zeta \in \mathbb{P}^1_{\text{Berk}}$, and let $E \subset \mathbb{P}^1_{\text{Berk}} \backslash \{\zeta\}$.

To define the transfinite diameter, for each $n = 2, 3, 4, \ldots$ put

$$d_n(E)_\zeta = \sup_{x_1,\ldots,x_n \in E} \left(\prod_{i \neq j} \delta(x_i, x_j)_\zeta \right)^{1/(n(n-1))}.$$

Note that unlike the classical case, $d_n(E)_\zeta$ can be nonzero even if some of the x_i coincide; this will happen, for example, if $E = \{a\}$ for a point of type II, III, or IV.

LEMMA 6.22. *The sequence $\{d_n(E)_\zeta\}$ is monotone decreasing.*

PROOF. Fix $n \geq 2$. If $d_{n+1}(E)_\zeta = 0$, then certainly $d_n(E)_\zeta \geq d_{n+1}(E)_\zeta$. Otherwise, take ε with $0 < \varepsilon < d_{n+1}(E)_\zeta$. Then there exist $z_1, \ldots, z_{n+1} \in E$ such that

$$\prod_{\substack{i,j=1 \\ i \neq j}}^{n+1} \delta(z_i, z_j)_\zeta \geq (d_{n+1}(E)_\zeta - \varepsilon)^{(n+1)n}.$$

By the definition of $d_n(E)_\zeta$, for each $k = 1, \ldots, n+1$, we have

$$(d_n(E)_\zeta)^{n(n-1)} \geq \prod_{\substack{i \neq j \\ i,j \neq k}} \delta(z_i, z_j)_\zeta.$$

Taking the product over all k, we see that

$$(d_n(E)_\zeta)^{(n+1)n(n-1)} \geq \left(\prod_{\substack{i,j=1 \\ i \neq j}}^{n+1} \delta(z_i, z_j)_\zeta \right)^{n-1} = (d_{n+1}(E)_\zeta - \varepsilon)^{(n+1)n(n-1)}.$$

Since $\varepsilon > 0$ is arbitrary, $d_n(E)_\zeta \geq d_{n+1}(E)_\zeta$. □

Set

$$d_\infty(E)_\zeta = \lim_{n \to \infty} d_n(E)_\zeta.$$

We call $d_\infty(E)_\zeta$ the *transfinite diameter* of E with respect to ζ.

We will define three variants of the Chebyshev constant. For each positive integer n and $a_1, \ldots, a_n \in \mathbb{P}^1_{\text{Berk}} \backslash \{\zeta\}$ (which need not be distinct), define the '*pseudo-polynomial*'

$$P_n(x; a_1, \ldots, a_n) = \prod_{i=1}^n \delta(x, a_i)_\zeta.$$

This definition is modeled on the absolute value of a usual polynomial, in factored form.

6.4. THE TRANSFINITE DIAMETER AND THE CHEBYSHEV CONSTANT

Put $\|P_n(x;a_1,\ldots,a_n)\|_E = \sup_{x\in E} P_n(x;a_1,\ldots,a_n)$ and let
$$\mathrm{CH}^*_n(E)_\zeta = \inf_{a_1,\ldots,a_n \in E} (\|P_n(x;a_1,\ldots,a_n)\|_E)^{1/n},$$
$$\mathrm{CH}^a_n(E)_\zeta = \inf_{a_1,\ldots,a_n \in \mathbb{P}^1(K)\backslash\{\zeta\}} (\|P_n(x;a_1,\ldots,a_n)\|_E)^{1/n},$$
$$\mathrm{CH}_n(E)_\zeta = \inf_{a_1,\ldots,a_n \in \mathbb{P}^1_{\mathrm{Berk}}\backslash\{\zeta\}} (\|P_n(x;a_1,\ldots,a_n)\|_E)^{1/n}.$$

We will show that the three quantities
$$\mathrm{CH}^*(E)_\zeta = \lim_{n\to\infty} \mathrm{CH}^*_n(E)_\zeta,$$
$$\mathrm{CH}^a(E)_\zeta = \lim_{n\to\infty} \mathrm{CH}^a_n(E)_\zeta,$$
$$\mathrm{CH}(E)_\zeta = \lim_{n\to\infty} \mathrm{CH}_n(E)_\zeta$$

exist and that $\mathrm{CH}^*(E)_\zeta \geq \mathrm{CH}^a(E)_\zeta = \mathrm{CH}(E)_\zeta$. They will be called the *restricted Chebyshev constant*, the *algebraic Chebyshev constant*, and the *unrestricted Chebyshev constant*, respectively.

The proofs that $\mathrm{CH}^*(E)_\zeta$, $\mathrm{CH}^a(E)_\zeta$, and $\mathrm{CH}(E)_\zeta$ are well-defined are all similar; we give the argument only for $\mathrm{CH}(E)_\zeta$. Put $\alpha = \liminf_n \mathrm{CH}_n(E)_\zeta$. If $\alpha = \infty$, then $\mathrm{CH}(E)_\zeta = \infty$ and there is nothing to prove. Otherwise, fix $\varepsilon > 0$. Then there are an integer N and points $a_1,\ldots,a_N \in \mathbb{P}^1_{\mathrm{Berk}}\backslash\{\zeta\}$ such that
$$M_N := \sup_{x\in E} P_N(x;a_1,\ldots,a_N) \leq (\alpha+\varepsilon)^N.$$

Let $\{b_j\} = \{a_1,\ldots,a_N,a_1,\ldots,a_N,\ldots\}$ be the sequence which cyclically repeats $\{a_1,\ldots,a_N\}$. Put $M_0 = 1$, and for each $r = 1,\ldots,N-1$, put $M_r = \|P_r(x;a_1,\ldots,a_r)\|_E$. Since $M_N \leq (\alpha+\varepsilon)^N$ is finite, each M_r is finite as well. Put $M = \max_{0\leq r \leq N} M_r/(\alpha+\varepsilon)^r$. For each n we can write $n = qN + r$ with $q, r \in \mathbb{Z}$ and $0 \leq r < N$. Then
$$\mathrm{CH}_n(E)_\zeta \leq (\|P_n(x;b_1,\ldots,b_n)\|)^{1/N}$$
$$\leq (M_N^q \cdot M_r)^{1/n} \leq M^{1/n} \cdot (\alpha+\varepsilon).$$

It follows that $\limsup_{n\to\infty} \mathrm{CH}_n(E)_\zeta \leq \alpha + \varepsilon$. Since $\varepsilon > 0$ was arbitrary,
$$\limsup_{n\to\infty} \mathrm{CH}_n(E)_\zeta \leq \alpha = \liminf_{n\to\infty} \mathrm{CH}_n(E)_\zeta$$
and $\mathrm{CH}(E)_\zeta = \lim_{n\to\infty} \mathrm{CH}_n(E)_\zeta$ exists.

Clearly $\mathrm{CH}(E)^*_\zeta \geq \mathrm{CH}(E)_\zeta$ and $\mathrm{CH}(E)^a_\zeta \geq \mathrm{CH}(E)_\zeta$. We will now show:

LEMMA 6.23. $\mathrm{CH}^a(E)_\zeta = \mathrm{CH}(E)_\zeta$.

PROOF. If $\mathrm{CH}(E)_\zeta = \infty$, there is nothing to prove. Otherwise put $\alpha = \mathrm{CH}(E)_\zeta$ and fix $\varepsilon > 0$. Then there are an N and $a_1,\ldots,a_N \in \mathbb{P}^1_{\mathrm{Berk}}$ such that $\|P_N(x;a_1,\ldots,a_N)\|_E \leq (\alpha+\varepsilon)^N$. Take $\eta > 0$. We claim that for each i, there is an a'_i of type I such that $\delta(x,a'_i)_\zeta \leq (1+\eta)\delta(x,a_i)_\zeta$ for all $x \in \mathbb{P}^1_{\mathrm{Berk}}\backslash\{\zeta\}$. If a_i of type I, put $a'_i = a_i$. If a_i is not of type I, let $r_i = \mathrm{diam}_\zeta(a_i)$; then there is an a'_i of type I in the ball $\mathcal{B}(a_i, r_i(1+\eta))_\zeta$. If

$x \notin \mathcal{B}(a_i, r_i(1+\eta))_\zeta$, then $\delta(x, a'_i)_\zeta = \delta(x, a_i)_\zeta$ by the ultrametric inequality. If $x \in \mathcal{B}(a_i, r_i(1+\eta))_\zeta$, note that $\delta(x, a_i)_\zeta \geq \delta(a_i, a_i)_\zeta = r_i$, so again by the ultrametric inequality,

$$\delta(x, a'_i)_\zeta \leq (1+\eta)r_i \leq (1+\eta)\delta(x, a_i)_\zeta ,$$

and in either case the claim is true. It follows that

$$\begin{aligned} \|P_N(x; a'_1, \ldots, a'_N)\|_E &\leq \|P_N(x; a_1, \ldots, a_N)\|_E \cdot (1+\eta)^N \\ &\leq (\alpha+\varepsilon)^N \cdot (1+\eta)^N . \end{aligned}$$

Since η is arbitrary, $\mathrm{CH}_N^a(E)_\zeta \leq \alpha + \varepsilon$, so $\mathrm{CH}^a(E)_\zeta = \mathrm{CH}(E)_\zeta$. □

When E is compact, all the capacitary functions above are equal:

THEOREM 6.24. *For any $\zeta \in \mathbb{P}^1_{\mathrm{Berk}}$ and any compact set $E \subset \mathbb{P}^1_{\mathrm{Berk}} \backslash \{\zeta\}$,*

$$\gamma_\zeta(E) = d_\infty(E)_\zeta = \mathrm{CH}(E)_\zeta = \mathrm{CH}^*(E)_\zeta = \mathrm{CH}^a(E)_\zeta .$$

PROOF. We have seen that $\mathrm{CH}^a(E)_\zeta = \mathrm{CH}(E)_\zeta \leq \mathrm{CH}^*(E)_\zeta$. We now show that $\gamma_\zeta(E) \leq \mathrm{CH}^a(E)_\zeta$, $\mathrm{CH}^*(E)_\zeta \leq d_\infty(E)_\zeta$, and $d_\infty(E)_\zeta \leq \gamma_\zeta(E)$.

I. $\gamma_\zeta(E) \leq \mathrm{CH}^a(E)_\zeta$.

If $\gamma_\zeta(E) = 0$, there is nothing to prove. Suppose $\gamma_\zeta(E) > 0$. We will show that $\gamma_\zeta(E) \leq \mathrm{CH}_n^a(E)_\zeta$ for each n. Fix n and take $\varepsilon > 0$; let $a_1, \ldots, a_n \in \mathbb{P}^1(K) \backslash \{\zeta\}$ be points such that

$$\sup_{z \in E} \prod_{i=1}^n \delta(z, a_i)_\zeta \leq (\mathrm{CH}_n^a(E)_\zeta + \varepsilon)^n .$$

Any finite set of type I points has capacity 0. By Corollary 6.21, replacing E by $E \cup \{a_1, \ldots, a_n\}$ does not change its capacity. Let ν be the probability measure on E with a point mass $1/n$ at each a_i. Then for each $z \in E$,

$$-\log_v(\mathrm{CH}_n^a(E)_\zeta + \varepsilon) \leq (1/n) \sum_{i=1}^n -\log_v \delta(z, a_i)_\zeta = u_\nu(z, \zeta) .$$

By Corollary 6.19, $\inf_{z \in E} u_\nu(z, \zeta) \leq V_\zeta(E)$. Thus $-\log_v(\mathrm{CH}_n^a(E)_\zeta + \varepsilon) \leq V_\zeta(E)$, which gives $\gamma_\zeta(E) \leq \mathrm{CH}_n^a(E)_\zeta + \varepsilon$. Since $\varepsilon > 0$ was arbitrary, we conclude by letting $n \to \infty$ that $\gamma_\zeta(E) \leq \mathrm{CH}^a(E)_\zeta$.

II. $\mathrm{CH}^*(E)_\zeta \leq d_\infty(E)_\zeta$.

We will show that $d_{n+1}(E)_\zeta \geq \mathrm{CH}_n^*(E)_\zeta$ for each n. Since E is compact, the sup defining $d_{n+1}(E)_\zeta$ is achieved: there are points $z_1, \ldots, z_{n+1} \in E$ for which

$$d_{n+1}(E)_\zeta^{(n+1)n} = \prod_{i \neq j} \delta(z_i, z_j)_\zeta .$$

If $z_1, \ldots, z_n$ are fixed, then by the definition of $d_{n+1}(E)_\zeta$,

$$H(z; z_1, \ldots, z_n) := \prod_{1 \leq i < j \leq n} \delta(z_i, z_j)_\zeta \cdot \prod_{i=1}^n \delta(z, z_i)_\zeta$$

achieves its maximum for $z \in E$ at $z = z_{n+1}$. But by the definition of $\mathrm{CH}_n^*(E)_\zeta$,

$$\max_{z \in E} \prod_{i=1}^{n} \delta(z, z_i)_\zeta \geq (\mathrm{CH}_n^*(E)_\zeta)^n .$$

Thus,

$$d_{n+1}(E)_\zeta^{(n+1)n/2} \geq \prod_{1 \leq i < j \leq n} \delta(z_i, z_j)_\zeta \cdot (\mathrm{CH}_n^*(E)_\zeta)^n .$$

A similar inequality holds if $\{z_1, \ldots, z_n\}$ is replaced by $\{z_1, \ldots, z_{n+1}\} \backslash \{z_k\}$ for each $k = 1, \ldots, n+1$. Multiplying these inequalities together, we find

$$\begin{aligned}
(d_{n+1}(E)_\zeta^{(n+1)n/2})^{n+1} &\geq (\mathrm{CH}_n^*(E)_\zeta)^{(n+1)n} \cdot \prod_{k=1}^{n+1} \Big(\prod_{\substack{1 \leq i < j \leq n+1 \\ i,j \neq k}} \delta(z_i, z_j)_\zeta \Big) \\
&= (\mathrm{CH}_n^*(E)_\zeta)^{(n+1)n} \cdot \Big(\prod_{1 \leq i < j \leq n+1} \delta(z_i, z_j)_\zeta \Big)^{n-1} \\
&= (\mathrm{CH}_n^*(E)_\zeta)^{(n+1)n} \cdot (d_{n+1}(E)_\zeta^{(n+1)n/2})^{n-1} .
\end{aligned}$$

Canceling terms and taking roots gives $d_{n+1}(E)_\zeta \geq \mathrm{CH}_n^*(E)_\zeta$.

III. $d_\infty(E)_\zeta \leq \gamma_\zeta(E)$.

We will show that $d_\infty(E)_\zeta \leq \gamma_\zeta(E) + \varepsilon$ for each $\varepsilon > 0$.

Fix $\varepsilon > 0$. By Corollary 6.9, there is a strict closed affinoid neighborhood W containing E such that $\gamma_\zeta(W) \leq \gamma_\zeta(E) + \varepsilon$. Then $\partial W = \{a_1, \ldots, a_m\}$, where each a_k is of type II. Put $r_k = \mathrm{diam}_\zeta(a_k)$ and put $r = \min(r_k) > 0$. Let $r_\zeta : E \to \partial W_\zeta$ be the retraction map which takes each x in E to the last point on the path from x to ζ which belongs to W; thus $r_\zeta(E) \subset \{a_1, \ldots, a_m\}$.

Note that if $x, y \in E$ and $r_\zeta(x) = a_k$, $r_\zeta(y) = a_\ell$, then $\delta(x, y)_\zeta \leq \delta(a_k, a_\ell)_\zeta$. Indeed, if $k = \ell$, then $x, y \in \mathcal{B}(a_k, r_k)_\zeta$, so $\delta(x, y)_\zeta \leq r_k = \delta(a_k, a_\ell)_\zeta$. If $k \neq \ell$, then $\delta(x, y)_\zeta = \delta(a_k, a_\ell)_\zeta$, since the paths from x and y to ζ encounter a_k, a_ℓ before they meet, so $\delta(x, y)_\zeta = \delta(a_k, a_\ell)_\zeta$.

Fix n. Since E is compact, there are points $z_1, \ldots, z_n \in E$ such that

$$(d_n(E)_\zeta)^{n(n-1)} = \prod_{\substack{i,j=1 \\ i \neq j}}^{n} \delta(z_i, z_j)_\zeta .$$

For each a_k, let m_k be the number of points z_i for which $r_\zeta(z_i) = a_k$, and let ν be the probability measure on W given by $\nu = \sum_{k=1}^{m} (m_k/n) \delta_{a_k}(z)$. Then

$I_\zeta(\nu) \geq V_\zeta(W)$. It follows that

$$-(1 - \frac{1}{n}) \log_v d_n(E)_\zeta = \frac{1}{n^2} \sum_{i \neq j} -\log_v \delta(z_i, z_j)_\zeta$$

$$\geq \frac{1}{n^2} \sum_{i \neq j} -\log_v \delta(r_\zeta(z_i), r_\zeta(z_j))_\zeta$$

$$= I_\zeta(\nu) + \frac{1}{n^2} \sum_{k=1}^m m_k \log_v(r_k)$$

$$\geq V_\zeta(W) + (1/n) \log_v(r) \ .$$

Letting $n \to \infty$, we see that $-\log_v d_\infty(E)_\zeta \geq V_\zeta(W)$. Thus, $d_\infty(E)_\zeta \leq \gamma_\zeta(W) \leq \gamma_\zeta(E) + \varepsilon$ as claimed. □

REMARK 6.25. The various capacitary functions need not be equal if E is not compact. For example, suppose $K = \mathbb{C}_p$ and ζ is of type I, and let $E = \mathbb{P}^1(\overline{\mathbb{Q}}) \backslash \{\zeta\}$ where $\overline{\mathbb{Q}}$ is the algebraic closure of $\mathbb{Q}$ in $\mathbb{C}_p$. Then $d_\zeta(E) = \text{CH}_\zeta(E) = \infty$ because $\delta(x,y)_\zeta$ is unbounded on E, while $\gamma_\zeta(E) = 0$ because E is countable.

We next prove the following result comparing our capacities with those defined in [**88**]. Suppose $K = \mathbb{C}_p$ and $\zeta \in \mathbb{P}^1(K)$. For each set $F \subset \mathbb{P}^1(K) \backslash \{\zeta\}$ which is bounded away from ζ, we let $\gamma_\zeta^R(F)$ be the *outer capacity* of F, in the sense of [**88**, §4.3]. Since the generalized Hsia kernel $\delta(x,y)_\zeta$ restricts to the canonical distance $[x,y]_\zeta$ on $\mathbb{P}^1(K) \times \mathbb{P}^1(K)$, it follows from the definitions in [**88**] that

$$\gamma_\zeta^R(F) = \lim_{n \to \infty} \inf_{a_1, \ldots, a_n \in \mathbb{P}^1(K) \backslash \{\zeta\}} (\|P_n(x; a_1, \ldots, a_n)\|_F)^{1/n} = \text{CH}^a(F)_\zeta \ ,$$

where as before

$$P_n(x; a_1, \ldots, a_n) = \prod_{i=1}^n \delta(x, a_i)_\zeta \ .$$

COROLLARY 6.26. *Suppose $K = \mathbb{C}_p$ and $\zeta \in \mathbb{P}^1(K)$. If E is a compact subset of $\mathbb{P}^1_{\text{Berk}} \backslash \{\zeta\}$ such that $E \cap \mathbb{P}^1(K)$ is dense in E, then*

$$\gamma_\zeta(E) = \gamma_\zeta^R(E \cap \mathbb{P}^1(K)) \ .$$

PROOF. Since $\gamma_\zeta(E) = \lim_{n \to \infty} \text{CH}_n^a(E)_\zeta$ by Theorem 6.24, it suffices to note that since $P_n(x; a_1, \ldots, a_n)$ is continuous in x on $\mathbb{P}^1 \backslash \{\zeta\}$ and $E' = E \cap \mathbb{P}^1(K)$ is dense in E, we have

$$(\|P_n(x; a_1, \ldots, a_n)\|_E)^{1/n} = (\|P_n(x; a_1, \ldots, a_n)\|_{E'})^{1/n}$$

for all n and all $a_1, \ldots, a_n \in \mathbb{P}^1(K) \backslash \{\zeta\}$. □

REMARK 6.27. In particular, the hypothesis of Corollary 6.26 is satisfied when E is a *strict closed affinoid* in $\mathbb{P}^1_{\text{Berk}}$, so that $E \cap \mathbb{P}^1(K)$ is an "RL-domain" in the sense of [**88**].

6.5. The Fekete-Szegö theorem

As an application of capacity theory on $\mathbb{P}^1_{\text{Berk}}$, in this section we will give Berkovich versions of the adelic Fekete and Fekete-Szegö theorems (Theorems 6.3.1 and 6.3.2 of [88]), in the case $\mathcal{C} = \mathbb{P}^1$ and $\mathfrak{X} = \{\infty\}$. Berkovich versions of the Fekete and Fekete-Szegö theorems for more general $\mathfrak{X}$ will be given in §7.8.

In the discussion that follows, we assume familiarity with basic concepts from algebraic number theory.

Let k be a global field (that is, a finite extension of $\mathbb{Q}$, or of $\mathbb{F}_p(T)$ for some finite field $\mathbb{F}_p$). For each place v of k, let k_v be the completion of k at v, let $\overline{k}_v$ be an algebraic closure of k_v, and let $\mathbb{C}_v$ denote the completion of $\overline{k}_v$. It is well known that $\mathbb{C}_v$ is algebraically closed. If v is non-Archimedean, let q_v be the order of the residue field of k_v and normalize the absolute value on $\mathbb{C}_v$ so that $|x|_v = q_v^{-\text{ord}_v(x)}$; put $N_v = 1$. If v is Archimedean, let $|x|_v = |x|$ be the usual absolute value on $\mathbb{C}_v \cong \mathbb{C}$. Put $N_v = 1$ if $k_v \cong \mathbb{R}$, and put $N_v = 2$ if $k_v \cong \mathbb{C}$. Then for $0 \neq \kappa \in k$, the product formula reads

$$\prod_v |\kappa|_v^{N_v} = 1 .$$

Let $\overline{k}$ be an algebraic closure of k. For each v, fix an embedding of $\overline{k}$ in $\mathbb{C}_v$ and view this embedding as an identification. Let $\mathbb{P}^1_{\text{Berk},v}$ denote the Berkovich projective line over $\mathbb{C}_v$ (which we take to mean $\mathbb{P}^1(\mathbb{C})$ if v is Archimedean). Using the identification above, view $\mathbb{P}^1(\overline{k})$ as a subset of $\mathbb{P}^1_{\text{Berk},v}$.

Fixing a coordinate system on $\mathbb{P}^1$, define a "compact Berkovich adelic set" (relative to the point $\zeta = \infty$) to be a set of the form

$$\mathbb{E} = \prod_v E_v ,$$

where E_v is a nonempty compact subset of $\mathbb{P}^1_{\text{Berk},v} \backslash \{\infty\}$ for each place v of k and where E_v is the closed disc $\mathcal{D}(0,1) = \mathbb{P}^1_{\text{Berk},v} \backslash \mathcal{B}(\infty,1)^-$ for all v outside a finite set of places S containing the Archimedean places. For each non-Archimedean v, let $\gamma_\infty(E_v)$ be the capacity defined in (6.1), using the weight q_v and absolute value $|x|_v$ above. If v is Archimedean and $k_v \cong \mathbb{R}$, let $\gamma_\infty(E_v)$ be the usual logarithmic capacity $\gamma(E_v)$; if $k_v \cong \mathbb{C}$, let $\gamma_\infty(E_v) = \gamma(E_v)^2$. The *logarithmic capacity* (relative to ∞) of $\mathbb{E}$, denoted $\gamma_\infty(\mathbb{E})$, is defined to be

$$\gamma_\infty(\mathbb{E}) = \prod_v \gamma_\infty(E_v) .$$

It is well-defined since $\gamma_\infty(E_v) = \gamma_\infty(\mathcal{D}(0,1)) = 1$ for each $v \notin S$.

Each continuous automorphism of $\mathbb{C}_v/k_v$ extends (e.g., through its evident action on nested sequences of discs) to an automorphism of $\mathbb{P}^1_{\text{Berk},v}$; we will say that E_v is k_v-*symmetric* if it is stable under all such automorphisms. We will say that $\mathbb{E}$ is k-symmetric if E_v is k_v-symmetric for all v.

A "Berkovich adelic neighborhood" of $\mathbb{E}$ is a set $\mathbb{U}$ of the form

$$\mathbb{U} = \prod_v U_v ,$$

where for each $v \in S$, U_v is an open neighborhood of E_v in $\mathbb{P}^1_{\text{Berk},v}$, and for each $v \notin S$, either U_v is an open neighborhood of E_v or $U_v = E_v = \mathcal{D}(0,1)$.

Our generalization of the Fekete and Fekete-Szegö theorems is as follows:

THEOREM 6.28. *Let $\mathbb{E}$ be a k-symmetric compact Berkovich adelic set (relative to ∞).*

(A) *If $\gamma_\infty(\mathbb{E}) < 1$, then there is a Berkovich adelic neighborhood $\mathbb{U}$ of $\mathbb{E}$ such that the set of points of $\mathbb{P}^1(\overline{k})$, all of whose conjugates are contained in $\mathbb{U}$, is finite.*

(B) *If $\gamma_\infty(\mathbb{E}) > 1$, then for every Berkovich adelic neighborhood $\mathbb{U}$ of $\mathbb{E}$, the set of points of $\mathbb{P}^1(\overline{k})$, all of whose conjugates are contained in $\mathbb{U}$, is infinite.*

This theorem is proved in [**88**, §6.3] with $\gamma_\infty(\mathbb{E}) = \prod_v \gamma_\infty(E_v)$ replaced by $\prod_v \gamma_\infty^R(E_v \cap \mathbb{P}^1(\mathbb{C}_v))$, assuming that $E_v \cap \mathbb{P}^1(\mathbb{C}_v)$ is *algebraically capacitable* for all non-Archimedean v (we refer to [**88**] for the definition of algebraic capacitability). It is enough for our purposes to know that for non-Archimedean v, the intersections of strict closed affinoids with $\mathbb{P}^1(\mathbb{C}_v)$ (which are called "RL-domains" in [**88**]) are algebraically capacitable (see [**88**, Theorem 4.3.3]) and are both open and closed in $\mathbb{P}^1(\mathbb{C}_v)$. We will reduce Theorem 6.28 to Theorems 6.3.1 and 6.3.2 of [**88**], using Corollary 6.9.

PROOF OF THEOREM 6.28. By the remark following Corollary 6.26, together with Theorems 6.3.1 and 6.3.2 of [**88**], the result is true when each non-Archimedean E_v is a strict closed affinoid. Let S be a finite set of places of k containing all the Archimedean places and all the places where $E_v \neq \mathcal{D}(0,1)$.

For part (A), for each non-Archimedean $v \in S$, Corollary 6.9 allows us to replace E_v by a strict closed affinoid F_v with $E_v \subseteq F_v$, for which $\gamma_\zeta(F_v)$ is arbitrarily close to $\gamma_\zeta(E_v)$. Since F_v has only finitely many k_v-conjugates, after replacing it by the intersection of those conjugates, we can assume without loss of generality that F_v is k_v-symmetric. Let $\mathbb{F}$ be the corresponding set, with the F_v chosen in such a way that $\gamma_\zeta(\mathbb{F})$ is strictly less than 1.

For each v, put $F_v^R = F_v \cap \mathbb{P}^1(\mathbb{C}_v)$, and let $\mathbb{F}^R = \prod_v F_v^R$. Then $\mathbb{F}^R$ is an adelic set of the type studied in [**88**]. Since F_v^R is dense in F_v for each v, Corollary 6.26 shows that

$$\gamma_\infty^R(\mathbb{F}^R) = \gamma_\infty(\mathbb{F}) < 1 .$$

By Theorem 6.3.1 of [**88**], there is an adelic neighborhood $\mathbb{U}^R = \prod_v U_v^R$ of $\mathbb{F}^R$ for which there are only finitely many $\alpha \in \mathbb{P}^1(\overline{k})$ whose $\text{Gal}(\overline{k}/k)$-conjugates all belong to U_v^R, for each v. (Here, $\text{Gal}(\overline{k}/k)$ denotes the group of all automorphisms of $\overline{k}$ fixing k). More precisely, the proof of Theorem 6.3.1 of [**88**] constructs a polynomial $f \in k[z]$ such that for each $v \in S$

$$\sup_{z \in F_v^R} |f(z)|_v < 1 , \qquad U_v^R = \{z \in \mathbb{P}^1(\mathbb{C}_v) : |f(z)|_v < 1\}$$

and for each $v \notin S$ (so $F_v = E_v = \mathcal{D}(0,1)$)

$$\sup_{z \in F_v^R} |f(z)|_v = 1, \qquad U_v^R = F_v^R = \{z \in \mathbb{P}^1(\mathbb{C}_v) : |f(z)|_v \le 1\} .$$

We can now conclude the proof as follows. For each $v \in S$, put $U_v = \{z \in \mathbb{P}^1_{\text{Berk},v} : [f]_z < 1\}$, and for each $v \notin S$, put $U_v = \{z \in \mathbb{P}^1_{\text{Berk},v} : [f]_z \le 1\}$. Then for each v, $U_v \cap \mathbb{P}^1(\mathbb{C}_v) = U_v^R$, and $F_v \subset U_v$ since F_v^R is dense in F_v. If $v \in S$, then U_v is an open neighborhood of F_v, and if $v \notin S$, then $U_v = F_v = \mathcal{D}(0,1)$. Thus $\mathbb{U} := \prod_v U_v$ is a Berkovich adelic neighborhood of $\mathbb{F}$, having the property required in the theorem.

In fact, the only points $\alpha \in \overline{k}$ whose conjugates all belong to $\mathbb{U}$ are the roots of $f(z)$. Suppose α and its conjugates belong to $\mathbb{U}$. If L is the normal closure of α over k and p^m is the degree of inseparability of L/k, then $\beta := N_{L/k}(f(\alpha))$ belongs to k and

$$\prod_v |\beta|_v^{N_v} = \prod_v \Big(\prod_{\sigma \in \text{Aut}(L/k)} |f(\sigma(\alpha))|_v^{N_v} \Big)^{p^m} < 1 ,$$

so $\beta = 0$ by the product formula. This means some $\sigma(\alpha)$ is a root of f, so α is a root as well.

For part (B), given an adelic neighborhood $\mathbb{U}$ of $\mathbb{E}$, Corollary 6.9 allows us to replace each E_v for $v \in S$ by a slightly larger strict closed affinoid F_v which is still contained in U_v. As before, we can assume without loss of generality that F_v is k_v-symmetric and hence that the corresponding adelic set $\mathbb{F}$ is k-symmetric. We now apply Theorem 6.3.2 of [**88**] to the set $\mathbb{F}^R = \prod_v (F_v \cap \mathbb{P}^1(\mathbb{C}_v))$ and its neighborhood $\mathbb{U}^R = \prod_v (U_v \cap \mathbb{P}^1(\mathbb{C}_v))$, noting that

$$\gamma_\infty^R(\mathbb{F}^R) = \gamma_\infty(\mathbb{F}) > 1 .$$

That theorem produces infinitely many $\alpha \in \mathbb{P}^1(\overline{k})$ whose conjugates all belong to $U_v \cap \mathbb{P}^1(\mathbb{C}_v)$ for each v. Since $\mathbb{U}^R \subset \mathbb{U}$, these α meet the requirements of Theorem 6.28(B). $\square$

REMARK 6.29. Compact subsets of $\mathbb{P}^1(\mathbb{C}_v)$ are algebraically capacitable. Note that a closed affinoid is not compact as a subset of $\mathbb{P}^1(\mathbb{C}_v)$, but both closed affinoids and compact subsets of $\mathbb{P}^1_{\text{Berk}}$ are intersections of compact subsets of $\mathbb{P}^1_{\text{Berk}}$ with $\mathbb{P}^1(\mathbb{C}_v)$. Thus Theorem 6.28 generalizes both the RL-domain and compact cases of the Fekete and Fekete-Szegö theorems from [**88**].

6.6. Notes and further references

An accessible reference for classical capacity theory over $\mathbb{C}$ is [**79**]; see also [**59**] and [**96**]. A non-Archimedean version of capacity theory (without Berkovich spaces) is developed in [**33**]; see [**88**] for a more systematic version. In his thesis, Thuillier [**94**] has developed parts of capacity theory in the context of Berkovich curves.

In [**88**], capacity theory is developed relative to an arbitrary finite set $\mathfrak{X}$ (instead of just relative to a point ζ). The Fekete and Fekete-Szegö theorems proved in [**88**] hold with $\mathbb{P}^1$ replaced by an arbitrary curve C/k and with ζ replaced by any k-symmetric finite set $\mathfrak{X} \subset C(\overline{k})$. A Berkovich space version of this multi-center Fekete-Szegö theorem is given in §7.8.

In [**44**] (which we learned about only after the present chapter was written), Bálint Farkas and Béla Nagy study the relationship between abstract notions of transfinite diameter, Chebyshev constant, and capacity on locally compact Hausdorff topological spaces with respect to a positive, symmetric, lower semicontinuous kernel function. Some of the results in the present chapter could be proved by referencing their general theory. Conversely, several of the arguments given in this chapter hold more or less without modification in the more general context considered in [**44**].

CHAPTER 7

Harmonic functions

In this chapter, we develop a theory of harmonic functions on $\mathbb{P}^1_{\text{Berk}}$. We first define what it means for a function to be harmonic on an open set. We then establish analogues of the Maximum Principle, Poisson's formula, Harnack's theorem, the Riemann Extension Theorem, and the theory of Green's functions. We characterize harmonic functions as limits of logarithms of norms of rational functions and show their stability under uniform limits and pullbacks by meromorphic functions.

As a byproduct of the theory, in §7.6 we obtain the uniqueness of the equilibrium measure, asserted in Chapter 6. As arithmetic applications, in §7.8 we give a Berkovich space version of the multi-center Fekete-Szegö theorem, and in §7.9 we give a Berkovich space generalization of Bilu's equidistribution theorem for a rather general class of adelic heights.

7.1. Harmonic functions

In this section we will define harmonic functions, give examples, and establish a basic result about their structure (Proposition 7.12).

Recall that a domain is a connected open subset U of $\mathbb{P}^1_{\text{Berk}}$ and that if U is a domain and $f \in \text{BDV}(U)$, then the Laplacian of f is the measure $\Delta_U(f) = \Delta_{\overline{U}}(f)|_U$ (Definition 5.16).

However, if V is open but not connected, we have not defined $\Delta_{\overline{V}}(f)$ or $\Delta_V(f)$, even if $f \in \text{BDV}(U)$ for each component U of V. This is because different components may share boundary points, creating unwanted interaction between the Laplacians on the components. For example, let $\zeta_{\text{Gauss}} \in \mathbb{P}^1_{\text{Berk}}$ be the Gauss point, and consider the open set $U = \mathbb{P}^1_{\text{Berk}} \backslash \{0, \zeta_{\text{Gauss}}\}$. It has one distinguished component $U_0 = \mathcal{B}(0,1)^- \backslash \{0\}$ and infinitely many other components $U_i = \mathcal{B}(a_i, 1)_0^-$. They all have ζ_{Gauss} as a boundary point. The function f which is $-\log_v(\delta(x,0)_\infty)$ in U_0 and is 0 elsewhere has Laplacian $\Delta_{\overline{U_0}}(f) = \delta_0 - \delta_{\zeta_{\text{Gauss}}}$, while $\Delta_{\overline{U_i}}(f) = 0$ for $i \neq 0$.

DEFINITION 7.1. If U is a domain, a function $f : U \to \mathbb{R}$ is *strongly harmonic* in U if it is continuous on U, belongs to $\text{BDV}(U)$, and satisfies
$$\Delta_U(f) \equiv 0\ .$$

DEFINITION 7.2. If U is an arbitrary open set, then $f : U \to \mathbb{R}$ is *harmonic* in U if for each $x \in U$, there is a domain $V_x \subset U$ with $x \in V_x$, on which f is strongly harmonic.

We denote the space of harmonic functions on U by $\mathcal{H}(U)$.

If f and g are harmonic (or strongly harmonic) in V, then so are $-f$ and $a \cdot f + b \cdot g$, for any $a, b \in \mathbb{R}$. In view of the Laplacians on $\mathbb{P}^1_{\text{Berk}}$ computed in §5.4, combined with Proposition 5.26, we have the following examples of strongly harmonic functions:

EXAMPLE 7.3. Each constant function $f(x) = C$ on a domain U is strongly harmonic.

EXAMPLE 7.4. For fixed $y, \zeta \in \mathbb{P}^1_{\text{Berk}}$, $-\log_v(\delta(x,y)_\zeta)$ is strongly harmonic on each component of $\mathbb{P}^1_{\text{Berk}} \backslash \{y, \zeta\}$.

EXAMPLE 7.5. If $0 \neq g \in K(T)$ has divisor $\text{div}(g) = \sum_{i=1}^m n_i(a_i)$, then $f(x) = -\log_v([g]_x)$ is strongly harmonic on $\mathbb{P}^1_{\text{Berk}} \backslash \{a_1, \ldots, a_m\}$.

EXAMPLE 7.6. Given a probability measure ν and a point $\zeta \notin \text{supp}(\nu)$, the potential function $u_\nu(z, \zeta)$ is strongly harmonic on each component of $\mathbb{P}^1_{\text{Berk}} \backslash (\text{supp}(\nu) \cup \{\zeta\})$.

In Example 7.15 below, we will give an example of a function which is harmonic on a domain U but not strongly harmonic. The following lemma clarifies the relation between being harmonic and strongly harmonic.

LEMMA 7.7.
- (A) If $U_1 \subset U_2$ are domains and f is strongly harmonic on U_2, then f is strongly harmonic on U_1.
- (B) If f is harmonic on an open set V and U is a subdomain of V with $\overline{U} \subset V$, then f is strongly harmonic on U.
- (C) If f is harmonic on V and $E \subset V$ is compact and connected, there is a subdomain $U \subset V$ containing E such that f is strongly harmonic on U.

In Corollary 7.11 below, we will see that if V is a domain, there is a countable exhaustion of V by subdomains on which f is strongly harmonic.

PROOF. For (A), note that $f \in \text{BDV}(U_1)$ and $\Delta_{U_1}(f) = \Delta_{U_2}(f)|_{U_1}$ by Proposition 5.26.

For (B), note that for each $x \in \overline{U}$ there is a connected neighborhood U_x of x contained in V such that f is strongly harmonic on U_x. Since $\overline{U}$ is compact, a finite number of the U_x cover $\overline{U}$, say $U_{x_1}, \ldots, U_{x_m}$. Put $W = \bigcup_{i=1}^m U_{x_i}$. Clearly W is open. It is also connected, since $\overline{U}$ and the U_{x_i} are connected, so it is a domain. By Proposition 5.27, $f|_W \in \text{BDV}(W)$ and $\Delta_W(f) \equiv 0$. In turn, by Proposition 5.26, $f|_U \in \text{BDV}(U)$ and $\Delta_U(f) \equiv 0$.

For (C), note that for each $x \in E$ there is a connected neighborhood $U_x \subset V$ of x on which f is strongly harmonic. Since E is compact, a finite number of the U_x cover E. Let U be the union of those neighborhoods. Then U is connected, and by the same argument as above, f is strongly harmonic on U. $\square$

A key observation is that the behavior of a harmonic function on a domain U is controlled by its behavior on a special subset.

DEFINITION 7.8. If U is a domain, the *main dendrite*[1] $D = D(U) \subset U$ is the set of all $x \in U$ belonging to paths between boundary points $y, z \in \partial U$.

Note that the main dendrite can be empty: that will be the case if and only if U has one or no boundary points, namely if U is a connected component of $\mathbb{P}^1_{\text{Berk}}\backslash\{\zeta\}$ for some $\zeta \in \mathbb{P}^1_{\text{Berk}}$ or if $U = \mathbb{P}^1_{\text{Berk}}$.

If the main dendrite is nonempty, it is an $\mathbb{R}$-tree which is finitely branched at each point. To see this, we first require the following lemma.

LEMMA 7.9. *Let $W \subset \mathbb{P}^1_{\text{Berk}}$ be a domain, let $x \in W$, and let $y \in \mathbb{P}^1_{\text{Berk}}\backslash W$. Then the unique path Γ from x to y contains some boundary point of W.*

PROOF. Let $W' = \mathbb{P}^1_{\text{Berk}}\backslash\overline{W}$. If $\Gamma \cap \partial W = \emptyset$, then $\Gamma = (\Gamma \cap W) \cup (\Gamma \cap W')$ is a union of two relatively open subsets, contradicting the fact that Γ is connected. Thus $\Gamma \cap \partial W \neq \emptyset$. □

PROPOSITION 7.10. *Let U be a domain in $\mathbb{P}^1_{\text{Berk}}$, and let D be the main dendrite of U. If D is nonempty, then:*

(A) *D is finitely branched at every point.*
(B) *D is a countable union of finite $\mathbb{R}$-trees, whose boundary points are all of type II.*

PROOF. For (A), fix $z \in D$, and choose a covering of ∂U by simple domains not containing z. Since ∂U is compact, a finite number of simple domains $W_1, \ldots, W_n$ will suffice. Let $W = W_1 \cup \cdots \cup W_n$, and let B be the finite set consisting of all boundary points of $W_1, \ldots, W_n$ in U. It follows from Lemma 7.9 that $D \cap (U\backslash W)$ coincides with the union P of all paths between z and points in B. This shows D is finitely branched at z, as desired.

For (B), we require a more "quantitative" argument[2]. Suppose D is nonempty. Fix $\zeta \in D$, and put $r_0 = \sup_{x \in \partial U}(\text{diam}_\zeta(x))$. Since ∂U is compact and $\text{diam}_\zeta(x) = \delta(x,x)_\zeta$ is upper semicontinuous (Proposition 4.10(A)), by [79, Theorem 2.1.2] the sup is achieved at some $x_0 \in \partial U$. Since $\text{diam}_\zeta(x) < \text{diam}_\zeta(\zeta)$ for each $x \in \mathbb{P}^1_{\text{Berk}}\backslash\{\zeta\}$, it follows that $r_0 < \text{diam}_\zeta(\zeta)$.

Let N be an integer large enough that $0 < 1/N < \text{diam}_\zeta(\zeta) - r_0$. For each integer $n \geq N$ and each $x \in \partial U$, let $r_{x,n}$ be a number which belongs to $|K^\times|$ and satisfies

$$\text{diam}_\zeta(x) + \frac{1}{n+1} < r_{x,n} < \text{diam}_\zeta(x) + \frac{1}{n}.$$

---

[1] The main dendrite of a domain is closely related to the notion of a *skeleton*, which plays a key role in the work of Berkovich [16, 20, 21, 19] and Thuillier [94]. (Note, however, that the skeleton of a formal model for $\mathbb{P}^1$ is a (compact) finite subgraph of $\mathbb{P}^1_{\text{Berk}}$, whereas the main dendrite of a simple domain U does not contain the boundary points of U and thus (when nonempty) is not compact.) The main dendrite of U is called the *arbre* (tree) associated to U in [81].

[2] Part (B) is closely related to the fact that any connected open subset of $\mathbb{P}^1_{\text{Berk}}$ is topologically "countable at infinity" (see [16]).

Since ∂U is compact, it can be covered by finitely many open balls $\mathcal{B}(x, r_{x,n})_\zeta^-$ (Proposition 4.10(D)), say the balls corresponding to $x_1, \ldots, x_{m_n}$. Since $\delta(x, y)_\zeta$ satisfies the ultrametric inequality (Proposition 4.10(C)), without loss of generality we can assume that these balls are pairwise disjoint. Let $V_n = \bigcup_{i=1}^{m_n} \mathcal{B}(x_i, r_{x_i,n})_\zeta^-$. Each ball $\mathcal{B}(x_i, r_{x_i,n})_\zeta^-$ has a single boundary point $z_{i,n}$, which is necessarily of type II and belongs to D if D is nonempty.

Let D_n be the smallest connected subset of $\mathbb{H}_{\text{Berk}}$ containing the points $z_{1,n}, \ldots, z_{m_n,n}$; then D_n is a finite $\mathbb{R}$-tree contained in U. If $y, z \in \partial U$ belong to the same ball $\mathcal{B}(x_i, r_{x_i,n})_\zeta^-$, then the path between them lies entirely in that ball, because a subset of $\mathbb{P}^1_{\text{Berk}}$ is connected if and only if it is path-connected and there is a unique path between any two points. If y and z belong to distinct balls $\mathcal{B}(x_i, r_{x_i,n})_\zeta^-$, $\mathcal{B}(x_j, r_{x_j,n})_\zeta^-$, the path between them is contained in those balls together with D_n. If we fix $x \in \partial U$ and let $n \to \infty$, then the balls $\mathcal{B}(x, r_{x,n})_\zeta^-$ will eventually omit any given point $z \in U$; this follows from Remark 4.12 and the fact that $\bigcap_n \mathcal{B}(x, r_{x,n})_\zeta^- = \mathcal{B}(x, \operatorname{diam}_\zeta(x))_\zeta$, since U is completely contained in $\mathbb{P}^1_{\text{Berk}} \backslash \mathcal{B}(x, \operatorname{diam}_\zeta(x))_\zeta$. If $n_2 > n_1$, then since $\delta(x, y)_\zeta$ satisfies the ultrametric inequality, our choice of the $r_{x,n}$ shows that V_{n_2} is contained in the interior of V_{n_1}. It follows that D_{n_1} is contained in the interior of D_{n_2}, and therefore as $n \to \infty$ the subgraphs D_n exhaust D. Thus

$$D = \bigcup_{n=N}^{\infty} D_n$$

and each D_n is a finite $\mathbb{R}$-tree whose boundary points are of type II. $\square$

Recall (Definition 2.27) that a strict simple domain is a simple domain whose boundary points are all of type II; strict simple domains coincide with strict open affinoid domains. From the proof of Proposition 7.10, we obtain the following useful result:

COROLLARY 7.11. *If $U \neq \mathbb{P}^1_{\text{Berk}}$ is a domain in $\mathbb{P}^1_{\text{Berk}}$, then U can be exhausted by a sequence $W_1 \subset W_2 \subset \cdots$ of strict simple domains with $\overline{W}_n \subset W_{n+1} \subset U$ for each n.*

PROOF. If the main dendrite D of U is nonempty, fix a point $\zeta \in D$; if D is empty, let $\zeta \in U$ be arbitrary. Let the sets V_n be as in the proof of Proposition 7.10(B). (Note that the construction of the V_n only requires that $\partial U \neq \emptyset$ and not that D is nonempty.)

Put $W_n = U \cap (\mathbb{P}^1_{\text{Berk}} \backslash \overline{V}_n)$. By construction $\overline{W}_n \subset W_{n+1}$, and the W_n exhaust U. Since W_n is a domain with finite boundary $\partial W_n = \partial V_n$ and each point of ∂W_n is of type II, it follows that W_n is a strict simple domain. $\square$

The connection between the main dendrite and harmonic functions is given by the following important result:

PROPOSITION 7.12. *Let f be harmonic on a domain U. If the main dendrite is empty, then f is constant on U; otherwise, f is constant along every path leading away from the main dendrite.*

PROOF. First suppose the main dendrite is nonempty, and let $x \in U \backslash D$. Fix a point $y_0 \in D$, and let w be the first point in D along the path from x to y_0. We claim that $f(x) = f(w)$.

To see this, let V be the connected component of $\mathbb{P}^1_{\text{Berk}} \backslash \{w\}$ containing x; then $V \subset U$ because x is not on D. By Lemma 7.7(B), f is strongly harmonic on V, and $\Delta_V(f) \equiv 0$. As $\partial V = \{w\}$, we have $\overline{V} = V \cup \{w\} \subset U$. By Proposition 5.25, $\Delta_{\overline{V}}(f)(\{w\}) = -\Delta_V(f)(V) = 0$. As $\Delta_V(f) = \Delta_{\overline{V}}(f)|_V$, it follows that $\Delta_{\overline{V}}(f) \equiv 0$. By Lemma 5.24, f is constant on $V \cap \mathbb{H}_{\text{Berk}}$. Since f is continuous on U, it must be constant on all of $\overline{V}$. Thus $f(x) = f(w)$.

Next suppose the main dendrite is empty. Then either $U = \mathbb{P}^1_{\text{Berk}}$ or U is a connected component of $\mathbb{P}^1_{\text{Berk}} \backslash \{\zeta\}$ for some $\zeta \in \mathbb{P}^1_{\text{Berk}}$. Fix $w \in U$, and let $x \in U$ be arbitrary. By the description of U above, there is a disc V with $\overline{V} \subset U$, for which both $x, w \in V$. The same argument as before shows that $f(x) = f(w)$. □

REMARK 7.13. Taken together, Propositions 7.12 and 7.10 show in particular that a harmonic function on a domain U is constant in all but finitely many tangent directions at every point $x \in U$.

REMARK 7.14. Proposition 7.12 shows that f is locally constant outside the main dendrite, for the weak topology.

EXAMPLE 7.15. Here is an example of a domain U and a function $f : U \to \mathbb{R}$ which is harmonic on U but not strongly harmonic.

Let $K = \mathbb{C}_p$, fix coordinates so that $\mathbb{P}^1_{\text{Berk}} = \mathbb{A}^1_{\text{Berk}} \cup \{\infty\}$, and take $U = \mathbb{P}^1_{\text{Berk}} \backslash \mathbb{Z}_p$. Then the main dendrite D is a rooted $\mathbb{R}$-tree whose root node is the Gauss point ζ_{Gauss} and whose other nodes are at the points corresponding to the discs $D(a, p^{-n})$ for $a, n \in \mathbb{Z}$, with $n \geq 1$ and $0 \leq a \leq p^n - 1$. The tree has p branches extending down from each node, and each edge has length 1.

To describe f, it suffices to give its values on the main dendrite. It will be continuous, with constant slope on each edge. Let $f(\zeta_{\text{Gauss}}) = 0$; we will recursively give its values on the other nodes. Suppose z_a is a node on which $f(z_a)$ has already been defined. If z_a is the root node, put $N_a = 0$; otherwise, let N_a be the slope of f on the edge entering z_a from above. Of the p edges extending down from z_a, choose two distinguished edges, and let $f(z)$ have slope $N_a + 1$ on one and slope -1 on the other, until the next node. On the $p - 2$ other edges, let $f(z)$ have the constant value $f(z_a)$ until the next node.

By construction, the sum of the slopes of f on the edges leading away from each node is 0, so f is harmonic on U. However, there are edges of D on which f has arbitrarily large slope; if Γ is an edge of D on which f has slope m_Γ, then $|\Delta_\Gamma(f)|(\Gamma) = 2|m_\Gamma|$. Thus $f \notin \text{BDV}(U)$, and f is not strongly harmonic on U.

7.2. The Maximum Principle

We will now show that harmonic functions on domains in $\mathbb{P}^1_{\text{Berk}}$ have many features in common with classical harmonic functions.

PROPOSITION 7.16 (Maximum Principle).
- (A) If f is a nonconstant harmonic function on a domain $U \subset \mathbb{P}^1_{\text{Berk}}$, then f does not achieve a maximum or a minimum value on U.
- (B) If f is harmonic on U and $\limsup_{x \to \partial U} f(x) \leq M$, then $f(x) \leq M$ for all $x \in U$; if $\liminf_{x \to \partial U} f(x) \geq m$, then $f(x) \geq m$ for all $x \in U$.

PROOF. It suffices to deal with the case of a maximum.

To prove (A), suppose f takes on its maximum at a point $x \in U$. Let $V \subset U$ be a subdomain containing x on which f is strongly harmonic. We will show that f is constant on V. It follows that $\{z \in U : f(z) = f(x)\}$ is both open and closed and hence equals U since U is connected, so f is constant on U.

By Proposition 7.12, we can assume that the main dendrite D of V is nonempty. Let T be the branch off of D containing x, and let w be the point where T attaches to D. By Proposition 7.12 again, we have $f(w) = f(x)$.

Let $\Gamma \subset D$ be an arbitrary finite subgraph with w in its interior. By the definition of the main dendrite, $r_{\overline{V},\Gamma}(\partial V)$ consists of the endpoints of Γ. Since $\Delta_{\overline{V}}(f)$ is supported on ∂V, $\Delta_\Gamma(f) = (r_{\overline{V},\Gamma})_*(\Delta_{\overline{V}}(f))$ is supported on the endpoints of Γ. By Proposition 3.11, $f|_\Gamma \in \text{CPA}(\Gamma)$. Let Γ_w be the connected component of $\{z \in \Gamma : f(z) = f(w)\}$ containing w. If $\Gamma_w \neq \Gamma$, let p be a boundary point of Γ_w in Γ. Since $f(p)$ is maximal, we must have $d_{\vec{v}} f(p) \leq 0$ for all tangent vectors $\vec{v} \in T_p(\Gamma)$. Since $f|_\Gamma \in \text{CPA}(\Gamma)$ and $f|_\Gamma$ is nonconstant near p, there must be some $\vec{v}$ with $d_{\vec{v}}(f)(p) < 0$. It follows that
$$\Delta_\Gamma(f)(p) = - \sum_{\vec{v} \in T_p(\Gamma)} d_{\vec{v}} f(p) > 0 \ .$$
This contradicts the fact that $\Delta_\Gamma(f)$ is supported on the endpoints of Γ, and we conclude that $\Gamma_w = \Gamma$.

Since Γ can be taken arbitrarily large, f must be constant on the whole main dendrite. By Proposition 7.12, it is constant everywhere on V.

Part (B) of the proposition is deduced as follows. Since the function $f^\# : \overline{U} \to \mathbb{R}$ defined by
$$f^\#(x) = \begin{cases} f(x) & \text{for } x \in U \ , \\ \limsup_{y \to x, y \in U} f(y) & \text{for } x \in \partial U \end{cases}$$
is upper semicontinuous, it attains its maximum value on $\overline{U}$. Since f is harmonic, the maximum value cannot occur on U, which implies that if $\limsup_{x \to \partial U} f(x) \leq M$, then $f^\#(x) \leq M$ for all $x \in \overline{U}$, and thus $f(x) \leq M$ for all $x \in U$. A similar argument using lower semicontinuity implies that if $\liminf_{x \to \partial U} f(x) \geq m$, then $f(x) \geq m$ for all $x \in U$. $\square$

7.2. THE MAXIMUM PRINCIPLE

There is an important strengthening of the Maximum Principle which allows us to ignore sets of capacity 0 in ∂U.

PROPOSITION 7.17 (Strong Maximum Principle). *Let f be harmonic on a domain $U \subset \mathbb{P}^1_{\text{Berk}}$. Assume either that f is nonconstant or that ∂U has positive capacity.*

If f is bounded above on U and there are a number M and a set $e \subset \partial U$ of capacity 0 such that $\limsup_{x \to z} f(x) \leq M$ for all $z \in \partial U \setminus e$, then $f(x) \leq M$ for all $x \in U$.

Similarly, if f is bounded below on U and there are a number m and a set $e \subset \partial U$ of capacity 0 such that $\liminf_{x \to z} f(x) \geq m$ for all $z \in \partial U \setminus e$, then $f(x) \geq m$ for all $x \in U$.

For the proof, we will need the existence of an 'Evans function'.

LEMMA 7.18. *Let $e \subset \mathbb{P}^1_{\text{Berk}}$ be a compact set of capacity 0, and let $\zeta \notin e$. Then there is a probability measure ν supported on e for which the potential function $u_\nu(z, \zeta)$ (which is defined on all of $\mathbb{P}^1_{\text{Berk}}$ as a function to $\mathbb{R} \cup \{\pm\infty\}$) satisfies*

$$\lim_{z \to x} u_\nu(z, \zeta) = \infty$$

for all $x \in e$. A function with this property is called an Evans function.

PROOF. By Theorem 6.24, the restricted Chebyshev constant $\text{CH}^*(e)_\zeta$ is 0. This means that for each $\epsilon > 0$, there are points $a_1, \ldots, a_N \in e$ for which the pseudo-polynomial

$$P_N(z; a_1, \ldots, a_N) = \prod_{i=1}^N \delta(z, a_i)_\zeta$$

satisfies $(\|P_N(z; a_1, \ldots, a_N)\|_e)^{1/N} < \varepsilon$.

Let $q = q_v$ be the base of the logarithm $\log_v(t)$, and take $\varepsilon = 1/q^{2^k}$ for $k = 1, 2, 3, \ldots$. For each k we obtain points $a_{k,1}, \ldots, a_{k,N_k}$ such that the corresponding pseudo-polynomial $P_{N_k}(z)$ satisfies $(\|P_{N_k}\|_e)^{1/N_k} < 1/q^{2^k}$. Put

$$\nu = \sum_{k=1}^\infty \frac{1}{2^k} \left(\frac{1}{N_k} \sum_{i=1}^{N_k} \delta_{a_{k,i}} \right).$$

Then ν is a probability measure supported on e, and

$$u_\nu(z, \zeta) = \sum_{k=1}^\infty \frac{1}{2^k} \left(-\frac{1}{N_k} \log_v(P_{N_k}(z)) \right).$$

Since $P_{N_k}(z)$ is continuous, there is a neighborhood U_k of e on which $P_{N_k}(z)^{1/N_k} < 1/q^{2^k}$. Without loss of generality, we can assume that we have $U_1 \supset U_2 \supset U_3 \cdots$. On U_m, for each $k = 1, \ldots, m$ we have

$$-\frac{1}{2^k} \cdot \frac{1}{N_k} \log_v(P_{N_k}(z)) \geq -\frac{1}{2^k} \log_v(1/q^{2^k}) = 1.$$

Hence $u_\nu(z,\zeta) \geq m$ on U_m. Letting $m \to \infty$, we see that $\lim_{z \to x} u_\nu(z,\zeta) = \infty$ for each $x \in e$. □

We can now prove Proposition 7.17, using the same argument as in the classical case.

PROOF. It suffices to deal with the case of a maximum.

Suppose that f is bounded above on U and that there are a number M and a set $e \subset \partial U$ of capacity 0 such that $\limsup_{x \to z} f(x) \leq M$ for all $z \in \partial U \backslash e$ but that for some $x_0 \in U$ we have

$$f(x_0) > M .$$

We are assuming that either f is nonconstant or that ∂U has positive capacity. If f is nonconstant, there is a point $\zeta \in U$ with $f(\zeta) \neq f(x_0)$, and after interchanging x_0 and ζ if necessary, we can assume $f(x_0) > f(\zeta)$. If ∂U has positive capacity, then for each $z \in \partial U \backslash e$ we have $\limsup_{x \to z} f(x) \leq M$, and since $U \backslash e$ is nonempty, there is again a point $\zeta \in U$ with $f(\zeta) < f(x_0)$.

Fix $M_1 > M$ with $f(x_0) > M_1 > f(\zeta)$, and let

$$W = \{x \in U : f(x) > M_1\} , \qquad W' = \{x \in U : f(x) < M_1\} .$$

Then W and W' are open, by the continuity of f on U. Furthermore, $\zeta \in W'$ and $W' \cap \overline{W} = \emptyset$.

Let V be the connected component of W containing x_0. Then V is open and is itself a domain in $\mathbb{P}^1_{\text{Berk}}$. Consider its closure $\overline{V}$. If $\overline{V} \subset U$, then each $y \in \partial V$ would have a neighborhood containing points of V and points of $U \backslash W$. By the continuity of f, this implies $f(y) = M_1$. However, $f(x_0) > M_1$. This violates the Maximum Principle for V.

Hence $e_1 = \overline{V} \cap \partial U$ is nonempty; clearly it is compact. By the definition of W, e_1 is contained in the exceptional set e where $\limsup_{x \to y} f(x) > M$. Since e has capacity 0, e_1 has capacity 0.

Our construction has assured that $\zeta \notin e_1$. Let h be an Evans function for e_1 with respect to ζ. Then h is harmonic on V, and $\lim_{x \to y} h(x) = \infty$ for each $y \in e_1$. Since $\zeta \notin \overline{V}$, h is bounded below on $\overline{V}$, say $h(x) \geq -B$ for all $x \in \overline{V}$.

For each $\eta > 0$, put

$$f_\eta(x) = f(x) - \eta \cdot h(x)$$

on V. Then f_η is harmonic on V. Since f is bounded above on U, for each $y \in e_1$ we have

$$\limsup_{x \to y} f_\eta(x) = -\infty .$$

On the other hand, for each $y \in \partial V \cap U$, f is continuous at y and $f(y) = M_1$, as before. Hence

$$\limsup_{x \to y} f_\eta(x) \leq M_1 + \eta \cdot B .$$

Since each $y \in \partial V$ belongs either to e_1 or to $\partial V \cap U$, the Maximum Principle shows that $f_\eta(x) \leq M_1 + \eta \cdot B$ on V, hence that
$$f(x) \leq M_1 + \eta \cdot (B + h(x)) \, .$$
Fixing x and letting $\eta \to 0$, we see that $f(x) \leq M_1$ for each $x \in V$. However, this contradicts $f(x_0) > M_1$.

Thus, $f(x) \leq M$ for all $x \in U$, as was to be shown. $\square$

PROPOSITION 7.19 (Riemann Extension Theorem). *Let U be a domain, and let $e \subset U$ be a compact set of capacity 0. Suppose f is harmonic and bounded in $U \backslash e$. Then f extends to a harmonic function on U.*

PROOF. Note that $U \backslash e$ is indeed a domain: it is open since e is closed, and it is connected since a set of capacity 0 is necessarily contained in $\mathbb{P}^1(K)$ and removing type I points cannot disconnect any connected set.

We claim that f is locally constant in a neighborhood of each point $a \in e$. To see this, given $a \in e$, let $\mathcal{B}$ be an open ball containing a whose closure is contained in U. (Such a ball exists, because $a \in \mathbb{P}^1(K)$.) Consider the restriction of f to the domain $V = \mathcal{B} \backslash e$. Since the intersection of ∂V with the complement of e is equal to $\partial \mathcal{B} = \{z\} \subset \mathbb{H}_{\text{Berk}}$ and since a harmonic function is continuous at each point of $\mathbb{H}_{\text{Berk}}$, it follows that
$$\lim_{\substack{x \to z \\ x \in V}} f(x) = f(z) \, .$$

Applying Proposition 7.17 to f on V, we see that for each $x \in V$, both $f(x) \leq f(z)$ and $f(x) \geq f(z)$. Thus $f(x) = f(z)$ for all $z \in V$.

If we extend f by putting $f(x) = f(z)$ for all $x \in \mathcal{B}$, it is easy to verify that the extended function is harmonic on U, as desired. $\square$

COROLLARY 7.20. *Let $\{a_1, \ldots, a_m\}$ be a finite set of points in $\mathbb{P}^1(K)$ (resp. in $\mathbb{P}^1(K) \cap \mathcal{B}(a,r)_\zeta^-$ for some ball $\mathcal{B}(a,r)_\zeta^-$). Then the only bounded harmonic functions in $\mathbb{P}^1_{\text{Berk}} \backslash \{a_1, \ldots, a_m\}$ (resp. $\mathcal{B}(a,r)_\zeta^- \backslash \{a_1, \ldots, a_m\}$) are the constant functions.*

PROOF. We have seen in Proposition 7.12 that if U is $\mathbb{P}^1_{\text{Berk}}$ or an open disc, the only harmonic functions on U are the constant functions. Since a finite set of type I points has capacity 0, Proposition 7.19 shows that a bounded harmonic function on $U \backslash \{a_1, \ldots, a_m\}$ extends to a function that is harmonic on U. $\square$

We can now prove the uniqueness of the equilibrium distribution, as promised in Chapter 6:

PROPOSITION 7.21. *Let $E \subset \mathbb{P}^1_{\text{Berk}}$ be a compact set with positive capacity, and let $\zeta \in \mathbb{P}^1_{\text{Berk}} \backslash E$. Then the equilibrium distribution μ_ζ of E with respect to ζ is unique.*

PROOF. Suppose μ_1 and μ_2 are two equilibrium distributions for E with respect to ζ, so that
$$I_\zeta(\mu_1) = I_\zeta(\mu_2) = V_\zeta(E).$$
Let $u_1(x) = u_{\mu_1}(x, \zeta)$ and $u_2(x) = u_{\mu_2}(x, \zeta)$ be the corresponding potential functions, and let U be the connected component of $\mathbb{P}^1_{\text{Berk}} \backslash E$ containing ζ.

Both u_1 and u_2 satisfy Frostman's theorem (Theorem 6.18): each is bounded above by $V_\zeta(E)$ for all z, each is equal to $V_\zeta(E)$ for all $z \in E$ except possibly on an F_σ set f_i of capacity 0, and each is continuous on E except on f_i. Furthermore, each is bounded in a neighborhood of E.

Consider the function $u = u_1 - u_2$. By Example 5.21, $u \in \text{BDV}(\mathbb{P}^1_{\text{Berk}})$ and
$$\Delta_{\mathbb{P}^1_{\text{Berk}}}(u) = (\mu_1 - \delta_\zeta) - (\mu_2 - \delta_\zeta) = \mu_1 - \mu_2.$$
By Proposition 6.8, μ_1 and μ_2 are both supported on $\partial E_\zeta = \partial \overline{U} \subset \overline{U}$. Since the retraction map $r_{\mathbb{P}^1_{\text{Berk}}, \overline{U}}$ fixes $\overline{U}$, it follows that

(7.1) $$\Delta_{\overline{U}}(u) = (r_{\mathbb{P}^1_{\text{Berk}}, \overline{U}})_*(\Delta_{\mathbb{P}^1_{\text{Berk}}}(u)) = \mu_1 - \mu_2.$$

Thus $\Delta_{\overline{U}}(u)$ is supported on $\partial \overline{U}$.

First suppose $\zeta \in \mathbb{H}_{\text{Berk}}$. By Proposition 6.8, both u_1 and u_2 are well-defined and bounded on all of $\mathbb{P}^1_{\text{Berk}}$ and are continuous on U. Hence u has these properties as well. By (7.1), u is strongly harmonic on U. Put $f = f_1 \cup f_2$; then f has capacity 0 by Corollary 6.21. By the discussion above, for each $x \in \partial U \backslash f$, u is continuous at x and $u(x) = 0$. In particular,

(7.2) $$\lim_{\substack{z \to x \\ z \in U}} u(z) = 0.$$

By the Strong Maximum Principle (Proposition 7.17), $u(z) \equiv 0$ on U. Hence $\Delta_{\overline{U}}(u) = 0$, so $\mu_1 = \mu_2$.

If $\zeta \in \mathbb{P}^1(K)$, there is a slight complication because $u_1(\zeta) - u_2(\zeta) = \infty - \infty$ is undefined. However, Proposition 6.12 shows that u is constant, and in fact is identically 0, in a deleted neighborhood of ζ. By Proposition 6.12 and Theorem 6.18, each u_i is bounded in a neighborhood of E and is continuous in $U \backslash \{\zeta\}$, so u is harmonic and bounded in $U \backslash \{\zeta\}$. The boundary limits (7.2) continue to hold for all $x \in \partial U \backslash f$, and in addition

(7.3) $$\lim_{\substack{z \to \zeta \\ z \in U}} u(z) = 0.$$

Applying the Strong Maximum Principle to u on the domain $U \backslash \{\zeta\}$, relative to the exceptional set $f \cup \{\zeta\}$ contained in its boundary $\partial U \cup \{\zeta\}$, we conclude that $u \equiv 0$ in $U \backslash \{\zeta\}$. Since the closure of $U \backslash \{\zeta\}$ is $\overline{U}$, it follows as before that $\Delta_{\overline{U}}(u) = 0$ and therefore that $\mu_1 = \mu_2$. □

7.3. The Poisson formula

In the classical theory of harmonic functions in the complex plane, if f is harmonic on an open disc $\mathcal{D}$ and has a continuous extension to the closure $\overline{\mathcal{D}}$, then the Poisson formula expresses the values of f on $\mathcal{D}$ in terms of its values on the boundary $\partial \mathcal{D}$.

Specifically, if $\mathcal{D} \subseteq \mathbb{C}$ is an open disc of radius r centered at z_0 and if f is harmonic on $\mathcal{D}$ and extends continuously to $\overline{\mathcal{D}}$, then $f(z_0) = \int_{\partial \mathcal{D}} f \, d\mu_{\mathcal{D}}$, where $\mu_{\mathcal{D}}$ is the uniform probability measure $d\theta/2\pi$ on the boundary circle $\partial \mathcal{D}$. More generally, for any $z \in \mathcal{D}$ there is a measure $\mu_{z,\mathcal{D}}$ depending only on z and $\mathcal{D}$, called the *Poisson-Jensen measure*, for which

$$f(z) = \int_{\partial \mathcal{D}} f \, d\mu_{z,\mathcal{D}}$$

for every harmonic function f on $\mathcal{D}$. We seek to generalize this formula to the Berkovich projective line.

Recall that a *finite-dendrite domain* (Definition 2.27) is a domain U which either has one boundary point (which belongs to $\mathbb{H}_{\text{Berk}}$) or is of the form $U = r_\Gamma^{-1}(\Gamma^0)$ for some finite subgraph $\Gamma \subset \mathbb{H}_{\text{Berk}}$, where $\Gamma^0 = \Gamma \backslash \partial \Gamma$ is the interior of Γ. In the latter case Γ^0 is the main dendrite of U. The class of finite-dendrite domains contains the class of *simple domains*, which we regard as the basic open neighborhoods in $\mathbb{P}^1_{\text{Berk}}$.

In this section we will show that every harmonic function f on a finite-dendrite domain V has a continuous extension to the closure of V, and we will give a formula describing the values of f on V in terms of its values on the boundary. This is the *Poisson formula*.

Conversely, given a prescribed set of values on ∂V, we will show that there is a function $f \in \mathcal{H}(V)$ with a continuous extension to $\overline{V}$, which takes the prescribed values on ∂V. We will also show that, given any discrete measure $\mu_{\partial V} = \sum_{i=1}^m c_i \delta_{x_i}$ of total mass 0 supported on ∂V, there is a function $f \in \mathcal{H}(V)$ for which $\Delta_{\partial V}(f) = \mu_{\partial V}$ and that such a function is unique up to an additive constant. This says that for a finite-dendrite domain, the *Dirichlet problem* and the *Neumann problem* are both solvable.

In one sense, the Berkovich Poisson formula is simpler than the classical one: it is a finite sum, obtained using linear algebra. However, in another sense, it is more complicated, because it depends on the structure of the main dendrite of V.

Let V be a finite-dendrite domain with boundary points $x_1, \ldots, x_m \in \mathbb{H}_{\text{Berk}}$. For $z \in \mathbb{P}^1_{\text{Berk}}$, define a real $(m+1) \times (m+1)$ matrix $M(z)$ as follows:

$$(7.4) \quad M(z) = \begin{pmatrix} 0 & 1 & \cdots & 1 \\ 1 & -\log_v(\delta(x_1, x_1)_z) & \cdots & -\log_v(\delta(x_1, x_m)_z) \\ \vdots & \vdots & \ddots & \vdots \\ 1 & -\log_v(\delta(x_m, x_1)_z) & \cdots & -\log_v(\delta(x_m, x_m)_z) \end{pmatrix}.$$

The matrix $M(z)$ first appeared (in a slightly different form) in D. Cantor's paper [**33**]; we will refer to it as the Cantor matrix (relative to z). Our first version of the Poisson formula for $\mathbb{P}^1_{\text{Berk}}$ relies on the following property of the Cantor matrix:

LEMMA 7.22. *For every $z \in \mathbb{P}^1_{\text{Berk}}$, the matrix $M(z)$ is nonsingular.*

PROOF. Let $M = M(z)$, and suppose that $M\vec{c} = 0$ for some $\vec{c} = [c_0, c_1, \ldots, c_m]^T \in \mathbb{R}^{m+1}$. Then $\sum_{i=1}^m c_i = 0$, and if

$$f(x) := c_0 + \sum_{i=1}^m c_i \cdot (-\log_v(\delta(x, x_i)_z)) \, ,$$

then $f \equiv 0$ on ∂V. Since $\Delta_{\mathbb{P}^1_{\text{Berk}}}(f) = \Delta_{\overline{V}}(f) = \sum_{i=1}^m c_i \delta_{x_i}$ by Example 5.19 and Proposition 5.26, the function f is harmonic on V, and therefore $f \equiv 0$ on V by the Maximum Principle (Proposition 7.16). Thus

$$0 = \Delta_{\overline{V}}(f) = \sum_{i=1}^m c_i \delta_{x_i} \, ,$$

which implies that $c_1 = \cdots = c_m = 0$, and therefore (by the definition of f) we have $c_0 = 0$ as well. Thus $\text{Ker}(M(z)) = \vec{0}$. □

PROPOSITION 7.23 (Poisson formula, Version I). *Let V be a finite-dendrite domain in $\mathbb{P}^1_{\text{Berk}}$ with boundary points $x_1, \ldots, x_m \in \mathbb{H}_{\text{Berk}}$. Then each harmonic function f on V belongs to $\text{BDV}(V)$ and has a continuous extension to $\overline{V}$. There is a unique such function with prescribed boundary values $A_1, \ldots, A_m$, which is given as follows. Fix $z \in \mathbb{P}^1_{\text{Berk}}$, and let $\vec{c} = [c_0, c_1, \ldots, c_m]^T \in \mathbb{R}^{m+1}$ be the unique solution to the linear equation $M(z)\vec{c} = [0, A_1, \ldots, A_m]^T$. Then*

$$(7.5) \qquad f(x) = c_0 + \sum_{i=1}^m c_i \cdot (-\log_v(\delta(x, x_i)_z)) \, .$$

(This should be understood as a limit, if z is of type I and $x = z \in V$.) Moreover,

$$\Delta_{\overline{V}}(f) = \sum_{i=1}^m c_i \delta_{x_i} \, .$$

PROOF. We first show that if f is harmonic on V, then f extends continuously to $\overline{V}$. If $m = 1$, then V is a generalized Berkovich disc, and by Proposition 7.12, the only harmonic functions on V are constant functions, which obviously extend to $\overline{V}$. (The Poisson formula is trivially satisfied in this case.) If $m \geq 2$, let Γ be the convex hull of $\partial V = \{x_1, \ldots, x_m\}$, so that $\Gamma^0 := \Gamma \backslash \{x_1, \ldots, x_m\}$ is the main dendrite of V. If f is harmonic on V, then $\Delta_\Gamma(f)(p) = -\sum_{\vec{v} \in T_p(\Gamma)} d_{\vec{v}} f(p) = 0$ for each $p \in \Gamma^0$. In particular, the restriction of f to each edge of Γ^0 is affine. Moreover, for every $x \in V$ we have $f(x) = f(r_\Gamma(x))$ by Proposition 7.12. It follows easily that f belongs to $\text{BDV}(V)$ and has a continuous extension to $\overline{V}$.

Suppose now that f_1, f_2 are harmonic functions on V whose extensions to $\overline{V}$ agree on ∂V. Then $f_1 - f_2 \equiv 0$ on ∂V and is harmonic on V, and therefore $f_1 - f_2 \equiv 0$ on V by the Maximum Principle. Thus there is at most one function f which is harmonic on V and whose continuous extension to ∂V satisfies $f(x_i) = A_i$ for $i = 1, \ldots, m$. We now check that such a function f is given by the explicit formula (7.5). By construction, $f(x_i) = A_i$ for $i = 1, \ldots, m$. By Proposition 4.10, f is continuous on $\overline{V}$, and by Example 5.19 and Proposition 5.26, f is harmonic on V. Thus f has all of the required properties.

The final assertion regarding $\Delta_{\overline{V}}(f)$ follows immediately from Example 5.19 and Proposition 5.26. $\square$

Note that by Cramer's rule, we can give an explicit formula for the coefficients c_i in (7.5) as follows: for each $i = 0, 1, \ldots m$, let $M_i(z, \vec{A})$ be the matrix obtained by replacing the ith column of $M(z)$ with $(0, A_1, \ldots, A_m)^T$. Then $c_i = \det(M_i(z, \vec{A})) / \det(M(z))$. Note also that taking $x = z$ in (7.5) shows that
$$f(z) \;=\; c_0 \;=\; \det(M_0(z, \vec{A})) / \det(M(z)) \;.$$

As a consequence of Proposition 7.23, we have

COROLLARY 7.24. *Let V be a finite-dendrite domain in $\mathbb{P}^1_{\text{Berk}}$, with boundary $\partial V = \{x_1, \ldots, x_m\}$. Then:*
(A) *(The Dirichlet problem for V is solvable): Given $A_1, \ldots, A_m \in \mathbb{R}$, there is a unique function f which is harmonic on V and continuous on $\overline{V}$, such that $f(x_i) = A_i$ for $i = 1, \ldots, m$.*
(B) *(The Neumann problem for V is solvable): Given $c_1, \ldots, c_m \in \mathbb{R}$ with $\sum_{i=1}^m c_i = 0$, there exists a function f which is harmonic on V and continuous on $\overline{V}$, for which*
$$\Delta_{\partial V}(f) \;=\; \Delta_{\overline{V}}(f) \;=\; \sum_{i=1}^m c_i \delta_{x_i} \;.$$

Moreover, such a function is unique up to addition of a constant.

PROOF. Part (A) is a reformulation of part of Proposition 7.23.

Part (B) follows from part (A) by linear algebra. Note that if $f \in \mathcal{H}(V)$, then Proposition 5.25 shows that $\Delta_{\partial V}(f) = \Delta_{\overline{V}}(f)$ has total mass 0, and hence $\Delta_{\overline{V}}(f) = \sum_{i=1}^m d_i \delta_{x_i}$ where $\sum_{i=1}^m d_i = 0$. Let $\vec{\partial}(f) = (d_1, \ldots, d_m)$ be the vector of coefficients in $\Delta_{\overline{V}}(f)$. By the continuity of f and Lemma 5.24, $\vec{\partial}(f) = 0$ if and only if f is a constant function.

Given $\vec{A} = (A_1, \ldots, A_m) \in \mathbb{R}^m$, let $f_{\vec{A}}$ be the solution to the Dirichlet problem for $A_1, \ldots, A_m$. Consider the linear map $L : \mathbb{R}^m \to \mathbb{R}^m$ defined by
$$L(\vec{A}) \;=\; \vec{\partial}(f_{\vec{A}}) \;.$$

By the Maximum Principle (Proposition 7.16), $f_{\vec{A}}$ is constant if and only if $A_1 = \cdots = A_m$. Hence $\text{Ker}(L)$ has dimension 1 and $\text{Im}(L)$ has dimension

$m - 1$. Since $\mathrm{Im}(L)$ is contained in the hyperplane $\sum_{i=1}^{m} d_i = 0$, it must be equal to that hyperplane.

Combining these facts shows that the Neumann problem is solvable for any $\vec{c} \in \mathbb{R}^m$ with $\sum_{i=1}^{m} c_i = 0$, and the solution is unique up to a constant. $\square$

We will now give another solution to the Dirichlet problem, which has a different flavor from the one in Proposition 7.23. For each $i = 1, \ldots, m$, let h_i be the unique function which is harmonic on V and continuous on $\overline{V}$, which takes the value 1 at x_i and 0 at x_j for $j \neq i$ (the existence of such a function is given by Proposition 7.23). Then h_i is a Berkovich space analogue of the classical *harmonic measure* for the boundary component x_i of U. By the maximum principle, $0 < h_i(z) < 1$ for each $z \in V$, and

$$(7.6) \qquad \sum_{i=1}^{m} h_i(z) \equiv 1 .$$

As a direct consequence of the uniqueness in Proposition 7.23, we obtain:

PROPOSITION 7.25 (Poisson formula, Version II). *Let V be a finite-dendrite domain in $\mathbb{P}^1_{\mathrm{Berk}}$ with boundary points $x_1, \ldots, x_m$. Let $A_1, \ldots, A_m \in \mathbb{R}$ be given. Then the unique function f which is harmonic on V, continuous on $\overline{V}$, and satisfies $f(x_i) = A_i$ for each $i = 1, \ldots, m$ is given by*

$$(7.7) \qquad f(z) = \sum_{i=1}^{m} A_i \cdot h_i(z)$$

for all $z \in \overline{V}$, where h_i is the harmonic measure for $x_i \in \partial U$.

By the remarks following Proposition 7.23, we have

$$(7.8) \qquad h_i(z) = \det(M_0(z, \hat{e}_i))/\det(M(z))$$

for each $i = 1, \ldots, m$, where $\hat{e}_i \in \mathbb{R}^{m+1}$ is the vector which is 1 in the $(i+1)^{\mathrm{st}}$ component and 0 elsewhere. We will give another formula for h_i below.

A useful reformulation of Proposition 7.25 is as follows (compare with [64, §4.2]). For $z \in \overline{V}$, define the *Poisson-Jensen measure* $\mu_{z,V}$ on $\overline{V}$ relative to the point z by

$$\mu_{z,V} = \sum_{i=1}^{m} h_i(z) \delta_{x_i} .$$

Then by Proposition 7.25, we have:

COROLLARY 7.26. *If V is a finite-dendrite domain in $\mathbb{P}^1_{\mathrm{Berk}}$, then a continuous function $f : \overline{V} \to \mathbb{R} \cup \{\pm\infty\}$ is harmonic on V if and only if*

$$f(z) = \int_{\partial V} f \, d\mu_{z,V}$$

for all $z \in V$.

7.3. THE POISSON FORMULA

Since the closures of simple domains form a fundamental system of compact neighborhoods for the topology on $\mathbb{P}^1_{\text{Berk}}$, it follows from Lemma 7.7 that a function f is harmonic on an open set $U \subseteq \mathbb{P}^1_{\text{Berk}}$ if and only if its restriction to every simple subdomain of U is harmonic, where a V is a *simple subdomain* of U if it is a simple domain whose *closure* $\overline{V}$ is contained in U (Definition 2.24). With this terminology, we have:

COROLLARY 7.27. *If U is a domain in $\mathbb{P}^1_{\text{Berk}}$ and $f : U \to \mathbb{R} \cup \{\pm\infty\}$ is a continuous function, then f is harmonic on U if and only if for every simple subdomain V of U we have*

$$f(z) = \int_{\partial V} f \, d\mu_{z,V}$$

for all $z \in V$.

We will now show that $\mu_{z,V}$ coincides with the equilibrium measure for ∂V relative to the point z, as defined in §6.2. Although we have already proved the uniqueness of the equilibrium measure, for the boundary of a finite-dendrite domain there is a simpler proof:

LEMMA 7.28. *If V is a finite-dendrite domain in $\mathbb{P}^1_{\text{Berk}}$ and $\zeta \in V$, then the following are equivalent for a probability measure ν supported on $E = \partial V$:*

(A) *$I_\zeta(\nu) = V_\zeta(E)$, i.e., the energy $I_\zeta(\nu)$ of ν is the minimum possible for a probability measure supported on ∂V.*

(B) *The potential function $u_\nu(x) = u_\nu(x, \zeta)$ defined by*

$$u_\nu(x) = \int -\log_v \delta(x,y)_\zeta \, d\nu(y)$$

is constant on ∂V.

Moreover, the equilibrium measure μ_ζ for ∂V relative to ζ is the unique probability measure satisfying these equivalent conditions.

PROOF. Let $E = \partial V = \{x_1, \ldots, x_m\} \subset \mathbb{H}_{\text{Berk}}$, and write $\nu = \sum_{i=1}^m \nu_i \delta_{x_i}$ with $\sum \nu_i = 1$ and $\nu_i \geq 0$ for all $i = 1, \ldots, m$. If ν satisfies (B), then by the Berkovich space analogue of Frostman's theorem (Theorem 6.18), we have $u_\nu(x) = V_\zeta(E)$ for all $x \in E$ (since every nonempty subset of E has positive capacity), so condition (A) is satisfied. The existence of a measure ν satisfying (A) follows from Proposition 6.6. So to prove that (B) implies (A) and to prove the uniqueness of the equilibrium measure, it suffices to show that there is a unique probability measure ν on ∂E satisfying (B). For this, note that the potential function u_ν is constant on ∂V if and only if

(7.9) $$M(\zeta)[\nu_0, \nu_1, \ldots, \nu_m]^T = [1, 0, 0, \ldots, 0]^T$$

for some $\nu_0 \in \mathbb{R}$, where $M(\zeta)$ is the Cantor matrix relative to ζ. Since $M(\zeta)$ is nonsingular by Lemma 7.22, there is a unique $\vec{\nu} = [\nu_0, \nu_1, \ldots, \nu_m]^T \in \mathbb{R}^{m+1}$ satisfying (7.9), which gives what we want. $\square$

We can now show that the Poisson-Jensen measure and the equilibrium measure coincide:

PROPOSITION 7.29. *Let $V \subseteq \mathbb{P}^1_{\text{Berk}}$ be a finite-dendrite domain, and let $\mu = \mu_{z,V}$ be the Poisson-Jensen measure for V relative to the point $z \in V$. Then μ satisfies condition* (B) *of Lemma 7.28 and is therefore the equilibrium measure for ∂V relative to z.*

PROOF. By Example 5.22, the potential function
$$u_\mu(x) = \int -\log_v(\delta(x,y)_z) \, d\mu(y)$$
belongs to $\text{BDV}(\mathbb{P}^1_{\text{Berk}})$, and by Proposition 5.26,
$$\Delta_{\overline{V}}(u_\mu) = \mu - \delta_z .$$

Let h be any strongly harmonic function on V, so that $h \in \text{BDV}(V)$ and $\Delta_{\overline{V}}(h)$ is a measure of total mass zero supported on ∂V. As in the proof of Proposition 7.23, h extends to a continuous function on $\overline{V}$. By Corollary 5.39, we have
$$0 = \left(\int_{\overline{V}} h \, d\mu \right) - h(z) = \int_{\overline{V}} h \, \Delta_{\overline{V}}(u_\mu) = \int_{\overline{V}} u_\mu \, \Delta_{\overline{V}}(h) ,$$
so that $\int_{\overline{V}} u_\mu \, \Delta_{\overline{V}}(h) = 0$ for every such h. By Proposition 5.28, for every measure ν of total mass zero supported on ∂V, there is a strongly harmonic function h on $\overline{V}$ (unique up to an additive constant) such that $\nu = \Delta_{\overline{V}}(h)$. Taking $\nu = \delta_{x_i} - \delta_{x_1}$ for $i = 2, \ldots, m$ shows that u_μ is constant on ∂V, as desired. □

A generalization of Proposition 7.29, with the finite-dendrite domain V replaced by an arbitrary domain U for which ∂U has positive capacity, will be proved in Proposition 7.43.

By Proposition 7.29 and the proof of Lemma 7.28, $\mu_{z,V}$ is the *unique* measure μ supported on ∂V such that

(7.10) $\qquad M(z)[\mu_0, \mu(x_1), \ldots, \mu(x_m)]^T = [1, 0, 0, \ldots, 0]^T$

for some $\mu_0 \in \mathbb{R}$, where $M(z)$ is the Cantor matrix relative to z. By Cramer's rule, we therefore obtain the following alternative explicit formula for the harmonic measure h_i, which is different from the one given in (7.8):

COROLLARY 7.30. *For $i = 1, \ldots, m$, the harmonic measure $h_i(z)$ for $x_i \in \partial V$ is given by the following formula. Let $M_i(z)$ be the matrix obtained by replacing the ith column of $M(z)$ by $[1, 0, \ldots, 0]^T$. Then*
$$h_i(z) = \det(M_i(z))/\det(M(z)) .$$

7.4. Uniform convergence

Poisson's formula implies that the limit of a sequence of harmonic functions is harmonic, under a much weaker condition than is required classically.

PROPOSITION 7.31. *Let U be an open subset of $\mathbb{P}^1_{\text{Berk}}$. Suppose $f_1, f_2, \ldots$ are harmonic on U and converge pointwise to a function $f : U \to \mathbb{R}$. Then f is harmonic on U, and the f_i converge uniformly to f on compact subsets of U.*

PROOF. Given $x \in U$, take a simple domain U_x containing x, with closure $\overline{U}_x \subset U$. If $\partial U_x = \{x_1, \ldots, x_m\}$, then by Proposition 7.25, for each $k \geq 1$ we have

$$f_k(z) = \sum_{i=1}^{m} f_k(x_i) h_i(z)$$

for all $z \in \overline{U}_x$. It follows that the functions f_k converge uniformly to f on $\overline{U}_x$ and $f(z) = \sum_{i=1}^{m} f(x_i) h_i(z)$ is strongly harmonic on U_x.

Thus f is harmonic on U. Since any compact set $E \subset U$ is covered by finitely many sets U_x, the sequence of functions f_i converges uniformly to f on E. □

Using Proposition 7.31, we can characterize harmonic functions as local uniform limits of suitably normalized logarithms of norms of rational functions.

COROLLARY 7.32. *If $U \subset \mathbb{P}^1_{\text{Berk}}$ is a domain and f is harmonic on U, there are rational functions $g_1(T), g_2(T), \ldots \in K(T)$ and rational numbers $R_1, R_2, \ldots \in \mathbb{Q}$ such that*

$$f(x) = \lim_{k \to \infty} R_k \cdot \log_v([g_k]_x)$$

uniformly on compact subsets of U.

PROOF. The assertion is trivial if the main dendrite of U is empty, since the only harmonic functions on U are constants. Hence we can assume that the main dendrite is nonempty. After a change of coordinates, if necessary, we can assume without loss of generality that $\infty \notin U$. By Corollary 7.11, we can choose an exhaustion of U by strict simple domains $\{U_k\}$ with closures $\overline{U}_k \subset U$. Fix k, and write $\partial U_k = \{x_{k,1}, \ldots, x_{k,m_k}\}$. Taking $z = \infty$ in Proposition 7.23, there are numbers $c_{k,i} \in \mathbb{R}$ with $\sum_{i=1}^{m_k} c_{k,i} = 0$ such that for all $x \in U_k$,

$$f(x) = c_{k,0} - \sum_{i=1}^{m_k} c_{k,i} \log_v(\delta(x, x_{k,i})_\infty) \ .$$

For each $i = 1, \ldots, m_k$, fix a type I point $a_{k,i}$ with $a_{k,i} \notin U_k$ such that the path $[a_{k,i}, \infty]$ passes through $x_{k,i}$. (Such points $a_{k,i}$ exist because each $x_{k,i}$ is of type II.) Then $\delta(x, x_{k,i})_\infty = \delta(x, a_{k,i})_\infty = [T - a_{k,i}]_x$ for each $x \in \overline{U}_k$.

Choose rational numbers $d_{k,i}$, with $\sum_{i=1}^{m_k} d_{k,i} = 0$, which are close enough to the $c_{k,i}$ that

$$f_k(x) := d_{k,0} - \sum_{i=1}^{m_k} d_{k,i} \log_v(\delta(x, a_{k,i})_\infty)$$

satisfies $|f_k(x) - f(x)| < 1/k$ on $\overline{U}_k$. Let N_k be a common denominator for the $d_{k,i}$, and put $n_{k,i} = N_k \cdot d_{k,i}$. Let $b_k \in K$ be a constant with $|b_k| = q_v^{-n_{k,0}}$, and put

(7.11) $$g_k(T) = b_k \cdot \prod_{i=1}^{m_k} (T - a_{k,i})^{n_{k,i}}.$$

Then $f_k(x) = -\frac{1}{N_k} \log_v([g_k]_x)$ on U_k. The result follows. $\square$

7.5. Harnack's principle

There are Berkovich space analogues of Harnack's inequality and Harnack's principle:

LEMMA 7.33 (Harnack's inequality). *Let $U \subseteq \mathbb{P}^1_{\text{Berk}}$ be a domain. Then for each $x_0 \in U$ and each compact set $X \subset U$, there is a constant $C = C(x_0, X)$ such that for any function h which is harmonic and nonnegative in U, each $x \in X$ satisfies*

(7.12) $$(1/C) \cdot h(x_0) \leq h(x) \leq C \cdot h(x_0) .$$

PROOF. It suffices to give the proof under the assumptions that $h(x_0) > 0$ and D is nonempty: if $h(x_0) = 0$, then $h(z) \equiv 0$ on U by Proposition 7.16; if the main dendrite D of U is empty, then h is constant by Proposition 7.12. In either case (7.12) holds for any $C \geq 1$. Since harmonic functions are constant on branches off D, we can also assume that x_0 belongs to D.

First consider the upper bound in (7.12). Let $\rho(x,y)$ be the logarithmic path distance on $\mathbb{P}^1_{\text{Berk}}$. Since the main dendrite is everywhere finitely branched, for each $p \in D$ there is an $\varepsilon > 0$ such that the closed neighborhood of p in D defined by $\Gamma(p,\varepsilon) = \{x \in D : \rho(x,p) \leq \varepsilon\}$ is a star, i.e., a union of n closed segments of length ε emanating from p, for some $n \geq 2$. Take ε as large as possible, subject to the conditions that $\varepsilon \leq 1$ and $\Gamma(p,\varepsilon)$ is a star, and write $q_i = p + \varepsilon \vec{v}_i$, $i = 1, \ldots, n$, for the endpoints of $\Gamma(p,\varepsilon)$.

The restriction of h to the finite subgraph $\Gamma(p,\varepsilon)$ is linear on each of the segments $[p, q_i]$, and $\Delta_{\Gamma(p,\varepsilon)}(h)(p) = 0$. Thus for each i and each $x = p + t\vec{v}_i \in [p, q_i]$, we have

$$0 = \Delta_{\Gamma(p,\varepsilon)}(h)(p) = \frac{h(x) - h(p)}{t} + \sum_{j \neq i} \frac{h(q_j) - h(p)}{\varepsilon} .$$

Since $h(q_j) \geq 0$ for each j, it follows that there is a constant C_p such that $h(x) \leq C_p \cdot h(p)$ for each $x \in \Gamma(p,\varepsilon)$ (indeed, we can take $C_p = n$).

Since X is compact, there is a finite subgraph Γ of D such that the retraction of X to the main dendrite is contained in the interior of Γ. Starting with $p = x_0$, then proceeding stepwise and using the compactness of Γ, it follows that there is a constant C_Γ such that for each $x \in \Gamma$, and hence for each $x \in X$,

(7.13) $$h(x) \leq C_\Gamma \cdot h(x_0) \,.$$

For the lower bound, let Γ^0 be the interior of Γ and put $U_\Gamma^0 = r_{\mathbb{P}^1_{\text{Berk}}, \Gamma}^{-1}(\Gamma^0)$. Then U_Γ^0 is a subdomain of U whose boundary ∂U_Γ^0 consists of the endpoints $x_1, \ldots, x_m$ of Γ. For each x_i, let $C_{\Gamma,i}$ be the constant constructed as above for x_i and Γ. Put $C_\Gamma' = \max(C_{\Gamma,1}, \ldots, C_{\Gamma,m})$. Then $h(x_0) \leq C_\Gamma' \cdot h(x_i)$ for each x_i.

Since h is harmonic on U_Γ^0 and continuous on $\overline{U}_\Gamma^0$, Proposition 7.16 shows that $\min(h(x_1), \ldots, h(x_m)) \leq h(x)$ for each $x \in X$. Hence

(7.14) $$h(x_0) \leq C_\Gamma' \cdot \min(h(x_i)) \leq C_\Gamma' \cdot h(x) \,.$$

Putting $C = \max(C_\Gamma, C_\Gamma')$ and combining (7.13), (7.14) yields the lemma. $\square$

PROPOSITION 7.34 (Harnack's principle). *Let $U \subseteq \mathbb{P}^1_{\text{Berk}}$ be a domain, and suppose $f_1, f_2, \ldots$ are harmonic on U with $0 \leq f_1 \leq f_2 \leq \cdots$. Then either*

(A) $\lim_{i \to \infty} f_i(x) = \infty$ *for each $x \in U$ or*
(B) $f(x) = \lim_{i \to \infty} f_i(x)$ *is finite for all x, the f_i converge uniformly to f on compact subsets of U, and f is harmonic on U.*

PROOF. Suppose there is some $x_0 \in U$ for which $f(x_0) = \lim_{i \to \infty} f_i(x_0)$ is finite. By Lemma 7.33, $f(x) = \lim_{i \to \infty} f_i(x)$ is finite for each x in U and the f_i converge pointwise to f. By Proposition 7.31, f is harmonic on U and the f_i converge locally uniformly to f. $\square$

7.6. Green's functions

In this section we will develop a theory of Green's functions for compact sets in $\mathbb{P}^1_{\text{Berk}}$. We first establish the basic properties of Green's functions (positivity, symmetry, monotonicity) and then prove an approximation theorem showing that Green's functions of arbitrary compact sets can be approximated by Green's functions of strict closed affinoids. As an application, we show that the equilibrium distribution plays the role of a reproducing kernel for harmonic functions.

The theory depends on the fact that the equilibrium distribution μ_ζ is unique, so the potential function $u_E(z, \zeta) = u_{\mu_\zeta}(z, \zeta)$ is well-defined. In the terminology of [**88**], the Green's functions defined here are *upper* Green's functions.

DEFINITION 7.35. If $E \subset \mathbb{P}^1_{\text{Berk}}$ is a compact set of positive capacity, then for each $\zeta \notin E$, the *Green's function of E relative to ζ* is
$$G(z, \zeta; E) = V_\zeta(E) - u_E(z, \zeta) \qquad \text{for all } z \in \mathbb{P}^1_{\text{Berk}} .$$

DEFINITION 7.36. If $U \subset \mathbb{P}^1_{\text{Berk}}$ is a domain for which ∂U has positive capacity, the *Green's function of U* is
$$G_U(z, \zeta) = G(z, \zeta; \mathbb{P}^1_{\text{Berk}} \backslash U) \qquad \text{for } z \in \mathbb{P}^1_{\text{Berk}}, \zeta \in U .$$

Note that throughout this section, we normalize the Hsia kernel as in (4.23), which imposes the normalization for the Robin constant $V_\zeta(E)$ and the capacity $\gamma_\zeta(E)$ given by (7.15) below. In certain arithmetic situations (see §7.8) it is natural to use other normalizations.

PROPOSITION 7.37. *Let $E \subset \mathbb{P}^1_{\text{Berk}}$ be a compact set of positive capacity. Then $G(z, \zeta; E)$ has the following properties:*

(A) *For each fixed $\zeta \notin E$, regarding $G(z, \zeta; E)$ as a function of z:*

(1) $G(z, \zeta; E) \geq 0$ *for all $z \in \mathbb{P}^1_{\text{Berk}}$.*

(2) $G(z, \zeta; E) > 0$ *for $z \in U_\zeta$, where U_ζ is the connected component of $\mathbb{P}^1_{\text{Berk}} \backslash E$ containing ζ.*

(3) $G(z, \zeta; E)$ *is finite on $\mathbb{P}^1_{\text{Berk}} \backslash \{\zeta\}$ and strongly harmonic on $U_\zeta \backslash \{\zeta\}$. For each $a \neq \zeta$, $G(z, \zeta; E) - \log_v(\delta(z, a)_\zeta)$ extends to a function harmonic on a neighborhood of ζ.*

(4) $G(z, \zeta; E) = 0$ *on $\mathbb{P}^1_{\text{Berk}} \backslash U_\zeta$, except on a (possibly empty) F_σ set $e \subset \partial U_\zeta$ of capacity 0 which depends only on the component $U = U_\zeta$ of $\mathbb{P}^1_{\text{Berk}} \backslash E$ containing ζ and not on ζ itself.*

(5) $G(z, \zeta; E)$ *is continuous on $\mathbb{P}^1_{\text{Berk}} \backslash e$ and is upper semicontinuous on $\mathbb{P}^1_{\text{Berk}} \backslash \{\zeta\}$.*

(6) *If $E' \subset \mathbb{P}^1_{\text{Berk}}$ is another compact subset for which the connected component U'_ζ of ζ in $\mathbb{P}^1_{\text{Berk}} \backslash E'$ coincides with U_ζ, then E' has positive capacity and $G(z, \zeta; E') = G(z, \zeta; E)$ for all $z \in \mathbb{P}^1_{\text{Berk}}$. In particular, $G(z, \zeta; E) = G(z, \zeta, \partial U_\zeta)$.*

(7) $G(z, \zeta; E)$ *is bounded if $\zeta \in \mathbb{H}_{\text{Berk}}$. It is unbounded if $\zeta \in \mathbb{P}^1(K)$. In either case*

(7.15) $$\lim_{z \to \zeta} \left(G(z, \zeta; E) + \log_v(\|z, \zeta\|) \right) = V_\zeta(E) \quad \text{and} \quad \gamma_\zeta(E) = q_v^{-V_\zeta(E)} .$$

If $\zeta \in \mathbb{P}^1(K)$, then there is a neighborhood of ζ on which

(7.16) $$G(z, \zeta; E) = V_\zeta(E) - \log_v(\|z, \zeta\|) .$$

(8) $G(z, \zeta; E) \in \text{BDV}(\mathbb{P}^1_{\text{Berk}})$, *with $\Delta_{\mathbb{P}^1_{\text{Berk}}}(G(z, \zeta; E)) = \delta_\zeta(z) - \mu_{E,\zeta}(z)$.*

(B) *As a function of two variables, $G(z_1, z_2; E) = G(z_2, z_1; E)$ for all $z_1, z_2 \notin E$.*

(C) *Suppose $E_1 \subseteq E_2$ are compact and have positive capacity. Then for all $\zeta \in \mathbb{P}^1_{\text{Berk}} \backslash E_2$, we have $V_\zeta(E_1) \geq V_\zeta(E_2)$, and for all $z \in \mathbb{P}^1_{\text{Berk}}$,*
$$G(z, \zeta; E_1) \geq G(z, \zeta; E_2) .$$

REMARK 7.38. In Chapter 8, when we have defined subharmonicity for functions on domains in $\mathbb{P}^1_{\text{Berk}}$, we will see that for each $\zeta \notin E$, $G(z,\zeta;E)$ is subharmonic on $\mathbb{P}^1_{\text{Berk}}\backslash\{\zeta\}$ (see Example 8.9). It then follows from Proposition 8.11 that for each $z \neq \zeta$,

$$(7.17) \qquad G(z,\zeta;E) \;=\; \limsup_{x \to z} G(x,\zeta;E) \;=\; \limsup_{\substack{x \to z \\ x \in \mathbb{H}_{\text{Berk}}}} G(x,\zeta;E) \,,$$

strengthening the lower semicontinuity in Proposition 7.37(A)(5).

PROOF. For (A), part (A)(1) is a consequence of Theorem 6.18 (Frostman's theorem). Parts (A)(4) and (A)(5) (except for the fact that e depends only on U_ζ, which we will prove later) follow from Proposition 6.12 and Frostman's theorem. These same facts show $G(\zeta,\zeta;E) > 0$. Part (A)(7) is a reformulation of assertions in Proposition 6.12, using the definition of $G(z,\zeta;E)$. Part (A)(8) is a reformulation of Example 5.21.

For the first part of (A)(3), we know that $G(z,\zeta;E)$ is finite and harmonic on $U_\zeta\backslash\{\zeta\}$ by Example 7.6. For the second part of (A)(3), fix $a \neq \zeta$ and consider the function $f(z) = G(z,\zeta;E) - \log_v(\delta(z,a)_\zeta)$. By Examples 5.19 and 5.21, f belongs to $\text{BDV}(\mathbb{P}^1_{\text{Berk}})$ and satisfies

$$\Delta_{\mathbb{P}^1_{\text{Berk}}}(f) \;=\; \delta_a - \mu_\zeta \,,$$

where we write μ_ζ for $\mu_{E,\zeta}$ when the set E is understood. Let V be a connected neighborhood of ζ with $\overline{V} \cap (E \cup \{a\}) = \emptyset$. If $\zeta \notin \mathbb{P}^1(K)$, then f is defined everywhere and is strongly harmonic on V. If $\zeta \in \mathbb{P}^1(K)$, then f is harmonic and bounded in $V\backslash\{\zeta\}$, so it extends to a strongly harmonic function in V by the Riemann Extension Theorem (Proposition 7.19).

For (A)(2), if $G(x,\zeta;E) = 0$ for some $x \in U_\zeta\backslash\{\zeta\}$, let W be the connected component of $U_\zeta\backslash\{\zeta\}$ containing x. Then $G(z,\zeta;E)$ achieves its minimum value 0 at an interior point of W and so is identically 0 on W by the Maximum Principle (Proposition 7.16). Thus

$$\lim_{\substack{z \to \zeta \\ z \in W}} G(z,\zeta;E) \;=\; 0 \,,$$

which contradicts the fact that $G(\zeta,\zeta;E) > 0$ and $G(z,\zeta;E)$ is continuous at ζ.

To see that the set e in (A)(4) depends only on U_ζ, we temporarily write e_ζ for the set e and take $\xi \neq \zeta$ in U_ζ. Let V be a neighborhood of ζ such that $\overline{V} \subset U_\zeta$ and $\xi \notin \overline{V}$. Since $\partial V \subset U_\zeta\backslash\{\zeta,\xi\}$, there is a constant $C > 0$ such that $G(z,\xi;E) > C \cdot G(z,\zeta;E)$ on ∂V. Let W be a connected component of $U_\zeta\backslash\overline{V}$. Then $\partial W \subset \partial U_\zeta \cup \partial V$. Consider $f(z) = G(z,\xi;E) - C \cdot G(z,\zeta;E)$, which is harmonic on W. For each $x \in \partial W \cap \partial V$, we have

$$\lim_{\substack{z \to x \\ z \in W}} f(z) \;>\; 0 \,,$$

while for each $x \in \partial W \cap (\partial U_\zeta \backslash e_\zeta)$, we have

$$\liminf_{\substack{z \to x \\ z \in W}} f(z) \geq 0 \ .$$

By the Strong Maximum Principle (Proposition 7.17), $f(z) \geq 0$ for all $z \in W$. Hence if $x \in \partial W \cap e_\zeta$, then

$$\liminf_{\substack{z \to x \\ z \in W}} G(z, \xi; E) \geq C \cdot \liminf_{\substack{z \to x \\ z \in W}} G(z, \zeta; E) > 0 \ .$$

Since each $x \in e_\zeta$ belongs to ∂W for some W, it follows that $e_\zeta \subset e_\xi$. By symmetry, $e_\zeta = e_\xi$.

Finally, we prove (A)(6). Since the equilibrium distribution μ_ζ of E is supported on $\partial U_\zeta = \partial U'_\zeta \subseteq E'$, necessarily E' has positive capacity. Put $u(z) = G(z, \zeta; E) - G(z, \zeta; E')$. By (A)(3), u extends to a function harmonic on U_ζ. By (A)(4), there are sets $e, e' \subset \partial U_\zeta$ of capacity 0 such that $u(z) = 0$ for all $z \in \mathbb{P}^1_{\text{Berk}} \backslash (U_\zeta \cup e \cup e')$. By (A)(5), u is continuous on $\mathbb{P}^1_{\text{Berk}} \backslash (e \cup e')$; in particular for each $z \in \partial U_\zeta \backslash (e \cup e')$,

$$\lim_{\substack{x \to z \\ x \in U_\zeta}} u(x) = 0 \ .$$

Since e and e' have capacity 0, Corollary 6.21 shows that $e \cup e'$ has capacity 0 as well. By the Strong Maximum Principle, $u(z) = 0$ for all $z \in U_\zeta$. Since any set of capacity 0 is contained in $\mathbb{P}^1(K)$, we have shown in particular that $u(z) = 0$ for all $z \in \mathbb{H}_{\text{Berk}}$.

Since the Laplacians $\Delta(G(z, \zeta; E))$ and $\Delta(G(z, \zeta; E'))$ depend only on the restrictions of $G(z, \zeta; E)$ and $G(z, \zeta; E')$ to $\mathbb{H}_{\text{Berk}}$, by (A)(8) the equilibrium distributions μ_ζ of E and μ'_ζ of E' coincide. Hence $G(z, \zeta; E) = G(z, \zeta; E')$ for all $z \in \mathbb{P}^1_{\text{Berk}}$.

For part (B), we can assume without loss of generality that z_1, z_2 belong to the same connected component U of $\mathbb{P}^1_{\text{Berk}} \backslash E$; otherwise by (A)(4) we have $G(z_1, z_2; E) = G(z_2, z_1; E) = 0$.

Let $\mathcal{S}_{\overline{U}}$ be a directed set indexing the finite subgraphs of $\overline{U}$, and let Γ_α be the finite subgraph of $\overline{U}$ corresponding to $\alpha \in \mathcal{S}_{\overline{U}}$. For $\alpha \in \mathcal{S}_{\overline{U}}$, let $r_\alpha := r_{\overline{U}, \Gamma_\alpha} : \overline{U} \to \Gamma_\alpha$ be the retraction map. Then $\lim_\alpha r_\alpha(z) = z$ for all $z \in \overline{U}$.

First, suppose both z_1 and z_2 are in $\mathbb{H}_{\text{Berk}}$. Put $g_1(x) = G(x, z_1; E)$, $g_2(x) = G(x, z_2; E)$, and write $\mu_1 = \mu_{z_1}$, $\mu_2 = \mu_{z_2}$. Then $g_1, g_2 \in \text{BDV}(\mathbb{P}^1_{\text{Berk}})$ by (A)(7).

For each finite subgraph Γ_α of $\overline{U}$, we have $g_1|_{\Gamma_\alpha}, g_2|_{\Gamma_\alpha} \in \text{CPA}(\Gamma_\alpha)$ since each $g_i(x)$ is harmonic on $U \backslash \{z_i\}$. By Proposition 3.2(D),

$$(7.18) \qquad \int_{\Gamma_\alpha} g_1 \, \Delta_{\Gamma_\alpha}(g_2) = \int_{\Gamma_\alpha} g_2 \, \Delta_{\Gamma_\alpha}(g_1) \ .$$

Assume α is chosen so that $z_1, z_2 \in \Gamma_\alpha$. Then for $\{i,j\} = \{1,2\}$, we have
$$\Delta_{\Gamma_\alpha}(g_i) = \delta_{r_\alpha(z_i)} - (r_\alpha)_*(\mu_i)$$
and
$$\int_{\Gamma_\alpha} g_i d((r_\alpha)_*(\mu_j)) = \int_{\overline{U}} (g_i \circ r_\alpha) d\mu_j .$$
Since μ_j is supported on ∂U, it follows that

(7.19) $\quad g_1(z_2) - \int_{\partial U} g_1(r_\alpha(x)) d\mu_2(x) = g_2(z_1) - \int_{\partial U} g_2(r_\alpha(x)) d\mu_1(x) .$

We claim that for each $\varepsilon > 0$, there is an $\alpha_0 \in \mathcal{S}_{\overline{U}}$ such that for each $\alpha \geq \alpha_0$,
$$\left| \int_{\partial U} g_i(r_\alpha(x)) d\mu_j(x) \right| < \varepsilon/2 .$$
Assuming the claim, it follows from (7.19) that $|g_1(z_2) - g_2(z_1)| < \varepsilon$ for every $\varepsilon > 0$ and thus that $G(z_1, z_2; E) = G(z_2, z_1; E)$ as desired.

To prove the claim, let e be the exceptional set of capacity 0 in ∂U given in (A)(4). By Lemma 6.16, $\mu_j(e) = 0$. Note that by (A)(7), $g_i(x) = G(x, z_i; E)$ is bounded on $E \supseteq \partial U$, say by $M > 0$. By the regularity of μ_j, there exists an open set V_e containing e such that $\mu_j(V_e) < \varepsilon/(4M)$. For each $x \in \partial U \backslash e$, since $\lim_\alpha r_\alpha(x) = x$ and $g_i(z)$ is continuous at x by (A)(5), we have
$$\lim_\alpha g_i(r_\alpha(x)) = g_i(x) = 0 .$$
Since $\partial U \backslash V_e$ is compact, there is an $\alpha_0 \in \mathcal{S}_{\overline{U}}$ such that if $\alpha \geq \alpha_0$, then $|g_i(r_\alpha(x))| < \varepsilon/4$ for all $x \in \partial U \backslash V_e$. For $\alpha \geq \alpha_0$, we therefore have
$$\left| \int_{\partial U} g_i(r_\alpha(x)) d\mu_j(x) \right| \leq \sup_{x \in \partial U \backslash V_e} |g_i(r_\alpha(x))| + M \cdot \mu_j(V_e)$$
$$< \varepsilon/4 + M \cdot \varepsilon/(4M) = \varepsilon/2$$
as claimed.

Now suppose $z_1 \in U$ is of type I, but $z_2 \in U$ is not of type I. Let $t \to z_1$ along a path in $\mathbb{H}_{\text{Berk}}$. Since $G(x, z_2; E)$ is continuous for $x \in U$,
$$G(z_1, z_2; E) = \lim_{t \to z_1} G(t, z_2; E) = \lim_{t \to z_1} G(z_2, t; E) .$$
We claim that $\lim_{t \to z_1} G(z_2, t; E) = G(z_2, z_1; E)$.

To see this, first suppose the main dendrite D of U is nonempty, and fix a point $z \in D$. By (4.26), there is a constant C_t such that
$$-\log_v(\delta(x,y)_t) = j_z(x,y) - j_z(x,t) - j_z(y,t) + C_t$$
for all $x, y \in \mathbb{P}^1_{\text{Berk}}$. Let w be the point where the path from z_1 to z first meets D. If t lies on the path from z_1 to w, then for all $x, y \in E$, $j_z(x,t) = j_z(x,w)$ and $j_z(y,t) = j_z(y,w)$. In other words, for $x, y \in E$, $\log_v(\delta(x,y)_t)$ depends

on t only through the constant C_t. Hence the same probability measure μ minimizes the energy integral

$$V_t(E) = \iint_{E \times E} -\log_v(\delta(x,y)_t) \, d\mu(x) d\mu(y)$$

for all t on the path from w to z_1, that is, $\mu_t = \mu_{z_1} = \mu$. Similarly, $-\log_v(\delta(z_2, y)_t)$ depends on t only through C_t, if $y \in E$ and t is sufficiently near z_1. Hence

$$G(z_2, t; E) = V_t(E) - \int_E -\log_v(\delta(z_2, u)_t) \, d\mu(u)$$

is a constant independent of t, for t sufficiently near z_1. Thus

$$\lim_{t \to z_1} G(z_2, t; E) = G(z_2, z_1; E).$$

If the main dendrite D of U is empty, then U is a disc with a single boundary point z. In this case, if we take $w = z$, then the same arguments as above carry through.

Finally, let $z_1, z_2 \in U$ be arbitrary. Let t approach z_1 through points of type I. Using continuity and an argument like the one above, we find that

$$G(z_1, z_2; E) = \lim_{t \to z_1} G(t, z_2; E)$$
$$= \lim_{t \to z_1} G(z_2, t; E) = G(z_2, z_1; E).$$

For part (C), fix $\zeta \notin E_2$, and let $U_{2,\zeta}$ be the connected component of $\mathbb{P}^1_{\text{Berk}} \backslash E_2$ containing ζ.

If $z \in \mathbb{P}^1_{\text{Berk}}$ is such that $G(z, \zeta; E_2) = 0$ (in particular if $z \notin \overline{U}_{2,\zeta}$ or if $z \notin \partial U_{2,\zeta} \backslash e$ where e is the exceptional set for $G(z, \zeta; E_2)$ given by part (A)(4)), then by part (A)(1) we trivially have $G(z, \zeta; E_1) \geq G(z, \zeta; E_2)$.

For $z \in U_{2,\zeta}$, consider the function on $U_{2,\zeta}$ given by

$$u(z) = \begin{cases} G(z, \zeta; E_1) - G(z, \zeta; E_2) & \text{if } z \neq \zeta, \\ V_\zeta(E_1) - V_\zeta(E_2) & \text{if } z = \zeta. \end{cases}$$

By parts (A)(3), (A)(5), and (A)(6), u is continuous and strongly harmonic on $U_{2,\zeta}$, and by parts (A)(1), (A)(4), and (A)(5), for each $x \in \partial U_{2,\zeta} \backslash e$

$$\liminf_{\substack{z \to x \\ z \in U_{2,\zeta}}} u(z) \geq 0.$$

By Proposition 7.17 (the Strong Maximum Principle), $u(z) \geq 0$ for all $z \in U_{2,\zeta}$. Hence $G(z, \zeta; E_1) \geq G(z, \zeta; E_2)$ for all $z \in U_{2,\zeta}$. In particular, by part (A)(6), $V_\zeta(E_1) \geq V_\zeta(E_2)$.

Finally, suppose $z \in e$. Since e has capacity 0, necessarily $e \subset \mathbb{P}^1(K)$. Thus z is of type I. Fix $y \in U_\zeta$, and let t approach z along the path $[y, z]$. By the final assertion in Proposition 6.12 and what has been shown above,

$$G(z, \zeta; E_1) = \lim_{t \to z} G(t, \zeta; E_1) \geq \lim_{t \to z} G(t, \zeta; E_2) = G(z, \zeta; E_2).$$

This completes the proof of (C). $\square$

If the exceptional set e in Proposition 7.37 is empty, then one has:

COROLLARY 7.39. *Let $E \subseteq \mathbb{P}^1_{\text{Berk}}$ be a compact set with positive capacity, let $\zeta \in \mathbb{P}^1_{\text{Berk}} \backslash E$, and let U_ζ denote the connected component of $\mathbb{P}^1_{\text{Berk}} \backslash E$ containing ζ. Assume that $G(z, \zeta; E)$ is identically zero on E. Then:*
- (A) *$G(z, \zeta; E)$ is continuous on all of $\mathbb{P}^1_{\text{Berk}} \backslash \{\zeta\}$.*
- (B) *$\{z \in \mathbb{P}^1_{\text{Berk}} : G(z, \zeta; E) > 0\} = U_\zeta$.*
- (C) *$\{z \in \mathbb{P}^1_{\text{Berk}} : G(z, \zeta; E) = 0\} = \mathbb{P}^1_{\text{Berk}} \backslash U_\zeta$.*
- (D) *The support of the equilibrium distribution $\mu_{E,\zeta}$ of E with respect to ζ is precisely ∂U_ζ.*

PROOF. Assertions (A), (B), and (C) all follow easily from Proposition 7.37.

For (D), we know by Proposition 6.8 that $\text{supp}(\mu_{E,\zeta}) \subseteq \partial U_\zeta$. Conversely, suppose for the sake of contradiction that $x_0 \in \partial U_\zeta$ but $x_0 \notin \text{supp}(\mu_{E,\zeta})$. Then there is a neighborhood U_0 of x_0 such that $G(x, \zeta; E)$ is harmonic on U_0. As $G(x_0, \zeta; E) = 0$ and $G(x, \zeta; E) \geq 0$ for all $x \in U_0$, the Maximum Principle (Proposition 7.16) implies that $G(x, \zeta; E) = 0$ for all $x \in U_0$. But then (B) implies that U_0 is disjoint from U_ζ, a contradiction. □

There is an alternate formula for the Green's function of a strict closed affinoid:

PROPOSITION 7.40. *Suppose $K = \mathbb{C}_p$. Let $W \subset \mathbb{P}^1_{\text{Berk}}$ be a strict closed affinoid, and let U be a connected component of $\mathbb{P}^1_{\text{Berk}} \backslash W$. Then $\partial U \cap \partial W = \{a_1, \ldots, a_m\}$ is a finite set of points of type II, and for each $\zeta \in \mathbb{P}^1(K) \cap U$:*
- (A) *$\mathbb{P}^1_{\text{Berk}} \backslash U$ can be decomposed as the disjoint union of closed balls*

$$\bigcup_{i=1}^{m} \mathcal{B}(a_i, r_i)_\zeta \ ,$$

  where $r_i = \text{diam}_\zeta(a_i)$ for each $i = 1, \ldots, m$.
- (B) *There is a rational function $f_\zeta(z) \in K(z)$, having poles only at ζ, for which*

$$\mathbb{P}^1_{\text{Berk}} \backslash U = \{z \in \mathbb{P}^1_{\text{Berk}} : [f_\zeta]_z \leq 1\} \ .$$

- (C) *For any function $f_\zeta(z)$ satisfying the conditions in part (B), letting $N = \deg(f_\zeta)$, we have*

$$G(z, \zeta; W) = \begin{cases} \frac{1}{N} \log_v([f_\zeta]_z) & \text{if } z \in U \ , \\ 0 & \text{if } z \in \mathbb{P}^1_{\text{Berk}} \backslash U \ . \end{cases}$$

PROOF. After an affine change of coordinates, we can assume $\zeta = \infty$. Introduce coordinates on $\mathbb{P}^1_{\text{Berk}}$ so that $\mathbb{P}^1_{\text{Berk}} = \mathbb{A}^1_{\text{Berk}} \cup \{\infty\}$.

For part (A), put $W' = \mathbb{P}^1_{\text{Berk}} \backslash U$. Then W' is a strict closed affinoid whose complement has one connected component, and $\partial W' = \{a_1, \ldots, a_m\}$. Let $r_i = \text{diam}_\infty(a_i)$. Each of the balls $\mathcal{B}(a_i, r_i)_\infty$ is connected and has a_i as its unique boundary point. We claim that each $\mathcal{B}(a_i, r_i)_\infty \subseteq W'$. If not, for some i there would be a point $z_i \in \mathcal{B}(a_i, r_i)_\infty \cap U$. Since U is uniquely path-connected, the geodesic path from z_i to ∞ would be contained in U. However, a_i lies on that path, so $a_i \in U$, a contradiction. The balls $\mathcal{B}(a_i, r_i)_\infty$ are pairwise disjoint, since otherwise by the ultrametric inequality one would be contained in another, contradicting the fact that each $\mathcal{B}(a_i, r_i)_\infty$ has exactly one boundary point.

To show that $W' = \bigcup_{i=1}^m \mathcal{B}(a_i, r_i)_\infty$, suppose to the contrary that there were some $z_0 \in W' \backslash \bigcup_{i=1}^m \mathcal{B}(a_i, r_i)_\infty$. Then the geodesic path from z_0 to ∞ would pass through some a_i. Since $\mathcal{B}(a_i, r_i)_\infty$ has a_i as its only boundary point, either $z_0 \in \mathcal{B}(a_i, r_i)_\infty$ or $\infty \in \mathcal{B}(a_i, r_i)_\infty$, both of which are false by assumption.

For part (B), we use the fact that since the balls $\mathcal{B}(a_i, r_i)_\infty$ are pairwise disjoint and have radii belonging to $|\mathbb{C}_p^\times|$, there is a polynomial $f \in K[z]$ for which
$$\bigcup_{i=1}^m (K \cap \mathcal{B}(a_i, r_i)_\infty) = \{z \in K : |f(z)| \leq 1\} .$$
This fact is originally due to Cantor [**33**, Lemma 3.2.3]; a self-contained proof is given in [**90**, Lemma 12.1]. Since $K \cap W'$ is dense in W' and K is dense in $\mathbb{P}^1_{\text{Berk}}$, it follows that
$$W' = \{z \in \mathbb{P}^1_{\text{Berk}} : [f]_z \leq 1\} .$$

For part (C), note that $g(z) := G(z, \infty; W) - (1/N) \log_v([f]_z)$ is harmonic on $U \backslash \{\infty\}$. It extends to a function harmonic on U, because it is bounded in a neighborhood of ∞. Clearly, for each of the points $a_i \in \partial U$,
$$\lim_{\substack{z \to a_i \\ z \in U}} \frac{1}{N} \log_v([f]_z) = 0 .$$
Furthermore, the singleton set $\{a_i\}$ has positive capacity because $a_i \in \mathbb{H}_{\text{Berk}}$. Hence by Proposition 7.37(A)(4) and (A)(5),
$$\lim_{\substack{z \to a_i \\ z \in U}} G(z, \zeta; W) = G(a_i, \zeta; W) = 0 .$$
By the Maximum Principle, $g(z) = 0$ on U. Proposition 7.37(A)(4) gives $G(z, \zeta; W) = 0$ for all $z \notin U$, so we are done. $\square$

As a special case of Proposition 7.40, we have the following analogue of a well-known fact over $\mathbb{C}$.

COROLLARY 7.41. *Let $g(z) \in K[z]$ be a nonconstant polynomial. Fix $0 < R \in |K^\times|$, and put $W_R = \{z \in \mathbb{P}^1_{\text{Berk}} : [g]_z \leq R\}$. Suppose $\deg(g) = N$. Then $\gamma_\infty(W_R) = R^{1/N}$, and*

(7.20) $$G(z, \infty; W_R) = \begin{cases} \frac{1}{N} \log_v([g]_z/R) & \text{if } z \notin W_R , \\ 0 & \text{if } z \in W_R . \end{cases}$$

PROOF. Let $C \in K^\times$ be such that $|C| = R$, and put $f = g/C$. Then $W_R = \{z \in \mathbb{P}^1_{\text{Berk}} : [f]_z \leq 1\}$. It is well known (see for example [**14**, Lemma 4.10]) that $W_R \cap K$ is a finite union of discs $D(a_i, R_i)$ with each $R_i \in |K^\times|$; it follows from this that W_R is a strict closed affinoid. Applying Proposition 7.40(C) to W_R, the function f, and the point $\zeta = \infty$, we obtain (7.20). By formula (7.15) in Proposition 7.37(A)(7), we have $V_\infty(W_R) = -\log_v(R)/N$, so $\gamma_\zeta(W_R) = R^{1/N}$. □

The following approximation property holds, generalizing Corollary 6.9:

PROPOSITION 7.42. *Let $E \subset \mathbb{P}^1_{\text{Berk}} \setminus \{\zeta\}$ be compact.*

If E has positive capacity, then for any $\varepsilon > 0$, any open neighborhood U of E, and any $\zeta \in \mathbb{P}^1_{\text{Berk}} \setminus E$, there is a strict closed affinoid $W \subset U$, containing E in its interior, satisfying

(7.21) $$0 < V_\zeta(E) - V_\zeta(W) < \varepsilon ,$$

and such that for all $z \in \mathbb{P}^1_{\text{Berk}} \setminus (U \cup \{\zeta\})$,

(7.22) $$0 < G(z, \zeta; E) - G(z, \zeta; W) < \varepsilon .$$

If E has capacity 0, then for any $M > 0$, any open neighborhood U of E, and any $\zeta \in \mathbb{P}^1_{\text{Berk}} \setminus E$, there is a strict closed affinoid $W \subset U$, containing E in its interior, satisfying

(7.23) $$V_\zeta(W) > M ,$$

and such that for all $z \in \mathbb{P}^1_{\text{Berk}} \setminus (U \cup \{\zeta\})$,

(7.24) $$G(z, \zeta; E) > M .$$

More generally, given a finite set $\mathfrak{X} = \{\zeta_1, \ldots, \zeta_n\} \subset \mathbb{P}^1_{\text{Berk}} \setminus E$, regardless of whether E has positive capacity or capacity 0, there is a strict closed affinoid $W \subset U$ such that the assertions above hold simultaneously for each $\zeta \in \mathfrak{X}$.

PROOF. Fix $\zeta \in \mathbb{P}^1_{\text{Berk}} \setminus E$.

First suppose E has positive capacity. After shrinking U, we can assume without loss of generality that $\zeta \notin \overline{U}$. Let D_ζ be the connected component of $\mathbb{P}^1_{\text{Berk}} \setminus E$ containing ζ, and put

$$X = (\mathbb{P}^1_{\text{Berk}} \setminus U) \cap \overline{D_\zeta} .$$

Clearly X is compact; since $\partial D_\zeta \subset U$, X is contained in D_ζ. Let B be a simple subdomain of D_ζ, containing X, for which $\overline{B} \subset D_\zeta$; such a subdomain exists by Corollary 7.11. Put $U' = U \cap (\mathbb{P}^1_{\text{Berk}} \setminus \overline{B}) \subseteq U$.

By Corollary 6.9, there is a sequence of strict closed affinoids W_k, contained in U' and containing E in their interior, for which $\lim_{k \to \infty} V_\zeta(W_k) = V_\zeta(E)$. We will show that for a suitable k, we can take $W = W_k$.

Since $V_\zeta(E) \geq V_\zeta(W_k)$ for all k, (7.21) holds for all sufficiently large k. For the second assertion, note that for any k, each connected component of $\mathbb{P}^1_{\text{Berk}} \setminus W_k$ is contained in a connected component of $\mathbb{P}^1_{\text{Berk}} \setminus E$. If $z \in \mathbb{P}^1_{\text{Berk}} \setminus U$

does not belong to X, then $z \notin \overline{D}_\zeta$, and Propositions 7.37(A)(4) and 7.40 show that $G(z, \zeta; E) = G(z, \zeta; W_k) = 0$.

Now consider points $z \in X$. Put $h_k(z) = G(z, \zeta; E) - G(z, \zeta; W_k)$. If $\zeta \in \mathbb{H}_{\text{Berk}}$, then by Proposition 7.37(A)(7), h_k is harmonic on B with

(7.25) $$h_k(\zeta) = V_\zeta(E) - V_\zeta(W_k) .$$

If $\zeta \in \mathbb{P}^1(K)$, then h_k is harmonic on $B \backslash \{\zeta\}$, and by Propositions 6.12 and 7.19 it extends to a function harmonic on B, still denoted h_k, which satisfies (7.25).

By Proposition 7.37(C), $h_k(z) \geq 0$ for all $z \in B$. As $\lim_{k \to \infty} h_k(\zeta) = 0$, Harnack's inequality (Lemma 7.33), applied to ζ and X, shows that

$$\lim_{k \to \infty} G(z, \zeta; W_k) = G(z, \zeta; E)$$

uniformly on X. This yields (7.22).

Next suppose E has capacity 0. Necessarily $E \subset \mathbb{P}^1(K)$, which means that neighborhoods of E of the form

$$U' = \bigcup_{i=1}^{m} \mathcal{B}(a_i, r_i)^-$$

are cofinal in the set of all neighborhoods of E. Fix such a U' for which $\overline{U}' \subseteq U$ and $\zeta \notin \overline{U}'$. Put $B = \mathbb{P}^1_{\text{Berk}} \backslash \overline{U}$. Then B is a simple domain containing ζ, and the compact set $X = \mathbb{P}^1_{\text{Berk}} \backslash U$ is contained in B.

By Corollary 6.9, there is a sequence of strict affinoids $W_n \subset U'$, each of which contains E in its interior, for which $\lim_{n \to \infty} \gamma_\zeta(W_n) = 0$. Equivalently,

$$\lim_{n \to \infty} V_\zeta(W_n) = \infty .$$

For each n, put $h_n(z) = G(z, \zeta; W_n) - G(z, \zeta; \overline{U}')$. As above, each h_n is harmonic and nonnegative in $B \backslash \{\zeta\}$ and extends to a function harmonic on B with $h_n(\zeta) = V_\zeta(W_n) - V_\zeta(\overline{U}')$.

Let $C = C(\zeta, X) \geq 1$ be the constant given by Harnack's inequality (Lemma 7.33) for ζ, X, and B. Fix $M > 0$, and take n large enough so that $V_\zeta(W_n) \geq M$ and

$$V_\zeta(W_n) - V_\zeta(\overline{U}') > C \cdot M .$$

Then $h_n(z) \geq M$ for all $z \in X$, and since $G(z, \zeta; \overline{U}') \geq 0$, we conclude that $G(z, \zeta; W_n) \geq M$ for all $z \in U \backslash \{\zeta\}$. This proves (7.23) and (7.24).

For the final assertion, let $W^{(i)}$ be the strict closed affinoid constructed above when $\zeta = \zeta_i$, and put $W = \bigcap_{i=1}^{n} W^{(i)}$. The monotonicity of Green's functions (Proposition 7.37(C)) shows that W meets the requirements of the proposition. $\square$

In a pleasing improvement over the theory of Green's functions developed in [88], the equilibrium measure μ_ζ has an interpretation as the reproducing kernel for harmonic functions. In light of Corollary 7.26, this generalizes Proposition 7.29:

7.6. GREEN'S FUNCTIONS

PROPOSITION 7.43. *Let $U \subset \mathbb{P}^1_{\text{Berk}}$ be a domain such that ∂U has positive capacity. For each $\zeta \in U$, let μ_ζ be the equilibrium measure on ∂U relative to ζ. Suppose f is harmonic on U and extends to a continuous function on $\overline{U}$. Then*

$$f(\zeta) = \int_{\partial U} f(z) \, d\mu_\zeta(z) \,.$$

PROOF. First suppose $\zeta \in U$ is not of type I. Using Corollary 7.11, we may choose an exhaustion of U by a sequence of simple subdomains $V_n \subset U$. Without loss of generality, we can assume that $\zeta \in V_n$ for all n. Put $g_n(x) = G(x, \zeta; V_n)$. Since ∂V_n consists of a finite number of points, none of which is of type I, $g_n(x)$ is continuous on $\overline{V}_n$ and $g_n(x) = 0$ for each $x \in \partial V_n$. Note that $g_n \in \text{BDV}(V_n)$, $g_n \in \mathcal{C}(\overline{V}_n)$ for each n, and f is continuous on V_n and strongly harmonic on V_n by Lemma 7.7; in particular $f \in \text{BDV}(V_n)$.

Fix n, and let $\Gamma \subset V_n$ be a finite subgraph. Then $f|_\Gamma, g|_\Gamma \in \text{CPA}(\Gamma)$. By Proposition 3.2(D),

$$\int_\Gamma f \, \Delta_\Gamma(g_n) = \int_\Gamma g_n \, \Delta_\Gamma(f) \,. \tag{7.26}$$

Taking a limit over finite subgraphs Γ, we find using Corollary 5.39 that

$$\int_{\overline{V}_n} f \, \Delta_{\overline{V}_n}(g_n) = \int_{\overline{V}_n} g_n \, \Delta_{\overline{V}_n}(f) \,. \tag{7.27}$$

Here $\Delta_{\overline{V}_n}(g_n) = \delta_\zeta - \mu_{\zeta,n}$, where $\mu_{\zeta,n}$ is the equilibrium measure of $E_n = \mathbb{P}^1_{\text{Berk}} \setminus V_n$ with respect to ζ. It is supported on ∂V_n. Since f is strongly harmonic on V_n, $\Delta_{\overline{V}_n}(f)$ is supported on ∂V_n, where $g_n(x) = 0$. Hence the integral on the right side of (7.27) is 0. It follows that

$$f(\zeta) = \int f(x) \, d\mu_{\zeta,n}(x) \,. \tag{7.28}$$

By Corollary 6.11 (applied to $W_n = \mathbb{P}^1_{\text{Berk}} \setminus V_n$), the sequence $\langle \mu_{\zeta,n} \rangle$ converges weakly to μ_ζ, the equilibrium measure of $E = \mathbb{P}^1_{\text{Berk}} \setminus U$. Since f is continuous on $\overline{U}$,

$$f(\zeta) = \lim_{n \to \infty} \int f(x) \, d\mu_{\zeta,n}(x) = \int f(x) \, d\mu_\zeta(x) \,,$$

yielding the result in this case.

If ζ is of type I, let t approach ζ along a path in $\mathbb{H}_{\text{Berk}}$. As in the proof of Proposition 7.37(B), $\mu_\zeta = \mu_t$ for t sufficiently near ζ. Since f is continuous, it follows from the previous case that $f(\zeta) = \int_{\partial U} f(x) \, d\mu_\zeta(x)$. □

In the classical theory over $\mathbb{C}$, for a domain U with a piecewise smooth boundary, the reproducing kernel is the inward normal derivative of the Green's function $G_U(z, \zeta)$ on ∂U,

$$\mu_\zeta = \frac{1}{2\pi} \frac{\partial}{\partial n} G_U(z, \zeta) \,.$$

On $\mathbb{P}^1_{\text{Berk}}$, if we take $E = \mathbb{P}^1_{\text{Berk}}\backslash U$, then Proposition 7.43, combined with part (A)(8) of Proposition 7.37, shows that the reproducing kernel is the negative of the boundary derivative of the Green's function:

$$\mu_\zeta = -\Delta_{\partial U}(G(z, \zeta; E)) := -\Delta(G(z, \zeta; E))|_{\partial U} \ .$$

This justifies our interpretation of the boundary derivative of a function in BDV(U) as an analogue of the classical outward normal derivative.

7.7. Pullbacks

In this section we show that harmonicity is preserved under pullbacks by meromorphic functions. As an application, we obtain a pullback formula for Green's functions under rational maps. We also show that pullbacks of sets of capacity 0 by rational maps have capacity 0.

We first need a lemma asserting that logarithms of norms of meromorphic functions are harmonic.

LEMMA 7.44. *Let U be a domain, and suppose g is a meromorphic function on U. Then $f(x) := \log_v([g]_x)$ is harmonic on $U \backslash \operatorname{supp}(\operatorname{div}(g))$.*

PROOF. By Corollary 7.11, we can exhaust U by an increasing sequence of simple domains $U_1 \subset U_2 \subset \cdots$ with $\overline{U}_k \subset U_{k+1}$ for each k. For each U_k, choose a strict closed affinoid V_k with $\overline{U}_k \subset V_k \subset U_{k+1}$. (Such a V_k can be constructed as follows: There are a finite subgraph Γ of $\mathbb{P}^1_{\text{Berk}}$ and connected open sets $W_k \subset W_{k+1} \subset \Gamma$ with $\overline{W}_k \subset W_{k+1}$ such that $U_i = r_\Gamma^{-1}(W_i)$ for $i = k, k+1$. Choose any subgraph Γ' of Γ with $W_k \subset \Gamma' \subset W_{k+1}$, and let $V_k = r_\Gamma^{-1}(\Gamma')$.)

Let $\mathcal{A}_k$ be an affinoid algebra for which $V_k = \mathcal{M}(\mathcal{A}_k)$. By the Weierstrass Preparation Theorem for affinoid algebras (see Theorem 2.2.9 and Proposition 2.5.10 of [**55**]), for each k there are coprime polynomials P_k and Q_k and a unit power series $u_k \in \mathcal{A}_k$, such that $g(z) = (P_k(z)/Q_k(z)) \cdot u_k(z)$ on $V_k \cap K$. A unit power series satisfies $[u_k]_x = 1$ for all x, so

$$\log_v([g]_x) = \log_v([P_k]_x) - \log_v([Q_k]_x)$$

is strongly harmonic on $V_k \backslash (\operatorname{div}(P_k/Q_k))$. $\square$

COROLLARY 7.45. *Let U and V be domains in $\mathbb{P}^1_{\text{Berk}}$. Suppose $\Phi(z)$ is meromorphic in U, with $\Phi(U) \subset V$. If f is harmonic on V, then $f \circ \Phi$ is harmonic on U.*

PROOF. Let x be the variable on U, and y the variable on V. By Corollary 7.32, there are rational functions g_k and rational numbers R_k such that

$$\lim_{k \to \infty} R_k \cdot \log_v([g_k]_y) = f(y)$$

uniformly on compact subsets of V. For each $y \in V$, let V_y be a neighborhood of y with $\overline{V}_y \subset V$; since f is harmonic, there is a constant C_y so that g_k has no zeros or poles in $\overline{V}_y$ if $k \geq C_y$.

Suppose $x \in U$ satisfies $\Phi(x) = y$. Choose a simple neighborhood U_x of x contained in $\Phi^{-1}(V_y)$. The functions $g_k \circ \Phi$ are meromorphic in U_x, and if $k \geq C_y$, they have no zeros or poles in U_x. The functions $R_k \cdot \log_v([g_k \circ \Phi]_z)$ converge to $f \circ \Phi(z)$ for each $z \in U_x$.

By Lemma 7.44 and Proposition 7.31, $f \circ \Phi$ is harmonic on U_x. By Proposition 7.23, $f \circ \Phi$ is strongly harmonic on U_x. Since $x \in U$ is arbitrary, $f \circ \Phi$ is harmonic on U. $\square$

LEMMA 7.46. *Let $\varphi : \mathbb{P}^1_{\text{Berk}} \to \mathbb{P}^1_{\text{Berk}}$ be induced by a nonconstant rational function $\varphi(T) \in K(T)$. If a set $e \subset \mathbb{P}^1_{\text{Berk}}$ has capacity 0, then $\varphi^{-1}(e)$ has capacity 0 as well.*

PROOF. Since e has capacity 0, necessarily $e \subset \mathbb{P}^1(K)$ (see the remarks before Proposition 6.1), so $\varphi^{-1}(e) \subset \mathbb{P}^1(K)$ as well (Proposition 2.15). Fix a coordinate system on $\mathbb{P}^1$ and use it on both the source and target. Removing a finite set of points from a set does not change its capacity (Corollary 6.21), so without loss of generality we can assume that e does not contain ∞ or $f(\infty)$. By formula (6.2), we can also assume that e is compact.

If $\text{char}(K) = 0$, then φ is automatically separable. If $\text{char}(K) = p > 0$, then $\varphi(T) = f(T^{p^m})$ for some m, where f is separable. By decomposing φ as a composition of a sequence of maps, there are two cases to consider: the case where φ is separable and the case where $\varphi(T) = T^p$ with $p = \text{char}(K)$.

First assume φ is separable. Using the same reasoning as before, we can arrange that e does not contain any of the finitely many ramification points of φ. Hence φ is locally one-to-one on $\varphi^{-1}(e)$ in $\mathbb{P}^1(K)$, and we can cover e by a finite number of (classical) discs $D(a_i, r_i)$ on which φ is one-to-one. Since φ is one-to-one on $D(a_i, r_i)$, there is a constant $C_i > 0$ such that for all $z, w \in D(a_i, r_i)$,

$$\frac{1}{C_i}|z - w| \leq |\varphi(z) - \varphi(w)| \leq C_i |z - w| \ .$$

Suppose $\varphi^{-1}(e)$ had positive capacity. A finite union of sets of capacity 0 has capacity 0 (Corollary 6.21), so for some i the set $A_i = \varphi^{-1}(e) \cap D(a_i, r_i)$ must have positive capacity. As $\varphi^{-1}(e)$ is compact and $D(a_i, r_i)$ is closed, A_i is compact. Put $B_i = f(A_i)$. Both A_i and B_i are Hausdorff, so $\varphi : A_i \to B_i$ is a homeomorphism.

Consider capacities relative to the point ∞. Proposition 6.1 shows that $\gamma_\infty(A_i) > 0$, and for $x, y \in \mathbb{A}^1(K)$ we have $\delta(x,y)_\infty = |x - y|$, so there is a probability measure μ supported on A_i for which

$$I(\mu) = \int -\log_v(|z - w|) \, d\mu(z) d\mu(w) \ < \ \infty \ .$$

Using φ, transport μ to a measure $\nu = \varphi_*(\mu)$ on B_i. Then

$$\begin{aligned}
I(\nu) &= \int -\log_v(|x-y|)\, d\nu(x)d\nu(y) \\
&= \int -\log_v(|\varphi(z) - \varphi(w)|)\, d\mu(z)d\mu(w) \\
&\leq \int -\log_v(|z-w|/C_i)\, d\mu(z)d\mu(w) < \infty.
\end{aligned}$$

(The second step can be justified either by viewing f as an identification or by considering a limit as $N \to \infty$ with the truncated kernels $k_N(x,y) = \min(N, -\log_v(|x-y|))$, which are continuous.) It follows that $\gamma_\infty(B_i) > 0$. Since $B_i \subset e$, this is a contradiction.

Now consider the case where $\varphi(T) = T^p$ is purely inseparable. In this case, φ is one-to-one on all of $\mathbb{P}^1(K)$ and

$$|\varphi(z) - \varphi(w)| = |z^p - w^p| = |(z-w)^p| = |z-w|^p.$$

Since $\varphi^{-1}(e)$ is compact, $\varphi : \varphi^{-1}(e) \to e$ is a homeomorphism. If $\varphi^{-1}(e)$ had positive capacity, then by an argument similar to the one earlier we would conclude that e had positive capacity. This is a contradiction, so we are done. $\square$

There is also a pullback formula for Green's functions. We prove the formula only when $\zeta \in \mathbb{P}^1(K)$, giving a version sufficient for the needs of Proposition 7.50 below. In Proposition 9.57, after we have developed a theory of subharmonic functions and a theory of multiplicities on $\mathbb{P}^1_{\text{Berk}}$, we will establish the formula for arbitrary $\zeta \in \mathbb{P}^1_{\text{Berk}} \backslash E$, without the exceptional set e.

PROPOSITION 7.47. *Suppose $E \subset \mathbb{P}^1_{\text{Berk}}$ is compact and has positive capacity, and let $\zeta \in \mathbb{P}^1(K) \backslash E$. Let $\varphi(T) \in K(T)$ be a nonconstant rational function. Write $\varphi^{-1}(\{\zeta\}) = \{\xi_1, \ldots, \xi_m\}$, and for the divisor (ζ) write*

$$\varphi^{-1}((\zeta)) = \sum_{i=1}^m n_i(\xi_i).$$

Then there is a set e of capacity 0, contained in $\varphi^{-1}(E)$, such that for each $z \in \mathbb{P}^1_{\text{Berk}} \backslash e$,

(7.29) $$G(\varphi(z), \zeta; E) = \sum_{i=1}^m n_i G(z, \xi_i; \varphi^{-1}(E)).$$

PROOF. By Proposition 7.37(A)(4) and (A)(5), there is a set $e_0 \subset E$ of capacity 0 such that $G(z, \zeta; E) = 0$ for each $z \in E \backslash e_0$ and $G(z, \zeta; E)$ is continuous on $z \in \mathbb{P}^1_{\text{Berk}} \backslash e_0$. Similarly, for each ξ_i there is a subset $e_i \subset \varphi^{-1}(E)$ of capacity 0 such that $G(z, \xi_i; \varphi^{-1}(E)) = 0$ for each $z \in \varphi^{-1}(E) \backslash e_0$

and $G(z, \xi_i; \varphi^{-1}(E))$ is continuous on $z \in \mathbb{P}^1_{\text{Berk}} \backslash e_0$. By Lemma 7.46, $\varphi^{-1}(e_0)$ has capacity 0. Put

$$e = \varphi^{-1}(e_0) \cup (\bigcup_{i=1}^{m} e_i) \ .$$

Then e has capacity 0 by Corollary 6.21.

For $z \in \mathbb{P}^1_{\text{Berk}} \backslash \{\xi_1, \ldots, \xi_m\}$, put

$$h(z) = G(\varphi(z), \zeta; E) - \sum_{i=1}^{m} n_i G(z, \xi_i; \varphi^{-1}(E)) \ .$$

On $\varphi^{-1}(E) \backslash e$, we have $h(z) = 0$ by the definition of e.

Let the zeros of $\varphi(z)$ be $a_1, \ldots, a_k$, with multiplicities $h_1, \ldots, h_k$, respectively. Since $\|z, w\|$ is the Hsia kernel relative to the Gauss point, by Corollary 4.14 there is a constant $C > 0$ such that for each $z \in \mathbb{P}^1_{\text{Berk}}$,

$$[\varphi]_z = C \cdot \prod_{j=1}^{k} \|z, a_j\|^{h_j} / \prod_{i=1}^{m} \|z, \xi_i\|^{n_i} \ .$$

Hence Proposition 7.37(A)(7) shows that h is bounded on a neighborhood of each ξ_i.

Now consider h on a connected component V of $\mathbb{P}^1_{\text{Berk}} \backslash \varphi^{-1}(E)$. By Proposition 7.37(A)(3) and Lemma 7.44, h is harmonic on $V \backslash \{\xi_1, \ldots, \xi_m\}$. By Proposition 7.19 and the remarks above, h extends to a function harmonic on V. By the definition of e, for each $x \in \partial V \backslash e$ we have

$$\lim_{\substack{z \to x \\ z \in V}} h(z) = 0 \ .$$

By the Strong Maximum Principle (Proposition 7.17), $h(z) \equiv 0$ on V.

Since this holds for each component of $\mathbb{P}^1_{\text{Berk}} \backslash \varphi^{-1}(E)$, we have $h(z) = 0$ for all $z \in \mathbb{P}^1_{\text{Berk}} \backslash (e \cup \{\xi_1, \ldots, \xi_m\})$. This yields (7.29), since both sides of (7.29) are ∞ at the points ξ_i. □

7.8. The multi-center Fekete-Szegö theorem

In this section we establish a Berkovich space version of the multi-center Fekete-Szegö theorem. The results of this section will not be used elsewhere in the book.

To provide context for the theorem, note that the Fekete-Szegö theorem proved in §6.5 concerns algebraic points whose conjugates satisfy two kinds of conditions: first, they lie *near* a specified set $E_v \subset \mathbb{P}^1_{\text{Berk},v} \backslash \{\infty\}$ at a finite number of places v; second, they *avoid* the point ∞ at all remaining places (since $E_v = \mathcal{D}(0,1) = \mathbb{P}^1_{\text{Berk},v} \backslash \mathcal{B}(\infty, 1)^-$ for all but finitely many v). The multi-center Fekete-Szegö theorem concerns points which avoid a finite set of points $\mathfrak{X} = \{x_1, \ldots, x_n\}$. For example, points which avoid ∞ at all finite places are algebraic integers, whereas points which avoid ∞ and 0 at all finite places are algebraic units.

We will use the same notations as in §6.5. Let k be a global field. For each place v of k, let k_v be the completion of k at v, let $\overline{k}_v$ be an algebraic closure of k_v, and let $\mathbb{C}_v$ denote the completion of $\overline{k}_v$. If v is non-Archimedean, let q_v be the order of the residue field of k_v and normalize the absolute value on $\mathbb{C}_v$ so that $|x|_v = q_v^{-\operatorname{ord}_v(x)}$. Write $\log_v(x)$ for the logarithm to the base q_v. If v is Archimedean, let $|x|_v = |x|$ be the usual absolute value on $\mathbb{C}_v \cong \mathbb{C}$, and let $\log_v(x) = \ln(x)$. Put $q_v = e$ if $k_v \cong \mathbb{R}$, and put $q_v = e^2$ if $k_v \cong \mathbb{C}$. For $0 \neq \kappa \in k$, the product formula reads

$$(7.30) \qquad \sum_v \log_v(|\kappa|_v) \ln(q_v) = 0 \ .$$

Let $\overline{k}$ be an algebraic closure of k. For each v, fix an embedding $\overline{k} \hookrightarrow \mathbb{C}_v$ and view it as an identification. Let $\mathbb{P}^1_{\text{Berk},v}$ be the Berkovich projective line over $\mathbb{C}_v$ (which we take to mean $\mathbb{P}^1(\mathbb{C})$ if v is Archimedean). Using the identification above, view $\mathbb{P}^1(\overline{k})$ as a subset of $\mathbb{P}^1_{\text{Berk},v}$.

Let $\mathfrak{X} = \{x_1, \ldots, x_n\}$ be a finite subset of $\mathbb{P}^1(\overline{k})$, stable under $\operatorname{Gal}(\overline{k}/k)$, the group of all automorphisms of $\overline{k}$ fixing k. Fix a coordinate system on $\mathbb{P}^1/k$ and use it to induce coordinates on $\mathbb{P}^1_{\text{Berk},v}$. Using the embedding of $\mathbb{P}^1(\overline{k})$ into $\mathbb{P}^1(\mathbb{C}_v)$ chosen above, we can identify $\mathfrak{X}$ with a subset of $\mathbb{P}^1_{\text{Berk},v}$, for each v. Since $\mathfrak{X}$ is stable under $\operatorname{Gal}(\overline{k}/k)$, its image is independent of the choice of the embedding. Define a *compact Berkovich adelic set relative to* $\mathfrak{X}$ to be a set of the form

$$\mathbb{E} = \prod_v E_v$$

where E_v is a compact nonempty subset of $\mathbb{P}^1_{\text{Berk},v} \backslash \mathfrak{X}$ for each place v of k and for all but finitely many v,

$$(7.31) \qquad E_v = \mathbb{P}^1_{\text{Berk},v} \backslash \bigcup_{i=1}^n \mathcal{B}(x_i, 1)^- .$$

If the balls $\mathcal{B}(x_i, 1)^-$ are pairwise disjoint, we call a set E_v of the form (7.31) $\mathfrak{X}$-*trivial*.

We will say that E_v is k_v-symmetric if it is stable under the group of continuous automorphisms $\operatorname{Gal}^c(\mathbb{C}_v/k_v)$, for the natural Galois action on $\mathbb{P}^1_{\text{Berk},v}$, and we will say that $\mathbb{E}$ is k-symmetric if each E_v is k_v-symmetric.

A *Berkovich adelic neighborhood of* $\mathbb{E}$ *relative to* $\mathfrak{X}$ is a set $\mathbb{U}$ of the form

$$\mathbb{U} = \prod_v U_v \ ,$$

where for each v, either U_v is an open neighborhood of E_v in $\mathbb{P}^1_{\text{Berk},v}$ or

$$U_v = E_v = \mathbb{P}^1_{\text{Berk},v} \backslash \bigcup_{i=1}^n \mathcal{B}(x_i, 1)^- \ .$$

Assuming that $\mathbb{E}$ is a k-symmetric compact Berkovich adelic set relative to $\mathfrak{X}$, we will now define the global capacity $\gamma(\mathbb{E}, \mathfrak{X})$.

7.8. THE MULTI-CENTER FEKETE-SZEGŐ THEOREM

If some E_v has capacity 0, put $\gamma(\mathbb{E}, \mathfrak{X}) = 0$. If each E_v has positive capacity, the definition uses the theory of Green's functions developed in §7.6. Note that in the Archimedean case, Green's functions satisfying the properties in Proposition 7.37 were constructed in [**88**, §3].

First suppose $\mathfrak{X} \subset \mathbb{P}^1(k)$. For each x_i, fix a k-rational uniformizing parameter $g_{x_i}(z)$ at x_i: a function $g_{x_i}(z) \in k(z)$ with a simple zero at x_i. For each v, define the $g_{x_i}(z)$-Robin constant

$$V_{g_{x_i}}(E_v) = \lim_{z \to x_i} G(z, x_i; E_v) + \log_v([g_{x_i}]_z) \ .$$

The limit exists by Proposition 7.37(A)(7) and its Archimedean counterpart: in terms of the Robin constant $V_{x_i}(E_v)$ defined earlier,

$$V_{g_{x_i}}(E_v) = V_{x_i}(E_v) + \lim_{z \to x_i} \log_v([g_{x_i}]_z / \|z, x_i\|_v) \ .$$

Define the *local Green's matrix*

$$\Gamma(E_v, \mathfrak{X}) = \begin{bmatrix} V_{g_{x_1}}(E_v) & G(x_1, x_2; E_v) & \cdots & G(x_1, x_n; E_v) \\ G(x_2, x_1; E_v) & V_{g_{x_2}}(E_v) & \cdots & G(x_2, x_n; E_v) \\ \vdots & \vdots & \ddots & \vdots \\ G(x_n, x_1; E_v) & G(x_n, x_2; E_v) & \cdots & V_{g_{x_n}}(E_v) \end{bmatrix} \ .$$

Then define the *global Green's matrix* by

$$\Gamma(\mathbb{E}, \mathfrak{X}) = \sum_v \Gamma(E_v, \mathfrak{X}) \ln(q_v) \ .$$

This is actually a finite sum; for each place v where E_v is $\mathfrak{X}$-trivial and where each $g_{x_i}(z)$ coincides with $\|z, x_i\|_v$ on a neighborhood of x_i, the matrix $\Gamma(E_v, \mathfrak{X})$ is the zero matrix; these conditions hold for all but finitely many v. The product formula (7.30) shows that $\Gamma(\mathbb{E}, \mathfrak{X})$ is independent of the choice of the $g_{x_i}(z)$.

By the symmetry and nonnegativity of the Green's functions (Proposition 7.37), $\Gamma(\mathbb{E}, \mathfrak{X})$ is symmetric and nonnegative off the diagonal. Let

$$\mathcal{P}^n = \{(s_1, \ldots, s_n) \in \mathbb{R}^n : \text{each } s_i \geq 0, s_1 + \cdots + s_n = 1\}$$

be the set of n-element probability vectors, viewed as column vectors. Define the *global Robin constant* by

(7.32) $$V(\mathbb{E}, \mathfrak{X}) = \max_{\vec{s} \in \mathcal{P}^n} \min_{\vec{t} \in \mathcal{P}^n} \vec{s}^T \Gamma(\mathbb{E}, \mathfrak{X}) \vec{t} \ .$$

Finally, define the *global capacity* by

(7.33) $$\gamma(\mathbb{E}, \mathfrak{X}) = e^{-V(\mathbb{E}, \mathfrak{X})} \ .$$

In the general case when $\mathfrak{X} \not\subset \mathbb{P}^1(k)$, let L/k be a finite normal extension such that $\mathfrak{X} \subset \mathbb{P}^1(L)$, and define a set $\mathbb{E}_L$ as follows. For each place v of k and each place w of L with $w|v$, fix an embedding $L_w \hookrightarrow \mathbb{C}_v$ over k_v. Extend it to an isomorphism $\mathbb{C}_w \cong \mathbb{C}_v$, which induces an isomorphism $\mathbb{P}^1_{\text{Berk},w} \cong \mathbb{P}^1_{\text{Berk},v}$; let E_w be the pullback of E_v under this isomorphism. Because E_v is stable under $\mathrm{Gal}^c(\mathbb{C}_v/k_v)$, the set E_w is independent of the embedding $L_w \hookrightarrow \mathbb{C}_v$ and is stable under $\mathrm{Gal}^c(\mathbb{C}_w/L_w)$. Put

$$\mathbb{E}_L = \prod_w E_w . \tag{7.34}$$

Using the Green's functions and numbers q_w obtained using the normalized absolute values for L, define

$$\Gamma(\mathbb{E},\mathfrak{X}) = \frac{1}{[L:k]}\Gamma(\mathbb{E}_L,\mathfrak{X}) = \frac{1}{[L:k]}\sum_{w \text{ of } L}\Gamma(E_w,\mathfrak{X})\ln(q_w) .$$

As in [**88**, §5.1], the Galois stability of the E_v and the local-global degree formula ($[L:k] = \sum_{w|v}[L_w:k_v]$ for each v of k; see [**88**, p. 321] for the proof in positive characteristic) can be used to show that $\Gamma(\mathbb{E},\mathfrak{X})$ is independent of the choice of L. We then define $V(\mathbb{E},\mathfrak{X})$ and $\gamma(\mathbb{E},\mathfrak{X})$ using (7.32), (7.33) as before.

This definition of the global capacity was originally proposed by D. Cantor in [**33**]. Its meaning is not immediately obvious. The best way to motivate it is to remark that $V(\mathbb{E},\mathfrak{X})$, which is the *value* $\mathrm{val}(\Gamma)$ *of* $\Gamma = \Gamma(\mathbb{E},\mathfrak{X})$ *as a matrix game*, is a numerical statistic of Green's matrices which has good functoriality properties under base change and pullbacks by rational maps and is negative if and only if $\Gamma(\mathbb{E},\mathfrak{X})$ is negative definite (see [**88**, §5.1] for more details). Thus, $\gamma(\mathbb{E},\mathfrak{X}) > 1$ if and only if $\Gamma(\mathbb{E},\mathfrak{X})$ is negative definite. Likewise, $\gamma(\mathbb{E},\mathfrak{X}) < 1$ if and only if when the rows and columns of $\Gamma(\mathbb{E},\mathfrak{X})$ are permuted so as to bring it into "minimal block-diagonal form", each "block" has a positive eigenvalue (see [**88**, p. 328]).

When $\mathfrak{X} = \{\infty\}$, the global capacity $\gamma(\mathbb{E},\{\infty\})$ coincides with the capacity $\gamma_\infty(\mathbb{E})$ defined in §6.5. To see this, note that if we take $g_\infty(z) = 1/z$, then $V_{g_\infty}(E_v) = V_\infty(E_v)$ for each v. Hence the global Green's matrix $\Gamma(\mathbb{E},\{\infty\})$ is the 1×1 matrix with unique entry

$$V(\mathbb{E},\{\infty\}) = \sum_v V_\infty(E_v)\ln(q_v) .$$

Since $\gamma_\infty(E_v) = q_v^{-V_\infty(E_v)}$ for each v, it follows that $\gamma(\mathbb{E},\{\infty\}) = \gamma_\infty(\mathbb{E})$.

We can now state the multi-center Fekete-Szegö theorem:

THEOREM 7.48. *Let k be a global field, let $\mathfrak{X} \subset \mathbb{P}^1(\overline{k})$ be a finite set of points stable under $\mathrm{Gal}(\overline{k}/k)$, and let $\mathbb{E}$ be a k-symmetric compact Berkovich adelic set relative to $\mathfrak{X}$.*

(A) *If $\gamma(\mathbb{E},\mathfrak{X}) < 1$, then there is a Berkovich adelic neighborhood $\mathbb{U}$ of $\mathbb{E}$ relative to $\mathfrak{X}$, such that the set of points of $\mathbb{P}^1(\overline{k})$, all of whose conjugates are contained in $\mathbb{U}$, is finite.*

(B) *If $\gamma(\mathbb{E},\mathfrak{X}) > 1$, then for every Berkovich adelic neighborhood $\mathbb{U}$ of $\mathbb{E}$ relative to $\mathfrak{X}$, the set of points of $\mathbb{P}^1(\overline{k})$, all of whose conjugates are contained in $\mathbb{U}$, is infinite.*

7.8. THE MULTI-CENTER FEKETE-SZEGÖ THEOREM

Theorem 7.48 will be proved by reducing it to Theorems 6.3.1 and 6.3.2 of [**88**]. To do so, we need to know that the generalized capacities $\gamma(\mathbb{E}, \mathfrak{X})$ defined here and in the book [**88**] are compatible. The definition of the generalized capacity in [**88**] is formally the same as that in (7.33), except that the Green's functions and Robin constants used are the ones from [**88**]. We now recall the definitions of those quantities, indicating them with a superscript R.

Let v be non-Archimedean and take $K = \mathbb{C}_v$. Given a point $\zeta \in \mathbb{P}^1(\mathbb{C}_v)$ and a set $F_v^R \subset \mathbb{P}^1(\mathbb{C}_v) \backslash \{\zeta\}$ which is bounded away from ζ, for each $z \in \mathbb{P}^1(\mathbb{C}_v)$ put

$$(7.35) \qquad G^R(z, \zeta; F_v^R) = \sup_{\substack{W \supset F_v^R, \ \zeta \notin W \\ W = \text{strict closed affinoid}}} G(z, \zeta; W) ,$$

and let

$$(7.36) \qquad V_\zeta^R(F_v^R) = \sup_{\substack{W \supset F_v^R, \ \zeta \notin W \\ W = \text{strict closed affinoid}}} V_\zeta(W) .$$

By Proposition 7.40 and Theorem 6.24,

$$V_\zeta^R(F_v^R) = -\log_v(\gamma_\zeta^R(F_v^R))$$

where $\gamma_\zeta^R(F_v^R)$ is the outer capacity defined in §6.4.

We claim that for *algebraically capacitable sets* F_v^R (see [**88**, §4.3]), which by [**88**, Theorem 4.3.11] include finite unions of compact sets and "RL-domains" (intersections of strict closed affinoids with $\mathbb{P}^1(\mathbb{C}_v)$), the function $G^R(z, \zeta; F_v^R)$ coincides with the Green's function defined in [**88**, §4.4].

For any F_v^R the definition of the *lower Green's function* $\underline{G}(z, \zeta; F_v^R)$ in [**88**, p. 282] is the same as the definition of $G^R(z, \zeta; F_v^R)$ in (7.35), but with the limit over strict closed affinoids W containing F_v^R replaced by a limit over "PL$_\zeta$-domains" containing F_v^R. By definition, a PL$_\zeta$-domain is a set of the form

$$D(f) = \{z \in \mathbb{P}^1(\mathbb{C}_v) : |f(z)|_v \leq 1\}$$

where $f(z) \in \mathbb{C}_v(z)$ is a function with poles only at ζ; the Green's function of $D(f)$ is defined to be $(1/\deg(f))\log_v(|f(z)|)$ for $z \notin D(f)$. Proposition 7.40 shows that the limit in (7.35) coincides with the limit described above.

For algebraically capacitable sets F_v^R, the Green's function in [**88**] is defined to be the lower Green's function: see [**88**, Definition 4.4.12]. Furthermore, by Proposition 6.12 and its proof,

$$V_\zeta^R(F_v^R) = \lim_{z \to \zeta} G^R(z, \zeta; F_v^R) + \log_v(\|z, \zeta\|_v) .$$

The right side is the definition of the Robin constant in [**88**], relative to the uniformizer $\|z, \zeta\|_v$: see [**88**, Theorem 4.4.15] and formula (37).

In [**88**] the global adelic capacity is defined for adelic sets of the form $\mathbb{F}^R = \prod_v F_v^R$, where F_v^R is an algebraically capacitable subset of $\mathbb{P}^1(\mathbb{C}_v)$ for each place v of k and for all but finitely many v,

$$F_v^R = \mathbb{P}^1(\mathbb{C}_v) \backslash (\bigcup_{i=1}^{n} B(x_i, 1)^-) \ .$$

For such sets, the local and global Green's matrices $\Gamma_v^R(F_v^R, \mathfrak{X})$ and $\Gamma^R(\mathbb{F}^R, \mathfrak{X})$, the global Robin constant $V^R(\mathbb{F}^R, \mathfrak{X})$, and the global capacity $\gamma^R(\mathbb{F}^R, \mathfrak{X})$ are defined by the same formalism as in (7.32), (7.33). See [**88**, §5.1] for details.

The following generalization of Corollary 6.26 holds:

COROLLARY 7.49. *Suppose $K = \mathbb{C}_v$ and $\zeta \in \mathbb{P}^1(K)$. If E is a compact subset of $\mathbb{P}^1_{\text{Berk}} \backslash \{\zeta\}$ of positive capacity and $E \cap \mathbb{P}^1(K)$ is dense in E, then*

$$V_\zeta^R(E \cap \mathbb{P}^1(K)) = V_\zeta(E) \ ,$$

and for each $z \in \mathbb{P}^1(K) \backslash (E \cup \{\zeta\})$

$$G^R(z, \zeta; E \cap \mathbb{P}^1(K)) = G(z, \zeta; E) \ .$$

PROOF. Since $E \cap \mathbb{P}^1(K)$ is dense in E, any strict closed affinoid W containing $E \cap \mathbb{P}^1(K)$ also contains E. Hence by Proposition 7.37(C),

$$V_\zeta^R(E \cap \mathbb{P}^1(K)) \leq V_\zeta^R(E) \quad \text{and} \quad G^R(z, \zeta; E \cap \mathbb{P}^1(K)) \leq G(z, \zeta; E)$$

for $z \in \mathbb{P}^1_{\text{Berk}} \backslash E$. The opposite inequalities follow from Proposition 7.42. □

PROOF OF THEOREM 7.48. We will show how to reduce the result to Theorems 6.3.1 and 6.3.2 of [**88**]. First note that the quantity

$$\text{val}(\Gamma) = \max_{\vec{s} \in \mathcal{P}^n} \min_{\vec{t} \in \mathcal{P}^n} \vec{s}^T \Gamma \vec{t}$$

is a continuous function of matrices $\Gamma \in M_n(\mathbb{R})$. By the definition (7.32) of $V(\mathbb{E}, \mathfrak{X})$, for each $\varepsilon > 0$, there is a $\delta > 0$ such that if $\Gamma \in M_n(\mathbb{R})$ is any matrix whose entries all satisfy

$$\left| \Gamma_{ij} - \Gamma(\mathbb{E}, \mathfrak{X})_{ij} \right| < \delta \ ,$$

then $|\text{val}(\Gamma) - V(\mathbb{E}, \mathfrak{X})| < \varepsilon$.

If $\gamma(\mathbb{E}, \mathfrak{X}) = 0$, then some E_v has capacity 0. Using Proposition 7.42 (or [**88**, Proposition 3.3.1] and its proof in the Archimedean case), we can replace each E_v with capacity 0 by a k_v-symmetric set of positive capacity containing it, in such a way that all the entries of $\Gamma(\mathbb{E}, \mathfrak{X})$ become positive. (The k_v-symmetry is obtained by intersecting the set W_v constructed there with its finitely many k_v-conjugates.) In this situation, trivially $V(\mathbb{E}, \mathfrak{X}) > 0$ and $0 < \gamma(\mathbb{E}, \mathfrak{X}) < 1$. Thus we can assume without loss of generality that $\gamma(\mathbb{E}, \mathfrak{X}) > 0$.

To prove (A), it suffices to establish the result after making a base extension to the field $L = k(\mathfrak{X})$, since if there are only finitely many $\alpha \in \mathbb{P}^1(\overline{k})$ whose $\text{Gal}(\overline{k}/L)$-conjugates lie near E_v for each v, then the same is true for the $\text{Gal}(\overline{k}/k)$-conjugates. Hence we can assume each x_i is rational over k.

Note that $0 < \gamma(\mathbb{E}, \mathfrak{X}) < 1$ iff $V(\mathbb{E}, \mathfrak{X}) > 0$. Let S be a finite set of places of k containing all the Archimedean places and all the places where E_v is not $\mathfrak{X}$-trivial. For each non-Archimedean $v \in S$, by applying Proposition 7.42 to E_v, $\mathfrak{X}$, and its neighborhood $\mathbb{P}^1_{\text{Berk},v} \backslash \mathfrak{X}$, we can construct a strict closed affinoid F_v containing E_v for which the local Green's matrix $\Gamma(F_v, \mathfrak{X})$ is arbitrarily close to $\Gamma(E_v, \mathfrak{X})$. By choosing the F_v suitably, we can arrange that the k-symmetric adelic set $\mathbb{F}$, obtained by replacing E_v with F_v for each non-Archimedean $v \in S$, satisfies $V(\mathbb{F}, \mathfrak{X}) > 0$. Since each adelic neighborhood of $\mathbb{F}$ is also an adelic neighborhood of $\mathbb{E}$, it suffices to prove the result for $\mathbb{F}$.

For each v, put $F_v^R = F_v \cap \mathbb{P}^1(\mathbb{C}_v)$, and let $\mathbb{F}^R = \prod_v F_v^R$. Then $\mathbb{F}^R$ is an adelic set of the type studied in [88]. Since F_v^R is dense in F_v for each v, Corollary 7.49 shows that

$$\gamma^R(\mathbb{F}^R, \mathfrak{X}) = \gamma(\mathbb{F}, \mathfrak{X}) < 1 .$$

By Theorem 6.3.1 of [88], there is an adelic neighborhood $\mathbb{U}^R = \prod_v U_v^R$ of $\mathbb{F}^R$ for which there are only finitely many $\alpha \in \mathbb{P}^1(\overline{k})$ whose $\text{Gal}(\overline{k}/k)$-conjugates all belong to U_v^R, for each v. More precisely, the proof of Theorem 6.3.1 of [88] constructs a rational function $f \in k(z)$, having poles supported on $\mathfrak{X}$, such that for each $v \in S$,

$$\sup_{z \in F_v^R} |f(z)|_v < 1 , \qquad U_v^R = \{z \in \mathbb{P}^1(\mathbb{C}_v) : |f(z)|_v < 1\} ,$$

and for each $v \notin S$ (so $F_v = E_v$ is $\mathfrak{X}$-trivial),

$$\sup_{z \in F_v^R} |f(z)|_v = 1 , \qquad U_v^R = F_v^R = \{z \in \mathbb{P}^1(\mathbb{C}_v) : |f(z)|_v \leq 1\} .$$

We can now conclude the proof as follows. For each $v \in S$, put $U_v = \{z \in \mathbb{P}^1_{\text{Berk},v} : [f]_z < 1\}$, and for each $v \notin S$, put $U_v = \{z \in \mathbb{P}^1_{\text{Berk},v} : [f]_z \leq 1\}$. Then for each v, $U_v \cap \mathbb{P}^1(\mathbb{C}_v) = U_v^R$, and $F_v \subset U_v$ since F_v^R is dense in F_v. If $v \in S$, then U_v is an open neighborhood of F_v, and if $v \notin S$, then $U_v = F_v$ is $\mathfrak{X}$-trivial. Thus $\mathbb{U} := \prod_v U_v$ is a Berkovich adelic neighborhood of $\mathbb{F}$ with respect to $\mathfrak{X}$, having the property required in the theorem.

For part (B), let S be as before. Given an adelic neighborhood $\mathbb{U} = \prod_v U_v$ of $\mathbb{E}$ with respect to $\mathfrak{X}$, after shrinking the U_v if necessary, we can assume that $\overline{U}_v$ is disjoint from $\mathfrak{X}$, for each v. Proposition 7.42 allows us to replace each E_v for $v \in S$ by a slightly larger strict closed affinoid F_v which is still contained in U_v, in such a way that the Robin constants $V_{x_i}(F_v)$ and Green's functions $G(z, x_i; F_v)$ are arbitrarily close to those of E_v, outside U_v. After replacing F_v with the intersection of its finitely many $\text{Gal}(\mathbb{C}_v/k_v)$-conjugates, we can assume that F_v is k_v-symmetric. Let $\mathbb{F}$ be the corresponding k-symmetric compact Berkovich adelic set, relative to $\mathfrak{X}$.

We have $\gamma(\mathbb{E}, \mathfrak{X}) > 1$ iff $V(\mathbb{E}, \mathfrak{X}) < 0$. By choosing the F_v appropriately, we can arrange that $\gamma(\mathbb{F}, \mathfrak{X}) > 1$. For each v, put $F_v^R = F_v \cap \mathbb{P}^1(\mathbb{C}_v)$, and let $\mathbb{F}^R = \prod_v F_v^R$. Likewise, put $U_v^R = U_v \cap \mathbb{P}^1(\mathbb{C}_v)$, and let $\mathbb{U}^R = \prod_v U_v^R$.

Since F_v^R is dense in F_v for each v, Corollary 7.49 shows that
$$\gamma^R(\mathbb{F}^R, \mathfrak{X}) = \gamma(\mathbb{F}, \mathfrak{X}) > 1 \,.$$
We now apply Theorem 6.3.2 of [**88**] to $\mathbb{F}^R$ and its neighborhood $\mathbb{U}^R$. That theorem shows the existence of infinitely many $\alpha \in \mathbb{P}^1(\overline{k})$ such that for each v, the $\text{Gal}(\overline{k}/k)$-conjugates of α all belong to U_v^R. Since $U_v^R \subset U_v$ for each v, the proof of (B) is complete. $\square$

As a complement, we note the following functoriality properties of the global capacity:

PROPOSITION 7.50. *Let k be a global field, let $\mathfrak{X} \subset \mathbb{P}^1(\overline{k})$ be a finite set of points stable under $\text{Gal}(\overline{k}/k)$, and let $\mathbb{E} = \prod_v E_v$ be a k-symmetric compact Berkovich adelic set relative to $\mathfrak{X}$. Then:*

(A) *For each finite extension L/k, let $\mathbb{E}_L$ be the L-symmetric compact Berkovich adelic set relative to $\mathfrak{X}$ defined in (7.34). Then*
$$\gamma(\mathbb{E}_L, \mathfrak{X}) = \gamma(\mathbb{E}, \mathfrak{X})^{[L:K]} \,.$$

(B) *Given a nonconstant rational function $\varphi(T) \in k(T)$, for each place v of k let $\varphi_v : \mathbb{P}^1_{\text{Berk},v} \to \mathbb{P}^1_{\text{Berk},v}$ be the induced map, and define $\varphi^{-1}(\mathbb{E}) = \prod_v \varphi_v^{-1}(E_v)$. Then $\varphi^{-1}(\mathbb{E})$ is a compact Berkovich adelic set relative to $\varphi^{-1}(\mathfrak{X})$, and*
$$\gamma(\varphi^{-1}(\mathbb{E}), \varphi^{-1}(\mathfrak{X})) = \gamma(\mathbb{E}, \mathfrak{X})^{1/\deg(\varphi)} \,.$$

PROOF. If $\gamma(\mathbb{E}, \mathfrak{X}) = 0$, then some E_v has capacity 0; in this situation (A) holds trivially, and (B) follows from Lemma 7.46 (and its well-known analogue in the Archimedean case). Hence we can assume that $\gamma(\mathbb{E}, \mathfrak{X}) > 0$.

For part (A), note that by the local-global sum formula for the degree, one has $\Gamma(\mathbb{E}_L, \mathfrak{X}) = [L:K]\Gamma(\mathbb{E}, \mathfrak{X})$; hence
$$V(\mathbb{E}_L, \mathfrak{X}) = [L:K] \cdot V(\mathbb{E}, \mathfrak{X}) \,.$$
This base change formula extends [**88**, Theorem 5.1.13].

Part (B) follows formally from the pullback formula for Green's functions (Proposition 7.47) by the same argument as in [**88**, Theorem 5.1.14]. $\square$

7.9. A Bilu-type equidistribution theorem

In this section, we give another global arithmetic application of capacity theory and the theory of Green's functions on $\mathbb{P}^1_{\text{Berk}}$: an equidistribution theorem for points of small height with respect to a compact Berkovich adelic set. The prototype for such a result is Bilu's celebrated equidistribution theorem [**24**], which asserts that if z_n is a sequence of distinct points of $\overline{\mathbb{Q}}$ whose logarithmic Weil height tends to zero and if δ_n denotes the discrete probability measure on $\mathbb{C}$ supported equally on the Galois conjugates of z_n, then δ_n converges weakly to the uniform probability measure on the boundary of the complex unit disc.

7.9. A BILU-TYPE EQUIDISTRIBUTION THEOREM

In [**89**, Theorem 1], an analogue of Bilu's theorem was proved with the complex unit disc replaced by an arbitrary compact set $E \subset \mathbb{C}$ with logarithmic capacity $\gamma_\infty(E)$ equal to 1 (and the Archimedean contribution $\log^+ |z|$ to the standard Weil height replaced by the Green's function for E relative to ∞). It was also remarked without proof that one could extend Theorem 1 of [**89**] to an adelic setting, using the machinery developed in [**88**]. In [**88**], adelic sets are restricted so that each E_v is either a compact subset of $\mathbb{C}_v$ or a PL-domain, but as we have seen in §6.5 and §7.8, a more general and more elegant notion of adelic set can be obtained by working with Berkovich spaces. In this section, we extend Theorem 1 of [**89**] to an adelic equidistribution theorem relative to a general compact Berkovich adelic set $\mathbb{E} = (E_v)$ (Theorem 7.52 below). Our principal motivation for doing this comes from dynamics: the Call-Silverman canonical height function $\hat{h}_\varphi$ attached to a polynomial map $\varphi(T) \in \mathbb{Q}[T]$ is the height function attached to a certain Berkovich compact adelic set (the "adelic Berkovich filled Julia set of φ"; see Remark 10.26), but in general the v-adic filled Julia set of φ is neither compact in $\mathbb{C}_v$ nor is it a PL-domain (see [**8**]). Another canonical height function of dynamical interest is the one associated to the Mandelbrot set (see [**8**, §8], [**47**, §6.5]).

Our result is also more general than Theorem 1 of [**89**] in that we work over an arbitrary product formula field rather than over a global field, and we deal with general Galois-stable subsets rather than just the set of Galois conjugates of a single point. The proof given in [**89**] can be easily adapted to this more general setting, but it is useful for certain applications to explicitly formulate the equidistribution theorem in this way.

We note that Theorem 7.52 is almost contained in a general equidistribution theorem due to Favre and Rivera-Letelier [**47**, Theorem 2], with the following differences:

(1) We work over an arbitrary product formula field, while Favre and Rivera-Letelier's theorem is stated for number fields. (However, their proof can be easily adapted to product formula fields.)

(2) The result of Favre and Rivera-Letelier, as stated, only applies when the Green's function associated to each E_v is continuous, while our E_v's can be arbitrary compact subsets of $\mathbb{A}^1_{\text{Berk}}$.

A Berkovich space generalization of the original Bilu equidistribution theorem was given by Chambert-Loir in [**35**]. Our equidistribution theorem generalizes his result.

In order to state our equidistribution theorem (Theorem 7.52), we first need some definitions and notation.

DEFINITION 7.51. A *product formula field*[3] is a field k, together with the following extra data:

---
[3]Note that the definition of a product formula field in [**1**, Chapter 12] is more restrictive than the definition we use here.

(1) a set $\mathcal{M}_k$ of nontrivial absolute values on k (which we may assume to be pairwise inequivalent),
(2) for each $v \in \mathcal{M}_k$, an integer $N_v \geq 1$

such that:

(3) For each $\alpha \in k^\times$, we have $|\alpha|_v = 1$ for all but finitely many $v \in \mathcal{M}_k$.
(4) Every $\alpha \in k^\times$ satisfies the *product formula*

$$\prod_{v \in \mathcal{M}_k} |\alpha|_v^{N_v} = 1 \ .$$

The most important examples of product formula fields are number fields (see §6.5) and function fields of normal projective varieties (see [**70**, §2.3] or [**25**, §1.4.6]). It is known (see [**1**, Chapter 12, Theorem 3]) that a product formula field for which at least one $v \in \mathcal{M}_k$ is Archimedean must be a number field.

Let $\overline{k}$ (resp. k^{sep}) denote a fixed algebraic closure (resp. separable closure) of k. For $v \in \mathcal{M}_k$, let k_v be the completion of k at v, let $\overline{k}_v$ be an algebraic closure of k_v, and let $\mathbb{C}_v$ denote the completion of $\overline{k}_v$. For each $v \in \mathcal{M}_k$, we fix an embedding of $\overline{k}$ in $\mathbb{C}_v$ extending the canonical embedding of k in k_v, and we view this embedding as an identification. By the discussion above, if v is Archimedean, then $\mathbb{C}_v \cong \mathbb{C}$; this also follows from the more general fact that the only complete Archimedean fields are $\mathbb{R}$ and $\mathbb{C}$ (see [**1**, Chapter 2, Theorem 4]). For each $v \in \mathcal{M}_k$, we let $\mathbb{P}^1_{\text{Berk},v}$ denote the Berkovich projective line over $\mathbb{C}_v$ (which we take to mean $\mathbb{P}^1(\mathbb{C})$ if v is Archimedean).

Recall from §6.5 that a *compact Berkovich adelic set* (relative to ∞) is a set of the form

$$\mathbb{E} = \prod_v E_v$$

where E_v is a nonempty compact subset of $\mathbb{P}^1_{\text{Berk},v} \setminus \{\infty\}$ for each $v \in \mathcal{M}_k$ and where E_v is the closed unit disc $\mathcal{D}(0,1) = \mathbb{P}^1_{\text{Berk},v} \setminus \mathcal{B}(\infty,1)^-$ for all but finitely many non-Archimedean $v \in \mathcal{M}_k$.

For each non-Archimedean $v \in \mathcal{M}_k$, let $\gamma_\infty(E_v)$ be the capacity defined in (6.1), using the weight $q_v = e$ and the absolute value $|x|_v$ corresponding to v. If v is Archimedean, let $\gamma_\infty(E_v)$ be the usual logarithmic capacity. The *logarithmic capacity* (relative to ∞) of $\mathbb{E}$, denoted $\gamma_\infty(\mathbb{E})$, is

$$\gamma_\infty(\mathbb{E}) = \prod_v \gamma_\infty(E_v)^{N_v} \ .$$

For each $v \in \mathcal{M}_k$, let $G_v : \mathbb{A}^1_{\text{Berk}} \to \mathbb{R}$ be the Green's function for E_v relative to ∞, i.e., $G_v(z) = G(z,\infty;E_v)$. If $S \subset k^{\text{sep}}$ is any finite set invariant under $\text{Gal}(k^{\text{sep}}/k)$, we define the *height of S relative to $\mathbb{E}$*, denoted $h_\mathbb{E}(S)$, by

$$(7.37) \qquad h_{\mathbb{E}}(S) = \sum_{v \in \mathcal{M}_k} N_v \left(\frac{1}{|S|} \sum_{z \in S} G_v(z) \right).$$

By Galois-invariance, the sum $\sum_{z \in S} G_v(z)$ does not depend on our choice of an embedding of $\overline{k}$ into $\mathbb{C}_v$.

If $z \in k^{\text{sep}}$, let $S_k(z) = \{z_1, \ldots, z_n\}$ denote the set of $\text{Gal}(k^{\text{sep}}/k)$-conjugates of z over k, where $n = [k(z) : k]$. We define a function $h_{\mathbb{E}} : k^{\text{sep}} \to \mathbb{R}_{\geq 0}$ by setting $h_{\mathbb{E}}(z) = h_{\mathbb{E}}(S_k(z))$. If $E_v = \mathcal{D}(0,1)$ for all $v \in \mathcal{M}_k$, then $G_v(z) = \log_v^+ |z|_v$ for all $v \in \mathcal{M}_k$ and all $z \in k^{\text{sep}}$, so $h_{\mathbb{E}}$ coincides with the standard logarithmic Weil height h.

If k is a number field or a function field in one variable over a finite field, then $h_{\mathbb{E}}$ satisfies the *Northcott finiteness property*: For any real number $M > 0$ and any integer $D \geq 1$, the set

$$(7.38) \qquad \{z \in k^{\text{sep}} : [k(z) : k] \leq D, \; h_{\mathbb{E}}(z) \leq M\}$$

is finite. This follows from the Northcott finiteness property for the standard logarithmic Weil height h (see [**93**, Theorem 3.12]), together with the easily verified fact that the difference $h_{\mathbb{E}} - h$ is bounded.

Finally, if $\gamma_\infty(E_v) \neq 0$, we let μ_v denote the equilibrium distribution for E_v relative to ∞. We can now state the main result of this section:

THEOREM 7.52. *Let k be a product formula field, and let $\mathbb{E}$ be a compact Berkovich adelic set with $\gamma_\infty(\mathbb{E}) = 1$. Suppose S_n is a sequence of $\text{Gal}(k^{\text{sep}}/k)$-invariant finite subsets of k^{sep} with $|S_n| \to \infty$ and $h_{\mathbb{E}}(S_n) \to 0$. Fix $v \in \mathcal{M}_k$, and for each n let δ_n be the discrete probability measure on $\mathbb{P}^1_{\text{Berk},v}$ supported equally on the elements of S_n. Then the sequence of measures $\{\delta_n\}$ converges weakly to μ_v on $\mathbb{P}^1_{\text{Berk},v}$.*

For concreteness, we explicitly state the special case of Theorem 7.52 in which k is a number field and S_n is the $\text{Gal}(\overline{k}/k)$-orbit of a point $z_n \in \overline{k}$:

COROLLARY 7.53. *Let k be a number field, and let $\mathbb{E}$ be a compact Berkovich adelic set with $\gamma_\infty(\mathbb{E}) = 1$. Suppose $\{z_n\}$ is a sequence of distinct points of $\overline{k}$ with $h_{\mathbb{E}}(z_n) \to 0$. Fix a place v of k, and for each n let δ_n be the discrete probability measure on $\mathbb{P}^1_{\text{Berk},v}$ supported equally on the $\text{Gal}(\overline{k}/k)$-conjugates of z_n. Then the sequence of measures $\{\delta_n\}$ converges weakly to μ_v on $\mathbb{P}^1_{\text{Berk},v}$.*

Note that if S_n is the set of $\text{Gal}(\overline{k}/k)$-conjugates of z_n in the statement of Corollary 7.53, then $|S_n| \to \infty$ by the Northcott finiteness property (7.38). Thus Corollary 7.53 is a special case of Theorem 7.52.

Before we can prove Theorem 7.52, we need the following two lemmas. The first is a corrected version of [**9**, Lemma 3.26]:

LEMMA 7.54. *Let δ_N be a sequence of discrete probability measures on a compact Hausdorff topological space X converging weakly to a probability measure ν, and assume that $|\mathrm{supp}(\delta_N)| \to \infty$. Let Δ be the diagonal in $X \times X$, and let $g : X \times X \to \mathbb{R} \cup \{+\infty\}$ be lower semicontinuous. Then*

$$\liminf_{N \to \infty} \iint_{X \times X \setminus \Delta} g(x,y)\, d\delta_N(x) d\delta_N(y) \geq \iint_{X \times X} g(x,y) d\nu(x) d\nu(y) \ .$$

PROOF. Since $X \times X$ is compact and g is lower semicontinuous, there is a constant $0 \leq M \in \mathbb{R}$ such that $g(x,y) \geq -M$ for all $x, y \in X$. By Proposition A.3 and Lemma 6.5, we have

$$\iint_{X \times X} g(x,y)\, d\nu(x) d\nu(y) = \sup_{\substack{h \in \mathcal{C}(X \times X) \\ -M \leq h \leq g}} \iint_{X \times X} h(x,y)\, d\nu(x) d\nu(y)$$

$$= \sup_{\substack{h \in \mathcal{C}(X \times X) \\ -M \leq h \leq g}} \lim_{N \to \infty} \iint_{X \times X} h(x,y)\, d\delta_N(x) d\delta_N(y).$$

Moreover, if $h \in \mathcal{C}(X \times X)$, then

$$\iint_{\Delta} h(x,y)\, d\delta_N(x) d\delta_N(y) = \frac{1}{|\mathrm{supp}(\delta_N)|} \int_X h(x,x)\, d\delta_N(x) \ ,$$

so since h is bounded (by compactness) and $|\mathrm{supp}(\delta_N)| \to \infty$ by assumption,

$$\lim_{N \to \infty} \iint_{\Delta} h(x,y)\, d\delta_N(x) d\delta_N(y) = 0 \ .$$

Combining these observations, it follows that

$$\liminf_{N \to \infty} \iint_{X \times X \setminus \Delta} g(x,y) d\delta_N(x) d\delta_N(y)$$

$$\geq \sup_{\substack{h \in \mathcal{C}(X \times X) \\ -M \leq h \leq g}} \liminf_{N \to \infty} \iint_{X \times X \setminus \Delta} h(x,y) d\delta_N(x) d\delta_N(y)$$

$$= \sup_{\substack{h \in \mathcal{C}(X \times X) \\ -M \leq h \leq g}} \liminf_{N \to \infty} \iint_{X \times X} h(x,y) d\delta_N(x) d\delta_N(y)$$

$$= \iint_{X \times X} g(x,y)\, d\nu(x) d\nu(y)$$

□

The following lemma is a variant of Lemmas 4.2 and 4.3 of [**9**].

LEMMA 7.55. *Let $\mathcal{M}$ be an indexing set, and suppose we are given real numbers $b_{w,n} \in \mathbb{R}$ for each $w \in \mathcal{M}$ and each natural number n such that:*
   (1) *For all but finitely many $w \in \mathcal{M}$, we have $b_{w,n} \geq 0$ for all n.*
   (2) *For all $w \in \mathcal{M}$, we have $\liminf_{n \to \infty} b_{w,n} \geq 0$.*
   (3) *For every n, we have $b_{w,n} = 0$ for all but finitely many $w \in \mathcal{M}$, and $\liminf_{n \to \infty} \sum_{w \in \mathcal{M}} b_{w,n} \leq 0$.*

Then $\liminf_{n \to \infty} b_{w,n} = 0$ for all $w \in \mathcal{M}$.

PROOF. Suppose for the sake of contradiction that there exists $v \in \mathcal{M}$ such that $\liminf_{n\to\infty} b_{v,n} = B_v > 0$. Then there is an N_v such that $b_{v,n} \geq \frac{B_v}{2} > 0$ for all $n \geq N_v$. Let T be the set of all $w \in \mathcal{M}\setminus\{v\}$ such that $b_{w,n} < 0$ for some n. By (1), the set T is finite. Let $\varepsilon = \frac{B_v}{4|T|} > 0$. By (2), for each $w \in T$ there is an N_w such that $b_{w,n} \geq -\varepsilon$ for all $n \geq N_w$. Let $N = \max\{N_v, \max_{w \in T} N_w\}$. Then for all $n \geq N$, we have

$$\sum_{w \in \mathcal{M}} b_{w,n} = b_{v,n} + \sum_{\substack{w \notin T \\ w \neq v}} b_{w,n} + \sum_{w \in T} b_{w,n} \geq \frac{B_v}{2} - |T|\varepsilon \geq \frac{B_v}{4}.$$

Thus

$$\liminf_{n \to \infty} \sum_{w \in \mathcal{M}} b_{w,n} \geq \frac{B_v}{4} > 0,$$

contradicting (3). $\square$

PROOF OF THEOREM 7.52. We give the proof assuming that v is non-Archimedean; the Archimedean case follows from the same argument using the analogous (well-known) facts from classical potential theory over $\mathbb{C}$. We may assume without loss of generality that $|S_n| \geq 2$. Extracting a subsequence of S_n if necessary, we may assume by Prohorov's theorem (Theorem A.11) that δ_n converges weakly to a probability measure ν on $\mathbb{P}^1_{\text{Berk},v}$. We want to show that $\nu = \mu_v$.

We will first show that ν is supported on the complement of U_v, where U_v is the connected component of ∞ in $\mathbb{P}^1_{\text{Berk},v} \setminus E_v$. To prove this, note that if $\text{supp}(\nu) \cap U_v \neq \emptyset$, then there is an open set A with $\overline{A} \subset U_v$ such that $\nu(A) > 0$. Since G_v is continuous and strictly positive on U_v by Proposition 7.37(A), there is a constant $C_A > 0$ such that $g_v(z) \geq C_A > 0$ for all $z \in \overline{A}$. By the Portmanteau theorem (Theorem A.13), applied with $\varepsilon = \nu(A)/2$, there is an N such that $\delta_n(A) \geq \nu(A)/2 > 0$ for all $n \geq N$. On the other hand, definition (7.37) shows that

$$h_\mathbb{E}(S_n) \geq N_v \cdot C_A \cdot \delta_n(A) \geq N_v \cdot C_A \cdot \frac{\nu(A)}{2} > 0$$

for all n, contradicting the assumption that $h_\mathbb{E}(S_n) \to 0$.

By Proposition 7.37(A), the Green's functions for the sets E_v and $E'_v := \mathbb{P}^1_{\text{Berk},v} \setminus U_v$ are the same, and these sets also have the same logarithmic capacity. Without loss of generality, we may therefore replace E_v by E'_v and assume that ν is supported on E_v.

For $w \in \mathcal{M}_k$ and $n \geq 1$, define

$$D_w(S_n) := \sum_{\substack{x,y \in S_n \\ x \neq y}} -\log(|x-y|_w).$$

By Galois theory, since S_n is $\mathrm{Gal}(k^{\mathrm{sep}}/k)$-stable, we have

$$\alpha_n := \prod_{\substack{x,y \in S_n \\ x \neq y}} (x-y) \in k^\times,$$

and therefore the product formula shows that

(7.39) $$\sum_{w \in \mathcal{M}_k} N_w D_w(S_n) = -\log\left(\prod_{w \in \mathcal{M}_k} |\alpha_n|_w^{N_w}\right) = 0$$

for all n.

We claim that

(7.40) $$\liminf_{n \to \infty} D_v(S_n) \geq V_\infty(E_v).$$

To see this, note that since $-\log \delta(x,y)_\infty$ is lower semicontinuous by Proposition 4.1, it follows from Lemma 7.54, the definition of the Robin constant $V_\infty(E_v)$, and the fact that ν is supported on E_v that

(7.41)
$$\liminf_{n \to \infty} D_v(S_n) = \liminf_{n \to \infty} \iint_{\mathbb{P}^1_{\mathrm{Berk},v} \times \mathbb{P}^1_{\mathrm{Berk},v} \setminus \Delta} -\log \delta(x,y)_\infty \, d\delta_n(x) \, d\delta_n(y)$$
$$\geq I_v(\nu) := \iint_{\mathbb{P}^1_{\mathrm{Berk},v} \times \mathbb{P}^1_{\mathrm{Berk},v}} -\log \delta(x,y)_\infty \, d\nu(x) \, d\nu(y)$$
$$\geq V_\infty(E_v),$$

proving the claim.

By the same argument, we have

(7.42) $$\liminf_{n \to \infty} D_w(S_n) \geq V_\infty(E_w)$$

for all $w \in \mathcal{M}_k$. (Note that by Prohorov's theorem and the definition of lim inf, it suffices to prove (7.42) under the assumption that δ_n converges weakly to some probability measure on $\mathbb{P}^1_{\mathrm{Berk},w}$.)

By (7.41), we have $I_v(\nu) \geq V_\infty(E_v)$. We claim that $I_v(\nu) \leq V_\infty(E_v)$ as well. To see this, for each $w \in \mathcal{M}_k$ and $n \geq 1$ we define

$$b_{w,n} = N_w \cdot \left(D_w(S_n) + 2G_w(S_n) - V_\infty(E_w)\right),$$

where $G_w(S_n) := \frac{1}{|S_n|} \sum_{z \in S_n} G_w(z)$. Note that since G_w is a nonnegative function for each $w \in \mathcal{M}_k$, the hypothesis $h_\mathbb{E}(S_n) \to 0$ and definition (7.37) imply that

(7.43) $$\lim_{n \to \infty} G_w(S_n) = 0$$

for all $w \in \mathcal{M}_k$.

We now make the following observations:

(1) If w is non-Archimedean and $E_w = \mathcal{D}(0,1)$ (which is the case for all but finitely many $w \in \mathcal{M}_k$), then $G_w(z) = \log^+(|z|_w)$ and (setting $N_n = |S_n|$)

$$b_{w,n} = N_w \cdot \left(\frac{1}{N_n(N_n-1)} \sum_{\substack{x,y \in S_n \\ x \neq y}} -\log(\|x,y\|_w) \right) \geq 0 \,.$$

(2) For all $w \in \mathcal{M}_k$, we have $\liminf_{n \to \infty} b_{w,n} \geq 0$ by (7.42) and (7.43).

(3) For each n, we have $b_{w,n} = 0$ for all but finitely many $w \in \mathcal{M}_k$, and by the product formula and the hypothesis $\gamma_\infty(\mathbb{E}) = 1$, we have $\sum_{w \in \mathcal{M}_k} b_{w,n} = 2h_\mathbb{E}(S_n)$. In particular, $\lim_{n \to \infty} \sum_{w \in \mathcal{M}_k} b_{w,n} = 0$.

Thus the hypotheses of Lemma 7.55 are satisfied, and we conclude that

$$\liminf_{n \to \infty} b_{w,n} = 0$$

for all $w \in \mathcal{M}_k$, which in turn implies that

$$\liminf_{n \to \infty} D_w(S_n) \leq V_\infty(E_w)$$

for all $w \in \mathcal{M}_k$. From (7.41), we obtain

$$I_v(\nu) \leq \liminf_{n \to \infty} D_v(S_n) \leq V_\infty(E_v) \,,$$

proving the claim.

Thus $I_v(\nu) = V_\infty(E_v)$. By the uniqueness of the equilibrium distribution μ_v for E_v (Proposition 7.21), we conclude that $\nu = \mu_v$ as desired. $\square$

7.10. Notes and further references

Harmonic functions on a domain $U \subseteq \mathbb{P}^1(K)$ were previously considered by Ernst Kani in [64]. Kani also defines Poisson-Jensen measures associated to an affinoid domain and gives a variant of the Poisson formula discussed above (but without ever mentioning Berkovich spaces).

A theory of harmonic functions, similar to the one which we have developed in this chapter, can be developed on an arbitrary Berkovich curve; see Thuillier's thesis [94].

Arakelov geometry provides an alternative approach to Bilu-type equidistribution theorems such as Theorem 7.52; see for example Autissier [2], Chambert-Loir [35], and Thuillier [94]. A major advantage of the Arakelov geometry point of view is that it can be generalized to higher dimensions, as in the recent work of Xinyi Yuan [99]. Even for $\mathbb{P}^1$, the Arakelov framework of adelically metrized line bundles is in several respects more general than the framework of compact Berkovich adelic sets used in §7.9. In order to translate between the Arakelov and capacity-theoretic points of view, however, one needs to relate an Arakelov self-intersection number $c_1(\mathcal{L})^2$ to global capacities; this can be done using the results in [94, §4.2.3] or [27, §5.E].

A useful feature of the approach taken in §7.9 is that the argument works over an arbitrary product formula field k; this level of generality is useful for applications to complex dynamics (cf. [**6**]).

CHAPTER 8

Subharmonic functions

In this chapter, we develop a theory of subharmonic functions on the Berkovich projective line. We begin by giving two definitions of subharmonicity, one in terms of nonpositivity of the Laplacian and upper semicontinuity, the other in terms of domination by harmonic functions. We prove their equivalence and then establish analogues of the main tools of the classical theory: stability properties, the Maximum Principle, the Domination Theorem, the Riesz Decomposition Theorem, Hartogs's lemma, conditions for convergence of Laplacians, and smoothing. We establish an exact sequence relating the harmonic functions, the subharmonic functions, and the positive measures on a domain. As an application, we introduce Arakelov-Green's functions as generalized potential kernels and prove an Energy Minimization Principle for them.

8.1. Subharmonic and strongly subharmonic functions

In this section, we define subharmonic functions and note some of their basic properties. The definition is motivated by the definition of subharmonicity over $\mathbb{C}$ and by Definitions 7.1 and 7.2.

Recall that a *domain* in $\mathbb{P}^1_{\text{Berk}}$ is a connected open subset of $\mathbb{P}^1_{\text{Berk}}$. Recall (Definition 2.27) that a domain $V \subset \mathbb{P}^1_{\text{Berk}}$ is called *simple* if ∂V is a nonempty finite set $\{x_1, \ldots, x_m\}$, where each x_i is of type II or III and V is called a *finite-dendrite domain* if each x_i is of type II, III, or IV. If $U \subset \mathbb{P}^1_{\text{Berk}}$ is open, then $V \subset U$ is a *subdomain* of U if V is a domain and $\overline{V} \subset U$. Combining these properties, we speak of simple subdomains and finite-dendrite subdomains.

DEFINITION 8.1. Let $U \subset \mathbb{P}^1_{\text{Berk}}$ be a domain. A function $f : U \to [-\infty, \infty)$ is *strongly subharmonic* on U if:
(A) $f \in \text{BDV}(U)$ and $\Delta_U(f) \leq 0$.
(B) f is upper semicontinuous on U, and for each $z \in U \cap \mathbb{P}^1(K)$,

$$(8.1) \qquad f(z) = \limsup_{\substack{t \to z \\ t \in U \cap \mathbb{H}_{\text{Berk}}}} f(t) .$$

DEFINITION 8.2. If $U \subset \mathbb{P}^1_{\text{Berk}}$ is an arbitrary open set, $f : U \to [-\infty, \infty)$ is *subharmonic* on V if for each $x \in U$, there is a domain $V_x \subset U$ with $x \in V_x$ on which f is strongly subharmonic.

We write $\mathcal{SH}(U)$ for the space of subharmonic functions on U.

Clearly subharmonicity is a local property.

So far we have only defined $\Delta_U(f)$ when U is a domain and $f \in \mathrm{BDV}(U)$. However, the coherence properties of the Laplacian enable us to define a generalized Laplacian $\Delta_U(f)$ as a negative signed measure, for subharmonic functions on arbitrary open sets U:

DEFINITION 8.3. Let $U \subset \mathbb{P}^1_{\mathrm{Berk}}$ be an open set, and let $f \in \mathcal{SH}(U)$. The Laplacian $\Delta_U(f)$ is the unique signed measure on U such that
$$\Delta_U(f)|_V = \Delta_V(f)$$
for each subdomain $V \subset U$.

Propositions 5.26 and 5.27 show that $\Delta_U(f)$ is well-defined. It may have finite or infinite total mass.

REMARK 8.4. In the definition of subharmonicity, some condition controlling f on $U \cap \mathbb{P}^1(K)$ is necessary, as is shown by the following example. Take $K = \mathbb{C}_p$, $U = \mathcal{B}(0,1)^-$, and put $f(z) = 0$ on $U \backslash \mathbb{Z}_p$, $f(z) = 1$ on $\mathbb{Z}_p$. Then f is upper semicontinuous on U, since for each $x \in U \backslash \mathbb{Z}_p$ there is a neighborhood V of x with $V \cap \mathbb{Z}_p = \emptyset$. It also belongs to $\mathrm{BDV}(U)$, with $\Delta_{\overline{U}}(f) \equiv 0$. However, f should not be considered subharmonic on U, since, for example, it does not satisfy the Maximum Principle (Proposition 8.14 below).

A function $f : U \to \mathbb{R} \cup \{\infty\}$ will be called *strongly superharmonic* on U if $-f$ is strongly subharmonic. It will be called *superharmonic* if $-f$ is subharmonic. Here are some examples of subharmonic and superharmonic functions.

EXAMPLE 8.5. The only functions which are subharmonic on all of $\mathbb{P}^1_{\mathrm{Berk}}$ are the constant functions.

Indeed, if f is subharmonic on $\mathbb{P}^1_{\mathrm{Berk}}$, then $\mathbb{P}^1_{\mathrm{Berk}}$ has a cover by subdomains U_α such that $\Delta_{U_\alpha}(f) \leq 0$ for each α. Since $\Delta_{\mathbb{P}^1_{\mathrm{Berk}}}|_{U_\alpha} = \Delta_{U_\alpha}(f)$ by Proposition 5.26, it follows that $\Delta_{\mathbb{P}^1_{\mathrm{Berk}}}(f) \leq 0$. Since $\Delta_{\mathbb{P}^1_{\mathrm{Berk}}}(f)$ has total mass 0, this means $\Delta_{\mathbb{P}^1_{\mathrm{Berk}}}(f) = 0$. For each finite subgraph $\Gamma \subset \mathbb{P}^1_{\mathrm{Berk}}$, we have
$$\Delta_\Gamma(f) = \left(r_{\mathbb{P}^1_{\mathrm{Berk}},\Gamma}\right)_* (\Delta_{\mathbb{P}^1_{\mathrm{Berk}}}(f)) \equiv 0,$$
so $f|_\Gamma$ is constant by Proposition 3.14(A). Thus f is constant on $\mathbb{H}_{\mathrm{Berk}}$. Since f is strongly subharmonic on each U_α, condition (B) in Definition 8.1 implies that f is constant on $\mathbb{P}^1_{\mathrm{Berk}}$.

EXAMPLE 8.6. For $a \neq \zeta \in \mathbb{P}^1_{\mathrm{Berk}}$, the function $f(x) = -\log_v(\delta(x,a)_\zeta)$ is strongly subharmonic on $\mathbb{P}^1_{\mathrm{Berk}} \backslash \{a\}$ and strongly superharmonic on $\mathbb{P}^1_{\mathrm{Berk}} \backslash \{\zeta\}$.

Indeed, $\delta(x,a)_\zeta$ is continuous by Proposition 4.10, and
$$\Delta_{\mathbb{P}^1_{\mathrm{Berk}}}(-\log_v(\delta(x,a)_\zeta)) = \delta_a(x) - \delta_\zeta(x)$$
by Example 5.19. Correspondingly, $\log_v(\delta(x,a)_\zeta)$ is strongly subharmonic on $\mathbb{P}^1_{\mathrm{Berk}} \backslash \{\zeta\}$ and strongly superharmonic on $\mathbb{P}^1_{\mathrm{Berk}} \backslash \{a\}$.

EXAMPLE 8.7. If $f \in K(T)$ is a nonzero rational function with divisor $\operatorname{div}(f) = \sum_{i=1}^{m} n_i \delta_{a_i}(x)$, let $\operatorname{supp}^-(\operatorname{div}(f))$, $\operatorname{supp}^+(\operatorname{div}(f))$ be its polar locus and zero locus, respectively. Then $-\log_v([f]_x)$ is strongly subharmonic on the complement of $\operatorname{supp}^+(\operatorname{div}(f))$, and it is strongly superharmonic on the complement of $\operatorname{supp}^-(\operatorname{div}(f))$. Likewise $\log_v([f]_x)$ is strongly subharmonic on the complement of $\operatorname{supp}^-(\operatorname{div}(f))$ and strongly superharmonic on the complement of $\operatorname{supp}^+(\operatorname{div}(f))$.

These assertions follow from the continuity of $[f]_x$ and Example 5.20.

EXAMPLE 8.8. If ν is a probability measure on $\mathbb{P}^1_{\text{Berk}}$ and $\zeta \notin \operatorname{supp}(\nu)$, then the potential function $u_\nu(x, \zeta)$ is strongly superharmonic on $\mathbb{P}^1_{\text{Berk}} \backslash \{\zeta\}$ and strongly subharmonic on $\mathbb{P}^1_{\text{Berk}} \backslash \operatorname{supp}(\nu)$. These assertions follow from Proposition 6.12 and Example 5.22.

EXAMPLE 8.9. If $E \subset \mathbb{P}^1_{\text{Berk}}$ is a compact set of positive capacity and $\zeta \notin E$, then the Green's function $G(z, \zeta; E)$ is strongly subharmonic on $\mathbb{P}^1_{\text{Berk}} \backslash \{\zeta\}$. Indeed, if μ_ζ is the equilibrium distribution of E for ζ, then $G(z, \zeta; E) = V_\zeta(E) - u_{\mu_\zeta}(z, \zeta)$, so the result follows from Example 8.8.

The proof of the following lemma is similar to that of Lemma 7.7, so we omit the details.

LEMMA 8.10.
- (A) If $U_1 \subset U_2$ are domains and f is strongly subharmonic on U_2, then f is strongly subharmonic on U_1.
- (B) If f is subharmonic on an open set U and V is a subdomain of U (so $\overline{V} \subset U$), then f is strongly subharmonic on V.
- (C) If f is subharmonic on U and $E \subset U$ is compact and connected, there is a simple subdomain V of U containing E such that f is strongly subharmonic on V. In particular, if U is a domain, there is an exhaustion of U by simple subdomains V on which f is strongly subharmonic.

The following observations, which include the fact that subharmonic functions are "strongly upper semicontinuous" in the sense of §A.2, are often useful. Recall that *convex* means *convex upwards* (or better, *not concave downwards*; see §A.1).

PROPOSITION 8.11. Let f be subharmonic on a domain U. Then:
- (A) If the main dendrite D of U is nonempty, then f is nonincreasing on paths leading away from D. If U is a disc, then f is nonincreasing on paths leading away from the boundary $\partial U = \{x\}$.
- (B) For each $p \in U$ and each path $[y, p] \subset U$,

$$(8.2) \qquad f(p) = \limsup_{z \to p} f(z) = \limsup_{\substack{z \to p \\ z \in U \cap \mathbb{H}_{\text{Berk}}}} f(z) = \lim_{\substack{z \to p \\ z \in [y,p]}} f(z).$$

- (C) f is continuous on each path $[p, q] \subset U$.
- (D) f is convex on each path $\Gamma = [p, q] \subset U$ with $r_\Gamma(\partial U) \subset \{p, q\}$.

PROOF. For (A), if the main dendrite D is nonempty, let q be a point on D; if U is a disc, let $q = x$ be its unique boundary point. Take a point $p \in U \backslash D$ lying on a path leading away from q. We claim that for each z between q and p, we have $f(q) \geq f(z) \geq f(p)$.

First assume that $p \in \mathbb{H}_{\text{Berk}}$. Then the path $[q, p]$ is a finite subgraph Γ of U. Since $f \in \text{BDV}(U)$, $f|_\Gamma \in \text{BDV}(\Gamma)$, and in particular $f|_\Gamma$ is continuous on Γ. Given $z \in \Gamma \backslash \{p, q\}$, let $\vec{v}$ be the unit tangent vector at z pointing towards p. We need to show that $d_{\vec{v}} f(z) \leq 0$. By construction of Γ, we have $r_{\overline{U}, \Gamma}(\partial U) = \{q\}$. Since f is subharmonic, the support of $\Delta_{\overline{U}}(f)^+$ is contained in ∂U. By the coherence property of Laplacians, we conclude that $\Delta_\Gamma(f)^+$ is supported on $\{q\}$, so that $\Delta_\Gamma(f) \leq 0$ on $\Gamma \backslash \{q\}$. On the other hand, formula (3.14) implies that

$$\Delta_\Gamma(f)\left((z, p]\right) = d_{\vec{v}} f(z),$$

so that $d_{\vec{v}} f(z) \leq 0$ as required.

If $p \in \mathbb{P}^1(K)$, the argument above shows that f is nonincreasing on $[q, p_1]$ for each p_1 in the interior of $[q, p]$. By condition (B) in the definition of subharmonicity, f is nonincreasing on $[q, p]$.

For (B), we can assume that $U \neq \mathbb{P}^1_{\text{Berk}}$, since otherwise, by Example 8.5, $f(z)$ is constant and the assertions are trivial.

First suppose $p \in U \cap \mathbb{P}^1(K)$. If the main dendrite D of U is nonempty, then after shrinking $[y, p]$ if necessary, we can assume that $[y, p] \cap D = \emptyset$; if D is empty, we can assume that y lies on the path from p to the unique boundary point x of U. By the definition of subharmonicity, we have

$$\limsup_{\substack{z \to p \\ z \in U \cap \mathbb{H}_{\text{Berk}}}} f(z) = f(p).$$

By part (B), f is nonincreasing on $[y, p]$. It follows that

$$\lim_{\substack{z \to p \\ z \in [y, p]}} f(z) = f(p).$$

Fix $\varepsilon > 0$. After further shrinking $[y, p]$ if necessary, we can assume that $f(y) \leq f(p) + \varepsilon$ and that for each $z \in [y, p)$, the component U_z of $\mathbb{P}^1_{\text{Berk}} \backslash \{z\}$ containing p is a disc contained in U. By part (A), for each $x \in U_z$ we have $f(x) \leq f(z) \leq f(p) + \varepsilon$. Since ε is arbitrary, it follows that

$$\limsup_{z \to p} f(z) = \limsup_{\substack{z \to p \\ z \in U \cap \mathbb{H}_{\text{Berk}}}} f(z) = f(p).$$

Next suppose $p \in U \cap \mathbb{H}_{\text{Berk}}$. Without loss of generality, we can assume that $y \in \mathbb{H}_{\text{Berk}}$ as well. Fix a subgraph $\Gamma \subset U$ for which $p \in \Gamma^0$ and $p \notin r_\Gamma(\partial U)$. (Such a subgraph exists because the main dendrite D of U is either empty or is finitely branched at each point.) If $p \in D$, we can assume that $\Gamma \subset D$. If $p \notin D$ and D is nonempty, we can assume that Γ is a segment with one of its endpoints in D. If D is empty, we can assume that Γ is a segment with one of its endpoints on the path from p to the unique boundary point x of U. Note that $f|_\Gamma$ is continuous, since $f|_\Gamma \in \text{BDV}(\Gamma)$. Similarly, $f|_{[y,p]}$ is continuous. Given $\varepsilon > 0$, choose a neighborhood $\Gamma(p, \delta)$ of p in Γ small enough that $|f(z) - f(p)| < \varepsilon$ for all $z \in \Gamma(p, \delta)$. By part (A), f is nonincreasing on paths off the main dendrite or on paths away from x if U is a disc; from this it is clear that

$$f(p) + \varepsilon \geq \limsup_{z \to p} f(z) \geq \limsup_{\substack{z \to p \\ z \in U \cap \mathbb{H}_{\text{Berk}}}} f(z) \geq \lim_{\substack{z \to p \\ z \in [y,p]}} f(z) = f(p).$$

Since ε is arbitrary, the proof of (B) is complete.

Part (C) follows immediately from part (B). For (D), note that since f is subharmonic and $r_{\overline{U},\Gamma}(\partial U) \subset \{p,q\}$, it follows that $\Delta_\Gamma(f) = r_\Gamma(\Delta_{\overline{U}}(f)) \leq 0$ on (p,q). By the same argument as in the proof of Proposition 3.14(C) (see the paragraph containing (3.34) and (3.35)), f is convex on (p,q). By part (B) it is continuous on $[p,q]$. Hence it is in fact convex on $[p,q]$. □

REMARK 8.12. According to Proposition 7.10, the main dendrite of a domain is finitely branched. Proposition 8.11 therefore implies that at any given point, there are only finitely many tangent directions in which a subharmonic function can be increasing.

EXAMPLE 8.13. A function $f : U \to \mathbb{R}$ is harmonic on an open set $U \subset \mathbb{P}^1_{\text{Berk}}$ if and only if it is both subharmonic and superharmonic on U. In Corollary 8.39 below, we will see that f is harmonic on U if and only if it is subharmonic and $\Delta_U(f) = 0$.

Subharmonic functions obey the following Maximum Principle:

PROPOSITION 8.14 (Maximum Principle). *Let $U \subseteq \mathbb{P}^1_{\text{Berk}}$ be a domain, and suppose that f is subharmonic on U. Then:*

(A) *If f attains a global maximum on U, then f is constant.*
(B) *If M is a bound such that for each $q \in \partial U$,*

$$\limsup_{\substack{z \to q \\ z \in U}} f(z) \leq M,$$

then $f(z) \leq M$ for all $z \in U$.

PROOF. (A) Suppose first that f attains its maximum value on U at a point $\zeta \in U \cap \mathbb{H}_{\text{Berk}}$. Let W be a simple subdomain of U containing ζ. Let Γ be a finite subgraph of W containing ζ such that $\zeta \notin r_{\overline{W},\Gamma}(\partial W)$. Since f is strongly subharmonic on W, it follows that $\Delta_{\overline{W}}(f)^+$ is supported on ∂W. By construction, we have $\Delta_\Gamma(f) = (r_{\overline{W},\Gamma})_*(\Delta_{\overline{W}}(f))$, so that $\Delta_\Gamma(f)^+$ is supported on $r_{\overline{W},\Gamma}(\partial W)$. But the maximum value of f on Γ is achieved at $\zeta \notin r_{\overline{W},\Gamma}(\partial W)$, contradicting Proposition 3.14(C).

Now suppose f attains its maximum value on U at a point $z \in U \cap \mathbb{P}^1(K)$. Let W be a simple subdomain of U containing z, which we may assume to be a disc. Let q be the unique boundary point of W. If $f(q) = f(z)$, then

we are done by the previous case. Hence we can assume that $f(q) < f(z)$. Since f is strongly subharmonic on W, we have
$$f(z) = \limsup_{\substack{p \to z \\ p \in W \cap \mathbb{H}_{\text{Berk}}}} f(p),$$
and thus there exists $p \in W \cap \mathbb{H}_{\text{Berk}}$ such that $f(q) < f(p)$.

Let Γ be the path from q to p; it is a finite subgraph of U contained in $\overline{W}$, and $r_{\overline{U},\Gamma}(\partial U) = \{q\}$. As above, we see that $\Delta_\Gamma(f)^+$ is supported on $\{q\}$, and thus the maximum value of f on Γ occurs at q by Proposition 3.14(C). This contradicts the fact that $f(q) < f(p)$.

(B) Define $f^* : \overline{U} \to [-\infty, \infty)$ by
$$f^*(q) = \limsup_{\substack{z \to q \\ z \in U}} f(z).$$
Then it is easy to verify that f^* is upper semicontinuous on $\overline{U}$ and $f^* = f$ on U. The result therefore follows from (A), together with the fact that an upper semicontinuous function on a compact set E achieves its maximum value on E. □

COROLLARY 8.15. *If f is subharmonic on U and V is a subdomain of U (so $\overline{V} \subset U$), then the maximum value of f on $\overline{V}$ is attained on ∂V.*

PROOF. The maximum value M of f on $\overline{V}$ is attained, since $\overline{V}$ is compact and f is upper semicontinuous on $\overline{V}$. By the Maximum Principle, either f is constant on V or $f(x) < M$ for all $x \in V$. If f is constant on V and $f(x) \equiv M$, then for each $q \in \overline{V}$, by semicontinuity $f(q) \geq M$ and therefore $f(q) = M$. Hence we can assume that $f(x) < M$ for all $x \in V$, which implies that $f(q) = M$ for some $q \in \partial V$. □

The following consequence of Proposition 8.11 will be used in §8.10.

PROPOSITION 8.16. *Suppose f is subharmonic on a domain $U \subsetneq \mathbb{P}^1_{\text{Berk}}$, and let $z_0 \in U \cap \mathbb{H}_{\text{Berk}}$. Then there exists a point $q \in \partial U$ and a path Λ from z_0 to q such that f is nondecreasing along Λ.*

PROOF. If U is a disc, the assertion is contained in Proposition 8.11. Hence we can assume that the main dendrite D of U is nonempty. By Proposition 8.11, we may assume without loss of generality that $z_0 \in D$.

Let S be the set of all paths $\Lambda = [z_0, p]$ from z_0 to a point $p \in \overline{D} = D \cup \partial U$ for which f is nondecreasing along the interior Λ^0 of Λ. (We allow the trivial path $\Lambda = \{z_0\}$ with $p = z_0$.) Then S is partially ordered by containment. Since $\overline{D}$ is uniquely path-connected, for each $\Lambda \in S$ we have $\Lambda \subseteq \overline{D}$. Every totally ordered subset of S has an upper bound in S, since the closure of the union of such paths is again a path in S. Also, $S \neq \emptyset$, since the trivial path $\{z_0\}$ is in S. By Zorn's lemma, there exists a maximal path $\Lambda_q = [z_0, q] \in S$.

If $q \in \partial U$, we are done. Assume for the sake of contradiction that $q \notin \partial U$. We must have $q \in \mathbb{H}_{\text{Berk}}$, since $\overline{D} \cap \mathbb{P}^1(K) \subseteq \partial U$. Thus Λ_q is a finite subgraph of D. Let Γ be a finite subgraph of D containing Λ_q for which $q \in \Gamma^0$ and $q \notin r_{\overline{U},\Gamma}(\partial U)$. (Such a subgraph Γ exists because D is *finitely branched* at q and because $q \in D$ implies that q is not of type IV.) By the subharmonicity of f on U, $\text{supp}(\Delta_\Gamma(f)^+) \subseteq r_{\overline{U},\Gamma}(\partial U)$. Hence there is an $\varepsilon > 0$ for which the path distance neighborhood $\Gamma(q,\varepsilon)$ is a star, with $\Gamma(q,\varepsilon) \cap r_{\overline{U},\Gamma}(\partial U) = \emptyset$. In particular, $\Delta_\Gamma(f) \leq 0$ on $\Gamma(q,\varepsilon)$.

Let $\vec{v}_1,\ldots,\vec{v}_n$ be the distinct unit tangent vectors to Γ at q. Here $n \geq 2$, since $q \in \Gamma^0$. Without loss of generality we can assume that $\vec{v}_1$ points toward z_0. Using the notation of Chapter 3, put $r_i = q + (\varepsilon/2)\vec{v}_i$ for each i.

Take $i \geq 2$ and let $[q,r_i] \subset \Gamma^0$ be the path from q to r_i. By Proposition 8.11(D), f is convex upward on $[q,r_i]$. If $d_{\vec{v}_i}(f)(q) \geq 0$, then f is nondecreasing on $[q,r_i]$ and therefore also on $\Lambda_q \cup [q,r_i]$, which contradicts the maximality of Λ_q. Hence $d_{\vec{v}_i}(f)(q) < 0$. Since f is nonincreasing on $[q,r_1]$ by the construction of Λ_q, necessarily $d_{\vec{v}_1}(f)(q) \leq 0$. Thus

$$\Delta_\Gamma(\{q\}) \;=\; -\sum_{i=1}^n d_{\vec{v}_i}(f)(q) \;>\; 0\;.$$

This contradicts the fact that $\Delta_\Gamma(f) \leq 0$ on $\Gamma(q,\varepsilon)$, so it must be that $q \in \partial U$. $\square$

8.2. Domination subharmonicity

In this section, we will show that subharmonic functions can be characterized in terms of domination by harmonic functions.

By Proposition 7.23, if V is a finite-dendrite domain, each harmonic function h on V extends to a continuous function on $\overline{V}$, and the Poisson formula expresses h on V in terms of its values on ∂V. We will often use this implicitly, and if h is harmonic on a finite-dendrite domain, we will speak of its values on ∂V.

DEFINITION 8.17. If V is a finite-dendrite domain, then a function $f : \overline{V} \to \mathbb{R} \cup \{-\infty\}$ is *majorized by harmonic functions on* V if for each harmonic function h on V, the inequality $h(x) \geq f(x)$ on ∂V implies that $h(x) \geq f(x)$ on V.

Recall that $f : U \to \mathbb{R} \cup \{\pm\infty\}$ is *upper semicontinuous* if for each $x \in U$,
$$\limsup_{z \to x} f(z) \;\leq\; f(x)\;.$$
The following definition is motivated by analogy with the classical theory over $\mathbb{C}$ and explains the name "subharmonic":

DEFINITION 8.18. Let $U \subset \mathbb{P}^1_{\text{Berk}}$ be an open set. A function $f : U \to \mathbb{R} \cup \{-\infty\}$ is *domination subharmonic* on U if

(A') f is upper semicontinuous, and $f(x) \not\equiv -\infty$ on any component of U.

(B') f is majorized by harmonic functions on each simple subdomain of U.

The following result shows that the notions of "subharmonic" and "domination subharmonic" are one and the same.

THEOREM 8.19. *Let $U \subset \mathbb{P}^1_{\text{Berk}}$ be an open set. Then a function $f : U \to \mathbb{R} \cup \{-\infty\}$ is subharmonic on U if and only if it is domination subharmonic on U. In that case, it is majorized by harmonic functions on each finite-dendrite subdomain of U.*

Before giving the proof of Theorem 8.19, we will need several lemmas. The hardest part of the proof is to show that if f is domination subharmonic on U, then every $x \in U$ has a neighborhood U_x with $f \in \text{BDV}(U_x)$.

If $\Gamma \subset \mathbb{H}_{\text{Berk}}$ is a finite subgraph whose set of endpoints is $\partial \Gamma$ and whose interior is $\Gamma^0 = \Gamma \backslash \partial \Gamma$, we will call $V = r_\Gamma^{-1}(\Gamma^0)$ the *domain associated to* Γ. Note that V is a finite-dendrite domain in the sense of Definition 2.27, with boundary $\partial V = \partial \Gamma$ and main dendrite Γ^0.

Recall that if V and U are domains, we call V a *subdomain* of U if $\overline{V} \subset U$ (Definition 2.24). Recall also that a Berkovich open disc (Definition 2.25) is a domain with a single boundary point, which is of type II or III.

LEMMA 8.20. *Let $U \subset \mathbb{P}^1_{\text{Berk}}$ be a domain.*

If $\Gamma \subset U$ is a finite subgraph, then the associated finite-dendrite domain $V = r_\Gamma^{-1}(\Gamma^0)$ is a subdomain of U if and only if $r_\Gamma(\partial U) \subset \partial \Gamma$.

If V is a Berkovich open disc, then V is a subdomain of U if and only if its boundary point x belongs to U and $\partial U \cap V = \emptyset$.

PROOF. Let $\Gamma \subset U$ be a finite subgraph. If $r_\Gamma(\partial U) \subset \partial \Gamma$, then $V = r_\Gamma^{-1}(\Gamma^0)$ is a domain with $\partial U \cap V = \emptyset$. Since U is connected and $\Gamma^0 \subset U$, necessarily $V \subset U$. Since $\Gamma \subset U$ and $\partial V = \partial \Gamma$, it follows that $\overline{V} \subset U$. Conversely, if $V = r_\Gamma^{-1}(\Gamma^0)$ is a subdomain of U, then $\overline{V} \subset U$. For each $x \in \partial U$ we have $x \notin V$, so $r_\Gamma(x) \notin \Gamma^0$. This means $r_\Gamma(x) \in \partial \Gamma$.

If a Berkovich open disc V is a subdomain of U, then by definition $\overline{V} \subset U$ so $\{x\} = \partial V \subset U$, and $\partial U \cap \overline{V} = \emptyset$ so $\partial U \cap V = \emptyset$. Conversely, if V is a Berkovich open disc with $\{x\} = \partial V \subset U$, then $V \cap U$ is nonempty. Since V is one of the connected components of $\mathbb{P}^1_{\text{Berk}} \backslash \{x\}$, if $\partial U \cap V = \emptyset$, necessarily $V \subset U$. It follows that $\overline{V} \subset U$. □

If U is a domain in $\mathbb{P}^1_{\text{Berk}}$ and $\Gamma \subset U$ is a finite subgraph, we will say that Γ is *well-oriented* relative to U if the domain $V = r_\Gamma^{-1}(\Gamma^0)$ is a subdomain of U. The first part of Lemma 8.20 can thus be rephrased by saying that Γ is a well-oriented relative to U if and only if $r_\Gamma(\partial U) \subset \partial \Gamma$.

LEMMA 8.21. *Let $U \subset \mathbb{P}^1_{\text{Berk}}$ be open. If V is a subdomain of U, then there is a simple subdomain W of U with $\overline{V} \subset W \subset U$.*

PROOF. Cover the compact set $\overline{V}$ with a finite number of simple domains whose closures are contained in U, and let W be their union. Then W is connected since $\overline{V}$ is connected and each simple domain is connected; $\overline{W} \subset U$; and ∂W is finite and consists of points of type II or III, since the boundary of each simple domain has those properties. □

LEMMA 8.22. *If f is domination subharmonic on an open set $U \subset \mathbb{P}^1_{\text{Berk}}$, then f is majorized by harmonic functions on each finite-dendrite subdomain of U.*

PROOF. Let V be a finite-dendrite domain with $\overline{V} \subset U$.

Let $h : V \to \mathbb{R}$ be a harmonic function such that $h(x) \geq f(x)$ on ∂V. Let $\varepsilon > 0$ be given. Since f is upper semicontinuous, for each $x \in \partial V$ there is a neighborhood U_x of x such that $f(z) \leq f(x) + \varepsilon$ for all $z \in U_x$. Since h is continuous on $\overline{V}$, after shrinking U_x if necessary, we can assume that $|h(z) - h(x)| < \varepsilon$ for each $z \in U_x \cap \overline{V}$.

Put $E = \overline{V} \backslash (\bigcup_{x \in \partial V} U_x)$. Then E is compact and $E \subset V$. By Corollary 7.11, there is an exhaustion of V by simple domains $V_1 \subset V_2 \subset \cdots \subset V$ with $\overline{V}_n \subset V_{n+1}$ for each n. By the compactness of E, there is an N such that $E \subset V_n$ for each $n \geq N$. Fix $n \geq N$, and take $p \in \partial V_n$. Then $p \in U_x$ for some x, and by the properties of U_x it follows that
$$f(p) \leq f(x) + \varepsilon \leq h(x) + \varepsilon \leq h(p) + 2\varepsilon \ .$$
Since f is majorized by harmonic functions on simple subdomains of U, it follows that $f(z) \leq h(z) + 2\varepsilon$ for all $z \in V_n$. Letting $n \to \infty$ and then letting $\varepsilon \to 0$, we see that $f(z) \leq h(z)$ for all $z \in V$. □

LEMMA 8.23. *Let $U \subset \mathbb{P}^1_{\text{Berk}}$ be a domain. Suppose $f : U \to \mathbb{R} \cup \{-\infty\}$ is majorized by harmonic functions on each finite-dendrite subdomain V of U. Then either $f(x) \equiv -\infty$ for all $x \in U$ or $f(x) \in \mathbb{R}$ for all $x \in U \cap \mathbb{H}_{\text{Berk}}$.*

PROOF. Suppose $f(x_1) = -\infty$ for some $x_1 \in U \cap \mathbb{H}_{\text{Berk}}$. We will show that $f(x) \equiv -\infty$ on U.

Let $x \in U$ be an arbitrary point with $x \neq x_1$, and take a simple subdomain V_1 of U which contains x and x_1. Let V be the connected component of $V_1 \backslash \{x_1\}$ which contains x. Then V is a finite-dendrite subdomain of U with x_1 as a boundary point, which contains x in its interior.

If x_1 is the only boundary point of V, then V is a disc. In that case, by Proposition 7.12, each harmonic function on V is constant and is determined by its value on $\partial V = \{x_1\}$. Since f is majorized by harmonic functions on V and $f(x_1) = -\infty$, we have $f(z) \leq C$ on V for each $C \in \mathbb{R}$. Thus $f(x) = -\infty$.

If x_1 is not the only boundary point of V, let $x_2, \ldots, x_m$ be the other boundary points of V. Fix numbers $A_2, \ldots, A_m$ with $A_i \geq f(x_i)$ for each i. Given $A_1 \in \mathbb{R}$, Poisson's formula (Proposition 7.25) constructs a harmonic function
$$h_{A_1}(z) = \sum_{i=1}^{m} A_i h_i(z)$$
on V, where $h_i(z)$ is the harmonic measure with $h_i(x_i) = 1$, $h_i(x_j) = 0$ for each $j \neq i$. By construction, $0 < h_i(z) < 1$ on V. Taking $z = x$ and letting $A_1 \to -\infty$, we see that $h_{A_1}(x) \to -\infty$. Since f is majorized by harmonic functions on V, again $f(x) = -\infty$.

Since $x \in U$ is arbitrary, we have shown that $f(x) \equiv -\infty$ on U. □

PROPOSITION 8.24. *Let $U \subset \mathbb{P}^1_{\text{Berk}}$ be a domain, and let $f : U \to \mathbb{R} \cup \{-\infty\}$ be a function which is majorized by harmonic functions on each finite-dendrite subdomain of U, with $f(x) \not\equiv -\infty$ on U. If Γ is a finite subgraph of U, then for each point $p \in \Gamma$ and each tangent direction $\vec{v} \in T_p(\Gamma)$, the directional derivative $d_{\vec{v}} f(p)$ exists and is finite. In particular, the restriction of f to Γ is continuous.*

PROOF. We first show that Γ can be enlarged to a finite subgraph which is well-oriented with respect to U. Let V be any simple subdomain of U containing Γ, and let $\partial V = \{x_1, \ldots, x_m\}$. Fix a point $a \in \Gamma$, and adjoin each of the segments $[a, x_i]$ to Γ. After the enlargement, each x_i is an endpoint of Γ. Since $r_{\overline{U},\overline{V}}(\partial U) \subset \partial V$, we trivially have $r_\Gamma(\partial U) \subset \partial \Gamma$.

Hence we can assume without loss of generality that Γ is well-oriented with respect to U. In the argument below, we will want to consider finite subgraphs $\Gamma' \subset \Gamma$ which are also well-oriented relative to U.

There are two types of Γ' for which this is assured. One is if $\Gamma' = [a, b]$ is a segment contained in an edge of Γ (that is, no branch point of Γ is contained in $(\Gamma')^0$). The other is if $\Gamma' = \overline{\Gamma_{x_0}(\varepsilon)} = \{x \in \Gamma : \rho(x, x_0) \leq \varepsilon\}$ is a closed neighborhood of a point $x_0 \in \Gamma$, for a sufficiently small $\varepsilon > 0$. If ε is small enough, then Γ' is a star, a union of closed segments $[x_0, x_i]$ for $i = 1, \ldots, m$. In either case, $r_{\Gamma, \Gamma'}(\partial \Gamma) \subset \partial \Gamma'$, so $r_{\Gamma'}(\partial U) \subset \partial \Gamma'$.

By Lemma 8.23, $f(x) \in \mathbb{R}$ for all $x \in \Gamma$.

By abuse of notation, write f for $f|_\Gamma$. We will first show that for each $p \in \Gamma$ and each $\vec{v} \in T_p(\Gamma)$, the derivative $d_{\vec{v}} f(p)$ exists in $\mathbb{R} \cup \{-\infty\}$. Fix p and $\vec{v}$. For each sufficiently small $T > 0$, the segment $\Gamma' = [p, p + T\vec{v}]$ corresponds to a finite-dendrite domain which is a subdomain of U. The harmonic function on $r_{\Gamma'}^{-1}((\Gamma')^0)$ whose values agree with those of f at p and $p + T\vec{v}$ is affine on Γ' and is given by

$$(8.3) \qquad h(p + t\vec{v}) = (1 - \frac{t}{T}) \cdot f(p) + \frac{t}{T} \cdot f(p + T\vec{v})$$

for $0 \leq t \leq T$. Since f is majorized by harmonic functions, $f(p + t\vec{v}) \leq h(p + t\vec{v})$ for all t. Using (8.3), this gives

$$(8.4) \qquad \frac{f(p + t\vec{v}) - f(p)}{t} \leq \frac{f(p + T\vec{v}) - f(p)}{T}.$$

Hence

$$(8.5) \qquad d_{\vec{v}} f(p) = \lim_{t \to 0^+} \frac{f(p + t\vec{v}) - f(p)}{t}$$

exists in $\mathbb{R} \cup \{-\infty\}$, since the limit on the right side is nonincreasing.

Next, we will show that if $p \notin \partial \Gamma$, then $d_{\vec{v}} f(p) \neq -\infty$. Since $p \notin \partial \Gamma$, for sufficiently small $T > 0$, the star $\Gamma' = \overline{\Gamma_p(T)}$ is a subgraph of Γ with $p \in (\Gamma')^0$, which is well-oriented relative to U. Let $x_1, \ldots, x_m$ be its endpoints, where $m \geq 2$, and let $\vec{v}_1, \ldots, \vec{v}_m$ be the direction vectors in $T_p(\Gamma)$, so $\Gamma' = \bigcup_{i=1}^m [p, p + T\vec{v}_i]$. Without loss of generality, suppose $\vec{v} = \vec{v}_1$. We claim that for each $0 < t < T$,

$$\text{(8.6)} \qquad \frac{f(p+t\vec{v}) - f(p)}{t} \geq -\sum_{i=2}^{m} \frac{f(x_i) - f(p)}{T}.$$

If this fails for some t, then

$$\text{(8.7)} \qquad f(p) > \frac{Tf(p+t\vec{v}) + t \cdot \sum_{i=2}^{m} f(x_i)}{T + (m-1) \cdot t}.$$

Consider the finite subgraph $\Gamma'' = [p, p+t\vec{v}] \cup (\bigcup_{i=2}^{m} [p, p+T\vec{v}_i])$, which corresponds to a finite-dendrite domain V'' which is a subdomain of U and has boundary points $x_1'' = p + t\vec{v}, x_2, \ldots, x_m$. The harmonic function h on V'' with
$$h(x_1'') = f(x_1''), \qquad h(x_i) = f(x_i) \quad \text{for } i \geq 2$$
satisfies
$$0 = \Delta_\Gamma h(p) = \frac{f(x_1'') - h(p)}{t} + \sum_{i=2}^{m} \frac{f(x_i) - h(p)}{T},$$
which implies that
$$h(p) = \frac{Tf(p+t\vec{v}) + t \cdot \sum_{i=2}^{m} f(x_i)}{T + (m-1) \cdot t} < f(p).$$

This contradicts the fact that f is majorized by harmonic functions on V''. By (8.6), the right side of
$$d_{\vec{v}} f(p) = \lim_{t \to 0^+} \frac{f(p+t\vec{v}) - f(p)}{t}$$
is bounded from below, so $d_{\vec{v}} f(p) > -\infty$.

It remains to consider points $p \in \partial \Gamma$.

First suppose $p \in \partial\Gamma \backslash r_\Gamma(\partial U)$. There is a single edge e of Γ emanating from p; let x approach p along this edge. For each such x, let V_x be the connected component of $\mathbb{P}^1_{\text{Berk}} \backslash \{x\}$ containing p. Then V_x is a Berkovich open disc with $\partial V_x = \{x\}$. Since $x \notin r_\Gamma(\partial U)$, it follows that $\overline{V}_x \subset U$, and so V_x is a subdomain of U. Let h be the harmonic function on V_x with $h(z) = f(x)$ for all $z \in V_x$. Since f is majorized by harmonic functions, $f(z) \leq f(x)$ for all $z \in V_x$. Thus, $f(x)$ is decreasing as $x \to p$ along e. If $\vec{v} \in T_p(\Gamma)$ is the unique direction vector at p, it follows that

$$\text{(8.8)} \qquad d_{\vec{v}} f(p) = \lim_{t \to 0^+} \frac{f(p+t\vec{v}) - f(p)}{t} \geq 0.$$

Finally, suppose $p \in r_\Gamma(\partial U) \subset \partial \Gamma$. Let $\vec{v} \in T_p(\Gamma)$ be the unique direction vector at p, and write $\partial \Gamma = \{x_1, \ldots, x_m\}$. Without loss of generality, suppose $p = x_1$. Let $W \subset U$ be a simple domain containing p but not $x_2, \ldots, x_m$, such that $\overline{W}$ is contained in U. Recalling that $V = r_\Gamma^{-1}(\Gamma^0)$, put $\tilde{V} = W \cup V$. Then $\tilde{V}$ is a connected open set whose closure is contained in U. Furthermore, $\partial \tilde{V}$ is a finite set contained in $\mathbb{H}_{\text{Berk}}$. Let $\tilde{\Gamma}$ be the finite subgraph spanned by $\partial \tilde{V}$. Then $\tilde{V}$ is the finite-dendrite domain associated to $\tilde{\Gamma}$ and is a subdomain of U. It follows that $\tilde{\Gamma}$ is well-oriented relative to U. Furthermore $\Gamma \subset \tilde{\Gamma}$, since $\{x_2, \ldots, x_m\}$ remain boundary points of $\tilde{V}$, and if $\tilde{x}$ is a boundary point of $\tilde{V}$ which is not contained in V, then the path from $\tilde{x}$ to each x_i, $i = 2, \ldots, m$, passes through p.

From this we see that p is an interior point of $\tilde{\Gamma}$. By what has been shown above, $d_{\vec{v}}f(p)$ is finite, when f is regarded as a function on $\tilde{\Gamma}$. However, $d_{\vec{v}}f(p)$ depends only on the restriction of f to Γ. Thus it is finite. □

PROPOSITION 8.25. *Let $U \subset \mathbb{P}^1_{\text{Berk}}$ be a domain, and let $f : U \to \mathbb{R} \cup \{-\infty\}$ be a function which is majorized by harmonic functions on each finite-dendrite subdomain of U, with $f(x) \not\equiv -\infty$ on U. If $\Gamma \subset U$ is well-oriented with respect to U, then $f|_\Gamma \in \text{BDV}(\Gamma)$ and $\text{supp}(\Delta_\Gamma(f)^+) \subseteq r_\Gamma(\partial U) \subseteq \partial \Gamma$.*

PROOF. We have seen in Lemma 8.23 that $f(x) \in \mathbb{R}$ for all $x \in \Gamma$, and in Proposition 8.24 that $d_{\vec{v}}f(p)$ exists and is finite for all $p \in \Gamma$ and all directions $\vec{v}$ at p. Thus, $\Delta_\Gamma(f)$ exists as a finitely additive set function on the Boolean algebra $\mathcal{A}(\Gamma)$ generated by the open, closed, and half-open segments in Γ.

We first claim that $\Delta_\Gamma(f)(p) \leq 0$ for each $p \in \Gamma^0$. Suppose not, and fix $p \in \Gamma^0$ with $\Delta_\Gamma(p) > \eta > 0$. Since $p \in \Gamma^0$, there are at least two edges emanating from p. Let $\varepsilon > 0$ be small enough that the closed neighborhood $\overline{\Gamma_p(\varepsilon)}$ is a star. If $\vec{v}_1, \ldots, \vec{v}_m \in T_p(\Gamma)$ are the tangent directions at p, then

$$(8.9) \qquad \sum_{i=1}^{m} d_{\vec{v}_i} f(p) \;=\; -\Delta_\Gamma(f)(p) \;<\; -\eta \;.$$

Since $d_{\vec{v}_i} f(p) = \lim_{t \to 0^+}(f(p+t\vec{v}_i) - f(p))/t \in \mathbb{R}$, for each i there is a number $0 < t_i < \varepsilon$ such that for $q_i = p + t_i \vec{v}_i$,

$$\left| \frac{f(q_i) - f(p)}{t_i} - d_{\vec{v}_i} f(p) \right| \;<\; \frac{\eta}{2m} \;.$$

Hence $\sum_{i=1}^{m} (f(q_i) - f(p))/t_i < 0$, which gives

$$(8.10) \qquad f(p) \;>\; \frac{\sum_i f(q_i)/t_i}{\sum_i 1/t_i} \;.$$

Put $\Gamma' = \bigcup_{i=1}^{m} [p, q_i]$. Then $V' = r_{\Gamma'}^{-1}((\Gamma')^0)$ is a subdomain of U. Let h be the harmonic function on V' with $h(q_i) = f(q_i)$. Since $\Delta_{\Gamma'}(h)(p) = 0$, we have

$$(8.11) \qquad h(p) \;=\; \frac{\sum_i f(q_i)/t_i}{\sum_i 1/t_i} \;.$$

Combining (8.10) and (8.11), we get a contradiction to the fact that f is majorized by harmonic functions on V'.

Thus for each $p \in \Gamma^0$,

$$(8.12) \qquad \Delta_\Gamma(f)(p) \;=\; -\sum_i d_{\vec{v}_i} f(p) \;\leq\; 0 \;.$$

If p is an endpoint of Γ which does not belong to $r_\Gamma(\partial U)$ and if $\vec{v} \in T_p(\Gamma)$ is the unique direction vector at p, it has already been shown in (8.8) that $d_{\vec{v}}f(p) \geq 0$. Hence for such points as well,

$$(8.13) \quad \Delta_\Gamma(f)(p) = -d_{\vec{v}}f(p) \leq 0.$$

Now consider an open segment (x, y) contained in an edge of Γ. Put $\Gamma'' = [x, y]$. The associated domain V'' is a subdomain of U with boundary points x, y. Let h be the harmonic function on V'' with $h(x) = f(x)$ and $h(y) = f(y)$. Let $T = \rho(x, y)$ be the length of Γ'', and put $\alpha = (f(y) - f(x))/T$. The restriction of h to Γ'' is affine, and $f(z) \leq h(z)$ for all $z \in \Gamma$. Let $\vec{v}_+$ be the direction vector at x pointing toward y, and let $\vec{v}_-$ be the direction vector at y pointing toward x. Then

$$d_{\vec{v}_+}f(x) = \lim_{t\to 0+} \frac{f(x + t\vec{v}_+) - f(x)}{t} \leq \lim_{t\to 0+} \frac{h(x + t\vec{v}_+) - h(x)}{t} = \alpha,$$

$$d_{\vec{v}_-}f(y) = \lim_{t\to 0+} \frac{f(y + t\vec{v}_-) - f(x)}{t} \leq \lim_{t\to 0+} \frac{h(x + t\vec{v}_-) - h(x)}{t} = -\alpha,$$

so by (3.15) the segment (x, y) has measure

$$(8.14) \quad \Delta_\Gamma(f)((x, y)) = d_{\vec{v}_+}f(x) + d_{\vec{v}_-}f(y) \leq 0.$$

We can now show that $f \in \mathrm{BDV}(\Gamma)$, i.e., that $\Delta_\Gamma(f)$ extends to a bounded Borel measure on Γ. By definition (3.18), we must show that there is a number B such that for any countable collection $\{T_i\}$ of pairwise disjoint sets in $\mathcal{A}(\Gamma)$,

$$(8.15) \quad \sum_{i=1}^{\infty} |\Delta_\Gamma(f)(T_i)| \leq B.$$

Since each T_i can be decomposed as a finite disjoint union of points and open intervals, it suffices to prove (8.15) under the assumption that each T_i is a point or an open interval. Since Γ has only finitely many edges, endpoints, and branch points, it also suffices to prove (8.15) assuming that all the T_i are contained in the interior of an edge $e = [a, b]$.

In this case, the fact that $\Delta_\Gamma(f)$ is finitely additive, with $\Delta_\Gamma(f)(p) \leq 0$ and $\Delta_\Gamma(f)((x, y)) \leq 0$ for each point p and open interval (x, y) contained in (a, b), means that for all n we have

$$\sum_{i=1}^{n} |\Delta_\Gamma(T_i)| = -\sum_{i=1}^{n} \Delta_\Gamma(T_i) \leq |\Delta_\Gamma((a, b))|.$$

Letting $n \to \infty$ gives (8.15). The argument also shows that $\Delta_\Gamma(f)$ is ≤ 0 on $\Gamma \backslash r_\Gamma(\partial U)$. Hence $\mathrm{supp}(\Delta_\Gamma(f)^+) \subset r_\Gamma(\partial U)$, and if $r_\Gamma(\partial U) = \{p_1, \ldots, p_m\}$, we can take $B = 2\sum_{i=1}^{m} |\Delta_\Gamma(p_i)|$. $\square$

We can now prove Theorem 8.19.

PROOF OF THEOREM 8.19. First suppose that f is subharmonic on U; we want to show that it is domination subharmonic. Let W be a connected component of U, and let V be a simple subdomain of W. Since $f \in \mathrm{BDV}(V)$ by Lemma 8.10(B), it follows that $f(x)$ is finite on each finite subgraph $\Gamma \subset V$; hence $f(x) \not\equiv -\infty$ on W. It remains to show that if V is a simple subdomain of W and if h is harmonic on V and satisfies $f(x) \leq h(x)$ on ∂V, then $f(z) \leq h(z)$ on V.

Let $g(x) = f(x) - h(x)$ for $x \in \overline{V}$. Then g is strongly subharmonic on V, so by the corollary to the Maximum Principle (Corollary 8.15), the maximum value of g on $\overline{V}$ is attained on ∂V. But $g(x) \leq 0$ on ∂V, so $g(x) \leq 0$ on V, or equivalently $f(x) \leq h(x)$ for all $x \in V$, as desired.

Thus if f is subharmonic on U, it is domination subharmonic on U. In this case, Lemma 8.22 shows that f is majorized by harmonic functions on each finite-dendrite subdomain of U.

Now suppose f is domination subharmonic on U. By Lemma 8.22, it is majorized by harmonic functions on each finite-dendrite subdomain of U, and by hypothesis $f(x) \not\equiv -\infty$ on any connected component of U. We will show that f is strongly subharmonic on each subdomain V of U.

By Lemma 8.10, it suffices to prove this for simple subdomains, since by Lemma 8.21 each subdomain V of U is contained in a simple subdomain V'.

So, let V be a simple subdomain contained in a component W of U. We will first show that $f \in \mathrm{BDV}(V)$. Let $\Gamma \subset V$ be an arbitrary finite subgraph. We must show that $f \in \mathrm{BDV}(\Gamma)$ and that there is a bound B independent of Γ such that $|\Delta_\Gamma(f)|(\Gamma) \leq B$. We will do this by enlarging Γ to a graph $\tilde{\Gamma}$ which contains all the boundary points of V and applying Proposition 8.25.

Let $\tilde{\Gamma}$ be the smallest finite subgraph of W containing both Γ and ∂V. Since $\Gamma \subset V$,

$$r_{\tilde{\Gamma}}(\partial W) \subseteq \partial V \subseteq \partial \tilde{\Gamma},$$

and thus $\tilde{\Gamma}$ is a well-oriented subgraph of W by Lemma 8.20.

Write $\partial V = \{x_1, \ldots, x_m\}$. Since each x_i is an endpoint of $\tilde{\Gamma}$, there is a single direction vector $\vec{v}_i \in T_{x_i}(\tilde{\Gamma})$, and

$$\Delta_{\tilde{\Gamma}}(f)(x_i) = -d_{\vec{v}_i} f(x_i) .$$

The x_i are the only points where $\Delta_{\tilde{\Gamma}}(f)$ can have positive mass. Since $\Delta_{\tilde{\Gamma}}(f)$ has total mass 0, we see that

(8.16) $$|\Delta_{\tilde{\Gamma}}(f)|(\tilde{\Gamma}) \leq 2 \cdot \sum_{i=1}^{m} |d_{\vec{v}_i} f(x_i)| .$$

The right side of (8.16) is a bound B independent of the graph $\tilde{\Gamma}$.

Taking the retraction to Γ, it follows that

$$\Delta_\Gamma(f) = (r_{\tilde{\Gamma},\Gamma})_*(\Delta_{\tilde{\Gamma}}(f))$$

has total mass at most B. Since Γ is arbitrary, $f \in \mathrm{BDV}(V)$. Furthermore $\Delta_\Gamma(f)^+$ is supported on $r_{\overline{V},\Gamma}(\partial V)$. Since

$$\Delta_\Gamma(f) = (r_{\overline{V},\Gamma})_*(\Delta_{\overline{V}}(f))$$

for each Γ, taking the limit over finite subgraphs Γ shows that $\Delta_{\overline{V}}(f)^+$ is supported on ∂V, that is, $\Delta_{\overline{V}}(f)|_V \leq 0$.

It remains to show that for each $p \in V \cap \mathbb{P}^1(K)$,

(8.17) $$f(p) = \limsup_{\substack{z \to p \\ z \in V \cap \mathbb{H}_{\mathrm{Berk}}}} f(z) .$$

Fix such a p. By assumption f is upper semicontinuous, so

(8.18) $$f(p) \geq \limsup_{z \to p} f(z) .$$

Fix $y \in V$, and let x approach p along the path $[y, p]$. If x is close enough to p, then the connected component V_x of $\mathbb{P}^1_{\mathrm{Berk}} \setminus \{x\}$ containing p is an open disc whose closure is $V_x \cup \{x\} \subset W$. Since f is subharmonic, $f(z) \leq f(x)$ for all $z \in V_x$. These discs V_x are a fundamental system of neighborhoods of p, so

$$f(p) \leq \lim_{\substack{x \to p \\ x \in [y,p)}} f(x) .$$

Combined with (8.18), this establishes (8.17) and in fact shows that

(8.19) $$f(p) = \lim_{\substack{x \to p \\ x \in [y,p)}} f(x) .$$

$\square$

8.3. Stability properties

We will now show that subharmonic functions on $\mathbb{P}^1_{\mathrm{Berk}}$ are stable under the same operations as classical subharmonic functions (see [67, p. 49]). This will enable us to give additional examples of subharmonic functions.

Recall that if $U \subset \mathbb{P}^1_{\mathrm{Berk}}$ is open, we write $\mathcal{SH}(U)$ for the space of functions which are subharmonic on U. Recall from §A.2 that if $f : U \to [\infty, \infty)$ is locally bounded above, the *upper semicontinuous regularization* f^* of f is

$$f^*(x) := \limsup_{y \to x} f(y).$$

PROPOSITION 8.26. *Let $U \subset \mathbb{P}^1_{\mathrm{Berk}}$ be an open set. Then:*
- (A) *$\mathcal{SH}(U)$ is a convex cone: if $0 \leq \alpha, \beta \in \mathbb{R}$ and if $f, g \in \mathcal{SH}(U)$, then $\alpha f + \beta g \in \mathcal{SH}(U)$.*
- (B) *If U is connected and if $\mathcal{F}$ is a family of functions in $\mathcal{SH}(U)$, put $f(x) = \inf_{g \in \mathcal{F}} g(x)$. Then either $f \in \mathcal{SH}(U)$ or $f(x) \equiv -\infty$ on U.*
- (C) *If $\langle f_\alpha \rangle_{\alpha \in A}$ is a net of functions in $\mathcal{SH}(U)$ which converges uniformly to a function $f : U \to \mathbb{R}$ on compact subsets of U, then $f \in \mathcal{SH}(U)$.*
- (D) *If $f, g \in \mathcal{SH}(U)$, then $\max(f, g) \in \mathcal{SH}(U)$.*

(E) If $\{f_\alpha\}_{\alpha \in A}$ is a net of functions in $\mathcal{SH}(U)$ which is locally bounded from above and if $f(x) = \sup_\alpha f_\alpha(x)$, then $f^*(x) \in \mathcal{SH}(U)$. Furthermore $f^*(x) = f(x)$ for all $x \in U \cap \mathbb{H}_{\text{Berk}}$.

(F) If U is connected and $\{f_\alpha\}_{\alpha \in A}$ is a net of functions in $\mathcal{SH}(U)$ which is locally bounded from above, put $f(x) = \limsup_\alpha f_\alpha(x)$. Then either $f(x) \equiv -\infty$ on U or $f^*(x) \in \mathcal{SH}(U)$. Furthermore $f^*(x) = f(x)$ for all $x \in U \cap \mathbb{H}_{\text{Berk}}$.

PROOF. Except for the final assertions in (E) and (F), the proofs of these are essentially the same as their classical counterparts and rely on general properties of semicontinuity and domination by harmonic functions.

(A) If $f, g \in \mathcal{SH}(U)$, then f and g are upper semicontinuous and $\alpha f + \beta g$ is upper semicontinuous since $\alpha, \beta \geq 0$. Neither f nor g is $-\infty$ on $U \cap \mathbb{H}_{\text{Berk}}$, so $\alpha f + \beta g \not\equiv -\infty$ on $U \cap \mathbb{H}_{\text{Berk}}$, and certainly $\alpha f + \beta g \not\equiv -\infty$ on any component of U. If V is a simple subdomain of U, let $\partial V = \{x_1, \ldots, x_m\}$. Suppose h is harmonic on V with $h(x_i) \geq \alpha f(x_i) + \beta g(x_i)$ for each i. Let h_1 be the harmonic function on V with $h_1(x_i) = f(x_i)$ on ∂V, and let h_2 be the harmonic function on V with $h_2(x_i) = g(x_i)$ on ∂V. Then $h_1(z) \geq f(z)$ on V, and $h_2(z) \geq g(z)$ on V. Put $H = h - \alpha h_1 - \beta h_2$. Then $H(x_i) \geq 0$ on ∂V, so by the Maximum Principle for harmonic functions, $H(z) \geq 0$ on V. Hence $\alpha f(z) + \beta g(z) \leq \alpha h_1(z) + \beta h_2(z) \leq h(z)$ on V.

(B) The fact that f is upper semicontinuous follows from Lemma A.2. By Lemma 8.23, either $f(x) \not\equiv -\infty$ on $U \cap \mathbb{H}_{\text{Berk}}$ or $f(x) \equiv -\infty$ on U.

Suppose $f \not\equiv -\infty$. If V is a simple subdomain of U and h is a harmonic function on V with $f(x_i) \leq h(x_i)$ for each x_i in the finite set ∂V, then for each $\varepsilon > 0$ there is a $g \in \mathcal{F}$ such that $g(x_i) \leq h(x_i) + \varepsilon$ and all $x_i \in \partial V$. It follows that $f(z) \leq g(z) \leq h(z) + \varepsilon$ on V. Since $\varepsilon > 0$ is arbitrary, $f(z) \leq h(z)$ for all $z \in V$.

(C) Suppose $\langle f_\alpha \rangle_{\alpha \in A}$ is a net of subharmonic functions converging uniformly to a function $f : U \to \mathbb{R}$ on compact subsets of U. Since each f_α is upper semicontinuous, f is upper semicontinuous. By hypothesis $f \not\equiv -\infty$ on any component of U.

Let V be a simple subdomain of U, and let h be a harmonic function on V with $f(x_i) \leq h(x_i)$ for all $x_i \in \partial V$. Take $\varepsilon > 0$. Since ∂V is finite, there is a β_1 so that $f_\alpha(x_i) \leq h(x_i) + \varepsilon$ on ∂V for all $\alpha \geq \beta_1$. Thus $f_\alpha(z) \leq h(z) + \varepsilon$ on V for each $\alpha \geq \beta_1$. Since the f_α converge uniformly to f on compact subsets, there is a β_2 such that $|f(z) - f_\alpha(z)| < \varepsilon$ on $\overline{V}$ for all $\alpha \geq \beta_2$. Taking $\alpha \in A$ such that $\alpha \geq \beta_1$ and $\alpha \geq \beta_2$, we see that $f(z) \leq h(z) + 2\varepsilon$ on V. Since $\varepsilon > 0$ is arbitrary, $f(z) \leq h(z)$ on V.

(D) Suppose $f, g \in \mathcal{SH}(U)$, and put $F(z) = \max(f(z), g(z))$. Since f and g are upper semicontinuous, so is F. Since neither f nor g is $\equiv -\infty$ on any component of U, the same is true for F. If V is a simple subdomain of U and h is a harmonic function on U with $h(x_i) \geq F(x_i)$ on ∂V, then $h(x_i) \geq f(x_i)$ and $h(x_i) \geq g(x_i)$ on ∂V, so $h(z) \geq f(z)$ and $h(z) \geq g(z)$ on V, which means that $h(z) \geq \max(f(z), g(z)) = F(z)$ on V.

(E) Let $\{f_\alpha\}_{\alpha \in A}$ be a family of subharmonic functions on U, and put $f(z) = \sup_\alpha(f_\alpha(z))$ on U. Then f^* is upper semicontinuous on U. Since no f_α is $\equiv -\infty$ on any component of U, the same is true for f and also for $f^* \geq f$. If V is a simple subdomain of U and h is harmonic on U with $h(x_i) \geq f^*(x_i)$ on ∂V, then $h(x_i) \geq f_\alpha(x_i)$ on ∂V for each $\alpha \in A$. It follows that $h(z) \geq f_\alpha(z)$ on V for each α, so $h(z) \geq f(z)$ on V. However, $h(z)$ is continuous on V, hence certainly upper semicontinuous, so $h(z) \geq f^*(z)$ on V by the properties of the upper semicontinuous regularization.

We will now show that for each $x \in U \cap \mathbb{H}_{\mathrm{Berk}}$,

$$(8.20) \qquad f(x) \geq \limsup_{z \to x} f(z) .$$

We first construct a simple subdomain V of U such that x lies on the main dendrite of V. If x is on the main dendrite of U, let V be any simple subdomain of U containing x; without loss of generality we may assume that V is not a disc. Then x is on the main dendrite of V and in particular is an interior point of the graph Γ with $V = r_\Gamma^{-1}(\Gamma^0)$. If x is not on the main dendrite of U, let V_0 be an open subdisc of U containing x. Let x_0 be the boundary point of V_0, and let $x_1 \in V_0 \cap \mathbb{H}_{\mathrm{Berk}}$ be a point such that x is in the interior of the path $\Gamma = [x_0, x_1]$. Put $V = r_\Gamma^{-1}(\Gamma^0)$; then x lies on the main dendrite of V.

Next, we claim that $f|_{\Gamma^0}$ is continuous. Indeed, for each subgraph $\Gamma' \subset \Gamma^0$ of the type considered in the proof of Proposition 8.24, let V' be the corresponding simple subdomain of V. Let h be the harmonic function on V' whose value at each $p \in \partial V' = \partial \Gamma'$ is $f(p)$. For each $\alpha \in A$ and each $p \in \partial \Gamma'$, $f_\alpha(p) \leq f(p)$. Hence $f_\alpha(z) \leq h(z)$ for all $z \in V'$, and in turn $f(z) = \sup_\alpha f_\alpha(z) \leq h(z)$. By the same argument as in the proof of Proposition 8.24, the directional derivative $d_{\vec{v}} f(p)$ exists for each $p \in \Gamma^0$ and each $\vec{v}$ at p. Thus f is continuous on Γ^0.

Now fix $\varepsilon > 0$ and let $\Gamma_x(\delta) = \{q \in \Gamma : \rho(q,x) < \delta\}$ be a neighborhood of x in Γ on which $f(q) < f(x) + \varepsilon$. Let V'' be the simple subdomain of V associated to $\overline{\Gamma_x(\delta)}$. For each $\alpha \in A$ and each $q \in \Gamma_x(\delta)$, $f_\alpha(q) \leq f(x) + \varepsilon$. By Proposition 8.11, f_α is nonincreasing on paths off the main dendrite $\Gamma_p(\delta)$ of V'', so $f_\alpha(z) \leq f(x) + \varepsilon$ on V''. Hence $f(z) = \sup_\alpha f_\alpha(z) \leq f(x) + \varepsilon$ for each $z \in V''$. Since ε is arbitrary, this gives (8.20) and shows that $f^*(x) = f(x)$.

(F) For each α, put $F_\alpha(z) = \sup_{\beta \geq \alpha} f_\beta(x)$ and let F_α^* be the upper semicontinuous regularization of F_α. By the final assertion in part (E), $F_\alpha^*(x) = F_\alpha(x)$ for each $x \in U \cap \mathbb{H}_{\mathrm{Berk}}$. Since $\langle F_\alpha^* \rangle$ is a decreasing net of subharmonic functions and $f(x) = \lim_\alpha F_\alpha^*(x)$, by part (B) we must have either $f \equiv -\infty$ or $f \in \mathcal{SH}(U)$. Furthermore, by the same argument as in (E), $f(x) = f^*(x)$ for each $x \in U \cap \mathbb{H}_{\mathrm{Berk}}$. $\square$

Subharmonic functions are also stable under integration over suitably bounded families on a parameter space. The following is a Berkovich space version of [**67**, Theorem 2.6.5, p. 51].

PROPOSITION 8.27. *Let μ be a positive, σ-finite measure on a measure space T, and let $U \subset \mathbb{P}^1_{\text{Berk}}$ be a domain. Suppose that $F : U \times T \to [-\infty, \infty)$ is a measurable function such that*

(A) *for each $t \in T$, the function $F_t(z) = F(z,t) : U \to [-\infty, \infty)$ is subharmonic on U and*

(B) *there is a function $G \in L^1(T, \mu)$ such that for each $z \in U$, the function $F^z(t) = F(z,t)$ satisfies $F^z(t) \le G(t)$ for each $t \in T$.*

Put
$$f(z) = \int_T F(z,t)\, d\mu(t) .$$
Then either f is subharmonic on U or $f(z) \equiv -\infty$ on U.

PROOF. We first show that f is majorized by harmonic functions on each finite-dendrite subdomain V of U.

Let V be a finite-dendrite subdomain of U and write $\partial V = \{x_1, \ldots, x_m\}$. Let $h_1, \ldots, h_m$ be the unique harmonic functions on V with $h_i(x_j) = \delta_{ij}$ (see the discussion before Proposition 7.25). Recall that $0 < h_i(z) < 1$ and $\sum_{i=1}^m h_i(z) = 1$ for each $z \in V$. For each $t \in T$ put $h_t(z) = \sum_{i=1}^m F_t(x_i) h_i(z)$, and put $h(z) = \sum_{i=1}^m f(x_i) h_i(z)$. Then h is the least function which is harmonic on V and dominates f on ∂V.

For each t the function F_t is subharmonic on U and hence is majorized by harmonic functions on V. It follows that for each $z \in V$,
$$F_t(z) \le \sum_{i=1}^m F_t(x_i) h_i(z) .$$
Integrating over μ shows that
$$f(z) \le \sum_{i=1}^m f(x_i) h_i(z) = h(z) .$$
Thus f is majorized by harmonic functions on V.

By Lemma 8.23, if $f(x) = -\infty$ for some $x \in U \cap \mathbb{H}_{\text{Berk}}$, then $f(z) \equiv -\infty$ on U. Henceforth we will assume that $f(x) \in \mathbb{R}$ for all $x \in U \cap \mathbb{H}_{\text{Berk}}$. To show that f is subharmonic on U, we will show that it is domination subharmonic. We have shown above that f satisfies condition (B') in Definition 8.18. It remains to establish condition (A'), that f is upper semicontinuous on U.

Fix $p \in U$. If $p \in U \cap \mathbb{H}_{\text{Berk}}$, the upper semicontinuity of f at p follows from the fact that f is majorized by harmonic functions on finite-dendrite subdomains. To see this, let Γ be a finite subgraph of U containing p. As in the proof of Proposition 8.24, we can enlarge p so that it is well-oriented with respect to U and so that $p \notin r_\Gamma(\partial U)$.

By Proposition 8.24, the restriction of f to Γ is continuous. Hence for each $\varepsilon > 0$, there is a connected open neighborhood W of p in Γ such that $|f(x) - f(p)| < \varepsilon$ for each $x \in W$. After shrinking W, if necessary, we can assume that W is connected and does not contain any points of $r_\Gamma(\partial U)$.

Put $V = r_\Gamma^{-1}(W)$; then V is a finite-dendrite subdomain of U containing p, whose boundary points are contained in $\overline{W}$. Let h be the constant function on $\overline{V}$ defined by $h(z) = f(p) + \varepsilon$. By the continuity of f on Γ, for each $x \in \partial V$
$$f(x) \leq f(p) + \varepsilon = h(x) \ .$$
Since f is majorized by harmonic functions, it follows that $f(z) \leq f(p) + \varepsilon$ on V. As $\varepsilon > 0$ is arbitrary, this gives $\limsup_{z \to p} f(z) \leq f(p)$. Since $f(x) \to f(p)$ as x approaches p along Γ, we actually have
$$(8.21) \qquad \limsup_{z \to p} f(z) \ = \ f(p) \ .$$

Next suppose $p \in U \cap \mathbb{P}^1(K)$. Fix a path $[a,p] \subset U$, and let $x_1, x_2, \ldots$ be a sequence of points approaching p along $[a, p)$. For each x_n, let V_n be the component of $U \backslash \{x_n\}$ containing p. Thus V_n is a Berkovich disc with boundary point x_n. Let h_n be the harmonic function on V_n taking the constant value $f(x_n)$ on V_n. Since f is majorized by harmonic functions on finite-dendrite subdomains of U, we see that $f(z) \leq f(x_n)$ on V_n.

On the other hand, Fatou's lemma asserts that for a sequence of functions $g_n : T \to \mathbb{R} \cup \{-\infty\}$ which are uniformly bounded above by a constant $B < \infty$, we have
$$\limsup_{n \to \infty} \int_T g_n \, d\mu \ \leq \ \int_T (\limsup_{n \to \infty} g_n) \, d\mu \ .$$
Applying this to the functions $g_n(t) = F(x_n, t) - G(t)$, we obtain
$$(8.22) \qquad \limsup_{n \to \infty} f(x_n) \ \leq \ f(p) \ .$$
Combining (8.22) with the fact that $f(z) \leq f(x_n)$ on V_n shows that
$$(8.23) \qquad \limsup_{z \to p} f(z) \ = \ f(p) \ .$$
This completes the proof of upper semicontinuity. $\square$

Here is still another way of getting new subharmonic functions from old ones (compare [**67**, Theorem 2.6.6, p. 51]):

LEMMA 8.28. *Let $U \subset \mathbb{P}^1_{\text{Berk}}$ be open. If f is subharmonic on U and $\varphi : \mathbb{R} \to \mathbb{R}$ is convex and nondecreasing, then $\varphi \circ f$ is subharmonic on U, provided $\varphi(-\infty)$ is interpreted as $\lim_{t \to -\infty} \varphi(t)$.*

PROOF. Note that $\varphi(t)$ can be written as
$$\varphi(t) \ = \ \sup(\{a \cdot t + b : a \geq 0, b \in \mathbb{R}, \text{ and } a \cdot t + b \leq \varphi(t) \text{ for all } t \in \mathbb{R}\}) \ .$$
Let A be the corresponding set of pairs $(a, b) \in \mathbb{R}^2$; then
$$\varphi \circ f(z) \ = \ \sup_{(a,b) \in A} a \cdot f(z) + b \ .$$
For each $(a, b) \in A$, we have $a \cdot f(z) + b \in \mathcal{SH}(U)$. By Proposition 8.26(E), if $F(z) = \varphi(f(z))$, then $F^*(z)$ is subharmonic.

A convex function on $\mathbb{R}$ is automatically continuous, since its one-sided derivatives exist at each point. Since φ is continuous and nondecreasing and f is upper semicontinuous, it follows that $F(z) = \varphi(f(z))$ is upper semicontinuous. Hence $F^*(z) = F(z)$. $\square$

COROLLARY 8.29. *Let $U \subset \mathbb{P}^1_{\text{Berk}}$ be open.*
(A) *If f is subharmonic on U and $q \geq 1$, then the function $F(z) = q^{f(z)}$ is subharmonic on U.*
(B) *If f is subharmonic and nonnegative on U, then for any $\alpha \geq 1$, the function $F(z) = f(z)^\alpha$ is subharmonic on U.*

PROOF. (See [**67**, Corollary 2.6.8, p. 52].) Note that $t \to q^t$ and $t \to t^\alpha$ are convex and nondecreasing, and apply Lemma 8.28. $\square$

We conclude this section by using the theory developed above to give more examples of subharmonic functions.

EXAMPLE 8.30. Fix $\zeta \in \mathbb{P}^1_{\text{Berk}}$. For each $\alpha > 0$ and each $a \neq \zeta$, the function $f(x) = \delta(x,a)^\alpha_\zeta$ is subharmonic on $\mathbb{P}^1_{\text{Berk}} \backslash \{\zeta\}$. In particular this applies to $\delta(x,a)_\zeta$. More generally, for each $\alpha_1, \ldots, \alpha_n \geq 0$ and each $a_1, \ldots, a_n \in \mathbb{P}^1_{\text{Berk}} \backslash \{\zeta\}$, the generalized pseudo-polynomial

$$P(x, \vec{\alpha}, \vec{a}) = \prod_{i=1}^{m} \delta(x, a_i)^{\alpha_i}_\zeta$$

is subharmonic on $\mathbb{P}^1_{\text{Berk}} \backslash \{\zeta\}$.

This follows from Corollary 8.29(A) and Proposition 8.26(A), taking $q = q_v$, since $\alpha_i \cdot \log_v(\delta(x, a_i)_\zeta)$ is subharmonic on $\mathbb{P}^1_{\text{Berk}} \backslash \{\zeta\}$.

In particular, consider $f(x) = \delta(x, 0)_\infty$ on $\mathbb{A}^1_{\text{Berk}}$. It is constant on branches off the path $[0, \infty]$. Give $(0, \infty)$ the arclength parametrization, so that x_t is the point corresponding to the disc $B(0, q_v^t)$ for $-\infty < t < \infty$. Then $f(x_t) = q_v^t$. For each disc $V_T := \mathcal{B}(0, q_v^T)^-$, the Laplacian $\Delta_{V_T}(f)$ is supported on $[0, x_T)$. Relative to the arclength parametrization,

$$\begin{aligned} \Delta_{V_T}(f) &= f'(x_T) \cdot \delta_T(t) - f''(x_t) dt \\ &= q_v^T \log(q_v) \cdot \delta_T(t) - q_v^t (\log(q_v))^2 dt \ . \end{aligned}$$

The total variation of these measures grows to ∞ as $T \to \infty$. Thus f is subharmonic, but not strongly subharmonic, on $\mathbb{A}^1_{\text{Berk}}$.

EXAMPLE 8.31. Consider the function $\psi(t) = \arcsin(q_v^t)$. It is bounded, increasing, and convex on $[-\infty, 0)$, with a vertical tangent at $t = 0$.

Put $f(x) = \psi(\log_v(\delta(x, 0)_\infty))$ on $U := \mathcal{B}(0,1)^-$. By Lemma 8.28, $f(x)$ is bounded and subharmonic on U. However, it is not strongly subharmonic, and it cannot be extended to a subharmonic function on any larger domain.

EXAMPLE 8.32. For a nonzero rational function $f \in K(T)$ with divisor $\text{div}(f)$, the function $F(x) = [f]_x$ is subharmonic on $\mathbb{P}^1_{\text{Berk}} \backslash \text{supp}(\text{div}(f)^-)$.

This follows from Corollary 8.29(A), taking $q = q_v$, since $\log_v([f]_x)$ is subharmonic on the complement of $\mathrm{supp}(\mathrm{div}(f)^-)$.

EXAMPLE 8.33. Let $E \subset \mathbb{P}^1_{\mathrm{Berk}}$ be a compact set of positive capacity, and take $\zeta \in \mathbb{P}^1_{\mathrm{Berk}} \backslash E$. Then for each $\alpha \geq 1$, the function $G(x,\zeta;E)^\alpha$ is subharmonic on $\mathbb{P}^1_{\mathrm{Berk}} \backslash \{\zeta\}$.

This follows from Corollary 8.29(B), taking $\varphi(t) = t^\alpha$, since $G(x,\zeta;E)$ is subharmonic and nonnegative on $\mathbb{P}^1_{\mathrm{Berk}} \backslash \{\zeta\}$.

EXAMPLE 8.34. Let $f_1, \ldots, f_m \in K(T)$ be nonzero rational functions with poles supported on $\{\zeta_1, \ldots, \zeta_d\}$, and let $N_1, \ldots, N_m$ be positive integers. Then

$$g(x) = \max\left(\frac{1}{N_1} \log_v([f_1]_x), \ldots, \frac{1}{N_m} \log_v([f_m]_x)\right)$$

is subharmonic on $\mathbb{P}^1_{\mathrm{Berk}} \backslash \{\zeta_1, \ldots, \zeta_d\}$.

This follows from Example 8.32 above and Proposition 8.26(D).

8.4. The Domination Theorem

If f and g are subharmonic functions on an open set U, one can ask for conditions which assure that $f(z) \leq g(z)$ on U. In the classical case, the desired result is called the Domination Theorem.

PROPOSITION 8.35 (Domination Theorem). *Let $U \subsetneq \mathbb{P}^1_{\mathrm{Berk}}$ be an open set. Suppose f and g are subharmonic on U and that*

(A) *for each $q \in \partial U$,* $\limsup\limits_{\substack{z \to q \\ z \in U}} (f(z) - g(z)) \leq 0$,

(B) $\Delta_U(f) \leq \Delta_U(g)$ *on U.*

Then $f(z) \leq g(z)$ on U.

PROOF. Consider the function $h = f - g$ on U. If we knew that h were subharmonic on U, the result would follow from the Maximum Principle. Indeed, for each component V of U, $\limsup_{z \to q} h(z) \leq 0$ for each $q \in \partial V$, and $\Delta_V(h) = \Delta_V(f) - \Delta_V(g) \leq 0$. Unfortunately, the difference of two upper semicontinuous functions need not be upper semicontinuous, so we must go back to first principles.

Fix a simple subdomain V of U, and fix $p \in V$. As noted above, $h \in \mathrm{BDV}(V)$ and $\Delta_{\overline{V}}(h)|_V \leq 0$. We claim that

(8.24) $$h(p) \leq \max_{q_i \in \partial V} h(q_i) \ .$$

To see this, let Γ be any finite subgraph of $\overline{V}$ with $\partial V \subseteq \partial \Gamma$. Then $h|_\Gamma \in \mathrm{BDV}(\Gamma)$ and the positive part of $\Delta_\Gamma(h) = (r_{\overline{V},\Gamma})_*(\Delta_{\overline{V}}(h))$ is supported on $\partial V \subset \partial \Gamma$. By Proposition 3.14(C), $h|_\Gamma$ achieves its maximum value on $r_{\overline{V},\Gamma}(\partial V) = \partial V$, so that $h(x) \leq \max_{q_i \in \partial V} h(q_i)$ for all $x \in \Gamma$.

If $p \in \mathbb{H}_{\mathrm{Berk}}$, we may choose Γ so that $p \in \Gamma$, and thus (8.24) is satisfied in this case.

If $p \in \mathbb{P}^1(K)$, let $y = r_{\overline{V},\Gamma}(p)$, and consider the path $[y,p] \subset \overline{V}$. We have just shown that $h(x) \le \max_{q_i \in \partial V} h(q_i)$ for all $x \in [y,p)$. But in (8.19), we saw that
$$f(p) = \lim_{\substack{x \to p \\ x \in [y,p)}} f(x), \qquad g(p) = \lim_{\substack{x \to p \\ x \in [y,p)}} g(x).$$
Thus $h(p) = \lim_{\substack{x \to p \\ x \in [y,p)}} h(x) \le \max_{q_i \in \partial V} h(q_i)$, which verifies (8.24) in all cases.

Now let $p \in U$, and let V_α be the directed system of all simple subdomains of U containing p. For each α, choose a point $x_\alpha \in \partial V_\alpha$ such that $h(p) \le h(x_\alpha)$. This defines a net $\langle x_\alpha \rangle$ in V. Since $\overline{U}$ is compact, there is a subnet $\langle y_\beta \rangle$ of $\langle x_\alpha \rangle$ converging to a point $q \in \overline{U}$. As $\bigcup_\alpha V_\alpha = U$ and $U \ne \mathbb{P}^1_{\text{Berk}}$, it follows that $q \in \partial U$. But then $h(p) \le \limsup_\beta h(y_\beta) \le \limsup_{z \to q} h(z) \le 0$, as desired. □

As a special case, the Domination Theorem gives:

COROLLARY 8.36. *Let $U \subset \mathbb{P}^1_{\text{Berk}}$ be an open set with nonempty boundary. Suppose f and g are subharmonic functions on U such that*

(A) *for each $q \in \partial U$, $-\infty < \limsup_{\substack{z \to q \\ z \in U}} f(z) \le \liminf_{\substack{z \to q \\ z \in U}} g(z) < \infty$ and*

(B) $\Delta_U(f) \le \Delta_U(g)$ *on U.*

Then $f(z) \le g(z)$ on U.

The proof of the Domination Theorem also yields the following useful criterion for equality of subharmonic functions:

COROLLARY 8.37. *Let $U \subset \mathbb{P}^1_{\text{Berk}}$ be a domain, and let V be a simple subdomain of U. Suppose f and g are subharmonic on U with $\Delta_V(f) = \Delta_V(g)$ and $f(x) = g(x)$ on ∂V. Then $f(z) \equiv g(z)$ on V.*

PROOF. By symmetry, it suffices to show that $f(z) \le g(z)$. Put $h(z) = f(z) - g(z)$. The proof of Proposition 8.35 shows that for all $z \in V$, we have
$$h(z) \le \max_{q_i \in \partial V} h(q_i) = 0.$$
Thus $f(z) \le g(z)$ as desired. □

8.5. The Riesz Decomposition Theorem

Let U be an open set in $\mathbb{P}^1_{\text{Berk}}$, and let $V \subset U$ be a simple subdomain. Suppose f is subharmonic on U. By Lemma 8.10, $f \in \text{BDV}(V)$. Put $\nu = -\Delta_V(f)$, i.e., $\nu = -\Delta_{\overline{V}}(f)|_V$. Then $\nu \ge 0$. Fix $\zeta \notin \overline{V}$, and consider the potential function
$$u_\nu(z,\zeta) = \int -\log_v(\delta(z,y)_\zeta) \, d\nu(y).$$

Note that if $\nu = 0$, then $u_\nu(z,\zeta) = 0$.

8.5. THE RIESZ DECOMPOSITION THEOREM

THEOREM 8.38 (Riesz Decomposition Theorem). *Let V be a simple subdomain of an open set $U \subset \mathbb{P}^1_{\text{Berk}}$. Fix $\zeta \in \mathbb{P}^1_{\text{Berk}} \backslash \overline{V}$. Suppose f is subharmonic on U, and let ν be the positive measure $\nu = -\Delta_V(f)$. Then there is a continuous function h_V on $\overline{V}$, whose restriction to V is harmonic, such that*
$$f(z) = h_V(z) - u_\nu(z, \zeta) \quad \text{for all } z \in \overline{V}.$$

PROOF. Define $F(z) = -u_\nu(z, \zeta)$. By Example 5.22, $F \in \text{BDV}(\mathbb{P}^1_{\text{Berk}})$ and
$$\Delta_{\mathbb{P}^1_{\text{Berk}}}(F) = \nu(\mathbb{P}^1_{\text{Berk}}) \delta_\zeta(x) - \nu.$$
If V is not a disc, let Γ be the finite graph which is the closure of the main dendrite of V. If V is a disc, let Γ be a segment having ∂V as one endpoint. By the retraction property of the Laplacian, $\Delta_\Gamma(f) = (r_{\overline{V},\Gamma})_*(\Delta_{\overline{V}}(f))$, and similarly for $\Delta_\Gamma(F)$. It follows that $\Delta_\Gamma(F) - \Delta_\Gamma(f)$ is a discrete measure supported on $\partial V \subset \Gamma$.

Define $h_V(x) = f(x) - F(x)$ on Γ, and extend h_V to $\overline{V}$ by setting $h_V(z) = h_V \circ r_{\overline{V},\Gamma}(z)$ for all $z \in \overline{V}$. Then h_V is harmonic on V, since it is continuous on $\overline{V}$ and $\Delta_{\overline{V}}(h_V) = \Delta_\Gamma(F) - \Delta_\Gamma(f)$ is supported on ∂V. Now consider the functions f and $h_V + F$. By construction, both functions are subharmonic on V and satisfy $\Delta_V(f) = \Delta_V(h_V + F) = -\nu$. In addition, they agree on $\partial V \subset \Gamma$. By Corollary 8.37, $f(z) \equiv h_V(z) + F(z)$. □

It follows from the Riesz Decomposition Theorem that if f is subharmonic on an open set U and its Laplacian is 0, then f is harmonic on U:

COROLLARY 8.39. *Let $U \subset \mathbb{P}^1_{\text{Berk}}$ be an open set, and suppose $f \in \mathcal{SH}(U)$. If $\Delta_U(f) = 0$, then $f \in \mathcal{H}(U)$.*

PROOF. Suppose $f \in \mathcal{SH}(U)$ and $\Delta_U(f) = 0$. Let V be any simple subdomain of U. By Lemma 8.10(B) and Proposition 5.26, we have $f \in \text{BDV}(V)$ and $\Delta_V(f) = 0$. By Theorem 8.38, there is a function h_V which is continuous on $\overline{V}$ and harmonic on V, such that $f(z) = h_V(z)$ for each $z \in V$. Thus f is strongly harmonic on V. Since harmonicity is a local property, the result follows. □

If f is subharmonic on U, we define its *polar set* to be
$$\mathcal{P}_U(f) = \{z \in U : f(z) = -\infty\}.$$
It follows from the Riesz Decomposition Theorem that the polar set of f has capacity 0:

COROLLARY 8.40. *Let U be an open set in $\mathbb{P}^1_{\text{Berk}}$, and let $f \in \mathcal{SH}(U)$. Then its polar set $\mathcal{P}_U(f)$ has capacity 0.*

PROOF. If $U = \mathbb{P}^1_{\text{Berk}}$, each $f \in \mathcal{SH}(U)$ is constant by Example 8.5, so $\mathcal{P}_U(f)$ is empty. Hence we can assume that $U \neq \mathbb{P}^1_{\text{Berk}}$.

To show that $\mathcal{P}_U(f)$ has capacity 0, we must show that each compact set $E \subset \mathcal{P}_U(f)$ has capacity 0. Fix such an E, and assume to the contrary that E has positive capacity. Since each compact subset of U is contained in finitely many components of U and the union of finitely many sets of capacity 0 has capacity 0 (Corollary 6.17), we can assume without loss of generality that E is contained in a single component of U. Thus we are reduced to the case where U is a domain.

By Corollary 7.11, there is an exhaustion of U by a sequence of simple subdomains $V_1 \subset V_2 \subset \cdots$ with $\overline{V}_n \subset V_{n+1}$ for each n. Let V be a simple subdomain of U which contains E, and put $\nu = \Delta_V(f)$. Since f is strongly subharmonic on V, ν has finite total mass. Put $M = |\nu(V)|$. If $M = 0$, then f is harmonic on V and E is empty. Hence we can assume that $M > 0$.

Since U is connected and $U \neq \mathbb{P}^1_{\text{Berk}}$, there is a point $\zeta \in \mathbb{P}^1(K) \backslash U$. Fix such a ζ. By Theorem 8.38, there is a function h_V which is harmonic on V and continuous on $\overline{V}$, such that $f(z) = h_V(z) - u_\nu(z, \zeta)$ for each $z \in \overline{V}$. Since h_V is bounded and $E \subset \mathcal{P}_U(f)$, it follows that

$$E \subseteq \{z \in V : u_\nu(z, \zeta) = \infty\} \ .$$

Consider the Green's function $G(z, \zeta; E)$. By Proposition 7.37(A)(7), there is a neighborhood of ζ on which

(8.25) $$G(z, \zeta; E) = V_\zeta(E) - \log_v(\|z, \zeta\|) \ .$$

For each $N \in \mathbb{N}$, put $W_{V,N} = \{z \in V : u_\nu(z, \zeta) > N\}$. By construction, $E \subset W_{V,N}$. Since $u_\nu(z, \zeta)$ is lower semicontinuous on V (Proposition 6.12), $W_{V,N}$ is open. Put $g_N(z) = \frac{1}{M}(N - u_\nu(z, \zeta))$, noting that $g_N(z)$ is subharmonic on $\mathbb{P}^1_{\text{Berk}} \backslash \{\zeta\}$. Note also that $\frac{1}{M} u_\nu(z, \zeta) = u_{\nu/M}(z, \zeta)$ is a potential function attached to a probability measure. By Example 5.22, $\Delta_{\mathbb{P}^1_{\text{Berk}}}(g_N) = \delta_\zeta - \nu/M$, and by Proposition 6.12, there is a neighborhood of ζ on which $g_N(z) = \frac{N}{M} - \log_v(\|z, \zeta\|)$.

Next put $G_N(z) = \max(0, g_N(z))$. By Proposition 8.26(D), G_N is subharmonic on $\mathbb{P}^1_{\text{Berk}} \backslash \{\zeta\}$. By the definition of $W_{V,N}$, $G_N(z) = 0$ on $W_{W,N}$, and $G_N(z)$ coincides with $g_N(z)$ outside $W_{V,N}$. In particular, since $\zeta \notin \overline{V}$, there is a neighborhood of ζ on which

(8.26) $$G_N(z) = \frac{N}{M} - \log_v(\|z, \zeta\|) \ .$$

Let Y be the connected component of $\mathbb{P}^1_{\text{Berk}} \backslash E$ which contains ζ, and consider the function given by $a_N(z) = G_N(z) - G(z, \zeta; E)$ on $Y \backslash \{\zeta\}$. By (8.25) and (8.26), a_N is constant on a punctured neighborhood of ζ, with value $N/M - V_\zeta(E)$, so it extends to a function which is harmonic at ζ. Thus, we can regard a_N as defined on all of Y, with $a_N(\zeta) = N/M - V_\zeta(E)$. Since $G_N(z)$ is subharmonic on $\mathbb{P}^1_{\text{Berk}} \backslash \{\zeta\}$ and $G(z, \zeta; E)$ is harmonic on $\mathbb{P}^1_{\text{Berk}} \backslash (E \cup \{\zeta\})$, a_N is subharmonic on Y. In addition, since $G_N(z) = 0$ on $W_{V,N}$ and $G(z, \zeta; E) \geq 0$ for all z, it follows that $a_N(z) \leq 0$ on $Y \cap W_{V,N}$.

On the other hand, ∂Y is contained in E, which in turn is contained in $W_{V,N}$. Thus, for each $x \in \partial Y$,
$$\limsup_{\substack{z \to x \\ z \in Y}} a_N(z) \leq 0 .$$
By the Maximum Principle (Proposition 8.14(B)), $a_N(z) \leq 0$ for all $z \in Y$. In particular, $a_N(\zeta) \leq 0$, so $V_\zeta(E) \geq N/M$.

This holds for each N, so $V_\zeta(E) = \infty$, contradicting our assumption that $\gamma_\zeta(E) > 0$. Thus, $\mathcal{P}_U(f)$ has capacity 0. $\square$

By an argument similar to the one in the proof of the Riesz Decomposition Theorem, we can obtain interesting information about the structure of functions in $\mathrm{BDV}(U)$. Recall that Definition 5.11 says that $f \in \mathrm{BDV}(U)$ iff $f|_\Gamma \in \mathrm{BDV}(\Gamma)$ for every finite subgraph $\Gamma \subset U$ and there is a uniform bound B for the measures $|\Delta_\Gamma(f)|(\Gamma)$ for all $\Gamma \subset U$. Nothing is said about the continuity of f on U or the behavior of f on $U \cap \mathbb{P}^1(K)$.

PROPOSITION 8.41. *Let $U \subsetneq \mathbb{P}^1_{\mathrm{Berk}}$ be a domain, and let $f \in \mathrm{BDV}(U)$. Then there are functions $g_1, g_2 \in \mathcal{SH}(U) \cap \mathrm{BDV}(U)$ such that $f(z) = g_1(z) - g_2(z)$ for each $z \in U \cap \mathbb{H}_{\mathrm{Berk}}$.*

In particular, there is a set $e = e(f) \subset \mathbb{P}^1(K) \cap U$ of capacity 0 such that for each $p \in (\mathbb{P}^1(K) \cap U) \setminus e$, as z approaches p along any path $[y, p]$, the limit
$$L(p) = \lim_{\substack{z \to p \\ z \in [y, p)}} f(z)$$
exists and is finite.

PROOF. Since $f \in \mathrm{BDV}(U)$, the measure $\nu = \Delta_U(f)$ has finite total mass. Let ν_1, ν_2 be the positive and negative measures in the Jordan decomposition of ν, so $\nu = \nu_1 - \nu_2$. Put $M_1 = \nu_1(U)$ and $M_2 = \nu_2(U)$. Fix $\zeta \in \mathbb{P}^1_{\mathrm{Berk}} \setminus U$, and let
$$g_1(z) = -M_1 \log_v(\|z, \zeta\|) - u_{\nu_1}(z, \zeta_{\mathrm{Gauss}}) ,$$
$$g_2(z) = -M_2 \log_v(\|z, \zeta\|) - u_{\nu_2}(z, \zeta_{\mathrm{Gauss}}) .$$
By Examples 5.19 and 5.21, g_1 and g_2 are subharmonic on $\mathbb{P}^1_{\mathrm{Berk}} \setminus \{\zeta\}$ and belong to $\mathrm{BDV}(\mathbb{P}^1_{\mathrm{Berk}})$, with $\Delta(g_1) = M_1 \delta_\zeta - \nu_1$ and $\Delta(g_2) = M_2 \delta_\zeta - \nu_2$. Thus
$$\Delta_U(g_1 - g_2) = \nu_1 - \nu_2 = \nu .$$

Set $H_0 = f - (g_1 - g_2)|_U$. Then $H_0 \in \mathrm{BDV}(U)$ and $\Delta_U(H_0) = 0$. By Corollary 7.11, there is an exhaustion of U by simple subdomains $V_1 \subset V_2 \subset \cdots$ with $\overline{V}_n \subset V_{n+1}$ for each n. By Proposition 5.26, for each n we have $\Delta_{V_n}(H_0) = \Delta_U(H_0)|_{V_n} = 0$, so $\Delta_{\overline{V}_n}(H_0)$ is supported on ∂V_n. By the solvability of the Neumann problem for V_n (Corollary 7.24(B)), there is a function $h_n \in \mathcal{H}(V_n)$ which satisfies $\Delta_{\overline{V}_n}(h_n) = \Delta_{\overline{V}_n}(H_0)$. By Lemma 5.24, there is a constant C_n such that $H_0(z) = h_n(z) + C_n$ for each $z \in V_n \cap \mathbb{H}_{\mathrm{Berk}}$. By replacing h_n with $h_n + C_n$, we can assume $h_n(z) = H_0(z)$ on $V_n \cap \mathbb{H}_{\mathrm{Berk}}$.

For each n, h_n is continuous on V_n, so $h_{n+1}|_{V_n} = h_n$. It follows that we can define a function $H \in \mathcal{H}(U)$, which coincides with H_0 on $U \cap \mathbb{H}_{\text{Berk}}$, by setting $H(z) = h_n(z)$ if $z \in V_n$. Since $H_0 \in \text{BDV}(U)$, we also have $H \in \text{BDV}(U)$.

By restricting g_1 and g_2 to U and replacing g_1 with $g_1 + H$, we obtain the first assertion. The second assertion now follows by applying Corollary 8.40 to g_1 and g_2 and using Proposition 8.11(B). □

For future applications, it is useful to know that if f is continuous and subharmonic on U and if $\nu = -\Delta_V(f)$, then the potential function $u_\nu(z, \zeta)$ is continuous everywhere.

PROPOSITION 8.42. *Suppose f is continuous and subharmonic on an open set $U \subset \mathbb{P}^1_{\text{Berk}}$. Let V be a simple subdomain of U, and fix $\zeta \in \mathbb{P}^1_{\text{Berk}} \backslash \overline{V}$. Put $\nu = -\Delta_{\overline{V}}(f)|_V = -\Delta_V(f)$. Then $u_\nu(z, \zeta)$ is continuous on all of $\mathbb{P}^1_{\text{Berk}}$.*

PROOF. By Theorem 8.38, there is a continuous function h on $\overline{V}$, harmonic on V, such that $u_\nu(z, \zeta) = h(z) - f(z)$ on V. In particular, $u_\nu(z, \zeta)$ is continuous on V. Since $\text{supp}(\nu) \subset \overline{V}$, $u_\nu(z, \zeta)$ is also continuous on $\mathbb{P}^1_{\text{Berk}} \backslash \overline{V}$ by Proposition 6.12 (which includes continuity at ζ). It remains to show that $u_\nu(z, \zeta)$ is continuous on the finite set ∂V.

Fix $x_i \in \partial V$. Since f and h are continuous on $\overline{V}$,
$$\lim_{\substack{z \to x_i \\ z \in V}} u_\nu(z, \zeta) = u_\nu(x_i, \zeta) .$$

Now consider the behavior of $u_\nu(z, \zeta)$ on each connected component of $\mathbb{P}^1_{\text{Berk}} \backslash \{x_i\}$, as $z \to x_i$. Each such component is an open disc with x_i as its boundary point. One component contains V; we have already dealt with it.

Suppose W is a component which does not contain V. If W contains ζ, choose a point p on the interior of the path from x_i to ζ; otherwise, let $p \in U \backslash \mathbb{P}^1(K)$ be arbitrary. Put $\Gamma = [x_i, p]$. Since $u_\nu(z, \zeta) \in \text{BDV}(\mathbb{P}^1_{\text{Berk}})$, its restriction to Γ belongs to $\text{BDV}(\Gamma)$, hence is continuous on Γ. Thus

(8.27) $$\lim_{\substack{z \to x_i \\ z \in \Gamma}} u_\nu(z, \zeta) = u_\nu(x_i, \zeta) .$$

If W is a component which does not contain ζ, then $u_\mu(z, \zeta)$ is harmonic on the open disc W, hence constant. By (8.27), $u_\nu(z, \zeta) \equiv u_\nu(x_i, \zeta)$ on W for such a component. If W is the component containing ζ, put $W_0 = r_\Gamma^{-1}(\Gamma^0)$; then W_0 is a simple subdomain of $\mathbb{P}^1_{\text{Berk}}$ contained in $W \backslash \{\zeta\}$, and $u_\nu(z, \zeta)$ is harmonic on W_0. The main dendrite of W_0 is Γ^0. Since harmonic functions are constant on branches off the main dendrite,

(8.28) $$\lim_{\substack{z \to x_i \\ z \in W}} u_\nu(z, \zeta) = \lim_{\substack{z \to x_i \\ z \in \Gamma}} u_\nu(z, \zeta) = u_\nu(x_i, \zeta) .$$

Combining all cases, we see that $u_\nu(z, \zeta)$ is continuous at x_i. □

8.6. The topological short exact sequence

Let $U \subsetneq \mathbb{P}^1_{\text{Berk}}$ be a domain. In this section we equip the space of subharmonic functions $\mathcal{SH}(U)$ with the *topology of pointwise convergence on* $U \cap \mathbb{H}_{\text{Berk}}$ and show that it is complete in this topology. We then present a short exact sequence, due to Thuillier [**94**], which describes the space of subharmonic functions $\mathcal{SH}(U)$ in terms of the space of harmonic functions $\mathcal{H}(U)$ and the space $\mathcal{M}^+(U)$ of positive, locally finite Borel measures on U. We conclude by giving related results for Arakelov-Green's functions, due to Favre and Jonsson [**45**], when $U = \mathbb{P}^1_{\text{Berk}}$.

The space $\mathcal{H}(U)$ is a real vector space which contains the constant functions. If $U = \mathbb{P}^1_{\text{Berk}}$, then by Lemma 5.24 and the fact that harmonic functions are continuous by definition, $\mathcal{H}(U)$ is precisely the set of constant functions. If U is a finite-dendrite domain, then by Proposition 7.25 (the Poisson formula), the dimension of $\mathcal{H}(U)$ is finite and is equal to the cardinality of ∂U. In general $\mathcal{H}(U)$ can be infinite dimensional.

The space $\mathcal{SH}(U)$ is a real cone; that is, if $0 \leq a, b \in \mathbb{R}$ and $f, g \in \mathcal{SH}(U)$, then $a \cdot f + b \cdot g \in \mathcal{SH}(U)$. It contains $\mathcal{H}(U)$ as a subspace. We give $\mathcal{SH}(U)$ and $\mathcal{H}(U)$ the *topology of pointwise convergence on* $U \cap \mathbb{H}_{\text{Berk}}$. That is, as f ranges over $\mathcal{SH}(U)$, x ranges over $U \cap \mathbb{H}_{\text{Berk}}$, and ε ranges over all positive reals, the topology on $\mathcal{SH}(U)$ is generated by the open sets

$$(8.29) \qquad U_{f,x}(\varepsilon) \;=\; \{g \in \mathcal{SH}(U) : |g(x) - f(x)| < \varepsilon\} \ .$$

The topology on $\mathcal{H}(U)$ is defined similarly and coincides with the subspace topology of $\mathcal{H}(U)$ in $\mathcal{SH}(U)$.

The following result asserts that $\mathcal{SH}(U)$ is complete, when given the topology of pointwise convergence on $U \cap \mathbb{H}_{\text{Berk}}$:

PROPOSITION 8.43. *Let $U \subset \mathbb{P}^1_{\text{Berk}}$ be a domain. If $\{f_\alpha\}_{\alpha \in A}$ is a net of functions in $\mathcal{SH}(U)$ which converges pointwise on $U \cap \mathbb{H}_{\text{Berk}}$ to a function $f : U \to \mathbb{R} \cup \{-\infty\}$ such that $f(x) \not\equiv -\infty$ on U, then f is finite on $U \cap \mathbb{H}_{\text{Berk}}$, $f^*(x) = f(x)$ for all $x \in U \cap \mathbb{H}_{\text{Berk}}$, and $f^* \in \mathcal{SH}(U)$.*

PROOF. The argument is similar to the one in Proposition 8.27. We first show that f is majorized by harmonic functions on each finite-dendrite subdomain V of U.

Let V be a finite-dendrite subdomain of U and write $\partial V = \{x_1, \ldots, x_m\}$. Let $h_1, \ldots, h_m$ be the unique harmonic functions on V with $h_i(x_j) = \delta_{ij}$. For each $\alpha \in A$ put $h_\alpha(z) = \sum_{i=1}^m f_\alpha(x_i) h_i(z)$, and put $h(z) = \sum_{i=1}^m f(x_i) h_i(z)$. Then h is the least function which is harmonic on V and dominates f on ∂V, and $\lim_\alpha h_\alpha(z) = h(z)$ for each $z \in V$.

For each α the function f_α is subharmonic on U and hence is majorized by h_α on V. It follows that for each $z \in V$,

$$f(z) \;=\; \lim_\alpha f_\alpha(z) \;\leq\; \lim_\alpha h_\alpha(z) \;=\; h(z) \ .$$

Thus f is majorized by harmonic functions on V.

As in the proof of Proposition 8.27, it follows from the fact that f is majorized by harmonic functions on finite-dendrite subdomains that either $f(x) \equiv -\infty$ on U or that $f(x) \in \mathbb{R}$ for all $x \in U \cap \mathbb{H}_{\text{Berk}}$.

Suppose $f \not\equiv -\infty$. Since f is majorized by harmonic functions on finite-dendrite subdomains, the same argument that gives (8.21) shows that for each $p \in U \cap \mathbb{H}_{\text{Berk}}$,

(8.30) $$\limsup_{z \to p} f(z) = f(p) \ .$$

Now consider $f^*(x) = \limsup_{z \to x} f(z)$. Since f is majorized by harmonic functions on finite-dendrite domains and each harmonic function is continuous, f^* is also majorized by harmonic functions on finite-dendrite domains. By definition it is upper semicontinuous, so it is subharmonic. Furthermore, by (8.30), it coincides with f on $U \cap \mathbb{H}_{\text{Berk}}$. $\square$

The space of positive, locally finite Borel measures $\mathcal{M}^+(U)$ is also a real cone. Let $\mathcal{C}_c(U)$ be the space of continuous functions $f : U \to \mathbb{R}$ for which there is a compact subset $X_f \subset U$ such that f vanishes outside X_f. We give $\mathcal{M}^+(U)$ the *topology of weak convergence relative to* $\mathcal{C}_c(U)$. That is, as μ ranges over $\mathcal{M}^+(U)$, f ranges over $\mathcal{C}_c(U)$, and ε ranges over all positive reals, the topology on $\mathcal{M}^+(U)$ is generated by the open sets

(8.31) $$U_{\mu,f}(\varepsilon) = \{\nu \in \mathcal{M}^+(U) : \left| \int f d\nu - \int f d\mu \right| < \varepsilon \} \ .$$

Thus, a net of measures $\langle \mu_\alpha \rangle_{\alpha \in A}$ in $\mathcal{M}^+(U)$ converges weakly to μ in $\mathcal{M}^+(U)$ iff $\int f d\mu_\alpha \to \int f d\mu$ for each $f \in \mathcal{C}_c(U)$.

THEOREM 8.44 (Thuillier). *Let $U \subsetneq \mathbb{P}^1_{\text{Berk}}$ be a domain. Equip $\mathcal{SH}(U)$ with the topology of pointwise convergence on $U \cap \mathbb{H}_{\text{Berk}}$ and $\mathcal{M}^+(U)$ with the topology of weak convergence relative to $\mathcal{C}_c(U)$. Then the map $-\Delta_U : \mathcal{SH}(U) \to \mathcal{M}^+(U)$ is continuous, and the following sequence is exact:*

(8.32) $$0 \longrightarrow \mathcal{H}(U) \longrightarrow \mathcal{SH}(U) \overset{-\Delta_U}{\longrightarrow} \mathcal{M}^+(U) \longrightarrow 0 \ .$$

The proof will be given after a series of lemmas. By Corollary 8.39, we know that if f is subharmonic on a domain U and its Laplacian is 0, then f is harmonic on U. This gives exactness at the middle joint in (8.32).

Our first two lemmas establish the surjectivity on the right in (8.32).

LEMMA 8.45. *Let $V, W \subset \mathbb{P}^1_{\text{Berk}}$ be simple domains with $\overline{V} \subset W$. Assume that each component of $\mathbb{P}^1_{\text{Berk}} \backslash V$ contains at least one boundary point of W. Then each function $h_V \in \mathcal{H}(V)$ has an extension to a function $h_W \in \mathcal{H}(W)$.*

PROOF. If V is a disc, then by Proposition 7.12, $h_V(z) = C$ is constant on V. Setting $h_W(z) = C$ for all $z \in W$, we are done.

If V is not a disc, write $\partial V = \{x_1, \ldots, x_\ell\}$, $\partial W = \{y_1, \ldots, y_m\}$. By hypothesis $\ell \geq 2$, so $m \geq 2$ as well. Let Γ_V be the finite graph with endpoints $x_1, \ldots, x_\ell$ and let Γ_W be the finite graph with endpoints $y_1, \ldots, y_m$. Then

$\Gamma_V^0 = \Gamma_V \backslash \{x_1, \ldots, x_\ell\}$ is the main dendrite of V, and $\Gamma_W^0 = \Gamma_W \backslash \{y_1, \ldots, y_m\}$ is the main dendrite of W.

Note that Γ_W can be obtained from Γ_V by adjoining certain trees T_{x_i}, where the endpoints of T_{x_i} are x_i and some of the $y_j \in \partial W$. By our hypothesis, for each $x_i \in \partial V$ there is at least one $y_j \in \partial W$ which is an endpoint of T_{x_i}; fix such a point and denote it y_{j_i}.

By Propositions 7.12 and 7.23, there is a function $H_V \in \text{CPA}(\Gamma_V)$ such that $h_V(z) = H_V \circ r_{\mathbb{P}^1_{\text{Berk}}, \Gamma_V}(z)$ for each $z \in V$. Let $\vec{v}_i$ be the unique tangent direction at x_i in Γ_V, and put $m_i = d_{\vec{v}_i}(H_V)(x_i)$. Let H_W be the function on Γ_W obtained by extending H_V to each T_{x_i} by making it affine with slope $-m_i$ on the segment $[x_i, y_{j_i}]$ and by making it constant on branches of T_{x_i} off $[x_i, y_{j_i}]$. Then $\Delta_{\Gamma_W}(H_W)$ is supported on ∂W.

Put $h = H_W \circ r_{\mathbb{P}^1_{\text{Berk}}, \Gamma_W}$. Then $h \in \text{BDV}(\mathbb{P}^1_{\text{Berk}})$, with $\Delta(h) = \Delta_{\Gamma_W}(H_W)$ by Example 5.18. It follows that $h_W := h|_W \in \mathcal{H}(W)$. By construction $h_W|_V = h_V$. □

LEMMA 8.46. *Let $U \subsetneq \mathbb{P}^1_{\text{Berk}}$ be a domain, and let $\mu \in \mathcal{M}^+(U)$. Then there is an $f \in \mathcal{SH}(U)$ for which $\Delta_U(f) = -\mu$.*

PROOF. By Corollary 7.11, there is an exhaustion of U by simple subdomains $V_1 \subset V_2 \subset \cdots$ with $\overline{V}_n \subset V_{n+1}$ for each n. We claim that we can assume that for each n, each component of $\mathbb{P}^1_{\text{Berk}} \backslash V_n$ contains at least one boundary point of U. If this fails, let N be the least index n for which it fails, and let $W_1, \ldots, W_r$ be the components of $\mathbb{P}^1_{\text{Berk}} \backslash V_N$ which do not contain points of ∂U (note that since V_N is simple, $\mathbb{P}^1_{\text{Berk}} \backslash V_N$ has finitely many components). Then each $W_i \subset U$, and by replacing V_n with $V_n \cup W_1 \cup \cdots \cup W_r$ for each $n \geq N$, we obtain a new sequence of simple domains exhausting U, with $\overline{V}_n \subset V_{n+1}$ for each n, and each component of $\mathbb{P}^1_{\text{Berk}} \backslash V_N$ contains a point of ∂U. The claim now follows by induction.

Note that since for each n, each component of $\mathbb{P}^1_{\text{Berk}} \backslash V_n$ contains a point of ∂U, it is also the case that for each n and each $m > n$, each component of $\mathbb{P}^1_{\text{Berk}} \backslash V_n$ contains at least one point of ∂V_m.

For each n, put $\mu_n = \mu|_{V_n}$: that is, μ_n is the positive Borel measure on $\mathbb{P}^1_{\text{Berk}}$ such that for each measurable set F, $\mu_n(F) = \mu(F \cap V_n)$. We will inductively construct functions $f_1, f_2, \ldots$, with each $f_n \in \mathcal{SH}(V_n)$, such that $\Delta_{V_n}(f_n) = -\mu_n$ and $f_{n+1}|_{V_n} = f_n$. By the compatibility of the Laplacian with restriction to subdomains (Proposition 5.26), we can then define $f \in \mathcal{SH}(U)$ with $\Delta_U(f) = -\mu$ by requiring that $f|_{V_n} = f_n$ for each n.

Fix a point $\zeta \in \mathbb{P}^1_{\text{Berk}} \backslash U$, and put $f_1(z) = -u_{\mu_1}(z, \zeta)|_{V_1}$. By Example 5.21, f_1 is subharmonic on V_1, and by Example 5.21 and Proposition 5.26, $\Delta_{V_1}(f_1) = -\mu_1$. Put $h_1(z) \equiv 0$.

Inductively, suppose that for some n, we have constructed a function $f_n \in \mathcal{SH}(V_n)$ with $\Delta_{V_n}(f_n) = -\mu_n$ and a function $h_n \in \mathcal{H}(V_n)$ such that $f_n(z) = h_n(z) - u_{\nu_n}(z, \zeta)$ on V_n. Put $g_{n+1}(z) = -u_{\mu_{n+1}}(z, \zeta)|_{V_{n+1}}$. By Example 5.21 and Proposition 5.26, $g_{n+1} \in \mathcal{SH}(V_{n+1})$ and $\Delta_{V_{n+1}}(g_{n+1}) =$

$-\mu_{n+1}$. Furthermore, by Proposition 5.26,
$$\Delta_{V_n}(g_{n+1}|_{V_n}) = -\mu_{n+1}|_{V_n} = -\mu_n.$$
By Theorem 8.38 (the Riesz Decomposition Theorem), there is a harmonic function $H_n \in \mathcal{H}(V_n)$ such that
$$g_{n+1}|_{V_n}(z) = H_n(z) - u_{\nu_n}(z, \zeta)$$
for all $z \in V_n$.

By Lemma 8.45, there is a function $h_{n+1} \in \mathcal{H}(V_{n+1})$ with $h_{n+1}|_{V_n} = h_n - H_n$. Put $f_{n+1}(z) = h_{n+1}(z) - u_{\nu_{n+1}}(z, \zeta)|_{V_{n+1}}$. An easy computation shows that $f_{n+1}|_{V_n} = f_n$. Furthermore $f_{n+1} \in \mathcal{SH}(V_{n+1})$, and $\Delta_{V_{n+1}}(f_{n+1}) = \Delta_{V_{n+1}}(g_{n+1}) = -\mu_{n+1}$. □

Let $\text{CPA}_c(U)$ be the space of functions $f \in \mathcal{C}_c(U)$ for which there are a compact subset $X_f \subset U$ and a finite subgraph $\Gamma \subset U$ such that $f = F \circ r_{U,\Gamma}$ for some $F \in \text{CPA}(\Gamma)$ and f vanishes outside X_f. We regard $\text{CPA}_c(U)$ as the space of "smooth test functions on U". We now show that $\text{CPA}_c(U)$ is dense in $\mathcal{C}_c(U)$ under the sup norm:

LEMMA 8.47. *Let $U \subset \mathbb{P}^1_{\text{Berk}}$ be a domain. Then for each $f \in \mathcal{C}_c(U)$, each simple subdomain V of U with $\text{supp}(f) \subset V$, and each $\varepsilon > 0$, there is a function $g \in \text{CPA}_c(U)$ such that $|f(z) - g(z)| < \varepsilon$ for each $z \in U$ and $g(z) = 0$ for each $z \notin \overline{V}$.*

PROOF. If $U = \mathbb{P}^1_{\text{Berk}}$, the result follows from Proposition 5.4. Henceforth, we will assume that $U \neq \mathbb{P}^1_{\text{Berk}}$.

Fix $f \in \mathcal{C}_c(U)$. If $f \equiv 0$, the result is trivial. Otherwise, put $X = \text{supp}(f)$. By hypothesis, X is nonempty, compact, and contained in U. Let V be a simple subdomain of U containing X. By Corollary 7.11 (applied to V), there is a simple subdomain V_1 of V which contains X. Noting that $\overline{V}_1 \subset V$, write $\partial V_1 = \{x_1, \ldots, x_\ell\}$.

Let $\varepsilon > 0$ be given. By Proposition 5.4, there are a finite subgraph $\tilde{\Gamma} \subset U$ and a function $\tilde{G} \in \text{CPA}(\tilde{\Gamma})$, such that the function $\tilde{g} = \tilde{G} \circ r_{U,\tilde{\Gamma}}$ satisfies $|f(z) - \tilde{g}(z)| < \varepsilon$ for all $z \in U$. We will construct g from $\tilde{g}$ in two steps.

We first construct a finite subgraph $\Gamma_1 \subset \overline{V}_1$ and a function $G_1 \in \text{CPA}(\Gamma_1)$, such that $g_1 = G_1 \circ r_{U,\Gamma_1}$ satisfies $|f(z) - g_1(z)| < \varepsilon$ for all $z \in U$. Fix a point $\zeta_1 \in \tilde{\Gamma}$, and put
$$\tilde{\Gamma}_1 = \tilde{\Gamma} \cup \left(\bigcup_{j=1}^{\ell} [\zeta_1, x_j] \right).$$

Since $\tilde{g}$ is constant on branches off $\tilde{\Gamma}$, the function $\widetilde{G}_1 = \tilde{g}|_{\tilde{\Gamma}_1}$ belongs to $\text{CPA}(\tilde{\Gamma}_1)$. As $r_{U,\tilde{\Gamma}} = r_{\tilde{\Gamma}_1,\tilde{\Gamma}} \circ r_{U,\tilde{\Gamma}_1}$, it follows that $\tilde{g} = \tilde{G}_1 \circ r_{U,\tilde{\Gamma}_1}$.

Now put $\Gamma_1 = \overline{V}_1 \cap \tilde{\Gamma}_1$. Since $\overline{V}_1$ is compact and path-connected, Γ_1 is a finite graph and $x_1, \ldots, x_\ell$ are among its endpoints. (If V_1 is a disc, it is possible that Γ_1 is a degenerate graph, reduced to a point.) Note that $\tilde{\Gamma}_1$ can be obtained from Γ_1 by adjoining certain trees $\tilde{T}_{x_i}$ at the points $x_i \in \partial V_1$. Put $G_1 = \tilde{g}|_{\Gamma_1}$ and put $g_1 = G_1 \circ r_{U,\Gamma_1}$. Then $G_1 \in \text{CPA}(\Gamma_1)$. To see that $|f(z) - g_1(z)| < \varepsilon$ for all $z \in U$, fix z. If $z \in V_1$, then $g_1(z) = \tilde{g}(z)$, since $r_{U,\tilde{\Gamma}_1}(z) = r_{U,\Gamma_1}(z) \in \Gamma_1 \subset \tilde{\Gamma}_1$. If $z \notin V_1$, then $r_{U,\Gamma_1}(z) = x_i$ for some $x_i \in \partial V_1$. Since g_1 is constant on branches off Γ_1 and coincides with $\tilde{g}$ on Γ_1, it follows that $g_1(z) = g_1(x_i) = \tilde{g}(x_i)$. However $f(z) = f(x_i) = 0$, as $\text{supp}(f) \subset V_1$. Thus $|\tilde{g}(x_i)| = |f(x_i) - \tilde{g}(x_i)| < \varepsilon$, so $|f(z) - g_1(z)| < \varepsilon$.

We will now modify g_1 to make it vanish outside $\overline{V}$. We will construct a finite graph $\Gamma \subset \overline{V}$ and a $G \in \text{CPA}(\Gamma)$ such that the function $g = G \circ r_{U,\Gamma}$ coincides with g_1 on $\overline{V}_1$, vanishes on ∂V, and satisfies $|f(z) - g(z)| < \varepsilon$ for all $z \in U$. Fix $\zeta \in \Gamma_1$, and write $\partial V = \{y_1, \ldots, y_m\}$. Define

$$\Gamma = \Gamma_1 \cup \left(\bigcup_{k=1}^{m} [\zeta, y_k] \right) .$$

Then $\Gamma \subset \overline{V}$, and Γ is obtained from Γ_1 by adjoining trees T_{x_i} at the points $x_i \in \partial V_1$ (it is possible that some of the T_{x_i} are reduced to a point). The endpoints of each T_{x_i} consist of x_i and some of the $y_k \in \partial V$.

Define G on Γ as follows. First, put $G = G_1$ on Γ_1. Then, extend G to each T_{x_i} in such a way that $G|_{T_{x_i}} \in \text{CPA}(T_{x_i})$ and $G(z)$ is between 0 and $g(x_i)$ on T_{x_i}, with $G(x_i) = G_1(x_i)$ and $G(y_k) = 0$ at each other endpoint y_k of T_{x_i}. To do this, for each branch point z_j of T_{x_i}, it suffices to choose a value for $G(z_j)$ between 0 and $g(x_i)$ and then to extend G to the edges of T_{x_i} by linearly interpolating between the specified values at the endpoints and branch points. In particular, $|G(z)| < \varepsilon$ for each $z \in T_{x_i}$.

Clearly $G \in \text{CPA}(\Gamma)$. Put $g = G \circ r_{U,\Gamma}$. Since $G(y_k) = 0$ for each $y_k \in \partial V$, g vanishes outside $\overline{V}$ and thus $g \in \text{CPA}_c(U)$. We claim that $|f(z) - g(z)| < \varepsilon$ for each $z \in U$. If $z \in \overline{V}_1$, then $r_{U,\Gamma}(z) = r_{U,\Gamma_1}(z) \in \Gamma_1$, so since g is constant on branches off Γ and g_1 is constant on branches off Γ_1, it follows that $g(z) = g_1(z)$. Thus $|f(z) - g(z)| = |f(z) - g_1(z)| < \varepsilon$. If $z \notin \overline{V}_1$, then $f(z) = 0$, while $r_{U,\Gamma}(z) \in T_{x_i}$ for some x_i, so $|g(z)| = |G(r_{U,\Gamma}(z))| < \varepsilon$. Thus $|f(z) - g(z)| < \varepsilon$ holds trivially. $\square$

The following lemma shows that in order to establish weak convergence in $\mathcal{M}^+(U)$ relative to functions in $\mathcal{C}_c(U)$, it suffices to check weak convergence relative to functions in $\text{CPA}_c(U)$:

LEMMA 8.48. *Let $U \subsetneq \mathbb{P}^1_{\text{Berk}}$ be a domain. If $\mu \in \mathcal{M}^+(U)$ and $\langle \mu_\alpha \rangle_{\alpha \in A}$ is a net in $\mathcal{M}^+(U)$, then $\langle \mu_\alpha \rangle_{\alpha \in A}$ converges weakly to μ if and only if $\int g \, d\mu_\alpha \to \int g \, d\mu$ for each $g \in \text{CPA}_c(U)$.*

PROOF. Let $f \in \mathcal{C}_c(U)$ be given. We must show that if $\int g \, d\mu_\alpha \to \int g \, d\mu$ for each $g \in \text{CPA}_c(U)$, then $\int f \, d\mu_\alpha \to \int f \, d\mu$.

If $f \equiv 0$, then there is nothing to prove. Otherwise, put $X = \text{supp}(f)$. By hypothesis, X is compact and contained in U. If $U \neq \mathbb{P}^1_{\text{Berk}}$, then by

Corollary 7.11, there a simple subdomain V of U with $X \subset V$. We claim that there are a bound B and an index α_0 such that $\mu_\alpha(\overline{V}) < B$ for all $\alpha \geq \alpha_0$ and such that $\mu(\overline{V}) < B$.

Since V is a simple subdomain of U, we have $\overline{V} \subset U$. By Corollary 7.11 again, there is a simple subdomain W of U with $\overline{V} \subset W$. Fix a point $\zeta \in V \cap \mathbb{H}_{\text{Berk}}$, and let $\partial V = \{x_1, \ldots, x_\ell\}$, $\partial W = \{y_1, \ldots, y_m\}$. Put

$$\Gamma_V = \bigcup_{i=1}^{\ell} [\zeta, x_i], \quad \Gamma_W = \bigcup_{k=1}^{m} [\zeta, y_k].$$

Then Γ_W can be obtained from Γ_V by adjoining a finite number of trees T_{x_i} at the points $x_i \in \partial V$; the endpoints of T_{x_i} consist of x_i and some of the $y_k \in \partial W$.

Define a function $G \in \text{CPA}(\Gamma_W)$ by setting $G(z) \equiv 1$ on Γ_V, and then extend G to each T_{x_i} in such a way that $G|_{T_{x_i}} \in \text{CPA}(T_{x_i})$, with $0 \leq G(z) \leq 1$ for each $z \in T_{x_i}$ and with $G(x_i) = 1$ and $G(y_k) = 0$ for each other endpoint of T_{x_i}. This can be done by a construction like that in the proof of Lemma 8.47. Put $g = G \circ r_{U, \Gamma_W}$. Then $g(z) = 1$ on $\overline{V}$, $g(z) = 0$ outside $\overline{W}$, and $0 \leq g(z) \leq 1$ for all $z \in U$. In particular, $g \in \text{CPA}_c(U)$.

Put $B = \int g\, d\mu + 1$. Clearly $\mu(\overline{V}) < B$. Since $g \in \text{CPA}_c(U)$, there is an α_0 with $|\int g\, d\mu_\alpha - \int g\, d\mu| < 1$ for each $\alpha \geq \alpha_0$. For such α, we have $\mu_\alpha < B$.

Now let $\varepsilon > 0$ be given. By Lemma 8.47, there is a $g_\varepsilon \in \text{CPA}_c(U)$ with $\text{supp}(g_\varepsilon) \subseteq \overline{V}$ such that $|f(z) - g_\varepsilon(z)| < \varepsilon$ for each $z \in U$. By hypothesis, there is an $\alpha(\varepsilon)$ such that $|\int g_\varepsilon\, d\mu_\alpha - \int g_\varepsilon\, d\mu| < \varepsilon$ for each $\alpha \geq \alpha(\varepsilon)$. Fix β with $\beta \geq \alpha_0$, $\beta \geq \alpha(\varepsilon)$. Then for each $\alpha \geq \beta$,

$$\left| \int f\, d\mu_\alpha - \int f\, d\mu \right| \leq \int |f - g_\varepsilon|\, d\mu_\alpha + \left| \int g_\varepsilon\, d\mu_\alpha - \int g_\varepsilon\, d\mu \right|$$
$$+ \int |g_\varepsilon - f|\, d\mu$$
$$(8.33) \qquad\qquad \leq (2B + 1)\varepsilon.$$

Since $\varepsilon > 0$ is arbitrary, it follows that $\langle \mu_\alpha \rangle$ converges weakly to μ. $\square$

PROOF OF THEOREM 8.44. Clearly $\mathcal{H}(U)$ is contained in $\text{Ker}(-\Delta_U)$, and by Corollary 8.39, $\text{Ker}(-\Delta_U) = \mathcal{H}(U)$. By Lemma 8.46, $\text{Im}(-\Delta_U) = \mathcal{M}^+(U)$. Thus, (8.32) is exact.

It remains to show that the map $-\Delta_U : \mathcal{SH}(U) \to \mathcal{M}^+(U)$ is continuous. Suppose $\langle f_\alpha \rangle_{\alpha \in A}$ is a net of functions in $\mathcal{SH}(U)$ which converges to $f \in \mathcal{SH}(U)$ pointwise on $U \cap \mathbb{H}_{\text{Berk}}$. Put $\mu_\alpha = -\Delta_U(f_\alpha)$ and $\mu = -\Delta_U(f)$. We must show that for each $g \in \mathcal{C}_c(U)$, the net $\langle \int g\, d\mu_\alpha \rangle_{\alpha \in A}$ converges to $\int g\, d\mu$.

By Lemma 8.48, it suffices to check this when $g \in \text{CPA}_c(U)$. Given such a g, put $-\Delta_U(g) = \sum_{i=1}^{m} c_i \delta_{x_i}$ with $x_1, \ldots, x_m \in U \cap \mathbb{H}_{\text{Berk}}$. By Corollary 5.39,

$$(8.34) \quad \int g\, d\mu_\alpha = -\int g\, \Delta_U(f_\alpha) = -\int f_\alpha\, \Delta_U(g) = \sum_{i=1}^{m} c_i f_\alpha(x_i).$$

Similarly,

$$\int g \, d\mu = -\int g \, \Delta_U(f) = -\int f \, \Delta_U(g) = \sum_{i=1}^{m} c_i f(x_i) \, . \tag{8.35}$$

Since $f_\alpha(x_i) \to f(x_i)$ for each x_i, the result follows. $\square$

When $U = \mathbb{P}^1_{\text{Berk}}$, the only subharmonic functions on U are the constant functions (Example 8.5). It is more interesting to fix a point $\zeta \in \mathbb{P}^1_{\text{Berk}}$, let μ range over $\mathcal{M}^+_1$ (the set of positive Borel measures on $\mathbb{P}^1_{\text{Berk}}$ with mass 1), and consider the collection of functions $f \in \text{BDV}(\mathbb{P}^1_{\text{Berk}})$ which belong to $\mathcal{SH}(\mathbb{P}^1_{\text{Berk}} \backslash \{\zeta\})$ and satisfy

$$\Delta(f) = \delta_\zeta - \mu \, . \tag{8.36}$$

Such functions are called *Arakelov-Green's functions* relative to ζ. The set of functions $f \in \text{BDV}(\mathbb{P}^1_{\text{Berk}}) \cap \mathcal{SH}(\mathbb{P}^1_{\text{Berk}} \backslash \{\zeta\})$ satisfying (8.36) for some μ will be denoted $\mathcal{AG}[\zeta]$.

We have already encountered examples of Arakelov-Green's functions: the Green's functions $G(z, \zeta; E)$ associated to a set E of positive capacity. In Chapter 10 we will study another class of examples, the functions $g_\varphi(z, \zeta)$ associated to a rational function $\varphi \in K(T)$ with $\deg(\varphi) \geq 2$.

By Example 5.21, for $\zeta \in \mathbb{H}_{\text{Berk}}$ each function of the form

$$f(z) = -u_\mu(z, \zeta) + C = \int \log_v([z, w]_\zeta) \, d\mu(w) + C \, , \tag{8.37}$$

with $\mu \in \mathcal{M}^+_1$, is in $\mathcal{AG}[\zeta]$. This also holds when $\zeta \in \mathbb{P}^1(K)$ and $\zeta \notin \text{supp}(\mu)$. For an arbitrary $\zeta \in \mathbb{P}^1_{\text{Berk}}$ and $\mu \in \mathcal{M}^+_1$, functions of the form

$$f(z) = -\log_v(\|z, \zeta\|) - u_\mu(z, \zeta_{\text{Gauss}}) + C \tag{8.38}$$

belong to $\mathcal{AG}[\zeta]$ by Examples 5.19 and 5.21 (recall that $\|z, \zeta\| = [z, \zeta]_{\zeta_{\text{Gauss}}}$). Note that when $\zeta \in \mathbb{P}^1(K)$, we have $f(\zeta) = \infty$ unless $\mu = \delta_\zeta$. In that case, $f(z) = C$ for all $z \neq \zeta$, while $f(\zeta)$ is indeterminate. When $\mu = \delta_\zeta$, we will by convention regard the function f in (8.38) as constant.

Let $\mathcal{M}^+_0[\zeta]$ be the set of all measures of the form $\mu - \delta_\zeta$, with $\mu \in \mathcal{M}^+_1$. The next two results, due to Favre and Jonsson [45, Theorem 7.64], assert that $\mathcal{AG}[\zeta]$ is complete for the topology of pointwise convergence on $\mathbb{H}_{\text{Berk}}$ and that all Arakelov-Green's functions have the form (8.38):

PROPOSITION 8.49 (Favre and Jonnson). *Fix $\zeta \in \mathbb{H}_{\text{Berk}}$. Let $\{f_\alpha\}_{\alpha \in A}$ be a net of functions in $\mathcal{AG}[\zeta]$ which converges pointwise on $\mathbb{H}_{\text{Berk}}$ to a function $f : \mathbb{P}^1_{\text{Berk}} \to \mathbb{R} \cup \{-\infty\}$ with $f \not\equiv -\infty$. Then f is finite on $\mathbb{H}_{\text{Berk}}$, $f^*(x) = f(x)$ for all $x \in \mathbb{H}_{\text{Berk}}$, and $f^* \in \mathcal{AG}[\zeta]$.*

If $\zeta \in \mathbb{P}^1(K)$, the same assertions hold, except that the domain of f should be taken as $\mathbb{P}^1_{\text{Berk}} \backslash \{\zeta\}$.

PROOF. By replacing each $f_\alpha(z)$ with $f_\alpha(z) - \log_v(\|z,\zeta\|)$, we can reduce to the case where $\zeta = \zeta_{\text{Gauss}} \in \mathbb{H}_{\text{Berk}}$.

For each $g \in \mathcal{AG}[\zeta]$, the fact that $\Delta(g) = \delta_\zeta - \mu$, where μ is a probability measure, means that g takes its maximum on $\mathbb{P}^1_{\text{Berk}}$ at ζ (Proposition 8.14) and that $|d_{\vec{v}} g(x)| \leq 1$ for each $x \in \mathbb{H}_{\text{Berk}}$ and each $\vec{v} \in T_x$. (To see this, let $\Gamma = [x,p]$ be a segment emanating from x in the direction $\vec{v}$. By formula (3.15), $|d_{\vec{v}} g(x)| = |\Delta_\Gamma(g)((x,p])| = |\Delta(g)(r_\Gamma^{-1}(x,p])| \leq 1$.) Consequently, for any two points $x, y \in \mathbb{H}_{\text{Berk}}$ we have $|g(x) - g(y)| \leq \rho(x,y)$. It follows that if $\lim_\alpha f_\alpha(a) = -\infty$ for one point $a \in \mathbb{H}_{\text{Berk}}$, then $\lim_\alpha f_\alpha(\zeta) = -\infty$. Since each f_α takes its maximum at ζ, necessarily $f(x) \equiv -\infty$ on $\mathbb{P}^1_{\text{Berk}}$.

Suppose $f \not\equiv -\infty$. By an argument similar to the one in the proof of Proposition 8.49, using domination subharmonicity on each component of $\mathbb{P}^1_{\text{Berk}} \backslash \{\zeta\}$, one concludes that f^* is subharmonic on $\mathbb{P}^1_{\text{Berk}} \backslash \{\zeta\}$.

Next, we claim that $f^* \in \text{BDV}(\mathbb{P}^1_{\text{Berk}})$. By the definition of subharmonicity, for each $x \in \mathbb{P}^1_{\text{Berk}} \backslash \{\zeta\}$, there is a neighborhood V_x of x such that f^* is strongly subharmonic on V_x; in particular $f^*|_{V_x} \in \text{BDV}(V_x)$. Now consider f^* near ζ. Let ξ be a point of $\mathbb{H}_{\text{Berk}}$ distinct from ζ. For each $\alpha \in A$, put $\tilde{f}_\alpha(x) = f_\alpha(x) - \log_v(\delta(x,\xi)_\zeta)$; then $\tilde{f}_\alpha \in \mathcal{AG}[\xi]$. The functions $\tilde{f}_\alpha$ converge pointwise to $\tilde{f}(x) = f(x) - \log_v(\delta(x,\xi)_\zeta)$ on $\mathbb{H}_{\text{Berk}}$, and by the discussion above, $\tilde{f}^*$ is subharmonic on $\mathbb{P}^1_{\text{Berk}} \backslash \{\xi\}$. Let V_ζ be a neighborhood of ζ on which $\tilde{f}^*$ is strongly subharmonic. Since $\Delta(-\log_v(\delta(x,\xi)_\zeta)) = \delta_\xi - \delta_\zeta$, it follows that $f^*|_{V_\zeta} \in \text{BDV}(V_\zeta)$ and that $\Delta_{V_\zeta}(f^*) = \Delta_{V_\zeta}(\tilde{f}^*) + \delta_\zeta$. Since $\mathbb{P}^1_{\text{Berk}}$ is compact, finitely many of the sets V_ζ and V_x cover it. Our claim that $f \in \text{BDV}(\mathbb{P}^1_{\text{Berk}})$ now follows from Proposition 5.27.

Finally, we claim that $f^* \in \mathcal{AG}[\zeta]$. The discussion above shows that $\Delta(f) = \delta_\zeta - \mu$ for some positive Borel measure μ. However, by Proposition 5.25, $\Delta(f)$ has total mass 0. Hence μ is a probability measure, and we are done. □

THEOREM 8.50 (Favre and Jonsson). *Fix $\zeta \in \mathbb{P}^1_{\text{Berk}}$. Equip $\mathcal{AG}[\zeta]$ with the topology of pointwise convergence on $\mathbb{H}_{\text{Berk}}$ and $\mathcal{M}_0^+[\zeta]$ with the topology of weak convergence relative to $C(\mathbb{P}^1_{\text{Berk}})$. Then the map $-\Delta : \mathcal{AG}[\zeta] \to \mathcal{M}_0^+[\zeta]$ is continuous, and the following sequence is exact:*

$$(8.39) \qquad 0 \longrightarrow \mathbb{R} \longrightarrow \mathcal{AG}[\zeta] \xrightarrow{-\Delta} \mathcal{M}_0^+[\zeta] \longrightarrow 0 \ .$$

PROOF. We first show that (8.39) is exact. Surjectivity on the right follows from (8.38) by Examples 5.19 and 5.21. Clearly each constant function belongs to $\text{Ker}(-\Delta)$. Conversely, if $f \in \mathcal{AG}[\zeta]$ satisfies $-\Delta(f) = 0$, then since $f \in \text{BDV}(\mathbb{P}^1_{\text{Berk}})$, Proposition 5.28 shows that $f = C$ is constant on $\mathbb{H}_{\text{Berk}}$. For each $\xi \in \mathbb{P}^1(K) \backslash \{\zeta\}$, Proposition 8.11(C) shows that $f(\xi) = C$ as well. In view of the convention made after (8.38), this gives exactness at the middle joint in (8.39).

We next show that $-\Delta : \mathcal{AG}[\zeta] \to \mathcal{M}_0^+[\zeta]$ is continuous for the given topologies. Since $\mathbb{P}^1_{\text{Berk}}$ is compact, $\text{CPA}_c(\mathbb{P}^1_{\text{Berk}})$ coincides with the space of all functions of the form $f = F \circ r_{\mathbb{P}^1_{\text{Berk}}, \Gamma}$, where Γ is a finite subgraph of $\mathbb{P}^1_{\text{Berk}}$ and $F \in \text{CPA}(\Gamma)$. By Proposition 5.4(B), such functions are dense in $\mathcal{C}(\mathbb{P}^1_{\text{Berk}})$. For each $\nu \in \mathcal{M}_0^+[\zeta]$, the positive and negative parts of the Jordan decomposition of ν have mass at most 1, so the argument in the proof of Lemma 8.48 shows that in order to prove that a net $\langle \nu_\alpha \rangle$ in $\mathcal{M}_0^+[\zeta]$ converges weakly to $\nu \in \mathcal{M}_0^+[\zeta]$ relative to functions in $\mathcal{C}(\mathbb{P}^1_{\text{Berk}})$, it suffices to prove it converges weakly relative to functions in $\text{CPA}_c(\mathbb{P}^1_{\text{Berk}})$. Now the continuity of $-\Delta$ in (8.39) follows from (8.34) and (8.35) by the same argument as in the proof of Theorem 8.44. $\square$

8.7. Convergence of Laplacians

Let U be a domain in $\mathbb{P}^1_{\text{Berk}}$, and let $\langle f_\alpha \rangle_\alpha$ be a net of subharmonic functions on U which converges to a subharmonic function F. One can ask under what conditions the net of measures $\langle \Delta_U(f_\alpha) \rangle$ converges to $\Delta_U(F)$.

PROPOSITION 8.51. *Let $U \subset \mathbb{P}^1_{\text{Berk}}$ be a domain, and let $\langle f_\alpha \rangle_{\alpha \in A}$ be a net of subharmonic functions on U. Suppose one of the following conditions holds:*

(A) *The f_α converge uniformly to F on each finite subgraph $\Gamma \subset U$.*
(B) *The f_α decrease monotonically to F on each finite subgraph $\Gamma \subset U$, and $F \not\equiv -\infty$ on U.*
(C) *The f_α increase monotonically on each finite subgraph $\Gamma \subset U$, and the family $\{f_\alpha\}$ is locally bounded above. Put $f(z) = \lim_\alpha f_\alpha(z)$ and let $F(z) = f^*(z)$ be the upper semicontinuous regularization of $f(z)$.*

Then F is subharmonic on U, and the measures $\Delta_{\overline{V}}(f_\alpha)$ converge weakly to $\Delta_{\overline{V}}(F)$ for each simple subdomain $V \subset U$.

PROOF. In all three cases, Proposition 8.26 shows that F is subharmonic. Only the convergence of the measures needs to be established.

Proof of (A). Let $V \subset U$ be a simple subdomain.

We first claim that the measures $|\Delta_{\overline{V}}(f_\alpha)|$ have uniformly bounded total mass. This depends on the fact that V can be enlarged within U.

Let W be a simple subdomain of U containing $\overline{V}$, and let $\Gamma \subset \tilde{\Gamma} \subset W$ be finite subgraphs with $\Gamma \subset \tilde{\Gamma}^0$, $\partial W \subseteq \partial \tilde{\Gamma}$, and $\partial V \subseteq \partial \Gamma$. Write $\partial V = \{x_1, \ldots, x_m\}$. When x_i is regarded as a point of Γ, it has a unique tangent vector $\vec{v}_i$ leading into Γ. As in (8.16) in the proof of Lemma 8.10, for each $\alpha \in A$ we have

$$(8.40) \quad |\Delta_\Gamma(f_\alpha)|(\overline{V}) = |\Delta_\Gamma(f_\alpha)|(\Gamma) = 2 \cdot \sum_{i=1}^m |d_{\vec{v}_i} f_\alpha(x_i)|.$$

Thus, to bound the $|\Delta_{\overline{V}}(f_\alpha)|(\overline{V})$, it suffices to show that the $|d_{\vec{v}_i} f_\alpha(x_i)|$ are uniformly bounded, for all i and α.

Fix $x_i \in \partial V$. When x_i is regarded as a point of $\tilde{\Gamma}$, it will have several tangent vectors $\vec{v}_{ij}$, $j = 1, \ldots, n_i$, with each $n_i \geq 2$. Without loss of generality, we can assume the $\vec{v}_{ij}$ are indexed so that $\vec{v}_i = \vec{v}_{i1}$ for each i. Fix $T > 0$ small enough that for each i, j, the point $q_{ij} = x_i + t\vec{v}_{ij}$ lies on the edge emanating from x_i in the direction $\vec{v}_{ij}$.

Consider the limit function F, and fix $\varepsilon > 0$. Since the f_α converge uniformly to F on $\tilde{\Gamma}$, there is an α_0 such that $|f_\alpha(q_{ij}) - F(q_{ij})| < \varepsilon$ and $|f_\alpha(x_i) - F(x_i)| < \varepsilon$ for all $\alpha \geq \alpha_0$ and all i, j. It follows that for such i, j, and α, we have

$$(8.41) \qquad \frac{f_\alpha(q_{ij}) - f_\alpha(x_i)}{T} \leq \frac{(F(q_{ij}) + \varepsilon) - (F(x_i) - \varepsilon)}{T}.$$

Since each f_α is convex on edges of $\tilde{\Gamma}$, for each i, j

$$(8.42) \qquad d_{\vec{v}_{ij}} f_\alpha(x_i) \leq \frac{f_\alpha(q_{ij}) - f_\alpha(x_i)}{T}.$$

Thus (8.41) provides an upper bound B_{ij} for each $d_{\vec{v}_{ij}} f_\alpha(x_i)$ which is independent of α.

Since x_i is an interior point of $\tilde{\Gamma}$ and $n_i \geq 2$, Proposition 8.25 gives $0 \geq \Delta_{\tilde{\Gamma}}(f_\alpha)(x_i) = -\sum_{j=1}^{n_i} d_{\vec{v}_{ij}} f_\alpha(x_i)$, so

$$(8.43) \qquad d_{\vec{v}_{i1}} f_\alpha(x_i) \geq -\sum_{j=2}^{n_i} d_{\vec{v}_{ij}} f_\alpha(x_i) \geq -\sum_{j=2}^{n_i} B_{ij}.$$

Since $d_{\vec{v}_i} f_\alpha(x_i) = d_{\vec{v}_{i1}} f_\alpha(x_i)$, we have shown that $|d_{\vec{v}_i} f_\alpha(x_i)|$ is uniformly bounded for all i and all $\alpha \geq \alpha_0$. In particular, there exists $B > 0$ such that $|\Delta_{\overline{V}}(f_\alpha)|(\overline{V}) \leq B$ for all $\alpha \geq \alpha_0$.

We will now show that the net of measures $\Delta_{\overline{V}}(f_\alpha)$ converges weakly to a measure μ on $\overline{V}$. Recall that by Proposition 5.4(B), as Γ ranges over all finite subgraphs of V and G ranges over all functions in CPA(Γ), the functions $g(x) = G \circ r_\Gamma(x)$ are dense in $\mathcal{C}(\overline{V})$. For each subgraph $\Gamma' \supset \Gamma$, each g, and each f_α, Lemma 3.14(D) gives

$$\int_{\Gamma'} g \, \Delta_{\Gamma'}(f_\alpha) = \int_{\Gamma'} f_\alpha \, \Delta_{\Gamma'}(g) = \int_\Gamma f_\alpha \, \Delta_\Gamma(G),$$

where the last equality holds because $\Delta_{\Gamma'}(g)|_\Gamma = \Delta_\Gamma(G)$. Passing to the limit over all finite subgraphs Γ' with $\Gamma \subset \Gamma' \subset V$, we find that

$$(8.44) \qquad \int_{\overline{V}} g \, \Delta_{\overline{V}}(f_\alpha) = \int_{\overline{V}} f_\alpha \, \Delta_{\overline{V}}(g) = \int_\Gamma f_\alpha \, \Delta_\Gamma(G).$$

Since the f_α converge uniformly to F on Γ, we can define a linear functional

$$(8.45) \qquad \Lambda_F(g) = \lim_\alpha \int_{\overline{V}} f_\alpha \, \Delta_{\overline{V}}(g) = \int_{\overline{V}} F \, \Delta_{\overline{V}}(g) = \int_\Gamma F \, \Delta_\Gamma(G)$$

on the dense space of functions $g(x) = G \circ r_\Gamma(x)$ in $\mathcal{C}(\overline{V})$.

On the other hand, by (8.44), we have

(8.46) $$\Lambda_F(g) = \lim_\alpha \int_{\overline{V}} g \, \Delta_{\overline{V}}(f_\alpha) \ .$$

Since the measures $|\Delta_{\overline{V}}(f_\alpha)|(\overline{V})$ are uniformly bounded, Λ_F extends to a bounded linear functional on $\mathcal{C}(\overline{V})$. By the Riesz Representation Theorem, there is a unique measure μ such that

$$\Lambda_F(g) = \int_{\overline{V}} g(x) \, d\mu(x)$$

for all $g \in \mathcal{C}(\overline{V})$. This measure is the weak limit of the $\Delta_{\overline{V}}(f_\alpha)$.

To complete the proof, we must show that $\mu = \Delta_{\overline{V}}(F)$. Since $\Delta_{\overline{V}}(F)$ is a bounded measure on $\overline{V}$, it suffices to check that

$$\Lambda_F(g) = \int_{\overline{V}} g(x) \, \Delta_{\overline{V}}(F)(x)$$

on the dense space of functions considered above. This follows immediately from (8.44), with f_α replaced by F. This proves (A).

For (B), fix a finite subgraph $\Gamma \subset U$. By part (A), it suffices to show that f_α converges uniformly to F on Γ. This follows from Dini's lemma (Lemma A.7), together with the observation that by Proposition 8.24, the restriction to Γ of a subharmonic function on U is continuous.

For (C), a similar argument shows that since $F(z) = f^*(z)$ is subharmonic on U, for each finite subgraph $\Gamma \subset U$ the restriction of F to Γ is continuous. By Proposition 8.26(E), $F(z)$ coincides with $f(z) = \lim_\alpha f_\alpha(z)$ on $U \cap \mathbb{H}_{\mathrm{Berk}}$. Hence $f|_\Gamma = F|_\Gamma$ is continuous, and the f_α converge uniformly to F on Γ by Dini's lemma. $\square$

We augment Proposition 8.51 with two results dealing with other cases in Proposition 8.26. The first follows from the linearity of the Laplacian.

PROPOSITION 8.52. *Let $U \subset \mathbb{P}^1_{\mathrm{Berk}}$ be open. Suppose $f, g \in \mathcal{SH}(U)$, and let $0 \leq \alpha, \beta \in \mathbb{R}$. Then $\Delta_U(\alpha f + \beta g) = \alpha \cdot \Delta_U(f) + \beta \cdot \Delta_U(g)$.*

PROPOSITION 8.53. *Let $U \subset \mathbb{P}^1_{\mathrm{Berk}}$ be open. Suppose f and g are continuous subharmonic functions on U, with $f(z) > g(z)$ outside a compact subset $Z \subset U$. Put $h(z) = \max(f(z), g(z))$. Then:*

(A) $\Delta_U(h)|_{U \setminus Z} = \Delta_U(f)|_{U \setminus Z}$, *and*
(B) $\Delta_U(h)(Z) = \Delta_U(f)(Z)$.

PROOF. We can assume without loss of generality that U is a domain, since the result holds if and only if it holds for each component of U.

The proof of (A) is easy. For each $x \in U \setminus Z$, there is a simple subdomain $W \subset U$ such that $x \in W$ and $\overline{W} \cap Z = \emptyset$. We have $h(z) = f(z)$ on W, so $\Delta_{\overline{W}}(h) = \Delta_{\overline{W}}(f)$, and $\Delta_U(f)|_W = \Delta_{\overline{W}}(f)|_W = \Delta_{\overline{W}}(h)|_W = \Delta_U(h)|_W$. Since $U \setminus Z$ can be covered by such W, the result follows.

For part (B), using that Z is compact, choose a simple subdomain $V \subset U$ with $Z \subset V$. By Lemma 8.10, the restrictions of f, g, and h to V belong to $\mathrm{BDV}(V)$. Note that ∂V is compact. Fix $\zeta \in Z$, and cover ∂V with a finite number of balls $\mathcal{B}(x_i, r_i)_\zeta^-$, where the closure of each $\mathcal{B}(x_i, r_i)_\zeta^-$ is disjoint from Z. Without loss, we can assume the $\mathcal{B}(x_i, r_i)_\zeta^-$ are pairwise disjoint. Let p_i be the unique boundary point of $\mathcal{B}(x_i, r_i)_\zeta^-$; it belongs to $V \backslash Z$. Put $V_i = V \cap \mathcal{B}(x_i, r_i)_\zeta^-$; then $h(z) = f(z)$ on V_i, and $\partial V_i \subset \partial V \cup \{p_i\}$.

By the retraction formula for Laplacians,
$$(r_{\overline{V}, \overline{V}_i})_*(\Delta_{\overline{V}}(f)) = \Delta_{\overline{V}_i}(f) = \Delta_{\overline{V}_i}(h) = (r_{\overline{V}, \overline{V}_i})_*(\Delta_{\overline{V}}(h)) \ .$$
Since $r_{\overline{V}, \overline{V}_i}$ is the identity map on $\overline{V}_i \backslash \{p_i\}$ and $r_{\overline{V}, \overline{V}_i}(\overline{V} \backslash V_i) = \{p_i\}$,
$$\Delta_{\overline{V}}(f)|_{\overline{V}_i \backslash \{p_i\}} = \Delta_{\overline{V}}(h)|_{\overline{V}_i \backslash \{p_i\}} \ .$$
Combined with the result from (A), this gives $\Delta_{\overline{V}}(f)|_{\overline{V} \backslash Z} = \Delta_{\overline{V}}(h)|_{\overline{V} \backslash Z}$. Since both $\Delta_{\overline{V}}(f)$ and $\Delta_{\overline{V}}(h)$ have total mass 0,
$$\Delta_{\overline{V}}(f)(Z) = -\Delta_{\overline{V}}(f)(\overline{V} \backslash Z) = -\Delta_{\overline{V}}(h)(\overline{V} \backslash Z) = \Delta_{\overline{V}}(h)(Z) \ .$$
□

8.8. Hartogs's lemma

In this section we establish a Berkovich space analogue of Hartogs's lemma. The proof is modeled on [**46**, Proposition 2.16] and uses the fact that the support of any finite signed Borel measure on $\mathbb{P}^1_{\mathrm{Berk}}$, equipped with the subspace topology, is a separable metric space (Lemma 5.7).

PROPOSITION 8.54 (Hartogs's lemma). *Let $U \subset \mathbb{P}^1_{\mathrm{Berk}}$ be a domain, and let $\{g_n\}$ be a sequence of functions which are subharmonic on U. Suppose the functions g_n are uniformly bounded above on U. Then either*

(A) *there is a subsequence $\{g_{n_k}\}$ which converges uniformly to $-\infty$ on each compact subset $X \subset U$ or*

(B) *there are a subsequence $\{g_{n_k}\}$ and a function G subharmonic on U, such that the g_{n_k} converge pointwise to G on $\mathbb{H}_{\mathrm{Berk}} \cap U$ and such that for each continuous function $f : U \to \mathbb{R}$ and each compact subset $X \subset U$,*

$$(8.47) \qquad \limsup_{k \to \infty} \left(\sup_{z \in X} (g_{n_k}(z) - f(z)) \right) \leq \sup_{z \in X} (G(z) - f(z)) \ .$$

The crux of the argument is contained in the following lemma:

LEMMA 8.55. *Let $U \subset \mathbb{P}^1_{\mathrm{Berk}}$ be a domain, and let $V \subset U$ be a simple subdomain (so that $\overline{V} \subset U$). Let $\{g_n\}$ be a sequence of functions in $\mathcal{SH}(U)$. Suppose that*

(A) *for each $x_\ell \in \partial V$, $\lim_{n \to \infty} g_n(x_\ell) = A_\ell$ exists and is finite and*

(B) *the sequence of positive measures $\nu_n = -\Delta_V(g_n)$ converges weakly to a finite measure ν.*

Fix a point $\zeta \in \mathbb{P}^1_{\text{Berk}} \backslash \overline{V}$, and let $u(z) = u_\nu(z, \zeta)$ be the potential function associated to ν and ζ. Let h be the unique function which is continuous on $\overline{V}$, harmonic on V, and satisfies $h(x_\ell) = A_\ell + u(x_\ell)$ for each ℓ. Define

$$G(z) = h(z) - u(z).$$

Then G is subharmonic on V, and the functions g_n converge pointwise to G on $\mathbb{H}_{\text{Berk}} \cap V$. Furthermore,

$$(8.48) \qquad \limsup_{n \to \infty} \left(\sup_{z \in V} g_n(z) \right) \leq \sup_{z \in V} G(z).$$

PROOF. Let G, h, and u be as in the lemma. Clearly G is subharmonic on V. The function h is continuous on $\overline{V}$, and u, which belongs to $\text{BDV}(\mathbb{P}^1_{\text{Berk}})$ since it is a potential function, is continuous when restricted to any finite subgraph. Hence for each $x_\ell \in \partial V$, G is continuous when restricted to any segment $[a, x_\ell] \subset \overline{V}$. It follows from the Maximum Principle (Proposition 7.16) that

$$(8.49) \qquad \sup_{z \in V} G(z) = \sup_{z \in \overline{V}} G(z) = \max_{x_\ell \in \partial V} G(x_\ell) = \max_{x_\ell \in \partial V} A_\ell.$$

For each n, put $u_n(z) = u_{\nu_n}(z, \zeta)$ and let h_n be the (unique) function given by the Riesz Decomposition Theorem (Theorem 8.38) such that $g_n(z) = h_n(z) - u_n(z)$ on $\overline{V}$. Then h_n is continuous on $\overline{V}$ and harmonic on V, and $h_n(x_\ell) = g_n(x_\ell) + u_n(x_\ell)$ for each $x_\ell \in \partial V$.

For each fixed $x \in \mathbb{H}_{\text{Berk}}$, the kernel $-\log_v([x, w]_\zeta)$ is a continuous function of $w \in \overline{V}$. Since the measures ν_n converge weakly to ν, for each $x \in \mathbb{H}_{\text{Berk}}$

$$(8.50) \quad u(x) = \int -\log_v([x, w]_\zeta) \, d\nu(w)$$
$$= \lim_{n \to \infty} \int -\log_v([x, w]_\zeta) \, d\nu_n(w) = \lim_{n \to \infty} u_n(x).$$

In particular, (8.50) holds for each $x_\ell \in \partial V$. Since $\lim_{n \to \infty} g_n(x_\ell) = A_\ell$, it follows that $h(x_\ell) = \lim_{n \to \infty} h_n(x_\ell)$. From this and the Poisson formula (Proposition 7.25), for each $z \in \overline{V}$ we have

$$(8.51) \qquad h(z) = \lim_{n \to \infty} h_n(z).$$

Combining (8.50) and (8.51), it follows that for each $x \in \mathbb{H}_{\text{Berk}} \cap \overline{V}$, we have

$$(8.52) \qquad G(x) = \lim_{n \to \infty} g_n(x).$$

By Corollary 8.15, each $g_n(z)$ achieves its maximum for $z \in \overline{V}$ at a point $x_\ell \in \partial V$. Combining (8.49) and (8.52), we obtain (8.48). □

PROOF OF PROPOSITION 8.54. If $U = \mathbb{P}^1_{\text{Berk}}$, then by Example 8.5, each g_n is a constant function and the assertions in the proposition are trivial. Hence we can assume that $U \neq \mathbb{P}^1_{\text{Berk}}$. Let M be a finite upper bound for the functions g_n on U. Without loss of generality, we can assume that $M \geq 0$.

First suppose there is a point $x_0 \in \mathbb{H}_{\text{Berk}} \cap U$ where $\inf_n g_n(x_0) = -\infty$. Choose a subsequence $\{g_{n_k}\}$ such that $g_{n_k}(x_0) < 0$ for each k and $\lim_{k \to \infty} g_{n_k}(x_0) = -\infty$.

Fix a compact set $X \subset U$, and choose a simple subdomain V of U with $X \cup \{x_0\} \subset V$. Let $x_1, \ldots, x_m$ be the boundary points of V. Applying Proposition 7.25 to each component of $V \backslash \{x_0\}$, one obtains a function h which is continuous on $\overline{V}$, harmonic on each component of $V \backslash \{x_0\}$, and satisfies $h(x_0) = 1$, $h(x_1) = \cdots = h(x_m) = 0$. Note that there are only finitely many components of $V \backslash \{x_0\}$ which are not discs, and on such components $h(z)$ is given by formula (7.7). On each component which is a disc, $h(z) \equiv 1$. The discussion before formula (7.7) shows that $h(z) > 0$ for all $z \in V$. Since X is compact, there is a number $\delta > 0$ such that $h(z) \geq \delta$ for all $z \in X$.

Since each g_{n_k} is domination subharmonic on V, with $g_{n_k}(x_0) < 0$ and $M \geq 0$, it follows that $g_{n_k}(z) \leq M + g_{n_k}(x_0) \cdot h(z)$ for all $z \in V$ and in turn that $g_{n_k}(z) \leq M + g_{n_k}(x_0) \cdot \delta$ for $z \in X$. Thus $\{g_{n_k}\}$ converges uniformly to $-\infty$ on X.

Next suppose that for each $x \in \mathbb{H}_{\text{Berk}} \cap U$, there is a finite lower bound L_x for the values $\{g_n(x)\}$. Using Corollary 7.11, choose an exhaustion of U by simple subdomains $V_1, V_2, \ldots$ with $\overline{V}_j \subset V_{j+1}$ for each j. Write $\partial V_j = \{x_{j,1}, \ldots, x_{j,m_j}\}$.

Fix j and put $\nu_{n,j} = -\Delta_{V_j}(g_n)$. Since each g_n is subharmonic on U, the measures $\nu_{n,j}$ are positive. Since for each $x \in \mathbb{H}_{\text{Berk}} \cap U$ the values $g_n(x)$ lie in the bounded interval $[L_x, M]$ for all n, the argument in the proof of Proposition 8.51(A) shows that there is a finite bound B_j such that $\nu_{n,j}(V_j) \leq B_j$ for each n. Indeed, for each $x_{j,\ell}$ and each tangent vector $\vec{v}$ at $x_{j,\ell}$, in place of (8.42) we have

$$(8.53) \qquad d_{\vec{v}} g_n(x_{j,\ell}) \leq \frac{M - L_{x_{j,\ell}}}{T} \;,$$

and by inserting this in (8.43) and using (8.40), we obtain the desired bound B_j. By Lemma 5.7, there is a separable closed set $Y_j \subset \overline{V}_j$ on which all the $\nu_{n,j}$ are supported. Hence for each j, any infinite subsequence of the measures $\{\nu_{n,j}\}_{n \geq 1}$ has a further subsequence which converges weakly.

As j varies, there are only countably many boundary points $x_{j,\ell}$, so by a diagonalization argument we can choose a subsequence $\{g_{n_k}\}$ such that

$$\lim_{k \to \infty} g_{n_k}(x_{j,\ell}) = A_{j,\ell}$$

exists for each $x_{j,\ell}$. Using a further diagonalization and the results in the previous paragraph, we can assume that for each j the measures $\nu_{n_k,j}$ converge weakly to a measure ν_j on $\overline{V}_j$.

By Lemma 8.55, for each j there is a function $G_j \in SH(V_j)$ such that the g_{n_k} converge pointwise to G_j on $\mathbb{H}_{\text{Berk}} \cap V_j$. If $j_1 < j_2$, it follows that the restrictions of G_{j_1} and G_{j_2} to V_{j_1} coincide on $\mathbb{H}_{\text{Berk}} \cap V_{j_1}$. By Proposition 8.11(C), they coincide on all of V_{j_1}. It follows that there is a function

$G \in SH(U)$ such that $G|_{V_j} = G_j$ for each V_j. In particular, $\lim_{k \to \infty} g_{n_k}(x) = G(x)$ for each $x \in \mathbb{H}_{\text{Berk}} \cap U$.

To complete the proof, it remains to show that for each continuous $f : U \to \mathbb{R}$ and each compact subset $X \subset U$, the inequality (8.47) holds. Fix $\varepsilon > 0$. For each $x \in X$, there is a neighborhood W_x of x such that $|f(z) - f(x)| < \varepsilon$ for all $z \in W_x$. After shrinking W_x if necessary, we can assume that W_x is a simple subdomain of U. By compactness, there are finitely many subsets $W_{x_1}, \ldots, W_{x_m}$ which cover X.

For each $z \in W_{x_i}$ we have
$$g_{n_k}(z) - f(z) \le g_{n_k}(z) - f(x_i) + \varepsilon, \quad G(z) - f(x_i) \le G(z) - f(z) + \varepsilon .$$
Clearly the measures $\Delta_{W_{x_i}}(g_{n_k})$ converge weakly to $\Delta_{W_{x_i}}(G)$. Applying Lemma 8.55 to the sequence $\{g_{n_k}\}$ and taking $V = W_{x_i}$, we see that
$$\begin{aligned} \limsup_{k \to \infty} \left(\sup_{z \in W_{x_i}} (g_{n_k}(z) - f(z)) \right) &\le \limsup_{k \to \infty} \left(\sup_{z \in W_{x_i}} (g_{n_k}(z) - f(x_i)) \right) + \varepsilon \\ &\le \sup_{z \in W_{x_i}} (G(z) - f(x_i)) + \varepsilon \\ &\le \sup_{z \in W_{x_i}} (G(z) - f(z)) + 2\varepsilon . \end{aligned}$$
Since the W_{x_i} cover X, we obtain (8.47). $\square$

REMARK 8.56. When part (B) of Hartogs's lemma applies, it need not be the case that $\{g_{n_k}\}$ converges pointwise to G on $U \cap \mathbb{P}^1(K)$. For example, take $U = \mathcal{D}(0,1)^-$, and let $g_n(z) = \frac{1}{n} \log_v([z,0]_\infty)$. Each g_n is subharmonic on U by Example 8.6, and trivially the g_n are uniformly bounded above by 0. Take $G(z) \equiv 0$. Then G is subharmonic (in fact harmonic) on U, and $\{g_n\}$ converges pointwise to G on $\mathbb{H}_{\text{Berk}} \cap U$. But $G(0) = 0$, while $g_n(0) = -\infty$ for each n.

Fix a point $\zeta \in \mathbb{H}_{\text{Berk}}$. Recall from §8.6 that $\mathcal{AG}[\zeta]$ is the set of Arakelov-Green's functions relative to ζ and that $\mathcal{M}_1^+$ denotes the set of positive Borel measures on $\mathbb{P}^1_{\text{Berk}}$ of total mass 1. Both $\mathcal{AG}[\zeta]$ and $\mathcal{M}_1^+$ are convex sets. Each $f \in \mathcal{AG}[\zeta]$ is subharmonic on $\mathbb{P}^1_{\text{Berk}} \backslash \{\zeta\}$, and $\Delta(f) = \delta_\zeta - \nu$ with $\nu \in \mathcal{M}_1^+$. By Proposition 7.16 (the Maximum Principle), $\max_{z \in \mathbb{P}^1_{\text{Berk}}} f(z) = f(\zeta)$.

In [**46**, Proposition 2.16], Favre and Rivera-Letelier prove the following version of Hartogs's lemma for $\mathcal{AG}[\zeta]$, which is the key result needed for their purely local proof of the equidistribution theorem for the periodic points of a rational function of degree ≥ 2 in characteristic 0 ([**46**, Theorem B]).

PROPOSITION 8.57 (Hartogs's lemma for Arakelov-Green's functions). *Fix $\zeta \in \mathbb{H}_{\text{Berk}}$, and let $\{g_n\}$ be a sequence of functions in $\mathcal{AG}[\zeta]$. Suppose the functions g_n are uniformly bounded above on $\mathbb{P}^1_{\text{Berk}}$. Then either*

(A) *there is a subsequence $\{g_{n_k}\}$ which converges uniformly to $-\infty$ on $\mathbb{P}^1_{\text{Berk}}$, or*

(B) *there are a subsequence* $\{g_{n_k}\}$ *and a function* $G \in \mathcal{AG}[\zeta]$ *such that the* g_{n_k} *converge pointwise to* G *on* $\mathbb{H}_{\text{Berk}}$ *and such that for each compact set* $X \subset \mathbb{P}^1_{\text{Berk}}$ *and each continuous* $f : X \to \mathbb{R}$,

$$(8.54) \qquad \limsup_{k \to \infty} \left(\sup_{z \in X} (g_{n_k}(z) - f(z)) \right) \leq \sup_{z \in X} (G(z) - f(z)).$$

PROOF. The proof is similar to that of Proposition 8.54, so we only sketch it. Fix n. Since g_n is subharmonic on each component of $\mathbb{P}^1_{\text{Berk}} \backslash \{\zeta\}$, it is nonincreasing on paths leading away from ζ (Proposition 8.11(A)). From this, one concludes that if there is a subsequence $\{g_{n_k}\}$ for which $\lim_{k \to \infty} g_{n_k}(\zeta) = -\infty$, then $\{g_{n_k}\}$ converges uniformly to $-\infty$ on $\mathbb{P}^1_{\text{Berk}}$.

Hence we can assume that the values $\{g_k(\zeta)\}$ are uniformly bounded. Write $\Delta(g_k) = \delta_\zeta - \nu_k$, where ν_k is a probability measure. Since $\mathcal{M}_1^+$ is sequentially compact in the weak-* topology, there are a sequence of indices $\{n_k\}$, a number L, and a measure $\nu \in \mathcal{M}_1^+$ such that

(1) $\lim_{k \to \infty} g_{n_k}(\zeta) = L$,
(2) the measures $\{\nu_{n_k}\}$ converge weakly to ν on $\mathbb{P}^1_{\text{Berk}}$.

Consider the potential function $u_\nu(z, \zeta) = \int -\log_v(\delta(z, w)_\zeta) \, d\nu(w)$. Let $C = L + u_\nu(\zeta, \zeta)$, and put $G(z) = C - u_\nu(z, \zeta)$. Then $G \in \mathcal{AG}[\zeta]$, $G(\zeta) = L$, and $\Delta(G) = \zeta - \nu$. By an argument similar to the one leading to (8.52), the g_{n_k} converge pointwise to G on $\mathbb{H}_{\text{Berk}}$.

Given a compact set $X \subset \mathbb{P}^1_{\text{Berk}}$ and a continuous function $f : X \to \mathbb{R}$, cover X by finitely many simple domains $W_1, \ldots, W_m$. For each W_i, choose a point $\zeta_i \in \mathbb{H}_{\text{Berk}} \backslash \overline{W}_i$. Let U_i be the component of $\mathbb{P}^1_{\text{Berk}} \backslash \{\zeta_i\}$ containing $\overline{W}_i$. Put $f_i(z) = f(z) - \log_v(\delta(z, \zeta_i)_\zeta)$ on $X \cap \overline{W}_i$, and put $G_i(z) = G(z) - \log_v(\delta(z, \zeta_i)_\zeta)$ and $g_{n_k,i}(z) - \log_v(\delta(z, \zeta_i)_\zeta)$ on U_i. Then $f_i : X \cap \overline{W}_i \to \mathbb{R}$ is continuous, and $g_{n_k,i} \in \mathcal{SH}(U_i)$ for each i. By applying the argument at the end of the proof of Proposition 8.54 to G_i, f_i, and $\{g_{n_k,i}\}$ for each i, we obtain (8.54). $\square$

8.9. Smoothing

In the classical theory over the complex numbers, each subharmonic function f is locally a decreasing limit of $\mathcal{C}^\infty$ subharmonic functions.

In the case of a domain $U \subset \mathbb{P}^1_{\text{Berk}}$, each subharmonic function f on U can locally be approximated by subharmonic functions of a special form. We will discuss two ways of doing this, first by Favre–Rivera-Letelier smoothing (see §5.7) and second by CPA-smoothing (defined below).

Favre–Rivera-Letelier smoothing is more canonical. It produces a sequence of smoothed functions which coincide with f on an increasing sequence of balls which exhaust $U \cap \mathbb{H}_{\text{Berk}}$. On the other hand, CPA-smoothing produces "nicer" functions, but in a noncanonical way, and the functions constructed converge uniformly to f on an increasing sequence of balls but do not in general coincide with it.

8.9. SMOOTHING

We first consider Favre–Rivera-Letelier smoothing. Let $U \subset \mathbb{P}^1_{\text{Berk}}$ be a domain. Fix a point $\zeta \in U \cap \mathbb{H}_{\text{Berk}}$, and let $\{\delta_n\}_{n \geq 1}$ be a decreasing sequence of numbers with $0 < \delta_n < 1/\|\zeta, \zeta\|$ such that $\lim_{n \to \infty} \delta_n = 0$. For each n, let $X(\zeta, \delta_n) = \{z \in \mathbb{P}^1_{\text{Berk}} : \text{diam}_\zeta(z) \geq \delta_n\}$, and let $r_{\zeta, \delta_n} : \mathbb{P}^1_{\text{Berk}} \to X(\zeta, \delta_n)$ be the retraction. Let $f_n = f_{\zeta, \delta_n} = f \circ r_{\zeta, \delta_n}|_{U_n}$ be the Favre–Rivera-Letelier smoothing of f.

If V is a simple subdomain of U, put $\delta(V) = \min_{x \in \partial V} \text{diam}_\zeta(x)$; then $\partial V \subset X(\zeta, \delta(V))$.

THEOREM 8.58. *Let $U \subset \mathbb{P}^1_{\text{Berk}}$ be a domain, take $\zeta \in U \cap \mathbb{H}_{\text{Berk}}$, and let f be subharmonic on U. Choose an exhaustion of U by an increasing sequence of simple subdomains $U_1 \subset U_2 \subset \cdots \subset U$, and let $\{\delta_n\}_{n \geq 1}$ be a decreasing sequence of numbers with $0 < \delta_n < 1/\|\zeta, \zeta\|$ such that $\lim_{n \to \infty} \delta_n = 0$. For each n, let $f_n = (f|_{U_n})_{\zeta, \delta_n} = f \circ r_{\zeta, \delta_n}|_{U_n}$ be the Favre–Rivera-Letelier smoothing of f on U_n relative to ζ and δ_n.*

Given a simple subdomain V of U, let $N = N(V)$ be large enough that $U_n \supset \overline{V}$ and $\delta_v < \delta(V)$ for all $n \geq N$. Then:
- (A) *If $n \geq N(V)$, then f_n is strongly subharmonic on V.*
- (B) *The sequence $\{f_n\}_{n \geq N(V)}$ decreases monotonically to f on $\overline{V}$, and for each $\delta > 0$, f_n coincides with f on $\overline{V} \cap X(\zeta, \delta)$ for all large enough n.*
- (C) *The measures $\Delta_{\overline{V}}(f_n)$ converge weakly to $\Delta_{\overline{V}}(f)$.*

PROOF. After replacing $\{f_n\}$ with a subsequence, we can assume that $N(V) = 1$.

We first show (A). By Theorem 5.44, f_n belongs to BDV(V) and
$$\Delta_{\overline{V}}(f_n) = (r_{\zeta, \delta_n})_*(\Delta_{\overline{V}}(f)) .$$
Write $\Delta_{\overline{V}}(f) = \Delta_{\partial V}(f) + \Delta_V(f)$, where $\Delta_{\partial V}(f) = \Delta_{\overline{V}}(f)|_{\partial V}$ is the boundary derivative and $\Delta_V(f) = \Delta_{\overline{V}}(f)|_V$. Since f is subharmonic on V, $\Delta_V(f) \leq 0$. Since r_{ζ, δ_n} fixes ∂V, it follows that $\Delta_V(f_n) = (r_{\zeta, \delta_n})_*(\Delta_V(f)) \leq 0$. Thus f_n is strongly subharmonic on V.

We next show (B). By construction, f_n coincides with f on $\overline{V} \cap X(\zeta, \delta_n)$ and is constant on branches off $X(\zeta, \delta_n)$ in V. If V is a disc, its unique boundary point is contained in $X(\zeta, \delta_n)$; if V is not a disc, then since $\partial V \subset X(\zeta, \delta_n)$, its main dendrite D is contained in $X(\zeta, \delta_n)$. By Proposition 8.11(A), if V is a disc, then f is nonincreasing on paths in V away from the boundary point of V; if V is not a disc, then f is nonincreasing on paths in V away from D. In either case, we conclude that f is nonincreasing on paths in V away from $X(\zeta, \delta_n)$. Since the sets $X(\zeta, \delta_n)$ are increasing with n, the functions $f_n|_{\overline{V}}$ are monotonically decreasing with n.

Since the sets $\overline{V} \cap X(\zeta, \delta_n)$ exhaust $\overline{V} \cap \mathbb{H}_{\text{Berk}}$, for each $x \in \overline{V} \cap \mathbb{H}_{\text{Berk}}$ we have $f_n(x) = f(x)$ for all sufficiently large n. In particular, for any $\delta > 0$, f_n coincides with f on $X(\zeta, \delta)$ as soon as $\delta_n < \delta$. If $x \in V \cap \mathbb{P}^1(K)$, put $x_n = r_{\zeta, \delta_n}(x)$; by the definition of f_n, we have $f_n(x) = f(x_n)$. As $n \to \infty$,

the points x_n approach x along the path $[\zeta, x]$. Since f is subharmonic, Proposition 8.11(B) shows that
$$\lim_{\substack{t \to x \\ t \in [\zeta, x)}} f(t) = f(x) \ .$$
Thus $\lim_{n \to \infty} f_n(x) = x$.

Part (C) follows immediately from Proposition 5.45(A). $\square$

The "nicest" functions on $\mathbb{P}^1_{\text{Berk}}$ we have encountered are functions of the form $f(z) = F \circ r_\Gamma(z)$, where Γ is a finite subgraph and $F \in \text{CPA}(\Gamma)$. We call the space of such functions $\text{CPA}(\mathbb{P}^1_{\text{Berk}})$ and deem them to be the smooth functions on $\mathbb{P}^1_{\text{Berk}}$. More generally, we make the following definition:

DEFINITION 8.59. Let $U \subseteq \mathbb{P}^1_{\text{Berk}}$ be a domain, and let $f : U \to \mathbb{R}$ be a function. We say that f is CPA-*smooth* on U if there are a finite subgraph $\Gamma \subset U$ and a function $F \in \text{CPA}(\Gamma)$ such that $f = F \circ r_{U,\Gamma}$. We call the space of all such functions $\text{CPA}(U)$.

We have the following smoothing theorem:[1]

THEOREM 8.60. *Let $U \subseteq \mathbb{P}^1_{\text{Berk}}$ be a domain, and take $f \in \mathcal{SH}(U)$. Let V be a simple subdomain of U, and let $\zeta \in V \cap \mathbb{H}_{\text{Berk}}$. Then there is a sequence of CPA-smooth functions $\{f_n\}_{n \geq 1}$ on U such that*

(A) *each f_n is strongly subharmonic on V,*
(B) *$\{f_n\}$ decreases monotonically to f on $\overline{V}$,*
(C) *for each $\delta > 0$, $\{f_n\}$ converges uniformly to f on $\overline{V} \cap X(\zeta, \delta)$,*
(D) *the sequence of measures $\{\Delta_{\overline{V}}(f_n)\}$ converges weakly to $\Delta_{\overline{V}}(f)$.*

Before giving the proof, we will need two lemmas:

LEMMA 8.61. *Let $U \subseteq \mathbb{P}^1_{\text{Berk}}$ be a domain. If f is continuous on U and $f \in \text{BDV}(U)$, then $f \in \text{CPA}(U)$ if and only if $\Delta_{\overline{U}}(f)$ is a discrete measure supported on $U \cap \mathbb{H}_{\text{Berk}}$.*

PROOF. If $f = F \circ r_{U,\Gamma} \in \text{CPA}(U)$, then $\Delta_{\overline{U}}(f) = \Delta_\Gamma(F)$ is a discrete measure supported on $\Gamma \subset U \cap \mathbb{H}_{\text{Berk}}$. Conversely, if $\Delta_{\overline{U}}(f)$ is a discrete measure supported on $U \cap \mathbb{H}_{\text{Berk}}$, choose a finite subgraph $\Gamma \subset U$ containing $\text{supp}(\Delta_{\overline{U}}(f))$. Then $\Delta_\Gamma(f) = (r_{\overline{U},\Gamma})_* \Delta_{\overline{U}}(f) = \Delta_{\overline{U}}(f)$ is a discrete measure supported on Γ. Put $F = f|_\Gamma$; then $F \in \text{CPA}(\Gamma)$. Since $\Delta_{\overline{U}}(F \circ r_{U,\Gamma}) = \Delta_{\overline{U}}(f)$, Lemma 5.24 shows that $f = F \circ r_{U,\Gamma}$ on $U \cap \mathbb{H}_{\text{Berk}}$. Since f is continuous on U, $f = F \circ r_{U,\Gamma}$ on all of U and therefore $f \in \text{CPA}(U)$. $\square$

LEMMA 8.62. *Let $U \subseteq \mathbb{P}^1_{\text{Berk}}$ be a domain, and take $f \in \mathcal{C}(\overline{U})$. Then as Γ ranges over all finite subgraphs of U, the net of functions $f \circ r_{\overline{U},\Gamma}$ converges uniformly to f on $\overline{U}$.*

---

[1] In an earlier version of Theorem 8.60, we showed only that there was a *net* $\langle f_\alpha \rangle$ of smooth functions converging to f. We thank Charles Favre for pointing out that Lemma 5.7 could be used to improve this to the existence of a *sequence* $\{f_n\}$.

PROOF. We must show that if $\varepsilon > 0$, there is a finite subgraph $\Gamma_0 \subset U$ such that if Γ is any finite subgraph of U containing Γ_0, then for all $x \in \overline{U}$,
$$|f(r_{\overline{U},\Gamma}(x)) - f(x)| < \varepsilon .$$

Fix a point $\zeta \in U \cap \mathbb{H}_{\text{Berk}}$. By continuity, for each $x \in \overline{U}$, there is a simple domain $V_x \subset \mathbb{P}^1_{\text{Berk}}$ containing x, such that $|f(x) - f(y)| < \varepsilon/2$ for all $y \in \overline{V}_x \cap \overline{U}$. If $x \in \partial U$, we can assume without loss of generality that $\zeta \notin V_x$. Then $V_x \cap U$ is nonempty and connected, and the unique path in $\overline{U}$ from x to ζ must pass through a boundary point of V_x belonging to U. By compactness, a finite number of the simple domains V_x cover $\overline{U}$, say $V_{x_1}, \ldots, V_{x_m}$, and each of these has a finite number of boundary points. Let Γ_0 be the convex hull of all points belonging to $\partial V_{x_i} \cap U$ for some $i = 1, \ldots, m$, so that Γ_0 is a finite subgraph of U. Let Γ be any finite subgraph of U containing Γ_0. Let $x \in \overline{U}$, let $\tilde{x} = r_\Gamma(x)$, and choose an i so that $x \in V_{x_i}$. Then $\tilde{x} \in \overline{V}_{x_i}$, since $\tilde{x}$ belongs to the path Λ from x to some point of $\partial V_{x_i} \cap U$ and Λ is contained in $\overline{V}_{x_i}$ by the unique path-connectedness of $\mathbb{P}^1_{\text{Berk}}$. Since $x, \tilde{x} \in \overline{V}_{x_i}$, the triangle inequality shows that $|f(x) - f(\tilde{x})| \leq |f(x) - f(x_i)| + |f(x_i) - f(\tilde{x})| < \varepsilon$, as desired. $\square$

PROOF OF THEOREM 8.60. Let $f \in \mathcal{SH}(U)$, fix a simple subdomain V of U, and take $\zeta \in V \cap \mathbb{H}_{\text{Berk}}$. By Lemma 8.10(B), $f|_V \in \text{BDV}(V)$. Put $\nu = \Delta_{\overline{V}}(f)$. By Lemma 5.7, there is a sequence of finite subgraphs $\Gamma_1 \subseteq \Gamma_2 \subseteq \cdots \subset V$ such that $\text{supp}(\nu)$ is contained in the closure of $\bigcup_{n=1}^\infty \Gamma_n$.

Choose a decreasing sequence $\{\delta_n\}_{n \geq 1}$ with $\lim_{n \to \infty} \delta_n = 0$, such that $0 < \delta_n < \delta(V)$ for each n. Let $g_n = f \circ r_{\zeta, \delta_n}|_V$ be the Favre–Rivera-Letelier smoothing of f on V; by Theorem 5.44, g_n coincides with f on $V \cap X(\zeta, \delta_n)$ and extends to a continuous function on $\overline{V}$. By Lemma 8.62 there is a finite graph $\tilde{\Gamma}_n \subset V$ such that for any Γ with $\tilde{\Gamma}_n \subset \Gamma \subset U$, we have $|g_n(z) - g_n \circ r_{V,\Gamma}(z)| < 1/n$ for all $z \in V$. After enlarging $\tilde{\Gamma}_n$ if necessary, we can assume that $\zeta \in \tilde{\Gamma}_n$. Since $V \cap X(\zeta, \delta_n)$ is uniquely path-connected, if $z \in V \cap X(\zeta, \delta_n)$, then $r_{V,\Gamma}(z) \in \Gamma \cap V \cap X(\zeta, \delta_n)$.

Since $f(z) = g_n(z)$ for each $z \in V \cap X(\zeta, \delta_n)$, we have shown that if $\tilde{\Gamma}_n \subset \Gamma$, then

(8.55) $\qquad |f(z) - f \circ r_{V,\Gamma}(z)| < 1/n$

for all $z \in V \cap X(\zeta, \delta_n)$.

Let $\partial V = \{a_1, \ldots, a_m\}$. After replacing Γ_n with $\Gamma_n \cup \left(\bigcup_{k=1}^n \tilde{\Gamma}_k\right) \cup \left(\bigcup_{k=1}^m [\zeta, a_k]\right)$ if necessary, we can arrange that the Γ_n are still increasing and that for each n, all of ζ, ∂V, and $\tilde{\Gamma}_n$ are contained in Γ_n. (Note that now $\Gamma_n \subset \overline{V}$, rather than $\Gamma_n \subset V$.) Let $X \subset \overline{V}$ be the closure of $\bigcup_{n=1}^\infty \Gamma_n$.

Let T_1 be the set consisting of the end points and branch points of Γ_1. Inductively define $T_{n+1} \subset \Gamma_{n+1}$ to be the smallest set which contains

(1) the end points and branch points of Γ_{n+1},
(2) T_n, and
(3) for each pair of distinct points $p, q \in T_n$, the midpoint of $[p, q]$.

Then each T_n is finite, and if $p, q \in T_n$, then for each $k \geq 1$, T_{n+k} contains the equally spaced 2^k-division points on $[p, q]$. By construction, $\partial V \subset T_n$.

Let F_n be the function on Γ_n such that $F_n(x) = f(x)$ for each $x \in T_n$ and which interpolates linearly between those values on segments of $\Gamma_n \backslash T_n$. Thus, $F_n \in \mathrm{CPA}(\Gamma_n)$. Let $f_n(z) = F_n \circ r_{U,\Gamma_n}$ be the CPA-smooth function on U which agrees with F_n on Γ_n and is constant on branches off Γ_n. Then $\Delta_{\overline{V}}(f_n) = \Delta_{\Gamma_n}(F_n)$ is supported on Γ_n.

Since ∂V is contained in T_n, for each pair of adjacent points $p, q \in T_n$, we have $r_{[p,q]}(\partial V) \subset \{p, q\}$. By Proposition 8.11(D), $f|_{\Gamma_n}$ is convex upwards on each segment in $\Gamma_n \backslash T_n$. This means that $f(z) \leq f_n(z)$ for each $z \in \Gamma_n$. Since f_n is constant on branches off Γ_n, while $f|_V$ is nonincreasing on branches off Γ_n by Proposition 8.11(A), it follows that $f(z) \leq f_n(z)$ for all $z \in \overline{V}$. Furthermore, the discrete measure

$$\Delta_V(f_n) = \Delta_{\overline{V}}(f_n)|_V = \Delta_{\Gamma_n}(F_n)|_V = \sum_{y \in T_n \backslash \partial V} \Delta_{\Gamma_n}(F_n)(y) \delta_y$$

is negative, because the subharmonicity of f, combined with the fact that $f(z) \leq f_n(z)$ for each $z \in \Gamma_n$, means that at each $y \in T_n \backslash \partial V$ we have

$$\Delta_{\Gamma_n}(F_n)(y) = -\sum_{\vec{v} \in T_y(\Gamma_n)} d_{\vec{v}} F_n(y) \leq -\sum_{\vec{v} \in T_y(\Gamma_n)} d_{\vec{v}} f|_{\Gamma_n}(y)$$
$$= \Delta_{\Gamma_n}(f|_{\Gamma_n})(y) = r_{\Gamma_n}(\Delta_V(f))(y) \leq 0.$$

Hence f_n is strongly subharmonic on V. This shows (A).

Since $T_n \subset T_{n+1}$, the function F_{n+1} coincides with F_n at each point in T_n. The convexity of f discussed above implies that $F_{n+1}(z) \leq F_n(z)$ on Γ_n. Since f_n is constant on branches off Γ_n and f_{n+1} is nonincreasing on branches off Γ_n by Proposition 8.11(A), it follows that $f_{n+1}(z) \leq f_n(z)$ for all $z \in U$.

Furthermore, for each N, the restriction of f to Γ_N is continuous, hence uniformly continuous. Since T_{n+1} contains the midpoint of each segment in $\Gamma_n \backslash T_n$, as $n \to \infty$ the points of $T_n \cap \Gamma_N$ become increasingly dense in Γ_N, and the functions f_n converge uniformly to f on Γ_N.

We will now show that for each $w \in \overline{V}$, the sequence $\{f_n(w)\}$ converges to $f(w)$. If $w \in \Gamma_n$ for some n, the desired convergence has been shown above. Suppose $w \notin \Gamma_n$ for any n, and consider the path $[\zeta, w]$. Since $\zeta \in X$ and X is connected (and hence path-connected), $[\zeta, w] \cap X$ is a segment $[\zeta, x]$, where x is the last point of $[\zeta, w]$ belonging to X. We claim that $f_n(w) = f_n(x)$ for each n and that $f(w) = f(x)$. For the f_n, this holds because each f_n is constant on branches off Γ_n. For f, it holds because $\nu = \Delta_{\overline{V}}(f)$ is supported on X and has total mass 0, while $[x, w] \cap X = \{x\}$. This means that $(r_{[x,w]})_*(\nu)$ is the zero measure, and so f is constant on $[x, w]$.

Hence it suffices to show convergence when $w \in X \backslash (\bigcup_{n=1}^\infty \Gamma_n)$. Let $x_k = r_{\Gamma_k}(w) \in \Gamma_k$ be the point such that $[\zeta, w] \cap \Gamma_k = [\zeta, x_k]$. Since X is the closure of $\bigcup_{k=1}^{\infty} \Gamma_k$ and each x_k belongs to $[\zeta, w]$, the points x_k approach w along $[\zeta, w]$. If V is a disc, then Γ_1 contains the unique boundary point of V, while if V is not a disc, then Γ_1 contains the main dendrite of V. Thus Proposition 8.11(A) shows that f is nonincreasing on $[x_1, w]$ and

$$(8.56) \qquad \lim_{k \to \infty} f(x_k) = f(w) .$$

Fix k. If x_k is one of the points in T_k, then $f_k(z)$ is constant on $[x_k, w]$ and $f_k(w) = f(x_w)$. If $x_k \notin T_k$, then x_k is not an endpoint of Γ_k. This means that $[x_k, w]$ is a branch off of Γ_k, and since $x_m \to w$ as $m \to \infty$, there is some $m > k$ for which $x_m \in (x_k, w]$. This means that x_k is a branch point of Γ_m and hence that $x_k \in T_m$. Now let $n \geq m$. Since $T_m \subset T_n$, it follows that $f_n(x_k) = f(x_k)$. On the other hand, by Proposition 8.11(A), f_n is nonincreasing on branches off Γ_k. Combining these facts, we see that for each $n \geq m$,

$$(8.57) \qquad f(w) \leq f_n(w) \leq f_n(x_k) = f(x_k) .$$

Since k can be chosen so that $f(x_k)$ is arbitrarily close to $f(w)$, it follows that $\lim_{n \to \infty} f_n(w) = w$. This completes the proof of (B).

We will now establish (C). Fix $0 < \delta < \delta(V)$. We must show that the functions f_n converge uniformly to f on $\overline{V} \cap X(\zeta, \delta)$. For each n, we have $\partial V \subset T_n$ by construction, so $f_n(x) = f(x)$ for each $x \in \partial V$. Hence it suffices to prove uniform convergence on $V \cap X(\zeta, \delta)$.

Fix $\varepsilon > 0$, and let N be large enough that $\delta_N < \delta$ and $1/N < \varepsilon$. As shown above, the functions f_n converge uniformly to f on Γ_N. Let $N_1 \geq N$ be such that $|f_n(x) - f(x)| < \varepsilon$ for all $n \geq N_1$ and all $x \in \Gamma_N$. Take $w \in V \cap X(\zeta, \delta)$, and put $x_n = r_{\Gamma_n}(w) \in \Gamma_n$ for each n. As noted above, since $\partial V \subset \Gamma_1$, by Proposition 8.11(A) each of the functions f and f_n is nonincreasing on $[x_1, w]$. On the other hand, since $\tilde{\Gamma}_N \subset \Gamma_N$, (8.55) shows that

$$(8.58) \qquad |f(w) - f(x_N)| < \varepsilon .$$

For each $n \geq N_1$, since $\Gamma_1 \subset \Gamma_N \subset \Gamma_n$, we have $x_n \in [x_N, w] \subset [x_1, w]$. Combining these facts, we see that

$$f(w) + 2\varepsilon > f(x_N) + \varepsilon > f_n(x_N) \geq f_n(w) \geq f(w) .$$

Thus $|f_n(w) - f(w)| < 2\varepsilon$ for all $w \in V \cap X(\zeta, \delta)$ and all $n \geq N_1$.

It remains to prove assertion (D) that the measures $\Delta_{\overline{V}}(f_n)$ converge weakly to $\Delta_{\overline{V}}(f)$. This can be shown directly using Proposition 5.32, but it is easier to finesse the issue by using Proposition 8.51(A). That proposition says that for each simple subdomain W of V, the measures $\Delta_{\overline{W}}(f_n)$ converge weakly to $\Delta_{\overline{W}}(f)$. Hence, if we had first enlarged V to a simple subdomain V' of U with $\overline{V} \subset V'$ and then carried out the above construction for f and V', then with the resulting sequence $\{f_n\}$ all the conditions in Theorem 8.60 would hold for V. $\square$

8.10. The Energy Minimization Principle

In this section we will study a class of generalized potential kernels which arise in the study of dynamics of rational functions on $\mathbb{P}^1_{\text{Berk}}$. We call them 'Arakelov-Green's functions' by analogy with the functions introduced by Arakelov in arithmetic geometry; we have met them briefly in §8.6 and §8.8. We show that the potential functions obtained from them satisfy analogues of Maria's and Frostman's theorems, as well as an Energy Minimization Principle which is one of the key results used in [9] to prove a non-Archimedean equidistribution theorem for points of small dynamical height[2].

We will give three forms of the Energy Minimization Principle. We provide a self-contained proof for the first and derive the last two from properties of the Dirichlet pairing.

Recall that for each $\zeta \in \mathbb{P}^1_{\text{Berk}}$, the generalized Hsia kernel is defined by

$$(8.59) \qquad \delta(x,y)_\zeta = \frac{\|x,y\|}{\|x,\zeta\|\,\|y,\zeta\|}$$

and that $\|x,y\| = \delta(x,y)_{\zeta_{\text{Gauss}}}$. By Definition 5.40, a finite signed measure μ on $\mathbb{P}^1_{\text{Berk}}$ has *continuous potentials* if for each $\zeta \in \mathbb{H}_{\text{Berk}}$ the potential function

$$u_\mu(z,\zeta) = \int -\log_v(\delta(z,w)_\zeta)\,d\mu(w)$$

is continuous (hence bounded) on $\mathbb{P}^1_{\text{Berk}}$. The discussion after Definition 5.40 shows that if $u_\mu(z,\xi)$ is continuous for some $\xi \in \mathbb{H}_{\text{Berk}}$, then $u_\mu(z,\zeta)$ is continuous for every $\zeta \in \mathbb{H}_{\text{Berk}}$.

For us, the most important example of a probability measure with continuous potentials will be the canonical measure attached to a rational function $\varphi(T) \in K(T)$, constructed in §10.1. Here are some other examples:

EXAMPLE 8.63. Let ν be a probability measure supported on a ball $\widehat{B}(a,R) \subset \mathbb{H}_{\text{Berk}}$. Then ν has continuous potentials by Proposition 5.42.

In particular, if $\mu = \delta_\zeta$ for some $\zeta \in \mathbb{H}_{\text{Berk}}$, then μ has continuous potentials. Similarly, if $\Gamma \subset \mathbb{P}^1_{\text{Berk}}$ is a finite subgraph and μ is a probability measure supported on Γ, then μ has continuous potentials.

EXAMPLE 8.64. Take $K = \mathbb{C}_p$ and introduce coordinates so that $\mathbb{P}^1(K) = \mathbb{C}_p \cup \{\infty\}$. Let $L/\mathbb{Q}_p$ be a finite extension and let $\mathcal{O}_L$ be its valuation ring. If μ is additive Haar measure on $\mathcal{O}_L$ and if $\mathcal{O}_L$ is viewed as a subset of $\mathbb{P}^1_{\text{Berk}}$, then μ has continuous potentials. This follows from [88, Example 4.1.24, p. 213], where an explicit formula is given for $u_\mu(z,\infty)$ on $\mathbb{P}^1(\mathbb{C}_p)$. This formula extends to $\mathbb{P}^1_{\text{Berk}}$ and shows that $u_\mu(z,\infty)$ is continuous as a function to $\mathbb{R} \cup \{-\infty\}$. An easy modification of the construction shows that $u_\mu(z,\zeta_{\text{Gauss}}) = \max(0, u_\mu(z,\infty))$, so $u_\mu(z,\zeta_{\text{Gauss}})$ is continuous and bounded.

---

[2]Favre and Rivera-Letelier use a related energy minimization principle in their proof of the non-Archimedean equidistribution theorem; see [47, Proposition 4.5]. Our Theorem 8.71 is an improved form of their Proposition 4.5.

8.10. THE ENERGY MINIMIZATION PRINCIPLE

PROPOSITION 8.65. *Let μ be a positive measure on $\mathbb{P}^1_{\text{Berk}}$ for which $-\mu$ is locally the Laplacian of a continuous subharmonic function. Then μ has continuous potentials.*

PROOF. Fix $x \in \mathbb{P}^1_{\text{Berk}}$, and let $V \subset \mathbb{P}^1_{\text{Berk}}$ be a neighborhood on which $-\mu|_V = \Delta_V(f)$ for some continuous subharmonic function f. After shrinking V if necessary, we can assume that V is a simple domain and that f is bounded and strongly subharmonic on V. Fix $\zeta \in \mathbb{H}_{\text{Berk}} \backslash \overline{V}$. Put $\mu_V = \mu|_V$, and consider the potential function

$$u_{\mu_V}(z, \zeta) = \int -\log_v(\delta(z,y)_\zeta) \, d\mu_V(y) \ .$$

By Proposition 8.42, $u_{\mu_V}(z, \zeta)$ is continuous on all of $\mathbb{P}^1_{\text{Berk}}$. It is bounded since $\zeta \notin \mathbb{P}^1(K)$.

Now replace ζ by ζ_{Gauss}. By (4.29), there is a constant $C > 0$ such that for all $x, y \in \mathbb{H}_{\text{Berk}}$,

$$\delta(x,y)_{\zeta_{\text{Gauss}}} = C \cdot \frac{\delta(x,y)_\zeta}{\delta(x, \zeta_{\text{Gauss}})_\zeta \delta(y, \zeta_{\text{Gauss}})_\zeta} \ .$$

Since $\log_v(\delta(x, \zeta_{\text{Gauss}})_\zeta)$ is a continuous, bounded function of x, it follows that $u_{\mu_V}(z, \zeta_{\text{Gauss}})$ is continuous and bounded as well.

As $\mathbb{P}^1_{\text{Berk}}$ is compact, a finite number of such simple domains V_i cover $\mathbb{P}^1_{\text{Berk}}$. It is easy to see that the assertions made for the restriction of μ to the V_i hold also for its restriction to their intersections $V_{i_1} \cap \cdots \cap V_{i_r}$. By inclusion-exclusion, $u_\mu(z, \zeta_{\text{Gauss}})$ is continuous on $\mathbb{P}^1_{\text{Berk}}$. $\square$

Given a probability measure μ with continuous potentials, define the *normalized Arakelov-Green's function* $g_\mu(x, y)$ by

$$(8.60) \qquad g_\mu(x,y) = \int_{\mathbb{P}^1_{\text{Berk}}} -\log_v(\delta(x,y)_\zeta) \, d\mu(\zeta) + C \ ,$$

where the normalizing constant C is chosen so that

$$(8.61) \qquad \iint g_\mu(x,y) \, d\mu(x) d\mu(y) = 0 \ .$$

Using (8.59), one sees that

$$(8.62) \qquad g_\mu(x,y) = -\log_v(\|x,y\|) - u_\mu(x, \zeta_{\text{Gauss}}) - u_\mu(y, \zeta_{\text{Gauss}}) + C$$

and that $g_\mu(x, y)$ is valued in $[M, \infty]$ for some $M \in \mathbb{R}$. When $\zeta \in \mathbb{H}_{\text{Berk}}$ and $\mu = \delta_\zeta$, then $g_\mu(x, y) = -\log_v(\delta(x,y)_\zeta) + C$. Thus Arakelov-Green's functions may be viewed as potential kernels whose 'poles' have been spread out over the support of μ.

It is a formal consequence of (8.61) that for each $y \in \mathbb{P}^1_{\text{Berk}}$, we have

$$(8.63) \qquad \int g_\mu(x,y) \, d\mu(x) = 0 \ .$$

Indeed, integrating (8.62) with respect to $d\mu(x)$ and using the definition of $u_\mu(y, \zeta_{\text{Gauss}})$, we have

$$\int g_\mu(x,y)\,d\mu(x) = \int -\log_v(\|x,y\|)\,d\mu(x) - \int u_\mu(x,\zeta_{\text{Gauss}})\,d\mu(x)$$
$$-u_\mu(y,\zeta_{\text{Gauss}}) + C$$
(8.64)
$$= -\int u_\mu(x,\zeta_{\text{Gauss}})\,d\mu(x) + C\,.$$

Then, integrating (8.64) with respect to $d\mu(y)$ and using the normalization (8.61), we see that

$$(8.65) \quad 0 = \iint g_\mu(x,y)\,d\mu(x)d\mu(y) = -\int u_\mu(x,\zeta_{\text{Gauss}})\,d\mu(x) + C\,.$$

Inserting (8.65) into (8.64) gives (8.63). Furthermore, it follows from (8.62) and (8.65) that $C = \iint \log_v(\|x,y\|)\,d\mu(x)d\mu(y)$.

Since $u_\mu(z, \zeta_{\text{Gauss}})$ is continuous, $g_\mu(x,y)$ inherits the following properties from $-\log_v(\|x,y\|)$ (see Proposition 4.7 and Examples 5.19 and 8.6):

PROPOSITION 8.66. *Let μ be a probability measure on $\mathbb{P}^1_{\text{Berk}}$ with continuous potentials. As a function of two variables, the Arakelov-Green's function $g_\mu(x,y)$ is symmetric, lower semicontinuous everywhere, continuous off the diagonal and on the type I diagonal, and bounded from below. For each fixed y, the function $G_y(x) = g_\mu(x,y)$ is continuous and belongs to $\text{BDV}(\mathbb{P}^1_{\text{Berk}})$; it is subharmonic on $\mathbb{P}^1_{\text{Berk}}\setminus\{y\}$ and satisfies $\Delta_{\mathbb{P}^1_{\text{Berk}}}(G_y(x)) = \delta_y(x) - \mu$.*

Given a probability measure ν on $\mathbb{P}^1_{\text{Berk}}$, define the generalized potential function

$$(8.66) \qquad u_\nu(x,\mu) = \int g_\mu(x,y)\,d\nu(y)$$

and the μ-energy integral

$$(8.67) \qquad I_\mu(\nu) = \iint g_\mu(x,y)\,d\nu(x)d\nu(y) = \int u_\nu(x,\mu)\,d\nu(x)\,.$$

We will now prove the following Energy Minimization Principle:

THEOREM 8.67. *Let μ be a probability measure on $\mathbb{P}^1_{\text{Berk}}$ with continuous potentials. Then for each probability measure ν on $\mathbb{P}^1_{\text{Berk}}$,*
  (A) $I_\mu(\nu) \geq 0$,
  (B) $I_\mu(\nu) = 0$ *if and only if $\nu = \mu$.*

The proof rests on analogues of Maria's theorem and Frostman's theorem for the function $f(x) = u_\nu(x,\mu)$. We begin by showing that generalized potential functions have the same properties as the ones studied in §6.3.

PROPOSITION 8.68. *Let μ be a probability measure on $\mathbb{P}^1_{\text{Berk}}$ with continuous potentials, and let ν be an arbitrary probability measure on $\mathbb{P}^1_{\text{Berk}}$.*

8.10. THE ENERGY MINIMIZATION PRINCIPLE

Then $u_\nu(z,\mu)$ is lower semicontinuous everywhere and is continuous at each $z \notin \mathrm{supp}(\nu)$. For each $z \in \mathbb{P}^1_{\mathrm{Berk}}$, as t approaches z along any path $[y, z]$,

$$\lim_{\substack{t \to z \\ t \in [y,z]}} u_\nu(t, \mu) = u_\nu(z, \mu) .$$

Moreover, $u_\nu(x, \mu)$ belongs to $\mathrm{BDV}(\mathbb{P}^1_{\mathrm{Berk}})$, and $\Delta_{\mathbb{P}^1_{\mathrm{Berk}}}(u_\nu(x, \mu)) = \nu - \mu$.

PROOF. Note that since $\|x, y\| = \delta(x, y)_{\zeta_{\mathrm{Gauss}}}$,

$$\begin{aligned}
u_\nu(x, \mu) &= \iint \Big(-\log_v(\|x, y\|) + \log_v(\|x, \zeta\|) \\
&\qquad\qquad + \log_v(\|y, \zeta\|) + C \Big) d\mu(\zeta) d\nu(y) \\
&= u_\nu(x, \zeta_{\mathrm{Gauss}}) - u_\mu(x, \zeta_{\mathrm{Gauss}}) + C'
\end{aligned}$$

where C' is a finite constant. Since $u_\mu(x, \zeta_{\mathrm{Gauss}})$ is bounded and continuous, the assertions about continuity, semicontinuity, and path limits follow from the analogous facts for $u_\nu(x, \zeta_{\mathrm{Gauss}})$ proved in Proposition 6.12.

We have also seen in Example 5.22 that $u_\nu(x, \zeta_{\mathrm{Gauss}})$ and $u_\mu(x, \zeta_{\mathrm{Gauss}})$ belong to $\mathrm{BDV}(\mathbb{P}^1_{\mathrm{Berk}})$. By the computations of the Laplacians there,

$$\Delta(u_\nu(x, \mu)) = (\nu - \delta_{\zeta_{\mathrm{Gauss}}}(x)) - (\mu - \delta_{\zeta_{\mathrm{Gauss}}}(x)) = \nu - \mu .$$

$\square$

The following result is an analogue of Maria's theorem from classical potential theory:

PROPOSITION 8.69. *Let μ be a probability measure on $\mathbb{P}^1_{\mathrm{Berk}}$ with continuous potentials, and let ν be an arbitrary probability measure on $\mathbb{P}^1_{\mathrm{Berk}}$. If there is a constant $M < \infty$ such that $u_\nu(z, \mu) \leq M$ on $\mathrm{supp}(\nu)$, then $u_\nu(z, \mu) \leq M$ for all $z \in \mathbb{P}^1_{\mathrm{Berk}}$.*

PROOF. Put $E = \mathrm{supp}(\nu)$, and let U be a component of $\mathbb{P}^1_{\mathrm{Berk}} \backslash E$. By Proposition 8.68, $u_\nu(z, \mu)$ is continuous on U and $\Delta_U(u_\nu(z, \mu)) \leq 0$. Hence $u_\nu(z, \mu)$ is subharmonic on U.

Let $f(z) = u_\nu(z, \mu)$. Suppose there exists $p_0 \in U$ with $f(p_0) > M$. As f is continuous on U and points of $\mathbb{H}_{\mathrm{Berk}}$ are dense in U, we may assume without loss of generality that $p_0 \in \mathbb{H}_{\mathrm{Berk}}$. Fix $\sigma > 0$ with $f(p_0) \geq M + \sigma$. By Proposition 8.16, there exist a point $q \in \partial U$ and a path Λ from p_0 to q such that f is nondecreasing along Λ. In particular, $f(z) \geq M + \sigma$ for all $z \in \Lambda^0$ and

$$(8.68) \qquad \lim_{\substack{x \to q \\ x \in \Lambda^0}} f(x) \geq M + \sigma .$$

By Proposition 8.68, the limit on the left-hand side of (8.68) equals $f(q)$. Therefore $f(q) \geq M + \sigma$. But $q \in \partial U \subseteq \mathrm{supp}(\nu)$, contradicting the assumption that $f(z) \leq M$ on $\mathrm{supp}(\nu)$. $\square$

Define the 'μ-Robin constant' to be

$$V_\mu(\mathbb{P}^1_{\text{Berk}}) \;=\; \inf_{\substack{\nu \\ \text{prob meas}}} I_\mu(\nu) \;=\; \inf_{\substack{\nu \\ \text{prob meas}}} \iint g_\mu(x,y)\,d\nu(x)d\nu(y)\;,$$

where ν runs over all probability measures supported on $\mathbb{P}^1_{\text{Berk}}$. For compactness of notation, we will write $V(\mu)$ for $V_\mu(\mathbb{P}^1_{\text{Berk}})$. Trivially $V(\mu) > -\infty$, since $g_\mu(x,y)$ is bounded below, and $V(\mu) \le 0$, since $I_\mu(\mu) = 0$ by the normalization of $g_\mu(x,y)$. Taking the weak limit of a sequence of measures $\nu_1, \nu_2, \ldots$ for which $I_\mu(\nu_n) \to V(\mu)$ and applying the same argument used to prove the existence of an equilibrium measure, one obtains a probability measure ω for which $I_\mu(\omega) = V(\mu)$.

PROPOSITION 8.70 (Frostman). *Let μ be a probability measure on $\mathbb{P}^1_{\text{Berk}}$ with continuous potentials, and let ω be a probability measure for which $I_\mu(\omega) = V(\mu)$. Then the generalized potential function $u_\omega(z, \mu)$ satisfies*

$$u_\omega(z, \mu) \equiv V(\mu) \qquad \text{on} \quad \mathbb{P}^1_{\text{Berk}}\;.$$

PROOF. In outline, the proof, which is modeled on the classical one, will be as follows. First, using a quadraticity argument, we will show that $u_\omega(z, \mu) \ge V(\mu)$ for all $z \in \mathbb{P}^1_{\text{Berk}}$ except possibly on a set f of capacity 0. Then, we will show that $u_\omega(z, \mu) \le V(\mu)$ on $\text{supp}(\omega)$. By Maria's theorem, $u_\omega(z, \mu) \le V(\mu)$ for all $z \in \mathbb{P}^1_{\text{Berk}}$. Since a set of capacity 0 is necessarily contained in $\mathbb{P}^1(K)$, it follows that $u_\omega(z, \mu) = V(\mu)$ on $\mathbb{H}_{\text{Berk}}$. Finally, if $p \in \mathbb{P}^1(K)$, let $z \to p$ along a path $[y, p]$. Proposition 8.68 gives $u_\omega(p, \mu) = V(\mu)$.

Put

$$\begin{aligned} f &= \{z \in \mathbb{P}^1_{\text{Berk}} : u_\omega(z, \mu) < V(\mu)\}\;, \\ f_n &= \{z \in \mathbb{P}^1_{\text{Berk}} : u_\omega(z, \mu) < V(\mu) - 1/n\}\;, \quad \text{for } n = 1, 2, 3, \ldots. \end{aligned}$$

Since $u_\omega(z, \mu)$ is lower semicontinuous, each f_n is closed, hence compact. By Corollary 6.17, f has capacity 0 if and only if each f_n has capacity 0.

Suppose f_n has positive capacity for some n; then there is a probability measure σ supported on f_n such that

$$I_{\zeta_{\text{Gauss}}}(\sigma) := \iint -\log_v(\|x, y\|)\,d\sigma(x)d\sigma(y) \;<\; \infty\;.$$

Since $g_\mu(x,y)$ differs from $-\log_v(\|x, y\|)$ by a bounded function, $I_\mu(\sigma) < \infty$ as well. Furthermore, since

$$V(\mu) \;=\; \iint g_\mu(x,y)\,d\omega(x)d\omega(y) \;=\; \int u_\omega(x, \mu)\,d\omega(x)\;,$$

there is a point $q \in \text{supp}(\omega)$ with $u_\omega(q, \mu) \ge V(\mu)$. Since $u_\omega(z, \mu)$ is lower semicontinuous, there is a neighborhood U of q on which $u_\omega(z, \mu) > V(\mu) - 1/(2n)$. After shrinking U if necessary, we can assume its closure $\overline{U}$ is disjoint from f_n. Since $q \in \text{supp}(\omega)$, it follows that $M := \omega(\overline{U}) > 0$. Define

8.10. THE ENERGY MINIMIZATION PRINCIPLE

a measure σ_1 of total mass 0 by

$$\sigma_1 = \begin{cases} M \cdot \sigma & \text{on } f_n, \\ -\omega & \text{on } \overline{U}, \\ 0 & \text{elsewhere}. \end{cases}$$

We claim that $I_\zeta(\sigma_1)$ is finite. Indeed

$$I_\zeta(\sigma_1) = M^2 \cdot \iint_{f_n \times f_n} g_\mu(x,y)\,d\sigma(x)d\sigma(y)$$
$$- 2M \cdot \iint_{f_n \times \overline{U}} g_\mu(x,y)\,d\sigma(x)d\omega(y)$$
$$+ \iint_{\overline{U} \times \overline{U}} g_\mu(x,y)\,d\omega(x)d\omega(y).$$

The first integral is finite by hypothesis. The second is finite because f_n and $\overline{U}$ are disjoint, so $g_\mu(x,y)$ is bounded on $f_n \times \overline{U}$. The third is finite because $I_\mu(\omega)$ is finite and $g_\mu(x,y)$ is bounded below.

For each $0 \le t \le 1$, $\omega_t := \omega + t\sigma_1$ is a probability measure. By an expansion like the one above,

$$I_\mu(\omega_t) - I_\mu(\omega) = 2t \cdot \int_{f_n \cup \overline{U}} u_\omega(z,\mu)\,d\sigma_1(z) + t^2 \cdot I_\mu(\sigma_1)$$
$$\le 2t \cdot ((V(\mu) - 1/n) - (V(\mu) - 1/(2n))) \cdot M + t^2 \cdot I_\mu(\sigma_1)$$
$$= (-M/n) \cdot t + I_\mu(\sigma_1) \cdot t^2.$$

For sufficiently small $t > 0$, the right side is negative. This contradicts the fact that ω minimizes the energy integral. It follows that f_n has capacity 0, and hence $f = \bigcup_{n=1}^\infty f_n$ has capacity 0 by Corollary 6.17.

Since a set of capacity 0 is necessarily contained in $\mathbb{P}^1(K)$, it follows that $u_\omega(z,\mu) \ge V(\mu)$ for all $z \in \mathbb{H}_{\text{Berk}}$.

The second part requires showing that $u_\omega(z,\mu) \le V(\mu)$ for all $z \in \text{supp}(\omega)$. If $u_\omega(q,\mu) > V(\mu)$ for some $q \in \text{supp}(\omega)$, let $\varepsilon > 0$ be small enough that $u_\omega(q,\mu) > V(\mu) + \varepsilon$. The lower semicontinuity of $u_\omega(z,\mu)$ shows that there is a neighborhood U of q on which $u_\omega(z,\mu) > V(\mu) + \varepsilon$. Then $T := \omega(U) > 0$, since $q \in \text{supp}(\omega)$. On the other hand, by Lemma 6.16, $\omega(f) = 0$ since $I_\mu(\omega) < \infty$ implies that $I_{\zeta_{\text{Gauss}}}(\omega) < \infty$. Since $u_\omega(z,\mu) \ge V(\mu)$ for all $z \notin f$,

$$V(\mu) = \int_U u_\omega(z,\mu)\,d\omega(z) + \int_{\mathbb{P}^1_{\text{Berk}} \setminus U} u_\omega(z,\mu)\,d\omega(z)$$
$$\ge T \cdot (V(\mu) + \varepsilon) + (1 - T) \cdot V(\mu) = V(\mu) + T\varepsilon,$$

which is impossible. Hence $u_\omega(z,\mu) \le V(\mu)$ on $\text{supp}(\omega)$, and Maria's theorem implies that $u_\omega(z,\mu) \le V(\mu)$ for all $z \in \mathbb{P}^1_{\text{Berk}}$.

Combining the first and second parts, we find that $u_\omega(z,\mu) = V(\mu)$ for all $z \in \mathbb{H}_{\text{Berk}}$. As noted at the beginning of the proof, the fact that $u_\omega(p,\mu) = V(\mu)$ for each $p \in \mathbb{P}^1(K)$ now follows from Proposition 8.68. $\square$

We can now complete the proof of Theorem 8.67:

PROOF. Suppose ω minimizes the energy integral $I_\mu(\nu)$. By Proposition 8.70, $u_\omega(z,\mu)$ is constant. It follows that $\Delta(u_\omega(z,\mu)) \equiv 0$. On the other hand, by Proposition 8.68, $\Delta(u_\omega(z,\mu)) = \omega - \mu$. Hence $\omega = \mu$.

However, $I_\mu(\mu) = 0$ by the normalization of $g_\mu(x,y)$, so $V(\mu) = 0$. It follows that $I_\mu(\nu) \geq 0$ for all ν, and if $I_\mu(\nu) = 0$, then $\nu = \mu$. □

There is an alternate approach to the Energy Minimization Principle based on the Dirichlet pairing, due to Favre and Rivera-Letelier.

In this approach, one only needs to assume that the probability measure μ has bounded potentials. Recall that by Definition 5.40, μ has bounded potentials if for each $\zeta \in \mathbb{H}_{\text{Berk}}$ the potential function $u_\mu(z,\zeta)$ is bounded on $\mathbb{P}^1_{\text{Berk}}$. A probability measure with bounded, but not continuous, potentials is given in [**88**, Example 4.1.29, p. 217]. The discussion after Definition 5.40 shows that if $u_\mu(z,\xi)$ is bounded for some $\xi \in \mathbb{H}_{\text{Berk}}$, then $u_\mu(z,\zeta)$ is bounded for every $\zeta \in \mathbb{H}_{\text{Berk}}$. Furthermore, as shown in the proof of Proposition 5.41, $u_\mu(z,\zeta)$ is Borel measurable.

If μ is a probability measure with bounded potentials, one defines the normalized Arakelov-Green's function $g_\mu(x,y)$ by (8.60) and (8.61) as before. Using (8.59), one sees that

$$(8.69) \quad g_\mu(x,y) = -\log_v(\|x,y\|) - u_\mu(x,\zeta_{\text{Gauss}}) - u_\mu(y,\zeta_{\text{Gauss}}) + C$$

and that $g_\mu(x,y)$ is valued in $[M,\infty]$ for some $M \in \mathbb{R}$. Furthermore, as in (8.63), one has

$$(8.70) \quad \int g_\mu(x,y)\,d\mu(x) = 0$$

for each $y \in \mathbb{P}^1_{\text{Berk}}$.

We will now give an Energy Minimization Principle for finite signed measures ν of total mass 0. For such measures, in order to assure that the integral defining the energy $I_\mu(\nu)$ is well-defined, it is necessary to impose conditions on ν.

If μ is a probability measure with bounded potentials and if ν is a finite signed Borel measure such that in the Jordan decomposition $\nu = \nu_1 - \nu_2$, the negative part ν_2 has bounded potentials, then $I_\mu(\nu)$ is well-defined. This is because

$$
\begin{aligned}
I_\mu(\nu) &= \iint g_\mu(x,y)\,d\nu(x)d\nu(y) \\
(8.71) \quad &= \sum_{i,j=1}^{2}(-1)^{i+j}\iint g_\mu(x,y)\,d\nu_i(x)\,d\nu_j(y),
\end{aligned}
$$

where all the integrals on the right are well-defined and those with $i = 2$ or $j = 2$ are finite. Indeed, by (8.69), for all (i, j)

$$
\begin{aligned}
&\iint g_\mu(x, y) \, d\nu_i(x) \, d\nu_j(y) \\
&= \iint -\log_v(\|x, y\|) \, d\nu_i(x) \, d\nu_j(y) - \nu_j(\mathbb{P}^1_{\text{Berk}}) \int u_\mu(x, \zeta_{\text{Gauss}}) \, d\nu_i(x) \\
&\quad - \nu_i(\mathbb{P}^1_{\text{Berk}}) \int u_\mu(y, \zeta_{\text{Gauss}}) \, d\nu_j(y) + C \cdot \nu_i(\mathbb{P}^1_{\text{Berk}}) \nu_j(\mathbb{P}^1_{\text{Berk}}) \ .
\end{aligned}
$$
(8.72)

The function $-\log_v(\|x, y\|)$ is Borel measurable since it is lower semicontinuous (Proposition 4.10).

Since $-\log_v(\|x, y\|) \geq 0$, the first term on the right is defined and belongs to $[0, \infty]$. If $i = 2$, then by Tonelli's theorem it equals $\int u_{\nu_2}(y, \zeta_{\text{Gauss}}) \, d\nu_j(y)$, which is finite because ν_2 has bounded potentials; similarly it is finite if $j = 2$. The second and third terms on the right are finite because $u_\mu(z, \zeta_{\text{Gauss}})$ is bounded; the fourth term is clearly finite.

Our second Energy Minimization Principle concerns measures of total mass zero.

THEOREM 8.71. *Let μ be a probability measure on $\mathbb{P}^1_{\text{Berk}}$ with bounded potentials. Then for each signed Borel measure ν on $\mathbb{P}^1_{\text{Berk}}$ with total mass 0 whose negative part has bounded potentials,*

(A) $I_\mu(\nu) \geq 0$,
(B) $I_\mu(\nu) = 0$ *if and only if* $\nu = 0$.

PROOF. We keep the notation from the discussion above. Substituting (8.72) into (8.71), collecting terms, and using that $\nu_1(\mathbb{P}^1_{\text{Berk}}) = \nu_2(\mathbb{P}^1_{\text{Berk}})$,

$$
\begin{aligned}
I_\mu(\nu) &= \sum_{i,j=1}^{2} (-1)^{i+j} \iint -\log_v(\|x, y\|) \, d\nu_i(x) \, d\nu_j(y) \\
&= \iint -\log_v(\|x, y\|) \, d\nu(x) \, d\nu(y) \ .
\end{aligned}
$$
(8.73)

Noting that $\int -\log_v(\|x, y\|) \, d\nu_j(y) = u_{\nu_j}(x, \zeta_{\text{Gauss}})$, put

$$f(x) = u_{\nu_1}(x, \zeta_{\text{Gauss}}) - u_{\nu_2}(x, \zeta_{\text{Gauss}}) \ .$$

Then f belongs to $\text{BDV}(\mathbb{P}^1_{\text{Berk}})$, and $\Delta(f) = \nu_1 - \nu_2 = \nu$. By our hypothesis on ν_2, using Proposition 5.41, we can rewrite (8.73) as

(8.74) $\qquad\qquad I_\mu(\nu) \;=\; \langle f, f \rangle_{\mathbb{P}^1_{\text{Berk}}, \text{Dir}} \ .$

By Proposition 5.34 $\langle f, f \rangle_{\mathbb{P}^1_{\text{Berk}}, \text{Dir}} \geq 0$, and $\langle f, f \rangle_{\mathbb{P}^1_{\text{Berk}}, \text{Dir}} = 0$ if and only if f is constant on $\mathbb{H}_{\text{Berk}}$. However, by Lemma 5.24, f is constant on $\mathbb{H}_{\text{Berk}}$ if and only if $\nu_1 - \nu_2 = 0$, and the result follows. $\square$

Our final Energy Minimization Principle simultaneously generalizes Theorems 8.67 and 8.71 and gives an independent proof of Theorem 8.67.

THEOREM 8.72. *Let μ be a probability measure on $\mathbb{P}^1_{\text{Berk}}$ with bounded potentials. Let ν be a finite signed Borel measure on $\mathbb{P}^1_{\text{Berk}}$ of arbitrary mass whose negative part has bounded potentials. Then:*

(A) $I_\mu(\nu) \geq 0$,
(B) $I_\mu(\nu) = 0$ if and only if $\nu = m(\nu) \cdot \mu$, where $m(\nu) = \nu(\mathbb{P}^1_{\text{Berk}})$.

PROOF. Put $\nu_0 = \nu - m(\nu) \cdot \mu$. Then ν_0 has total mass 0, and by our hypotheses on ν and μ, the negative part of ν_0 has bounded potentials. By Theorem 8.71, $I_\mu(\nu_0) \geq 0$ and $I_\mu(\nu_0) = 0$ if and only if $\nu_0 = 0$, that is, $I_\mu(\nu_0) = 0$ if and only if $\nu = m(\nu) \cdot \mu$.

On the other hand, by (8.70) for each y we have $\int g_\mu(x, y) \, d\mu(x) = 0$. If the Jordan decomposition of ν is $\nu = \nu_1 - \nu_2$, then

$$\iint g_\mu(x, y) \, d\mu(x) d\nu(y)$$
$$= \iint g_\mu(x, y) \, d\mu(x) d\nu_1(y) - \iint g_\mu(x, y) \, d\mu(x) d\nu_2(y) = 0$$

using Fubini-Tonelli. Expanding $I_\mu(\nu_0)$ by linearity, we see that

$$I_\mu(\nu_0) = \iint g_\mu(x, y) \, d\nu(x) d\nu(y) - 2m(\nu) \cdot \iint g_\mu(x, y) \, d\mu(x) d\nu(y)$$
$$+ m(\nu)^2 \cdot \iint g_\mu(x, y) \, d\mu(x) d\mu(y)$$
$$= I_\mu(\nu) \, .$$

This yields the theorem. $\square$

8.11. Notes and further references

Thuillier [94] has defined subharmonicity on arbitrary Berkovich curves. He uses subharmonic functions in the context of non-Archimedean Arakelov theory. The heart of Thuillier's approach involves CPA-smoothing.

The smoothing of subharmonic functions is also fundamental to the approach to higher-dimensional non-Archimedean potential theory developed recently by Sebastien Boucksom, Charles Favre, and Mattias Jonsson in [28].

CHAPTER 9

Multiplicities

Given a nonconstant rational map $\varphi \in K(T)$, in this chapter we develop a theory of multiplicities for φ at points of $\mathbb{P}^1_{\text{Berk}}$, generalizing the classical algebraic multiplicities at points of $\mathbb{P}^1(K)$. We prove a Berkovich space analogue of the open mapping theorem, and we establish several formulas relating multiplicities to geometric properties of φ. We give a proof of Rivera-Letelier's theorem that the image under φ of a finite graph is again a finite graph, and we give conditions for the image under φ of a disc or an annulus to be a disc or annulus. We then define pullback and pushforward measures under φ and establish some of their functorial properties. We conclude by showing that the pullback by φ of a subharmonic function is again subharmonic, and we use this to derive pullback formulas for arbitrary functions in $\text{BDV}(U)$ and for Green's functions.

9.1. An analytic construction of multiplicities

Let $\varphi \in K(T)$ be a nonconstant rational function of degree d. For classical points $a, b \in \mathbb{P}^1(K)$ with $\varphi(a) = b$, write $m_\varphi(a) = m_{\varphi,b}(a)$ for the (classical algebraic) multiplicity of φ at a. For a, b with $\varphi(a) \neq b$, put $m_{\varphi,b}(a) = 0$. Thus for each $b \in K$,

$$\text{div}(\varphi(T) - b) = \sum_{\varphi(a)=b} m_{\varphi,b}(a)\delta_a - \sum_{\varphi(a)=\infty} m_{\varphi,\infty}(a)\delta_a ,$$

and for each $b \in \mathbb{P}^1(K)$,

$$\sum_{a \in \mathbb{P}^1(K)} m_{\varphi,b}(a) = d .$$

Using the theory of subharmonic functions, we will give an analytic construction of multiplicities on $\mathbb{P}^1_{\text{Berk}}$ which share these properties. Favre and Rivera-Letelier [48, §2.2] and Thuillier [94, §3.2.3] have given algebraic constructions of multiplicities using ranks of modules[1]. Rivera-Letelier [81, §2]

---

[1]The constructions of multiplicities given in [48] and [94] generalize to higher dimensions and are based on the following fact: if $f : Y \to X$ is a finite flat map between smooth, separable Berkovich spaces and if $y \in Y$, then there are an affinoid neighborhood W of y and an affinoid neighborhood V of $f(y)$ such that $f(W) = V$ and y is the only preimage of $f(y)$ in W, such that the affinoid algebra $\mathcal{A}_W$ is free of finite rank over $\mathcal{A}_V$. This rank is the multiplicity.

and [**84**, §12] has given a different construction based on local mapping properties of $\varphi(T)$; he calls the multiplicity the "local degree". This terminology is explained by Corollary 9.17 below. In [**48**], Favre and Rivera-Letelier give a characterization of multiplicities which shows that all the constructions yield the same result.

After constructing multiplicities, we establish formulas expressing them in terms of various geometric properties of φ. We then apply these formulas to give insight into the action of φ on $\mathbb{P}^1_{\text{Berk}}$, proving an open mapping theorem and related results.

For each $b, \zeta \in \mathbb{P}^1_{\text{Berk}}$ with $b \neq \zeta$, put
$$h_{b,\zeta}(x) = \log_v(\delta(\varphi(x), b)_\zeta) \, .$$
This is a measure of the proximity of $\varphi(x)$ to b. The motivation for our definition of multiplicities is the following computation:

LEMMA 9.1. *If $b, \zeta \in \mathbb{P}^1(K)$, then $h_{b,\zeta} \in \text{BDV}(\mathbb{P}^1_{\text{Berk}})$ and*
$$\Delta_{\mathbb{P}^1_{\text{Berk}}}(h_{b,\zeta}) = \sum_{\varphi(a)=\zeta} m_{\varphi,\zeta}(a)\delta_a - \sum_{\varphi(a)=b} m_{\varphi,b}(a)\delta_a \, . \tag{9.1}$$

PROOF. After making a change of coordinates, we can assume without loss of generality that $\zeta = \infty$.

Write $\mathbb{P}^1_{\text{Berk}} = \mathbb{A}^1_{\text{Berk}} \cup \{\infty\}$, and as in §2.1, identify $\mathbb{A}^1_{\text{Berk}}$ with the set of multiplicative seminorms on $K[T]$ which extend the absolute value on K. Extend each such seminorm to its local ring $\mathcal{R}_x$. By Corollary 4.2, for each $y \in \mathbb{A}^1_{\text{Berk}}$ we have $\delta(y, b)_\infty = [T - b]_y$. By formula (2.11), for each $f \in K(T)$ belonging to the local ring $\mathcal{R}_{\varphi(x)}$, we have $f \circ \varphi \in \mathcal{R}_x$ and $[f]_{\varphi(x)} = [f \circ \varphi]_x$. If $\varphi(x) \neq \infty$, it follows that
$$\delta(\varphi(x), b)_\infty = [T - b]_{\varphi(x)} = [\varphi(T) - b]_x \, . \tag{9.2}$$
By Example 5.20, $h_{b,\infty} \in \text{BDV}(\mathbb{P}^1_{\text{Berk}})$, and formula (9.1) holds. □

In Proposition 9.5 below, we will see that a similar result holds for each $b \in \mathbb{P}^1_{\text{Berk}}$, and we will use this to define multiplicities for φ.

LEMMA 9.2. *For each $b \in \mathbb{P}^1_{\text{Berk}}$ and each $\zeta \in \mathbb{P}^1(K)$ with $b \neq \zeta$, we have $h_{b,\zeta} \in \text{BDV}(\mathbb{P}^1_{\text{Berk}})$.*

PROOF. As before, by making a change of coordinates, we can assume that $\zeta = \infty$. Put $U = \mathbb{P}^1_{\text{Berk}} \setminus \varphi^{-1}(\{\infty\})$. When $b \in \mathbb{P}^1(K)$, formula (9.1) shows that $h_{b,\infty}(x) \in \mathcal{SH}(U)$.

We start by showing that $h_{b,\infty}(x) \in \mathcal{SH}(U)$ for all b.

First assume b is of type II or type III. Identifying $\mathbb{P}^1_{\text{Berk}}$ with $\mathbb{A}^1_{\text{Berk}} \cup \{\infty\}$, let b correspond to a disc $D(\beta, r)$ under Berkovich's classification theorem (Theorem 2.2), with $\beta \in K$ and $r > 0$. Recall that we write $D(\beta, r) = \{z \in K : |z - \beta| \leq r\}$ for the 'classical' disc and $\mathcal{D}(\beta, r) = \{z \in \mathbb{A}^1_{\text{Berk}} : \delta(z, \beta)_\infty \leq r\}$ for the corresponding Berkovich disc. Let $\preceq$ be the partial order on $\mathbb{A}^1_{\text{Berk}}$ such that $p \preceq q$ iff q lies on the path from p to ∞. Recall from §4.1 that $p \vee_\infty q$, the point where the paths $[p, \infty]$ and $[q, \infty]$ first meet, is the least element for which $p, q \preceq p \vee_\infty q$. For each $y \in \mathbb{A}^1_{\text{Berk}}$, we have $\delta(y, b)_\infty = \text{diam}_\infty(y \vee_\infty b)$. If $y \in \mathcal{D}(\beta, r)$, then $y \vee_\infty b = b$ and $\delta(y, b)_\infty = r$. If $y \notin \mathcal{D}(\beta, r)$, then $y \vee_\infty \beta = y \vee_\infty b \succ b$ so $\delta(y, b)_\infty = \delta(y, \beta)_\infty > r$. Combining these, we find that $\delta(y, b)_\infty = \max(\delta(y, \beta)_\infty, r)$. Hence if $\varphi(x) \neq \infty$, then $\delta(\varphi(x), b)_\infty = \max(\delta(\varphi(x), \beta)_\infty, r)$ and

$$(9.3) \qquad h_{b,\infty}(x) = \max(h_{\beta,\infty}(x), \log_v(r)) \ .$$

It follows from Proposition 8.26(D) that $h_{b,\infty} \in \mathcal{SH}(U)$.

If b is of type IV, let b correspond to a sequence of nested discs $\mathcal{D}(\beta_1, r_1) \supset \mathcal{D}(\beta_2, r_2) \supset \cdots$ with empty intersection, under Berkovich's classification theorem. Here $\beta_i \in K$, and $r_i > 0$. For each i, let $b_i \in \mathbb{A}^1_{\text{Berk}}$ be the point corresponding to $\mathcal{D}(\beta_i, r_i)$; note that each b_i is of type II or III, $\lim_{i \to \infty} b_i = b$, and the $r_i = \text{diam}_\infty(b_i)$ decrease monotonically to $r = \text{diam}_\infty(b) > 0$. We claim that the $h_{b_i,\infty}$ converge uniformly to $h_{b,\infty}$ on U. To see this, fix $\varepsilon > 0$ and let N be large enough that $\log_v(r_N) < \log_v(r) + \varepsilon$. Take $i, j > N$ with $i < j$. Since $b_j \in \mathcal{D}(\beta_i, r_i)$, for each $y \notin \mathcal{D}(\beta_i, r_i)$ we have $\delta(y, b_i) = \delta(y, b_j) > r_i$. Hence if $x \in U$ is such that $\varphi(x) \notin \mathcal{D}(\beta_i, r_i)$, then $h_{b_i,\infty}(x) = h_{b_j,\infty}(x)$. On the other hand, if $\varphi(x) \in \mathcal{D}(\beta_i, r_i)$, then $\log_v(r_N) > \log_v(r_i) = h_{b_i,\infty}(x)$ and

$$\log_v(r_i) \geq \log_v(\max(\delta(\varphi(x), b_j)_\infty, r_j)) = h_{b_j,\infty}(x) > \log_v(r) \ .$$

In either case we have $|h_{b_i,\infty}(x) - h_{b_j,\infty}(x)| < \varepsilon$. It follows from Proposition 8.26(C) that $h_{b,\infty} \in \mathcal{SH}(U)$.

We can now show that $h_{b,\infty} \in \text{BDV}(\mathbb{P}^1_{\text{Berk}})$. Write

$$\partial U = \varphi^{-1}(\{\infty\}) = \{a_1, \ldots, a_r\} \subset \mathbb{P}^1(K) \ .$$

Fixing b, for each a_i choose a ball $U_i = \mathcal{B}(a_i, \delta_i)^-$ which contains a_i but does not meet the closed set $\varphi^{-1}(\{b\})$. After shrinking the U_i if necessary, we can assume that $U_1, \ldots, U_r$ are pairwise disjoint. The construction above shows there is a $\beta \in \mathbb{P}^1(K)$ such that $h_{b,\infty} = h_{\beta,\infty}$ on each U_i. Since $h_{\beta,\infty} \in \text{BDV}(\mathbb{P}^1_{\text{Berk}})$, it follows from Proposition 5.26 that for each i we have $h_{b,\infty}|_{U_i} \in \text{BDV}(U_i)$ and

$$(9.4) \qquad \Delta_{U_i}(h_{b,\infty}) = \Delta_{U_i}(h_{\beta,\infty}) = m_\varphi(a_i) \delta_{a_i} \ .$$

Next, for each a_i choose a closed ball $W_i = \mathcal{B}(a_i, \delta'_i) \subseteq U_i$ containing a_i, and put $U_0 = \mathbb{P}^1_{\text{Berk}} \backslash (\bigcup_{i=1}^r W_i)$. Then U_0 is a simple subdomain of U. Since $h_{b,\infty} \in \mathcal{SH}(U)$, Lemma 8.10(B) shows that $h_{b,\infty} \in \text{BDV}(U_0)$.

Since $U_0, U_1, \ldots, U_r$ form a finite open cover of $\mathbb{P}^1_{\text{Berk}}$, Proposition 5.27 gives $h_{b,\infty} \in \text{BDV}(\mathbb{P}^1_{\text{Berk}})$. $\square$

We next show that for each $b \in \mathbb{P}^1_{\text{Berk}}$, there is a measure $\omega_{\varphi,b}$ supported on $\varphi^{-1}(b)$ which behaves as if it were the polar divisor of $h_{b,\zeta}$. In Proposition 9.5 we will see that it is discrete. (We have not yet proved that the set $\varphi^{-1}(b)$ is finite, though we will see that it is.)

LEMMA 9.3. *For each $b \in \mathbb{P}^1_{\text{Berk}}$, there is a positive measure $\omega_{\varphi,b}$ of total mass d, supported on $\varphi^{-1}(\{b\})$ and independent of $\zeta \in \mathbb{P}^1(K)$ with $\zeta \neq b$, for which*

$$\Delta_{\mathbb{P}^1_{\text{Berk}}}(h_{b,\zeta}) = \sum_{\varphi(a)=\zeta} m_{\varphi,\zeta}(a)\delta_a - \omega_{\varphi,b} . \tag{9.5}$$

When $b \in \mathbb{P}^1(K)$, then $\omega_{\varphi,b} = \sum_{\varphi(a)=b} m_{\varphi,b}(a)\delta_a$.

PROOF. As in the preceding two proofs, we can assume without loss of generality that $\zeta = \infty$. Write $U = \mathbb{P}^1_{\text{Berk}} \backslash \varphi^{-1}(\{\infty\})$. In the proof of Lemma 9.2, it was shown that $h_{b,\zeta} \in \mathcal{SH}(U)$.

We first show that $\Delta_U(h_{b,\infty})$ is supported on $\varphi^{-1}(\{b\})$. When $b \in K$, this is trivial.

If b is of type II or type III, let b correspond to $D(\beta,r)$ under Berkovich's classification theorem, and write $h_{b,\zeta}(x) = \max(h_{\beta,\zeta}(x), \log_v(r))$ as in (9.3). Let V be a connected component of $U \backslash \varphi^{-1}(\{b\})$. Then $\varphi(V)$ is connected and disjoint from $\{b, \infty\}$. As b is the unique boundary point of $\mathcal{D}(\beta, r)$, we must have either $V \subseteq \mathcal{D}(\beta,r) \backslash \{b\}$ or $V \subseteq \mathbb{A}^1_{\text{Berk}} \backslash \mathcal{D}(\beta,r)$. If $\varphi(V) \subseteq \mathcal{D}(\beta,r)$, then $h_{b,\infty} \equiv \log_v(r)$ on V, and Proposition 5.26 shows that $\Delta_U(h_{b,\infty})|_V = \Delta_V(h_{b,\infty}) = 0$. If $\varphi(V) \subset \mathbb{A}^1_{\text{Berk}} \backslash \mathcal{D}(\beta,r)$, then $h_{b,\infty} = h_{\beta,\infty}$ on V, and since $h_{\beta,\infty}$ is harmonic on V, we again have $\Delta_U(h_{b,\infty})|_V = 0$. Thus $\Delta_U(h_{b,\infty})$ is supported on $\varphi^{-1}(\{b\})$ in this case.

Finally, suppose b is of type IV, and let the b_i and $D(\beta_i, r_i)$ be as in the proof of Lemma 9.2. As we have just seen, each $h_{b_i,\infty}(x)$ is harmonic on the open set $U \backslash \varphi^{-1}(\mathcal{D}(\beta_i, r_i))$. Since these sets form an exhaustion of $U \backslash \varphi^{-1}(\{b\})$ and the $h_{b_i,\infty}$ converge uniformly to $h_{b,\infty}$ on U, it follows that $h_{b,\infty}$ is harmonic on $U \backslash \varphi^{-1}(\{b\})$. Thus again $\Delta_U(h_{b,\infty})$ is supported on $\varphi^{-1}(\{b\})$.

Since $\Delta_U(h_{b,\infty}) \leq 0$, we can write

$$\Delta_U(h_{b,\infty}) = -\omega_{\varphi,b} \tag{9.6}$$

where $\omega_{\varphi,b}$ is nonnegative (but might depend on the point $\zeta = \infty$; we will see below that this dependence does not occur). Let $U_0, U_1, \ldots, U_r$ be as in the proof of Lemma 9.2. By (9.4), (9.6), and the restriction formulas in Proposition 5.27, we have

$$\Delta_{\mathbb{P}^1_{\text{Berk}}}(h_{b,\infty}) = \sum_{i=1}^{r} m_\varphi(a_i)\delta_{a_i} - \omega_{\varphi,b} . \tag{9.7}$$

Since $\Delta_{\mathbb{P}^1_{\text{Berk}}}(h_{b,\infty})$ has total mass 0, $\omega_{\varphi,b}$ necessarily has total mass d.

Lastly, we show that $\omega_{\varphi,b}$ is independent of $\zeta \in \mathbb{P}^1(K)$. For this part of the proof, we drop our simplifying assumption that $\zeta = \infty$ and temporarily write $\omega_{\varphi,b,\zeta}$ for the measure associated to b and ζ by the construction above.

9.1. AN ANALYTIC CONSTRUCTION OF MULTIPLICITIES

The key is the identity

$$(9.8) \qquad \delta(x,y)_\zeta \;=\; C_{\zeta,\infty} \cdot \frac{\delta(x,y)_\infty}{\delta(x,\zeta)_\infty \delta(y,\zeta)_\infty} \,,$$

which holds for each $\zeta \neq \infty$ and all $x,y \in \mathbb{P}^1_{\text{Berk}} \backslash \{\infty\}$, with $C_{\zeta,\infty} = \|\zeta,\infty\|^{-2}$. (This is a special case of formula (4.29).) Recall that the Laplacian of a function in $\text{BDV}(\mathbb{P}^1_{\text{Berk}})$ depends only on the function's restriction to $\mathbb{H}_{\text{Berk}}$.

Replacing x by $\varphi(x)$ and y by b in (9.8) and then taking logarithms, we see that

$$h_{b,\zeta}(x) \;=\; h_{b,\infty}(x) - h_{\zeta,\infty}(x) + C' \,.$$

Taking Laplacians and using (9.1), (9.7), we obtain

$$\sum_{\varphi(a)=\zeta} m_\varphi(a)\delta_a - \omega_{\varphi,b,\zeta} \;=\; \sum_{\varphi(a)=\zeta} m_\varphi(a)\delta_a - \omega_{\varphi,b,\infty} \,.$$

Thus $\omega_{\varphi,b,\zeta} = \omega_{\varphi,b,\infty}$.

The formula for $\omega_{\varphi,b}$ when $b \in \mathbb{P}^1(K)$ follows from Lemma 9.1. □

In order to show that $\omega_{\varphi,b}$ is a discrete measure for each $b \in \mathbb{P}^1_{\text{Berk}}$, which we will deduce from the known case where $b \in \mathbb{P}^1(K)$, we need the following 'continuity' lemma:

LEMMA 9.4. *Given $b \in \mathbb{P}^1_{\text{Berk}}$, if $\langle b_\alpha \rangle$ is a net in $\mathbb{P}^1_{\text{Berk}}$ converging to b, then the net of measures $\langle \omega_{\varphi,b_\alpha} \rangle$ converges weakly to $\omega_{\varphi,b}$.*

PROOF. Without loss of generality, we can assume that there is a simple domain V containing b and all the b_α. Fix a point $\zeta \in \mathbb{P}^1(K) \backslash V$, and write

$$h_\alpha(x) = h_{b_\alpha,\zeta}(x), \qquad h(x) = h_{b,\zeta}(x) \,.$$

Let $\nu_\alpha = \Delta_{\mathbb{P}^1_{\text{Berk}}}(h_\alpha)$ and $\nu = \Delta_{\mathbb{P}^1_{\text{Berk}}}(h)$. By Lemma 9.3, to show that the net $\langle \omega_{\varphi,b_\alpha} \rangle$ converges weakly to $\omega_{\varphi,b}$, it suffices to show that the net $\langle \nu_\alpha \rangle$ converges weakly to ν. Since $|\nu_\alpha|(\mathbb{P}^1_{\text{Berk}}) = 2d$ for each α, by Proposition 5.32 it suffices in turn to show that for each sufficiently large finite subgraph Γ of $\mathbb{P}^1_{\text{Berk}}$,

$$h_\alpha|_\Gamma \;\to\; h|_\Gamma$$

uniformly on Γ.

By formula (4.26), for each $\xi \in \mathbb{H}_{\text{Berk}}$ there is a constant C_ξ such that for all $y,z \in \mathbb{P}^1_{\text{Berk}}$, we have

$$\log_v(\delta(y,z)_\zeta) \;=\; -j_\xi(y,z) + j_\xi(y,\zeta) + j_\xi(z,\zeta) + C_\xi \,.$$

Expanding $h_\alpha(x) = \log_v(\delta(\varphi(x),b_\alpha)_\zeta)$ and $h(x) = \log_v(\delta(\varphi(x),b)_\zeta)$, we find that

$$(9.9) \quad |h_\alpha(x) - h(x)| \;\leq\; |j_\xi(\varphi(x),b_\alpha) - j_\xi(\varphi(x),b)| + |j_\xi(b_\alpha,\zeta) - j_\xi(b,\zeta)| \,.$$

Let Γ be any finite subgraph of $\mathbb{P}^1_{\text{Berk}}$. There are now two cases to consider, according as $b \notin \varphi(\Gamma)$ or $b \in \varphi(\Gamma)$:

If $b \notin \varphi(\Gamma)$, then since $\varphi(\Gamma)$ is compact and connected, we can choose $\xi \in \mathbb{H}_{\text{Berk}}$ so that $\varphi(\Gamma)$ and b belong to different components of $\mathbb{P}^1_{\text{Berk}} \backslash \{\xi\}$.

In this situation, the first term on the right in (9.9) is 0, and the second can be made as small as we wish by taking b_α sufficiently near b, since $j_\xi(z, \zeta)$ is a continuous function of z (Proposition 4.10(A)). Thus, given $\varepsilon > 0$, there is a neighborhood W of b such that $|h_\alpha(x) - h(x)| < \varepsilon$ for all $x \in \Gamma$ and all $b_\alpha \in W$.

If $b \in \varphi(\Gamma)$, take $\xi = b$. Since $j_b(y, z) \geq 0$ and $j_b(z, b) = 0$ for all y, z, (9.9) becomes

$$(9.10) \qquad |h_\alpha(x) - h(x)| \leq j_b(\varphi(x), b_\alpha) + j_b(b_\alpha, \zeta) \ .$$

We now make use of the fact, proved in Proposition 9.32 in §9.2 below, that the topology on $\varphi(\Gamma)$ induced by the path distance metric $\rho(z, w)$ coincides with the relative topology induced from $\mathbb{P}^1_{\text{Berk}}$. This implies that given $\varepsilon > 0$, there exists a simple domain W_0 containing b for which $W_0 \cap \varphi(\Gamma) \subseteq \widehat{\mathcal{B}}_{\varphi(\Gamma)}(b, \varepsilon)$, where $\widehat{\mathcal{B}}_{\varphi(\Gamma)}(b, \varepsilon) = \{z \in \varphi(\Gamma) : \rho(z, b) < \varepsilon\}$.

For each $x \in \Gamma$, if c is the point where the paths $[\varphi(x), b]$ and $[b_\alpha, b]$ first meet, then

$$j_b(\varphi(x), b_\alpha) \ = \ \rho(b, c) \ .$$

If $b_\alpha \in W_0$, then $c \in W_0 \cap \varphi(\Gamma)$ since W_0 and $\varphi(\Gamma)$ are uniquely path-connected. Hence $j_b(\varphi(x), b_\alpha) = \rho(b, c) < \varepsilon$. By the continuity of $j_b(z, \zeta)$, there is also a neighborhood W_1 of b such that $j_b(z, \zeta) < \varepsilon$ for all $z \in W_1$. It follows that $|h_\alpha(x) - h(x)| < 2\varepsilon$ for all $x \in \Gamma$ and all α such that $b_\alpha \in W_0 \cap W_1$. □

We can now show that $\omega_{\varphi, b}$ is a discrete measure:

PROPOSITION 9.5. *For each $b \in \mathbb{P}^1_{\text{Berk}}$, the measure $\omega_{\varphi, b}$ is positive and discrete, with total mass $d = \deg(\varphi)$. Furthermore, $\mathrm{supp}(\omega_{\varphi, b}) = \varphi^{-1}(\{b\})$ is finite, and there are integers $m_{\varphi, b}(a) > 0$ such that*

$$(9.11) \qquad \omega_{\varphi, b} \ = \ \sum_{\varphi(a) = b} m_{\varphi, b}(a) \delta_a \ .$$

For each $b, \zeta \in \mathbb{P}^1_{\text{Berk}}$ with $b \neq \zeta$, we have $h_{b, \zeta} \in \mathrm{BDV}(\mathbb{P}^1_{\text{Berk}})$ and

$$(9.12) \qquad \Delta_{\mathbb{P}^1_{\text{Berk}}}(h_{b, \zeta}) \ = \ \sum_{\varphi(a) = \zeta} m_{\varphi, \zeta}(a) \delta_a - \sum_{\varphi(a) = b} m_{\varphi, b}(a) \delta_a \ .$$

For the proof, we will need the "Portmanteau theorem" about convergence of nets of measures. It is proved as Theorem A.13 in §A.7:

THEOREM 9.6 (Portmanteau theorem). *Let X be a locally compact Hausdorff space, and let $\langle \omega_\alpha \rangle_{\alpha \in I}$ be a net of positive Radon measures, each of total mass $M > 0$, converging weakly to a positive Radon measure ω on X. Then:*

(A) $\omega_\alpha(A) \to \omega(A)$ *for every Borel set $A \subseteq X$ with $\omega(\partial A) = 0$.*

(B) *More generally, for every Borel set $A \subseteq X$ and every $\varepsilon > 0$, there exists $\alpha_0 \in I$ such that*

$$\omega(A^\circ) - \varepsilon \ \leq \ \omega_\alpha(A) \ \leq \ \omega(\overline{A}) + \varepsilon$$

for all $\alpha \geq \alpha_0$.

9.1. AN ANALYTIC CONSTRUCTION OF MULTIPLICITIES

PROOF OF PROPOSITION 9.5. Fix b, and write $\omega = \omega_{\varphi,b}$. We already know from Lemma 9.3 that ω is a positive measure of total mass d.

We first show that ω is a discrete measure. Choose a net $\langle b_\alpha \rangle$ in $\mathbb{P}^1(K)$ converging to b. If we set $\omega_\alpha = \omega_{\varphi,b_\alpha}$, then by Lemma 9.3, ω_α is a discrete, positive measure with integer coefficients and total mass d. Lemma 9.4 shows that ω_α converges weakly to ω.

To see that ω is discrete, we proceed as follows. Fix $a \in \mathrm{supp}(\omega)$. By assumption, for each open neighborhood U of a we have $\omega(U) > 0$. Choose ε with $0 < \varepsilon < \omega(U)$. The Portmanteau theorem shows that there exists α_0 such that for $\alpha \geq \alpha_0$, we have

$$\omega_\alpha(U) \geq \omega(U) - \varepsilon > 0 .$$

Since $\omega_\alpha(U) \in \mathbb{Z}$ for all α, we must in fact have $\omega_\alpha(U) \geq 1$ for all $\alpha \geq \alpha_0$. Another application of the Portmanteau theorem shows that for all $\varepsilon > 0$ and α sufficiently large, we have

$$\omega(\overline{U}) \geq \omega_\alpha(U) - \varepsilon \geq 1 - \varepsilon .$$

Therefore $\omega(\overline{U}) \geq 1$. Since this is true for every open neighborhood U of a, we conclude that $\omega(\{a\}) \geq 1$. Since ω is positive with total mass d and $\omega(\{a\}) \geq 1$ for each $a \in \mathrm{supp}(\omega)$, it follows that $\mathrm{supp}(\omega)$ is finite. Thus ω is a discrete measure as claimed.

Now that we know that ω is discrete, it follows that for each $x \in \mathbb{P}^1_{\mathrm{Berk}}$, there exists an open neighborhood U of x such that $\omega(\{x\}) = \omega(U)$ and $\omega(\partial U) = 0$. Put $m_{\varphi,b}(x) = \omega(\{x\})$. By the Portmanteau theorem, we have $\omega_\alpha(U) \to \omega(U)$, and therefore $\omega(U)$ is a nonnegative integer. Hence $0 \leq m_{\varphi,b}(x) \in \mathbb{Z}$.

We next show that for $a, b \in \mathbb{P}^1_{\mathrm{Berk}}$, $m_{\varphi,b}(a) > 0$ if and only if $\varphi(a) = b$. For this, choose a net $\langle a_\alpha \rangle$ in $\mathbb{P}^1(K)$ converging to a. Also, note that everything we have said so far applies to *any* net $\langle b_\alpha \rangle$ in $\mathbb{P}^1(K)$ which converges to b.

Suppose that $\varphi(a) = b$. Then by continuity, $\varphi(a_\alpha) \to b$. Taking $b_\alpha = \varphi(a_\alpha)$, we see that for any sufficiently small neighborhood U of a, we have $\omega_\alpha(U) \to \omega(U) = m_{\varphi,b}(a)$. But $\omega_\alpha(U)$ is the number of preimages of b_α lying in U, which by construction is at least 1 for α sufficiently large, since $a_\alpha \to a \in U$. Thus $m_{\varphi,b}(a) \geq 1$.

Suppose conversely that $\varphi(a) \neq b$. Then there exist open neighborhoods U of a and V of b such that $U \cap \varphi^{-1}(V) = \emptyset$. (To see this, note that since $\varphi(a) \neq b$ and $\mathbb{P}^1_{\mathrm{Berk}}$ is a compact Hausdorff space, there exists an open neighborhood V of b such that $\varphi(a) \notin \overline{V}$. Since a is in the complement of the closed set $\varphi^{-1}(\overline{V})$, there is an open neighborhood U of a disjoint from $\varphi^{-1}(\overline{V})$.) For this open neighborhood U of a and for any net $\langle b_\alpha \rangle$ in $\mathbb{P}^1(K)$ converging to b, there exists α_0 such that $b_\alpha \in V$ for $\alpha \geq \alpha_0$. By construction, there are no preimages of b_α lying in U when $b_\alpha \in V$. It follows that $\omega_\alpha(U) = 0$ for $\alpha \geq \alpha_0$, and thus $\omega(U) = 0$ as well, i.e., $m_{\varphi,b}(a) = 0$.

Finally, we show (9.12). Take $b, \zeta \in \mathbb{P}^1_{\text{Berk}}$ with $b \neq \zeta$, and choose $\xi \in \mathbb{P}^1(K)$ with $\xi \neq b, \zeta$. By (4.29), the identity

$$(9.13) \qquad \delta(x,y)_\zeta = C_{\zeta,\xi} \cdot \frac{\delta(x,y)_\xi}{\delta(x,\zeta)_\xi \delta(y,\zeta)_\xi}$$

holds for each $\zeta \neq \xi$ and all $x, y \in \mathbb{P}^1_{\text{Berk}} \backslash \{\xi\}$, with $C_{\zeta,\xi} = \|\zeta, \xi\|^{-2}$. Replacing x by $\varphi(x)$ and y by b in (9.13) and then taking logarithms, we see that

$$h_{b,\zeta}(x) = h_{b,\xi}(x) - h_{\zeta,\xi}(x) + C .$$

Since $h_{b,\xi}, h_{\zeta,\xi} \in \text{BDV}(\mathbb{P}^1_{\text{Berk}})$, it follows that $h_{b,\zeta} \in \text{BDV}(\mathbb{P}^1_{\text{Berk}})$. Taking Laplacians and using Lemma 9.3, we obtain (9.12). $\square$

DEFINITION 9.7. For $a, b \in \mathbb{P}^1_{\text{Berk}}$, if $\varphi(a) = b$, we define the *multiplicity* $m_{\varphi,b}(a)$ to be the coefficient of δ_a in (9.11), and if $\varphi(a) \neq b$, we set $m_{\varphi,b}(a) = 0$. Thus, $m_{\varphi,b}(a)$ is the weight of a in the Laplacian of the proximity function $h_{b,\zeta}(x)$, for any $\zeta \neq b$.

If $b = \varphi(a)$, we call $m_\varphi(a) := m_{\varphi,b}(a)$ the multiplicity of φ at a.

The following theorem summarizes the properties of multiplicities established above[2]:

THEOREM 9.8. *Let $\varphi(T) \in K(T)$ be a nonconstant rational function. Then for all $a, b \in \mathbb{P}^1_{\text{Berk}}$, the multiplicities $m_\varphi(a), m_{\varphi,b}(a)$ satisfy the following properties:*

(A) $m_\varphi(a) \in \mathbb{Z}$, with $1 \leq m_\varphi(a) \leq \deg(\varphi)$.
(B) $m_{\varphi,b}(a) > 0$ if and only if $\varphi(a) = b$, in which case $m_{\varphi,b}(a) = m_\varphi(a)$.
(C) $\sum_{\varphi(a)=b} m_\varphi(a) = \deg(\varphi)$.
(D) *For points a of type I, $m_\varphi(a)$ coincides with the usual algebraic multiplicity.*

Later in this section we will give formulas for our multiplicities which explain their geometric meaning. However, we first note some properties of φ which follow from their existence.

As an immediate consequence of Theorem 9.8, we find:

COROLLARY 9.9. *Let $\varphi(T) \in K(T)$ be a nonconstant rational function. Then $\varphi : \mathbb{P}^1_{\text{Berk}} \to \mathbb{P}^1_{\text{Berk}}$ is surjective, and for each $b \in \mathbb{P}^1_{\text{Berk}}$ there are at most $d = \deg(\varphi)$ points $a \in \mathbb{P}^1_{\text{Berk}}$ with $\varphi(a) = b$.*

We also obtain a Berkovich space analogue of the "Open Mapping Theorem" from complex analysis[3]:

---

[2] A number of results which occur in our development of the theory have counterparts in the work of Rivera-Letelier and Favre–Rivera-Letelier. In particular, Theorem 9.8 is analogous to Proposition-Definition 2.1 and Proposition 2.2 of [48], and Corollary 9.9 is contained in Proposition 2.2 of [48].

[3] It is in fact the case that any finite morphism between irreducible Berkovich k-analytic spaces of the same dimension is both open and closed; see [16, Lemma 3.2.4] (note that finite morphisms are proper and, therefore, closed).

COROLLARY 9.10 (Open Mapping Theorem). *Let $\varphi(T) \in K(T)$ be a nonconstant rational function. Then $\varphi : \mathbb{P}^1_{\text{Berk}} \to \mathbb{P}^1_{\text{Berk}}$ is open in the Berkovich topology.*

PROOF. Suppose $U \subset \mathbb{P}^1_{\text{Berk}}$ is an open set for which $\varphi(U)$ is not open. Then there exists $a \in U$ and a net $\langle b_\alpha \rangle$ converging to $b = \varphi(a)$ such that $b_\alpha \notin \varphi(U)$ for all α. By Lemma 9.4, the measures $\omega_\alpha = \omega_{\varphi, b_\alpha}$ converge weakly to $\omega = \omega_{\varphi, b}$. Also, by Theorem 9.8, $a \in \text{supp}(\omega)$. But each ω_α is supported on the compact set $U^c = \mathbb{P}^1_{\text{Berk}} \backslash U$, and therefore ω is supported on U^c as well. This contradicts the fact that $a \in U \cap \text{supp}(\omega)$. □

From the Open Mapping Theorem, we obtain the following corollary:

COROLLARY 9.11. *Let $\varphi(T) \in K(T)$ be a nonconstant rational function. If $U \subset \mathbb{P}^1_{\text{Berk}}$ is a domain and $V = \varphi(U)$, then V is a domain and $\partial V \subseteq \varphi(\partial U)$. If U is a simple domain, then V is also a simple domain.*

PROOF. If $U \subset \mathbb{P}^1_{\text{Berk}}$ is a domain, then by the Open Mapping Theorem, $V = \varphi(U)$ is connected and open, so it is a domain. If y is a boundary point of V, take a net $\langle y_\alpha \rangle_{\alpha \in A}$ in V converging to y. For each y_α choose an $x_\alpha \in U$ with $\varphi(x_\alpha) = y_\alpha$. Since $\overline{U}$ is compact, the generalized Bolzano-Weierstrass theorem (Proposition A.6) shows that the net $\langle x_\alpha \rangle_{\alpha \in A}$ has a (cofinal) subnet $\langle x_\alpha \rangle_{\alpha \in B}$ converging to a point $x \in \overline{U}$. By continuity $\varphi(x) = y$. If $x \in U$, then $y \in V$, which contradicts the fact that V is open. Hence $x \in \partial U$, so $\partial V \subseteq \varphi(\partial U)$.

Finally, by Proposition 2.15, φ takes $\mathbb{H}^{\mathbb{R}}_{\text{Berk}}$ to itself. If U is a simple domain, it follows that $\varphi(U)$ is a domain with finitely many boundary points and whose boundary points belong to $\mathbb{H}^{\mathbb{R}}_{\text{Berk}}$, so it is a simple domain. □

We also note the following complement to Corollary 9.11:

LEMMA 9.12. *Let $\varphi(T) \in K(T)$ be a nonconstant rational function of degree $d = \deg(\varphi)$. If $U \subset \mathbb{P}^1_{\text{Berk}}$ is a domain, then each component W of $\varphi^{-1}(U)$ is a domain with $\partial W \subseteq \varphi^{-1}(\partial U)$, and φ maps W surjectively onto U and ∂W surjectively onto ∂U. In particular, $\varphi^{-1}(U)$ has at most d components. Furthermore, if U is a simple domain, then each component of $\varphi^{-1}(U)$ is a simple domain.*

PROOF. As φ is continuous, $\varphi^{-1}(U)$ is open. Let W be a component of $\varphi^{-1}(U)$ and take $x \in \partial W$. Since W is open, $x \notin W$. We claim that $\varphi(x) \in \partial U$. By continuity, $\varphi(x) \in \overline{U}$. If $\varphi(x) \in U$, then there would be a neighborhood V of x with $\varphi(V) \subset U$. We can assume that V is connected, since the connected neighborhoods of a point in $\mathbb{P}^1_{\text{Berk}}$ are cofinal in all neighborhoods. Therefore $V \cup W$ is connected, and $\varphi(V \cup W) \subseteq U$. From the maximality of W, we conclude that $V \subseteq W$, so $x \in W$, a contradiction. Thus $\varphi(\partial W) \subseteq \partial U$. Since $\partial U \subseteq \varphi(\partial W)$ by Corollary 9.11, we see that $\varphi(\partial W) = \partial U$.

Next, we claim that W maps surjectively onto U. Suppose not; then there exists a point $z \in U \backslash \varphi(W)$. Fix $t \in \varphi(W)$, and consider the path $[t, z]$.

Since U is uniquely path-connected, $[t,z] \subset U$. Since $[t,z]$ (for its induced topology) is homeomorphic to a real interval, with $t \in \varphi(W)$, $z \notin \varphi(W)$, there is a point $y \in [t,z]$ such that each neighborhood of y contains points of $\varphi(W)$ and $U\backslash\varphi(W)$. Thus, $y \in \partial(\varphi(W))$. By Corollary 9.10, $\varphi(W)$ is open, so $y \notin \varphi(W)$.

Choose a net $\langle y_\alpha \rangle_{\alpha \in A}$ in $\varphi(W)$ converging to y. For each y_α, take an $x_\alpha \in W$ with $\varphi(x_\alpha) = y_\alpha$. Since $\overline{W}$ is compact, the generalized Bolzano-Weierstrass theorem shows that the net $\langle x_\alpha \rangle_{\alpha \in A}$ has a subnet $\langle x_\alpha \rangle_{\alpha \in B}$ converging to a point $x \in \overline{W}$. By continuity, $\varphi(x) = y$. If $x \in W$, then $y \in \varphi(W)$, a contradiction. Thus $x \notin W$. On the other hand, since $\varphi(x) = y \in U$, by continuity there is a neighborhood V of x such that $\varphi(V) \in U$. Without loss of generality we can assume that V is connected, so $W \cup V$ is connected. Since $\varphi(W \cup V) \subset U$, the maximality of W shows that $V \subseteq W$. This contradicts the fact that $x \notin W$. We conclude that $\varphi(W) = U$.

By Corollary 9.9, each $y \in U$ has at most d preimages, so $\varphi^{-1}(U)$ has at most d components.

Finally, if U is a simple domain, then ∂U is finite and is contained in $\mathbb{H}^{\mathbb{R}}_{\text{Berk}}$. By Corollary 9.9, $\varphi^{-1}(\partial U)$ is finite, and by Proposition 2.15 it is contained in $\mathbb{H}^{\mathbb{R}}_{\text{Berk}}$. Hence each component of $\varphi^{-1}(U)$ is a simple domain. $\square$

A useful consequence of Lemma 9.4 and Theorem 9.8 is the following[4]:

PROPOSITION 9.13. *If $f : \mathbb{P}^1_{\text{Berk}} \to \mathbb{R}$ is continuous, then so is the function $\varphi_* f : \mathbb{P}^1_{\text{Berk}} \to \mathbb{R}$ defined by*

$$(\varphi_* f)(y) = \sum_{\varphi(z)=y} m_{\varphi,y}(z) f(z) \ .$$

PROOF. If $\langle y_\alpha \rangle$ is a net converging to y, we need to show that the net $(\varphi_* f)(y_\alpha)$ converges to $(\varphi_* f)(y)$. Since

$$(\varphi_* f)(y_\alpha) = \int_{\mathbb{P}^1_{\text{Berk}}} f \, d\omega_{\varphi,y_\alpha} \quad \text{and} \quad (\varphi_* f)(y) = \int_{\mathbb{P}^1_{\text{Berk}}} f \, d\omega_{\varphi,y} \ ,$$

the result follows from the fact that the net $\langle \omega_{\varphi,y_\alpha} \rangle$ converges weakly to $\omega_{\varphi,y}$ by Lemma 9.4. $\square$

As a consequence, we can show that if $U \subset \mathbb{P}^1_{\text{Berk}}$ is a domain, then φ has a well-defined *local degree* on each connected component of $\varphi^{-1}(U)$:[5]

COROLLARY 9.14. *Let $\varphi(T) \in K(T)$ be a nonconstant rational function of degree $d = \deg(\varphi)$. If U is a domain in $\mathbb{P}^1_{\text{Berk}}$ and W is a connected component of $\varphi^{-1}(U)$, then there exists an integer $n = \deg_W(\varphi)$ with $1 \le n \le \deg(\varphi)$, called the* local degree *of φ on W, such that every point in U has exactly n preimages in W, counting multiplicities.*

---

[4]Compare with Proposition-Definition 2.3 of [**48**].

[5]Compare with Proposition-Definition 2.1 of [**48**].

9.1. AN ANALYTIC CONSTRUCTION OF MULTIPLICITIES

PROOF. By Lemma 9.12, $\varphi^{-1}(U)$ has only finitely many components $W_1, \ldots, W_r$, each of which maps surjectively onto U. Let them be labeled in such a way that $W = W_1$.

We first give the proof under the assumption that $\overline{W}_1, \ldots, \overline{W}_r$ are pairwise disjoint. Since $\mathbb{P}^1_{\text{Berk}}$ is a compact Hausdorff space, it is normal and there is a continuous function $f_W : \mathbb{P}^1_{\text{Berk}} \to [0,1]$ satisfying $(f_W)|_{\overline{W}_1} = 1$ and $(f_W)|_{\overline{W}_i} = 0$ for $i = 2, \ldots, r$. Define a function $n_W : U \to \mathbb{R}$ by

$$n_W(b) = \sum_{a \in W} m_{\varphi, b}(a) \ .$$

Then $n_W = (\varphi_* f_W)|_U$. By Proposition 9.13, $\varphi_* f_W$ is continuous, so n_W is continuous. Since n_W takes positive integer values and U is connected, we conclude that n_W is constant on U.

In the general case, using Corollary 7.11, choose an exhaustion of U by a sequence of simple subdomains U_j with $\overline{U}_j \subset U_{j+1}$ for each j. If $j < k$, then each connected component of $\varphi^{-1}(U_j)$ is contained in a connected component of $\varphi^{-1}(U_k)$, which in turn is contained in a connected component of $\varphi^{-1}(U)$. On the other hand, by Lemma 9.12 each component of $\varphi^{-1}(U_k)$ surjects onto U_k, so it contains at least one component of $\varphi^{-1}(U_j)$. Since each $\varphi^{-1}(U_j)$ has finitely many components, the number of components of $\varphi^{-1}(U_j)$ decreases monotonically with j and eventually stabilizes at the number of components of $\varphi^{-1}(U)$. Thus without loss of generality we can assume that each $\varphi^{-1}(U_j)$ has the same number of components as $\varphi^{-1}(U)$ and that for each j, each component of $\varphi^{-1}(U)$ contains a unique component of $\varphi^{-1}(U_j)$.

Fix j and let $W_{j,1}, \ldots, W_{j,r}$ be the components of $\varphi^{-1}(U_j)$, labeled in such a way that $W_{j,i} \subset W_i$ for $i = 1, \ldots, r$. Since $\overline{U}_j \subset U$, it follows that $\overline{W}_{j,i} \subset W_i$, so $\overline{W}_{j,1}, \ldots, \overline{W}_{j,r}$ are pairwise disjoint. By what has been shown above, φ has a well-defined local degree on each $W_{j,i}$.

The $W_{j,1}$ exhaust W, so φ has a well-defined local degree on W. $\square$

We will say that a domain U is φ-saturated if U is a connected component of $\varphi^{-1}(\varphi(U))$. Recall that if X and Y are topological spaces, a map $\psi : X \to Y$ is closed if it takes closed sets to closed sets.

PROPOSITION 9.15. *Let $\varphi(T) \in K(T)$ be a nonconstant rational function. Let $U \subset \mathbb{P}^1_{\text{Berk}}$ be a domain, and put $V = \varphi(U)$. Then the following are equivalent:*

(A) *U is φ-saturated.*
(B) *$\varphi(\partial U) = \partial V$.*
(C) *$\varphi(\partial U) \subseteq \partial V$.*
(D) *The map $\varphi|_U : U \to V$ is closed.*

Under these conditions, the local degree $\deg_U(\varphi)$ is well-defined.

PROOF. (A) $\Rightarrow$ (B) is part of Lemma 9.12. (B) $\Rightarrow$ (C) is trivial.

(C) ⇒ (D). Let $X \subset U$ be relatively closed in U, and let $\overline{X}$ be the closure of X in $\mathbb{P}^1_{\text{Berk}}$. Then $\varphi(\overline{X})$ is compact, hence closed in $\mathbb{P}^1_{\text{Berk}}$. On the other hand, $\overline{X} \subset X \cup \partial U$, so $\varphi(\overline{X}) \subset \varphi(X) \cup \varphi(\partial U)$. Since $\varphi(\partial U) \subseteq \partial V$ by hypothesis, it follows that $\varphi(\overline{X}) \cap V = \varphi(X)$. Thus $\varphi(X)$ is relatively closed in V, and $\varphi|_U : U \to V$ is closed.

(D) ⇒ (A). Suppose $\varphi|_U : U \to V$ is closed, and let U_0 be the connected component of $\mathbb{P}^1_{\text{Berk}} \backslash \varphi^{-1}(\partial V)$ containing U. If $U \neq U_0$, there is a point $a \in \partial U \cap U_0$. Put $b = \varphi(a)$, and choose a closed neighborhood X of a in $\mathbb{P}^1_{\text{Berk}}$ for which $X \cap \varphi^{-1}(\{b\}) = \{a\}$. This can be done because $\varphi^{-1}(\{b\})$ is finite. Then $X \cap U$ is closed in U, but $\varphi(X \cap U)$ is not closed in V because $b \notin \varphi(X \cap U)$, yet $b \in \overline{\varphi(X \cap U)} \cap V$. This contradiction shows that $U = U_0$. □

EXAMPLE 9.16. Let $\varphi(T) = T^2 - T$. A Newton polygon argument shows that $U = \mathcal{D}(0,p)^-$ is φ-saturated, since $\varphi(U) = \mathcal{D}(0,p^2)^-$ and $\varphi^{-1}(\varphi(U)) = U$, but $U' = \mathcal{D}(0,p)^- \backslash \mathcal{D}(0,\frac{1}{p})$ is not φ-saturated, since $\varphi(U') = \varphi(U) = \mathcal{D}(0,p^2)^-$. Note that $\varphi(\partial U) = \partial\varphi(U) = \{\zeta_{0,p^2}\}$, but

$$\varphi(\partial U') = \varphi(\{\zeta_{0,p}, \zeta_{0,1/p}\}) = \{\zeta_{0,p^2}, \zeta_{0,1/p}\} \supsetneq \partial\varphi(U') = \{\zeta_{0,p^2}\} ,$$

which illustrates Proposition 9.15 in this case.

What is happening is that φ maps $\mathcal{D}(0,p)^-$ onto $\mathcal{D}(0,p^2)^-$ with multiplicity 2 and maps each of the subdiscs $\mathcal{D}(0,\frac{1}{p})$ and $\mathcal{D}(1,\frac{1}{p})$ onto $\mathcal{D}(0,\frac{1}{p})$ with multiplicity 1. So the restriction of φ to the open annulus $U' = \mathcal{D}(0,p)^- \backslash \mathcal{D}(0,\frac{1}{p})$ still surjects onto $\mathcal{D}(0,p^2)^-$, and φ does not have a well-defined local degree on U'.

For each $a \in \mathbb{P}^1_{\text{Berk}}$, we claim that the φ-saturated open neighborhoods are cofinal in the set of all neighborhoods of a. To see this, let W be any neighborhood of a, and let $V \subset W$ be a simple domain containing a, none of whose boundary points belongs to the finite set $\varphi^{-1}(\{\varphi(a)\})$. Put $S = \varphi^{-1}(\varphi(\partial V))$; then S is a finite set of points of type II or III. Let U be the connected component of $\mathbb{P}^1_{\text{Berk}} \backslash S$ containing a. Clearly $U \subset V \subset W$. Let $\tilde{U}$ be the connected component of $\varphi^{-1}(\varphi(U))$ containing U; we will show $\tilde{U} = U$. Since the boundary points of U belong to S, the boundary points of $\varphi(U)$ belong to $\varphi(S)$, and hence the boundary points of $\tilde{U}$ belong to $\varphi^{-1}(\varphi(S)) = S$. Since U is a connected component of $\mathbb{P}^1_{\text{Berk}} \backslash S$, necessarily $\tilde{U} = U$. Thus U is a φ-saturated open neighborhood of a contained in W.

We will say that a neighborhood W of a is φ-*small relative to* a if

$$W \cap \varphi^{-1}(\{\varphi(a)\}) = \{a\} .$$

Since $\varphi^{-1}(\{\varphi(a)\})$ is finite, φ-small neighborhoods of a certainly exist. Clearly any subneighborhood of a φ-small neighborhood is itself φ-small.

If U is a φ-saturated, φ-small neighborhood of a and if $b = \varphi(a)$, the local degree $\deg_U(\varphi)$ of φ on U coincides with the multiplicity $m_\varphi(a) =$

$m_{\varphi,b}(a)$ of φ at a. We therefore obtain a purely topological interpretation of multiplicities, which explains their intrinsic meaning:

COROLLARY 9.17 (Topological characterization of multiplicities). *For each $a \in \mathbb{P}^1_{\text{Berk}}$ and for each sufficiently small φ-saturated neighborhood U of a, the multiplicity $m_\varphi(a)$ is equal to the number of preimages in U (counting multiplicities) of each $\alpha \in \varphi(U) \cap \mathbb{P}^1(K)$.*

REMARK 9.18. If K has characteristic zero, then φ has only finitely many critical values in $\mathbb{P}^1(K)$, and therefore all but finitely many type I points of $\varphi(U)$ will have exactly $m_\varphi(a)$ distinct preimages in U for any φ-small, φ-saturated neighborhood U of a.

Corollary 9.17 shows that our multiplicities coincide with the multiplicities defined by Rivera-Letelier (see for example [**81**]).

We will now give formulas relating multiplicities to other quantities associated to $\varphi(T)$ and note some consequences of these formulas.

Take $a \in \mathbb{P}^1_{\text{Berk}}$, and put $b = \varphi(a)$. By Proposition 2.15, b has the same type (I, II, III, or IV) as a. Recall from §B.6 that we can associate to a the set T_a of formal *tangent vectors* (or tangent directions) $\vec{v}$ at a, whose elements are the equivalence classes of paths starting at a which share a common initial segment. If a is of type I or IV, then T_a has one element; if a is of type III, then T_a has two elements. If a is of type II, then T_a has infinitely many elements, which are in one-to-one correspondence with the elements of $\mathbb{P}^1(k)$, where k is the residue field of K. If $\Gamma \subset \mathbb{P}^1_{\text{Berk}}$ is a subgraph and $a \in \Gamma$, we identify the set $T_a(\Gamma)$ of tangent directions to a in Γ (see §3.1) with a subset of T_a.

The tangent directions $\vec{v}$ at a are in one-to-one correspondence with the components of $\mathbb{P}^1_{\text{Berk}} \backslash \{a\}$. Recall that we write $\mathcal{B}_a(\vec{v})^-$ for the component of $\mathbb{P}^1_{\text{Berk}} \backslash \{a\}$ corresponding to $\vec{v}$.

Given a tangent direction $\vec{v}$ at a, consider the component $V = \mathcal{B}_a(\vec{v})^-$. Since $\varphi^{-1}(\{b\})$ is finite, for each $x \in V$ sufficiently near a we have $\varphi(x) \neq \varphi(a)$. If $x_1, x_2 \in V$ and $\varphi(x_1), \varphi(x_2)$ belong to different components of $\mathbb{P}^1_{\text{Berk}} \backslash \{b\}$, then since φ is continuous and $\mathbb{P}^1_{\text{Berk}}$ is uniquely path-connected, the segment $[x_1, x_2]$ must contain a point a' with $\varphi(a') = b$. Hence there is a unique component W of $\mathbb{P}^1_{\text{Berk}} \backslash \{b\}$ such that $\varphi(x) \in W$ for all $x \in V$ sufficiently near a. If $\vec{w}$ is the tangent direction at b corresponding to W, we will write $\varphi_*(\vec{v}) = \vec{w}$. We can think of the map $\varphi_* : T_a \to T_{\varphi(a)}$ as the "projectivized derivative" of φ at a.

Let $[a, c]$ be a segment with initial tangent vector $\vec{v}$, and as usual let $\rho(x, y)$ denote the path distance metric on $\mathbb{H}_{\text{Berk}}$. If $a \in \mathbb{H}_{\text{Berk}}$, we define

$$(9.14) \qquad r_\varphi(a, \vec{v}) \;=\; d_{\vec{v}}(\rho(\varphi(x), b))(a) \;=\; \lim_{\substack{x \to a \\ x \in [a,c]}} \frac{\rho(\varphi(x), b)}{\rho(x, a)} ,$$

while if $a \in \mathbb{P}^1(K)$, we define

$$(9.15) \quad r_\varphi(a, \vec{v}) = d_{\vec{v}}(\rho(\varphi(x), b))(a) = \lim_{\substack{x,y \to a \\ x,y \in [a,c] \\ x \neq y}} \frac{\rho(\varphi(x), \varphi(y))}{\rho(x, y)},$$

provided the limits (9.14), (9.15) exist. In the following proposition, we will see that the limits exist and are integers in the range $1 \leq m \leq \deg(\varphi)$; we call $r_\varphi(a, \vec{v})$ the *rate of repulsion* of φ at a, in the direction $\vec{v}$. Since any two segments with the same initial point and tangent direction share an initial segment, they depend only on φ, a, and $\vec{v}$ (and not on c).

The following formula, due to Rivera-Letelier [**84**, Proposition 3.1], expresses $m_\varphi(a)$ as a sum of the rates of repulsion of φ in the tangent directions at a that map to a given direction at b.

THEOREM 9.19 (Repulsion formula). *Let $\varphi(T) \in K(T)$ be a nonconstant rational function. Let $a \in \mathbb{P}^1_{\text{Berk}}$, and put $b = \varphi(a)$. Then:*

(A) *For each tangent vector $\vec{v}$ at a, the rate of repulsion $r_\varphi(a, \vec{v})$ is a well-defined integer m in the range $1 \leq m \leq \deg(\varphi)$.*

(B) *For each tangent direction $\vec{w}$ at b,*

$$(9.16) \quad m_\varphi(a) = \sum_{\substack{\vec{v} \in T_a \\ \varphi_*(\vec{v}) = \vec{w}}} r_\varphi(a, \vec{v}) .$$

PROOF. Let $\vec{w}$ be a tangent direction at b, and let $W = \mathcal{B}_b(\vec{w})^-$ be the corresponding component of $\mathbb{P}^1_{\text{Berk}} \setminus \{b\}$.

Fix $\zeta \in W$, and consider the function $h_{b,\zeta}(x) = \log_v(\delta(\varphi(x), b)_\zeta)$. By Proposition 9.5, $h_{b,\zeta} \in \text{BDV}(\mathbb{P}^1_{\text{Berk}})$. Let $a_1, \ldots, a_d, \zeta_1, \ldots, \zeta_d \in \mathbb{P}^1_{\text{Berk}}$ be the solutions to $\varphi(a_i) = b$ and $\varphi(\zeta_i) = \zeta$, listed with their multiplicities; let the a_i be labeled so that $a = a_1$. Put

$$(9.17) \quad f(x) = \sum_{i=1}^d \log_v(\delta(x, a_i)_{\zeta_i}) .$$

By Example 5.19, $f(x) \in \text{BDV}(\mathbb{P}^1_{\text{Berk}})$ and

$$(9.18) \quad \Delta_{\mathbb{P}^1_{\text{Berk}}}(f) = \sum_{i=1}^d \delta_{\zeta_i} - \sum_{i=1}^d \delta_{a_i} = \Delta_{\mathbb{P}^1_{\text{Berk}}}(h_{\beta,\zeta}) .$$

Both $h_{b,\zeta}$ and f are continuous and belong to $\text{BDV}(\mathbb{P}^1_{\text{Berk}})$. Since their difference is a function whose Laplacian is identically 0, Lemma 5.24 shows that $h_{\beta,\zeta}(x) = f(x) + C$, for some constant C.

First suppose $a \in \mathbb{H}_{\text{Berk}}$, and let Γ be a finite subgraph of $\mathbb{P}^1_{\text{Berk}}$ which contains the points a_i and is large enough that the retractions of distinct ζ_j to Γ are different from each other and from the a_i. Since the restriction to Γ of each term $\log_v(\delta(x, a_i)_{\zeta_i})$ in (9.17) belongs to $\text{CPA}(\Gamma)$, the restriction of $h_{b,\zeta}(x) = f(x) + C$ to Γ belongs to $\text{CPA}(\Gamma)$. Moreover, since each term

$\log_v(\delta(x,a_i)_{\zeta_i})$ has slope 1, 0, or -1 in every tangent direction at every point of $\mathbb{H}_{\text{Berk}}$, formula (9.17) shows that the slopes of $h_{b,\zeta}|_\Gamma$ are integers in the interval $[-d, d]$.

The retraction map r_Γ fixes the a_i, and
$$\Delta_\Gamma(h_{b,\zeta}) = (r_\Gamma)_*(\Delta_{\mathbb{P}^1_{\text{Berk}}}(h_{b,\zeta})),$$
so we can compute $m_{\varphi,b}(a)$ using the formula
$$(9.19) \quad m_{\varphi,b}(a) = \omega_{\varphi,b}(\{a\}) = -\Delta_\Gamma(h_{b,\zeta})(\{a\}) = \sum_{\vec{v} \in T_a(\Gamma)} (d_{\vec{v}} h_{b,\zeta})(a),$$
where the sum ranges over all tangent directions at a in the finite metrized graph Γ.

We claim that the set of tangent directions $\vec{v}$ in T_a for which $\varphi_*(\vec{v}) = \vec{w}$ is contained in the set of directions at a in Γ. To see this, note that $h_{b,\zeta}$ is constant on branches of $\mathbb{P}^1_{\text{Berk}}$ off Γ at a, because f has this property. Thus $d_{\vec{v}}(h_{b,\zeta})(a) = 0$ for each $\vec{v} \in T_a$ which is not a tangent direction in Γ. On the other hand, if $\vec{v}$ is a direction a for which $\varphi_*(\vec{v}) = \vec{w}$ and if V is the component of $\mathbb{P}^1_{\text{Berk}} \backslash \{a\}$ corresponding to $\vec{v}$, then $\varphi(x)$ lies in the component of $\mathbb{P}^1_{\text{Berk}} \backslash \{b\}$ containing ζ for each $x \in V$ sufficiently near a, so if $\varphi(x) \vee_\zeta b$ is the point where the paths $[\varphi(x), \zeta]$ and $[b, \zeta]$ first meet, then
$$(9.20) \quad \delta(\varphi(x), b)_\zeta = \text{diam}_\zeta(\varphi(x) \vee_\zeta b) > \text{diam}_\zeta(b) = \delta(\varphi(a), b)_\zeta.$$
Since
$$h_{b,\zeta}(x) = \log_v(\delta(\varphi(x), b)_\zeta)$$
and in any case $d_{\vec{v}}(h_{b,\zeta})(a)$ is an integer, it follows that $d_{\vec{v}}(h_{b,\zeta})(a) > 0$. Thus $\vec{v}$ must be a direction at a in Γ.

In fact (9.20) shows that $d_{\vec{v}}(h_{b,\zeta})(a) > 0$ if and only if $\varphi_*(\vec{v}) = \vec{w}$. We have just seen that if $\varphi_*(\vec{v}) = \vec{w}$, then $d_{\vec{v}}(h_{b,\zeta})(a) > 0$. Conversely, if $d_{\vec{v}}(h_{b,\zeta})(a) > 0$, then by the previous paragraph $\vec{v}$ must be a tangent direction to a in Γ; let $[a, c]$ be the edge of Γ emanating from a in the direction $\vec{v}$. For $x \in [a, c]$ sufficiently near a, we have $h_{b,\zeta}(x) > h_{b,\zeta}(a)$, so by (9.20), $\delta(\varphi(x), b)_\zeta > \text{diam}_\zeta(b)$. This means that $\varphi(x)$ belongs to the component $W = \mathcal{B}_b(\vec{w})^-$ of $\mathbb{P}^1_{\text{Berk}} \backslash \{b\}$ containing ζ, so $\varphi_*(\vec{v}) = \vec{w}$.

Let $\vec{v}$ be a direction with $\varphi_*(\vec{v}) = \vec{w}$, and let e be the edge of Γ with initial point a and direction $\vec{v}$. Put $m = d_{\vec{v}}(h_{b,\zeta})(a)$. By the discussion above, m is an integer in the range $[1, d]$. Since $h_{b,\zeta}|_\Gamma \in \text{CPA}(\Gamma)$, (9.20) shows that $\delta(\varphi(x), b)_\zeta$ is monotonically increasing on an initial segment $[a, c'] \subset e$. It follows that $\varphi([a, c'])$ is a segment with initial point $b = \varphi(a)$ and direction $\vec{w}$. Since any two paths with the same initial point and direction must share an initial segment, after shrinking $[a, c']$, we can assume that $\varphi([a, c']) \subset [b, \zeta]$. Hence for each $x \in [a, c']$ we have $\varphi(x) \vee_\zeta b = \varphi(x)$, which means that
$$h_{b,\zeta}(x) = \log_v(\text{diam}_\zeta(\varphi(x))) = \rho(\varphi(x), b) + \text{diam}_\zeta(b).$$
Thus
$$(9.21) \quad m = d_{\vec{v}}(h_{b,\zeta})(a) = d_{\vec{v}}(\rho(\varphi(x), b))(a),$$

and (9.16) follows from (9.19). By the linearity of the path distance on segments, we have $\varphi([a,c']) = [\varphi(a), \varphi(c')]$ and for all $x, y \in [a, c']$,
$$\rho(\varphi(x), \varphi(y)) = m \cdot \rho(x, y) \ .$$

The arguments above show that $r_\varphi(a, \vec{v}) = d_{\vec{v}}(\rho(\varphi(x), b))(a)$ is a well-defined integer in the range $1 \leq m \leq d$ for each $\vec{v}$ with $\varphi_*(\vec{v}) = \vec{w}$, proving the existence of the limit (9.14). Fixing $\vec{v}$ and putting $\vec{w} = \varphi_*(\vec{v})$, we see that it is well-defined for each $\vec{v}$. As noted before, the quantity $r_\varphi(a, \vec{v})$ depends only on φ, a, and $\vec{v}$, since any two paths with initial tangent direction $\vec{v}$ share an initial segment. This establishes the theorem in this case.

Next suppose $a \in \mathbb{P}^1(K)$. Then there is a neighborhood V of a and a point $c \in \mathbb{H}_{\text{Berk}}^{\mathbb{R}}$ such that the restriction of $h_{b,\zeta}$ to $[a, c]$ has constant slope $m = m_\varphi(a)$ on $[a, c] \cap V$. By an argument similar to the one above, for all $x, y \in [a, c] \cap V$ with $x \neq y$,
$$\frac{\rho(\varphi(x), \varphi(y))}{\rho(x, y)} = m \ .$$

This clearly shows the existence of the limit (9.15), and since there is only one tangent vector $\vec{v}$ at a, it also yields the assertions in the theorem. $\square$

COROLLARY 9.20. *Let $\varphi(T) \in K(T)$ be a nonconstant rational function. Suppose $a \in \mathbb{H}_{\text{Berk}}$, and put $b = \varphi(a)$. Then φ_* maps the tangent space T_a surjectively onto the tangent space T_b, and there are at most $d = \deg(\varphi)$ tangent directions at a which are taken to a given tangent direction at b.*

If a is of type I, III, or IV, then φ induces a one-to-one correspondence between the tangent directions at a and b, and $m_\varphi(a) = r_\varphi(a, \vec{v})$ for each tangent direction $\vec{v}$ at a.[6]

PROOF. The first assertion is immediate from Theorem 9.19; the second follows because there are exactly two tangent directions at a point of type III and exactly one at a point of type I or IV. $\square$

We also record for future use the following fact[7], established in the proof of Theorem 9.19:

COROLLARY 9.21. *Let $\varphi(T) \in K(T)$ be a nonconstant rational function. Take $a \in \mathbb{P}^1_{\text{Berk}}$, and put $b = \varphi(a)$. Let $\vec{v}$ be a tangent direction at a. Then there is a segment $[a, c']$ with initial point a, representing the tangent direction $\vec{v}$, for which $\varphi([a, c'])$ is the segment $[\varphi(a), \varphi(c')]$ and such that for all $x, y \in [a, c']$,*
$$\rho(\varphi(x), \varphi(y)) = r_\varphi(a, \vec{v}) \cdot \rho(x, y) \ .$$

Fix $a \in \mathbb{P}^1_{\text{Berk}}$ and let $\vec{v}$ be a tangent vector at a. We will now define an integer $m_\varphi(a, \vec{v})$ called the *multiplicity of φ at a in the direction $\vec{v}$*, and we will show that $m_\varphi(a, \vec{v}) = r_\varphi(a, \vec{v})$.

---

[6]When $K = \mathbb{C}_p$, Corollary 9.20 and Theorem 9.22 summarize results in [**82**, §4].
[7]When $K = \mathbb{C}_p$, Corollary 9.21 corresponds to [**82**, Proposition 4.6].

Let $[a, c]$ be a segment with initial tangent direction $\vec{v}$. Parametrize $[a, c]$ by a segment $[0, L]$ and for each $0 \le t \le L$, let b_t be the point of $[a, c]$ corresponding to t. Recall that a *(generalized open) annulus* in $\mathbb{P}^1_{\text{Berk}}$ is a domain with exactly two boundary points. If $t > 0$, let $\mathcal{A}_t(\vec{v}) = r^{-1}_{[a, b_t]}((a, b_t))$ be the annulus with boundary points a, b_t. By Corollary 9.11, the image $\varphi(\mathcal{A}_t(\vec{v}))$ has 0, 1, or 2 boundary points and hence is either $\mathbb{P}^1_{\text{Berk}}$, an open disc, or another generalized open annulus. We claim that $\varphi(\mathcal{A}_t(\vec{v}))$ is an annulus for all sufficiently small t. If not, then since the image is decreasing with t, for sufficiently small t the image would either be $\mathbb{P}^1_{\text{Berk}}$ or a fixed disc. But this would violate the continuity of φ, since for each neighborhood V of a, there is a $\delta > 0$ such that $\mathcal{A}_t(\vec{v}) \subseteq V$ when $0 < t < \delta$.

If $\varphi(\mathcal{A}_t(\vec{v}))$ is an annulus, then $\mathcal{A}_t(\vec{v})$ is φ-saturated by Proposition 9.15(B), and the local degree $\deg_{\mathcal{A}_t(\vec{v})}(\varphi)$ is a well-defined positive integer. It is nonincreasing as $t \to 0$, so it eventually stabilizes at an integer $1 \le m \le d$. We define this integer to be $m_\varphi(a, \vec{v})$. It is independent of the segment $[a, c]$, since any two segments with the same initial point and tangent vector must share an initial segment.

THEOREM 9.22. *Let $\varphi(T) \in K(T)$ be a nonconstant rational function. Fix $a \in \mathbb{P}^1_{\text{Berk}}$, and put $b = \varphi(a)$. Then:*[8]

(A) *For each tangent direction $\vec{v}$ at a,*

(9.22) $$m_\varphi(a, \vec{v}) \;=\; r_\varphi(a, \vec{v}) \;.$$

(B) *For each tangent direction $\vec{v}$ at a, there is a segment $[a, c'']$ with initial point a and tangent direction $\vec{v}$, such that $m_\varphi(z) = m_\varphi(a, \vec{v})$ for each $z \in (a, c'')$ and such that if $\mathcal{A}$ is the annulus with boundary points a, c'', then $\varphi(\mathcal{A})$ is the annulus with boundary points $\varphi(a)$, $\varphi(c'')$.*

(C) *(Directional multiplicity formula) For each tangent vector $\vec{w}$ at b,*

(9.23) $$m_\varphi(a) \;=\; \sum_{\substack{\vec{v} \in T_a \\ \varphi_*(\vec{v}) = \vec{w}}} m_\varphi(a, \vec{v}) \;.$$

PROOF. Fix a φ-saturated neighborhood V of a with $m_\varphi(V) = m_\varphi(a)$.

Let $\vec{w}$ be a tangent vector at b, and consider the tangent vectors $\vec{v}_i$ at a for which $\varphi_*(\vec{v}_i) = \vec{w}$. By Corollary 9.21, for each such $\vec{v}_i$ we can find a segment $[a, c'_i]$ with initial tangent vector $\vec{v}_i$, such that

(9.24) $$\rho(\varphi(x), \varphi(y)) \;=\; m_i \cdot \rho(x, y)$$

for all $x, y \in [a, c'_i]$, where $m_i = r_\varphi(a, \vec{v}_i)$. After replacing $[a, c'_i]$ with an initial subsegment of itself if necessary, we can assume that the annulus $\mathcal{A}_i$ with boundary points a, c'_i is φ-saturated, contained in V, and small enough that $m_\varphi(\mathcal{A}_i) = m_\varphi(a, \vec{v}_i)$.

---

[8]See footnote 6 above.

We claim that for each $z \in (a, c_i')$,

(9.25) $$m_\varphi(z) \geq d_{\vec{v}_i}(\rho(\varphi(x), b))(a) .$$

Indeed, fix such a z and put $b_z = \varphi(z)$. Let $\vec{v}$ be one of the tangent vectors to z in $[a, c_i']$. Equation (9.24) shows that $d_{\vec{v}}(\rho(\varphi(x), b_z))(z) = m_i$. Hence (9.25) follows from Theorem 9.19.

For each $\vec{v}_i$, (9.24) also shows that $\varphi([a, c_i'])$ is a segment $[b, q_i]$ with initial tangent vector $\vec{w}$. All of these segments share a common initial segment $[b, q]$. For each i, let c_i'' be the point in $[a, c_i']$ for which $\varphi(c_i'') = q$. Fix $y \in (b, q)$, and let $z_i \in [a, c_i'']$ be the point with $\varphi(z_i) = y$.

The equation $\varphi(x) = y$ has $\deg_{\mathcal{A}_i}(\varphi) \geq m_\varphi(z_i)$ solutions in $\mathcal{A}_i$, and it has $\deg_V(\varphi) = m_\varphi(a)$ solutions in V. Since the $\mathcal{A}_i$ are pairwise disjoint and contained in V, Theorem 9.19 and the discussion above show that

(9.26) $$\sum_i d_{\vec{v}_i}(\rho(\varphi(x), b))(a) = m_\varphi(a) \geq \sum_i \deg_{\mathcal{A}_i}(\varphi)$$
$$\geq \sum_i m_\varphi(z_i) \geq \sum_i d_{\vec{v}_i}(\rho(\varphi(x), b))(a) .$$

Hence equality must hold throughout.

Formula (9.26) implies, first, that for each $\vec{v}_i$,
$$m_\varphi(a, \vec{v}_i) = \deg_{\mathcal{A}_i}(\varphi) = d_{\vec{v}_i}(\rho(\varphi(x), b))(a) = r_\varphi(a, \vec{v}_i) .$$

It also shows that $m_\varphi(z_i) = m_\varphi(a, \vec{v}_i)$ for each z_i. As y runs over (b, q), the z_i run over (a, c_i''). Hence for each $\vec{v}_i$, we conclude that $m_\varphi(z) = m_\varphi(a, \vec{v}_i)$ for all $z \in (a, c_i'')$. Finally, formula (9.26) shows that $m_\varphi(a) = \sum_i m_\varphi(a, \vec{v}_i)$.

Letting $\vec{w}$ vary over all tangent directions at b, we obtain the theorem. □

Next suppose a is of type II or III, and put $b = \varphi(a)$. Then $\mathbb{P}^1_{\text{Berk}} \setminus \{b\}$ has at least two components. Let $\beta, \zeta \in \mathbb{P}^1_{\text{Berk}}$ be points in different components of $\mathbb{P}^1_{\text{Berk}} \setminus \{b\}$, and let $\beta_1, \ldots, \beta_d, \zeta_1, \ldots, \zeta_d$ be the solutions to $\varphi(\beta_i) = \beta$, $\varphi(\zeta_i) = \zeta$, listed with their multiplicities. For each component V of $\mathbb{P}^1_{\text{Berk}} \setminus \{a\}$, let $N_\beta(V)$ be the number of β_i in V (counted with multiplicities), and let $N_\zeta(V)$ be the number of ζ_i in V (counted with multiplicities). Put $N^+_{\zeta,\beta}(V) = \max(0, N_\zeta(V) - N_\beta(V))$.

The following formula relates multiplicities to global solvability of equations:

THEOREM 9.23. *Let $\varphi(T) \in K(T)$ be a nonconstant rational function. Suppose $a \in \mathbb{P}^1_{\text{Berk}}$ is of type II or type III, and put $\varphi(a) = b$. Let $\zeta, \beta \in \mathbb{P}^1_{\text{Berk}}$ belong to different components of $\mathbb{P}^1_{\text{Berk}} \setminus \{b\}$. Then:*

(A) (Imbalance formula)

(9.27) $$m_\varphi(a) = \sum_{\substack{\text{components } V \\ \text{of } \mathbb{P}^1_{\text{Berk}} \setminus \{a\}}} N^+_{\zeta,\beta}(V) .$$

(B) (Directional imbalance formula) *If $\vec{w}$ is the tangent direction at b for which $\zeta \in \mathcal{B}_b(\vec{w})^-$, then for each tangent direction $\vec{v}$ at a with $\varphi_*(\vec{v}) = \vec{w}$,*

$$(9.28) \quad m_\varphi(a, \vec{v}) \;=\; N_\zeta(\mathcal{B}_a(\vec{v})^-) - N_\beta(\mathcal{B}_a(\vec{v})^-) \;=\; N^+_{\zeta,\beta}(\mathcal{B}_a(\vec{v})^-) \;.$$

REMARK 9.24. If we take $\zeta, \beta \in \mathbb{P}^1(K)$, then (9.27) gives another characterization of $m_\varphi(a)$ for points of type II or III.

PROOF. The proof is based on further study of the formula (9.3) for $h_{b,\infty}(x)$ obtained in Lemma 9.2.

Since b lies on the path $[\beta, \zeta]$, if $r = \delta(b,\beta)_\zeta$, then by a mild generalization of formula (9.3), the function $h_{b,\zeta}(x) = \log_v(\delta(\varphi(x), b)_\zeta)$ satisfies

$$(9.29) \quad h_{b,\zeta}(x) \;=\; \max(h_{\beta,\zeta}(x), \log_v(r)) \;.$$

Since $\varphi(a) = b$, we have $h_{b,\zeta}(a) = \log_v(r)$. Let $\beta_1, \ldots, \beta_d, \zeta_1, \ldots, \zeta_d \in \mathbb{P}^1_{\text{Berk}}$ be the solutions to $\varphi(\beta_i) = \beta$ and $\varphi(\zeta_i) = \zeta$, listed with multiplicities. Put

$$(9.30) \quad f(x) \;=\; \sum_{i=1}^d \log_v(\delta(x,\beta_i)_{\zeta_i}) \;.$$

By the same argument as in the proof of Theorem 9.19, $h_{\beta,\zeta}(x) = f(x) + C$ for some constant C.

As in the proof of Theorem 9.19, there is a finite subgraph Γ of $\mathbb{P}^1_{\text{Berk}}$ containing $\varphi^{-1}(\{b\})$ for which the retractions of the ζ_i to Γ are distinct from the points $a_i \in \varphi^{-1}(\{b\})$. In particular, $a \in \Gamma$.

Since the restriction of each term in (9.30) to Γ belongs to CPA(Γ), the restrictions of $h_{\beta,\zeta}$ and $h_{b,\zeta}$ to Γ belong to CPA(Γ). As in the proof of Theorem 9.19,

$$m_\varphi(a) \;=\; \omega_{\varphi,b}(\{a\}) \;=\; -\Delta_\Gamma(h_{b,\zeta})(\{a\}) \;=\; \sum_{\vec{v} \in T_a(\Gamma)} d_{\vec{v}} h_{b,\zeta}(a) \;,$$

where $\Delta_{\mathbb{P}^1_{\text{Berk}}}(h_{b,\zeta}) = \omega_{\varphi,\zeta} - \omega_{\varphi,b}$, $d_{\vec{v}} h_{b,\zeta}(a) \geq 0$ for all $\vec{v}$, and $d_{\vec{v}} h_{b,\zeta}(a) > 0$ if and only if $\varphi_*(\vec{v}) = \vec{w}$.

Given a tangent direction $\vec{v}$ at a in Γ, let $V = \mathcal{B}_a(\vec{v})^-$ be the corresponding component of $\mathbb{P}^1_{\text{Berk}} \backslash \{a\}$. The directional derivative $d_{\vec{v}} h_{\beta,\zeta}(a)$ is the sum of the terms $d_{\vec{v}} \log_v(\delta(x,\beta_i)_{\zeta_i})(a)$ for $i = 1, \ldots, d$. Such a term is $+1$ if $\zeta_i \in V$, -1 if $\beta_i \in V$, and 0 if neither β_i nor ζ_i belongs to V. It follows that $d_{\vec{v}} h_{\beta,\zeta}(a) = N_\zeta(V) - N_\beta(V)$. Since $h_{b,\zeta}(x) = \max(h_{\beta,\zeta}(x), \log_v(r))$ and $h_{b,\zeta}(a) = \log_v(r)$, if $d_{\vec{v}} h_{\beta,\zeta}(a) > 0$, then $d_{\vec{v}} h_{b,\zeta}(a) = d_{\vec{v}} h_{\beta,\zeta}(a)$; otherwise $d_{\vec{v}} h_{b,\zeta}(a) = 0$. Thus $d_{\vec{v}} h_{b,\zeta}(a) = N^+_{\zeta,\beta}(V)$, and if $\varphi_*(\vec{v}) = \vec{w}$, then $d_{\vec{v}} h_{b,\zeta}(a) = N_\zeta(V) - N_\beta(V)$. Comparing this with (9.21) and (9.22) gives (9.28).

For the components V of $\mathbb{P}^1_{\text{Berk}} \backslash \{a\}$ not meeting Γ, we automatically have $N_\beta(V) = N_\zeta(V) = 0$. Hence

$$(9.31) \quad m_\varphi(a) \;=\; \sum_{\vec{v} \in T_a} d_{\vec{v}} h_{b,\zeta}(a) \;=\; \sum_V N^+_{\zeta,\beta}(V) \;. \qquad \square$$

For points a of type II, formulas (9.27) and (9.28) have an algebraic interpretation. Assume first that $a = \zeta_{\text{Gauss}}$ and that $\varphi(\zeta_{\text{Gauss}}) = \zeta_{\text{Gauss}}$. In this case, Lemma 2.17 shows that φ has a well-defined and nonconstant reduction $\tilde{\varphi}$ in $\tilde{K}(T)$. Explicitly, the rational map $\varphi(T)$ can be written as a quotient of polynomials $P(T), Q(T) \in \mathcal{O}[T]$, having no common factors, such that at least one coefficient of each belongs to $\mathcal{O}^\times$, and $\tilde{\varphi}(T) \in \tilde{K}(T)$ is the rational function of degree $1 \leq \tilde{d} \leq d$ obtained by canceling common factors in the reductions $\tilde{P}(T), \tilde{Q}(T)$.

There is a one-to-one correspondence between tangent directions at ζ_{Gauss} and points $\alpha \in \mathbb{P}^1(\tilde{K})$. Let $\vec{v}_\alpha$ be the tangent direction corresponding to α, and let $m_{\tilde{\varphi}}(\alpha)$ be the (classical algebraic) multiplicity of $\tilde{\varphi}(T)$ at α.

The polynomials $\tilde{P}(T)$ and $\tilde{Q}(T)$ have a common factor exactly when $\varphi(T)$ has a zero and a pole belonging to the same component of V of $\mathbb{P}^1_{\text{Berk}} \backslash \{\zeta_{\text{Gauss}}\}$. If we take $\zeta = 0$ and $\beta = \infty$ in (9.27), then $\sum_V N^+_{0,\infty}(V)$ coincides with the degree of the positive part of the divisor of $\tilde{\varphi}(T)$, which is the same as the degree of $\tilde{\varphi}(T)$. A corresponding local assertion holds for each directional multiplicity $m_\varphi(\zeta_{\text{Gauss}}, \vec{v}_\alpha)$. Thus $m_\varphi(\zeta_{\text{Gauss}}) = \deg(\tilde{\varphi}(T))$, and for each $\alpha \in \mathbb{P}^1(\tilde{K})$,
$$m_\varphi(\zeta_{\text{Gauss}}, \vec{v}_\alpha) = m_{\tilde{\varphi}}(\alpha) \ .$$

Now assume that a is an arbitrary point of type II, and let $b = \varphi(a)$. Applying Corollary 2.13 to the map $\varphi : \mathbb{P}^1_{\text{Berk}} \to \mathbb{P}^1_{\text{Berk}}$, there exist $\psi, \eta \in \text{PGL}_2(K)$ such that $\psi(a) = \eta(b) = \zeta_{\text{Gauss}}$. Letting $\varphi' = \eta \circ \varphi \circ \psi^{-1}$, we have $\varphi'(\zeta_{\text{Gauss}}) = \zeta_{\text{Gauss}}$, so $\varphi'(T)$ has a well-defined nonconstant reduction $\tilde{\varphi}'(T)$, which we call the *reduction of φ with respect to ψ and η*. When there is no possibility of confusion, we will write $\tilde{\varphi}(T)$ instead of $\tilde{\varphi}'(T)$.

By the above discussion, $m_{\varphi'}(\zeta_{\text{Gauss}}) = \deg(\tilde{\varphi}'(T))$ and
$$m_{\varphi'}(\zeta_{\text{Gauss}}, \vec{v}_\alpha) = m_{\tilde{\varphi}'}(\alpha)$$
for each $\alpha \in \mathbb{P}^1(\tilde{K})$. On the other hand, it follows easily from Proposition 9.28(C) below that $m_{\varphi'}(\zeta_{\text{Gauss}}) = m_\varphi(a)$, and by a similar computation
$$m_{\varphi'}(\zeta_{\text{Gauss}}, \psi_* \vec{v}) = m_\varphi(a, \vec{v}) \ .$$

Thus we obtain the following result, originally proved by Rivera-Letelier [81, Proposition 3.3]:

COROLLARY 9.25 (Algebraic Reduction Formula). *Let $\varphi(T) \in K(T)$ be a nonconstant rational function, and let $a \in \mathbb{P}^1_{\text{Berk}}$ be a point of type II. Put $b = \varphi(a)$, choose $\psi, \eta \in \text{PGL}_2(K)$ such that $\psi(a) = \eta(b) = \zeta_{\text{Gauss}}$, and let $\tilde{\varphi}(T)$ be the reduction of $\varphi(T)$ relative to ψ and η. Then*

(9.32) $$m_\varphi(a) = \deg(\tilde{\varphi}(T)) \ ,$$

and for each $\alpha \in \mathbb{P}^1(\tilde{K})$, if $\vec{v}_\alpha \in T_a$ is the associated tangent direction under the bijection between T_a and $\mathbb{P}^1(\tilde{K})$ afforded by ψ_*, we have

(9.33) $$m_\varphi(a, \vec{v}_\alpha) = m_{\tilde{\varphi}}(\alpha) \ .$$

For purposes of reference, we summarize Theorems 9.19, 9.22, 9.23 and Corollaries 9.20 and 9.25 as follows:

THEOREM 9.26. *Let $\varphi(T) \in K(T)$ have degree $\deg(\varphi) \geq 2$, let $a \in \mathbb{P}^1_{\text{Berk}}$, and let $b = \varphi(a)$. If a is of type II, choose $\psi, \eta \in \text{PGL}_2(K)$ such that $\psi(a) = \eta(b) = \zeta_{\text{Gauss}}$. Then:*

(A) *For each $\vec{v} \in T_a$, $m_\varphi(a, \vec{v}) = r_\varphi(a, \vec{v})$. If a is of type II or III, if $\zeta \in \mathcal{B}_{\varphi(a)}(\varphi_*(\vec{v}))^-$, and if $\beta \in \mathcal{B}_{\varphi(a)}(\vec{w})^-$ for some $\vec{w} \in T_{\varphi_*(a)}$ with $\vec{w} \neq \varphi_*(\vec{v})$, then*

$$m_\varphi(a, \vec{v}) = N_\zeta(\mathcal{B}_a(\vec{v})^-) - N_\beta(\mathcal{B}_a(\vec{v})^-) = N^+_{\zeta, \beta}(\mathcal{B}_a(\vec{v})^-).$$

If a is of type II, if $\tilde{\varphi}(T)$ is the reduction of $\varphi(T)$ relative to ψ and η, and if $\psi_*(\vec{v}) = \vec{v}_\alpha$ with $\alpha \in \mathbb{P}^1(\tilde{K})$, then $m_\varphi(a, \vec{v}) = m_{\tilde{\varphi}}(\alpha)$.

(B) *For each $\vec{w} \in T_{\varphi(a)}$,*

$$m_\varphi(a) = \sum_{\varphi_*(\vec{v}) = \vec{w}} r_\varphi(a, \vec{v}) = \sum_{\varphi_*(\vec{v}) = \vec{w}} m_\varphi(a, \vec{v}).$$

If a is of type II or III and ζ and β belong to different components of $\mathbb{P}^1_{\text{Berk}} \setminus \{\varphi(a)\}$, then $m_\varphi(a) = \sum_V N^+_{\zeta, \beta}(V)$, where V runs over all components of $\mathbb{P}^1_{\text{Berk}} \setminus \{a\}$. If a is of type II and if $\tilde{\varphi}(T)$ is the reduction of $\varphi(T)$ relative to ψ and η, then $m_\varphi(a) = \deg(\tilde{\varphi})$. If a is of type I, III, or IV, then $m_\varphi(a) = m_\varphi(a, \vec{v})$ for each $\vec{v} \in T_a$.

Combining Corollary 9.25, Lemma 2.17, and Corollary 9.14, we obtain the following result due to Rivera-Letelier:

COROLLARY 9.27. *Let $\varphi(T) \in K(T)$ be a nonconstant rational function. Then:*

(A) *φ has a well-defined reduction $\tilde{\varphi}$ iff $\varphi(\zeta_{\text{Gauss}}) \in \mathcal{D}(0, 1)$.*

(B) *φ has nonconstant reduction (i.e., $\deg(\tilde{\varphi}) \geq 1$) iff $\varphi(\zeta_{\text{Gauss}}) = \zeta_{\text{Gauss}}$.*

(C) *φ has good reduction (i.e., $\deg(\tilde{\varphi}) = \deg(\varphi)$) iff $\varphi^{-1}(\zeta_{\text{Gauss}}) = \{\zeta_{\text{Gauss}}\}$.*

We close this section by studying $m_\varphi(x)$ as x varies. The following properties of $m_\varphi(x)$ are due to Favre and Rivera-Letelier [**48**]:

PROPOSITION 9.28. *Let $\varphi(T) \in K(T)$ be a nonconstant rational function. Then $m_\varphi(x)$ takes integer values in the range $1 \leq m \leq \deg(\varphi)$, and*

(A) *m_φ is upper semicontinuous;*

(B) *if $m_\varphi(a) = 1$, then φ is locally injective at a, and if $\text{char}(K) = 0$, the converse holds;*

(C) *if $\psi(T) \in K(T)$ is another nonconstant rational function, then*

(9.34) $$m_{\psi \circ \varphi}(x) = m_\varphi(x) \cdot m_\psi(\varphi(x)).$$

PROOF. Theorem 9.8 shows that $m_\varphi(x)$ takes integer values in the range $1 \leq m \leq \deg(\varphi)$. Assertions (A), (B), and (C) all follow from the existence of the local degree (Corollary 9.14) and the topological characterization of multiplicities (Corollary 9.17).

For (A), fix $a \in \mathbb{P}^1_{\text{Berk}}$ and let V be a φ-small, φ-saturated neighborhood of a. Then V is a component of $\varphi^{-1}(\varphi(V))$. By Corollary 9.14, for each $x \in V$,
$$\sum_{\substack{y \in V \\ \varphi(y) = \varphi(x)}} m_\varphi(y) = m_\varphi(a) .$$
In particular, $m_\varphi(x) \leq m_\varphi(a)$ for each $x \in V$. Thus m_φ is upper semi-continuous.

For (B), note that if $m_\varphi(a) = 1$, then by Corollary 9.14, for any φ-small, φ-saturated neighborhood V of a we have $m_\varphi(x) = m_\varphi(a) = 1$ for all $x \in V$, which means that for each $x \in V$ there are no other points x' in U with $\varphi(x') = \varphi(x)$. Conversely, if φ is one-to-one on a neighborhood V of a and $\text{char}(K) = 0$, then the type I points in V all have multiplicity 1. By Corollary 9.17, $m_\varphi(a) = 1$ for all $a \in V$.

For (C), fix a, let V be a $\psi \circ \varphi$-small, $\psi \circ \varphi$-saturated neighborhood of a, and put $W = \varphi(V)$. Then W is a ψ-small, ψ-saturated neighborhood of $\varphi(a)$, and $\psi(W) = \psi \circ \varphi(V)$. By Corollary 9.17, each $\alpha \in \psi(W) \cap \mathbb{P}^1(K)$ has precisely $m_\psi(\varphi(a))$ preimages in W (counted with multiplicities), and each of these has $m_\varphi(a)$ preimages in V (counted with multiplicities). Hence (9.34) follows from another application of Corollary 9.17 and the multiplicativity of the usual algebraic multiplicities. $\square$

Define the "ramification function" $R_\varphi : \mathbb{P}^1_{\text{Berk}} \to \mathbb{Z}_{\geq 0}$ by
$$(9.35) \qquad R_\varphi(x) = m_\varphi(x) - 1 .$$

COROLLARY 9.29. *Let $\varphi(T) \in K(T)$ be a nonconstant rational function. The ramification function $R_\varphi(x)$ is upper semicontinuous and takes integer values in the range $0 \leq n \leq \deg(\varphi) - 1$. Furthermore, if $\text{char}(K) = 0$, then $R_\varphi(a) = 0$ if and only if φ is locally injective at a.*

EXAMPLE 9.30. Suppose K has characteristic 0. Then it is well known that there are only finitely many points in $\mathbb{P}^1(K)$ where $\varphi(T)$ is ramified. However, the situation is very different on Berkovich space.

Consider $\varphi(T) = T^2$. For a point $x \in \mathbb{A}^1_{\text{Berk}}$ corresponding to a disc $D(a,r)$ (i.e., for a point of type II or III), φ is ramified at x (that is, $R_\varphi(x) \geq 1$) if and only if there are distinct points $a_1, a_2 \in D(a,r)$ for which $\varphi(a_1) = \varphi(a_2)$, that is, $a_1^2 = a_2^2$, so $a_1 = -a_2$. It follows that x is ramified if and only if $r \geq |a_1 - (-a_1)| = |2a_1|$ for some $a_1 \in B(a,r)$. If $r \geq |2a_1|$ for some $a_1 \in D(a,r)$, then $r \geq |a - a_1| \geq |2| \cdot |a - a_1| = |2a - 2a_1|$, so $r \geq |2a|$ by the ultrametric inequality. Conversely, if $r \geq |2a|$, then $-a \in D(a,r)$. Thus φ is ramified at x if and only if $r \geq |2a|$.

If the residue characteristic of K is not 2, then $|2a| = |a|$, so φ is ramified at the point corresponding to $D(a,r)$ if and only if $r \geq |a|$, or, equivalently, if and only if $D(a,r) = D(0,r)$. These are the points corresponding to the interior of the path $[0, \infty]$ in $\mathbb{P}^1_{\text{Berk}}$. One sees easily that φ is not ramified at any points of type IV. On the other hand, the classical ramification points of φ in $\mathbb{P}^1(K)$ are 0 and ∞. Thus, the ramification locus of $\varphi(T)$ is $[0, \infty]$.

If the residue characteristic of K is 2, then $|2a| = |a|/2$ (assuming $|x|$ is normalized so that it extends the usual absolute value on $\mathbb{Q}_2$). In this case, φ is ramified at a point x corresponding to a disc $D(a,r)$ if and only if $r \geq |a|/2$. A little thought shows that this is a much larger set than $[0, \infty]$: for each $a \in K^\times$, it contains the line of discs $\{D(a,t) : |a|/2 \leq t \leq |a|\}$, which is a path leading off $[0, \infty]$. The union of these paths gives an infinitely branched dendritic structure whose interior is open in the path metric topology. If one views $\mathbb{P}^1_{\text{Berk}}$ as $\mathbb{A}^1_{\text{Berk}} \cup \{\infty\}$, the ramification locus of φ can be visualized as an "inverted Christmas tree".

9.2. Images of segments and finite graphs

In this section, we investigate the effect of a nonconstant rational map $\varphi \in K(T)$ on paths and finite subgraphs in $\mathbb{H}_{\text{Berk}}$.

We first establish the assertions concerning the path distance topology on $\varphi(\Gamma)$ needed to complete the proof of Lemma 9.4 in the construction of multiplicities. These are contained in Proposition 9.32 and will be a consequence of continuity properties of φ on $\mathbb{H}_{\text{Berk}}$ which are of independent interest. (The reader will note that nothing beyond Lemma 9.1 is used in the proof of Proposition 9.32.) Then, using the theory of multiplicities, we establish the stronger result, due to Rivera-Letelier [**82**, §4], that the image $\varphi(\Gamma)$ of a finite subgraph Γ of $\mathbb{P}^1_{\text{Berk}}$ is itself a finite graph. We also use these ideas to show that φ is Lipschitz continuous with respect to the metric $d(x,y)$ discussed in §2.7, and we prove Rivera-Letelier's "Incompressibility Lemma", which shows that $\varphi : \mathbb{H}_{\text{Berk}} \to \mathbb{H}_{\text{Berk}}$ is an open map for the path distance topology.

Consider $\mathbb{H}_{\text{Berk}}$, equipped with the path distance topology. Let $\rho(x,y)$ be the path distance metric; it is finite for all $x, y \in \mathbb{H}_{\text{Berk}}$.

LEMMA 9.31. *Let $\varphi(T) \in K(T)$ be a nonconstant rational function of degree $d = \deg(\varphi)$. Then φ acts on $\mathbb{H}_{\text{Berk}}$ and for all $a, c \in \mathbb{H}_{\text{Berk}}$,*

$$\rho(\varphi(a), \varphi(c)) \leq d \cdot \rho(a,c) .$$

In particular, φ is Lipschitz continuous on $\mathbb{H}_{\text{Berk}}$ relative to the path distance metric.[9]

---

[9] When $K = \mathbb{C}_p$, Lemma 9.31 is proved in [**82**, Corollary 4.7].

PROOF. We know that φ takes $\mathbb{H}_{\text{Berk}}$ to itself by Proposition 2.15. For the assertion about path distances, we use ideas from §9.1.

As $\rho(x,y)$ is continuous on $\mathbb{H}_{\text{Berk}}$, it suffices to show that $d(\varphi(a), \varphi(c)) \leq d \cdot \rho(a,c)$ when a, c are of type II or III. Fix $a, c \in \mathbb{H}_{\text{Berk}}^{\mathbb{R}}$ and consider the finite subgraph $\Gamma = [\varphi(a), \varphi(c)] \subset \mathbb{H}_{\text{Berk}}$. Fix $\beta, \zeta \in \mathbb{P}^1(K)$ with $r_\Gamma(\beta) = \varphi(a)$, $r_\Gamma(\zeta) = \varphi(c)$, so $\varphi(a)$ and $\varphi(c)$ lie on the path $[\beta, \zeta]$. It follows that

$$(9.36) \qquad \rho(\varphi(a), \varphi(c)) = |\log_v(\delta(\varphi(a), \beta)_\zeta) - \log_v(\delta(\varphi(c), \beta)_\zeta)| .$$

However, by Proposition 5.28 and Lemma 9.1, $h_{\beta,\zeta}(x) = \log_v(\delta(\varphi(x), \beta)_\zeta)$ is also given by

$$h_{\beta,\zeta}(x) = \sum_{i=1}^d \log_v(\delta(x, \beta_i)_{\zeta_i}) + C$$

where the β_i and ζ_i are the preimages of β and ζ, listed with multiplicities. Inserting this in (9.36) gives

$$(9.37) \qquad \rho(\varphi(a), \varphi(c)) \leq \sum_{i=1}^d |\log_v(\delta(a, \beta_i)_{\zeta_i}) - \log_v(\delta(c, \beta_i)_{\zeta_i})| .$$

For each $\xi \in \mathbb{H}_{\text{Berk}}$ the representation (4.26) of the Hsia kernel in terms of $j_\xi(z,w)$ gives

$$\begin{aligned}
\log_v(\delta(a, \beta_i)_{\zeta_i}) &- \log_v(\delta(c, \beta_i)_{\zeta_i}) \\
&= (-j_\xi(a, \beta_i) + j_\xi(a, \zeta_i) + j_\xi(\beta_i, \zeta_i) + C_\xi) \\
&\quad -(-j_\xi(c, \beta_i) + j_\xi(c, \zeta_i) + j_\xi(\beta_i, \zeta_i) + C_\xi) \\
&= j_\xi(a, \zeta_i) - j_\xi(a, \beta_i) + j_\xi(c, \beta_i) - j_\xi(c, \zeta_i) .
\end{aligned}$$

Now take $\xi = a$; then $j_a(a, \beta_i) = j_a(a, \zeta_i) = 0$, so

$$|\log_v(\delta(a, \beta_i)_{\zeta_i}) - \log_v(\delta(c, \beta_i)_{\zeta_i})| = |j_a(c, \beta_i) - j_a(c, \zeta_i)| .$$

Here $j_a(c, \zeta_i) = \rho(a, t)$, where $t \in [a, c]$ is the point where the paths $[c, a]$ and $[\zeta_i, a]$ first meet. Likewise, $j_a(c, \beta_i) = \rho(a, u)$, where $u \in [a, c]$ is the point where the paths $[c, a]$ and $[\beta_i, a]$ first meet. Since $t, u \in [a, c]$,

$$|\log_v(\delta(a, \beta_i)_{\zeta_i}) - \log_v(\delta(c, \beta_i)_{\zeta_i})| = \rho(t, u) \leq \rho(a, c) .$$

Inserting this in (9.37) gives the result. $\square$

The bound in the lemma is sharp, as shown by the example $\varphi(T) = T^d$, which takes the point $x \in [0, \infty]$ corresponding to $D(0, r)$ (under Berkovich's classification theorem) to the point $\varphi(x) \in [0, \infty]$ corresponding to $D(0, r^d)$. Using the definition of the path length, we see that $\rho(\varphi(a), \varphi(b)) = d \cdot \rho(a, b)$ for each $a, b \in (0, \infty)$.

Write $\ell(\Gamma)$ for the total path length of a finite subgraph $\Gamma \subset \mathbb{P}^1_{\text{Berk}}$. For any connected subset $Z \subset \mathbb{P}^1_{\text{Berk}}$, define its length to be

$$\ell(Z) = \sup_{\substack{\text{finite subgraphs} \\ \Gamma \subseteq Z}} \ell(\Gamma) .$$

9.2. IMAGES OF SEGMENTS AND FINITE GRAPHS

PROPOSITION 9.32. *Let $\varphi(T) \in K(T)$ be a nonconstant rational function of degree $d = \deg(\varphi)$, and let $\Gamma \subset \mathbb{P}^1_{\text{Berk}}$ be a finite subgraph. Then:*

(A) *$\varphi(\Gamma)$ is compact and connected for the path distance topology on $\mathbb{H}_{\text{Berk}}$.*
(B) *$\ell(\varphi(\Gamma)) \leq d \cdot \ell(\Gamma)$.*
(C) *The path distance topology on $\varphi(\Gamma)$ coincides with the induced topology from the Berkovich topology on $\mathbb{P}^1_{\text{Berk}}$.*

PROOF. Assertion (A) is trivial, since φ is continuous for the path distance topology, and Γ is compact and connected in the path distance topology.

For (B), let $\tilde{\Gamma}$ be an arbitrary finite subgraph of $\varphi(\Gamma)$. Choose a finite set of points $X = \{x_1, \ldots, x_m\} \subset \Gamma$ such that all the endpoints and branch points of Γ are among the x_i and such that for each endpoint or branch point p of $\tilde{\Gamma}$, there is an $x_i \in X$ with $\varphi(x_i) = p$. Let A be the set of pairs (i,j) such that x_i and x_j are adjacent in Γ, and put

$$\Gamma' = \bigcup_{(i,j) \in A} [\varphi(x_i), \varphi(x_j)] \, .$$

(Note that we are taking the segments $[\varphi(x_i), \varphi(x_j)]$, not the path images $\varphi([x_i, x_j])$.) Then Γ' is a finite subgraph of $\varphi(\Gamma)$ which contains all the endpoints and branch points of $\tilde{\Gamma}$ and hence contains $\tilde{\Gamma}$. For each $(i,j) \in A$, Lemma 9.31 gives $\rho(\varphi(x_i), \varphi(x_j)) \leq d \cdot \rho(x_i, x_j)$. Hence

$$\ell(\tilde{\Gamma}) \leq \ell(\Gamma') \leq \sum_{(i,j) \in A} \rho(\varphi(x_i), \varphi(x_j)) \leq d \cdot \ell(\Gamma) \, .$$

Taking the sup over all finite subgraphs shows that $\ell(\varphi(\Gamma)) \leq d \cdot \ell(\Gamma)$.

For (C), it suffices to note that $\varphi(\Gamma)$ is compact in the topology induced from the Berkovich (weak) topology on $\mathbb{P}^1_{\text{Berk}}$, since Γ is compact and φ is continuous in the weak topology. By part (A), $\varphi(\Gamma)$ is compact in the path distance (strong) topology. Hence part (C) follows from the fact that any continuous bijection between compact Hausdorff topological spaces is a homeomorphism (apply this to the identity map from $\varphi(\Gamma)$ to itself, giving the first copy the strong topology and the second copy the weak topology). □

In the results above, we have not used the theory of multiplicities, so the proof of Lemma 9.4 is complete and our construction of multiplicities is valid.

We will now use the theory of multiplicities to strengthen Proposition 9.32 and show that $\varphi(\Gamma)$, which is compact, connected, and has finite length, but could conceivably have infinitely many branch points, is in fact a finite subgraph.

For this, we need a lemma originally proved by Rivera-Letelier ([**81**, Proposition 3.1]) using the theory of Newton polygons.

LEMMA 9.33. *Let $\varphi(T) \in K(T)$ be a nonconstant rational function of degree $d = \deg(\varphi)$, and let $[a, c] \subset \mathbb{P}^1_{\text{Berk}}$ be a segment with initial tangent vector $\vec{v}$. Put $m = m_\varphi(a, \vec{v})$. Then $1 \leq m \leq d$, and for each $a' \in (a, c]$ sufficiently near a:*

(A) *$\varphi([a, a'])$ is the segment $[\varphi(a), \varphi(a')]$ with initial tangent vector $\varphi_*(\vec{v})$.*
(B) *$m_\varphi(x) = m$ for each $x \in (a, a')$.*
(C) *$\rho(\varphi(x), \varphi(y)) = m \cdot \rho(x, y)$ for all $x, y \in [a, a']$.*
(D) *If $\mathcal{A}$ is the annulus with boundary points a, a', then $\varphi(\mathcal{A})$ is the annulus with boundary points $\varphi(a)$, $\varphi(a')$.*

PROOF. Note that (A) follows from (C). Since any two segments with the same initial point and tangent direction have a common initial segment, the result follows by combining Corollary 9.21 and Theorem 9.22. □

REMARK 9.34. In Theorem 9.46 below, we will see that (A), (B), and (C) all follow from (D). For this reason, we will call a segment $[a, a']$ with the properties in Lemma 9.33 *annular*.

The following important result is due to Rivera-Letelier ([**84**, Corollaries 4.7 and 4.8]):

THEOREM 9.35. *Let $\varphi(T) \in K(T)$ be a nonconstant rational function of degree $d = \deg(\varphi)$, and let $[a, c] \subset \mathbb{P}^1_{\text{Berk}}$ be a segment. Then:*

(A) *There is a finite partition $\{a_0, \ldots, a_n\}$ of $[a, c]$, with $a_0 = a$, $a_n = c$ and with $a_1, \ldots, a_{n-1}$ of type II, such that for each $i = 1, \ldots, n$ the segment $[a_{i-1}, a_i]$ is annular. In particular, for each $i = 1, \ldots, n$ there is an integer $1 \leq m_i \leq d$ such that*
  (1) *$\varphi([a_{i-1}, a_i])$ is the segment $[\varphi(a_{i-1}), \varphi(a_i)]$,*
  (2) *$m_\varphi(x) = m_i$ is constant on (a_{i-1}, a_i),*
  (3) *$\rho(\varphi(x), \varphi(y)) = m_i \cdot \rho(x, y)$ for all $x, y \in [a_{i-1}, a_i]$.*
(B) *If $a, c \in \mathbb{H}_{\text{Berk}}$, then*

$$\rho(\varphi(a), \varphi(c)) \leq \int_{[a,c]} m_\varphi(t)\, dt \, , \tag{9.38}$$

where the integral is taken relative to the path distance. Equality holds if and only if φ is injective on $[a, c]$.

PROOF. We need only prove (A), since (B) follows trivially from (A). Using a compactness argument, we will first establish (A) without the condition that $a_1, \ldots, a_{n-1}$ are of type II. We will then eliminate points of type III from the partition.

To each $x \in [a, c]$, we associate a (relatively) open neighborhood of x in $[a, c]$ as follows. If $x = a$, then by Lemma 9.33 there is an annular segment $[a, a''] \subset [a, c]$; associate the half-open interval $[a, a'')$ to a. If x is an interior point of $[a, c]$, then by Lemma 9.33 (applied first to the reversed interval $[x, a]$ and then to the interval $[a, c]$), there are points $x' \in [a, x)$ and $x'' \in (x, c]$

such that each of the segments $[x', x]$ and $[x, x']$ is annular. Associate to x the open interval (x', x''). Likewise, if $x = c$, there is an annular segment $[c', c]$; associate to c the half-open interval $(c', c]$.

By compactness, a finite number of these intervals cover $[a, c]$. Let $\{a_0, \ldots, a_n\}$ be the corresponding endpoints. Since the intersection or union of two overlapping annuli is again an annulus, we obtain a partition with the required properties, except that some of $a_1, \ldots, a_{n-1}$ might be of type III.

If a_i is of type III, let $\vec{v}_1$ and $\vec{v}_2$ be the two tangent directions at a_i. By Corollary 9.20,
$$m_\varphi(a_i, \vec{v}_1) = m_\varphi(a_i) = m_\varphi(a_i, \vec{v}_2),$$
so $m_{i-1} = m_i$. The linearity of the path distance on segments shows that properties (1)–(3) hold for $[a_{i-1}, a_{i+1}]$. Furthermore, the union of a_i and the annuli associated to (a_{i-1}, a_i) and (a_i, a_{i+1}) is the annulus associated to (a_{i-1}, a_{i+1}). Its image under φ is the annulus associated to $(\varphi(a_{i-1}), \varphi(a_{i+1}))$, since $\varphi(a_i)$ is of type III, and by Corollary 9.20 the tangent vectors $\varphi_*(\vec{v}_1)$, $\varphi_*(\vec{v}_2)$ are distinct. Hence a_i can be eliminated from the partition. $\square$

As an easy consequence, we obtain the promised strengthening of Proposition 9.32:

COROLLARY 9.36. *Let $\varphi(T) \in K(T)$ be a nonconstant rational function, and let $\Gamma \subset \mathbb{H}_{\text{Berk}}$ be a finite subgraph. Then the image $\varphi(\Gamma)$ is itself a finite subgraph.*

In the rest of this section we will give applications of the results above. We first show that the action of φ on $\mathbb{P}^1_{\text{Berk}}$ is Lipschitz continuous relative to the metric $d(x, y)$ from §2.7. This was used without proof in [**47, 48**].

PROPOSITION 9.37. *Let $\varphi(T) \in K(T)$ have degree $d \geq 1$. Then there is a constant $M > 0$, depending only on φ, such that for all $x, y \in \mathbb{P}^1_{\text{Berk}}$,*
$$(9.39) \qquad d(\varphi(x), \varphi(y)) \leq M \cdot d(x, y).$$

PROOF. Fix a coordinate system on $\mathbb{P}^1/K$, and let $\|x, y\|$ be the associated spherical metric. Let $\alpha_1, \ldots, \alpha_m \in \mathbb{P}^1(K)$ be the zeros and poles of $\varphi(T)$. Fix a number R_0 satisfying
$$0 < R_0 < \min_{i \neq j} \|\alpha_i, \alpha_j\|.$$
In particular, $R_0 < 1$. Consider the rational functions $\varphi_1(T) = \varphi(T)$, $\varphi_2(T) = \varphi(1/T)$, $\varphi_3(T) = 1/\varphi(T)$, and $\varphi_4(T) = 1/\varphi(1/T)$. We can write each $\varphi_i(T)$ as a quotient of polynomials $P_i(T), Q_i(T) \in K(T)$ having no common roots, such that $Q_i(T)$ is monic. Let B be the maximum of the absolute values of the coefficients of the $P_i(T)$, $Q_i(T)$. Fix a number $0 < R_1 \leq R_0$ small enough that $BR_1/R_0^d < 1$.

Recall from (2.21), (4.33) that

$$d(x,y) = (\text{diam}(x \vee y) - \text{diam}(x)) + (\text{diam}(x \vee y) - \text{diam}(y)),$$
$$\rho(x,y) = -\log_v\left(\frac{\text{diam}(x \vee y)}{\text{diam}(x)}\right) - \log_v\left(\frac{\text{diam}(x \vee y)}{\text{diam}(y)}\right),$$

where $x \vee y$ is the point where the paths $[x, \zeta_{\text{Gauss}}]$ and $[y, \zeta_{\text{Gauss}}]$ first meet. For each $0 < t \leq u \leq 1$ we have $-\log_v(u/t) \geq (u-t)/\log(q_v)$, so

(9.40) $$d(x,y) \leq \log(q_v) \cdot \rho(x,y).$$

On the other hand, if $R > 0$ and $C(R) = 1/(R\log(q_v))$ is the Lipschitz constant for $-\log_v(t)$ on the interval $[R, 1]$, then for x, y with $\text{diam}(x), \text{diam}(y) \geq R$ we have

(9.41) $$\rho(x,y) \leq C(R) \cdot d(x,y).$$

To prove the proposition, we will consider three cases.

First suppose that there is a ball $\mathcal{B}(a, R_1) = \{z \in \mathbb{P}^1_{\text{Berk}} : \|z, a\| \leq R_1\}$ such that $x, y \in \mathcal{B}(a, R_1)$, where $a \in \mathbb{P}^1(K)$ and $\|z, a\| = \delta(x,y)_{\zeta_{\text{Gauss}}}$ is the extension of the spherical distance to $\mathbb{P}^1_{\text{Berk}}$ discussed in §4.3. Let $B(a, R_1) = \{z \in \mathbb{P}^1(K) : \|z, a\| \leq R_1\}$ be the associated classical ball.

Since $R_1 \leq R_0 < 1$, after replacing $\varphi(T)$ by $\varphi_2(T)$, $\varphi_3(T)$, or $\varphi_4(T)$ if necessary, we can assume that $B(a, R_1) = D(a, R_1) \subseteq D(a, R_0) \subseteq D(0, 1)$ and that $\varphi(T)$ has no poles in $D(a, R_0)$. Since points of type II are dense in $\mathbb{P}^1_{\text{Berk}}$ for the strong topology, we can assume without loss of generality that x and y are of type II. Noting that $d(x,y) = d(x, x \vee y) + d(x \vee y, y)$ and replacing y by $x \vee y$ if necessary, we can assume that y lies on the path $[x, \zeta_{\text{Gauss}}]$. Thus x and y correspond to discs $D(b, r)$, $D(b, R)$ under Berkovich's classification, and $0 < r \leq R \leq R_1$. Finally, after changing the center of $D(a, R_1)$ if necessary, we can assume that $b = a$. Since $\varphi(T)$ has no poles in $D(a, R_0)$, Corollary A.18 shows that the images $\varphi(D(a, r))$ and $\varphi(D(a, R))$ are discs, with $\varphi(D(a, r)) \subseteq \varphi(D(a, R))$. We will now estimate their radii.

Write $\varphi(T) = P(T)/Q(T)$ where $P(T), Q(T) \in K[T]$ have no common factors and $Q(T)$ is monic. Expand

$$P(T) = b_0 + b_1(T-a) + \cdots + b_m(T-a)^m,$$
$$Q(T) = c_0 + c_1(T-a) + \cdots + (T-a)^n.$$

Since $|a| \leq 1$ and $P(T) = P_i(T)$, $Q(T) = Q_i(T)$ for some i, we must have $\max(|b_j|) \leq B$. Since $Q(T)$ has no zeros in $D(a, R_0)$, we also have $|c_0| > R_0^d$. By Corollary A.18, for each $0 < s \leq R_1$, the image of $D(a, s)$ under $\varphi(T)$ is a disc of radius $f(s)$ where $f(s) = \max_{j \geq 1} |b_j|s^j/|c_0|$. Here $\max_{j \geq 1} |b_j|s^j \leq Bs \leq BR_1$ and $|c_0| > R_0^d$, so $f(s) < BR_1/R_0^d < 1$.

Thus $\varphi(D(a, r))$ has radius $f(r)$ and $\varphi(D(a, R))$ has radius $f(R)$, and both discs are contained in $D(0, 1)$. It follows that

$$d(\varphi(x), \varphi(y)) = \text{diam}(\varphi(y)) - \text{diam}(\varphi(x)) = f(R) - f(r).$$

Let J be an index with $|b_J|R^J = \max_{j\geq 1}|b_j|R^j$. Then $|b_J|r^J \leq \max_{j\geq 1}|b_j|r^j$ and
$$R^J - r^J = (R-r)\Big(\sum_{k+\ell=J-1} R^k r^\ell\Big) \leq (R-r)\cdot d\,,$$
giving
$$f(R) - f(r) \leq (|b_J|R^J - |b_J|r^J)/|c_0| \leq B\cdot(R-r)\cdot d/R_0^d\,.$$
Here $d(x,y) = R - r$, so $d(\varphi(x),\varphi(y)) \leq M_1\cdot d(x,y)$, where $M_1 = Bd/R_0^d$.

Second, suppose that $\text{diam}(x), \text{diam}(y) \geq R_1^2$. Then by Lemma 9.31, (9.40), and (9.41),
$$\begin{aligned} d(\varphi(x),\varphi(y)) &< \log(q_v)\cdot\rho(\varphi(x),\varphi(y)) \\ &\leq \log(q_v)\cdot d\cdot\rho(x,y) \leq \log(q_v)\cdot d\cdot C(R_1^2)\cdot d(x,y)\,. \end{aligned}$$
Put $M_2 = \log(q_v)\cdot d\cdot C(R_1^2) = d/R_1^2$.

Third, suppose that neither case above applies. Then $\text{diam}(x\vee y) > R_1$, since x and y do not belong to any ball $\mathcal{B}(a,R_1)$. In addition, either $\text{diam}(x) < R_1^2$ or $\text{diam}(y) < R_1^2$. It follows that $d(x,y) > R_1 - R_1^2$. On the other hand, trivially $d(\varphi(x),\varphi(y)) \leq 2$. Thus $d(\varphi(x),\varphi(y)) \leq M_3\cdot d(x,y)$, where $M_3 = 2/(R_1 - R_1^2)$.

Taking $M = \max(M_1, M_2, M_3)$, we obtain the proposition. $\square$

Next we will show that $\varphi : \mathbb{H}_{\text{Berk}} \to \mathbb{H}_{\text{Berk}}$ is an open map in the strong topology. For this, we will need a lemma, which is also used later in §10.7.

LEMMA 9.38. *Let $\varphi(T) \in K(T)$ be nonconstant, and let $U \subseteq \mathbb{P}^1_{\text{Berk}}$ be a φ-saturated domain. Take $a \in U \cap \mathbb{H}_{\text{Berk}}$, and put $b = \varphi(a)$. Let $b' \in \varphi(U) \cap \mathbb{H}_{\text{Berk}}$ be a point with $b' \neq b$. Let $\vec{w}$ be the tangent vector to $[b,b']$ at b, and let $\vec{v} \in T_a$ be a tangent vector with $\varphi_*(\vec{v}) = \vec{w}$.*

Then there is a segment $[a,a'] \subset U$, with initial tangent vector $\vec{v}$, such that φ maps $[a,a']$ homeomorphically onto $[b,b']$. Furthermore, $\rho(a,a') \leq \rho(b,b')$.

PROOF. Put $a_0 = a$, $b_0 = b$. We will construct $[a,a']$ by transfinite induction (see [**63**, §IV.7]).

Suppose that for some ordinal ω, we have constructed a_ω, b_ω with $a_\omega \in U$, $b_\omega = \varphi(a_\omega) \in [b,b']$, and if $\omega > 0$, then $[a,a_\omega] \subset U$ has initial tangent vector $\vec{v}$ and φ maps $[a,a_\omega]$ homeomorphically onto $[b,b_\omega] \subseteq [b,b']$. If $b_\omega = b'$, then, taking $a' = a_\omega$, we are done.

Otherwise, we proceed as follows.

If $\omega = 0$, put $\vec{v}_0 = \vec{v}$ and $\vec{w}_0 = \vec{w}$. If $\omega > 0$, let $\vec{w}_\omega$ be the tangent vector to b_ω in $[b,b']$ in the direction of b', and let $\vec{w}_\omega^0$ be the tangent vector to b_ω in $[b,b']$ in the direction of b. By Corollary 9.20 there is a tangent vector $\vec{v}_\omega$ at a_ω for which $\varphi_*(\vec{v}_\omega) = \vec{w}_\omega$. Let $\vec{v}_\omega^0$ be the tangent vector to $[a,a_\omega]$ at a_ω in the direction of a. Since $\vec{w}_\omega \neq \vec{w}_\omega^0$ and $\varphi_*(\vec{v}_\omega^0) = \vec{w}_\omega^0$, it follows that $\vec{v}_\omega \neq \vec{v}_\omega^0$.

Let $\omega + 1$ be the successor to ω. By Corollary 9.21, there is a segment $[a_\omega, a_{\omega+1}]$ with initial tangent vector $\vec{v}_\omega$ such that φ maps $[a_\omega, a_{\omega+1}]$ homeomorphically onto a segment $[b_\omega, b_{\omega+1}]$ with initial tangent vector $\vec{w}_\omega$. Since $[b_\omega, b_{\omega+1}]$ and $[b_\omega, b']$ both have initial tangent vector $\vec{w}_\omega$, they share a common initial segment. By moving $a_{\omega+1}$ closer to a_ω, if necessary, we can assume that $[b_\omega, b_{\omega+1}] \subseteq [b_\omega, b']$ and that $[a_\omega, a_{\omega+1}] \subset U$. As $\vec{v}_\omega$ and $\vec{v}_\omega^0$ are distinct, $[a, a_{\omega+1}] = [a, a_\omega] \cup [a_\omega, a_{\omega+1}]$. Similarly $[b, b_{\omega+1}] = [b, b_\omega] \cup [b_\omega, b_{\omega+1}]$, and by induction φ maps $[a, a_{\omega+1}]$ homeomorphically onto $[b, b_{\omega+1}]$.

Next suppose that λ is a limit ordinal and that for each $\omega < \lambda$ we have constructed a_ω, b_ω with the properties above. Since $[b, b']$ has finite total length, the sequence $\{b_\omega\}_{\omega < \lambda}$ is a Cauchy net relative to the path distance metric. For each $\omega, \tau < \lambda$, Theorem 9.35 shows that $\rho(a_\omega, a_\tau) \leq \rho(b_\omega, b_\tau)$. Thus $\{a_\omega\}_{\omega < \lambda}$ is also a Cauchy net relative to the path distance metric. By Proposition 2.29, it has a unique limit point $a_\lambda \in \mathbb{H}_{\text{Berk}}$. Put $b_\lambda = \varphi(a_\lambda)$.

Clearly $a_\lambda \in \overline{U}$. We claim that $a_\lambda \in U$. Suppose to the contrary that $a_\lambda \in \partial U$. Then $b_\lambda \in \partial(\varphi(U))$, since U is φ-saturated. However, by continuity $b_\lambda = \lim_\omega b_\omega$. Since each $b_\omega \in [b, b']$ and $[b, b']$ is closed, it follows that $b_\lambda \in [b, b']$. Furthermore, $[b, b'] \subset \varphi(U)$, since $\varphi(U)$ is connected. In particular, $b_\lambda \in \varphi(U)$ which is open (Corollary 9.10), which means that $b_\lambda \notin \partial(\varphi(U))$. Thus $a_\lambda \in U$. Since the $[a, a_\omega]$ for $\omega < \lambda$ are strictly increasing and since φ maps each $[a, a_\omega]$ homeomorphically onto $[b, b_\omega]$, it maps $[a, a_\lambda]$ homeomorphically onto $[b, b_\lambda]$.

Since there are ordinals of arbitrarily high cardinality, but the map $\omega \mapsto b_\omega \in [b, b']$ is injective on the ordinals with $b_\omega \neq b'$, it must be that $b_\omega = b'$ for some ω. Finally, Theorem 9.35(B) shows that $\rho(a, a') \leq \rho(b, b')$. □

Given $a \in \mathbb{H}_{\text{Berk}}$ and $r > 0$, write
$$\widehat{\mathcal{B}}(a, r)^- = \{x \in \mathbb{H}_{\text{Berk}} : \rho(x, a) < r\}.$$

The following important result from [**84**], due to Rivera-Letelier, shows that φ is an open map on $\mathbb{H}_{\text{Berk}}$ relative to the path distance topology.

COROLLARY 9.39 (Incompressibility Lemma). *Let $\varphi(T) \in K(T)$ be nonconstant, and let $a \in \mathbb{H}_{\text{Berk}}$. Then for each $r > 0$,*
$$\widehat{\mathcal{B}}(\varphi(a), r)^- \subseteq \varphi(\widehat{\mathcal{B}}(a, r)^-).$$

PROOF. Put $b = \varphi(a)$, and take $U = \mathbb{P}^1_{\text{Berk}}$ in Lemma 9.38. By that lemma, for each b' with $\rho(b, b') < r$, there is an a' with $\rho(a, a') < r$ such that $\varphi(a') = b'$. □

9.3. Images of discs and annuli

In this section, we will study the images of discs and annuli under a rational map $\varphi(T) \in K(T)$. We give new proofs for the key mapping properties of discs and annuli used by Rivera-Letelier in his theory of dynamics on $\mathbb{P}^1_{\text{Berk}}$, and we characterize when discs and annuli are φ-saturated, in terms of multiplicities.

Recall that by a *generalized open Berkovich disc* we mean a domain $\mathcal{B} \subset \mathbb{P}^1_{\text{Berk}}$ with exactly one boundary point ζ (Definition 2.25). Thus, $\mathcal{B}$ is a connected component of $\mathbb{P}^1_{\text{Berk}} \backslash \{\zeta\}$. Explicitly, if $\vec{v}$ is the tangent direction corresponding to $\mathcal{B}$, then $\mathcal{B} = \mathcal{B}_\zeta(\vec{v})^-$. If $r = \text{diam}_\zeta(\zeta)$, then $\mathcal{B} = \mathcal{B}(\alpha, r)_\zeta$ for each $\alpha \in \mathcal{B}$. Here r is finite if $\zeta \in \mathbb{H}_{\text{Berk}}$ and $r = \infty$ if $\zeta \in \mathbb{P}^1(K)$. We say that $\mathcal{B}$ is *capped* (or is simply an open Berkovich disc) if ζ is of type II or III and that $\mathcal{B}$ is *strict* if ζ is of type II.

PROPOSITION 9.40. *Let $\varphi(T) \in K(T)$ be a nonconstant rational function. If $\mathcal{B} \subset \mathbb{P}^1_{\text{Berk}}$ is a generalized open Berkovich disc, then $\varphi(\mathcal{B})$ is either another generalized open Berkovich disc or it is all of $\mathbb{P}^1_{\text{Berk}}$.*

Furthermore, $\mathcal{B}$ is φ-saturated iff $\varphi(\mathcal{B})$ is a generalized open Berkovich disc. If $\mathcal{B}$ is φ-saturated, then $\varphi(\mathcal{B})$ is capped (resp. strict) if and only if $\mathcal{B}$ is capped (resp. strict).

PROOF. By Corollary 9.11, for any domain V, $\partial(\varphi(V)) \subseteq \varphi(\partial V)$, and by Proposition 9.15, V is φ-saturated if and only if $\varphi(\partial V) = \partial(\varphi(V))$. Hence $\varphi(\mathcal{B})$ has 0 or 1 boundary points, and $\mathcal{B}$ is φ-saturated if and only if $\varphi(\mathcal{B})$ is a disc. If $\varphi(\mathcal{B})$ is a disc, then by Proposition 2.15 its boundary point has the same type as the boundary point of $\mathcal{B}$. □

The following fundamental fact is due to Rivera-Letelier ([**81**, Lemma 2.1]):

PROPOSITION 9.41. *Let $\varphi(T) \in K(T)$ be a nonconstant rational function of degree $d = \deg(\varphi)$. Let $\mathcal{B}$ be a generalized open Berkovich disc with boundary point ζ, and let $\vec{v}$ be the tangent vector at ζ which points into $\mathcal{B}$. Put $\xi = \varphi(\zeta)$, and put $\vec{w} = \varphi_*(\vec{v})$. Then $\varphi(\mathcal{B})$ always contains the generalized open Berkovich disc $\mathcal{B}' = \mathcal{B}_\xi(\vec{w})^-$, and either $\varphi(\mathcal{B}) = \mathcal{B}'$ or $\varphi(\mathcal{B}) = \mathbb{P}^1_{\text{Berk}}$. Put $m = m_\varphi(\zeta, \vec{v})$.*

(A) *If $\varphi(\mathcal{B}) = \mathcal{B}'$, then for each $y \in \mathcal{B}'$ there are $\deg_\mathcal{B}(\varphi) = m$ solutions to $\varphi(x) = y$ in $\mathcal{B}$ (counted with multiplicities).*

(B) *If $\varphi(\mathcal{B}) = \mathbb{P}^1_{\text{Berk}}$, there is an integer $N \geq m$ (with $N > m$ if $\mathcal{B}$ is capped), such that for each $y \in \mathcal{B}'$, there are N solutions to $\varphi(x) = y$ in $\mathcal{B}$ (counted with multiplicities), and for each $y \in \mathbb{P}^1_{\text{Berk}} \backslash \overline{\mathcal{B}'}$ there are $N - m$ solutions to $\varphi(x) = y$ in $\mathcal{B}$ (counted with multiplicities).*

PROOF. Part (A) follows from Corollary 9.14 and the fact that $\mathcal{B}$ is φ-saturated.

Part (B) is trivial when ζ is of type I or IV: if we take $N = d$, then the assertions about points in $\mathcal{B}'$ are trivial, and the assertions about points in $\mathbb{P}^1_{\text{Berk}} \backslash \overline{\mathcal{B}'}$ are vacuous. Suppose ζ is of type II or III. Fix $\tau \in \mathcal{B}'$, and let $N = N_\tau(\mathcal{B})$ be the number of solutions to $\varphi(x) = \tau$ in $\mathcal{B}$, counted with multiplicities. Let $\beta \in \mathbb{P}^1_{\text{Berk}} \backslash \overline{\mathcal{B}'}$ be arbitrary. By Theorem 9.23(B),

(9.42) $$N_\tau(\mathcal{B}) - N_\beta(\mathcal{B}) = m_\varphi(\zeta, \vec{v}) = m,$$

so there are $N - m$ solutions to $\varphi(x) = \beta$ in $\mathcal{B}$. Fixing β, let τ vary over $\mathcal{B}'$. By (9.42), for each τ there are N solutions to $\varphi(x) = \tau$ in $\mathcal{B}$. □

The following theorem illustrates the strong influence that multiplicities have on the local mapping properties of φ.

THEOREM 9.42. *Let $\varphi(T) \in K(T)$ be a nonconstant rational function, and let $\mathcal{B}$ be a generalized open Berkovich disc with boundary point ζ. Then $\varphi(\mathcal{B})$ is a generalized open Berkovich disc if and only if for each $c \in \mathcal{B}$, the multiplicity function $m_\varphi(x)$ is nonincreasing on $[\zeta, c]$.*

For the proof, we will need a lemma:

LEMMA 9.43. *Let $a \in \mathbb{H}_{\text{Berk}}$, and suppose $\vec{v}$ is a tangent direction at a for which $m_\varphi(a, \vec{v}) < m_\varphi(a)$. Then $\varphi(\mathbb{P}^1_{\text{Berk}} \backslash \mathcal{B}_a(\vec{v})^-) = \mathbb{P}^1_{\text{Berk}}$.*

PROOF. Put $b = \varphi(a)$. By Corollary 9.20, φ maps the tangent directions at a surjectively onto the tangent directions at b. Since $m_\varphi(a, \vec{v}) < m_\varphi(a)$, it follows from Theorem 9.22 that there is a tangent vector $\vec{v}' \neq \vec{v}$ at a for which $\varphi_*(\vec{v}') = \varphi_*(\vec{v})$. Hence φ induces a surjection from the tangent vectors at a, distinct from $\vec{v}$, onto the tangent vectors at b.

By Proposition 9.41, if $\vec{v}$ is a tangent vector at a and $\varphi_*(\vec{v}) = \vec{w}$, then $\varphi(\mathcal{B}_a(\vec{v})^-)$ contains $\mathcal{B}_b(\vec{w})^-$. Since $\varphi(a) = b$ and

$$\mathbb{P}^1_{\text{Berk}} = \{b\} \cup \bigcup_{\vec{w} \in T_b} \mathcal{B}_b(\vec{w})^-,$$

the image under φ of $\mathbb{P}^1_{\text{Berk}} \backslash \mathcal{B}_a(\vec{v})^-$ is all of $\mathbb{P}^1_{\text{Berk}}$. □

PROOF OF THEOREM 9.42. Fix $c \in \mathcal{B}$ and consider the segment $[\zeta, c] \subset \overline{\mathcal{B}}$. By Theorem 9.35, there is a finite partition $\{a_0, \ldots, a_n\}$ of $[\zeta, c]$, with $a_0 = \zeta$ and $a_n = c$, such that $m_\varphi(x)$ takes a constant value m_i on each open segment (a_{i-1}, a_i) and φ stretches distances on $[a_{i-1}, a_i]$ by the factor m_i.

First suppose $\varphi(\mathcal{B})$ is a generalized open Berkovich disc. Since $m_\varphi(x)$ is upper semicontinuous (Proposition 9.28(A)), for each subsegment $[a_{i-1}, a_i]$ we have $m_\varphi(a_{i-1}) \geq m_i$ and $m_\varphi(a_i) \geq m_i$. Suppose $m_\varphi(a_i) > m_i$, and let $\vec{v}_{i,-}$ be the tangent vector at a_i in the direction of a_{i-1}. By Lemma 9.43, $\varphi(\mathbb{P}^1_{\text{Berk}} \backslash \mathcal{B}_{a_i}(\vec{v}_{i,-})^-) = \mathbb{P}^1_{\text{Berk}}$. Since $\mathbb{P}^1_{\text{Berk}} \backslash \mathcal{B}_{a_i}(\vec{v}_{i,-})^- \subseteq \mathcal{B}$, this contradicts the assumption that $\varphi(\mathcal{B})$ is a disc. Hence $m_\varphi(a_i) = m_i$, and it follows inductively that $m_\varphi(x)$ is nonincreasing on $[\zeta, c]$.

Conversely, suppose $m_\varphi(x)$ is nonincreasing on $[\zeta, c]$. We claim that φ is injective on $[\zeta, c]$. If not, then since $\mathbb{P}^1_{\text{Berk}}$ is uniquely path-connected, it would have to backtrack. Since φ is injective on each $[a_{i-1}, a_i]$, the backtracking would take place at some a_i, $1 \leq i \leq n-1$. Let $\vec{v}_{i,-}$ and $\vec{v}_{i,+}$ be the tangent directions to a_i in $[a, c]$. Then $\varphi_*(\vec{v}_{i,-}) = \varphi_*(\vec{v}_{i,+})$. Since $m_\varphi(a_i, \vec{v}_{i,-}) = m_{i-1}$ and $m_\varphi(a_i, \vec{v}_{i,+}) = m_i$, Theorem 9.22 implies that $m_\varphi(a_i) \geq m_{i-1} + m_i > m_i$. This contradicts that $m_\varphi(x)$ is nonincreasing.

Let $\vec{v}$ be the tangent vector at ζ for which $\mathcal{B} = \mathcal{B}_\zeta(\vec{v})^-$, and put $b = \varphi(\zeta)$ and $\vec{w} = \varphi_*(\vec{v})$. Since φ is injective on $[\zeta, c]$ and $\varphi([\zeta, c])$ has initial tangent vector $\vec{w}$, we conclude that $c \in \mathcal{B}_b(\vec{w})^-$. This holds for each $c \in \mathcal{B} \cap \mathbb{H}_{\text{Berk}}$, and $\varphi(\mathcal{B})$ is either a disc or all of $\mathbb{P}^1_{\text{Berk}}$, so by continuity $\varphi(\mathcal{B}) = \mathcal{B}_b(\vec{w})^-$. □

9.3. IMAGES OF DISCS AND ANNULI

Recall that by a generalized open Berkovich annulus in $\mathbb{P}^1_{\text{Berk}}$ we mean a domain $\mathcal{A} = \mathcal{A}_{a,c} \subset \mathbb{P}^1_{\text{Berk}}$ with two boundary points a, c (Definition 2.26). Thus, $\mathcal{A}$ is the connected component of $\mathbb{P}^1_{\text{Berk}}\backslash\{a,c\}$ containing the open segment (a, c). The *modulus* of $\mathcal{A}$ is defined to be

$$\text{Mod}(\mathcal{A}) = \rho(a, c) \ .$$

If $a, c \in \mathbb{H}_{\text{Berk}}$, then $\text{Mod}(\mathcal{A}) < \infty$; if either $a \in \mathbb{P}^1(K)$ or $c \in \mathbb{P}^1(K)$, then $\text{Mod}(\mathcal{A}) = \infty$. We will say that $\mathcal{A}$ is *capped* (or is simply an open Berkovich annulus) if a and c are of type II or III (they need not be of the same type), and we will say that $\mathcal{A}$ is *strict* if a and c are both of type II.

PROPOSITION 9.44. *Let $\varphi(T) \in K(T)$ be a nonconstant rational function. If $\mathcal{A} \subset \mathbb{P}^1_{\text{Berk}}$ is a generalized open Berkovich annulus, then $\varphi(\mathcal{A})$ is either another generalized open Berkovich annulus or it is a generalized open Berkovich disc or it is all of $\mathbb{P}^1_{\text{Berk}}$.*

Furthermore, $\mathcal{A}$ is φ-saturated if $\varphi(\mathcal{A})$ is a generalized open Berkovich annulus or it is a generalized open Berkovich disc and both boundary points of $\mathcal{A}$ are mapped to the unique boundary point of the disc. If $\mathcal{A}$ is φ-saturated, then $\varphi(\mathcal{A})$ is capped (resp. strict) if and only if $\mathcal{A}$ is capped (resp. strict).

PROOF. By Corollary 9.11, for any domain V, $\partial(\varphi(V)) \subseteq \varphi(\partial V)$. Hence if $V = \mathcal{A}$ is a generalized open Berkovich annulus, $\varphi(\mathcal{A})$ has 0, 1, or 2 boundary points.

By Proposition 9.15, V is φ-saturated if and only if $\varphi(\partial V) = \partial(\varphi(V))$. Thus $\mathcal{A}$ is φ-saturated if and only if $\varphi(\mathcal{A})$ is an annulus or if $\varphi(\mathcal{A}) = \mathcal{B}$ is a disc and both boundary points of $\mathcal{A}$ are mapped to the boundary point of $\mathcal{B}$. If $\mathcal{A}$ is φ-saturated, Proposition 2.15 shows that $\varphi(\mathcal{A})$ is strict if and only if $\mathcal{A}$ is strict. $\square$

If $\mathcal{A}$ is a generalized open Berkovich annulus and a is one of its boundary points, then $\mathcal{A}$ has an "initial subannulus" with boundary point a whose image under φ is a generalized open Berkovich annulus. This basic property of annuli is due to Rivera-Letelier and was stated as a part of Lemma 9.33. It was proved in the discussion preceding Theorem 9.22 as a consequence of the Open Mapping Theorem and the continuity of φ.

LEMMA 9.45. *Let $\varphi(T) \in K(T)$ be a nonconstant rational function, and let $\mathcal{A} = \mathcal{A}_{a,c}$ be any generalized open Berkovich annulus. Then there is a point $a' \in (a, c)$ such that the image of the generalized open Berkovich annulus $\mathcal{A}_{a,a'}$ under φ is another generalized open Berkovich annulus.*

The following theorem characterizes when $\varphi(\mathcal{A})$ is a generalized open Berkovich annulus and shows that if $\varphi(\mathcal{A})$ is a generalized open Berkovich annulus, then assertions (A)–(C) of Lemma 9.33 hold automatically:

THEOREM 9.46. *Let $\varphi(T) \in K(T)$ be a nonconstant rational function, and let $\mathcal{A} = \mathcal{A}_{a,c}$ be a generalized open Berkovich annulus. Then $\varphi(\mathcal{A})$ is a generalized open Berkovich annulus iff the multiplicity function $m_\varphi(x)$ is constant on the open segment (a, c) and nonincreasing on paths off (a, c).*

In this situation, if $m_\varphi(x) = m$ on (a, c), then $\mathcal{A}$ is φ-saturated. Also:
- (A) $\varphi(\mathcal{A}) = \mathcal{A}_{\varphi(a), \varphi(c)}$.
- (B) *If $\vec{v}_a$ (resp. $\vec{v}_c$) is the tangent vector to a (resp. c) in $[a, c]$, then* $\deg_\mathcal{A}(\varphi) = m_\varphi(a, \vec{v}_a) = m_\varphi(c, \vec{v}_c) = m$.
- (C) $\varphi([a, c]) = [\varphi(a), \varphi(c)]$, $\rho(\varphi(x), \varphi(y)) = m \cdot \rho(x, y)$ *for all $x, y \in [a, c]$, and* $\mathrm{Mod}(\varphi(\mathcal{A})) = m \cdot \mathrm{Mod}(\mathcal{A})$.

PROOF. ($\Rightarrow$) Suppose $\varphi(\mathcal{A})$ is an annulus. Then $\mathcal{A}$ is φ-saturated, and $\varphi(\mathcal{A})$ has two boundary points. Since $\partial(\varphi(\mathcal{A})) = \varphi(\partial \mathcal{A})$ by Proposition 9.15, it follows that $\varphi(\mathcal{A}) = \mathcal{A}_{\varphi(a), \varphi(c)}$.

We first show that φ is injective on $[a, c]$ and that $\varphi([a, c]) = [\varphi(a), \varphi(c)]$. Parametrize $[a, c]$ by an interval $[0, L]$, and for each $0 \leq t \leq L$ let $b_t \in [a, c]$ be the point corresponding to t. Consider the set

$$S = \{t \in [0, L] : \varphi \text{ is injective on } [0, b_t] \text{ and } \varphi([0, b_t]) \subset [\varphi(a), \varphi(c)]\} .$$

Put $T = \sup_{t \in S} t$. Since $\varphi((a, c)) \subset \mathcal{A}_{\varphi(a), \varphi(c)}$, Lemma 9.33 shows that $T > 0$. By continuity, $T \in S$. Suppose $T < L$, and put $b = b_T$. Let $\vec{v}_-$ and $\vec{v}_+$ be the tangent directions to b in $[a, c]$ pointing toward a and c, respectively. Likewise, let $\vec{w}_-$ and $\vec{w}_+$ be the tangent directions to $\varphi(b)$ in $[\varphi(a), \varphi(c)]$ pointing toward $\varphi(a)$ and $\varphi(c)$, respectively. Since $T \in S$, $\varphi_*(\vec{v}_-) = \vec{w}_-$. However, $\varphi_*(\vec{v}_+) \neq \vec{w}_+$; otherwise, by Lemma 9.33, there would be a $T_1 > T$ with $T_1 \in S$. Since φ gives a surjection from the tangent directions at b onto the tangent directions at $\varphi(b)$, there is a tangent direction $\vec{v}$ at b, with $\vec{v} \neq \vec{v}_+$, $\vec{v} \neq \vec{v}_-$, for which $\varphi_*(\vec{v}) = \vec{w}_+$. By Proposition 9.41, $\varphi(\mathcal{B}_b(\vec{v})^-)$ contains $\mathcal{B}_{\varphi(b)}(\vec{w}_+)^-$. However, this is a contradiction, since $\mathcal{B}_b(\vec{v})^- \subset \mathcal{A}$ and $\varphi(c) \in \mathcal{B}_{\varphi(b)}(\vec{w}_+)^-$, while $\varphi(c) \notin \varphi(\mathcal{A})$.

By Theorem 9.35, there is a finite partition $\{a_0, \ldots, a_n\}$ of $[a, c]$, with $a_0 = a$, $a_n = c$, with the property that for each subsegment $[a_{i-1}, a_i]$ there is an integer $1 \leq m_i \leq \deg(\varphi)$ such that φ stretches distances on $[a_{i-1}, a_i]$ by the factor m_i and such that $m_\varphi(x) = m_i$ for all $x \in (a_{i-1}, a_i)$. Consider a_i, $1 \leq i \leq n-1$. Let $\vec{v}_{i,-}$ be the tangent vector to a_i in $[a, c]$ pointing toward a, and let $\vec{v}_{i,+}$ be the tangent vector pointing toward c. Let $\vec{w}_{i,-}, \vec{w}_{i,+}$ be the corresponding tangent vectors at $\varphi(b)$. By the stretching property and Theorem 9.22(A), $m_\varphi(a_i, \vec{v}_{i,-}) = m_{i-1}$. By Theorem 9.22(C), $m_\varphi(a_i) \geq m_\varphi(a_i, \vec{v}_{i,-})$. Suppose $m_\varphi(a_i) > m_\varphi(a_i, \vec{v}_{i,-})$. Then there would be another tangent direction $\vec{v}$ at a_i with $\varphi_*(\vec{v}) = \varphi_*(\vec{v}_{i,-})$. By Proposition 9.41, $\varphi(\mathcal{B}_{a_i}(\vec{v})^-)$ contains $\mathcal{B}_{\varphi(a_i)}(\vec{w}_{i,-})^-$. However, the latter contains $\varphi(a)$, which is impossible because $\mathcal{B}_{a_i}(\vec{v})^- \subset \mathcal{A}$, while $\varphi(a) \notin \varphi(\mathcal{A})$. Thus $m_\varphi(a_i) = m_\varphi(a_i, \vec{v}_{i,-}) = m_{i-1}$.

In a similar way, we find $m_\varphi(a_i) = m_\varphi(a_i, \vec{v}_{i,+}) = m_i$. Setting $m = m_1$ and proceeding inductively, we find that $m_i = m$ for all i and $\varphi(x) = m$ for all $x \in (a, c)$. From the distance-stretching property of φ on each $[a_{i-1}, a_i]$ and the additivity of the path distance, we see that $\rho(\varphi(x), \varphi(y)) = m \cdot \rho(x, y)$ for all $x, y \in [a, c]$. In particular, $\mathrm{Mod}(\varphi(\mathcal{A})) = m \cdot \mathrm{Mod}(\mathcal{A})$ and $m_\varphi(\mathcal{A}) = m_\varphi(a, \vec{v}_a) = m_\varphi(c, \vec{v}_c) = m$.

Finally, consider an arbitrary $b \in (a, c)$. Let $\vec{v}_-, \vec{v}_+$ be the two tangent vectors to b in $[a, c]$ and let $\vec{v}$ be any other tangent vector at b. Since $\mathcal{B}_b(\vec{v})^- \subset \mathcal{A}$ and $\varphi(\mathcal{A})$ is an annulus, $\varphi(\mathcal{B}_b(\vec{v})^-)$ must be a disc and not all of $\mathbb{P}^1_{\text{Berk}}$. By Theorem 9.42, $m_\varphi(x)$ is nonincreasing on paths off b in $\mathcal{B}_b(\vec{v})^-$.

($\Leftarrow$) Suppose $m_\varphi(x) = m$ is constant on (a, c) and $m_\varphi(x)$ is nonincreasing on paths off (a, c) in $\mathcal{A}$. We wish to show that $\varphi(\mathcal{A})$ is the annulus $\mathcal{A}_{\varphi(a),\varphi(c)}$.

Considering a partition $\{a_0, \ldots, a_n\}$ of $[a, c]$ with the properties in Theorem 9.35 and arguing as before, we see that φ stretches distances on $[a, c]$ by a factor of m and consequently that φ is injective on $[a, c]$ and $\varphi([a, c]) = [\varphi(a), \varphi(c)]$.

Clearly $\varphi((a, c)) \subset \mathcal{A}_{\varphi(a),\varphi(c)}$; let $b \in (a, c)$ be arbitrary, and let $\vec{v}_+, \vec{v}_-$ be the tangent directions at b in $[a, c]$. Put $\vec{w}_+ = \varphi_*(\vec{v}_+)$ and $\vec{w}_- = \varphi_*(\vec{v}_-)$. Then $\vec{w}_+$ and $\vec{w}_-$ are the tangent directions at $\varphi(b)$ in $[\varphi(a), \varphi(b)]$. If b is of type II, let $\vec{v}$ be a tangent direction at b which is different from $\vec{v}_+$, $\vec{v}_-$. Since $m_\varphi(b) = m_\varphi(b, \vec{v}_+) = m_\varphi(b, \vec{v}_-)$, it must be that $\vec{w} := \varphi_*(\vec{v})$ is different from $\vec{w}_+$, $\vec{w}_-$. Since $m_\varphi(x)$ is nonincreasing on paths off b in $\mathcal{B}_b(\vec{v})^-$, Theorem 9.42 shows that $\varphi(\mathcal{B}_b(\vec{v})^-)$ is the disc $\mathcal{B}_{\varphi(b)}(\vec{w})^-$, which is contained in $\mathcal{A}_{\varphi(a),\varphi(c)}$.

Thus $\varphi(\mathcal{A}) \subseteq \mathcal{A}_{\varphi(a),\varphi(c)}$. Since both $\varphi(a), \varphi(c) \in \partial(\varphi(\mathcal{A}))$, it must be that $\varphi(\mathcal{A}) = \mathcal{A}_{\varphi(a),\varphi(c)}$. $\square$

Similarly, the following result characterizes when $\mathcal{A}$ is φ-saturated and $\varphi(\mathcal{A})$ is a disc. We omit the proof, which is similar to the previous one:

THEOREM 9.47. *Let $\varphi(T) \in K(T)$ be a nonconstant rational function, and let $\mathcal{A} = \mathcal{A}_{a,c}$ be a generalized open Berkovich annulus. Suppose $\mathcal{A}$ is φ-saturated. Then $\varphi(\mathcal{A})$ is a generalized open Berkovich disc if and only if*

(A) *a and c both have the same type (I, II, III, or IV) and $\varphi(a) = \varphi(c)$;*
(B) *there are a point $b \in (a, c)$ (which is the midpoint of (a, c) if $\rho(a, c)$ is finite) and an integer $1 \leq m \leq \deg(\varphi)/2$, such that $m_\varphi(x) = m$ for all $x \in (a, b) \cup (b, c)$ and $m_\varphi(b) = 2m$;*
(C) *$m_\varphi(x)$ is nonincreasing on paths off (a, c).*

We now interpret the results above in terms of "classical" annuli in $\mathbb{P}^1(K)$.

DEFINITION 9.48. An *open annulus* A in $\mathbb{P}^1(K)$ is any set obtained by taking a proper open disc in $\mathbb{P}^1(K)$ and removing a proper closed subdisc from it. The annulus A is *rational* if the radii of both discs belong to $|K^\times|$.

After a change of coordinates by a suitable Möbius transformation, any open annulus can be put in the form

(9.43) $\quad A_{r,R}(\alpha) = \{z \in K : r < |z - \alpha| < R\} = D(\alpha, R)^- \backslash D(\alpha, r)$

where $\alpha \in K$ and $0 < r < R < \infty$. In particular, for $0 < r < 1$, we define the *standard open annulus* A_r *of height* r to be

$$A_r = \{z \in K : r < |z| < 1\} = D(0, 1)^- \backslash D(0, r).$$

If $A \subset \mathbb{P}^1(K)$ is an open annulus, there is a unique open Berkovich annulus $\mathcal{A} = \mathcal{A}_{a,c}$ for which $A = \mathcal{A} \cap \mathbb{P}^1(K)$; it is obtained by taking the closure of A in $\mathbb{P}^1_{\text{Berk}}$ and removing its boundary points a, c, which are necessarily of type II or III. When $A = D(\alpha, R)^- \backslash D(\alpha, r)$, then $a = \zeta_{\alpha, r}$ and $c = \zeta_{\alpha, R}$ are the points corresponding to $D(a, r)$ and $D(a, R)$ under Berkovich's classification theorem. In that case, $\text{Mod}(\mathcal{A}) = \rho(a, c) = \log_v(R/r)$.

Since the path distance is invariant under change of coordinates (Proposition 2.30), the ratio R/r in (9.43) is independent of the choice of coordinates. The *modulus* of an open annulus A in $\mathbb{P}^1(K)$ is defined to be
$$\text{Mod}(A) = \log_v(R/r)$$
for any representation (9.43). In particular, for the standard open annulus A_r of height r, we have $\text{Mod}(A_r) = -\log_v r$.

If A is a *rational* open annulus (i.e., if the boundary points a, c of $\mathcal{A}$ are type II), then it is an elementary exercise to show that there exists a Möbius transformation taking A to a standard open annulus A_r, whose height r is uniquely determined:

PROPOSITION 9.49. *Let A, A' be rational open annuli in $\mathbb{P}^1(K)$. Then there exists a Möbius transformation taking A to A' iff $\text{Mod}(A) = \text{Mod}(A')$.*

PROOF. By Corollary 2.13(B) there is a Möbius transformation taking any rational annulus to a standard open annulus. Hence it suffices to show that if φ is a Möbius transformation taking A_r to A_s, then $r = s$. Since φ extends to a function $\mathcal{A}_r \to \mathcal{A}_s$, where $\mathcal{A}_r = \mathcal{D}(0,1)^- \backslash \mathcal{D}(0,r)$ denotes the corresponding open Berkovich annulus, it follows easily that the function $f = -\log_v |\varphi| : \mathcal{A}_r \to \mathbb{R}$ is harmonic and extends continuously to $\overline{\mathcal{A}}_r = \mathcal{A}_r \cup \{\zeta_{0,1}, \zeta_{0,r}\}$. Consequently, $\Delta_{\overline{\mathcal{A}}_r}(f) = d\delta_{\zeta_{0,1}} - d\delta_{\zeta_{0,r}}$ for some integer $d \in \mathbb{Z}$. Since $\deg(\varphi) = 1$, we have $d = \pm 1$. Applying a Möbius transformation if necessary, we may assume without loss of generality that $d = 1$. As $g(x) = -\log_v([T]_x)$ has the same property, it follows that $f(x) = g(x) + C'$ for some constant C' and therefore that $|\varphi(x)| = C \cdot |x|$ for all $x \in \mathcal{A}_r$. Letting $|x| \to 1$ shows that $C = 1$, and then letting $|x| \to r$ shows that $r = s$. $\square$

In terms of open annuli in K, Lemma 9.45 translates as follows:

LEMMA 9.50. *Let $\varphi(T) \in K(T)$ be a nonconstant rational function. If $A = \{z \in K : r < |z| < R\}$ is an open annulus in K with $0 < r < R$, then there is a real number $s \in (r, R)$ for which the image of the annulus $A' = \{z \in K : r < |z| < s\}$ under φ is also an open annulus in K.*

Similarly, Theorem 9.46 translates in part as:

PROPOSITION 9.51. *Let A be an open annulus in K for which $\varphi(A)$ is also an open annulus. Then $\text{Mod}(\varphi(A)) = m \cdot \text{Mod}(A)$ for some integer $1 \leq m \leq \deg(\varphi)$, and the map $\varphi : A \to \varphi(A)$ has degree m.*

9.4. The pushforward and pullback measures

Let $\varphi(T) \in K(T)$ be a nonconstant rational function of degree d.

If ν is a bounded Borel measure on $\mathbb{P}^1_{\text{Berk}}$, the *pushforward measure* $\varphi_*(\nu)$ is the Borel measure defined by

$$\varphi_*(\nu)(U) \;=\; \nu(\varphi^{-1}(U))$$

for all Borel subsets $U \subset \mathbb{P}^1_{\text{Berk}}$. Here $\varphi^{-1}(U)$ is a Borel set because φ is continuous for the Berkovich topology.

The pullback of a function $f : \mathbb{P}^1_{\text{Berk}} \to \mathbb{R}$ is defined by

$$\varphi^*(f)(x) \;=\; f(\varphi(x)) \,.$$

When ν is positive, then for each nonnegative Borel measurable f, one has the usual identity

$$(9.44) \qquad \int f(y) \, d(\varphi_*(\nu))(y) \;=\; \int \varphi^*(f)(x) \, d\nu(x) \;=\; \int f(\varphi(x)) \, d\nu(x) \,.$$

This follows from the definition of $\varphi_*(\nu)$ when f is the characteristic function of a Borel set, and the general case follows by taking limits.

If ν is a bounded signed measure, then (9.44) holds for all Borel measurable functions f for which the integrals in (9.44) are finite. In particular, it holds for continuous functions.

The pullback measure $\varphi^*(\nu)$ is more complicated to define. If g is a continuous real-valued function on $\mathbb{P}^1_{\text{Berk}}$, then as in the statement of Proposition 9.13, we define $\varphi_*(g)$ by

$$\varphi_*(g)(y) \;=\; \sum_{\varphi(x)=y} m_\varphi(x) \cdot g(x) \,.$$

By Proposition 9.13, the function $\varphi_*(g)$ is continuous. The definition of $\varphi_*(g)$ shows that $\|\varphi_*(g)\| \leq d \cdot \|g\|$, where $\|g\| = \sup_{x \in \mathbb{P}^1_{\text{Berk}}} |g(x)|$.

The *pullback measure* $\varphi^*(\nu)$ is the measure representing the bounded linear functional $\Lambda : \mathcal{C}(\mathbb{P}^1_{\text{Berk}}) \to \mathbb{R}$ defined by $\Lambda(g) = \int \varphi_*(g)(y) \, d\nu(y)$. Thus

$$(9.45) \qquad \int g(x) \, d\varphi^*(\nu)(x) \;=\; \int \varphi_*(g)(y) \, d\nu(y) \,.$$

In particular, the measures $\omega_{\varphi,b}(x)$ defined in Lemma 9.3 have an interpretation as pullbacks:

PROPOSITION 9.52. *Let $\varphi(T) \in K(T)$ be a nonconstant rational function. Then for each $b \in \mathbb{P}^1_{\text{Berk}}$,*

$$\varphi^*(\delta_b) \;=\; \omega_{\varphi,b} \,.$$

PROOF. For any continuous function g on $\mathbb{P}^1_{\text{Berk}}$,

$$\int g(x)\, d\omega_{\varphi,b}(x) = \sum_{\varphi(a)=b} m_{\varphi,b}(a) g(a)$$
$$= \int \varphi_*(g)(x)\, \delta_b(x) = \int g(x)\, d\varphi^*(b)(x) \,.$$

$\square$

The pushforward and pullback measures satisfy the expected functorial properties.

THEOREM 9.53. *Let $\varphi(T) \in K(T)$ be a nonconstant rational function of degree d, and suppose ν is a signed Borel measure on a compact set $F \subseteq \mathbb{P}^1_{\text{Berk}}$.*

(A) *Let $\langle \nu_\alpha \rangle$ be a net of signed Borel measures on F which converges weakly to ν. Then the net $\langle \varphi^*(\nu_\alpha) \rangle$ converges weakly to $\varphi^*(\nu)$ on F.*

If $U = \varphi^{-1}(\varphi(U))$, then $\langle \varphi_*(\nu_\alpha) \rangle$ converges weakly to $\varphi_*(\nu)$ on F.

(B) $\varphi_*(\varphi^*(\nu)) = d \cdot \nu$.

PROOF. The first assertion in (A) holds because for each $g \in \mathcal{C}(\varphi^{-1}(F))$,

$$\int g(x)\, d\varphi^*(\nu)(x) = \int \varphi_*(g)(y)\, d\nu(y)$$
$$= \lim_\alpha \int \varphi_*(g)(y)\, d\nu_\alpha(y) = \lim_\alpha \int g(x)\, d\varphi^*(\nu_\alpha)(x) \,.$$

The second assertion holds because for each open subset $V \subset \varphi(U)$, the Portmanteau theorem (Theorem A.13) implies that

$$\nu_*(V) = \nu(\varphi^{-1}(V))$$
$$= \lim_\alpha \nu_\alpha(\varphi^{-1}(V)) = \lim_\alpha \varphi_*(\nu_\alpha)(V) \,,$$

and equality of two regular Borel measures on open sets implies their equality for all Borel sets.

Part (B) also follows from a simple computation. Formula (9.45) extends to characteristic functions of Borel sets. For each Borel set $E \subset \mathbb{P}^1_{\text{Berk}}$, the identity $\sum_{\varphi(x)=y} m_{\varphi,y}(x) = d$ means that $\varphi_*(\chi_{\varphi^{-1}(E)}) = d \cdot \chi_E$. Hence

$$\varphi_*(\varphi^*(\nu))(E) = \varphi^*(\nu)(\varphi^{-1}(E))$$
$$= \int \chi_{\varphi^{-1}(E)}(x)\, d\varphi^*(\nu)(x) = \int \varphi_*(\chi_{\varphi^{-1}(E)})(y)\, d\nu(y)$$
$$= \int d \cdot \chi_E(y)\, d\nu(y) = d \cdot \nu(E) \,.$$

$\square$

9.5. The pullback formula for subharmonic functions

In this section we prove a pullback formula for subharmonic functions and then use it to derive pullback formulas for arbitrary functions in BDV(U) and for Green's functions.

Recall from Lemma 9.12 that if $U \subset \mathbb{P}^1_{\text{Berk}}$ is a domain and $\varphi(T) \in K(T)$ is nonconstant, then $\varphi^{-1}(U)$ is a finite disjoint union of domains $V_1, \ldots, V_m$. By Definition 8.3, the Laplacian operator for subharmonic functions on $\varphi^{-1}(U)$ is given by

$$\Delta_{\varphi^{-1}(U)} = \sum_{i=1}^{m} \Delta_{V_i} .$$

PROPOSITION 9.54. *Let $\varphi(T) \in K(T)$ be a nonconstant rational map of degree d. Let $U \subset \mathbb{P}^1_{\text{Berk}}$ be a domain, and let f be subharmonic on U. Then the function $\varphi^*(f) = f \circ \varphi$ is subharmonic on $\varphi^{-1}(U)$, and*

$$\Delta_{\varphi^{-1}(U)}(\varphi^*(f)) = \varphi^*(\Delta_U(f)) .$$

REMARK 9.55. If $U \subset \mathbb{P}^1_{\text{Berk}}$ is an open set, then each of its components is a domain. Since subharmonicity is a local property and the Laplacian of a subharmonic function is a negative signed measure, Proposition 9.54 holds for arbitrary open sets U.

PROOF OF PROPOSITION 9.54. To see that $f \circ \varphi$ is subharmonic, we use the Riesz Decomposition Theorem (Theorem 8.38). Let V be a simple subdomain of U, and put $\nu = -\Delta_V(f)$. Let ζ be an arbitrary point in $\mathbb{P}^1_{\text{Berk}} \backslash \overline{V}$. Theorem 8.38 shows there is a harmonic function h on V such that for all $y \in V$,

$$f(y) = h(y) + \int_V \log_v(\delta(y,z)_\zeta) \, d\nu(z) .$$

Composing with φ gives, for all $x \in \varphi^{-1}(V)$,

$$f(\varphi(x)) = h(\varphi(x)) + \int_V \log_v(\delta(\varphi(x),z)_\zeta) \, d\nu(z) .$$

By Corollary 7.45, $h \circ \varphi$ is harmonic in each component W of $\varphi^{-1}(V)$. By Lemma 9.3, for each $z \in V$ the function $\log_v(\delta(\varphi(x),z)_\zeta)$ is subharmonic in $\mathbb{P}^1_{\text{Berk}} \backslash \varphi^{-1}(\{\zeta\})$, hence in particular in $\varphi^{-1}(V)$. Since $f \circ \varphi - h \circ \varphi$ is upper semicontinuous, it follows from Proposition 8.27 that

$$F(x) := \int_V \log_v(\delta(\varphi(x),z)_\zeta) \, d\nu(z)$$

is subharmonic on $\varphi^{-1}(V)$. (The function $F(x)$ is not identically $-\infty$ on any component of $\varphi^{-1}(V)$, since $F(x) = f(\varphi(x)) - h(\varphi(x))$.) It follows that $f \circ \varphi = h \circ \varphi + F$ is subharmonic on $\varphi^{-1}(V)$. Since subharmonicity is local, it follows from the Open Mapping Theorem (Corollary 9.10) that $f \circ \varphi$ is subharmonic on $\varphi^{-1}(U)$.

We now establish the pullback formula for Laplacians. By the smoothing theorem (Theorem 8.60), there is a sequence f_n of smooth subharmonic functions on W such that $\lim f_n(z) = f(z)$ for all $z \in V$ and where the convergence is uniform on each finite subgraph $\Gamma \subset V$. By smoothness, each $\Delta_V(f_n)$ is a discrete measure supported on a finite set of points. We will first prove the pullback formula for the f_n.

Fix $n \in \mathbb{N}$. Since f_n is smooth, there exist constants $c_{ni} \in \mathbb{R}$ and points $p_{ni}, \zeta_{ni} \in V$ such that
$$f_n(y) = c_{n0} + \sum_{i=1}^{M_n} c_{ni} \log_v(\delta(y, p_{ni})_{\zeta_{ni}}) \ .$$
Composing with φ gives
$$f_n(\varphi(x)) = c_{n0} + \sum_{i=1}^{M_n} c_{ni} \log_v(\delta(\varphi(x), p_{ni})_{\zeta_{ni}}) \ .$$
By Proposition 9.5, each function $F_{ni}(x) = \log_v(\delta(\varphi(x), p_{ni})_{\zeta_{ni}})$ belongs to $\mathrm{BDV}(\mathbb{P}^1_{\mathrm{Berk}})$, and by Lemma 9.3 and Proposition 9.52,
$$\Delta_{\mathbb{P}^1_{\mathrm{Berk}}}(F_{ni}) = \varphi^*(\delta_{\zeta_{ni}} - \delta_{p_{ni}}) \ .$$
Summing over all i, we find that
$$\Delta_{\varphi^{-1}(V)}(f_n \circ \varphi) = \sum_{i=1}^{M_n} c_{ni} \varphi^*(\delta_{\zeta_{ni}} - \delta_{p_{ni}}) = \varphi^*(\Delta_V(f_n)) \ .$$

The functions $f_n \circ \varphi$ converge uniformly to $f \circ \varphi$ on each finite subgraph $\Gamma' \subset \varphi^{-1}(V)$, since $\varphi(\Gamma')$ is a finite subgraph of V by Corollary 9.36. Therefore by Proposition 8.51(A), the measures $\Delta_{\varphi^{-1}(V)}(f_n \circ \varphi)$ converge weakly to $\Delta_{\varphi^{-1}(V)}(f \circ \varphi)$. Proposition 8.51(A) also shows that the measures $\Delta_{\overline{V}}(f_n)$ converge weakly to $\Delta_{\overline{V}}(f)$. Hence, by Theorem 9.53, the measures $\varphi^*(\Delta_{\overline{V}}(f_n))$ converge weakly to $\varphi^*(\Delta_{\overline{V}}(f))$. Thus
$$\Delta_{\varphi^{-1}(V)}(f \circ \varphi) = \varphi^*(\Delta_V(f)) \ .$$
$\square$

Let U be a domain. We next give a pullback formula for functions in $\mathrm{BDV}(U)$. If $V_1, \ldots, V_m$ are the components of $\varphi^{-1}(U)$ and if $g|_{V_i} \in \mathrm{BDV}(V_i)$ for each i, we define
$$\Delta_{\varphi^{-1}(U)}(g) = \sum_{i=1}^{m} \Delta_{V_i}(g) \ .$$

PROPOSITION 9.56. *Let $\varphi(T) \in K(T)$ be a nonconstant rational map. Let $U \subset \mathbb{P}^1_{\mathrm{Berk}}$ be a domain, and let $V_1, \ldots, V_m$ be the components of $\varphi^{-1}(U)$. Take $f \in \mathrm{BDV}(U)$. Then $\varphi^*(f)|_{V_i} \in \mathrm{BDV}(V_i)$ for each $i = 1, \ldots, m$, and*

(9.46) $$\Delta_{\varphi^{-1}(U)}(\varphi^*(f)) = \varphi^*(\Delta_U(f)).$$

PROOF. First assume $U \neq \mathbb{P}^1_{\text{Berk}}$. Let $Y_1 \subset Y_2 \subset \cdots \subset U$ be an exhaustion of U by simple domains Y_i with $\overline{Y}_i \subset Y_{i+1}$ for each i, as given by Corollary 7.11.

By Proposition 8.41, for each i, there are subharmonic functions $g_i, h_i \in \mathcal{SH}(Y_{i+1})$ such that $f(x) = g_i(x) - h_i(x)$ for all $x \in Y_{i+1} \cap \mathbb{H}_{\text{Berk}}$. By Proposition 9.54, $\varphi^*(g_i)$ and $\varphi^*(h_i)$ are subharmonic on $\varphi^{-1}(Y_{i+1})$, and
$$\Delta_{\varphi^{-1}(Y_{i+1})}(\varphi^*(g_i)) = \varphi^*(\Delta_{Y_{i+1}}(g_i)) , \quad \Delta_{\varphi^{-1}(Y_{i+1})}(\varphi^*(h_i)) = \varphi^*(\Delta_{Y_{i+1}}(h_i)) .$$

Since φ is an open mapping and $\overline{Y}_i \subset Y_{i+1}$, it follows that $\overline{\varphi^{-1}(Y_i)} = \varphi^{-1}(\overline{Y}_i) \subset \varphi^{-1}(Y_{i+1})$. Hence for each component $W_{i,j}$ of $\varphi^{-1}(Y_i)$ we have $\overline{W}_{i,j} \subset \varphi^{-1}(Y_{i+1})$. By Lemma 8.10, $\varphi^*(g_i)|_{W_{i,j}}$ and $\varphi^*(h_i)|_{W_{i,j}}$ are strongly subharmonic on $W_{i,j}$ and hence belong to $\text{BDV}(W_{i,j})$. Since the Laplacian of a function depends only on its restriction to $\mathbb{H}_{\text{Berk}}$, it follows that $\varphi^*(f)|_{W_{i,j}}$ belongs to $\text{BDV}(W_{i,j})$. This shows that $\Delta_{\varphi^{-1}(Y_i)}(\varphi^*(f))$ is well-defined; from Proposition 9.54 we obtain
$$\begin{aligned} \Delta_{\varphi^{-1}(Y_i)}(\varphi^*(f)) &= \Delta_{\varphi^{-1}(Y_i)}(\varphi^*(g_i)) - \Delta_{\varphi^{-1}(Y_i)}(\varphi^*(h_i)) \\ &= \varphi^*(\Delta_{Y_i}(g_i)) - \varphi^*(\Delta_{Y_i}(h_i)) = \varphi^*(\Delta_{Y_i}(f)) . \end{aligned}$$

On the other hand, let ν_i^+ and ν_i^- be the measures in the Jordan decomposition of $\Delta_{Y_i}(f)$. Then $\varphi^*(\Delta_{Y_i}(f)) = \varphi^*(\nu_i^+) - \varphi^*(\nu_i^-)$ is the Jordan decomposition of $\varphi^*(\Delta_{Y_i}(f))$. Since $f \in \text{BDV}(U)$, there is a bound B, independent of Y_i, such that ν_i^+ and ν_i^- have mass at most B. By Theorem 9.53(B), $\varphi^*(\nu_i^+)$ and $\varphi^*(\nu_i^-)$ have mass at most $\deg(\varphi) \cdot B$. Hence $\Delta_{\varphi^{-1}(Y_i)}(\varphi^*(f))$ has total mass at most $2\deg(\varphi) \cdot B$, independent of i.

Since the $\varphi^{-1}(Y_i)$ exhaust $\varphi^{-1}(U)$, it follows that $\varphi^*(f)|_{V_i} \in \text{BDV}(V_i)$ for each component V_i of $\varphi^{-1}(U)$. By the compatibility of Laplacians with restriction (Proposition 5.26), we obtain (9.46).

Now suppose $U = \mathbb{P}^1_{\text{Berk}}$. Take a finite cover of $\mathbb{P}^1_{\text{Berk}}$ by simple subdomains $U_1, \ldots, U_n$, and apply the case above to each U_i. The components of the $\varphi^{-1}(U_i)$ form a finite cover of $\mathbb{P}^1_{\text{Berk}}$, so Proposition 5.27 shows that $\varphi^*(f) \in \text{BDV}(\mathbb{P}^1_{\text{Berk}})$ and that (9.46) holds. $\square$

Finally, we give a definitive pullback formula for Green's functions, improving Proposition 7.47:

PROPOSITION 9.57. *Suppose $E \subset \mathbb{P}^1_{\text{Berk}}$ is compact and has positive capacity, and take $\zeta \in \mathbb{P}^1_{\text{Berk}} \backslash E$. Let $\varphi(T) \in K(T)$ be a nonconstant rational function. Write $\varphi^{-1}(\{\zeta\}) = \{\xi_1, \ldots, \xi_m\}$, and let*
$$\varphi^*(\delta_\zeta) = \sum_{i=1}^m n_i \delta_{\xi_i} ,$$
where $n_i = m_\varphi(\xi_i)$. Then for each $z \in \mathbb{P}^1_{\text{Berk}}$,

(9.47) $$G(\varphi(z), \zeta; E) = \sum_{i=1}^m n_i G(z, \xi_i; \varphi^{-1}(E)) .$$

PROOF. We first claim that for each $\zeta \in \mathbb{P}^1_{\text{Berk}} \setminus E$, there is a set $e \subset \varphi^{-1}(E)$ of capacity 0 such that (9.47) holds for all $z \in \mathbb{P}^1_{\text{Berk}} \setminus e$ and fails for $z \in e$. This has already been shown for $\zeta \in \mathbb{P}^1(K)$ in Proposition 7.47, so we can assume $\zeta \in \mathbb{H}_{\text{Berk}}$.

By Proposition 7.37(A)(8), $G(z, \zeta; E) \in \text{BDV}(\mathbb{P}^1_{\text{Berk}})$ and
$$\Delta_{\mathbb{P}^1_{\text{Berk}}}(G(z, \zeta; E)) = \delta_\zeta(z) - \omega_\zeta(z) ,$$
where $\omega_\zeta(z)$ is a probability measure supported on E. By Proposition 9.56,
$$\Delta_{\mathbb{P}^1_{\text{Berk}}}(G(\varphi(z), \zeta; E)) = \sum_{i=1}^{m} n_i \delta_{\xi_i}(z) - \varphi^*(\omega_\zeta) .$$
Put
$$h(z) = G(\varphi(z), \zeta; E) - \sum_{i=1}^{m} n_i G(z, \xi_i; \varphi^{-1}(E)) .$$
Since ζ and the ξ_i belong to $\mathbb{H}_{\text{Berk}}$, Propositions 9.56 and 7.37(A)(7) and (A)(8) imply that h is harmonic on each component of $\mathbb{P}^1_{\text{Berk}} \setminus \varphi^{-1}(E)$. Now the same argument as in Proposition 7.47 shows that (9.47) holds.

We will next show that the exceptional set e is empty. Since e has capacity 0, it is contained in $\mathbb{P}^1(K)$. In particular, (9.47) holds for all $z \in \mathbb{H}_{\text{Berk}}$. Suppose $p \in e$, and let $[p, q]$ be a segment. By Lemma 9.33, we can assume that $\varphi([p, q])$ is a segment $[\varphi(p), \varphi(q)]$ and that φ is injective on $[p, q]$. Let $z \to p$ in (p, q); then $\varphi(z) \to \varphi(p)$ in $(\varphi(p), \varphi(q))$. By Proposition 9.54, $G(\varphi(z), \zeta; E)$ is subharmonic on $U = \mathbb{P}^1_{\text{Berk}} \setminus \varphi^{-1}(\{\zeta\})$; by Example 8.9 and Proposition 8.26(A), the same is true for $\sum_{i=1}^{m} n_i G(z, \xi_i; \varphi^{-1}(E))$. Since (p, q) and $(\varphi(p), \varphi(q))$ are contained in $\mathbb{H}_{\text{Berk}}$, Proposition 8.11(C) and the case treated above show that
$$\begin{aligned} G(\varphi(p), \zeta; E) &= \lim_{\substack{z \to p \\ z \in (p,q)}} G(\varphi(z), \zeta; E) = \lim_{\substack{z \to p \\ z \in (p,q)}} \sum_{i=1}^{m} n_i G(z, \xi_i; \varphi^{-1}(E)) \\ &= \sum_{i=1}^{m} n_i G(p, \xi_i; \varphi^{-1}(E)) . \end{aligned}$$
This contradicts the assumption that $p \in e$, so e is empty. $\square$

9.6. Notes and further references

As should be clear by now, this chapter has been greatly influenced by the work of Rivera-Letelier. Many of the results in §9.1–§9.3 are due to him, though we have given new proofs. Our analytic construction of multiplicities using proximity functions is new, as are the characterizations of images of discs and annuli in terms of multiplicities.

We have already remarked that Rivera-Letelier [81, 84], Favre and Rivera-Letelier [48], and Thuillier [94] have given other constructions of multiplicities. A theory of multiplicities for étale morphisms of analytic spaces is discussed in §6.3 of Berkovich's paper [17].

CHAPTER 10

Applications to the dynamics of rational maps

In this chapter, we study the dynamics of a rational map $\varphi \in K(T)$ of degree $d \geq 2$ acting on $\mathbb{P}^1_{\text{Berk}}$, where K is a complete and algebraically closed non-Archimedean field.

We begin by showing, in §10.1, that there is a probability measure μ_φ on $\mathbb{P}^1_{\text{Berk}}$ which satisfies $\varphi_*(\mu_\varphi) = \mu_\varphi$ and $\varphi^*(\mu_\varphi) = d \cdot \mu_\varphi$. By analogy with the classical situation for $\mathbb{P}^1(\mathbb{C})$ (see [54], [72]), we call μ_φ the *canonical measure* associated to φ. Loosely speaking, the canonical measure is the negative of the Laplacian of the Call-Silverman local height function from [32] (extended to $\mathbb{P}^1_{\text{Berk}}$ in a natural way).

We also show that the associated Arakelov-Green's function $g_{\mu_\varphi}(x,y)$ satisfies a certain energy minimization principle. This was used in [5] and [9] to investigate properties of algebraic points of small dynamical height.

In §10.2, we give an explicit formula for the Arakelov-Green's function $g_{\mu_\varphi}(x,y)$, and we establish some functorial properties satisfied by $g_{\mu_\varphi}(x,y)$, including the important functional equation in Theorem 10.18.

In §10.3, we prove an adelic equidistribution theorem (due independently to Baker and Rumely, to Chambert-Loir, and to Favre and Rivera-Letelier) for sequences of distinct points in $\mathbb{P}^1$ over a number field k whose dynamical heights tend to zero. Our proof is a simplified version of that in [9], and it applies not just to number fields but to arbitrary product formula fields.

In §10.4, following Favre and Rivera-Letelier [46, 48], we prove an equidistribution theorem for the preimages under φ of any nonexceptional point of $\mathbb{P}^1_{\text{Berk}}$. This is the Berkovich space analogue of a classical result due independently to Lyubich [72] and Freire, Lopes, and Mañé [54]. Our proof, which differs from the one in [46, 48], makes use of properties of $g_{\mu_\varphi}(x,y)$. However, our proof is valid only when $\text{char}(K) = 0$, whereas Favre and Rivera-Letelier's proof holds in positive characteristic as well.

In §10.5–§10.8, we develop Fatou-Julia theory on $\mathbb{P}^1_{\text{Berk}}$ when $\text{char}(K) = 0$. We recover most of the basic facts known classically over $\mathbb{C}$, as presented for example in Milnor's book [73].

The modern approach to complex dynamics was initiated by Fatou and Julia in the early twentieth century. The best understood and most classical part of the theory concerns the iteration of a rational function $\varphi \in \mathbb{C}(T)$ on the complex projective line $\mathbb{P}^1(\mathbb{C})$. Its primary goal is to understand the structure of the Fatou set and its complement, the Julia set.

In p-adic dynamics, there is an analogous, more recent theory for a rational function $\varphi \in K(T)$. The theory was initially developed when $K = \mathbb{C}_p$,[1] for $\varphi \in \mathbb{C}_p(T)$ acting on $\mathbb{P}^1(\mathbb{C}_p)$, by Silverman, Benedetto, Hsia, and others (see e.g. [**14**], [**15**], [**61**], [**74**], [**81**]). One finds both striking similarities and striking differences between rational dynamics on $\mathbb{P}^1(\mathbb{C})$ and $\mathbb{P}^1(\mathbb{C}_p)$. (For an overview of the "classical" results in this field, see [**93**].) The differences arise for the most part from the fact that the topology on $\mathbb{C}_p$ is totally disconnected and not locally compact. Even the most basic topological questions about the Fatou and Julia sets have a completely different flavor when working over $\mathbb{C}_p$ versus $\mathbb{C}$. In fact, it was not *a priori* clear what the Fatou and Julia sets in $\mathbb{P}^1(\mathbb{C}_p)$ should be, since (for example) the notions of equicontinuity and normality do not coincide. Definitions which seemed reasonable (such as defining the Fatou set in terms of equicontinuity of the family of iterates $\{\varphi^{(n)}\}$) turned out to behave differently than expected. For example, in $\mathbb{P}^1(\mathbb{C}_p)$ the Julia set of any map with good reduction is empty, in sharp contrast to the situation over $\mathbb{C}$, where one of the most fundamental facts is that the Julia set is always nonempty. It took considerable effort to give a satisfactory definition for Fatou components in $\mathbb{P}^1(\mathbb{C}_p)$ [**83**].

However, as Rivera-Letelier's thesis [**81**] and subsequent works [**82, 83, 84, 46, 47, 48**] have made abundantly clear, a rational map $\varphi \in K(T)$ should be thought of as acting not just on $\mathbb{P}^1(K)$, but on $\mathbb{P}^1_{\text{Berk}}$. One advantage of working with $\mathbb{P}^1_{\text{Berk}}$ is evident in the existence of the canonical probability measure μ_φ. In contrast to $\mathbb{P}^1(K)$, the compact connected Hausdorff space $\mathbb{P}^1_{\text{Berk}}$ serves naturally as a support for measures.

In §10.5 we define the Berkovich Julia set J_φ as the support of the canonical measure and the Berkovich Fatou set F_φ as the complement of J_φ.[2] A direct consequence of this definition is that the Berkovich Julia set is always nonempty.

In Theorem 10.56, we show that the Berkovich Fatou set of φ coincides with the set of all points in $\mathbb{P}^1_{\text{Berk}}$ having a neighborhood V whose forward iterates under φ omit at least one nonexceptional point of $\mathbb{P}^1_{\text{Berk}}$ (or, equivalently, at least three points of $\mathbb{P}^1(K)$). This was Rivera-Letelier's definition of the Berkovich Fatou set for $K = \mathbb{C}_p$. We use this result to deduce information about the topological structure of the Berkovich Julia set, in the style of [**73**].

In §10.6, we compare our Berkovich Fatou and Julia sets with the corresponding sets defined in terms of equicontinuity, and we use this comparison to obtain structural information about the Berkovich Fatou and Julia sets.

---

[1]There are some subtle but important differences between dynamics over $\mathbb{C}_p$ and dynamics over an arbitrary complete and algebraically closed non-Archimedean field K of characteristic 0; see e.g. Example 10.70 below.

[2]As Rivera-Letelier pointed out to us (with examples like Example 10.70 below), over a general complete and algebraically closed non-Archimedean ground field K, pathologies arise if one tries to define the Berkovich Julia and Fatou sets in terms of equicontinuity or normality.

We also show (Theorem 10.67) that for any complete and algebraically closed non-Archimedean ground field K of characteristic 0, the intersection of the Berkovich Fatou set with $\mathbb{P}^1(K)$ coincides with the classical Fatou set of φ in $\mathbb{P}^1(K)$ considered by Silverman, Benedetto, Hsia, and others.

Using deep results of Rivera-Letelier, we prove (Theorem 10.72) that over $\mathbb{C}_p$, the Berkovich Fatou set consists of all points having a neighborhood on which the set of iterates $\{\varphi^{(n)}\}$ is equicontinuous (in the sense of uniform spaces; cf. §A.9). However, over a general complete and algebraically closed non-Archimedean ground field K of characteristic 0, this is no longer true, as shown by Example 10.70.

In §10.7, we give a proof of Rivera-Letelier's theorem that a rational map $\varphi(T) \in K(T)$ of degree at least two has at least one repelling fixed point in $\mathbb{P}^1_{\text{Berk}}$. We also show that the repelling periodic points belong to and are dense in the Berkovich Julia set. The complex analogue of this result is well known; indeed, the complex Julia set of a rational map $\varphi(T) \in \mathbb{C}(T)$ is sometimes defined as the closure of the repelling periodic points.

In §10.8, we show that when $\varphi(T) \in K[T]$ is a polynomial, the Berkovich Julia set J_φ coincides with the boundary of the Berkovich filled Julia set K_φ. This is the analogue of another well-known result over $\mathbb{C}$.

In §10.9, we consider the dynamics of rational functions over $\mathbb{C}_p$, which are much better understood than over arbitrary K. We recall the main results of Rivera-Letelier from [**81, 82, 83, 84**] and translate them into the language of $\mathbb{P}^1_{\text{Berk}}$.

Finally, in §10.10 we give examples illustrating the theory.

10.1. Construction of the canonical measure

Let $\varphi(T) \in K(T)$ be a rational function of degree $d \geq 2$. In this section we construct the canonical measure μ_φ, a non-Archimedean analogue of the measure constructed by Brolin [**31**] for a polynomial in $\mathbb{C}[T]$ and by Lyubich [**72**] and Freire, Lopes, and Mañé [**54**] for a rational function in $\mathbb{C}(T)$. Our construction follows [**9**]; other constructions have been given by Chambert-Loir [**35**] and by Favre and Rivera-Letelier [**46, 47, 48**].

Let $f_1(T), f_2(T) \in K[T]$ be coprime polynomials with

$$\varphi(T) = f_2(T)/f_1(T),$$

such that $\max(\deg(f_1), \deg(f_2)) = d$. Homogenizing $f_1(T)$ and $f_2(T)$, we obtain homogeneous polynomials $F_1(X, Y), F_2(X, Y) \in K[X, Y]$ of degree d such that $f_1(T) = F_1(1, T)$, $f_2(T) = F_2(1, T)$, and where $\text{Res}(F_1, F_2) \neq 0$. Let $F(X, Y) = (F_1(X, Y), F_2(X, Y))$.

We will be interested in the dynamics of φ on $\mathbb{P}^1_{\text{Berk}}$, so for each $n \geq 1$, let $\varphi^{(n)} = \varphi \circ \varphi \circ \cdots \circ \varphi$ (n times). Similarly, let $F^{(n)}$ be the n-fold composition of F with itself, and write $F^{(n)}(X, Y) = (F_1^{(n)}(X, Y), F_2^{(n)}(X, Y))$, where $F_1^{(n)}(X, Y), F_2^{(n)}(X, Y) \in K[X, Y]$ are homogeneous of degree d^n.

For the moment, regard $F_1(X,Y)$ and $F_2(X,Y)$ as functions on K^2, so that for $(x,y) \in K^2$ we have $F(x,y) = (F_1(x,y), F_2(x,y))$. Let
$$\|(x,y)\| = \max(|x|,|y|)$$
be the sup norm on K^2.

LEMMA 10.1. *There are numbers $0 < B_1 < B_2$, depending on F, such that for all $(x,y) \in K^2$,*

(10.1) $\qquad B_1 \cdot \|(x,y)\|^d \leq \|F(x,y)\| \leq B_2 \cdot \|(x,y)\|^d$.

Furthermore, if F_1 and F_2 have coefficients in the valuation ring of K, then we may take $B_1 = |\operatorname{Res}(F_1, F_2)|$ and $B_2 = 1$.

PROOF. The upper bound is trivial, since if $F_1(X,Y) = \sum c_{1,ij} X^i Y^j$, $F_2(X,Y) = \sum c_{2,ij} X^i Y^j$, and if we let $B_2 = \max(|c_{1,ij}|, |c_{2,ij}|)$, then by the ultrametric inequality
$$|F_1(x,y)|, |F_2(x,y)| \leq \max_{i,j}(|c_{1,ij}||x|^i|y|^j, |c_{2,ij}||x|^i|y|^j) \leq B_2 \|(x,y)\|^d .$$

Note that if F_1 and F_2 have coefficients in the valuation ring of K, then we may take $B_2 = 1$.

For the lower bound, note that since $\operatorname{Res}(F_1, F_2) \neq 0$, Lemma 2.11 shows that there are homogenous polynomials $G_1(X,Y), G_2(X,Y) \in K[X,Y]$ of degree $d-1$ for which

(10.2) $\quad G_1(X,Y) F_1(X,Y) + G_2(X,Y) F_2(X,Y) = \operatorname{Res}(F_1, F_2) X^{2d-1}$,

and there are homogeneous polynomials $H_1(X,Y), H_2(X,Y) \in K[X,Y]$ of degree $d-1$ such that

(10.3) $\quad H_1(X,Y) F_1(X,Y) + H_2(X,Y) F_2(X,Y) = \operatorname{Res}(F_1, F_2) Y^{2d-1}$.

By the upper bound argument applied to $G = (G_1, G_2)$ and $H = (H_1, H_2)$, there is an $A_2 > 0$ such that $\|G(x,y)\|, \|H(x,y)\| \leq A_2 \|(x,y)\|^{d-1}$ for all $(x,y) \in K^2$. By (10.2), (10.3), and the ultrametric inequality,
$$|\operatorname{Res}(F_1, F_2)||x|^{2d-1} \leq A_2 \|(x,y)\|^{d-1} \|F(x,y)\| ,$$
$$|\operatorname{Res}(F_1, F_2)||y|^{2d-1} \leq A_2 \|(x,y)\|^{d-1} \|F(x,y)\| .$$

Writing $A_1 = |\operatorname{Res}(F_1, F_2)|$, it follows that
$$A_1 \|(x,y)\|^{2d-1} \leq A_2 \|(x,y)\|^{d-1} \|F(x,y)\| .$$

Thus, taking $B_1 = A_1/A_2$, we have $\|F(x,y)\| \geq B_1 \|(x,y)\|^d$.

If F_1 and F_2 have coefficients in the valuation ring of K, then since the coefficients of G_1, G_2, H_1, H_2 are integral polynomials in the coefficients of F_1 and F_2 (see, for example, Proposition 2.13 in [**93**]), we find that G_1, G_2, H_1, H_2 have coefficients in the valuation ring of K as well. We may thus take $A_2 = 1$, in which case $B_1 = |\operatorname{Res}(F_1, F_2)|$. □

10.1. CONSTRUCTION OF THE CANONICAL MEASURE

Taking logarithms in (10.1), for $C_1 = \max(|\log_v(B_1)|, |\log_v(B_2)|)$ we have

(10.4) $$\left|\frac{1}{d}\log_v \|F(x,y)\| - \log_v \|(x,y)\|\right| \leq \frac{C_1}{d}$$

for all $(x,y) \in K^2 \backslash \{(0,0)\}$. Inserting $F^{(n-1)}(x,y)$ for (x,y) in (10.4) and dividing by d^{n-1}, we find that for each $n \geq 1$,

(10.5) $$\left|\frac{1}{d^n}\log_v \|F^{(n)}(x,y)\| - \frac{1}{d^{n-1}}\log_v \|F^{(n-1)}(x,y)\|\right| \leq \frac{C_1}{d^n} .$$

Put
$$\begin{aligned} h^{(n)}_{\varphi,v,(\infty)}(x) &= \frac{1}{d^n}\log_v \|F^{(n)}(1,x)\| \\ &= \frac{1}{d^n}\log_v\left(\max(|F_1^{(n)}(1,x)|, |F_2^{(n)}(1,x)|)\right) . \end{aligned}$$

The *Call-Silverman local height* for φ on $\mathbb{P}^1(K)$ (relative to the point ∞ and the dehomogenization $F_1(1,T), F_2(1,T)$) is defined for $x \in \mathbb{A}^1(K)$ by

(10.6) $$h_{\varphi,v,(\infty)}(x) = \lim_{n \to \infty} h^{(n)}_{\varphi,v,(\infty)}(x) .$$

The fact that the limit exists follows from (10.5) using a standard telescoping sum argument, as does the bound

(10.7) $$|h_{\varphi,v,(\infty)}(x) - \log_v(\max(1,|x|))| \leq \sum_{n=1}^{\infty} \frac{C_1}{d^n} = \frac{C_1}{d-1} .$$

We will denote the constant on the right side by C. Taking logarithms in the identity
$$F^{(n-1)}(1,\varphi(x)) = F^{(n-1)}\left(1, \frac{F_2(1,x)}{F_1(1,x)}\right) = \frac{F^{(n)}(1,x)}{F_1(1,x)^{d^{n-1}}} ,$$

then dividing by d^{n-1} and letting $n \to \infty$, gives the functional equation

(10.8) $$h_{\varphi,v,(\infty)}(\varphi(x)) = d \cdot h_{\varphi,v,(\infty)}(x) - \log_v(|F_1(1,x)|) ,$$

valid on $\mathbb{P}^1(K)\backslash(\{\infty\} \cup \varphi^{-1}(\{\infty\}))$.

Properties (10.7) and (10.8) characterize the Call-Silverman local height: in general, given a divisor D on $\mathbb{P}^1(K)$, a function $h_{\varphi,v,D} : \mathbb{P}^1(K)\backslash \mathrm{supp}(D) \to \mathbb{R}$ is a Call-Silverman local height for D if it is a Weil local height associated to D and if there exists a rational function $f \in K(T)$ with $\mathrm{div}(f) = \varphi^*(D) - d \cdot D$, such that
$$h_{\varphi,v,D}(\varphi(z)) = d \cdot h_{\varphi,v,D}(z) - \log_v(|f(z)|_v)$$

for all $z \in \mathbb{P}^1(K)\backslash(\mathrm{supp}(D) \cup \mathrm{supp}(\varphi^*(D)))$. If f is replaced by cf for some $c \neq 0$, then $h_{\varphi,v,D}$ is changed by an additive constant.

We will now "Berkovich-ize" the local height. For each n and each $x \in \mathbb{P}^1_{\mathrm{Berk}}$, put

(10.9) $$\hat{h}^{(n)}_{\varphi,v,(\infty)}(x) = \frac{1}{d^n}\max\left(\log_v[F_1^{(n)}(1,T)]_x, \log_v[F_2^{(n)}(1,T)]_x\right) .$$

(By convention, we set $\hat{h}^{(n)}_{\varphi,v,\infty}(\infty) = \infty$.) Then $\hat{h}^{(n)}_{\varphi,v,(\infty)}$ coincides with $h^{(n)}_{\varphi,v,(\infty)}$ on $\mathbb{A}^1(K)$ and is continuous and strongly subharmonic on $\mathbb{A}^1_{\text{Berk}}$. We claim that for all $x \in \mathbb{A}^1_{\text{Berk}}$,

$$(10.10) \qquad \left|\hat{h}^{(n)}_{\varphi,v,(\infty)}(x) - \hat{h}^{(n-1)}_{\varphi,v,(\infty)}(x)\right| \leq \frac{C_1}{d^n}.$$

Indeed, this holds for all type I points in $\mathbb{A}^1(K)$; such points are dense in $\mathbb{A}^1_{\text{Berk}}$ and the functions involved are continuous, so it holds for all $x \in \mathbb{A}^1_{\text{Berk}}$.

It follows that the functions $\hat{h}^{(n)}_{\varphi,v,(\infty)}$ converge uniformly to a continuous subharmonic function $\hat{h}_{\varphi,v,(\infty)}$ on $\mathbb{A}^1_{\text{Berk}}$ which extends the Call-Silverman local height $h_{\varphi,v,(\infty)}$. By the same arguments as before, for all $x \in \mathbb{A}^1_{\text{Berk}}$ we have

$$(10.11) \qquad |\hat{h}_{\varphi,v,(\infty)}(x) - \log_v(\max(1, [T]_x))| \leq C,$$

and for all $x \in \mathbb{P}^1_{\text{Berk}} \backslash (\infty \cup \varphi^{-1}(\infty))$,

$$(10.12) \qquad \hat{h}_{\varphi,v,(\infty)}(\varphi(x)) = d \cdot \hat{h}_{\varphi,v,(\infty)}(x) - \log_v([F_1(1,T)]_x).$$

Actually, (10.12) can be viewed as an identity for all $x \in \mathbb{P}^1_{\text{Berk}}$, if at $x = \infty$ one takes the right side to be given by its limit as $x \to \infty$.

Let $V_1 = \mathbb{A}^1_{\text{Berk}} = \mathbb{P}^1_{\text{Berk}} \backslash \{\infty\}$. Since $\hat{h}_{\varphi,v,(\infty)}$ is subharmonic on V_1, there is a nonnegative measure μ_1 on V_1 such that $\Delta_{V_1}(\hat{h}_{\varphi,v,(\infty)}) = -\mu_1$. On the other hand, each $\hat{h}^{(n)}_{\varphi,v,(\infty)}$ belongs to $\text{BDV}(\mathbb{P}^1_{\text{Berk}})$ and satisfies

$$\Delta_{\mathbb{P}^1_{\text{Berk}}}(\hat{h}^{(n)}_{\varphi,v,(\infty)}) = \delta_\infty - \mu^{(n)}_1,$$

where $\mu^{(n)}_1 \geq 0$ has total mass 1. Since the $\hat{h}^{(n)}_{\varphi,v,(\infty)}$ converge uniformly to $\hat{h}_{\varphi,v,(\infty)}$ on V_1, it follows from Proposition 5.32 that $\Delta_{\mathbb{P}^1_{\text{Berk}}}(\hat{h}^{(n)}_{\varphi,v,(\infty)}) \to \Delta_{\mathbb{P}^1_{\text{Berk}}}(\hat{h}_{\varphi,v,(\infty)})$. From this, one concludes from Proposition 8.51 that the $\mu^{(n)}_1$ converge weakly to μ_1 on simple subdomains of V_1, that $\hat{h}_{\varphi,v,(\infty)}$ belongs to $\text{BDV}(\mathbb{P}^1_{\text{Berk}})$, and that there is a nonnegative measure μ_φ on $\mathbb{P}^1_{\text{Berk}}$ of total mass 1 such that

$$(10.13) \qquad \Delta_{\mathbb{P}^1_{\text{Berk}}}(\hat{h}_{\varphi,v,(\infty)}) = \delta_\infty - \mu_\varphi.$$

(Note that functions in $\text{BDV}(\mathbb{P}^1_{\text{Berk}})$ are allowed to take the values $\pm\infty$ on points of $\mathbb{P}^1(K)$; the definition of the Laplacian only involves their restriction to $\mathbb{H}_{\text{Berk}}$.)

In the affine patch $V_2 := \mathbb{P}^1_{\text{Berk}} \backslash \{0\}$, relative to the coordinate function $U = 1/T$, the map φ is given by $F_1(U,1)/F_2(U,1)$. By a construction similar to the one above, using the functions $F_1(U,1)$ and $F_2(U,1)$, we obtain a function $\hat{h}_{\varphi,v,(0)}(x) \in \text{BDV}(\mathbb{P}^1_{\text{Berk}})$ which is continuous and subharmonic on

10.1. CONSTRUCTION OF THE CANONICAL MEASURE

V_2 and extends the Call-Silverman local height relative to the point 0. For all $x \in V_2$, it satisfies

(10.14) $\quad |\hat{h}_{\varphi,v,(0)}(x) - \log_v(\max(1, [1/T]_x))| \leq C$,

(10.15) $\hat{h}_{\varphi,v,(0)}(\varphi(x)) = d \cdot \hat{h}_{\varphi,v,(0)}(x) - \log_v([F_2(1/T, 1)]_x)$,

where (10.15) holds in the same sense as (10.12). Using $F^{(n)}(U, 1) = F^{(n)}(1, T)/T^{d^n}$, taking logarithms, dividing by d^n, and letting $n \to \infty$ gives

(10.16) $\quad \hat{h}_{\varphi,v,(0)}(x) = \hat{h}_{\varphi,v,(\infty)}(x) - \log_v([T]_x)$.

Applying the Laplacian and using (10.13) shows that

(10.17) $\quad \Delta_{\mathbb{P}^1_{\text{Berk}}}(\hat{h}_{\varphi,v,(0)}) = \delta_0 - \mu_\varphi$.

We will refer to the probability measure μ_φ on $\mathbb{P}^1_{\text{Berk}}$ appearing in (10.13) and (10.17) as the *canonical measure* associated to φ.

THEOREM 10.2. *Let $\varphi(T) \in K(T)$ have degree $d \geq 2$. Then the Call-Silverman local height $\hat{h}_{\varphi,v,(\infty)}$ belongs to* $\text{BDV}(\mathbb{P}^1_{\text{Berk}})$ *and* $\mathcal{SH}(\mathbb{P}^1_{\text{Berk}}\backslash\{\infty\})$, *with*

(10.18) $\quad \Delta_{\mathbb{P}^1_{\text{Berk}}}(\hat{h}_{\varphi,v,(\infty)}) = \delta_\infty - \mu_\varphi$.

The canonical measure μ_φ is nonnegative and has total mass 1. It satisfies the functional equations

(10.19) $\quad \varphi^*(\mu_\varphi) = d \cdot \mu_\varphi$, $\quad \varphi_*(\mu_\varphi) = \mu_\varphi$.

PROOF. The assertions about $\hat{h}_{\varphi,v,(\infty)}$ have been established earlier in this section.

Put $\mu = \mu_\varphi$. To show that $\varphi^*(\mu) = d \cdot \mu$, we use (10.12) and (10.15). Let $x_1, \ldots, x_d$ be the zeros (not necessarily distinct) of $F_1(1, T)$, and put $U_1 = \varphi^{-1}(V_1) = \mathbb{P}^1_{\text{Berk}}\backslash\{x_1, \ldots, x_d\}$. By the pullback formula for subharmonic functions (Proposition 9.54), $H(x) := \hat{h}_{\varphi,v,(\infty)}(\varphi(x))$ is subharmonic on U_1, and

$$\Delta_{U_1}(H) = \varphi^*(\Delta_{V_1}(\hat{h}_{\varphi,v,(\infty)}))$$.

Taking Laplacians in (10.12) gives $\varphi^*(\mu|_{V_1}) = d \cdot \mu|_{U_1}$. Likewise, put $U_2 = \varphi^{-1}(V_2) = \mathbb{P}^1_{\text{Berk}}\backslash\{y_1, \ldots, y_d\}$, where $y_1, \ldots, y_d$ are the zeros of $F_2(1/T, 1) = F_2(1, T)/T^d$. Since $F_1(1, T)$ and $F_2(1, T)$ are coprime, the sets $\{x_1, \ldots, x_d\}$ and $\{y_1, \ldots, y_d\}$ are disjoint. Then $G(x) := \hat{h}_{\varphi,v,(0)}(\varphi(x))$ is subharmonic on U_2 and satisfies

$$\Delta_{U_2}(G) = \varphi^*(\Delta_{V_2}(\hat{h}_{\varphi,v,(0)}))$$.

Taking Laplacians in (10.15) gives $\varphi^*(\mu|_{V_2}) = d \cdot \mu|_{U_2}$. Since $V_1 \cup V_2 = U_1 \cup U_2 = \mathbb{P}^1_{\text{Berk}}$, it follows that $\varphi^*(\mu) = d \cdot \mu$.

The identity $\varphi_*(\mu) = \mu$ follows formally from $\varphi^*(\mu) = d \cdot \mu$. By Theorem 9.53(B), $\varphi_*(\varphi^*(\mu)) = d \cdot \mu$. Since $\varphi^*(\mu) = d \cdot \mu$, this gives $\varphi_*(\mu) = \mu$. $\square$

We give the following definition (cf. Corollary 9.27):

DEFINITION 10.3. A rational function $\varphi(T) \in K(T)$ of degree $d \geq 1$ has *good reduction* if it can be written as $\varphi(T) = F_2(1,T)/F_1(1,T)$ where $F_1, F_2 \in K[X,Y]$ are homogeneous polynomials of degree d whose coefficients belong to the valuation ring $\mathcal{O}$ of K and where $\text{Res}(F_1, F_2)$ is a unit in $\mathcal{O}$.

A rational function $\varphi(T) \in K(T)$ of degree $d \geq 1$ has *simple reduction*[3] if there is a Möbius transformation $M \in \text{PGL}(2,K)$ such that the conjugate $M \circ \varphi \circ M^{-1}$ has good reduction.

EXAMPLE 10.4. Suppose $\varphi(T)$ has good reduction and degree $d \geq 2$. We claim that the canonical measure μ_φ coincides with the Dirac measure $\delta_{\zeta_{\text{Gauss}}}$, where ζ_{Gauss} is the Gauss point of $\mathbb{P}^1_{\text{Berk}}$.

To see this, recall formula (10.1) and note that under our hypotheses, we may take $B_1 = B_2 = 1$. Thus for all $(x,y) \in K^2$,
$$\|F(x,y)\| = \|(x,y)\|^d .$$
By iteration, for each n, $\|F^{(n)}(x,y)\| = \|(x,y)\|^{d^n}$. Examining the construction of $\hat{h}_{\varphi,v,(\infty)}$, one finds that
$$\hat{h}_{\varphi,v,(\infty)}(x) = \log_v \max(1, [T]_x) = \log_v(\delta(x, \zeta_{\text{Gauss}})_\infty) ,$$
so that $\Delta(\hat{h}_{\varphi,v,(\infty)}) = \delta_\infty - \delta_{\zeta_{\text{Gauss}}}$ by Example 5.19. Hence $\mu_\varphi = \delta_{\zeta_{\text{Gauss}}}$.

More generally, we have:

PROPOSITION 10.5 (Rivera-Letelier[4]). *A rational map $\varphi(T) \in K(T)$ of degree at least 2 has good reduction if and only if $\mu_\varphi = \delta_{\zeta_{\text{Gauss}}}$.*

PROOF. We have already shown in Example 10.4 that if φ has good reduction, then $\mu_\varphi = \delta_{\zeta_{\text{Gauss}}}$. Conversely, suppose that $\mu_\varphi = \delta_{\zeta_{\text{Gauss}}}$. Then using the identities $\varphi_*(\mu_\varphi) = \mu_\varphi$ and $\varphi^*(\mu_\varphi) = \deg(\varphi) \cdot \mu_\varphi$ proved in Theorem 10.2, it follows that $\varphi(\zeta_{\text{Gauss}}) = \zeta_{\text{Gauss}}$ and $m_\varphi(\zeta_{\text{Gauss}}) = \deg(\varphi)$. In particular, $\varphi^{-1}(\zeta_{\text{Gauss}}) = \{\zeta_{\text{Gauss}}\}$. By Corollary 9.27, this means that φ has good reduction. $\square$

REMARK 10.6. Later on (in Proposition 10.45), we will see that $\varphi(T)$ has simple reduction if and only if there is a point ζ (necessarily of type II) such that $\mu_\varphi = \delta_\zeta$. We will also see (in Corollary 10.47) that this is the only case in which μ_φ has point masses: μ_φ can never charge points of $\mathbb{P}^1(K)$, and it cannot charge a point of $\mathbb{H}_{\text{Berk}}$ unless φ has simple reduction.

We will now show that the canonical measure is better behaved than an arbitrary measure, in that it has continuous potentials (Definition 5.40): for each $\zeta \in \mathbb{H}_{\text{Berk}}$, the potential function $u_{\mu_\varphi}(z, \zeta) = \int -\log_v(\delta(z,y)_\zeta) \, d\mu_\varphi(y)$ is continuous (hence bounded).

---

[3]This is Rivera-Letelier's terminology.
[4]See [46, Proposition 0.1] and [48, Theorem E].

PROPOSITION 10.7. *For any rational function $\varphi(T) \in K(T)$ of degree $d \geq 2$, the canonical measure μ_φ has continuous potentials.*

PROOF. Apply Proposition 8.65, noting that in $V_1 = \mathbb{P}^1_{\text{Berk}} \setminus \{\infty\}$, the Laplacian of $\hat{h}_{\varphi,v,(\infty)}$ is $\mu_\varphi|_{V_1}$ and that in $V_2 = \mathbb{P}^1_{\text{Berk}} \setminus \{0\}$ the Laplacian of $\hat{h}_{\varphi,v,(0)}$ is $\mu_\varphi|_{V_2}$. □

Since μ_φ has continuous potentials, the theory of Arakelov-Green's functions developed in §8.10 applies to it. For $x, y \in \mathbb{P}^1_{\text{Berk}}$, we define the *normalized Arakelov-Green's function attached to μ_φ* by

$$(10.20) \qquad g_{\mu_\varphi}(x,y) = \int -\log_v(\delta(x,y)_\zeta) \, d\mu_\varphi(\zeta) + C ,$$

where the constant C is chosen so that

$$\iint g_{\mu_\varphi}(x,y) \, d\mu_\varphi(x) d\mu_\varphi(y) = 0 .$$

(In §10.2, we will determine the constant C explicitly.) By Proposition 8.66, $g_{\mu_\varphi}(x,y)$ is continuous in each variable separately (as a function from $\mathbb{P}^1_{\text{Berk}}$ to $\mathbb{R} \cup \{\infty\}$). As a function of two variables, it is symmetric, bounded below, and continuous off the diagonal and on the type I diagonal, but at points on the diagonal in $\mathbb{H}_{\text{Berk}} \times \mathbb{H}_{\text{Berk}}$ it is only lower semicontinuous.

Given a probability measure ν on $\mathbb{P}^1_{\text{Berk}}$, define the μ_φ-energy integral

$$I_{\mu_\varphi}(\nu) = \iint g_{\mu_\varphi}(x,y) \, d\nu(x) d\nu(y) .$$

Theorem 8.67 yields the following energy minimization principle, which will be used in §10.3 below to prove an equidistribution theorem for algebraic points of small dynamical height:

THEOREM 10.8 (μ_φ-Energy Minimization Principle).
(A) $I_{\mu_\varphi}(\nu) \geq 0$ *for each probability measure ν on $\mathbb{P}^1_{\text{Berk}}$.*
(B) $I_{\mu_\varphi}(\nu) = 0$ *if and only if $\nu = \mu_\varphi$.*

Combining Proposition 10.5, Theorem 10.8, and Corollary 9.27, we obtain:

COROLLARY 10.9. *We have $g_{\mu_\varphi}(\xi, \xi) \geq 0$ for all $\xi \in \mathbb{H}_{\text{Berk}}$, with equality iff $\mu_\varphi = \delta_\xi$. In particular, $g_{\mu_\varphi}(\zeta_{\text{Gauss}}, \zeta_{\text{Gauss}}) \geq 0$, with equality iff φ has good reduction.*

10.2. The Arakelov-Green's function $g_{\mu_\varphi}(x,y)$

Our goal in this section is to provide an explicit formula for the normalized Arakelov-Green's function $g_{\mu_\varphi}(x,y)$: we will show that $g_{\mu_\varphi}(x,y)$ coincides with the function $g_\varphi(x,y)$ given by (10.21) or (10.23) below. We also establish an important functional equation satisfied by $g_{\mu_\varphi}(x,y)$.

We continue the notation from §10.1. In particular, $\varphi(T) \in K(T)$ is a rational function of degree $d \geq 2$, and $F_1(X,Y), F_2(X,Y)$ are homogeneous polynomials of degree d such that $\varphi(T) = F_2(1,T)/F_1(1,T)$ and $\text{Res}(F_1, F_2) \neq 0$. The polynomials F_1, F_2 are determined by φ up to multiplication by a common nonzero scalar $c \in K^\times$.

With the conventions that $-\log_v(\delta(x,x)_\infty) = \infty$ for all $x \in \mathbb{P}^1(K)$ and that $\hat{h}_{\varphi,v,(\infty)}(\infty) = \infty$, we make the following definition:

For $x, y \in \mathbb{P}^1_{\text{Berk}}$, let $R = |\text{Res}(F_1, F_2)|^{-\frac{1}{d(d-1)}}$ and define

$$(10.21) \quad g_\varphi(x,y) = \begin{cases} -\log_v(\delta(x,y)_\infty) + \hat{h}_{\varphi,v,(\infty)}(x) \\ \quad + \hat{h}_{\varphi,v,(\infty)}(y) + \log_v R & \text{if } x, y \neq \infty, \\ \hat{h}_{\varphi,v,(\infty)}(x) + \hat{h}_{\varphi,v,(0)}(\infty) + \log_v R & \text{if } y = \infty, \\ \hat{h}_{\varphi,v,(\infty)}(y) + \hat{h}_{\varphi,v,(0)}(\infty) + \log_v R & \text{if } x = \infty. \end{cases}$$

For any fixed $y \neq \infty$, taking $a = 0$ in Corollary 4.2 and using Proposition 4.1(C) with $z = 0$ shows that $\delta(x,y)_\infty = \log_v([T]_x)$ for all x sufficiently near ∞. Hence by (10.16)

$$(10.22) \quad \lim_{x \to \infty} \hat{h}_{\varphi,v,(\infty)}(x) - \log_v(\delta(x,y)_\infty) = \hat{h}_{\varphi,v,(0)}(\infty).$$

From this, it follows that for each fixed $y \in \mathbb{P}^1_{\text{Berk}}$, $g_\varphi(x,y)$ is continuous and belongs to $\text{BDV}(\mathbb{P}^1_{\text{Berk}})$ as a function of x. It is also clear from the definition that $g_\varphi(x,y)$ is symmetric in x and y.

As we will see in Lemma 10.10, the reason for the choice of the normalizing constant $\log_v R$ in (10.21) is to make g_φ independent of the lift $F = (F_1, F_2)$ of φ. This has the magical consequence that the functional equation in Theorem 10.18 holds and that in (10.25) the normalizing constant C is equal to 0.

It is possible to give a more elegant description of $g_\varphi(x,y)$ when $x, y \in \mathbb{P}^1(K)$. To do this, given $\tilde{x} = (x_1, x_2)$, $\tilde{y} = (y_1, y_2) \in K^2$, define $\tilde{x} \wedge \tilde{y}$ to be

$$\tilde{x} \wedge \tilde{y} = x_1 y_2 - x_2 y_1,$$

and let $\|\tilde{x}\| = \max(|x_1|, |x_2|)$.

Recall that $F = (F_1, F_2) : K^2 \to K^2$ is a lifting of φ to K^2 and that we defined $F^{(n)} : K^2 \to K^2$ to be the nth iterate of F. For $\tilde{z} \in K^2 \setminus \{0\}$, define the *homogeneous dynamical height* $H_F : K^2 \setminus \{0\} \to \mathbb{R}$ by

$$H_F(\tilde{z}) := \lim_{n \to \infty} \frac{1}{d^n} \log_v \|F^{(n)}(\tilde{z})\|.$$

(By convention, we put $H_F(0,0) := -\infty$.)

By a standard telescoping series argument (see [9]), it follows that the limit $\lim_{n \to \infty} \frac{1}{d^n} \log_v \|F^{(n)}(\tilde{z})\|$ exists for all $\tilde{z} \in K^2 \setminus \{0\}$ and that the sequence $\frac{1}{d^n} \log_v \|F^{(n)}(\tilde{z})\|$ converges uniformly on $K^2 \setminus \{0\}$ to $H_F(\tilde{z})$. The definition of H_F is independent of the norm used to define it; this follows easily from the equivalence of norms on K^2. Moreover, it is easy to check that $H_F(F(z)) = dH_F(z)$ for all $z \in K^2 \setminus \{0\}$.

10.2. THE ARAKELOV-GREEN'S FUNCTION $g_{\mu_\varphi}(x,y)$

LEMMA 10.10. *Let $x, y \in \mathbb{P}^1(K)$, and let $\tilde{x}, \tilde{y}$ be arbitrary lifts of x, y to $K^2 \backslash \{0\}$. Then*

(10.23) $\qquad g_\varphi(x,y) = -\log_v |\tilde{x} \wedge \tilde{y}| + H_F(\tilde{x}) + H_F(\tilde{y}) + \log_v R$.

Moreover, the right-hand side of (10.23) is independent of the choice of the lifts F_1, F_2 of F.

PROOF. This is a straightforward computation (see [**9**, §3.4]). □

We will now show that $g_\varphi(x,y) = g_{\mu_\varphi}(x,y)$ for all $x, y \in \mathbb{P}^1_{\text{Berk}}$. The proof is based on the following observation.

LEMMA 10.11. *For each $y \in \mathbb{P}^1_{\text{Berk}}$, we have $\Delta_x g_\varphi(x,y) = \delta_y - \mu_\varphi$.*

PROOF. This follows from the definition of $g_\varphi(x,y)$, together with the identities $\Delta_x(-\log_v(\delta(x,y)_\infty)) = \delta_y - \delta_\infty$ and $\Delta_x(\hat{h}_{\varphi,v,(\infty)}) = \delta_\infty - \mu_\varphi$ (see Theorem 10.2). □

It follows from Proposition 8.66 and Lemma 10.11 that for each $y \in \mathbb{P}^1_{\text{Berk}}$,

$$\Delta_x(g_\varphi(x,y)) = \delta_y - \mu_\varphi = \Delta_x(g_{\mu_\varphi}(x,y)) .$$

By Proposition 5.28, there is a constant $C(y) \in \mathbb{R}$ such that

(10.24) $\qquad g_\varphi(x,y) = g_{\mu_\varphi}(x,y) + C(y)$

for all $x \in \mathbb{H}_{\text{Berk}}$. The continuity of $g_\varphi(x,y)$ and $g_{\mu_\varphi}(x,y)$ in x show that (10.24) in fact holds for all $x \in \mathbb{P}^1_{\text{Berk}}$.

Now fix x, and note that by symmetry, $g_\varphi(x,y)$ and $g_{\mu_\varphi}(x,y)$ are continuous and belong to $\text{BDV}(\mathbb{P}^1_{\text{Berk}})$ as functions of y. Hence $C(y)$ is continuous and $C(y) \in \text{BDV}(\mathbb{P}^1_{\text{Berk}})$. Taking Laplacians with respect to y of both sides of (10.24), we find that $\Delta_y(C(y)) = 0$. By the same argument as before, $C(y) = C$ is constant on $\mathbb{P}^1_{\text{Berk}}$, so

(10.25) $\qquad g_\varphi(x,y) = g_{\mu_\varphi}(x,y) + C$

for all $x, y \in \mathbb{P}^1_{\text{Berk}}$.

We claim that $C = 0$. To see this, it suffices to show that for some y,

$$\int g_\varphi(x,y) \, d\mu_\varphi(x) = \int g_{\mu_\varphi}(x,y) \, d\mu_\varphi(x) = 0 .$$

Below, we will show that both integrals are 0 for each y.

For $g_{\mu_\varphi}(x,y)$, this follows easily from known facts:

PROPOSITION 10.12. *For each $y \in \mathbb{P}^1_{\text{Berk}}$, we have*

$$\int_{\mathbb{P}^1_{\text{Berk}}} g_{\mu_\varphi}(x,y) \, d\mu_\varphi(x) = 0 .$$

PROOF. By Proposition 8.68, the potential function

$$u_{\mu_\varphi}(y, \mu_\varphi) = \int g_{\mu_\varphi}(x,y) \, d\mu_\varphi(x)$$

has Laplacian equal to $\mu_\varphi - \mu_\varphi = 0$. By continuity, $\int g_{\mu_\varphi}(x,y)\, d\mu_\varphi(x)$ is a constant independent of y. Integrating against μ_φ with respect to y and using the normalization in the definition of $g_{\mu_\varphi}(x,y)$ shows that this constant is 0. □

It remains to show that for each fixed $y \in \mathbb{P}^1_{\text{Berk}}$,
$$\int_{\mathbb{P}^1_{\text{Berk}}} g_\varphi(x,y)\, d\mu_\varphi(x) = 0\ .$$
We will prove this in Corollary 10.20 after first establishing a useful functional equation for $g_\varphi(x,y)$ (Theorem 10.18 below). For this we need several preliminary results.

LEMMA 10.13. *If $M \in \operatorname{GL}_2(K)$, let $F' = M^{-1} \circ F \circ M$, and let $\varphi' = M^{-1} \circ \varphi \circ M$. Then for all $z, w \in K^2 \backslash \{0\}$,*

(10.26) $$H_F(M(z)) = H_{F'}(z)$$

and

(10.27) $$g_\varphi(M(z), M(w)) = g_{\varphi'}(z, w)\ .$$

PROOF. First of all note that, given M, there exist constants $C_1, C_2 > 0$ such that

(10.28) $$C_1 \|z\| \leq \|M(z)\| \leq C_2 \|z\|\ .$$

Indeed, if we take C_2 to be the maximum of the absolute values of the entries of M, then clearly $\|M(z)\| \leq C_2 \|z\|$ for all z. By the same reasoning, if we let C_1^{-1} be the maximum of the absolute values of the entries of M^{-1}, then
$$\|M^{-1}(M(z))\| \leq C_1^{-1} \|M(z)\|\ ,$$
which gives the other inequality. (Alternatively, one can use the argument from Lemma 10.1.)

By the definition of H_F, we have
$$H_F(M(z)) = \lim_{n \to \infty} \frac{1}{d^n} \log_v \|F^{(n)}(M(z))\|$$
$$= \lim_{n \to \infty} \frac{1}{d^n} \log_v \|MF'^{(n)}(z)\| = H_{F'}(z)\ ,$$
where the last equality follows from (10.28). This proves (10.26).

Since $|M(z) \wedge M(w)| = |\det(M)| \cdot |z \wedge w|$, we have

(10.29) $$-\log_v |M(z) \wedge M(w)| = -\log_v |\det(M)| - \log_v |z \wedge w|\ .$$

On the other hand, by (10.31) below and the fact that $|\operatorname{Res}(M)| = |\det(M)|$, it follows that

(10.30) $$-\frac{1}{d(d-1)} \log_v |\operatorname{Res}(F')| = -\frac{1}{d(d-1)} \log_v |\operatorname{Res}(F)| - \log_v |\det(M)|\ .$$

Combining (10.26), (10.29), and (10.30) gives (10.27). □

COROLLARY 10.14. *Let $M \in \mathrm{PGL}_2(K)$ be a Möbius transformation. Then for all $z, w \in \mathbb{P}^1(K)$, we have*
$$g_\varphi(M(z), M(w)) = g_{M^{-1} \circ \varphi \circ M}(z, w) .$$

COROLLARY 10.15. *Let $M \in \mathrm{PGL}_2(K)$ be a Möbius transformation. Then $\mu_{M^{-1} \circ \varphi \circ M} = M^* \mu_\varphi$.*

PROOF. This follows from Corollary 10.14, together with Lemma 10.11 and Proposition 9.56. □

Let k be a field. By [**71**, Theorem IX.3.13], if $F = (F_1, F_2)$ and $G = (G_1, G_2)$ where $F_1, F_2 \in k[X, Y]$ are homogeneous of degree d and $G_1, G_2 \in k[X, Y]$ are homogeneous of degree e, then

$$(10.31) \qquad \mathrm{Res}(F \circ G) = \mathrm{Res}(F)^e \mathrm{Res}(G)^{d^2} .$$

In particular, it follows by induction on n that

$$(10.32) \qquad \mathrm{Res}(F^{(n)}) = \mathrm{Res}(F)^{\frac{d^{n-1}(d^n - 1)}{d-1}} .$$

LEMMA 10.16. *For every positive integer n, we have $\mu_\varphi = \mu_{\varphi^{(n)}}$ and*
$$g_\varphi(z, w) = g_{\varphi^{(n)}}(z, w)$$
for all $z, w \in \mathbb{P}^1_{\mathrm{Berk}}$.

PROOF. The second assertion follows easily from the definition of $g_\varphi(z, w)$ using the fact that
$$|\mathrm{Res}(F)|^{-\frac{1}{d(d-1)}} = |\mathrm{Res}(F^{(n)})|^{-\frac{1}{d^n(d^n-1)}}$$
for any homogeneous lifting F of φ, by (10.32). The first assertion follows from this by Lemma 10.11. □

As a consequence of Lemma 10.16 and Proposition 10.5, we obtain the following fact, originally proved by Benedetto [**14**] by a different method (see also [**81**, §7]):

COROLLARY 10.17. *For a rational map $\varphi \in K(T)$ of degree at least 2, the following are equivalent:*

(A) *φ has good reduction.*
(B) *$\varphi^{(n)}$ has good reduction for some integer $n \geq 2$.*
(C) *$\varphi^{(n)}$ has good reduction for every integer $n \geq 2$.*

For each $x, y \in \mathbb{P}^1_{\mathrm{Berk}}$, define

$$(10.33) \qquad g_\varphi(x, \varphi^*(y)) := \sum_{i=1}^{d} g_\varphi(x, y_i) ,$$

where $y_1, \ldots, y_d$ are the preimages of y under φ, counting multiplicities.

The function $g_\varphi(x, y)$ satisfies the following functional equation:

THEOREM 10.18. *Let $\varphi(T) \in K(T)$ be a rational function of degree $d \geq 2$. Then for all $x, y \in \mathbb{P}^1_{\text{Berk}}$,*

$$g_\varphi(\varphi(x), y) = g_\varphi(x, \varphi^*(y)) . \tag{10.34}$$

We first establish this in the special case where both $x, y \in \mathbb{P}^1(K)$:

LEMMA 10.19. *For all $x, y \in \mathbb{P}^1(K)$, we have*

$$g_\varphi(\varphi(x), y) = g_\varphi(x, \varphi^*(y)) .$$

PROOF. Using Corollary 10.14, we may assume without loss of generality that $y = \infty$. Let F be a homogeneous lifting of φ, and let $R = |\operatorname{Res}(F)|^{-\frac{1}{d(d-1)}}$. For $\tilde{z} \in K^2 \backslash \{0\}$, let $[\tilde{z}]$ denote the class of $\tilde{z}$ in $\mathbb{P}^1(K)$. If $\tilde{w} = (1, 0)$, then $[\tilde{w}] = \infty$. Let $w_1, \ldots, w_d$ be the preimages of ∞ under φ (counting multiplicities), and let $\tilde{w}_1, \ldots, \tilde{w}_d$ be any solutions to $F(\tilde{w}_i) = \tilde{w}$ with $[\tilde{w}_i] = w_i$.

It suffices to show that for all $\tilde{z} \in K^2 \backslash \{0\}$, we have

$$g_\varphi(\varphi([\tilde{z}]), [\tilde{w}]) = \sum_{i=1}^d g_\varphi([\tilde{z}], [\tilde{w}_i]) . \tag{10.35}$$

Since $H_F(F(\tilde{z})) = d H_F(\tilde{z})$ and $H_F(\tilde{w}_i) = \frac{1}{d} H_F(\tilde{w})$ for all i, (10.35) is equivalent to

$$-\log_v |F(\tilde{z}) \wedge \tilde{w}| = (d-1) \log_v R - \sum_{i=1}^d \log_v |\tilde{z} \wedge \tilde{w}_i| ,$$

which itself is equivalent to

$$|F(\tilde{z}) \wedge \tilde{w}| = |\operatorname{Res}(F)|^{1/d} \prod_{i=1}^d |\tilde{z} \wedge \tilde{w}_i| . \tag{10.36}$$

We verify (10.36) by an explicit calculation (compare with [**41**, Lemma 6.5]). Write $F_1(\tilde{z}) = \prod_{i=1}^d \tilde{z} \wedge \tilde{a}_i$, $F_2(\tilde{z}) = \prod_{i=1}^d \tilde{z} \wedge \tilde{b}_i$ with $\tilde{a}_i, \tilde{b}_i \in K^2$. Since $F(\tilde{w}_j) = \tilde{w} = (1, 0)$, it follows that $\prod_{i=1}^d \tilde{w}_j \wedge \tilde{a}_i = 1$, $\prod_{i=1}^d \tilde{w}_i \wedge \tilde{b}_i = 0$. Thus for each $j = 1, \ldots, d$, we can assume $\tilde{w}_j$ has been chosen so that

$$\tilde{w}_j = \frac{\tilde{b}_j}{(\prod_i \tilde{a}_i \wedge \tilde{b}_j)^{1/d}} ,$$

where for each j we fix some d^{th} root of $\prod_i \tilde{a}_i \wedge \tilde{b}_j$. Note that $|\tilde{w}_j|$ is independent of which d^{th} root we pick. It follows that

$$\prod_{j=1}^d |\tilde{z} \wedge \tilde{w}_j| = \frac{\prod_j |\tilde{z} \wedge \tilde{b}_j|}{\prod_{i,j} |\tilde{a}_i \wedge \tilde{b}_j|^{1/d}} = \frac{|F(\tilde{z}) \wedge \tilde{w}|}{|\operatorname{Res}(F)|^{1/d}} ,$$

which gives (10.36). □

Recall that by Proposition 9.13, for any continuous function $f : \mathbb{P}^1_{\text{Berk}} \to \mathbb{R}$, the function $\varphi_*(f)$ defined by

$$\varphi_*(f)(y) = \sum_{\varphi(x)=y} m_\varphi(x) f(x) \tag{10.37}$$

is itself continuous. (Here $m_\varphi(x)$ denotes the analytic multiplicity of φ at x, as defined in §9.1.)

Using this, we can show that Theorem 10.18 holds for all $x, y \in \mathbb{P}^1_{\text{Berk}}$:

PROOF OF THEOREM 10.18. First, fix $y \in \mathbb{P}^1(K)$. Then as functions of x, $g_\varphi(\varphi(x), y)$ and $g_\varphi(x, \varphi^*(y))$ are continuous from $\mathbb{P}^1_{\text{Berk}}$ to the extended reals. By Lemma 10.19, they agree on the dense subset $\mathbb{P}^1(K)$ of $\mathbb{P}^1_{\text{Berk}}$, so (10.34) holds for all $x \in \mathbb{P}^1_{\text{Berk}}$, $y \in \mathbb{P}^1(K)$.

Next, fix $x \in \mathbb{H}^1_{\text{Berk}}$. Then as a function of y, $g_\varphi(x, y)$ is bounded and continuous, and $g_\varphi(x, \varphi^*(y))$ coincides with the pushforward of the function $f(y) = g_\varphi(x, y)$ as in (10.37). By Proposition 9.13 it is continuous for all $y \in \mathbb{P}^1_{\text{Berk}}$. Thus both $g_\varphi(\varphi(x), y)$ and $g_\varphi(x, \varphi^*(y))$ are continuous and real-valued for all $y \in \mathbb{P}^1_{\text{Berk}}$. By what has been shown above, they agree for $y \in \mathbb{P}^1(K)$, so (10.34) holds for all $x \in \mathbb{H}_{\text{Berk}}$ and $y \in \mathbb{P}^1_{\text{Berk}}$.

Finally, let $y \in \mathbb{H}_{\text{Berk}}$. Then $g_\varphi(\varphi(x), y)$ and $g_\varphi(x, \varphi^*(y))$ are continuous and real-valued functions of $x \in \mathbb{P}^1_{\text{Berk}}$, and they agree on the dense subset $\mathbb{H}_{\text{Berk}}$, so they agree for all $x \in \mathbb{P}^1_{\text{Berk}}$.

The cases above cover all possibilities for $x, y \in \mathbb{P}^1_{\text{Berk}}$. $\square$

Using Theorem 10.18, we deduce the following:

COROLLARY 10.20. *For all $y \in \mathbb{P}^1_{\text{Berk}}$, we have*

$$\int_{\mathbb{P}^1_{\text{Berk}}} g_\varphi(x, y) \, d\mu_\varphi(x) = 0 . \tag{10.38}$$

PROOF. By (10.25) and Proposition 10.12, there is a constant C such that $\int g_\varphi(x, y) \, d\mu_\varphi(x) = C$ for each $y \in \mathbb{P}^1_{\text{Berk}}$, and we need to show that $C = 0$.

Fix $y \in \mathbb{P}^1_{\text{Berk}}$. Since $\varphi_* \mu_\varphi = \mu_\varphi$, it follows from Theorems 10.2 and 10.18 and formula (9.44) that

$$\begin{aligned} C &= \int g_\varphi(x, y) \, d\mu_\varphi(x) = \int g_\varphi(x, y) \, d(\varphi_* \mu_\varphi)(x) \\ &= \int g_\varphi(\varphi(x), y) \, d\mu_\varphi(x) = \int g_\varphi(x, \varphi^*(y)) \, d\mu_\varphi(x) = d \cdot C , \end{aligned}$$

and thus $C = 0$ as desired. $\square$

From (10.25), using Proposition 10.12 and Corollary 10.20, we finally conclude:

THEOREM 10.21. $g_\varphi(x, y) = g_{\mu_\varphi}(x, y)$ *for all $x, y \in \mathbb{P}^1_{\text{Berk}}$.*

We close this section by remarking that Theorem 10.18 shows that $g_\varphi(x, y)$ is a one-parameter family of Call-Silverman local heights, normalized by the condition (10.38). Indeed, for each fixed $y \in \mathbb{P}^1(K)$,

$$\Delta_x\big(g_\varphi(x, y) - \log_v(\|x, y\|)\big) \;=\; \delta_{\zeta_{\text{Gauss}}} - \mu_\varphi \;=\; \Delta_x(g_\varphi(x, \zeta_{\text{Gauss}})) \;,$$

so since $g_\varphi(x, y) - \log_v(\|x, y\|)$ and $g_\varphi(x, \zeta_{\text{Gauss}})$ both have continuous extensions to $\mathbb{P}^1_{\text{Berk}}$, there is a constant $C(y)$ such that

$$g_\varphi(x, y) - \log_v(\|x, y\|) \;=\; g_\varphi(x, \zeta_{\text{Gauss}}) + C(y) \;.$$

Here $g_\varphi(x, \zeta_{\text{Gauss}})$ is continuous and bounded since μ_φ has continuous potentials (Proposition 10.7). Hence $|g_\varphi(x, y) - \log_v(\|x, y\|)|$ is bounded, and $g_\varphi(x, y)$ is a Weil height with respect to the divisor (y). Furthermore, if $f_y(T) \in K(T)$ is a function with $\text{div}(f_y) = d(y) - \varphi^*((y))$, then

$$\begin{aligned}\Delta_x(g_\varphi(\varphi(x), y)) \;=\; \Delta_x(g_\varphi(x, \varphi^*(y))) \;&=\; \sum_{\varphi(y_i)=y} m_\varphi(y_i)\delta_{y_i}(x) - d \cdot \mu_\varphi \\ &=\; \Delta_x(d \cdot g_\varphi(x, y) - \log_v([f_y]_x)) \;.\end{aligned}$$

After replacing f_y by a constant multiple, if necessary, we obtain

$$g_\varphi(\varphi(x), y) \;=\; d \cdot g_\varphi(x, y) - \log_v([f_y]_x) \;.$$

This holds for all $x \in \mathbb{P}^1_{\text{Berk}} \setminus (\text{supp}(D) \cup \text{supp}(\varphi^*(D)))$.

Thus $g_\varphi(x, y)$ satisfies both conditions in the definition of a Call-Silverman local height for (y).

10.3. Adelic equidistribution of dynamically small points[5]

Let k be a number field. In this section, we prove an adelic equidistribution theorem (Theorem 10.24) for sequences of distinct points $P_n \in \mathbb{P}^1(\overline{k})$ for which $\widehat{h}_\varphi(P_n) \to 0$, where $\widehat{h}_\varphi$ is the canonical height associated to a rational function $\varphi(T) \in k(T)$. This result is due (independently) to Baker and Rumely [9], to Chambert-Loir [35], and to Favre and Rivera-Letelier [47]. Here we follow the proof from [9], with simplifications. Since it is not much more difficult to prove, we actually formulate our equidistribution theorem in a more general context, with Galois invariant subsets instead of Galois orbits of points and with k an arbitrary product formula field, of any characteristic. Our exposition has been influenced by the treatment of this material in [34] and [47]. We quote without proof the necessary results from potential theory over $\mathbb{C}$ in the Archimedean case, but otherwise we provide a complete and self-contained argument.

Our proof of the adelic equidistribution theorem depends on (part (A) of) the following purely local result:

---

[5]Parts of §10.3 are taken from "A lower bound for average values of dynamical Green's functions", Math. Res. Lett. 13(2):245–257, 2006. Copyright International Press, 2006. Used with permission.

THEOREM 10.22. *Let K be a field with a nontrivial (Archimedean or non-Archimedean) absolute value. Let $\varphi \in K(T)$ be a rational function of degree $d \geq 2$, and let g_φ be the normalized Arakelov-Green's function associated to φ. For $z_1, \ldots, z_N$ in $\mathbb{P}^1(K)$ ($N \geq 2$), define the* discrepancy *of $z_1, \ldots, z_N$ relative to φ to be*

$$D_\varphi(z_1, \ldots, z_N) \;=\; \frac{1}{N(N-1)} \sum_{\substack{1 \leq i,j \leq N \\ i \neq j}} g_\varphi(z_i, z_j) \;.$$

(By convention, we set $D_\varphi(z_1, \ldots, z_N) = +\infty$ if $z_i = z_j$ for some $i \neq j$.) Then:

(A) *The sequence $D_N := \inf_{z_1, \ldots, z_N \in \mathbb{P}^1(K)} D_\varphi(z_1, \ldots, z_N)$ is nondecreasing, and $\lim_{N \to \infty} D_N \geq 0$.*
(B) *If K is complete and algebraically closed, then*

$$\lim_{N \to \infty} D_N = 0 \;.$$

We will prove Theorem 10.22 at the end of this section. Following [**3**], we will also prove the following quantitative version of the nonnegativity assertion in Theorem 10.22(A).

THEOREM 10.23. *Let K be a field with a nontrivial (Archimedean or non-Archimedean) absolute value, and let $\varphi \in K(T)$ be a rational function of degree $d \geq 2$. Then there is an effective constant $C > 0$, depending only on φ and K, such that*

(10.39) $$D_\varphi(z_1, \ldots, z_N) \;\geq\; -C \frac{\log N}{N-1}$$

for all $z_1, \ldots, z_N \in \mathbb{P}^1(K)$ with $N \geq 2$.

Theorem 10.23 is a dynamical analogue of classical results due to Mahler and Elkies and can be used to make the equidistribution in Theorem 10.24 "quantitative".

In order to state the adelic equidistribution theorem for points of small dynamical height, we first need some definitions and notation. Let k be a product formula field (Definition 7.51), and let $\overline{k}$ (resp. k^{sep}) denote a fixed algebraic closure (resp. separable closure) of k. For $v \in \mathcal{M}_k$, let k_v be the completion of k at v, let $\overline{k}_v$ be an algebraic closure of k_v, and let $\mathbb{C}_v$ denote the completion of $\overline{k}_v$. For each $v \in \mathcal{M}_k$, we fix an embedding of $\overline{k}$ in $\mathbb{C}_v$ extending the canonical embedding of k in k_v and view this embedding as an identification. As discussed in §7.9, if v is Archimedean, then $\mathbb{C}_v \cong \mathbb{C}$.

Let $\varphi \in k(T)$ be a rational map of degree $d \geq 2$ defined over k, so that φ acts on $\mathbb{P}^1(\overline{k})$ in the usual way. As before, we define a *homogeneous lifting* of φ to be a choice of homogeneous polynomials $F_1, F_2 \in k[X, Y]$ of degree $d \geq 2$ having no common linear factor in $\overline{k}[X, Y]$ such that

$$\varphi([z_0 : z_1]) \;=\; [F_1(z_0, z_1) : F_2(z_0, z_1)]$$

for all $[z_0 : z_1] \in \mathbb{P}^1(\overline{k})$. The polynomials F_1, F_2 are uniquely determined by φ up to multiplication by a common scalar $c \in k^\times$, and the condition that F_1, F_2 have no common linear factor in $\overline{k}[X, Y]$ is equivalent to saying that the homogeneous resultant $\mathrm{Res}(F) = \mathrm{Res}(F_1, F_2) \in k$ is nonzero.

The mapping
$$F = (F_1, F_2) : \overline{k}^2 \to \overline{k}^2$$
is a lifting of φ to $\overline{k}^2$, and we denote by $F^{(n)} : \overline{k}^2 \to \overline{k}^2$ the iterated map $F \circ F \circ \ldots \circ F$ (n times).

The global canonical height function $\widehat{H}_F$ is defined for $P \in k^2 \setminus \{0\}$ by
$$\widehat{H}_F(P) = \lim_{n \to \infty} \frac{1}{d^n} \sum_{v \in \mathcal{M}_k} N_v \log \|F^{(n)}(P)\|_v ,$$
where $\|(x, y)\|_v = \max\{|x|_v, |y|_v\}$. Equivalently, in the terminology of §10.2, we have
$$\widehat{H}_F(P) = \sum_{v \in \mathcal{M}_k} N_v H_{F,v}(P) ,$$
where $H_{F,v} : \mathbb{C}_v^2 \setminus \{0\} \to \mathbb{R}$ is the homogeneous dynamical height function associated to F, relative to the ground field $K = \mathbb{C}_v$. By convention, we set $\widehat{H}_F(0) = 0$.

One sees easily from the product formula that $\widehat{H}_F(P) \geq 0$ for all P, that $\widehat{H}_F(P)$ depends only on the class of P in $\mathbb{P}^1(k)$, and that $\widehat{H}_{cF} = \widehat{H}_F$ for $c \in k^\times$. Thus $\widehat{H}_F$ descends to a well-defined global canonical height function $\widehat{h}_\varphi : \mathbb{P}^1(k) \to \mathbb{R}_{\geq 0}$ depending only on φ.

More generally, if $S \subset \mathbb{P}^1(k^{\mathrm{sep}})$ is any finite set which is invariant under $\mathrm{Gal}(k^{\mathrm{sep}}/k)$, we define $\widehat{h}_\varphi(S)$ by

(10.40) $$\widehat{h}_\varphi(S) = \sum_{v \in \mathcal{M}_k} N_v \left(\frac{1}{|S|} \sum_{P \in S} H_{F,v}(\tilde{P}) \right) ,$$

where $\tilde{P}$ is an arbitrary lifting of P to $\overline{k}^2 \setminus \{0\}$. Using Galois-invariance, one sees easily that $\sum_{P \in S} H_{F,v}(\tilde{P})$ does not depend on the choice of an embedding of $\overline{k}$ into $\mathbb{C}_v$. By the product formula, $\widehat{h}_\varphi(S)$ does not depend on the choice of $\tilde{P}$ in (10.40).

If $P \in \mathbb{P}^1(k^{\mathrm{sep}})$, let $S_k(P) = \{P_1, \ldots, P_n\}$ denote the set of $\mathrm{Gal}(k^{\mathrm{sep}}/k)$-conjugates of P over k, where $n = [k(P) : k]$. We define $\widehat{h}_\varphi(P) = \widehat{h}_\varphi(S_k(P))$. This extends $\widehat{h}_\varphi : \mathbb{P}^1(k) \to \mathbb{R}_{\geq 0}$ in a canonical way to a function $\widehat{h}_\varphi : \mathbb{P}^1(k^{\mathrm{sep}}) \to \mathbb{R}_{\geq 0}$.

For all $P \in \mathbb{P}^1(k^{\mathrm{sep}})$, $\widehat{h}_\varphi$ satisfies the functional equation

(10.41) $$\widehat{h}_\varphi(\varphi(P)) = d\, \widehat{h}_\varphi(P) ,$$

which is easily verified using the fact that $H_{F,v} \circ F = d\, H_{F,v}$ for each $v \in \mathcal{M}_k$.

If $h : \mathbb{P}^1(\bar{k}) \to \mathbb{R}_{\geq 0}$ denotes the standard logarithmic Weil height on $\mathbb{P}^1(\bar{k})$, defined for $P = [x : y] \in \mathbb{P}^1(k)$ by

$$h(P) = \sum_{v \in \mathcal{M}_k} N_v \log \max\{|x|_v, |y|_v\}$$

and extended to $\mathbb{P}^1(k^{\mathrm{sep}})$ as above, then there is a constant $C > 0$ such that

(10.42) $\qquad |\widehat{h}_\varphi(P) - h(P)| \leq C \quad \text{for all } P \in \mathbb{P}^1(k^{\mathrm{sep}}).$

This follows from the easily verified fact that

$$\widehat{h}_\varphi(P) = \lim_{n \to \infty} \frac{1}{d^n} h(\varphi^{(n)}(P))$$

for all $P \in \mathbb{P}^1(k^{\mathrm{sep}})$. If k is a global field (i.e., a number field or a function field in one variable over a finite field), then from the corresponding fact for the standard logarithmic Weil height h, it follows from (10.42) that $\widehat{h}_\varphi$ satisfies the *Northcott finiteness property*: For any real number $M > 0$ and any integer $D \geq 1$, the set

(10.43) $\qquad \{P \in \mathbb{P}^1(k^{\mathrm{sep}}) : [k(P):k] \leq D, \widehat{h}_\varphi(P) \leq M\}$

is finite (see [**93**, Theorem 3.12]).

It is easy to see using (10.41) and the Northcott finiteness property that if k is a global field and $P \in \mathbb{P}^1(k)$, then $\widehat{h}_\varphi(P) = 0$ if and only if P is *preperiodic* for φ (that is, the orbit of P under iteration of φ is finite).

For $v \in \mathcal{M}_k$, we let $\mathbb{P}^1_{\mathrm{Berk},v}$ denote the Berkovich projective line over $\mathbb{C}_v$ (which we take to mean $\mathbb{P}^1(\mathbb{C})$ if v is Archimedean). We let $\mu_{\varphi,v}$ be the canonical measure on $\mathbb{P}^1_{\mathrm{Berk},v}$ associated to the iteration of φ over $\mathbb{C}_v$, as defined in §10.1. Finally, we denote by $g_{\varphi,v}$ the corresponding normalized Arakelov-Green's function on $\mathbb{P}^1_{\mathrm{Berk},v} \times \mathbb{P}^1_{\mathrm{Berk},v}$. Our main result is:

THEOREM 10.24 (Adelic Equidistribution of Small Points). *Let k be a product formula field, and let $\varphi \in k(T)$ be a rational function of degree $d \geq 2$. Suppose S_n is a sequence of $\mathrm{Gal}(k^{\mathrm{sep}}/k)$-invariant finite subsets of $\mathbb{P}^1(k^{\mathrm{sep}})$ with $|S_n| \to \infty$ and $\widehat{h}_\varphi(S_n) \to 0$. Fix $v \in \mathcal{M}_k$, and for each n let δ_n be the discrete probability measure on $\mathbb{P}^1_{\mathrm{Berk},v}$ supported equally on the elements of S_n. Then the sequence of measures $\{\delta_n\}$ converges weakly to the canonical measure $\mu_{\varphi,v}$ on $\mathbb{P}^1_{\mathrm{Berk},v}$ associated to φ.*

For concreteness, we note the special case of Theorem 10.24 where k is a number field and S_n is the set of Galois conjugates of a point $P_n \in \mathbb{P}^1(\bar{k})$:

COROLLARY 10.25. *Let k be a number field, and let $\varphi \in k(T)$ be a rational function of degree $d \geq 2$. Suppose $\{P_n\}$ is a sequence of distinct points of $\mathbb{P}^1(\bar{k})$ with $\widehat{h}_\varphi(P_n) \to 0$. Fix a place v of k, and for each n let δ_n be the discrete probability measure on $\mathbb{P}^1_{\mathrm{Berk},v}$ supported equally on the $\mathrm{Gal}(\bar{k}/k)$-conjugates of P_n. Then the sequence of measures $\{\delta_n\}$ converges weakly to the canonical measure $\mu_{\varphi,v}$ on $\mathbb{P}^1_{\mathrm{Berk},v}$ associated to φ.*

Note that if S_n is the set of $\mathrm{Gal}(\bar{k}/k)$-conjugates of P_n in the statement of Corollary 10.25, then $|S_n| \to \infty$ by the Northcott finiteness property (10.43). Thus Corollary 10.25 is a special case of Theorem 10.24.

REMARK 10.26. If $\varphi(T)$ is a polynomial, then Theorem 10.24 is a special case of Theorem 7.52. Indeed, for $v \in \mathcal{M}_k$ let E_v be the Berkovich filled Julia set of $\varphi(T)$ over $\mathbb{C}_v$ (Definition 10.90), and let $\mathbb{E}$ be the corresponding compact Berkovich adelic set. Then by Theorem 10.91 below, $\widehat{h}_\varphi = h_\mathbb{E}$ and $\mu_{\varphi,v}$ coincides with the equilibrium distribution μ_v for E_v.

Before giving the proof of Theorem 10.24, we need some preliminary results. Let $S = \{P_1, \ldots, P_n\}$ be a finite subset of $\mathbb{P}^1(k^{\mathrm{sep}})$, with $n \geq 2$. For each $v \in \mathcal{M}_k$, define the *v-adic discrepancy* of S relative to φ by

$$D_v(S) = D_{\varphi,v}(S) = \frac{1}{n(n-1)} \sum_{\substack{i,j=1 \\ i \neq j}}^{n} g_{\varphi,v}(P_i, P_j) .$$

LEMMA 10.27. *For each* $\mathrm{Gal}(k^{\mathrm{sep}}/k)$*-invariant finite subset* S *of* $\mathbb{P}^1(k^{\mathrm{sep}})$, *we have*

$$\widehat{h}_\varphi(S) = \frac{1}{2} \sum_{v \in \mathcal{M}_k} N_v D_v(S) .$$

PROOF. Assume first that $\infty \notin S$. Write $S = \{P_1, \ldots, P_n\}$, and choose a homogeneous lifting $\tilde{P}_i = (1, z_i)$ of each P_i with $z_i \in k^{\mathrm{sep}}$. By definition of g_φ, we have

$$D_v(S) = -\frac{1}{n(n-1)} \sum_{i \neq j} \log |z_i - z_j|_v$$
$$+ \frac{2}{n} \sum_{i=1}^{n} H_{F,v}(\tilde{P}_i) - \frac{1}{d(d-1)} \log |\mathrm{Res}(F)|_v .$$

By Galois theory, $\prod_{i \neq j} (z_i - z_j) \in k^\times$, and also $\mathrm{Res}(F) \in k^\times$ by construction. The product formula therefore implies that

$$\frac{1}{2} \sum_{v \in \mathcal{M}_k} N_v D_v(S) = \sum_{v \in \mathcal{M}_k} N_v \left(\frac{1}{n} \sum_{i=1}^{n} H_{F,v}(\tilde{P}_i) \right) = \widehat{h}_\varphi(S) .$$

This proves the result when $\infty \notin S$. If $\infty \in S$, the result follows from a similar calculation (taking $\tilde{\infty} = (0,1)$). $\square$

PROPOSITION 10.28. *Let* $\{S_n\}$ *be a sequence of* $\mathrm{Gal}(k^{\mathrm{sep}}/k)$*-invariant finite subsets of* $\mathbb{P}^1(k^{\mathrm{sep}})$ *such that* $|S_n| \to \infty$ *and* $\lim_{n \to \infty} \widehat{h}_\varphi(S_n) = 0$. *Then for each* $w \in \mathcal{M}_k$, *we have*

$$\liminf_{n \to \infty} D_w(S_n) = 0 .$$

PROOF. For each $w \in \mathcal{M}_k$ and $n \geq 1$, define $b_{w,n} = N_w D_w(S_n)$. We make the following observations:

(1) If w is non-Archimedean and φ has good reduction at w (which is the case for all but finitely many $w \in \mathcal{M}_k$), then by Example 10.4 and (10.20), we have (setting $N_n = |S_n|$)

$$b_{w,n} = N_w \left(\frac{1}{N_n(N_n-1)} \sum_{\substack{x,y \in S_n \\ x \neq y}} -\log \|x,y\|_w \right) \geq 0 \ .$$

(2) For all $w \in \mathcal{M}_k$, we have $\liminf_{n \to \infty} b_{w,n} \geq 0$ by Theorem 10.22(A).
(3) For each n, we have $b_{w,n} = 0$ for all but finitely many $w \in \mathcal{M}_k$, and by Lemma 10.27 we have $\sum_{w \in \mathcal{M}_k} b_{w,n} = 2\widehat{h}_\varphi(S_n)$. In particular, $\lim_{n \to \infty} \sum_{w \in \mathcal{M}_k} b_{w,n} = 2 \lim_{n \to \infty} \widehat{h}_\varphi(S_n) = 0.$

Thus the hypotheses of Lemma 7.55 are satisfied, and we conclude that $\liminf_{n \to \infty} D_w(S_n) = 0$ for every $w \in \mathcal{M}_k$. $\square$

We can now give the proof of the Adelic Equidistribution of Small Points theorem:

PROOF OF THEOREM 10.24. We may assume without loss of generality that $|S_n| \geq 2$. Extracting a subsequence of S_n if necessary, we may assume by Prohorov's theorem (Theorem A.11) that δ_n converges weakly to a probability measure ν on $\mathbb{P}^1_{\text{Berk},v}$. We want to show that $\nu = \mu_{\varphi,v}$. To accomplish this, consider the energy of ν relative to $g_{\varphi,v}$, defined by

$$I_v(\nu) = \iint_{\mathbb{P}^1_{\text{Berk},v} \times \mathbb{P}^1_{\text{Berk},v}} g_{\varphi,v}(x,y) \, d\nu(x) \, d\nu(y) \ .$$

Recall from Theorem 10.8 and its Archimedean counterpart (proved in [**9**, §5]; see also [**34**, §3.2/2]) that $I_v(\nu) \geq 0$, with equality if and only if $\nu = \mu_{\varphi,v}$. On the other hand, we claim that $I_v(\nu) \leq 0$, which will immediately yield the desired result.

To prove the claim, let Δ be the diagonal in $\mathbb{P}^1_{\text{Berk},v} \times \mathbb{P}^1_{\text{Berk},v}$, and note that by definition of the v-adic discrepancy, we have

$$D_v(S_n) = \iint_{\mathbb{P}^1_{\text{Berk},v} \times \mathbb{P}^1_{\text{Berk},v} \setminus \Delta} g_{\varphi,v}(x,y) \, d\delta_n(x) \, d\delta_n(y) \ .$$

Since $g_{\varphi,v}$ is lower semicontinuous, Lemma 7.54 and Proposition 10.28 show that

$$I_v(\nu) \leq \liminf_{n \to \infty} D_v(S_n) = 0$$

as claimed. $\square$

We now give a proof of the local result, Theorem 10.22, which was used in our proof of Theorem 10.24. As in the statement of that result, we let K be a field endowed with a nontrivial (Archimedean or non-Archimedean) absolute value, and we let $\varphi \in K(T)$ be a rational function of degree $d \geq 2$. We begin with the following lemma [**9**, Lemma 3.48]:

LEMMA 10.29. *For $N \geq 2$, define*
$$D_N := \inf_{z_1,\ldots,z_N \in \mathbb{P}^1(K)} D_\varphi(z_1,\ldots,z_N) .$$
Then the sequence D_N is nondecreasing.

PROOF. If $z_0,\ldots,z_N \in \mathbb{P}^1(K)$, we have
$$\sum_{\substack{i,j=0 \\ i\neq j}}^N g_\varphi(z_i,z_j) = \frac{1}{N-1} \sum_{k=0}^N \left(\sum_{\substack{i\neq j \\ i,j\neq k}} g_\varphi(z_i,z_j) \right)$$
$$\geq \frac{1}{N-1} \sum_{k=0}^N N(N-1)D_N = N(N+1)D_N ,$$
from which we obtain $D_{N+1} \geq D_N$. □

The following result is taken from [**9**, Proof of Theorem 3.28]:

LEMMA 10.30. *Fix $N \geq 2$, and assume that K is complete and algebraically closed. Then for all $z_1,\ldots,z_N \in \mathbb{P}^1_{\text{Berk},K}$ we have*

(10.44) $$\frac{1}{N(N-1)} \sum_{i\neq j} g_\varphi(z_i,z_j) \geq D_N .$$

PROOF. If K is Archimedean, then this is trivial, since $\mathbb{P}^1_{\text{Berk},K} = \mathbb{P}^1(K)$ in that case. We may therefore assume that K is non-Archimedean. By (8.62) and Proposition 4.7, we have
$$g_\varphi(z,w) = \liminf_{\substack{(x,y)\to(z,w) \\ x,y\in\mathbb{P}^1(K)}} g_\varphi(x,y)$$
for all $z,w \in \mathbb{P}^1_{\text{Berk}}$. Thus for any $\varepsilon > 0$, there are points $x_1,\ldots,x_N \in \mathbb{P}^1(K)$ such that
$$|g_\varphi(z_i,z_j) - g_\varphi(x_i,x_j)| < \varepsilon$$
for all $i \neq j$. By definition, we have
$$\frac{1}{N(N-1)} \sum_{i\neq j} g_\varphi(x_i,x_j) \geq D_N ,$$
so letting $\varepsilon \to 0$ gives (10.44). □

We can now prove Theorem 10.22:

PROOF OF THEOREM 10.22. We will assume in the proof which follows that K is non-Archimedean. The Archimedean case follows from the same argument if we apply the complex counterpart of Theorem 10.8 (proved in [**9**, §5]; see also [**34**, §3.2/2]).

The sequence D_N is nondecreasing by Lemma 10.29, and in particular the limit $D_\infty := \lim_{N\to\infty} D_N$ exists. Replacing K by the completion of its algebraic closure if necessary, it therefore suffices to prove assertion (B). Let

$\mu = \mu_\varphi$ be the canonical measure for φ supported on $\mathbb{P}^1_{\text{Berk},K}$. We already know by (10.38) that $I_\mu(\mu) := \iint g_\varphi(x,y) d\mu(x) d\mu(y) = 0$, so it suffices to prove that $\lim_{N \to \infty} D_N = \iint g_\varphi(x,y) d\mu(x) d\mu(y)$. We do this following the argument of [**9**, Theorem 3.28].

By (10.44), for $N \geq 2$ we have
$$\frac{1}{N(N-1)} \sum_{i \neq j} g_\varphi(z_i, z_j) \geq D_N$$
for all $z_1, \ldots, z_N \in \mathbb{P}^1_{\text{Berk},K}$. Integrating this against $d\mu(z_1) \cdots d\mu(z_N)$ gives
$$I_\mu(\mu) = \frac{1}{N(N-1)} \sum_{i \neq j} I_\mu(\mu)$$
$$= \frac{1}{N(N-1)} \sum_{i \neq j} \iint g_\varphi(z_i, z_j) d\mu(z_i) d\mu(z_j) \geq D_N$$
for all $N \geq 2$. It follows that $I_\mu(\mu) \geq D_\infty$.

For the other direction, for each $N \geq 2$ choose distinct points $w_1, \ldots, w_N$ in $\mathbb{P}^1(K)$ such that
$$\frac{1}{N(N-1)} \sum_{i \neq j} g_\varphi(w_i, w_j) \leq D_N + \frac{1}{N},$$
and let $\nu_N := \frac{1}{N} \sum \delta_{w_i}$ be the discrete probability measure supported equally on each of the points w_i. Passing to a subsequence if necessary, we may assume by Prohorov's theorem (Theorem A.11) that the ν_N converge weakly to some probability measure ν on $\mathbb{P}^1_{\text{Berk},K}$. Since
$$\frac{N-1}{N}\left(D_N + \frac{1}{N}\right) \geq \frac{1}{N^2} \sum_{i \neq j} g_\varphi(w_i, w_j)$$
$$= \iint_{\mathbb{P}^1_{\text{Berk}} \times \mathbb{P}^1_{\text{Berk}} \setminus (\text{Diag})} g_\varphi(x,y) d\nu_N(x) d\nu_N(y) ,$$
it follows from Theorem 10.8 and Lemma 7.54 that
$$D_\infty = \lim_{N \to \infty} D_N \geq I_\mu(\nu) \geq I_\mu(\mu) . \qquad \square$$

Finally, we give a proof of Theorem 10.23, a strengthening of Theorem 10.22(A) which can be used to make the equidistribution in Theorem 10.24 quantitative.[6]

By Lemma 10.29, in order to prove Theorem 10.23, it will suffice to give lower bounds for D_N as N runs through some conveniently chosen subsequence of natural numbers. In the argument which follows, we will assume that N belongs to the subset Σ of $\mathbb{N}$ consisting of those positive integers which can be written as $N = td^k$ with $k \geq 0$ and $2 \leq t \leq 2d-1$.

---
[6]A different quantitative version of Theorem 10.24 is given in [**47**, Corollary 1.6].

We now recall a definition from [**9**]. The *homogeneous filled Julia set* K_F of F is defined to be the set of all $z \in K^2$ for which the iterates $F^{(n)}(z)$ remain bounded. Clearly $F^{-1}(K_F) = K_F$, and the same is true for each $F^{(-n)}$. Since all norms on K^2 are equivalent, the set K_F is independent of which norm is used to define it.

The filled Julia set K_F can be thought of as the 'unit ball' with respect to the homogeneous dynamical height H_F:

LEMMA 10.31. *We have*
$$K_F = \{z \in K^2 \ : \ H_F(z) \le 0\} \ .$$

PROOF. If $z \in K_F$, then there exists $M > 0$ such that $\|F^{(n)}(z)\| \le M$ for all n, and therefore $H_F(z) \le \lim_{n \to \infty} \frac{1}{d^n} \log M = 0$.

Conversely, suppose $z \notin K_F$. Then by Lemma 10.1, there exist $\alpha, \beta > 1$ and n_0 such that
$$\|F^{(n+n_0)}(z)\|_v > \beta \cdot \alpha^{d^n - 1}$$
for all $n \ge 0$. Therefore
$$H_F(z) \ge \lim_{n \to \infty} \frac{1}{d^{n+n_0}} \left((d^n - 1) \log \alpha + \log \beta\right) = \frac{1}{d^{n_0}} \log \alpha > 0 \ .$$
□

Let ϵ_K be zero if the absolute value on K is non-Archimedean and 1 if it is Archimedean. We now prove the following somewhat technical result, which will easily imply Theorem 10.23:

THEOREM 10.32. *Let $N = td^k \in \Sigma$, and let $z_1, \ldots, z_N$ be nonzero elements of the filled Julia set K_F whose images in $\mathbb{P}^1(K)$ are all distinct. Choose a real number $R(F) > 0$ so that $K_F \subseteq \{z \in K^2 \ : \ \|z\| \le R(F)\}$. Then*
$$\begin{aligned}\sum_{i \ne j} -\log|z_i \wedge z_j| \ \ge \ & r(F)N^2 - \epsilon_K N \log N \\ & -2\left(\log R(F)\right) \alpha N - r(F)(1 + \alpha)N \ ,\end{aligned}$$
where $r(F) = \frac{1}{d(d-1)} \log|\mathrm{Res}(F)|$ and $\alpha = t - 1 + (d-1)k > 0$ satisfies $2 \le \alpha \le (d-1)(\log_d N + 2)$.

An outline of the proof of Theorem 10.32 is as follows. First, we express $\prod_{i \ne j} |(x_i, y_i) \wedge (x_j, y_j)|$ as the determinant of a Vandermonde matrix S. We then replace this matrix with a new matrix H whose entries involve $F_1^{(k)}(x_i, y_i)$ and $F_2^{(k)}(x_i, y_i)$ for various $k \ge 0$, rather than the standard monomials $x_i^a y_i^b$. Replacing S by H amounts to choosing a different basis for the space of homogeneous polynomials in x and y of degree $N-1$, and we are able to explicitly calculate the determinant of the change of basis matrix under the assumption that $N \in \Sigma$. We then use Hadamard's inequality to estimate the determinant of H, using the fact that $\|F^{(k)}(x_i, y_i)\| \le R(F)$ for all $k \ge 0$.

Let $\Gamma^0(m)$ denote the vector space of homogeneous polynomials of degree m in $K[x, y]$, which has dimension $N = m + 1$ over K. If $N \in \Sigma$, that is, if $m = td^k - 1$ with $2 \leq t \leq 2d - 1$ and $k \geq 1$, we consider the collection $H(m)$ of polynomials in $\Gamma^0(m)$ given by

$$H(m) = \{x^{a_0} y^{b_0} F_1(x,y)^{a_1} F_2(x,y)^{b_1} \cdots F_1^{(k)}(x,y)^{a_k} F_2^{(k)}(x,y)^{b_k} : \\ a_i + b_i = d - 1 \text{ for } 0 \leq i \leq k-1 \text{ and } a_k + b_k = t - 1\}.$$

The cardinality of $H(m)$ is easily seen to be $N = \dim \Gamma^0(m)$. The following proposition shows that $H(m)$ forms a basis for $\Gamma^0(m)$ and explicitly calculates the determinant of the change of basis matrix between $H(m)$ and the standard monomial basis $S(m)$ given by

$$S(m) = \{x^a y^b \mid a + b = m\}.$$

PROPOSITION 10.33. *Let A be the matrix expressing the polynomials $H(m)$ (in some order) in terms of some ordering of the standard basis $S(m)$. Then $\det(A) = \pm \operatorname{Res}(F)^r$, where*

$$r = \frac{N^2}{2d(d-1)} - \frac{N}{2d(d-1)}(t + k(d-1)).$$

In particular, since $\operatorname{Res}(F) \neq 0$, $H(m)$ is a basis for $\Gamma^0(m)$.

PROOF. Let $H_1, \ldots, H_N$ be an ordering of the elements of $H(m)$, and let $S_1, \ldots, S_N$ be an ordering of the elements of $S(m)$, so that A is the $N \times N$ matrix whose $(i,j)^{\text{th}}$ entry is the coefficient of the monomial S_i in the expansion of H_j as a polynomial in x and y. We have $\det(A) = 0$ if and only if some nontrivial linear combination of the elements of $H(m)$ is zero.

Suppose $\det(A) = 0$. Then there exist homogeneous polynomials $h_1 \in H(d^k - 1)$ and $h_2 \in H((t-1)d^k - 1)$, not both zero, such that

$$h_1 \left(F_1^{(k)}(x,y)\right)^{t-1} + h_2 F_2^{(k)}(x,y) = 0.$$

We may assume that neither of h_1, h_2 is the zero polynomial. Thus $F_2^{(k)}$ divides $h_1 \left(F_1^{(k)}\right)^{t-1}$. Since $\deg(h_1) < \deg(F_2^{(k)})$, it follows that $F_1^{(k)}$ and $F_2^{(k)}$ have a common irreducible factor. Thus $\operatorname{Res}(F^{(k)}) = 0$. But $\operatorname{Res}(F^{(k)})$ is a power of $\operatorname{Res}(F)$ by (10.32), so $\operatorname{Res}(F) = 0$ as well.

Conversely, suppose $\operatorname{Res}(F) = 0$. By Lemma 2.11, there is a nontrivial relation of the form

(10.45) $$h_1 F_1 + h_2 F_2 = 0$$

with $h_1, h_2 \in K[x, y]$ homogeneous of degree $d - 1$. If $k = 1$, this already implies that $\det(A) = 0$. If $k \geq 2$, then multiplying both sides of (10.45) by $G(x, y)$, where

$$G(x,y) = F_1^{d-2}(F_1^{(2)})^{d-1} \cdots (F_1^{(k)})^{t-1},$$

gives a linear relation which shows that $\det(A) = 0$.

Now expand both $\det(A)$ and $\operatorname{Res}(F)$ as polynomials in the coefficients of F_1 and F_2. Since $\operatorname{Res}(F)$ is irreducible (see [**98**, §5.9]), we find that

$$\det(A) = C \cdot \operatorname{Res}(F)^r$$

for some $C \in K^\times$ and some natural number r. Now $\operatorname{Res}(F)$ is homogeneous of degree $2d$ in the coefficients of F_1 and F_2, and a straightforward calculation shows that $F_1^{(j)}$ and $F_2^{(j)}$ are each homogeneous of degree $(d^j - 1)/(d-1)$ in the coefficients of F_1 and F_2. It follows that $\det(A)$ is homogeneous of degree

$$r' = N\left(\sum_{j=1}^{k-1}(d^j - 1) + (t-1)\frac{d^k - 1}{d-1}\right)$$

in the coefficients of F_1 and F_2. Comparing degrees and performing some straightforward algebraic manipulations, we find that

$$r = \frac{r'}{2d} = \frac{N^2}{2d(d-1)} - \frac{N}{2d}\left(\frac{t}{d-1} + k\right).$$

Finally, to compute C, we set $F_1 = x^d$ and $F_2 = y^d$, in which case $H(m)$ is just a permutation of the standard monomial basis $S(m)$. It follows that $C = \pm 1$. □

If $v = (v_1, \ldots, v_N)^\mathrm{T} \in K^N$, define $\|v\|$ to be the L^2-norm $\|v\| = (\sum |v_i|^2)^{1/2}$ if K is Archimedean and to be the sup-norm $\|v\| = \sup |v_i|$ if K is non-Archimedean. If H is a matrix with columns $h_1, \ldots, h_N \in K^N$, then *Hadamard's inequality*[7] states that

(10.46) $$|\det(H)| \leq \prod_{i=1}^{N} \|h_i\|.$$

PROOF OF THEOREM 10.32. Let $\{S_1, \ldots, S_N\}$ and $\{H_1, \ldots, H_N\}$ be as in the proof of Proposition 10.33. Using the homogeneous version of the standard formula for the determinant of a Vandermonde matrix, we have

(10.47) $$\prod_{i \neq j} |x_i y_j - x_j y_i| = |\det(S)|^2,$$

where S is the matrix whose $(i,j)^{\mathrm{th}}$ entry is $S_j(x_i, y_i)$. If H is the matrix whose $(i,j)^{\mathrm{th}}$ entry is $H_j(x_i, y_i)$, then $H = SA$, so that by Proposition 10.33, we have

(10.48) $$\det(S)^2 = \det(H)^2 (\det(A))^{-2} = \det(H)^2 \operatorname{Res}(F)^{-2r}.$$

---

[7]See the discussion in [**3**] and the references therein.

On the other hand, we can estimate $|\det(H)|^2$ using Hadamard's inequality (10.46). Letting h_i be the ith column of H, we obtain

$$
\begin{aligned}
(10.49) \quad |\det(H)|^2 &\leq \prod_{i=1}^N \|h_i\| \\
&\leq N^{\epsilon_K N} \prod_i R(F)^{2(k(d-1)+(t-1))} \\
&= N^{\epsilon_K N} R(F)^{(2t-2+k(2d-2))N} .
\end{aligned}
$$

Putting together (10.47), (10.48), and (10.49) gives

$$
\begin{aligned}
(10.50) \quad \prod_{i\neq j} |x_i y_j - x_j y_i| &= |\det(S)|^2 \\
&= |\det(H)|^2 \cdot |\operatorname{Res}(F)|^{-2r} \\
&\leq N^{\epsilon_K N} R(F)^{2\alpha N} |\operatorname{Res}(F)|^{-2r} ,
\end{aligned}
$$

where $-2r = -\frac{N^2}{d(d-1)} + \frac{N}{d(d-1)}(t + k(d-1))$.

Taking the negative logarithm of both sides of (10.50) gives the desired lower bound

$$
\begin{aligned}
\sum_{i\neq j} -\log|x_i y_j - x_j y_i| \geq\ & -\epsilon_K N \log N - 2N\alpha \log R(F) \\
& + \tfrac{N^2}{d(d-1)} \log|\operatorname{Res}(F)| \\
& - \tfrac{N}{d(d-1)}(t + k(d-1)) \log|\operatorname{Res}(F)| .
\end{aligned}
$$

$\square$

We can now give the proof of Theorem 10.23:

PROOF OF THEOREM 10.23. Without loss of generality, we may assume that $N \geq 2d$. Let N' be the smallest integer less than or equal to N which belongs to Σ. One deduces easily from the definition of Σ that

$$
(10.51) \quad \frac{N-1}{2} \leq N' - 1 \leq N - 1 .
$$

Using the fact that $N' \in \Sigma$, we claim that there is a constant $C' > 0$ (independent of N and N') such that

$$
(10.52) \quad D_{N'} \geq -C' \frac{\log N'}{N' - 1} .
$$

From this, Lemma 10.29, and (10.51), it follows that

$$
D_\varphi(z_1, \ldots, z_N) \geq D_{N'} \geq -C \frac{\log N}{N - 1}
$$

as desired, where $C = 2C'$.

It remains to prove the claim. Replacing K by $\overline{K}$ if necessary, we may assume without loss of generality that the value group $|K^\times|$ is dense in the group $\mathbb{R}_{>0}$ of nonnegative reals. Since $H_F(cz) = H_F(z) + \log|c|$ for all $z \in K^2 \setminus \{0\}$ and all $c \in K^\times$, given $\varepsilon > 0$, we can choose coordinates (x_i, y_i) for z_i so that $-\varepsilon \leq H_F(x_i, y_i) \leq 0$. Then $(x_i, y_i) \in K_F$ by Lemma 10.31, and we can apply Theorem 10.32 to $(x_1, y_1), \ldots, (x_N, y_N)$. Simplifying the resulting expression and letting $\varepsilon \to 0$ gives the desired inequality (10.52). $\square$

REMARK 10.34. As an example of an application of Theorem 10.23, we note the following global result (see [**3**, §3] for a proof) concerning fields of definition for preperiodic points:

THEOREM 10.35. *Let k be a number field, and let $\varphi \in k(T)$ be a rational map of degree at least 2. Then there is a constant $C = C(\varphi, k)$ such that if $P_1, \ldots, P_N \in \mathbb{P}^1(\overline{k})$ are distinct preperiodic points of φ, then*

$$[k(P_1, \ldots, P_N) : k] \geq C \frac{N}{\log N} .$$

10.4. Equidistribution of preimages

In this and the following three sections, we develop the fundamental facts of Fatou-Julia theory when K has *characteristic 0*. This theory is very similar to the classical theory over $\mathbb{C}$, whereas the theory over fields of positive characteristic is different in some respects.

For the remainder of this section, we assume that K is a complete and algebraically closed non-Archimedean field of characteristic 0. Let $\varphi \in K(T)$ be a rational function of degree $d \geq 2$. Our goal is to prove an equidistribution theorem for the preimages under φ of any nonexceptional point of $\mathbb{P}^1_{\text{Berk}}$. This result is originally due to Favre and Rivera-Letelier [**46, 48**] and is the Berkovich space analogue of a classical result due independently to Lyubich [**72**] and Freire, Lopes, and Mañé [**54**].

A point $x \in \mathbb{P}^1_{\text{Berk}}$ is called *exceptional* for φ if the set of all forward and backward iterates of x is finite. In other words, if the *grand orbit of x* is

$$GO(x) = \{y \in \mathbb{P}^1_{\text{Berk}} : \varphi^{(m)}(x) = \varphi^{(n)}(y) \text{ for some } m, n \geq 0\} ,$$

then x is exceptional if and only if $GO(x)$ is finite. It is easy to see that if x is exceptional, then $GO(x)$ forms a finite cyclic orbit under φ and each $y \in GO(x)$ has multiplicity $m_\varphi(y) = d$. By analogy with the situation over $\mathbb{C}$, we define the *exceptional locus* E_φ of φ to be the set of all exceptional points of $\mathbb{P}^1_{\text{Berk}}$. We write $E_\varphi(K)$ for $E_\varphi \cap \mathbb{P}^1(K)$.

It is well known that if K has characteristic 0, then $|E_\varphi(K)| \leq 2$ (see Lemma 10.40 below). We will see below (see Proposition 10.45) that φ has at most one exceptional point in $\mathbb{H}_{\text{Berk}}$.

We will now prove the following result of Favre and Rivera-Letelier when $\text{char}(K) = 0$. In [**48**], it is established in positive characteristic as well.

THEOREM 10.36 (Favre, Rivera-Letelier). *Let $\varphi \in K(T)$ be a rational function of degree $d \geq 2$, and let y be any point in $\mathbb{P}^1_{\text{Berk}} \backslash E_\varphi(K)$. For each integer $n \geq 1$, consider the discrete probability measure*

$$\mu^y_{\varphi^{(n)}} = \frac{1}{d^n} \sum_{\varphi^{(n)}(y_i) = y} m_{\varphi^{(n)}}(y_i) \delta_{y_i} .$$

Then the sequence of measures $\{\mu^y_{\varphi^{(n)}}\}_{1 \leq n < \infty}$ converges weakly to the canonical measure μ_φ for φ.

In other words, the preimages under φ of any $y \in \mathbb{P}^1_{\text{Berk}} \backslash E_\varphi(K)$ are equidistributed with respect to the canonical measure μ_φ.

REMARK 10.37. Using the version of Theorem 10.36 proved by Favre and Rivera-Letelier in [48], one can establish characteristic p analogues of the results about Berkovich Fatou and Julia sets proved in later sections of this chapter. However, it should be noted that those results are not always identical with the ones in characteristic 0. For example, in characteristic p the exceptional set $E_\varphi(K)$ need not have at most two elements but can be countably infinite.

We begin with the following lemma:

LEMMA 10.38. *Suppose $g_n \in \mathrm{BDV}(\mathbb{P}^1_{\text{Berk}})$ and $g_n \to 0$ pointwise on $\mathbb{H}_{\text{Berk}}$. Then $\Delta_{\mathbb{P}^1_{\text{Berk}}}(g_n) \to 0$ weakly.*

PROOF. It suffices to prove that $\int f \Delta(g_n) \to 0$ for every simple function f on $\mathbb{P}^1_{\text{Berk}}$. Write $\Delta(f) = \sum_{i=1}^{t} a_i \delta_{\zeta_i}$ with $\zeta_i \in \mathbb{H}^{\mathbb{R}}_{\text{Berk}}$. By Corollary 5.39,

$$\lim_{n \to \infty} \int f \Delta(g_n) = \lim_{n \to \infty} \int g_n \Delta(f) = \lim_{n \to \infty} \sum_{i=1}^{t} a_i g_n(\zeta_i)$$

$$= \sum_{i=1}^{t} a_i (\lim_{n \to \infty} g_n(\zeta_i)) = 0,$$

as desired. $\square$

LEMMA 10.39. *The function $g_\varphi(x,y) = g_{\mu_\varphi}(x,y)$ differs from the function $-\log_v(\delta(x,y)_{\zeta_{\text{Gauss}}})$ by a bounded function on $\mathbb{P}^1_{\text{Berk}} \times \mathbb{P}^1_{\text{Berk}}$.*

PROOF. Since μ_φ has continuous potentials (cf. Definition 5.40 and Proposition 10.7), this follows directly from formula (8.62). $\square$

Recall that a point $z \in \mathbb{P}^1(K)$ is a critical point of φ iff $m_\varphi(z) > 1$. The following result is an immediate consequence of the classical Riemann-Hurwitz formula (which depends on our assumption that $\mathrm{char}(K) = 0$):

LEMMA 10.40. *Assume $\mathrm{char}(K) = 0$. Then the map φ has only finitely many critical points in $\mathbb{P}^1(K)$. More precisely,*

$$\sum_{z \in \mathbb{P}^1(K)} (m_\varphi(z) - 1) = 2d - 2 \ .$$

Let $E_\varphi(K)$ be the exceptional locus of φ in $\mathbb{P}^1(K)$. By Lemma 10.40, we have $|E_\varphi(K)| \leq 2$. It follows easily that if $E_\varphi(K)$ is nonempty, then it is a union of periodic cycles having length 1 or 2.

A periodic point $w \in \mathbb{P}^1(K)$ of period k for φ is called *attracting* if its multiplier $|(\varphi^{(k)})'(w)|$ (computed in any coordinate where $w \neq \infty$) is less than 1. It is *superattracting* if its multiplier is 0. A periodic cycle C of period k in $\mathbb{P}^1(K)$ is attracting if every point of C is an attracting periodic point. It is a well-known consequence of the chain rule that the multiplier is the same for every $w \in C$, so C is attracting if and only if some point of C is attracting. A periodic cycle composed of exceptional points for φ is necessarily attracting, since the multiplier is 0 at such points.

The *attracting basin* of an attracting periodic cycle $C \subseteq \mathbb{P}^1(K)$ for φ is defined to be the set of all $x \in \mathbb{P}^1_{\text{Berk}}$ for which, given any open neighborhood U of C, there exists an N such that $\varphi^{(n)}(x) \in U$ for $n \geq N$. The attracting basin of $E_\varphi(K)$ is defined to be empty if $E_\varphi(K)$ is empty and to be the union of the attracting basins of the cycles comprising $E_\varphi(K)$ otherwise.

LEMMA 10.41. *The attracting basin of any attracting periodic cycle $C \subseteq \mathbb{P}^1(K)$ is open in $\mathbb{P}^1_{\text{Berk}}$. In particular, the attracting basin of $E_\varphi(K)$ is open in $\mathbb{P}^1_{\text{Berk}}$.*

PROOF. Let $C \subseteq \mathbb{P}^1(K)$ be an attracting periodic cycle of period k. Replacing φ by $\varphi^{(k)}$, we may assume without loss of generality that $C = \{w\}$ with w a fixed point of φ. After a change of coordinates, we can assume that $w \neq \infty$. By classical non-Archimedean dynamics, there is a descending chain of open discs $D_0 \supseteq D_1 \supseteq D_2 \supseteq \cdots$ in $\mathbb{P}^1(K)$ with $\varphi(D_i) \subseteq D_{i+1}$ for all i, such that $\bigcap D_i = \{w\}$. Indeed, if $\varphi(T)$ has multiplicity m at w, then by expanding $\varphi(T)$ as a power series and using the theory of Newton polygons (see Corollary A.17 in §A.10), one finds an $r_0 > 0$ and a constant $B > 0$ (with $B = |\varphi'(w)| < 1$ if $m = 1$) such that $\varphi(D(w, r)^-) = D(w, Br^m)^-$ for each $0 < r \leq r_0$. Letting $\mathcal{D}_i$ be the Berkovich open disc corresponding to D_i, we will still have $\mathcal{D}_0 \supseteq \mathcal{D}_1 \supseteq \cdots$, $\varphi(\mathcal{D}_i) \subseteq \mathcal{D}_{i+1}$ for all i and $\bigcap \mathcal{D}_i = \{w\}$. Let $\mathcal{A}$ be the attracting basin of w. If $x \in \mathcal{A}$, then $\varphi^{(N)}(x) \in \mathcal{D}_0$ for some N. By continuity, there is an open neighborhood V of x such that $\varphi^{(N)}(V) \subseteq \mathcal{D}_0$. Then for $n \geq N$, we have $\varphi^{(n)}(V) \subseteq \varphi^{n-N}(\mathcal{D}_0) \subseteq \mathcal{D}_{n-N}$. Since the Berkovich open discs $\mathcal{D}_i$ form a fundamental system of open neighborhoods for $\{w\}$, this shows that $V \subset \mathcal{A}$, and thus $\mathcal{A}$ is open. □

PROOF OF THEOREM 10.36. Recall that μ_φ satisfies the functional equation
$$\Delta_x g_\varphi(x, y) = \delta_y - \mu_\varphi$$
for all $y \in \mathbb{P}^1_{\text{Berk}}$. Pulling back by $\varphi^{(n)}$ and applying Proposition 9.56 and Theorems 10.2 and 10.18, we obtain
$$\mu^y_{\varphi^{(n)}} - \mu_\varphi = \frac{1}{d^n} \Delta_x g_\varphi(x, (\varphi^{(n)})^*(y)) = \frac{1}{d^n} \Delta_x g_\varphi(\varphi^{(n)}(x), y) \ .$$
By Lemma 10.38, to prove Theorem 10.36 it suffices to show that
$$\frac{1}{d^n} g_\varphi(\varphi^{(n)}(x), y) \to 0$$
as $n \to \infty$, for each fixed $x \in \mathbb{H}_{\text{Berk}}$ and each $y \in \mathbb{P}^1_{\text{Berk}} \backslash E_\varphi(K)$.

Since g_φ is bounded below, we automatically have
$$\liminf \frac{1}{d^n} g_\varphi(\varphi^{(n)}(x), y) \geq 0 \ ,$$

so it suffices to prove that

(10.53) $$\limsup \frac{1}{d^n} g_\varphi(\varphi^{(n)}(x), y) \leq 0 \ .$$

We first dispose of some easy cases. If $y \in \mathbb{H}_{\text{Berk}}$, then $g_\varphi(x,y)$ is bounded as a function of x, and (10.53) holds trivially. Let $\mathcal{A}$ be the attracting basin in $\mathbb{P}^1_{\text{Berk}}$ of $E_\varphi(K)$. By Lemma 10.41, $\mathcal{A}$ is an open subset of $\mathbb{P}^1_{\text{Berk}}$. If $x \in \mathcal{A}$ and $y \in \mathbb{P}^1(K) \backslash \mathcal{A}$, then $\varphi^{(n)}(x)$ accumulates at points of $E_\varphi(K)$ for large n. This means $g_\varphi(\varphi^{(n)}(x), y)$ is close to $g_\varphi(z, y)$ for some $z \in E_\varphi(K)$, so (10.53) holds. Likewise, for $x \in \mathbb{H}_{\text{Berk}} \backslash \mathcal{A}$ and $y \in \mathbb{P}^1(K) \cap \mathcal{A}$, the iterates $\varphi^{(n)}(x)$ remain bounded away from y, and again (10.53) holds.

Thus it suffices to prove (10.53) for $x \in \mathbb{H}_{\text{Berk}} \backslash \mathcal{A}$ and $y \in \mathbb{P}^1(K) \backslash \mathcal{A}$. For such x, y, a potential problem arises when the sequence $\{\varphi^{(n)}(x)\}$ has y as an accumulation point.

In order to deal with this, we first make a reduction. Note that if k is any positive integer and the analogue of (10.53) holds with φ replaced by $\varphi^{(k)}$ and d replaced by d^k, then (10.53) holds for φ. (This follows easily from Lemma 10.16.) Suppose $z \in \mathbb{P}^1(K)$ is not exceptional. Consider the points $z_0 = z$, $z_1 = \varphi(z)$, and $z_2 = \varphi^{(2)}(z)$. We claim that $m_\varphi(z_i) < d$ for some i. Otherwise, by Lemma 10.40, two of them would have to coincide. If $z_0 = z_1$ or $z_0 = z_2$, then z would be exceptional. If $z_1 = z_2$, then z_1 would be exceptional, and then z, which belongs to $GO(z_1)$, would also be exceptional. Hence the formula $m_{\varphi \circ \psi}(z) = m_\varphi(\psi(z)) \cdot m_\psi(z)$ (see Proposition 9.28(C)) shows that if $z \in \mathbb{P}^1(K)$ is not an exceptional point for φ, then $m_{\varphi^{(3)}}(z) \leq d^3 - 1$. So in proving (10.53), after replacing φ by $\varphi^{(3)}$ if necessary, we can assume that $m_\varphi(z) \leq d - 1$ whenever $z \in \mathbb{P}^1(K)$ is not exceptional.

We now require the following estimate, whose proof is given after Corollary 10.44 below:

LEMMA 10.42. *Suppose $m_\varphi(z) \leq d-1$ for all $z \in \mathbb{P}^1(K) \backslash E_\varphi(K)$. Let $\mathcal{A}$ be the attracting basin of $E_\varphi(K)$. Then there is a constant $C > 0$ such that for all $x \in \mathbb{P}^1_{\text{Berk}} \backslash \mathcal{A}$ and all $y \in \mathbb{P}^1(K) \backslash \mathcal{A}$,*

(10.54) $$g_\varphi(\varphi(x), y) \leq C + (d-1) \max_{\varphi(z) = y} g_\varphi(x, z) \ .$$

For $x \in \mathbb{H}_{\text{Berk}} \backslash \mathcal{A}$ and $y \in \mathbb{P}^1(K) \backslash \mathcal{A}$, it follows from (10.54) that

$$g_\varphi(\varphi^{(n)}(x), y) \leq C(n) + (d-1)^n \max_{\varphi^n(z) = y} g_\varphi(x, z)$$

for all $n \geq 1$, where

$$\begin{aligned} C(n) &= C + (d-1)C + (d-1)^2 C + \cdots + (d-1)^{n-1} C \\ &= O\big(\max(n, (d-1)^n)\big) \ . \end{aligned}$$

Furthermore, the function $G(z) := g_\varphi(x,z)$ is bounded on $\mathbb{P}^1_{\text{Berk}}$, since $x \in \mathbb{H}_{\text{Berk}}$. Hence

$$\limsup \frac{1}{d^n} g_\varphi(\varphi^{(n)}(x), y)$$
$$\leq \limsup \frac{1}{d^n} \left(C(n) + (d-1)^n \max_{z \in \mathbb{P}^1_{\text{Berk}}} g_\varphi(x,z) \right) \leq 0$$

as desired. $\square$

To prove Lemma 10.42, we will need the following result about power series:

LEMMA 10.43 (Non-Archimedean Rolle's theorem). *Let K be a complete and algebraically closed non-Archimedean field of characteristic 0. Put $\rho_K = 1$ if K has residue characteristic 0, and put $\rho_K = |p|^{1/(p-1)} < 1$ if K has residue characteristic $p > 0$.*

Suppose $f(T) \in K[[T]]$ converges in $D(0,r)^-$. If $f(T)$ has two or more zeros in $D(0, \rho_K r)^-$, then $f'(T)$ has at least one zero in $D(0,r)^-$.

This is proved in [**85**, p. 316] when $K = \mathbb{C}_p$. The proof is carried over to the general case in Proposition A.20. The hypothesis that $\text{char}(K) = 0$ is used to assure that $|p| \neq 0$ when K has residue characteristic $p > 0$.

Let $\|x,y\|$ denote the spherical distance on $\mathbb{P}^1(K)$ (see §4.3). Recall that for $a \in \mathbb{P}^1(K)$ and $r > 0$, we write

$$B(a,r)^- = \{z \in \mathbb{P}^1(K) : \|z,a\| < r\} \ .$$

COROLLARY 10.44 (Uniform Injectivity Criterion). *Let K be a complete and algebraically closed non-Archimedean field of characteristic 0, and let $\varphi(T) \in K(T)$ be a nonconstant rational function. Then there is an $r_0 > 0$ (depending only on φ) such that if $a \in \mathbb{P}^1(K)$, $0 < r \leq r_0$, and $B(a,r)^-$ does not contain any critical points of φ, then φ is injective on $B(a, \rho_K r)^-$.*

PROOF. Let X be the set of all zeros and poles of $\varphi(T)$ in $\mathbb{P}^1(K)$. Put

$$r_0 = \min\left(1, \min_{\substack{\alpha,\beta \in X \\ \alpha \neq \beta}} (\|\alpha,\beta\|)\right) \ .$$

Let $a \in \mathbb{P}^1(K)$ and $0 < r \leq r_0$ be given. By replacing $\varphi(T)$ with $1/\varphi(T)$ if necessary, we can assume that $B(a,r)^-$ does not contain any poles of φ. If we identify $\mathbb{P}^1(K)$ with $K \cup \{\infty\}$, then after changing coordinates by an appropriate element of $\text{GL}_2(\mathcal{O}_K)$ (where $\mathcal{O}_K$ is the valuation ring of K), we can assume that $a = 0$. Observe that such a change of coordinates preserves the spherical distance. In this situation $B(a,r)^- = D(0,r)^- \subset D(0,1) \subset K$. Note that $D(0,r)^-$ does not contain any poles of $\varphi(T)$ and that $\|x,y\| = |x-y|$ for $x,y \in D(0,r)^-$.

Expand $\varphi(T)$ as a power series $f(T) \in K[[T]]$. Then $f(T)$ converges in $D(0,r)^-$. By Lemma 10.43, if $D(0,r)^-$ does not contain any critical points of $\varphi(T)$, then $\varphi(T)$ is injective on $D(0, \rho_K r)^-$. $\square$

10.4. EQUIDISTRIBUTION OF PREIMAGES

PROOF OF LEMMA 10.42. Let $x \in \mathbb{P}^1_{\text{Berk}} \backslash \mathcal{A}$ and $y \in \mathbb{P}^1(K) \backslash \mathcal{A}$, and assume that $m_\varphi(z) \leq d - 1$ for all $z \in \mathbb{P}^1(K) \backslash E_\varphi(K)$.

By Theorem 10.18, proving (10.54) is equivalent to showing that

$$(10.55) \qquad g_\varphi(x, \varphi^*(y)) \leq C + (d-1) \max_{\varphi(z)=y} g_\varphi(x, z) .$$

Note that for each fixed $y \in \mathbb{P}^1(K) \backslash \mathcal{A}$, both sides of (10.55) are continuous in x, as functions to the extended reals. The left side is a finite sum of continuous functions, and the right side concerns the maximum of those same functions. Since $\mathbb{P}^1(K) \backslash \mathcal{A}$ is dense in $\mathbb{P}^1_{\text{Berk}} \backslash \mathcal{A}$, it therefore suffices to prove (10.55) for $x \in \mathbb{P}^1(K) \backslash \mathcal{A}$.

If $z_1, \ldots, z_d$ are the points with $\varphi(z_i) = y$ (listed with multiplicities), then

$$g_\varphi(x, \varphi^*(y)) = \sum_{i=1}^d g_\varphi(x, z_i) .$$

Hence to prove (10.55) it is enough to show that there is a constant C such that for each $x, y \in \mathbb{P}^1(K) \backslash \mathcal{A}$, there is some z_i with $\varphi(z_i) = y$ for which $g_\varphi(x, z_i) \leq C$.

By Lemma 10.39, the Arakelov-Green's function $g_\varphi(x, y)$ differs from $-\log_v(\delta(x,y)_{\zeta_{\text{Gauss}}})$ by a bounded function on $\mathbb{P}^1_{\text{Berk}} \times \mathbb{P}^1_{\text{Berk}}$. Recall that for $x, y \in \mathbb{P}^1(K)$, $\delta(x,y)_{\zeta_{\text{Gauss}}}$ coincides with the spherical distance $\|x, y\|$. It therefore suffices to show that there is an $\varepsilon > 0$ such that for each $x, y \in \mathbb{P}^1(K) \backslash \mathcal{A}$, there is some z_i with $\varphi(z_i) = y$ for which

$$(10.56) \qquad \|x, z_i\| \geq \varepsilon .$$

For each fixed $w \in \mathbb{P}^1(K) \backslash E_\varphi(K)$, Corollary 9.17 and the fact that $m_\varphi(z) \leq d-1$ for all $z \in \varphi^{-1}(w)$ imply that one can find an open neighborhood U_w of w in $\mathbb{P}^1_{\text{Berk}}$ and disjoint connected components V, V' of $\varphi^{-1}(U_w)$. After shrinking U_w if necessary, we can assume that V and V' have disjoint closures. Hence, there is an $\varepsilon_w > 0$ such that $\|z, z'\| \geq \varepsilon_w$ for all $z \in \mathbb{P}^1(K) \cap V$, $z' \in \mathbb{P}^1(K) \cap V'$. For any $x \in \mathbb{P}^1(K)$, either $x \notin V$ or $x \notin V'$; it follows from this and the ultrametric inequality $\max(\|x, z\|, \|x, z'\|) \geq \|z, z'\|$ that for each $x \in \mathbb{P}^1(K) \backslash \mathcal{A}$ and each $y \in \mathbb{P}^1(K) \cap U_w$, there is some z_i with $\varphi(z_i) = y$ for which $\|x, z_i\| \geq \varepsilon_w$.

For each of the finitely many nonexceptional critical points $y_1, \ldots, y_t$ of φ, put $w_i = \varphi(y_i)$, choose a neighborhood $U_i = U_{w_i}$ as above, and let $\varepsilon_i := \varepsilon_{w_i}$ be the corresponding bound. For each $y \in \mathbb{P}^1(K) \cap (U_1 \cup \cdots \cup U_r)$, (10.56) holds with $\varepsilon = \min(\varepsilon_1, \ldots, \varepsilon_t)$.

Finally, put

$$\mathcal{B} = \mathcal{A} \cup U_1 \cup \cdots \cup U_t .$$

By construction, $\varphi^{-1}(\mathcal{B})$ contains all the critical points of φ. Since there are finitely many critical points, there is an $r > 0$ such that for each critical y, the ball $B(y, r)^-$ is contained in $\varphi^{-1}(\mathcal{B})$. After reducing r if necessary, we can assume that $r \leq r_0$, where r_0 is the number from Corollary 10.44. By the ultrametric inequality for the spherical distance, for each $z \in \mathbb{P}^1(K)\backslash \varphi^{-1}(\mathcal{B})$, the ball $B(z,r)^-$ does not contain any critical points. Hence Corollary 10.44 shows that φ is injective on $B(z, \rho_K r)^-$.

In particular, each $z \in \mathbb{P}^1(K)\backslash \varphi^{-1}(\mathcal{B})$ has multiplicity 1. Thus, if $y \in \mathbb{P}^1(K)\backslash \mathcal{B}$, then y has d distinct preimages $z_1, \ldots, z_d \in \mathbb{P}^1(K)\backslash \varphi^{-1}(\mathcal{B})$ under φ, and since $\varphi(z_i) = \varphi(z_j) = y$, necessarily $\|z_i, z_j\| \geq \rho_K r$ for all $i \neq j$. By the ultrametric inequality for the spherical distance, for each $x \in \mathbb{P}^1(K)\backslash \mathcal{A}$ there is at least one z_i with $\|x, z_i\| \geq \rho_K r$. Hence, if we take

$$\varepsilon = \min(\rho_K r, \varepsilon_1, \ldots, \varepsilon_t) ,$$

then (10.56) holds for all $x, y \in \mathbb{P}^1(K)\backslash \mathcal{A}$. □

As an application of Theorem 10.36, we give a characterization of the exceptional locus due to Rivera-Letelier.[8]

PROPOSITION 10.45. *Let K be a complete and algebraically closed non-Archimedean field of characteristic 0, and let $\varphi(T) \in K(T)$ have degree $d \geq 2$.*
  (A) *The exceptional locus E_φ of φ contains at most 3 points. At most two exceptional points are in $\mathbb{P}^1(K)$, and at most one is in $\mathbb{H}_{\mathrm{Berk}}$.*
  (B) *There is an exceptional point in $\mathbb{H}_{\mathrm{Berk}}$ (necessarily of type II) if and only if φ has simple reduction.*

PROOF. For (A), note that since $\varphi : \mathbb{P}^1_{\mathrm{Berk}} \to \mathbb{P}^1_{\mathrm{Berk}}$ is surjective by Corollary 9.9 (see also [81, Lemma 2.5]), φ must map each grand orbit onto itself. So φ maps each finite grand orbit bijectively onto itself, and thus each finite grand orbit is a periodic cycle. The fact that there are at most 2 exceptional points in $\mathbb{P}^1(K)$ is well known and follows from Lemma 10.40, as was noted earlier.

The fact that there is at most one exceptional point in $\mathbb{H}_{\mathrm{Berk}}$ follows from Theorem 10.36. Indeed, suppose $y \in \mathbb{H}_{\mathrm{Berk}}$ is exceptional. After replacing φ by $\varphi^{(k)}$ for an appropriate k, we can assume that y is a fixed point of φ. (Note that $\mu_{\varphi^{(k)}} = \mu_\varphi$ by Lemma 10.16.) For each $n \geq 1$, it follows that

$$\mu^y_{\varphi^{(n)}} = \frac{1}{d^n}(d^n \delta_y) = \delta_y .$$

Since the $\mu^y_{\varphi^{(n)}}$ converge weakly to μ_φ, we must have $\mu_\varphi = \delta_y$. Thus y is unique.

For assertion (B), suppose first that $y \in \mathbb{H}_{\mathrm{Berk}}$ is exceptional. We have just proved that $\mu_\varphi = \delta_y$ and that y is a fixed point of $\varphi^{(k)}$ for some $k \geq 1$. Moreover, $m_{\varphi^{(k)}}(y) = d^k$ by Theorem 10.2. By Lemma 10.80 below (whose proof uses different ideas than the ones at hand, but only depends on results from Chapter 9), y must be a point of type II. By Corollary 2.13, there is a Möbius transformation $M \in \mathrm{PGL}(2, K)$ such that $M(y) = \zeta_{\mathrm{Gauss}}$. Let $\psi = M \circ \varphi \circ M^{-1}$. Then $\psi^{(k)}(\zeta_{\mathrm{Gauss}}) = \zeta_{\mathrm{Gauss}}$, and $\mu_{\psi^{(k)}} = \delta_{\zeta_{\mathrm{Gauss}}}$ by

---

[8]The proof of [81, Theorem 3], written with $K = \mathbb{C}_p$, remains valid for arbitrary K.

Corollary 10.15. Thus $\psi^{(k)}$ has good reduction by Proposition 10.5. By Corollary 10.17, ψ itself has good reduction. Hence φ has simple reduction.

Conversely, suppose φ has simple reduction. Then by definition, there is a Möbius transformation $M \in \mathrm{PGL}(2,K)$ such that $\psi = M \circ \varphi \circ M^{-1}$ has good reduction. Thus $\mu_\psi = \delta_{\zeta_\mathrm{Gauss}}$, so by Corollary 10.15 we have $\mu_\varphi = \delta_y$, where $y = M^{-1}(\zeta_\mathrm{Gauss}) \in \mathbb{H}_\mathrm{Berk}$ is a point of type II. It follows from Theorem 10.2 that y is exceptional for φ. $\square$

REMARK 10.46. An alternate proof of Proposition 10.45(B) can be found in [**5**, Corollary 3.26]. The proof there is also based on properties of the canonical measure μ_φ and its associated Arakelov-Green's function $g_\varphi(x,y)$, especially Theorem 10.18.

As a second application of Theorem 10.36, we show that the canonical measure has point masses only when its support is a single point. This fact is also due to Favre and Rivera-Letelier [**48**]. With their version of Theorem 10.36, the proof remains valid in arbitrary characteristic.

COROLLARY 10.47. *Let $\varphi(T) \in K(T)$ have degree $d \geq 2$. Then the canonical measure μ_φ never charges points of $\mathbb{P}^1(K)$, and unless φ has simple reduction (so that J_φ consists of a single point, necessarily of type II), it does not charge points of $\mathbb{H}_\mathrm{Berk}$.*

PROOF. Since μ_φ has continuous potentials (Proposition 10.7), it cannot charge points of $\mathbb{P}^1(K)$.

Suppose that it charges some $\zeta \in \mathbb{H}_\mathrm{Berk}$. Put $\xi = \varphi(\zeta)$, and let $\zeta_1, \ldots, \zeta_n$ be the points in $\varphi^{-1}(\{\xi\})$, labeled in such a way that $\zeta = \zeta_1$. Let their multiplicities be $m_1, \ldots, m_n$ respectively, where $m_1 + \cdots + m_n = d$. By Proposition 9.52,
$$\varphi^*(\delta_\xi) = \sum_{i=1}^n m_i \delta_{\zeta_i}.$$
On the other hand, by Theorem 10.2, $\mu_\varphi = (1/d)\varphi^*(\mu_\varphi)$, so $\mu_\varphi(\{\zeta\}) = (m_1/d) \cdot \mu_\varphi(\{\xi\})$; that is
$$\mu_\varphi(\{\varphi(\zeta)\}) = \frac{d}{m_1}\mu_\varphi(\{\zeta\}) \geq \mu_\varphi(\{\zeta\}),$$
with strict inequality unless $m_1 = d$.

Inductively, we see that $\mu_\varphi(\{\varphi^{(k)}(\zeta)\}) \geq \mu_\varphi(\{\zeta\})$ for each $k \geq 0$. Since μ_φ has finite total mass, this is impossible unless there are integers $k_2 > k_1$ for which $\varphi^{(k_2)}(\zeta) = \varphi^{(k_1)}(\zeta)$. In this case, there must be a $\beta > 0$ such that $\mu_\varphi(\{\varphi^{(k)}(\zeta)\}) = \beta$ for each $k \geq k_1$. It follows from the discussion above that $\varphi^{(k)}(\zeta)$ has multiplicity d for each $k \geq k_1$. In particular, $\varphi^{(k_2-1)}(\zeta)$ is the only preimage of $\varphi^{(k_1)}(\zeta)$; hence $\varphi^{(k_1-1)}(\zeta) = \varphi^{(k_2-1)}(\zeta)$. Proceeding inductively backwards, we see that the grand orbit of ζ is finite. Thus ζ is an exceptional point.

By Proposition 10.45, we conclude that ζ is of type II and that φ has simple reduction. $\square$

We conclude this section with a third application of Theorem 10.36: for each $y \in \mathbb{H}_{\text{Berk}}$, the Arakelov-Green's function $g_\varphi(x,y)$ is not only continuous, but Hölder continuous. The Hölder continuity of dynamical Green's functions is well known in the classical case and is due to Favre and Rivera-Letelier in the Berkovich case (see [**47**, Proposition 6.5]). With their version of Theorem 10.36, the result holds in arbitrary characteristic.

We begin by recalling some definitions. For $x, y \in \mathbb{P}^1_{\text{Berk}}$, let $d(x,y) = 2\operatorname{diam}(x \vee y) - \operatorname{diam}(x) - \operatorname{diam}(y)$ be the metric introduced in §2.7, which induces the strong topology on $\mathbb{P}^1_{\text{Berk}}$. Here, $x \vee y$ denotes the point where the paths $[x, \zeta_{\text{Gauss}}]$ and $[y, \zeta_{\text{Gauss}}]$ first meet, and $\operatorname{diam}(x)$ is the diameter relative to ζ_{Gauss}.

We will say that a function $G : \mathbb{P}^1_{\text{Berk}} \to \mathbb{R}$ is *Lipschitz continuous* if there is a constant $C > 0$ such that for all $x, y \in \mathbb{P}^1_{\text{Berk}}$,

(10.57) $$|G(x) - G(y)| \leq C \cdot d(x,y) \;.$$

We will say that G is *Hölder continuous* if there are constants $C > 0$ and $0 < \kappa \leq 1$ such that

(10.58) $$|G(x) - G(y)| \leq C \cdot d(x,y)^\kappa \;.$$

If κ is given and (10.58) holds, we will say that G is κ-Hölder continuous. In particular, G is Lipschitz continuous iff it is 1-Hölder continuous.

EXAMPLE 10.48. For any pair of points $w, \zeta \in \mathbb{H}_{\text{Berk}}$, there is a continuous function $G \in \operatorname{BDV}(\mathbb{P}^1_{\text{Berk}})$, uniquely determined up to an additive constant, which satisfies $\Delta G = \delta_w - \delta_\zeta$. Explicitly, $G(z) = -\log_v(\delta(z,w)_\zeta) + C$ for each $z \in \mathbb{P}^1_{\text{Berk}}$ (see Example 5.19 and Lemma 5.24). We claim that G is Lipschitz continuous.

To show this, it suffices to establish (10.57) when x, y are points with $y \in [x, \zeta_{\text{Gauss}}]$. Indeed, if there is a constant C such that (10.57) holds in this case and if arbitrary $x, y \in \mathbb{P}^1_{\text{Berk}}$ are given, put $z = x \vee y$. Then
$$\begin{aligned} |G(x) - G(y)| &\leq |G(x) - G(z)| + |G(z) - G(y)| \\ &\leq C \cdot d(x,z) + C \cdot d(z,y) = C \cdot d(x,y) \;. \end{aligned}$$

Let $r = \min(\operatorname{diam}(w), \operatorname{diam}(\zeta)) > 0$. Note that G is constant on branches off $\Gamma = [\zeta, w]$, that its restriction to Γ has constant slope 1 (relative to the path distance metric), and that $G(\zeta) \leq G(z) \leq G(w)$ for each $z \in \mathbb{P}^1_{\text{Berk}}$. Assuming $y \in [x, \zeta_{\text{Gauss}}]$, we now consider three possibilities:

First, if $\operatorname{diam}(x), \operatorname{diam}(y) \leq r$, then x and y belong to a single branch off Γ, so $G(x) = G(y)$ and (10.57) holds trivially for any C. Next, if $\operatorname{diam}(x) \leq r/2$ but $\operatorname{diam}(y) \geq r$, let $M = G(w) - G(\zeta)$; then (10.57) holds trivially for any $C \geq 2M/r$. Finally, if $\operatorname{diam}(x), \operatorname{diam}(y) \geq r/2$, let $C \geq 2/r$ be any Lipschitz constant for $\log_v(t)$ on the interval $[r/2, 1] \subset \mathbb{R}$. Then since $G|_{[x,y]}$ is piecewise differentiable and $|G'(z)| \leq 1$ for each $z \in [x,y]$, we have
$$\begin{aligned} |G(x) - G(y)| &\leq \rho(x,y) = \log_v(\operatorname{diam}(y)) - \log_v(\operatorname{diam}(x)) \\ &\leq C \cdot (\operatorname{diam}(y) - \operatorname{diam}(x)) = C \cdot d(x,y) \;. \end{aligned}$$ □

10.4. EQUIDISTRIBUTION OF PREIMAGES

Let μ be a probability measure on $\mathbb{P}^1_{\text{Berk}}$ with continuous potentials, and let $g_\mu(x,y)$ be its normalized Arakelov-Green's function. We will say that μ has *Hölder continuous potentials* if $g_\mu(x, \zeta_{\text{Gauss}})$ is Hölder continuous.

To justify this terminology, we will show that if $G(x) := g_\mu(x, \zeta_{\text{Gauss}})$ is Hölder continuous, with Hölder exponent κ, then for each $y \in \mathbb{H}_{\text{Berk}}$ the function $G_y(x) = g_\mu(x,y)$ is κ-Hölder continuous as well. To see this, note that by (8.62), there is a constant B such that for all $x, y \in \mathbb{P}^1_{\text{Berk}}$

$$(10.59) \qquad g_\mu(x,y) \;=\; -\log_v(\|x,y\|) + G(x) + G(y) + B \;.$$

Fix $y \in \mathbb{H}_{\text{Berk}}$. By Example 10.48, $-\log_v(\|x,y\|) = -\log_v(\delta(x,y)_{\zeta_{\text{Gauss}}})$ is Lipschitz continuous, which means that $G_y(x) = g_\mu(x,y)$ is κ-Hölder continuous. (However, the Hölder constant C for G_y may depend on y.)

THEOREM 10.49 (Favre, Rivera-Letelier). *Let $\varphi \in K(T)$ be a rational function with $\deg(\varphi) \geq 2$. Then the canonical measure μ_φ has Hölder continuous potentials.*

PROOF. Put $D = \deg(\varphi)$, write $\zeta = \zeta_{\text{Gauss}}$, and let $x_1, \ldots, x_D$ be the preimages of ζ under φ, listed with multiplicities. Let $g \in \text{BDV}(\mathbb{P}^1_{\text{Berk}})$ be a continuous function satisfying

$$\Delta g \;=\; \delta_\zeta - \frac{1}{D}\varphi^*(\delta_\zeta) \;=\; \sum_{k=1}^{D} \frac{1}{D}(\delta_\zeta - \delta_{x_k}).$$

By Example 10.48, such a function exists and is bounded and Lipschitz continuous. Let C_1 be the Lipschitz constant of g, and let C_2 be a bound such that $|g(x)| \leq C_2$ for all $x \in \mathbb{P}^1_{\text{Berk}}$.

For each $n \geq 1$, put

$$g_n \;=\; \sum_{k=0}^{n-1} \frac{1}{D^k} g \circ \varphi^{(k)} \;.$$

Then $g_n \in \text{BDV}(\mathbb{P}^1_{\text{Berk}})$, and $\Delta g_n = \delta_\zeta - (1/D^n)(\varphi^{(n)})^*(\delta_\zeta)$. The functions g_n converge uniformly on $\mathbb{P}^1_{\text{Berk}}$ to

$$(10.60) \qquad g_\infty \;=\; \sum_{k=0}^{\infty} \frac{1}{D^k} g \circ \varphi^{(k)} \;.$$

By Theorem 10.36, the measures $(1/D^n)(\varphi^{(n)})^*(\delta_\zeta)$ converge weakly to μ_φ. Hence Proposition 5.32 shows that $g_\infty \in \text{BDV}(\mathbb{P}^1_{\text{Berk}})$ and that

$$\Delta(g_\infty) \;=\; \delta_\zeta - \mu_\varphi \;.$$

However, $\Delta(g_\mu(x,\zeta)) = \delta_\zeta - \mu_\varphi$ as well, and both $g_{\mu_\varphi}(x,\zeta)$ and $g_\infty(x)$ are continuous functions of x. By Lemma 5.24, there is a constant C_3 such that for each $x \in \mathbb{P}^1_{\text{Berk}}$,

$$g_{\mu_\varphi}(x,\zeta) \;=\; g_\infty(x) + C_3 \;.$$

We will now use the representation (10.60) to show that $g_{\mu_\varphi}(x,\zeta)$ is Hölder continuous. Let $N \geq 0$ be an integer to be specified later. By (10.60), for any $x, y \in \mathbb{P}^1_{\text{Berk}}$,

$$(10.61) \quad |g_\infty(x) - g_\infty(y)| \leq \sum_{k=0}^{N-1} \frac{1}{D^k} \left| g(\varphi^{(k)}(x)) - g(\varphi^{(k)}(y)) \right| + \sum_{k=N}^{\infty} \frac{2C_2}{D^k} .$$

By Proposition 9.37, there is a constant M such that for all $z, w \in \mathbb{P}^1_{\text{Berk}}$,

$$(10.62) \quad d(\varphi(z), \varphi(w)) \leq M \cdot d(z, w) .$$

Iterating (10.62), we see that

$$(10.63) \quad \sum_{k=0}^{N-1} \frac{1}{D_k} \left| g(\varphi^{(k)}(x)) - g(\varphi^{(k)}(y)) \right| \leq \left(\sum_{k=0}^{N-1} \frac{C_1 M^k}{D^k} \right) d(x,y) .$$

Without loss of generality, we can assume that $M > D$ and that $d(x, y) < 1$. It follows from (10.61) and (10.63) that

$$(10.64) \quad |g_\infty(x) - g_\infty(y)| \leq \frac{C_1}{\frac{M}{D} - 1} \cdot \left(\frac{M}{D}\right)^N \cdot d(x,y) + \frac{2C_2}{D-1} \cdot \left(\frac{1}{D}\right)^N .$$

If we take $N = \lceil -\log(d(x,y))/\log(M) \rceil$, the two terms on the right side of (10.64) are of the same order of magnitude, and simple algebraic manipulations show that with $\kappa = \log(D)/\log(M)$, there is a constant C independent of x, y such that

$$(10.65) \quad |g_\infty(x) - g_\infty(y)| \leq C \cdot d(x,y)^\kappa .$$

Thus $g_{\mu_\varphi}(x, \zeta)$ is Hölder continuous. □

10.5. The Berkovich Fatou and Julia sets

In this section, we define the Berkovich Fatou and Julia sets, give some characterizations of those sets, and note some of their structural properties. As in §10.4, we assume throughout this section that $\text{char}(K) = 0$.

DEFINITION 10.50. Let $\varphi \in K(T)$ be a rational map of degree $d \geq 2$. The *Berkovich Julia set* J_φ of φ is the support of the canonical measure μ_φ, and the *Berkovich Fatou set* F_φ is the complement in $\mathbb{P}^1_{\text{Berk}}$ of J_φ.

Note that since the Berkovich Julia set of φ is the support of the probability measure μ_φ, it is by definition always nonempty. We will eventually give four characterizations of J_φ:

(1) J_φ is the support of the canonical measure μ_φ (Definition 10.50).
(2) J_φ is the locus of points $x \in \mathbb{P}^1_{\text{Berk}}$ having the property that for every Berkovich neighborhood V of x, the union of the forward images $\varphi^{(n)}(V)$ contains $\mathbb{P}^1_{\text{Berk}} \backslash E_\varphi(K)$ (and in particular contains all of $\mathbb{H}_{\text{Berk}}$ and omits at most two points of $\mathbb{P}^1(K)$) (Theorem 10.56).

(3) J_φ is the smallest nonempty closed set in $\mathbb{P}^1_{\text{Berk}}$, disjoint from the classical exceptional locus $E_\varphi(K)$, which is both forward and backward invariant under φ (Corollary 10.57).

(4) J_φ is the closure of the set of repelling periodic points in $\mathbb{P}^1_{\text{Berk}}$ (Theorem 10.88).

The equivalence between these properties is due to Rivera-Letelier. For $K = \mathbb{C}_p$, he discussed (2) $\Leftrightarrow$ (3) $\Leftrightarrow$ (4) with us during a visit in 2005, and the general case is contained in [**80**]. The equivalence (1) $\Leftrightarrow$ (3) is stated in his note with Favre [**47**] and proved in [**48**, Theorem A]. The equivalence (2) $\Leftrightarrow$ (3) is stated in [**48**, Proposition 2.7]. In §10.6, we will discuss the relation between (1) and a more 'classical' definition of J_φ in terms of equicontinuity which holds when $K = \mathbb{C}_p$ but not in general; this fact is also due to Rivera-Letelier (unpublished).

Our first observation is that the Berkovich Julia sets (resp. Berkovich Fatou sets) of φ and $\varphi^{(n)}$ coincide for any integer $n \geq 1$:

LEMMA 10.51. *For any $n \geq 1$, we have $J_\varphi = J_{\varphi^{(n)}}$ and $F_\varphi = F_{\varphi^{(n)}}$.*

PROOF. This is an immediate consequence of Lemma 10.16. □

A subset $S \subseteq \mathbb{P}^1_{\text{Berk}}$ is called *completely invariant* for φ if $z \in S$ if and only if $\varphi(z) \in S$. We next observe:

LEMMA 10.52. *The Berkovich Julia and Fatou sets J_φ, F_φ are completely invariant for φ.*

PROOF. Write $\mu = \mu_\varphi$. If $\varphi(x) \in F_\varphi$, then there is an open neighborhood V of $\varphi(x)$ such that $\mu(V) = 0$. But then $\mu(\varphi^{-1}(V)) = (\varphi_*\mu)(V) = \mu(V) = 0$. As x belongs to the open set $\varphi^{-1}(V)$, it follows that $x \notin \text{supp}(\mu)$, that is, $x \in F_\varphi$. Thus F_φ is backward invariant for φ.

Conversely, if $x \in F_\varphi = \text{supp}(\mu)^c$, then there are an open neighborhood U of x and a continuous function f on $\mathbb{P}^1_{\text{Berk}}$ such that $f \equiv 1$ on U and $f \equiv 0$ on $\text{supp}(\mu)$. Let $V = \varphi(U)$, which is an open neighborhood of $\varphi(x)$ by the Open Mapping Theorem. Then $(\varphi_*f)(z) \geq 1$ for all $z \in V$, and $(\varphi_*f)(z) \geq 0$ for all $z \in \mathbb{P}^1_{\text{Berk}}$. Thus

$$0 \leq \mu(V) \leq \int_{\mathbb{P}^1_{\text{Berk}}} (\varphi_*f)\,d\mu = \int_{\mathbb{P}^1_{\text{Berk}}} f\,d\varphi^*\mu = \deg(\varphi) \cdot \int_{\mathbb{P}^1_{\text{Berk}}} f\,d\mu = 0$$

since $f \equiv 0$ on $\text{supp}(\mu)$ and $\varphi^*\mu = \deg(\varphi) \cdot \mu$. It follows that $\mu(V) = 0$, and therefore $\varphi(x) \notin \text{supp}(\mu)$, that is, $\varphi(x) \in F_\varphi$. □

We now use Proposition 10.45 to characterize maps with simple reduction in terms of their Julia sets. Recall that φ has *simple reduction* if, after a change of coordinates, it has good reduction.

LEMMA 10.53. *The Julia set J_φ consists of a single point (which is necessarily a point of $\mathbb{H}_{\text{Berk}}$ of type II) if and only if φ has simple reduction.*

PROOF. This is an immediate consequence of Corollary 10.47, but the following self-contained argument is easier.

If $J_\varphi = \{\zeta\}$ consists of a single point, then the result follows from Proposition 10.45, since ζ is then exceptional by the complete invariance of the Berkovich Julia set.

On the other hand, suppose φ has simple reduction. If φ has *good* reduction, then Example 10.4 shows that μ_φ is a point mass supported at the Gauss point. For the general case, if φ has good reduction after conjugating by a Möbius transformation $M \in \mathrm{PGL}(2, K)$, then by Corollary 10.15 the measure μ_φ is a point mass supported at the type II point $M(\zeta_{\mathrm{Gauss}})$. □

The following result and its corollary will be used several times below:

LEMMA 10.54. *Every attracting periodic point $x \in \mathbb{P}^1(K)$ for φ is contained in the Berkovich Fatou set.*

PROOF. Let $x \in \mathbb{P}^1(K)$ be an attracting periodic point of period n. Replacing φ by $\varphi^{(n)}$, we may assume that x is a fixed point of φ. Let U be a small open disc in $\mathbb{P}^1(K)$ centered at x which is contained in the (classical) attracting basin of x, and let $y \in U$ be a nonexceptional point of $\mathbb{P}^1(K)$. By Theorem 10.36, J_φ is contained in the closure of the set of inverse iterates of y. Since $\varphi(U) \subseteq U$, none of the inverse iterates of y is contained in U, and therefore U is disjoint from the closure of the set of inverse iterates of y. In particular, it follows that $x \notin J_\varphi$. □

COROLLARY 10.55. *Every point in $E_\varphi(K)$ is contained in F_φ.*

PROOF. Every exceptional point in $\mathbb{P}^1(K)$ is a (super)attracting fixed point, so the result follows from Lemma 10.54. □

The following result is another important consequence of the equidistribution theorem for preimages of nonexceptional points (Theorem 10.36):

THEOREM 10.56. *The Berkovich Fatou set F_φ coincides with*

(A) *the set of all points $x \in \mathbb{P}^1_{\mathrm{Berk}}$ having a neighborhood V whose forward iterates under φ omit at least one point of $\mathbb{P}^1_{\mathrm{Berk}} \backslash E_\varphi(K)$,*

(B) *the set of all points $x \in \mathbb{P}^1_{\mathrm{Berk}}$ having a neighborhood V whose forward iterates under φ omit at least three points of $\mathbb{P}^1(K)$.*

Equivalently, the Berkovich Julia set J_φ coincides with

(A′) *the set of all points in $x \in \mathbb{P}^1_{\mathrm{Berk}}$ such that for each neighborhood V of x, the union $\bigcup_{n=1}^\infty \varphi^{(n)}(V)$ contains $\mathbb{P}^1_{\mathrm{Berk}} \backslash E_\varphi(K)$,*

(B′) *the set of all points in $x \in \mathbb{P}^1_{\mathrm{Berk}}$ such that for each neighborhood V of x, the union $\bigcup_{n=1}^\infty \varphi^{(n)}(V)$ omits at most two points of $\mathbb{P}^1(K)$.*

PROOF. It suffices to prove the assertions concerning F_φ.

We first prove (A). Let F' be the set of all points in $\mathbb{P}^1_{\mathrm{Berk}}$ having a neighborhood V whose forward iterates under φ omit at least one point of $\mathbb{P}^1_{\mathrm{Berk}} \backslash E_\varphi(K)$. We want to show that $F' = F_\varphi$. First, suppose that $z \in F'$;

then one can find a neighborhood V of z and a point $y \in \mathbb{P}^1_{\text{Berk}} \backslash E_\varphi(K)$ not belonging to $\varphi^{(n)}(V)$ for any $n \geq 1$. If $\mu^y_{\varphi^{(n)}}$ denotes the probability measure $d^{-n}(\varphi^{(n)})^*(\delta_y)$, then $V \cap \text{supp}(\mu^y_{\varphi^{(n)}}) = \emptyset$ for all n, and thus $V \cap \text{supp}(\mu_\varphi) = \emptyset$ by Theorem 10.36. It follows that $F' \subseteq F_\varphi$.

Conversely, let J' be the complement of F'. We claim that J' is the smallest completely invariant nonempty compact subset of $\mathbb{P}^1_{\text{Berk}}$ disjoint from $E_\varphi(K) := E_\varphi \cap \mathbb{P}^1(K)$. Assuming the claim, since $J_\varphi = \text{supp}(\mu_\varphi)$ is completely invariant under φ, we obtain $J' \subseteq J_\varphi$, and hence $F_\varphi \subseteq F'$ as desired. (Note that $J_\varphi \cap E_\varphi(K) = \emptyset$ by Corollary 10.55.)

To prove the claim, we must show that if Z is any nonempty completely invariant compact subset of $\mathbb{P}^1_{\text{Berk}}$ which is disjoint from $E_\varphi(K)$, then Z contains J'. If U denotes the complement of Z, then U is also completely invariant under φ. In particular, $\varphi^{(n)}(U) \subseteq U$ for all $n \geq 1$. If φ does not have simple reduction, then it follows from Proposition 10.45 that $Z = U^c$ contains a nonexceptional point, and thus U is contained in F', which is equivalent to the assertion of the claim.

It remains to prove the claim in the case where φ has simple reduction and $Z = \{\zeta\}$ with $\zeta \in \mathbb{H}_{\text{Berk}}$ an exceptional point of φ. Conjugating by a Möbius transformation, we may assume without loss of generality that φ has *good* reduction and that $Z = \{\zeta_{\text{Gauss}}\}$. In this case, since we already know that $J' \neq \emptyset$, it suffices to prove that every point of $U = \mathbb{P}^1_{\text{Berk}} \backslash \{\zeta_{\text{Gauss}}\}$ belongs to F'. Define the *chordal diameter* of an open set $V \subseteq \mathbb{P}^1_{\text{Berk}}$ to be $\sup_{x,y \in V \cap \mathbb{P}^1(K)} \|x, y\|$. If x is any point of U and V is any sufficiently small connected open neighborhood of x contained in U, then the chordal diameter of V is some positive real number $\delta < 1$. Since φ has good reduction, it follows from [**93**, Theorem 2.17(a)] that φ is everywhere nonexpanding with respect to the spherical metric on $\mathbb{P}^1(K)$; this implies that the open set $\varphi^{(n)}(V)$ has chordal diameter at most δ for all $n \geq 1$. If y is any point of $\mathbb{H}_{\text{Berk}}$ such that $0 < \rho(y, \zeta_{\text{Gauss}}) < 1 - \delta$, then since the generalized Hsia kernel $\delta(\cdot, \cdot)_{\zeta_{\text{Gauss}}}$ extends the spherical metric on $\mathbb{P}^1(K)$, one verifies easily that y cannot belong to $\varphi^{(n)}(V)$ for any $n \geq 0$. Since y is automatically nonexceptional, it follows that x belongs to F' as desired. This completes the proof of (A).

The proof of (B) is similar; the only additional thing which needs to be proved is that if φ has good reduction and $x \in \mathbb{P}^1_{\text{Berk}} \backslash \{\zeta_{\text{Gauss}}\}$, then with V chosen as above, there exists a nonexceptional point in $\mathbb{P}^1(K)$ which does not belong to $\bigcup_{n \geq 0} \varphi^{(n)}(V)$. With y chosen as above, any (nonexceptional) point $z \in \mathbb{P}^1(K)$ whose retraction to the segment $[y, \zeta_{\text{Gauss}}]$ lies in the open interval $(y, \zeta_{\text{Gauss}})$ will work. $\square$

From the proof of Theorem 10.56, we obtain:

COROLLARY 10.57. *The Berkovich Julia set J_φ is the smallest completely invariant nonempty compact subset of $\mathbb{P}^1_{\text{Berk}}$ disjoint from $E_\varphi(K)$.*

The following is an immediate consequence of the definition of J_φ and Theorem 10.36:

COROLLARY 10.58. *If z_0 is any point of $\mathbb{P}^1_{\text{Berk}}\backslash E_\varphi(K)$, the closure of the backward orbit of z_0 contains J_φ. In particular, the backward orbit of any point in J_φ is dense in J_φ.*

The following consequence of Theorem 10.56 is analogous to Theorem 4.7 of [73].

THEOREM 10.59 (Transitivity Theorem). *Let $x \in J_\varphi$ and let V be an open neighborhood of x. Let U be the union of the forward images $\varphi^{(n)}(V)$ for $n \geq 0$. Then:*

(A) *U contains $\mathbb{P}^1_{\text{Berk}}\backslash E_\varphi(K)$: in particular, if $\text{char}(K) = 0$, it contains all of $\mathbb{H}_{\text{Berk}}$ and all but at most two points of $\mathbb{P}^1(K)$.*
(B) *U contains all of J_φ.*
(C) *If V is sufficiently small, then $U = \mathbb{P}^1_{\text{Berk}}\backslash E_\varphi(K)$.*

PROOF. (A) We first show that $\mathbb{P}^1(K)\backslash E_\varphi(K) \subset U$. Note that the set $\mathbb{P}^1(K)\backslash U$ can contain at most two points; otherwise, since $\varphi(U) \subseteq U$, it would follow from Theorem 10.56(B) that U is contained in the Fatou set, contradicting the fact that x belongs to both J_φ and U. Let $z \in \mathbb{P}^1(K)\backslash U$. Since $\varphi(U) \subseteq U$, each iterated preimage of z must belong to the finite set $\mathbb{P}^1(K)\backslash U$. It follows that z is exceptional, that is, $z \in E_\varphi(K)$.

The fact that $\mathbb{H}_{\text{Berk}}$ is contained in U follows similarly. If $y \in \mathbb{H}_{\text{Berk}}\backslash U$, then certainly $y \notin E_\varphi(K)$. By Corollary 10.58, the set $\{\varphi^{(n)*}(y)\}_{n\geq 1}$ has J_φ in its closure. On the other hand, $\varphi(U) \subset U$, so these preimages belong to $\mathbb{P}^1_{\text{Berk}}\backslash U$, meaning that $J_\varphi \subset \mathbb{P}^1_{\text{Berk}}\backslash U$. However this is a contradiction, since $x \in J_\varphi$ and $x \in U$.

(B) This is immediate from part (A) and Corollary 10.57.
(C) If $V \subseteq \mathbb{P}^1_{\text{Berk}}\backslash E_\varphi$, it follows from the proof of (A) that $U = \mathbb{P}^1_{\text{Berk}}\backslash E_\varphi$. $\square$

COROLLARY 10.60. *If φ does not have simple reduction, then J_φ is perfect (that is, contains no isolated points). In particular, J_φ either consists of a single point or it is uncountable.*

PROOF. It suffices to prove that J_φ contains a dense set of nonisolated points. By Proposition 10.45, we see that $E_\varphi \cap \mathbb{H}_{\text{Berk}} = \emptyset$. Furthermore, J_φ is infinite, since if it were finite, it would consist of exceptional points in $\mathbb{P}^1(K)$, which are necessarily in the Fatou set by Corollary 10.55. So J_φ contains at least one limit point z_0, and z_0 is nonexceptional. Since the measures $(1/d^n)\varphi^{(n)*}(\delta_{z_0})$ converge weakly to μ_φ, the iterated preimages of z_0 form a dense set of points in J_φ. If $w \in (\varphi^{(n)})^{-1}(\{z_0\})$, then for any neighborhood U of w, the image $\varphi^{(n)}(U)$ is a neighborhood of z_0 (Corollary 9.10). Since J_φ is backwards invariant, U contains points of J_φ distinct from z_0. As U is arbitrary, it follows that w is also a limit point of J_φ. $\square$

COROLLARY 10.61. *The Julia set J_φ is either connected or else has uncountably many connected components.*

PROOF. In the case where J_φ is perfect, the proof is identical to its classical counterpart (see [**73**, Corollary 4.15]). The only other possibility is that J_φ consists of a single point, in which case it is trivially connected. $\square$

10.6. Equicontinuity

In classical non-Archimedean dynamics, one usually defines the Fatou set of a rational map in terms of equicontinuity. Here we explore the relation between our definition and definitions in terms of equicontinuity. We assume throughout that $\text{char}(K) = 0$ and that $\varphi(T) \in K(T)$.

For $x, y \in \mathbb{P}^1(K)$, we write $\|x, y\|$ for the spherical distance (see §4.3), and for $0 < r \in \mathbb{R}$ we write $B(x, r)^- = \{y \in \mathbb{P}^1(K) : \|x, y\| < r\}$.

DEFINITION 10.62 (Equicontinuity on $\mathbb{P}^1(K)$). Let U be an open subset of $\mathbb{P}^1(K)$, and let $\mathcal{F}$ be a collection of functions $f : U \to \mathbb{P}^1(K)$. We say that $\mathcal{F}$ is *equicontinuous* on U if for every $x \in U$ and every $\varepsilon > 0$, there exists a $\delta > 0$ such that

$$f(B(x, \delta)^-) \subseteq B(f(x), \varepsilon)^-$$

for every $f \in \mathcal{F}$.

A point $x \in \mathbb{P}^1(K)$ is in the *classical Fatou set* for φ iff the collection $\{\varphi^{(n)}\}$ of iterates of φ is equicontinuous on some open neighborhood of x in $\mathbb{P}^1(K)$. The *classical Julia set* is defined to be the complement of the classical Fatou set in $\mathbb{P}^1(K)$.

We now seek to extend the definition of equicontinuity so that it works in the context of $\mathbb{P}^1_{\text{Berk}}$, which need not be metrizable. Fortunately, since $\mathbb{P}^1_{\text{Berk}}$ is a compact Hausdorff space, it has a canonical *uniform structure* (cf. §A.9), and the theory of uniform spaces provides for us a natural notion of equicontinuity. Concretely, this yields the following definition:

DEFINITION 10.63 (Equicontinuity on $\mathbb{P}^1_{\text{Berk}}$). If U is an open subset of $\mathbb{P}^1_{\text{Berk}}$ and $\mathcal{F}$ is a collection of functions $f : U \to \mathbb{P}^1_{\text{Berk}}$, we say that $\mathcal{F}$ is *equicontinuous* on U if for every $x \in U$ and every finite covering $\mathcal{C} = V_1 \cup \cdots \cup V_t$ of $\mathbb{P}^1_{\text{Berk}}$ by simple domains, there exists an open neighborhood U_x of x in $\mathbb{P}^1_{\text{Berk}}$ such that $f(U_x)$ is contained in some V_i for every $f \in \mathcal{F}$.

We define the *Berkovich equicontinuity locus* for φ to be the set of all $x \in \mathbb{P}^1_{\text{Berk}}$ for which the collection $\{\varphi^{(n)}\}$ of iterates of φ is equicontinuous on some Berkovich open neighborhood of x.

THEOREM 10.64. *The Berkovich equicontinuity locus and the classical Fatou sets are both contained in the Berkovich Fatou set.*

PROOF. Let x_0 be a point in either the Berkovich equicontinuity locus or the classical Fatou set. We will show that if U is a sufficiently small open neighborhood of x_0, then μ_φ restricted to U is the zero measure.

First assume that $x_0 \in \mathbb{A}^1_{\text{Berk}}$. By Theorem 10.2, it suffices to show that $\hat{h}_{\varphi,v,(\infty)}$ is harmonic on some neighborhood $U \subset \mathbb{A}^1_{\text{Berk}}$ of x_0. To do this, we follow an argument of Fornaess and Sibony [**52**], in outline. Define

$$O = \{x \in \mathbb{A}^1_{\text{Berk}} : [T]_x < 2\},$$
$$E = \{x \in \mathbb{A}^1_{\text{Berk}} : [T]_x \leq \frac{1}{2}\}.$$

We claim that there exists a neighborhood U of x_0 such that for each $n \geq 1$, either $\varphi^{(n)}(U) \subseteq O$ or $\varphi^{(n)}(U) \subseteq \mathbb{P}^1_{\text{Berk}} \backslash E$. Indeed, if x_0 is in the Berkovich equicontinuity locus, then this follows directly from the fact that $\{O, E^c\}$ is a finite covering of $\mathbb{P}^1_{\text{Berk}}$ by simple domains, while if x_0 is in the classical Fatou set, it follows easily from the ultrametric inequality. Passing to a subsequence $\varphi^{(n_k)}$ and replacing φ by $1/\varphi$ if necessary, we may therefore assume without loss of generality that $\varphi^{(n_k)}(U) \subseteq O$ for all k, or, equivalently, that $[\varphi^{(n_k)}(T)]_x < 2$ for all $x \in U$ and all k. In particular, in the notation of §10.1, $F_1^{(n_k)}(1,T) \neq 0$ on U for all k.

For $x \in U$ we have

$$\hat{h}^{(n_k)}_{\varphi,v,(\infty)}(x) = \frac{1}{d^{n_k}} \max(\log_v([F_1^{(n_k)}(1,T)]_x), \log_v([F_2^{(n_k)}(1,T)]_x))$$

(10.66)
$$= \frac{1}{d^{n_k}} \log_v([F_1^{(n_k)}(1,T)]_x) + \frac{1}{d^{n_k}} \max(1, \log_v([\varphi^{(n_k)}(T)]_x)).$$

The last term in (10.66) converges uniformly to 0, since the quantity $[\varphi^{(n_k)}(T)]_x$ is uniformly bounded as k varies. Moreover, the term $\frac{1}{d^{n_k}} \log_v([F_1^{(n_k)}(1,T)]_x)$ is harmonic on U for all k. Since $\hat{h}^{(n_k)}_{\varphi,v,(\infty)}$ converges uniformly to $\hat{h}_{\varphi,v,(\infty)}$, it follows that $\frac{1}{d^{n_k}} \log_v([F_1^{(n_k)}(1,T)]_x)$ converges uniformly to $\hat{h}_{\varphi,v,(\infty)}$ on U. Therefore $\hat{h}_{\varphi,v,(\infty)}$ is harmonic on U as desired.

Finally, for the case $x_0 = \infty$, apply a similar argument to the function $\hat{h}_{\varphi,v,(0)}$, using the coordinate function $1/T$ instead of T. $\square$

We now prove that the intersection of F_φ with $\mathbb{P}^1(K)$ is precisely the classical Fatou set. The proof will make use of the following non-Archimedean Montel theorem due to Hsia [**61**]:

THEOREM 10.65 (Hsia's theorem). *Let $D \subset \mathbb{P}^1(K)$ be a closed disc, and let $\mathcal{F}$ be a family of meromorphic functions on D. Suppose that $\bigcup_{f \in \mathcal{F}} f(D)$ omits two distinct points of $\mathbb{P}^1(K)$. Then $\mathcal{F}$ is equicontinuous on D.*

REMARK 10.66. There does not seem to be a satisfactory way to generalize Hsia's theorem to $\mathbb{P}^1_{\text{Berk}}$. This is one of the reasons why we have defined the Berkovich Julia set as the support of μ_φ and not in terms of equicontinuity. For example, let $\{\alpha_n\}$ be a set of coset representatives for the different residue classes in $D(0,1)$ over $\mathbb{C}_p$ and let $f_n = z + \alpha_n$. Then the sequence of functions $f_n = z + \alpha_n$ takes $\mathcal{D}(0,1)$ to itself (so $\bigcup_{n \geq 1} f_n(\mathcal{D}(0,1))$ omits an entire open disc in $\mathbb{P}^1_{\text{Berk}}$) but it is not equicontinuous at the Gauss point of $\mathbb{P}^1_{\text{Berk}}$. See Example 10.70 and Remark 10.71 below for related considerations.

THEOREM 10.67. *If $x \in \mathbb{P}^1(K)$, then x is in the Berkovich Fatou set of φ if and only if x is in the classical Fatou set of φ.*

PROOF. If x is in the classical Fatou set of φ, then we have just seen that x belongs to the Berkovich Fatou set of φ.

Conversely, suppose $x \in \mathbb{P}^1(K)$ belongs to the Berkovich Fatou set of φ. By Theorem 10.56(B), there is a Berkovich disc V containing x such that the union U of the forward orbits of V omits at least three points y_1, y_2, y_3 of $\mathbb{P}^1(K)$. At least one of these, say y_1, must be nonexceptional. Since $U \cap \mathbb{P}^1(K)$ omits the infinitely many preimages of y_1, Hsia's theorem shows x is contained in the classical Fatou set of φ. Thus the intersection of the Berkovich Fatou set with $\mathbb{P}^1(K)$ is contained in the classical Fatou set. □

By a theorem of Benedetto (see [**13**, Corollary 1.3]), the classical Fatou set of φ is always nonempty (in contrast to the situation over the complex numbers). We therefore conclude from Theorem 10.67:

COROLLARY 10.68. *The Berkovich Fatou set F_φ is always nonempty.*

The following corollary is stronger than what one obtains classically over $\mathbb{C}$, due to the absence in the non-Archimedean setting of rational maps with empty Fatou set:

COROLLARY 10.69. *The Berkovich Julia set J_φ always has empty interior.*

PROOF. If J_φ has an interior point z_1, then choosing a neighborhood $V \subset J_\varphi$ of z_1, the union U of forward images of V under φ is both contained in J_φ (by the forward invariance of J) and everywhere dense in $\mathbb{P}^1_{\text{Berk}}$ (since $\mathbb{P}^1_{\text{Berk}} \backslash U$ contains at most 2 points by the Transitivity Theorem). Since J_φ is closed, this implies that $J_\varphi = \mathbb{P}^1_{\text{Berk}}$ and thus that F_φ is empty, contradicting Corollary 10.68. □

Theorem 10.64 asserts that the Berkovich equicontinuity locus is always contained in the Berkovich Fatou set. When $K = \mathbb{C}_p$, we will see in Theorem 10.72 that the reverse inclusion holds as well, so that the Berkovich Fatou set coincides with the Berkovich equicontinuity locus. This relies on deep results of Rivera-Letelier concerning the structure of the Berkovich Fatou set over $\mathbb{C}_p$. However, for a general complete and algebraically closed non-Archimedean field K of characteristic 0, the Berkovich Fatou set need not be contained in the Berkovich equicontinuity locus, as we now show.

EXAMPLE 10.70 (A rational map whose Berkovich Fatou set is not contained in the Berkovich equicontinuity locus). Let p be an odd prime, and let K be a complete and algebraically closed non-Archimedean field of characteristic 0 and residue characteristic $p > 0$, having an element c satisfying $|c| = 1$ whose image in the residue field $\tilde{K}$ is transcendental over the prime field $\mathbb{F}_p$. Consider the polynomial map $\varphi(z) = pz^2 + cz$.

Then φ is conjugate to the map $\varphi'(w) = w^2 + cw$, which has good reduction, so J_φ is a point mass supported at a point $\zeta \in \mathbb{H}_{\text{Berk},K}$ with $\zeta \neq \zeta_{\text{Gauss}}$. In particular, the Gauss point of $\mathbb{P}^1_{\text{Berk},K}$ belongs to the Berkovich Fatou set of φ.

However, we claim that $\{\varphi^{(n)}\}$ is not equicontinuous at the Gauss point. To see this, first note that if $|a|, r < p$, then $\varphi(D(a,r)) = D(pa^2 + ca, r)$ and $\varphi(D(a,r)^-) = D(pa^2 + ca, r)^-$. In particular, $\varphi(D(0,1)) = D(0,1)$, so φ fixes the Gauss point. Moreover, $\varphi(D(0,1)^-) = D(0,1)^-$ and if $|a| = 1$, then $\varphi(D(a,1)^-) = D(ca,1)^-$. In particular, the open disc $V' = D(1,1)^-$ has infinitely many distinct preimages under φ, by our assumption on c. Let $\hat{V}' = \mathcal{D}(1,1)^-$ be the Berkovich open disc corresponding to V'. Let $\hat{V}$ be any simple Berkovich neighborhood of the Gauss point containing the complement of $\hat{V}'$ but not $\hat{V}'$ itself. Then $\{\hat{V}, \hat{V}'\}$ is a finite covering of $\mathbb{P}^1_{\text{Berk}}$ by simple domains. Suppose that $\{\varphi^{(n)}\}$ is equicontinuous at ζ_{Gauss}. By the definition of equicontinuity, since $\zeta_{\text{Gauss}} \notin \hat{V}'$, there must exist a simple domain $\hat{U}$ containing ζ_{Gauss} such that $\varphi^{(n)}(\hat{U}) \subseteq \hat{V}$ for all $n \geq 1$. However, this means that $\hat{U}$ must omit some element of the residue class $\varphi^{-n}(\hat{V}') = \mathcal{D}(c^{-n}a, 1)^-$, for each n. By hypothesis, this means that $\hat{U}$ is missing elements from infinitely many residue classes. On the other hand, a simple domain containing ζ_{Gauss} automatically contains all but finitely many residue classes. This contradiction shows that $\{\varphi^{(n)}\}$ is not equicontinuous at the Gauss point, as claimed.

REMARK 10.71. Philosophically, the point ζ_{Gauss} in Example 10.70 should be thought of as belonging to the Fatou set, since the iteration of φ in a small neighborhood of ζ_{Gauss} is quite understandable and predictable, and in some sense nearby points stay close together under iteration. However, there does not seem to be a notion of equicontinuity on $\mathbb{P}^1_{\text{Berk}}$ which captures this philosophical sentiment in a precise way.

For this reason we have defined the Berkovich Fatou and Julia sets without using equicontinuity (contrary to our initial inclination). We would like to thank Rivera-Letelier for bringing examples like Example 10.70 to our attention and for his patient insistence that for arbitrary complete algebraically closed fields K, equicontinuity should not be thought of as the basis for the fundamental dichotomy between order and chaos in the context of iteration on $\mathbb{P}^1_{\text{Berk}}$.

We conclude this section by sketching a proof that when $K = \mathbb{C}_p$, the Berkovich Fatou set coincides with the Berkovich equicontinuity locus.

THEOREM 10.72 (Rivera-Letelier). *Let* $\varphi \in \mathbb{C}_p(T)$ *be a rational function with* $\deg(\varphi) \geq 2$. *Take* $x_0 \in \mathbb{P}^1_{\text{Berk}}$, *and suppose there exists a neighborhood* V *of* x_0 *in* $\mathbb{P}^1_{\text{Berk}}$ *such that* $\bigcup_{n=1}^\infty \varphi^{(n)}(V)$ *omits at least three points of* $\mathbb{P}^1(\mathbb{C}_p)$ *or at least one point of* $\mathbb{H}_{\text{Berk}}$. *Then* x *is in the Berkovich equicontinuity locus of* φ.

Using Theorem 10.56 and Theorem 10.64, Theorem 10.72 implies:

COROLLARY 10.73. *Suppose $\varphi \in \mathbb{C}_p(T)$, and let $\deg(\varphi) \geq 2$. Then the Berkovich Fatou set of φ equals the Berkovich equicontinuity locus of φ.*

The proof of Theorem 10.72 is based on Rivera-Letelier's classification of the periodic components of F_φ, which we now recall[9] (cf. [**81**, **84**]). Let $\varphi \in \mathbb{C}_p(T)$ be a rational map of degree $d \geq 2$, and define F_φ^{RL} to be the set of $x \in \mathbb{P}^1_{\mathrm{Berk}}$ for which there exists a neighborhood V of x in $\mathbb{P}^1_{\mathrm{Berk}}$ such that $\bigcup_{n=1}^\infty \varphi^{(n)}(V)$ omits at least three points of $\mathbb{P}^1(\mathbb{C}_p)$. By Theorem 10.56, F_φ^{RL} coincides with F_φ. Rivera-Letelier defines the following two types of components of F_φ^{RL}:

Immediate basins of attraction: If $z_0 \in \mathbb{P}^1(\mathbb{C}_p)$ is an attracting periodic point of period n, its *basin of attraction* is the set
$$\mathcal{A}_{z_0}(\varphi) = \{z \in \mathbb{P}^1_{\mathrm{Berk}} : \varphi^{(nk)}(z) \to z_0 \text{ when } k \to \infty\}.$$
One easily shows that $\mathcal{A}_{z_0}(\varphi)$ is open and invariant under $\varphi^{(n)}$. The connected component $\mathcal{A}^0_{z_0}(\varphi)$ of $\mathcal{A}_{z_0}(\varphi)$ containing z_0 is called the *immediate basin of attraction* of z_0; it is also open and invariant under $\varphi^{(n)}$. Rivera-Letelier has shown that $\mathcal{A}_{z_0}(\varphi)$ is contained in F_φ^{RL} and that $\mathcal{A}^0_{z_0}(\varphi)$ is a connected component of F_φ^{RL}. He gives the following description of the action of $\varphi(T)$ on $\mathcal{A}^0_{z_0}(\varphi)$ [**81**, pp. 199–200]:

PROPOSITION 10.74. *Let $\varphi(T) \in \mathbb{C}_p(T)$ have degree $d \geq 2$, and let $\mathcal{A}^0_{z_0}(\varphi) \subset \mathbb{P}^1_{\mathrm{Berk}}$ be the immediate basin of attraction of an attracting periodic point z_0 of period n. Then there is a decreasing set of open neighborhoods $\{X_k\}_{k \in \mathbb{Z}}$ of z_0, cofinal in the collection of all neighborhoods of z_0, such that X_k is a disc for all $k \geq 0$, $\bigcup_k X_k = \mathcal{A}^0_{z_0}(\varphi)$, $\bigcap_k X_k = \{z_0\}$, and $\varphi^{(n)}(X_k) = X_{k+1}$ for each $k \in \mathbb{Z}$.*

The domain of quasi-periodicity: The *domain of quasi-periodicity* of φ, denoted $\mathcal{E}(\varphi)$, is defined to be the interior of the set of points in $\mathbb{P}^1_{\mathrm{Berk}}$ which are recurrent under φ. (A point is called *recurrent* if it lies in the closure of its forward orbit.) By definition, $\mathcal{E}(\varphi)$ is open and invariant under φ and is disjoint from each basin of attraction. Rivera-Letelier has shown that $\mathcal{E}(\varphi)$ is contained in F_φ^{RL} and that each connected component of $\mathcal{E}(\varphi)$ is in fact a connected component of F_φ^{RL}. The action of $\varphi(T)$ on $\mathcal{E}(\varphi)$ has the following property [**81**, Proposition 4.14]; see our Proposition 10.117 below and the discussion preceding it:

PROPOSITION 10.75. *Let $\varphi(T) \in \mathbb{C}_p(T)$ have degree $d \geq 2$. For each $x \in \mathcal{E}(\varphi) \cap \mathbb{H}_{\mathrm{Berk}}$, there is an $n \geq 1$ (depending on x) such that $\varphi^{(n)}(x) = x$. For each connected open affinoid X with $\overline{X} \subset \mathcal{E}(\varphi)$, there is an $N \geq 1$ (depending on X) such that $\varphi^{(N)}(X) = X$.*

---
[9]Technically Rivera-Letelier only works with $F_\varphi \cap \mathbb{P}^1(\mathbb{C}_p)$; in passing to F_φ, we are glossing over some subtleties which will be dealt with in §10.9.

Rivera-Letelier's classification of Fatou components says that immediate basins of attraction and components of the domain of quasi-periodicity are the only periodic Fatou components [**83**, Theorem A]:

THEOREM 10.76 (Rivera-Letelier). *Let $\varphi(T) \in \mathbb{C}_p(T)$ have degree at least 2. Then every periodic connected component of F_φ^{RL} is either an immediate basin of attraction or a connected component of the domain of quasi-periodicity of φ.*

REMARK 10.77. The proof of Theorem 10.76 uses the fact that the residue field of $\mathbb{C}_p$ is a union of finite fields, and in particular the proof does not extend to rational functions defined over an arbitrary K.

A component of F_φ^{RL} which is not preperiodic is called a *wandering component*. One deduces from Theorem 10.76 the following result:

COROLLARY 10.78 (Rivera-Letelier). *For each connected component U of F_φ^{RL}, exactly one of the following holds:*

(1) *$\varphi^{(m)}(U)$ is an immediate basin of attraction for some $m \geq 0$.*
(2) *$\varphi^{(m)}(U)$ is a connected component of the domain of quasi-periodicity for some $m \geq 0$.*
(3) *U is a wandering component.*

We can now prove Theorem 10.72.

PROOF OF THEOREM 10.72. Take $x_0 \in F_\varphi^{RL}$. Then the connected component U of F_φ^{RL} containing x_0 is one of the three types in the statement of Corollary 10.78. We will prove the result separately for each case. Fix an open covering $V_1, \ldots, V_t$ of $\mathbb{P}^1_{\mathrm{Berk}}$ by finitely many simple domains.

First suppose $\varphi^{(m)}(U)$ is the immediate basin of attraction of an attracting periodic point z_0.

We begin with the case where U itself is the immediate basin of attraction. Let n be the period of z_0, and let W be a neighborhood of z_0 small enough so that for each $\ell = 0, \ldots, n-1$, we have $\varphi^{(\ell)}(W) \subset V_i$ for some i.

Take an affinoid neighborhood Z of x_0 whose closure $\overline{Z}$ is contained in U. By Proposition 10.74, there is an exhaustion of U by connected open affinoids $\{X_k\}_{k \in \mathbb{Z}}$, with $X_k \subset W$ for all large k, such that $\varphi^{(n)}(X_k) = X_{k+1}$ for each k. Since $\overline{Z}$ is compact, it is contained in X_{k_0} for some k_0. Hence there is a k_1 such that $\varphi^{(nk_1)}(Z) \subset W$. By our choice of W, for each $k \geq k_1$ there is some V_i with $\varphi^{(k)}(Z) \subset V_i$. After shrinking Z if necessary, we can assume this holds for $0 \leq k < k_1$ as well. Thus the $\varphi^{(k)}(T)$ are equicontinuous at x_0.

More generally, if there is an $m \geq 1$ such that $\varphi^{(m)}(U)$ is an immediate basin of attraction and if Z is the neighborhood constructed above for $\varphi^{(m)}(x_0)$, then we can choose a neighborhood Z_0 of x_0 small enough so that $\varphi^{(m)}(Z_0) \subset Z$, and for each $\ell = 0, 1, \ldots, m-1$ there is some i for which $\varphi^{(\ell)}(Z_0) \subset V_i$. So once again the family $\varphi^{(k)}(T)$ is equicontinuous at x_0.

Next suppose that $\varphi^{(m)}(U)$ is a connected component of $\mathcal{E}(\varphi)$. We begin with the case where U itself is a component of $\mathcal{E}(\varphi)$. Take $x_0 \in U$.

First assume $x_0 \in \mathbb{P}^1(\mathbb{C}_p)$. By Theorem 10.67, the iterates $\varphi^{(k)}(T)$ are equicontinuous at x_0 relative to the spherical distance $\|x, y\|$. Let $\zeta_1, \ldots, \zeta_s$ be the finitely many points in $\partial V_1 \cup \cdots \cup \partial V_t$, and put $\varepsilon = \min_{1 \leq j \leq s} \operatorname{diam}(\zeta_j)$. (See (2.20) for the definition of diam.) By the definition of equicontinuity in a metric space, there is a $\delta > 0$ such that if $x \in \mathbb{P}^1(K)$ and $\|x, x_0\| < \delta$, then $\|\varphi(x), \varphi(x_0)\| < \varepsilon$. Let $B(x_0, \delta)^- = \{z \in \mathbb{P}^1(K) : \|z, x_0\| < \delta\}$, and let $Z = \mathcal{B}(x_0, \delta)^-$ be the associated Berkovich open ball.

Fix $k \in \mathbb{N}$. By construction, $\varphi^{(k)}(B(x_0, \delta)^-) \subset B(\varphi^{(k)}(x_0), \varepsilon)^-$, so by continuity, $\varphi^{(k)}(Z) \subset \mathcal{B}(\varphi^{(k)}(x_0), \varepsilon)^-$. Since $\operatorname{diam}(z) < \varepsilon$ for each $z \in \mathcal{B}(\varphi^{(k)}(x_0), \varepsilon)^-$, no ζ_j can belong to $\varphi^{(k)}(Z)$, and hence there is some V_i with $\varphi^{(k)}(Z) \subset V_i$.

Next assume $x_0 \in \mathbb{H}_{\text{Berk}}$. By Proposition 10.75, there is an $n \geq 1$ such that $\varphi^{(n)}(x_0) = x_0$. We claim that the neighborhoods Z of x_0 with $\varphi^{(n)}(Z) = Z$ are cofinal in all neighborhoods of x_0. To see this, let V be a neighborhood of x_0. After shrinking V if necessary, we can assume that V is a connected open affinoid with $\overline{V} \subset U$. By Proposition 10.75, there is an $N \geq 1$ such that $\varphi^{(N)}(V) = V$. For each $\ell = 1, \ldots, N$, $\varphi^{(n\ell)}(V)$ is open and $x_0 \in \varphi^{(n\ell)}(V)$. Put

$$Z = \bigcap_{\ell=1}^{N} \varphi^{(n\ell)}(V) .$$

Then Z is open, $\varphi^{(n)}(Z) = Z$, and $x_0 \in Z \subseteq V$.

For each $k = 1, \ldots, n$, choose a neighborhood W_k of x_0 small enough so that $\varphi^{(k)}(W_k) \subset V_i$ for some i depending on k. Put $W = \bigcap_{k=1}^{n} W_k$, and let $Z \subset W$ be a neighborhood of x_0 such that $\varphi^{(n)}(Z) = Z$. Then for each $k \geq 0$, there is some V_i with $\varphi^{(k)}(Z) \subset V_i$. Hence the $\varphi^{(k)}(T)$ are equicontinuous at x_0.

More generally, if $\varphi^{(m)}(U)$ belongs to $\mathcal{E}(\varphi)$ and if $x_0 \in U$, one can show that the $\varphi^{(k)}(T)$ are equicontinuous at x_0 by an argument much like the one in the case of immediate basins of attraction.

Lastly, suppose U is a wandering component. Let $\zeta_1, \ldots, \zeta_s$ be the finitely many points belonging to $\partial V_1 \cup \cdots \cup \partial V_t$. Since U is wandering, each ζ_j belongs to $\varphi^{(n)}(U)$ for at most one n. Thus there is an N such that for each $1 \leq j \leq s$, we have $\zeta_j \notin \varphi^{(k)}(U)$ for all $k > N$. Now fix $x_0 \in U$, and choose a subdomain $Z \subset U$ containing x_0 small enough so that for each $k \leq N$, $\varphi^{(k)}(Z)$ is contained in some V_i. For $k > N$, $\varphi^{(k)}(Z)$ is a domain which does not contain any boundary point of any V_i, so it must belong to some V_i as well. It follows that $\{\varphi^{(k)}(T)\}$ is equicontinuous at x_0. $\square$

REMARK 10.79. The proof which we have just given (including equicontinuity on the wandering components) is due to Rivera-Letelier (unpublished). We thank him for allowing us to reproduce his argument here.

10.7. Fixed point theorems and their applications

In this section, we obtain further information about the Berkovich Julia and Fatou sets. We continue our blanket assumption that char$(K) = 0$, but most of the arguments do not use this.

Let $\varphi(T) \in K(T)$ be nonconstant. Recall that a fixed point of $\varphi(T)$ in $\mathbb{P}^1(K)$ is called *superattracting* if its multiplier $\lambda = 0$, *attracting* if $|\lambda| < 1$, *indifferent* if $|\lambda| = 1$ and *repelling* if $|\lambda| > 1$. Following Rivera-Letelier, we call a fixed point x of $\varphi(T)$ in $\mathbb{H}_{\text{Berk}}$ *indifferent* if it has multiplicity $m_\varphi(x) = 1$, and *repelling* if $m_\varphi(x) > 1$. The reason for this terminology will be explained below. A periodic point for $\varphi(T)$, with period n, will be called indifferent or repelling according to whether the corresponding fixed point of $\varphi^{(n)}(T)$ is indifferent or repelling.

We first show that repelling periodic points are necessarily of type I or II, and that all repelling periodic points belong to the Berkovich Julia set J_φ. We next prove a fixed point theorem for expanding domains and show that if $\deg(\varphi) \geq 2$, then $\varphi(T)$ always has at least one repelling fixed point. We apply this to show that the topological connected components of the Berkovich Fatou set coincide with Rivera-Letelier's Fatou components. We also prove fixed point theorems for sets stabilized by or contracted by $\varphi(T)$. Finally, using a classical argument of Fatou involving homoclinic orbits (see [73, §14]), we show that repelling periodic points are dense in J_φ.

Our first lemma, which combines [82, Lemmas 5.3 and 5.4], shows that repelling fixed points in $\mathbb{H}_{\text{Berk}}$ must be of type II:

LEMMA 10.80 (Rivera-Letelier). *Let $\varphi(T) \in K(T)$ have $\deg(\varphi) \geq 2$. If x is a fixed point of $\varphi(T)$ of type III or IV, then:*

(A) *x is an indifferent fixed point.*
(B) *$\varphi_*(\vec{v}) = \vec{v}$ for each $\vec{v} \in T_x$.*

PROOF. First suppose x is a fixed point of type III, and put $m = m_\varphi(x)$. Let $\vec{v}_1$ and $\vec{v}_2$ be the tangent vectors at x. By Lemma 9.33, $m_\varphi(\vec{v}_1) = m_\varphi(\vec{v}_2) = m$, and there is a segment $[x,c]$ with initial tangent vector $\vec{v}_1$ such that $\rho(x, \varphi(z)) = m \cdot \rho(x,z)$ for each $z \in [x,c]$. Without loss of generality we can assume that c is of type II. Since x is of type III, it follows that $\rho(x,c) \notin \log_v(|K^\times|)$. Note that $\log_v(|K^\times|)$ is a divisible group, since K is algebraically closed. If $\varphi_*(\vec{v}_1) = \vec{v}_2$, then

$$\rho(c, \varphi(c)) = \rho(c, x) + \rho(x, \varphi(c)) = (1+m)\rho(x,c) \ .$$

However $\rho(c, \varphi(c)) \in \log_v(|K^\times|)$ since c and $\varphi(c)$ are of type II. This contradicts $\rho(x,c) \notin \log_v(|K^\times|)$, so it must be that $\varphi_*(\vec{v}_1) = \vec{v}_1$. Hence $[x,c]$ and $\varphi([x,c]) = [x, \varphi(c)]$ share a common initial segment. Let $y \in [x,c]$ be a type II point close enough to x that $\varphi(y) \in [x,c]$, and put $r = \rho(x,y)$. Then

$$mr = \rho(x, \varphi(y)) = r + \rho(y, \varphi(y)) \ .$$

Since $\rho(y, \varphi(y))$ belongs to $\log_v(|K^\times|)$ but r does not, this is a contradiction unless $m = 1$.

Next suppose x is of type IV. If we identify $\mathbb{P}^1_{\text{Berk}}$ with $\mathbb{A}^1_{\text{Berk}} \cup \{\infty\}$, then under Berkovich's classification theorem (Theorem 1.2), x corresponds to a nested sequence of closed discs in K with empty intersection:

$$D(a_1, r_1) \supset D(a_2, r_2) \supset D(a_3, r_3) \supset \cdots .$$

Since $\varphi(T)$ has only finitely many fixed points and finitely many poles in $\mathbb{P}^1(K)$, we can assume without loss of generality that none of the $D(a_i, r_i)$ contains a fixed point or a pole. We can also assume that each $r_i \in |K^\times|$. Let $z_i \in \mathbb{P}^1_{\text{Berk}}$ be the point corresponding to $D(a_i, r_i)$. The points z_i all belong to the path $[x, \infty]$.

Put $m = m_\varphi(x)$ and let $\vec{v}$ be the unique tangent vector at x. Clearly $\varphi_*(\vec{v}) = \vec{v}$. By Lemma 9.33, $m_\varphi(\vec{v}) = m$, and there is an initial segment $[x, c] \subset [x, \infty]$ such that $\rho(x, \varphi(z)) = m \cdot \rho(x, z)$ for each $z \in [x, c]$. Since $D(a_i, r_i)$ contains no poles of $\varphi(T)$, by Corollary A.18 its image $\varphi(D(a_i, r_i))$ is a disc $D(a'_i, R_i)$. This disc corresponds to $\varphi(z_i)$ under Berkovich's classification theorem. Since $[x, c]$ and $\varphi([x, c]) = [x, \varphi(c)]$ have the same initial tangent vector (the unique $\vec{v} \in T_x$), they share an initial segment. Hence for sufficiently large i, both z_i and $\varphi(z_i)$ belong to $[x, c]$. Suppose $m > 1$. Then

$$R_i := \text{diam}_\infty(\varphi(z_i)) = \text{diam}_\infty(x) + m \cdot \rho(x, z_i) > \text{diam}_\infty(z_i) = r_i .$$

Since the associated Berkovich discs $\mathcal{D}(a_i, r_i)$ and $\mathcal{D}(a'_i, R_i)$ both contain x and since $R_i > r_i$, it follows that $D(a_i, r_i)$ is properly contained in $D(a'_i, R_i)$.

We claim that $D(a_i, r_i)$ contains a fixed point of $\varphi(T)$ (in K), contradicting our assumption. After changing coordinates if necessary, we can assume that $D(a_i, r_i) = D(0, 1)$ and $D(a'_i, R_i) = D(0, R)$ where $R > 1$. Expand $\varphi(T)$ as a power series $f(T) = \sum_{k=0}^{\infty} a_k T^k$ converging on $D(0, 1)$. Then $\lim_{k \to \infty} |a_k| = 0$, and by the Maximum Principle $\max_k(|a_k|) = R$. By the ultrametric inequality, the coefficients of the power series $g(T) = f(T) - T$ have the same properties, and the theory of Newton polygons shows that $g(T) = 0$ has a solution in $D(0, 1)$ (see Proposition A.16). This point is a fixed point of $\varphi(T)$. $\square$

Let $x_0 \in \mathbb{P}^1_{\text{Berk}}$ be a fixed point of $\varphi(T)$. We now give a description of the local action of $\varphi(T)$ at x_0, which sheds light on the terminology "attracting", "indifferent", or "repelling".

First suppose $x_0 \in \mathbb{P}^1(K)$. After a change of coordinates, we can assume that $x_0 \neq \infty$. In this setting, if $\lambda = \varphi'(x_0)$ is the multiplier of x_0, we can expand $\varphi(T)$ as a power series which converges in a neighborhood of x_0:

$$(10.67) \qquad \varphi(T) = x_0 + \lambda(T - x_0) + \sum_{k=2}^{\infty} a_k (T - x_0)^k .$$

If $\lambda \neq 0$, then there is a disc $D(x_0, R)$ such that $|\varphi(x) - x_0| = |\lambda||x - x_0|$ for each $x \in D(x_0, R)$. Using the theory of Newton polygons, it is easy to show that for each $0 < r \leq R$, we have $\varphi(D(x_0, r)) = D(x_0, |\lambda|r)$ (see Corollary A.17). By continuity, $\varphi(T)$ maps the closed Berkovich disc $\mathcal{D}(x_0, r)$ to $\mathcal{D}(x_0, |\lambda|r)$ and the open Berkovich disc $\mathcal{D}(x_0, r)^-$ to $\mathcal{D}(x_0, |\lambda|r)^-$.

Furthermore, if $x_r \in \mathbb{P}^1_{\text{Berk}}$ is the point corresponding to $D(x_0, r)$ under Berkovich's classification theorem, the fact that $\varphi(D(x_0, r)) = D(x_0, |\lambda|r)$ means that $\varphi(x_r) = x_{|\lambda|r}$. Thus, if $|\lambda| < 1$, the points x_r approach x_0 along the segment $[x_0, x_R]$; if $|\lambda| = 1$, the segment $[x_0, R]$ is fixed; and if $|\lambda| > 1$, the x_r recede from x_0.

If $\lambda = 0$, so that x_0 is superattracting, let a_m be the first nonzero coefficient in the expansion (10.67). Then $m \geq 2$, and Corollary A.17 shows there is a disc $D(x_0, R)$ such that for $0 < r \leq R$, $\varphi(D(x_0, r)) = D(x_0, |a_m|r^m)$. As above, $\varphi(\mathcal{D}(x_0, r)) = \mathcal{D}(x_0, |a_m|r^m)$, $\varphi(\mathcal{D}(x_0, r)^-) = \mathcal{D}(x_0, |a_m|r^m)^-$, and $\varphi(x_r) = x_{|a_m|r^m}$. If r is small enough, then $|a_m|r^m < r$, and the points x_r approach x_0 along $[x_0, x_R]$ as $r \to 0$.

Next suppose x_0 is a fixed point of $\varphi(T)$ in $\mathbb{H}_{\text{Berk}}$ of type II. By Corollary 9.25, $m_\varphi(x_0) = \deg(\tilde{\varphi})$, where $\tilde{\varphi}(T) \in k(T)$ is the reduction of $\varphi(T)$ in suitable coordinates with respect to x_0 and $k = \tilde{K}$ is the residue field of K (see the discussion preceding Corollary 9.25). Furthermore, the tangent vectors $\vec{v}_\alpha \in T_{x_0}$ correspond bijectively to points $\alpha \in \mathbb{P}^1(k)$. Under this correspondence $\varphi_*(\vec{v}_\alpha) = \vec{v}_{\tilde{\varphi}(\alpha)}$. By Corollary 9.25, $m_\varphi(x_0, \vec{v}_\alpha) = m_\alpha$ where $m_\alpha = m_{\tilde{\varphi}}(a) \geq 1$ is the usual (algebraic) multiplicity of $\tilde{\varphi}(T)$ at α. Furthermore, by Theorem 9.22, for each $\vec{v}_\alpha \in T_{x_0}$,

$$(10.68) \qquad r_\varphi(x_0, \vec{v}_\alpha) := d_{\vec{v}_\alpha}(\rho(\varphi(x), x_0))(x_0) = m_\alpha .$$

Loosely speaking, (10.68) says that under the action of $\varphi(T)$, points sufficiently near x_0 in the direction $\vec{v}_\alpha$ recede from x_0 in the direction $\vec{v}_{\tilde{\varphi}(\alpha)}$ at the rate m_α. More precisely, by Lemma 9.33, there is a segment $[x_0, x_\alpha]$ with initial tangent vector $\vec{v}_\alpha$ such that $\varphi([x_0, x_\alpha]) = [x_0, \varphi(x_\alpha)]$ is a segment with initial tangent vector $\vec{v}_{\tilde{\varphi}(\alpha)}$ and for each $x \in [x_0, x_\alpha]$,

$$(10.69) \qquad \rho(\varphi(x), x_0) = m_\alpha \cdot \rho(x, x_0) .$$

Furthermore, if $\mathcal{A}_{x_0, x_\alpha}$ is the (generalized Berkovich open) annulus associated to the segment (x_0, x_α), then $\varphi(\mathcal{A}_{x_0, x_\alpha}) = \mathcal{A}_{x_0, \varphi(x_\alpha)}$ is the annulus associated to the segment $(x_0, \varphi(x_\alpha))$, and $\text{Mod}(\mathcal{A}_{x_0, \varphi(x_\alpha)}) = m_\alpha \cdot \text{Mod}(\mathcal{A}_{x_0, x_\alpha})$.

If $m_\varphi(x_0) = 1$, then $\deg(\tilde{\varphi}(T)) = 1$ and $m_{\tilde{\varphi}}(\alpha) = 1$ for each $\alpha \in \mathbb{P}^1(k)$. This means that φ_* permutes the tangent directions at x_0 and $\varphi(T)$ locally preserves path distances at x_0: $\rho(\varphi(x), x_0) = \rho(x, x_0)$ in (10.69). In this sense, the action of $\varphi(T)$ at x_0 is "indifferent".

By contrast, if $m_\varphi(x_0) > 1$, then $\deg(\tilde{\varphi}(T)) \geq 2$, so $\tilde{\varphi}(T)$ has at least one critical point α_0. Since $m_{\tilde{\varphi}}(\alpha_0) \geq 2$, points in the tangent direction $\vec{v}_{\alpha_0}$ on the segment $[x_0, x_{\alpha_0}]$ are "repelled" from x_0 in the direction $\vec{v}_{\tilde{\varphi}(\alpha_0)}$ at a rate $m_{\alpha_0} \geq 2$. In physical terms, one can visualize this action as being like an accretion disc swirling around a quasar, with one or more "jets" emanating from it. In this sense, the action of $\varphi(T)$ at x_0 is "repelling".

We will now show that, as in the classical complex case, repelling periodic points of φ always belong to the Berkovich Julia set $\mathcal{J}_\varphi$. The proof is the same as [**83**, Proposition 5.1].

10.7. FIXED POINT THEOREMS AND THEIR APPLICATIONS

THEOREM 10.81 (Rivera-Letelier). *Let $\varphi(T) \in K(T)$ have $\deg(\varphi) \geq 2$. Then each repelling periodic point for $\varphi(T)$ in $\mathbb{P}^1_{\text{Berk}}$ belongs to $\mathcal{J}_\varphi$.*

PROOF. Let x be a repelling periodic point of $\varphi(T)$ of period n. Since $\mathcal{J}_{\varphi^{(n)}} = \mathcal{J}_\varphi$ by Lemma 10.51, we can assume that x is a repelling fixed point.

First suppose $x \in \mathbb{P}^1(K)$. After a change of coordinates we can assume that $x \in K$ and $|x| \leq 1$. Let λ be the multiplier of x; by hypothesis $|\lambda| > 1$. There is a disc $D(x,r) \subset K$ such that $\|\varphi(z), x\| = |\varphi(z) - x| = |\lambda||z-x| = |\lambda| \cdot \|z, x\|$ for all $z \in D(x,r)$; without loss of generality, we can assume that $r < 1$. If $|z-x| \leq r/|\lambda|^n$, then $\|\varphi^{(n)}(z), x\| = |\lambda|^n \cdot \|z, x\|$. Thus the iterates $\{\varphi^{(n)}\}$ are not equicontinuous at x, and so $x \in \mathcal{J}_\varphi$ by Theorem 10.67.

Next suppose $x \in \mathbb{H}_{\text{Berk}}$. By Lemma 10.80, x is of type II. Let k be the residue field of K, and let $\tilde{\varphi}(T) \in k(T)$ be the nonconstant reduction of $\varphi(T)$ at x, in suitable coordinates. By hypothesis, $m_\varphi(x) \geq 2$; by Corollary 9.25, $m_\varphi(x) = \deg(\tilde{\varphi}(T))$.

As noted earlier, after identifying T_x with $\mathbb{P}^1(k)$, we have $\varphi_*(\vec{v}_\alpha) = \vec{v}_{\tilde{\varphi}(\alpha)}$ and $m_\varphi(x, \vec{v}_\alpha) = m_{\tilde{\varphi}}(\alpha)$. Recall that we write $\mathcal{B}_x(\vec{v})^-$ for the component of $\mathbb{P}^1_{\text{Berk}} \backslash \{x\}$ in the direction $\vec{v}$.

Fix a neighborhood U of x. There is a finite subset $S \subset \mathbb{P}^1(k)$ such that for each $\alpha \notin S$, the ball $\mathcal{B}_x(\vec{v}_\alpha)^-$ is contained in U. By Proposition 9.41, $\varphi(\mathcal{B}_x(\vec{v}_\alpha)^-)$ contains $\mathcal{B}_x(\vec{v}_{\tilde{\varphi}(\alpha)})^-$. Let S_0 be the (possibly empty) finite subset of S consisting of the points $\alpha \in S$ which are exceptional for $\tilde{\varphi}(T)$. If $\alpha \notin S_0$, then α has infinitely many preimages under the iterates $\tilde{\varphi}^{(n)}(T)$, so there are an $\alpha_0 \notin S$ and an n_0 such that $\tilde{\varphi}^{(n_0)}(\alpha_0) = \alpha$. It follows that $\mathcal{B}_x(\vec{v}_\alpha)^- \subset \varphi^{(n_0)}(\mathcal{B}_x(\vec{v}_{\alpha_0})^-)$, and so

$$\bigcup_{\alpha \notin S_0} \mathcal{B}_x(\vec{v}_\alpha)^- \subset \bigcup_{n=0}^\infty \varphi^{(n)}(U) \ .$$

Now consider the balls $\mathcal{B}_x(\vec{v}_\alpha)^-$ for $\alpha \in S_0$. After replacing φ by an appropriate iterate $\varphi^{(m)}$, we can assume that $\tilde{\varphi}$ fixes each $\alpha \in S_0$ and hence (since α is exceptional) that

$$m_\varphi(x, \vec{v}_\alpha) \ = \ m_{\tilde{\varphi}}(\alpha) \ = \ \deg(\tilde{\varphi}) \ \geq \ 2 \ .$$

For each $\alpha \in S_0$, consider an annulus $\mathcal{A}_x(\vec{v}_\alpha) := \mathcal{A}_{x,c} \subset U$, where $c \in \mathcal{B}_x(\vec{v}_\alpha)^-$ and $\mathcal{A}_{x,c}$ is the annulus with boundary points x, c and main dendrite (x,c), as in §9.3. After replacing $[x,c]$ by an initial segment of itself if necessary, we can assume that $[x,c] \subset \varphi([x,c])$, so that $\mathcal{A}_x(\vec{v}_\alpha) \subset \varphi(\mathcal{A}_x(\vec{v}_\alpha))$. By Proposition 9.44, for each n the image $\varphi^{(n)}(\mathcal{A}_x(\vec{v}_\alpha))$ is either another annulus, a disc, or all of $\mathbb{P}^1_{\text{Berk}}$. Since $\varphi^{(n)}_*(\vec{v}_\alpha) = \vec{v}_\alpha$, if the image under any iterate is a disc or $\mathbb{P}^1_{\text{Berk}}$, then $\mathcal{B}_x(\vec{v}_\alpha)^- \subset \varphi^{(n)}(U)$. Hence we can assume the image under each iterate is an annulus. By Theorem 9.46, the modulus of $\varphi^{(n)}(\mathcal{A}_x(\vec{v}_\alpha))$ is $\deg(\tilde{\varphi})^n \cdot \text{Mod}(\mathcal{A}_x(\vec{v}_\alpha))$. Since these annuli are increasing and their moduli grow to ∞, their union omits at most one point z_α in $\mathcal{B}_x(\vec{v}_\alpha)^-$, which is necessarily of type I.

Put $\mathcal{E}_0(K) = \{z_\alpha\}_{\alpha \in S_0}$. Then $\mathcal{E}_0(K)$ is a finite subset of $\mathbb{P}^1(K)$ and
$$\mathbb{P}^1_{\text{Berk}} \setminus \mathcal{E}_0(K) \subset \bigcup_{n=0}^{\infty} \varphi^{(n)}(U) \ .$$
Now take any $z \in \mathcal{E}_0(K) \setminus E_\varphi(K)$. Since z is not exceptional for φ, it has infinitely many distinct preimages under the $\varphi^{(n)}(T)$, so some preimage belongs to $\bigcup_{n=0}^{\infty} \varphi^{(n)}(U)$. It follows that z belongs to that set as well. Thus
$$\mathbb{P}^1_{\text{Berk}} \setminus E_\varphi(K) \subset \bigcup_{n=0}^{\infty} \varphi^{(n)}(U) \ .$$
Since U is arbitrary, Theorem 10.56 shows that $x \in J_\varphi$. $\square$

We now show that in repelling fixed points always exist.

THEOREM 10.82 (Rivera-Letelier[10]). *Let $\varphi(T) \in K(T)$ have $\deg(\varphi) \geq 2$. Then $\varphi(T)$ has at least one repelling fixed point in $\mathbb{P}^1_{\text{Berk}}$.*

PROOF. This follows immediately from Theorem 10.83 below. $\square$

Recall that a *simple domain* $U \subset \mathbb{P}^1_{\text{Berk}}$ is a connected open set whose boundary ∂U is a finite nonempty set, and each $x_i \in \partial U$ is of type II or III. Recall also that U is called φ-*saturated* if U is a connected component of $\varphi^{-1}(\varphi(U))$, or, equivalently, if $\varphi(\partial U) = \partial \varphi(U)$ (cf. Proposition 9.15).

The following result strengthens a fixed point theorem proved by Rivera-Letelier [**84**, Proposition 9.3] (see also [**82**, Lemma 6.2]).

THEOREM 10.83 (Repelling Fixed Point Criterion). *Let $\varphi(T) \in K(T)$ have $\deg(\varphi) \geq 2$. If either $U \subseteq \mathbb{P}^1_{\text{Berk}}$ is a φ-saturated simple domain with $\overline{U} \subset \varphi(U)$, or $U = \mathbb{P}^1_{\text{Berk}}$, then U contains a repelling fixed point for φ.*

PROOF. Given a point $x \in \mathbb{P}^1_{\text{Berk}}$, we will say that x is *strongly involutive* for $\varphi(T)$ if either

(1) x is fixed by $\varphi(T)$ or
(2) $x \neq \varphi(x)$ and the following condition holds: consider the path $[x, \varphi(x)]$; let $\vec{v}$ be its (inward-directed) tangent vector at x, and let $\vec{w}$ be its (inward-directed) tangent vector at $\varphi(x)$. Then $\varphi_*(\vec{v}) = \vec{w}$, and $\vec{v}$ is the only tangent vector $\vec{v}' \in T_x$ for which $\varphi_*(\vec{v}') = \vec{w}$.

We will consider two cases, according as each $x \in U \cap \mathbb{H}_{\text{Berk}}$ is strongly involutive or not.

First suppose some $x_0 \in U \cap \mathbb{H}_{\text{Berk}}$ is not strongly involutive. By definition, x_0 is not fixed by $\varphi(T)$. Set $x_{-1} = \varphi(x_0)$, and consider the segment $[x_{-1}, x_0]$. Let $\vec{v}_0$ be its tangent vector at x_0, and let $\vec{w}_0$ be its tangent vector at x_{-1}. Recall that the map $\varphi_* : T_{x_0} \to T_{x_{-1}}$ is surjective (Corollary 9.20). Since x_0 is not strongly involutive, there is a tangent vector $\vec{w}_1$ at x_0, with $\vec{w}_1 \neq \vec{v}_0$, for which $\varphi_*(\vec{w}_1) = \vec{w}_0$.

---

[10] When $K = \mathbb{C}_p$, this is proved in [**82**, Theorem B]. The argument carries through in the general case with some modifications; see [**80**]. Here we give a different proof.

By Lemma 9.38, there is a segment $[x_0, x_1] \subset U$ with initial tangent vector $\vec{w}_1$ such that $\varphi(T)$ maps $[x_0, x_1]$ homeomorphically onto $[x_{-1}, x_0]$, with $\varphi(x_0) = x_{-1}$ and $\varphi(x_1) = x_0$. Let $\vec{v}_1$ be the tangent vector to $[x_0, x_1]$ at x_1, noting that $\varphi_*(\vec{v}_1) = \vec{v}_0$. Since φ_* maps the tangent space T_{x_1} surjectively onto T_{x_0}, there is a $\vec{w}_2 \in T_{x_1}$ for which $\varphi_*(\vec{w}_2) = \vec{w}_1$. Clearly $\vec{w}_2 \neq \vec{v}_1$, since $\varphi_*(\vec{v}_1) = \vec{v}_0$.

We now iterate this process, inductively constructing segments $[x_1, x_2]$, $[x_2, x_3], \ldots$ contained in U such that $\varphi(T)$ maps $[x_n, x_{n+1}]$ homeomorphically onto $[x_{n-1}, x_n]$, with $\varphi(x_n) = x_{n-1}$ and $\varphi(x_{n+1}) = x_n$, such that the tangent vector $\vec{w}_{n+1}$ to x_n in $[x_n, x_{n+1}]$ is different from the tangent vector $\vec{v}_n$ to x_n in $[x_{n-1}, x_n]$. This means that

$$P = \bigcup_{n=0}^{\infty} [x_{n-1}, x_n]$$

is an arc.

If P has finite length, then since $\mathbb{H}_{\text{Berk}}$ is complete under $\rho(x, y)$ (Proposition 2.29), the points $\{x_n\}$ converge to a point $x \in \mathbb{H}_{\text{Berk}}$ which is an endpoint of P. Furthermore,

$$\varphi(x) = \lim_{n \to \infty} \varphi(x_n) = \lim_{n \to \infty} x_{n-1} = x \; ,$$

so x is a fixed point of $\varphi(T)$. Clearly $x \in \overline{U}$. We claim that $x \in U$. If $U = \mathbb{P}^1_{\text{Berk}}$, this is trivial. If $U \neq \mathbb{P}^1_{\text{Berk}}$, then since U is φ-saturated and $\overline{U} \subset \varphi(U)$, the boundary points of U are not fixed by $\varphi(T)$. Thus $x \in U$. Let $\vec{v}$ be the tangent vector to P at x. Since $\rho(x, x_n) < \rho(x, x_{n-1}) = \rho(x, \varphi(x_n))$, necessarily $d_{\vec{v}}(\rho(\varphi(z), x))(x) > 1$, so $m_\varphi(x) > 1$ (Theorem 9.19), and x is a repelling fixed point.

On the other hand, if P has infinite length, then we claim that the x_n converge to a point $x \in \mathbb{P}^1(K)$ and that P is a geodesic ray emanating from x. Indeed, we can choose coordinates so that ζ_{Gauss} lies on P and all but a finite initial segment of P belongs to $\mathcal{D}(0, 1)$. In this situation, if $x_n \in \mathcal{D}(0, 1)$ and $\rho(\zeta_{\text{Gauss}}, x_n) = t_n$, then under Berkovich's classification theorem (Theorem 1.2), x_n corresponds to a disc $D(a_n, r_n)$ with $r_n = q_v^{-t_n}$. These discs are nested and their radii go to 0. Since K is complete, their intersection is a point $x \in K$, which is the limit of the x_n in the strong topology. By the same argument as before, $x \in \overline{U}$ and x is a fixed point of $\varphi(T)$. If $U = \mathbb{P}^1_{\text{Berk}}$, trivially $x \in U$. Otherwise, U is a simple domain, so none of its boundary points belongs to $\mathbb{P}^1(K)$, and again $x \in U$. Under the action of $\varphi(T)$, the points x_n recede from x along P. If λ is the multiplier of $\varphi(T)$ at x, this means that $|\lambda| > 1$, so x is a repelling fixed point.

Next assume that each $x \in U \cap \mathbb{H}_{\text{Berk}}$ is strongly involutive.

If $\varphi(T)$ fixes each $x \in U \cap \mathbb{H}_{\text{Berk}}$, then by continuity it fixes each $z \in U$, and in particular $\varphi(z) = z$ for all $z \in \mathbb{P}^1(K) \cap U$. This means that $\varphi(T)$ has infinitely many fixed points in $\mathbb{P}^1(K)$, so $\varphi(T) = T$, which contradicts $\deg(\varphi) \geq 2$. Hence not every $x \in U \cap \mathbb{H}_{\text{Berk}}$ is a fixed point of $\varphi(T)$.

However, we claim that $\varphi(T)$ has at least one fixed point $c \in U \cap \mathbb{H}_{\text{Berk}}$, which could be an indifferent fixed point. Take an arbitrary $a \in U \cap \mathbb{H}_{\text{Berk}}$. If $\varphi(a) = a$, put $c = a$. If $\varphi(a) \neq a$, then by Lemma 10.84 below, the segment $[a, \varphi(a)]$ contains a fixed point c which belongs to U.

If $m_\varphi(c) > 1$, then c is a repelling fixed point and we are done. If $m_\varphi(c) = 1$, we consider two subcases, according as U is a φ-saturated simple domain with $\overline{U} \subset \varphi(U)$ or $U = \mathbb{P}^1_{\text{Berk}}$.

First, let U be a φ-saturated simple domain with $\overline{U} \subset \varphi(U)$. Write $\partial U = \{x_1, \ldots, x_m\}$, where each x_i is of type II or III. Consider the segments $[c, x_i]$, for $i = 1, \ldots, m$. Since $\overline{U} \subset \varphi(U)$ and $\varphi(\partial U) = \partial \varphi(U)$, we have $\varphi(x_i) \notin \overline{U}$ for each i. Since ∂U is finite, there is an i so that $\rho(c, x_i) \leq \rho(c, x_j)$ for all $1 \leq j \leq m$. For this choice of i we must have $\rho(c, x_i) < \rho(c, \varphi(x_i))$. Since $\varphi(c) = c$, Theorem 9.35(B) gives
$$\rho(c, x_i) < \rho(c, \varphi(x_i)) \leq \int_{[c, x_i]} m_\varphi(x) \, dx \ .$$
Hence there is a point $b \in (c, x_i)$ with $m_\varphi(b) > 1$. If $\varphi(b) = b$, then b is a repelling fixed point. If $\varphi(b) \neq b$, then by Lemma 10.84 there is a fixed point $d \in [b, \varphi(b)]$ which belongs to U and satisfies $m_\varphi(d) \geq m_\varphi(b)$. Thus d is a repelling fixed point.

Finally, suppose $U = \mathbb{P}^1_{\text{Berk}}$. We consider two more subcases.

First suppose there is a point $a \in \mathbb{H}_{\text{Berk}}$ with $m_\varphi(a) > 1$. If a is fixed by $\varphi(T)$, then a is a repelling fixed point, and we are done. Otherwise, by Lemma 10.84, there is a fixed point $d \in [a, \varphi(a)]$ with $m_\varphi(d) \geq m_\varphi(a)$, so d is a repelling fixed point.

Next suppose that $m_\varphi(a) = 1$ for each $a \in \mathbb{H}_{\text{Berk}}$. We will see that this leads to a contradiction. As shown above, $\varphi(T)$ has at least one fixed point $c \in \mathbb{H}_{\text{Berk}}$. Since $m_\varphi(c) = 1$ and c has $\deg(\varphi) \geq 2$ preimages under $\varphi(T)$, counting multiplicities (Theorem 9.8(C)), there is a point $a \in \mathbb{H}_{\text{Berk}}$ with $a \neq c$ and $\varphi(a) = c$. By Lemma 10.84, there is a fixed point $d \in [a, c]$, and $\varphi(T)$ maps $[a, d]$ homeomorphically onto $[c, d]$. In particular, if $\vec{v}_1$ is the tangent vector to d in the direction of a and $\vec{v}_2$ is the tangent vector at d in the direction of c, then $\varphi_*(\vec{v}_1) = \vec{v}_2$.

We claim that $\varphi(T)$ fixes the entire segment $[c, d]$. Take any $t \in [c, d]$. Since $m_\varphi(x) = 1$ for each $x \in [c, t]$, Theorem 9.35(B) gives
$$\rho(c, \varphi(t)) \leq \int_{[c, t]} m_\varphi(x) \, dx \leq \rho(c, t) \ .$$
Similarly, $\rho(d, \varphi(t)) \leq \rho(d, t)$. Since $t \in [c, d]$, the only point $z \in \mathbb{P}^1_{\text{Berk}}$ satisfying $\rho(c, z) \leq \rho(c, t)$ and $\rho(d, z) \leq \rho(d, t)$ is t itself. Thus $\varphi(t) = t$.

Since $\varphi(T)$ fixes $[c, d]$, it follows that $\varphi_*(\vec{v}_2) = \vec{v}_2$. By Theorem 9.22(C),
$$m_\varphi(d) = \sum_{\varphi(\vec{w}) = \vec{v}_2} m_\varphi(d, \vec{w}) \geq m_\varphi(d, \vec{v}_1) + m_\varphi(d, \vec{v}_2) \geq 2 \ .$$

Thus d is a repelling fixed point for φ. $\square$

10.7. FIXED POINT THEOREMS AND THEIR APPLICATIONS

The lemma below is needed to complete the proof of Theorem 10.83.

LEMMA 10.84. *Let $\varphi(T) \in K(T)$ be nonconstant, and let $U \subseteq \mathbb{P}^1_{\text{Berk}}$ be a φ-saturated domain with $U \subseteq \varphi(U)$. Suppose that each $x \in U \cap \mathbb{H}_{\text{Berk}}$ is strongly involutive. Let $a \in U \cap \mathbb{H}_{\text{Berk}}$, and put $b = \varphi(a)$.*

If $a \neq b$, then there is a point $c \in (a,b)$ which is fixed by $\varphi(T)$, and $\varphi(T)$ maps $[a,c]$ homeomorphically onto $[b,c]$. Furthermore, $c \in U$ and $m_\varphi(c) \geq m_\varphi(a)$.

PROOF. The proof is by transfinite induction. For readers unfamiliar with transfinite induction, we recommend [**63**, §IV.7]. Recall that each ordinal represents an isomorphism class of well-ordered sets. Transfinite induction generalizes induction on $\mathbb{N}$ to arbitrary well-ordered sets, by adding an additional case to deal with the fact that each ordinal which is not 0 or a successor ordinal must be a limit ordinal. The idea is to start at a and "push" towards b, with each push consisting of either a single step or passage to a limit of steps, until a fixed point is reached. Transfinite induction is needed since we have no control of the step size, so the fixed point might not be reached in a sequence of steps indexed by the natural numbers. We are assured that the process must eventually stop by cardinality considerations.

Put $a_0 = a$, $b_0 = b$. Suppose that for some ordinal ω, we have constructed points $a_\omega, b_\omega \in [a,b]$ with $a_\omega \in U$, $b_\omega = \varphi(a_\omega)$, and $m_\varphi(a_\omega) \geq m_\varphi(a)$, such that $[a, a_\omega) \cap (b_\omega, b] = \emptyset$ and $\varphi(T)$ maps $[a, a_\omega]$ homeomorphically onto $[b, b_\omega]$. If $a_\omega = b_\omega$, then taking $c = a_\omega$, we are done.

If not, let $\vec{v}_\omega$ (resp. $\vec{w}_\omega$) be the tangent vector to $[a_\omega, b_\omega]$ at a_ω (resp. b_ω). Since a_ω is strongly involutive, $\vec{v}_\omega$ is the only tangent vector at a_ω for which $\varphi_*(\vec{v}_\omega) = \vec{w}_\omega$, and Theorem 9.22 shows that $m_\varphi(a_\omega, \vec{v}_\omega) = m_\varphi(a_\omega)$. Put $m_\omega = m_\varphi(a_\omega)$.

By Lemma 9.33, there is an initial segment $[a_\omega, a_{\omega+1}] \subset [a_\omega, b_\omega]$ such that $\varphi(T)$ maps $[a_\omega, a_{\omega+1}]$ homeomorphically onto a segment $[b_\omega, b_{\omega+1}] \subset [b_\omega, a_\omega]$, with $m_\varphi(x) = m_\omega$ for all $x \in [a_\omega, a_{\omega+1}]$. After moving $a_{\omega+1}$ nearer to a_ω if necessary, we can assume that $a_{\omega+1} \in U$ and $[a, a_{\omega+1}) \cap (b_{\omega+1}, b_\omega] = \emptyset$.

Let $\vec{v}'_{\omega+1}$ be the tangent vector to $[a_\omega, a_{\omega+1}]$ at $a_{\omega+1}$. Since $m_\varphi(x) = m_\omega$ for each $x \in (a_\omega, a_{\omega+1})$, Theorem 9.22 shows that $m_\varphi(a_{\omega+1}, \vec{v}'_{\omega+1}) = m_\omega$ and in turn that $m_\varphi(a_{\omega+1}) \geq m_\omega$. Thus $m_\varphi(a_{\omega+1}) \geq m_\varphi(a_\omega) \geq m_\varphi(a)$. By induction, $\varphi(T)$ maps $[a, a_{\omega+1}]$ homeomorphically onto $[b, b_{\omega+1}]$ and $[a, a_{\omega+1}) \cap (b_{\omega+1}, b] = \emptyset$.

If λ is a limit ordinal and for each $\omega < \lambda$ we have constructed a_ω, b_ω with the properties above, then $\{a_\omega\}_{\omega < \lambda}$ is a Cauchy net in $\mathbb{H}_{\text{Berk}}$. By Proposition 2.29, it has a unique limit point $a_\lambda \in \mathbb{H}_{\text{Berk}}$. Put $b_\lambda = \varphi(a_\lambda)$.

Clearly $a_\lambda \in \overline{U}$ and $a_\lambda, b_\lambda \in [a,b]$. Since U is φ-saturated and $[a,b] \subset \varphi(U)$, the same argument as in the proof of Lemma 9.38 shows that $a_\lambda \in U$. By induction, $\varphi(T)$ maps $[a, a_\lambda]$ homeomorphically onto $[b, b_\lambda]$. Since $[a, a_\omega] \cap [b_\omega, b] = \emptyset$ for each $\omega < \lambda$, it follows that $[a, a_\lambda) \cap (b_\lambda, b] = \emptyset$. Since a_λ is the limit of the points a_ω, Theorem 9.22 shows that $m_\varphi(a_\lambda) \geq \limsup_{\omega < \lambda} m_\omega \geq m_\varphi(a)$.

Finally, since there are ordinals of arbitrarily high cardinality and the map $\omega \mapsto a_\omega$ is injective on ordinals with $a_\omega \neq b_\omega$, it must be that $a_\omega = b_\omega$ for some ω. $\square$

Rivera-Letelier [83] has also proved a fixed point theorem for closed, connected sets which are mapped into themselves by $\varphi(T)$.

THEOREM 10.85 (Stable Set Fixed Point Property). *Let $\varphi(T) \in K(T)$ have degree $d \geq 1$. Let $X \subseteq \mathbb{P}^1_{\text{Berk}}$ be a nonempty connected, closed set (for either the weak topology or the strong topology) which satisfies $\varphi(X) \subseteq X$. Then X contains a point fixed by φ.*

Furthermore, if X is not reduced to a point, then X contains either

(A) *an attracting fixed point of φ of type* I *or*

(B) *a repelling or indifferent fixed point of φ of type* II.

We remark that a set $X \subset \mathbb{P}^1_{\text{Berk}}$ is connected and closed in the weak topology if and only if it is connected and closed in the strong topology. To see this, first note that by Lemma B.18, X is connected in the weak topology if and only if it is connected in the strong topology. If X is closed in the weak topology, then trivially it is closed in the strong topology. Hence it suffices to observe that if X is connected and closed in the strong topology, then its complement is a union of open discs (for the weak topology), so X is closed in the weak topology.

PROOF. If X consists of a single point, that point is a fixed point of φ, but nothing can be said about its nature. Henceforth we will assume that X contains at least two (hence infinitely many) points.

We first show that there is at least one fixed point in X. As in Lemma 10.84, the idea is to start at an arbitrary point and use transfinite induction to "push" towards a fixed point, with each "push" being either a single step or passage to a limit of steps, until a fixed point is finally reached.

We then use this fixed point to deduce the existence of a fixed point with the properties described.

Since X is connected and contains at least two points, it has points in $\mathbb{H}_{\text{Berk}}$. Fix a point $x_0 \in X \cap \mathbb{H}_{\text{Berk}}$. If $\varphi(x_0) = x_0$, we are done. Otherwise, since $\varphi(x_0) \in X$ and X is connected, the path $[x_0, \varphi(x_0)]$ is contained in X. We claim that there is a point $x_1 \in (x_0, \varphi(x_0))$ such that for each $z \in [x_0, x_1)$ the path $[z, \varphi(z)]$ contains x_1 in its interior. By Lemma 9.31, for all $a, c \in \mathbb{H}_{\text{Berk}}$ the path distance satisfies $\rho(\varphi(a), \varphi(c)) \leq d \cdot \rho(a, c)$. Hence if x_1 is the point in $(x_0, \varphi(x_0))$ for which $\rho(x_0, x_1) = \rho(x_0, \varphi(x_0))/(d+1)$, then for each $z \in [x_0, x_1)$

$$\rho(\varphi(z), \varphi(x_0)) \leq d \cdot \rho(z, x_0) < \rho(x_1, \varphi(x_0)).$$

Since $\mathbb{H}_{\text{Berk}}$ is uniquely path-connected, the path from z to $\varphi(z)$ must pass through x_1.

Inductively, suppose $\omega \geq 1$ is an ordinal and that we have constructed points $\{x_\kappa\}_{\kappa \leq \omega} \subset X \cap \mathbb{H}_{\text{Berk}}$ such that for each $\kappa < \omega$,

(1) $[x_0, x_\kappa]$ is a proper initial segment of $[x_0, x_{\kappa+1}]$,
(2) $\varphi(x_\kappa) \neq x_\kappa$, and $[x_\kappa, x_{\kappa+1}]$ is a proper initial segment of $[x_\kappa, \varphi(x_\kappa)]$, such that for each $z \in [x_\kappa, x_{\kappa+1}]$ we have $x_{\kappa+1} \in [z, \varphi(z)]$,
(3) if ω is a limit ordinal, then $x_\omega = \lim_{\substack{\kappa \to \omega \\ \kappa < \omega}} x_\kappa$.

If $\varphi(x_\omega) = x_\omega$, we are done. Otherwise, by the same argument as above, there is a point $x_{\omega+1} \in (x_\omega, \varphi(x_\omega))$ such that $[x_\omega, x_{\omega+1}] \subset X$ and for each $z \in [x_\omega, x_{\omega+1})$ the path from z to $\varphi(z)$ passes through $x_{\omega+1}$. We claim that the path $[x_0, x_{\omega+1}]$ extends $[x_0, x_\omega]$.

If ω is a successor ordinal, say $\omega = \kappa+1$, then by hypothesis for each $z \in [x_\kappa, x_\omega)$ the path from z to $\varphi(z)$ contains x_ω in its interior. This means that $\varphi([x_\kappa, x_\omega))$ is contained in a single connected component V of $\mathbb{P}^1_{\text{Berk}} \setminus \{x_\omega\}$. By continuity, $\varphi(x_\omega)$ belongs to the closure of this component, which is $V \cup \{x_\omega\}$. Since $\varphi(x_\omega) \neq x_\omega$, it must be that $\varphi(x_\omega) \in V$. On the other hand, for each $z \in [x_\kappa, x_\omega]$ the path from z to $\varphi(z)$ contains x_ω, so $z \notin V$. It follows that $[x_\kappa, x_\omega] \cap [x_\omega, \varphi(x_\omega)] = \{x_\omega\}$, so $[x_0, x_{\omega+1}]$ extends $[x_0, x_\omega]$.

On the other hand, if ω is a limit ordinal and if $[x_\omega, x_{\omega+1}]$ does not extend $[x_0, x_\omega]$, then $[x_\omega, \varphi(x_\omega)]$ and $[x_\omega, x_0]$ have a common initial segment $[x_\omega, b]$, where $b \in (x_\omega, \varphi(x_\omega))$. There is an η_0 such that if $\eta_0 \leq \eta < \omega$, the point x_η belongs to $[x_\omega, b]$. By continuity, as $\eta \to \omega$, then $\varphi(x_\eta) \to \varphi(x_\omega)$, so there is an $\eta_1 > \eta_0$ such that if $\eta_1 < \eta < \omega$, the path from x_η to $\varphi(x_\eta)$ contains b. By construction, that path also contains $x_{\eta+1}$, which lies in $[x_\eta, x_\omega] \subset [b, x_\omega]$. This is a contradiction. Hence $[x_0, x_{\omega+1}]$ extends $[x_0, x_\omega]$.

Finally, suppose λ is a limit ordinal and that we have constructed points $\{x_\kappa\}_{\kappa < \lambda}$ satisfying properties (1) and (2) above. Then $P_\lambda := \bigcup_{\kappa < \lambda} [x_0, x_\kappa]$ is isomorphic to a half-open segment.

If P_λ has finite path length, then by the completeness of $\mathbb{H}_{\text{Berk}}$ under $\rho(x, y)$ (Proposition 2.29), the points $\{x_\kappa\}_{\kappa < \lambda}$ converge to a point $x_\lambda \in \mathbb{H}_{\text{Berk}}$. Since X is closed and x_λ is also the limit of the x_κ in the weak topology, $x_\lambda \in X$. Clearly $\{x_\kappa\}_{\kappa \leq \lambda}$ satisfies conditions (1), (2), and (3), so the induction can continue.

If P_λ has infinite path length, we claim that $\{x_\kappa\}_{\kappa < \lambda}$ converges to a point $x_\kappa \in X \cap \mathbb{P}^1(K)$ which is fixed by φ. For each $\kappa < \lambda$, let $\vec{v}_\kappa \in T_{x_\kappa}$ be the tangent vector to P_λ at x_κ in the direction of $x_{\kappa+1}$, and put $\mathcal{B}_\kappa^- = \mathcal{B}_{x_\kappa}(\vec{v}_\kappa)^-$. Then the $\mathcal{B}_\kappa^-$ form a nested collection of balls whose radii $r_\kappa = \text{diam}_{x_0}(\mathcal{B}_\kappa^-)$ approach 0. By the completeness of $\mathbb{P}^1_{\text{Berk}}$ in either the weak topology or the strong topology (Proposition 2.29), their intersection is a point

$$x_\lambda = \lim_{\kappa \to \lambda} x_\kappa \in \mathbb{P}^1(K).$$

Since X is closed in the weak topology, $x_\lambda \in X$. Moreover, for each κ we have $\varphi(x_\kappa) \in \mathcal{B}_\kappa^-$, so by continuity $\varphi(x_\lambda) = x_\lambda$.

The induction will eventually terminate, since there are ordinals of cardinality greater than the cardinality of $\mathbb{R}$. When it terminates, it has produced a fixed point in either $X \cap \mathbb{P}^1(K)$ or $X \cap \mathbb{H}_{\text{Berk}}$.

We will now show that the existence of the fixed point constructed above implies the existence of a fixed point with the specified properties.

First suppose the fixed point x_ω belongs to $X \cap \mathbb{P}^1(K)$. The path $[x_0, x_\omega]$ contains points x_κ with $\kappa < \omega$ which are arbitrarily near x_ω in the metric $d(x,y)$. By the description of the action of φ near a fixed point in $\mathbb{P}^1(K)$ given at the beginning of this section, if x_ω were repelling, there would be points x_κ arbitrarily near x_ω which are moved away from x_ω along the path $[x_0, x_\omega]$. If x_ω were indifferent, there would be points x_κ arbitrarily near x_ω which are fixed by φ. Neither of these is the case, so x_ω must be attracting.

Next suppose the fixed point x_ω belongs to $X \cap \mathbb{H}_{\text{Berk}}$. If x_ω is of type II, it may be either repelling or indifferent, but it cannot be attracting since there are no attracting fixed points in $\mathbb{H}_{\text{Berk}}$. If x_ω is of type III or IV, then by Lemma 10.80 it is an indifferent fixed point. Let $\vec{v} \in T_{x_\omega}$ be the tangent vector to the path $[x_0, x_\omega]$ at x_ω. Since $m_\varphi(x_\omega, \vec{v}) = 1$, there is a segment $[x_\omega, a]$ with initial tangent vector $\vec{v}$ which is fixed by φ. As the segments $[x_\omega, x_0]$ and $[x_\omega, a]$ have the same initial tangent vector, they must share a nontrivial initial segment $[x_\omega, b]$. This segment contains infinitely many points of type II which are fixed by φ. □

We can now prove the following counterpart to Theorem 10.83:

THEOREM 10.86 (Attracting Fixed Point Criterion). *Let $\varphi(T) \in K(T)$ have $\deg(\varphi) \geq 1$. Let $U \subset \mathbb{P}^1_{\text{Berk}}$ be a simple domain for which $\overline{\varphi(U)} \subset U$. Then U contains an attracting fixed point of φ, necessarily of type I.*

PROOF. Since $\varphi(\overline{U}) = \overline{\varphi(U)}$, the set $X = \overline{U}$ satisfies the conditions of Theorem 10.85. Thus $\overline{U}$ contains a type I or type II point x fixed by φ.

Since φ maps ∂U into U, the points of ∂U are not fixed, so x belongs to U. We claim that x must be of type I, in which case Theorem 10.85 shows that it is an attracting fixed point.

Suppose to the contrary that x were of type II. Let $x_1, \ldots, x_m$ be the finitely many boundary points of U; each is of type II or III. Put $r = \min_{1 \leq i \leq m} \rho(x, x_i)$. Then the ball

$$\widehat{\mathcal{B}}(x,r)^- = \{z \in \mathbb{H}_{\text{Berk}} : \rho(x,z) < r\}$$

is contained in U. By the Incompressibility Lemma (Corollary 9.39), $\widehat{\mathcal{B}}(x,r)^-$ is also contained in $\varphi(U)$. Hence the boundary point of U nearest to x, say x_j, is a point of $\overline{\varphi(U)}$. This contradicts the assumption that $\overline{\varphi(U)} \subset U$. □

As an application of Theorem 10.82, we show that each connected component U of the Berkovich Fatou set has the property that $\bigcup_{n=0}^{\infty} \varphi^{(n)}(U)$ omits at least three points of $\mathbb{P}^1(K)$. By Theorem 10.56, U is a maximal connected open set with respect to this property, so $U \cap \mathbb{P}^1(K)$ is a Fatou component of $F_\varphi \cap \mathbb{P}^1(K)$, in the sense of Rivera-Letelier [**83**].

COROLLARY 10.87. *Let $\varphi(T) \in K(T)$ have degree $\deg(\varphi) \geq 2$. Then for each connected component U of $F_\varphi = \mathbb{P}^1_{\text{Berk}} \backslash J_\varphi$, the union $\bigcup_{n=0}^{\infty} \varphi^{(n)}(U)$ omits at least three points of $\mathbb{P}^1(K)$.*

PROOF. First note that if U is a component of F_φ, then $\varphi(U)$ is also a component. Indeed, since $\varphi(U)$ is connected and F_φ is completely invariant under $\varphi(T)$ (Lemma 10.52), $\varphi(U)$ is contained in a component V of F_φ. Let U' be the component of $\varphi^{-1}(V)$ containing U. Since $\varphi^{-1}(V) \subset F_\varphi$ and U is a component of F_φ, it must be that $U' = U$. Hence $\varphi(U) = \varphi(U') = V$.

Thus $\varphi(T)$ acts on the set of components of F_φ.

If there is a component W of F_φ which is preperiodic but not periodic for this action, then every component U has the property in the corollary. Indeed, W has infinitely many distinct preimages W_n under the iterates $\varphi^{(n)}(T)$, and the set of forward images $\{\varphi^{(n)}(U)\}$ can contain only finitely many of the W_n. Hence there is some W_N which is disjoint from $\bigcup_{n=0}^{\infty} \varphi^{(n)}(U)$, and $W_N \cap \mathbb{P}^1(K)$ is infinite.

Similarly, if there is a wandering component (that is, a component that is not preperiodic) or if there are at least two periodic components W, W' with disjoint orbits under $\varphi(T)$, then every component U has the property in the corollary.

Thus we need only consider the case where the components form a single, necessarily finite, periodic orbit. By Theorems 10.81 and 10.82, J_φ contains a repelling fixed point x. According to Lemma 10.80, x is either of type I or type II. If x were of type II, then since $\mathbb{P}^1_{\text{Berk}} \backslash \{x\}$ has infinitely many components and J_φ has empty interior (Corollary 10.69), F_φ would have infinitely many components, a contradiction. Hence $x \in \mathbb{P}^1(K)$. Since $E_\varphi(K) \subset F_\varphi$ (Corollary 10.55), x is not exceptional. Thus x has infinitely many distinct preimages under the $\varphi^{(n)}(T)$, all of which belong to J_φ. For any component U of F_φ, the forward images $\varphi^{(n)}(U)$ omit these points. □

Finally, we show that periodic repelling points are dense in the Berkovich Julia set. (This fact is originally due to Rivera-Letelier [**80**].) The proof is a generalization of a classical argument of Fatou.

THEOREM 10.88 (Rivera-Letelier). *Let $\varphi(T) \in K(T)$ have $\deg(\varphi) \geq 2$. Then J_φ is the closure of the set of repelling periodic points of φ in $\mathbb{P}^1_{\text{Berk}}$.*

PROOF. By Theorem 10.81, all the repelling periodic points of φ belong to J_φ, so it suffices to show that those points are dense in J_φ.

By Theorem 10.82, there exists a repelling fixed point $z_0 \in \mathbb{P}^1_{\text{Berk}}$ of φ, and by Theorem 10.81, it is in J_φ. If $J_\varphi = \{z_0\}$, the theorem is clearly true.

Otherwise, let $V \subset \mathbb{P}^1_{\text{Berk}}$ be an open set, disjoint from z_0, which meets J_φ. We will construct a repelling periodic point whose orbit meets V, by first constructing a *homoclinic orbit*: a sequence of points $z_1, z_2, \ldots$ with $\varphi(z_k) = z_{k-1}$ for each $k \geq 1$, such that $\lim_{k \to \infty} z_k = z_0$ and some z_ℓ belongs to V. Thus the sequence $\{z_k\}$ abuts at z_0 under the action of $\varphi(T)$, but its "tail" loops back to z_0 and forms a sequence of points repelled by z_0.

By Theorem 10.59, the iterates $\varphi^{(k)}(V)$ cover J_φ, so there is an $\ell \geq 1$ for which $z_0 \in \varphi^{(\ell)}(V)$. Let $z_\ell \in V$ be such that $\varphi^{(\ell)}(z_\ell) = z_0$, and put $z_{\ell-k} = \varphi^{(k)}(z_\ell)$ for $k = 1, \ldots, \ell - 1$. We now consider two cases, according to whether $z_0 \in \mathbb{P}^1(K)$ or $z_0 \in \mathbb{H}_{\text{Berk}}$.

First suppose $z_0 \in \mathbb{P}^1(K)$. After a change of coordinates, we can assume that $z_0 \neq \infty$. Let λ be the multiplier for z_0, so $|\lambda| > 1$. By the discussion earlier in this section, there is an $R > 0$ such that $\varphi(\mathcal{D}(z_0, r)^-) = \mathcal{D}(z_0, |\lambda|r)^-$ for each $0 < r \leq R$. Put $Y = \mathcal{D}(z_0, R)^-$. After taking a smaller R if necessary, we can assume that $z_\ell \notin Y$. By Theorem 10.59, the iterates $\varphi^{(k)}(Y)$ cover J_φ, so there is an $m \geq 1$ for which $z_\ell \in \varphi^{(m)}(Y)$. Let $z_{\ell+m} \in Y$ be such that $\varphi^{(m)}(z_{\ell+m}) = z_\ell$. Set $z_{\ell+m-k} = \varphi^{(k)}(z_{\ell+m})$ for $k = 1, \ldots, m-1$.

Since $\varphi(\mathcal{D}(z_0, R/|\lambda|^k)^-) = \mathcal{D}(z_0, R/|\lambda|^{k-1})^-$ for each $k \geq 1$, we can inductively find points $z_{\ell+m+k} \in \mathcal{D}(z_0, R/|\lambda|^k)$ such that $\varphi(z_{\ell+m+k}) = z_{\ell+m+k-1}$ for each k. Clearly $\lim_{k\to\infty} z_k = z_0$.

By the Open Mapping Theorem (Corollary 9.10), $\varphi^{(m)}(Y)$ is an open neighborhood of z_ℓ. Since $\varphi^{(\ell)}$-saturated simple domains are cofinal in the neighborhoods of z_ℓ (see the discussion after Proposition 9.15), we can find a $\varphi^{(\ell)}$-saturated simple domain W_ℓ which contains z_ℓ and whose closure is contained in $V \cap \varphi^{(m)}(Y)$. Let $W_{\ell+m}$ be the connected component of $(\varphi^{(m)})^{-1}(W_\ell)$ which contains $z_{\ell+m}$. Clearly $\overline{W}_{\ell+m} \subset Y = \mathcal{D}(x_0, R)^-$. Inductively, for each $k \geq 1$ let $W_{\ell+m+k}$ be the connected component of $\varphi^{-1}(W_{\ell+m+k-1})$ which contains $z_{\ell+m+k}$; then $\overline{W}_{\ell+m+k} \subset \mathcal{D}(z_0, R/|\lambda|^k)^-$. Likewise, put $W_0 = \varphi^{(\ell)}(W_\ell)$; then W_0 is a neighborhood of z_0, and since W_ℓ is $\varphi^{(\ell)}$-saturated, W_ℓ is a connected component of $(\varphi^{(\ell)})^{-1}(W_0)$.

For some sufficiently large n, we will have $\mathcal{D}(z_0, R/|\lambda|^n) \subset W_0$. Put $q = \ell + m + n$, and put $U = W_{\ell+m+n}$. By Lemma 9.12, U is a simple domain. Furthermore, $W_0 = \varphi^{(q)}(U)$ and U is $\varphi^{(q)}$-saturated.

Since
$$\overline{U} \subset \mathcal{D}(z_0, R/|\lambda|^n)^- \subset W_0 = \varphi^{(q)}(U),$$
Theorem 10.83 shows that there is a repelling fixed point for $\varphi^{(q)}(T)$ which belongs to U. It is a repelling periodic point for $\varphi(T)$ whose orbit passes through W_ℓ, hence through V.

Next suppose $z_0 \in \mathbb{H}_{\text{Berk}}$. Let Y be an arbitrary neighborhood of z_0 which does not contain z_ℓ. As before, the iterates $\varphi^{(k)}(Y)$ cover J_φ, so there is an $m \geq 1$ for which $z_\ell \in \varphi^{(m)}(Y)$. Let $z_{\ell+m} \in Y$ be such that $\varphi^{(m)}(z_{\ell+m}) = z_\ell$.

Since the $\varphi^{(\ell)}$-saturated simple domains are cofinal in all neighborhoods of z_ℓ, we can find a $\varphi^{(\ell)}$-saturated simple domain W_ℓ containing z_ℓ, such that $\overline{W}_\ell \subset (V \cap \varphi^{(m)}(Y))$. Set $W_0 = \varphi^{(\ell)}(W_\ell)$; it is an open neighborhood of z_0, and W_ℓ is a component of $(\varphi^{(\ell)})^{-1}(W_0)$. Let $W_{\ell+m}$ be the component of $(\varphi^{(m)})^{-1}(W_\ell)$ containing $z_{\ell+m}$.

Put $Z = W_0 \cap Y$. Let $\tilde{\varphi}(T) \in k(T)$ be the reduction of $\varphi(T)$ at z_0 in suitable coordinates, and parametrize the tangent directions in T_{z_0} by

10.7. FIXED POINT THEOREMS AND THEIR APPLICATIONS

$\mathbb{P}^1(k)$ in the usual way. Let $S \subset \mathbb{P}^1(k)$ be the finite set of points such that $\mathcal{B}_{z_0}(\vec{v}_\alpha)^- \subset Z$ if $\alpha \notin S$. Let $S_0 \subset S$ be the (possibly empty) set of $\alpha \in S$ which are exceptional for $\tilde{\varphi}(T)$. Let $\beta \in \mathbb{P}^1(k)$ be the point for which $z_{\ell+m} \in \mathcal{B}_{z_0}(\vec{v}_\beta)^-$.

We now consider two subcases, according to whether $\beta \in S_0$ or not.

First suppose $\beta \notin S_0$. Then there are an $\alpha \notin S$ and an $n \geq 1$ with $\tilde{\varphi}^{(n)}(\alpha) = \beta$. By hypothesis, $\mathcal{B}_{z_0}(\vec{v}_\alpha)^- \subset Z$. Since $\varphi^{(n)}(\mathcal{B}_{z_0}(\vec{v}_\alpha)^-)$ contains $\mathcal{B}_{z_0}(\vec{v}_\beta)^-$ (Proposition 9.40), there is a point $z_{\ell+m+n} \in \mathcal{B}_{z_0}(\vec{v}_\alpha)^-$ with $\varphi^{(n)}(z_{\ell+m+n}) = z_{\ell+m}$, which means that $\varphi^{(m+n)}(z_{\ell+m+n}) = z_\ell$. Let $W_{\ell+m+n}$ be the component of $(\varphi^{(m+n)})^{-1}(W_\ell)$ which contains $z_{\ell+m+n}$. Since $W_\ell \subset V$ and $\varphi^{(\ell+m)}(z_0) = z_0$ but $z_0 \notin V$, clearly $z_0 \notin W_{\ell+m+n}$. However, $\mathcal{B}_{z_0}(\vec{v}_\alpha)^-$ is a component of $\mathbb{P}^1_{\text{Berk}} \backslash \{z_0\}$, this implies that $W_{\ell+m+n} \subseteq \mathcal{B}_{z_0}(\vec{v}_\alpha)^-$. Put $q = \ell + m + n$, and put $U = W_{\ell+m+n}$. As before, U is a simple domain, $W_0 = \varphi^{(q)}(U)$, and U is $\varphi^{(q)}$-saturated. Since

$$\overline{U} \subseteq (\mathcal{B}_{z_0}(\vec{v}_\alpha)^- \cup \{z_0\}) \subset Z \subseteq W_0 = \varphi^{(q)}(U) \, ,$$

we can conclude the proof by using Theorem 10.83.

Next suppose $\beta \in S_0$. Since β is exceptional for $\tilde{\varphi}(T)$, there is an $s \geq 1$ such that $\tilde{\varphi}^{(s)}(\beta) = \beta$. Since β is the only point $\alpha \in \mathbb{P}^1(k)$ for which $\tilde{\varphi}^{(s)}(\alpha) = \beta$, it follows that $\vec{v}_\beta$ is the only tangent vector $\vec{v} \in T_{z_0}$ for which $\varphi_*^{(s)}(\vec{v}) = \vec{v}_\beta$. Thus

$$M := m_{\varphi^{(s)}}(z_0, \vec{v}_\beta) = \deg(\tilde{\varphi})^s > 1 \, .$$

By Corollary 9.21 and Theorem 9.22, there is a segment $[z_0, t]$ such that $\varphi^{(s)}([z_0, t])$ is the segment $[z_0, \varphi^{(s)}(t)]$, and both segments have the initial tangent vector $\vec{v}_\beta$. Furthermore $\rho(z_0, \varphi^{(s)}(x)) = M \cdot \rho(z, x)$ for each $x \in [z_0, t]$, and if $\mathcal{A} = \mathcal{A}_{z_0, t}$ is the annulus associated to (z_0, t), then $\varphi^{(s)}(\mathcal{A})$ is the annulus associated to $(z_0, \varphi^{(s)}(t))$. Since any two segments with the same initial tangent vector share a common initial segment, after shrinking $[z_0, t]$ if necessary, we can assume that $[z_0, t] \subset [z_0, \varphi^{(s)}(t)]$.

Clearly $\mathcal{A} \subset \mathcal{B}_{z_0}(\vec{v}_\beta)^-$. After further shrinking $[z_0, t]$, we can arrange that $\mathcal{A} \subset W_0$. Put $r = \text{Mod}(\mathcal{A})$, and consider the iterates $\varphi^{(ks)}(\mathcal{A})$ for $k = 1, 2, \ldots$. By induction, $\varphi^{(ks)}(\mathcal{A}) \subset \varphi^{((k+1)s)}(\mathcal{A})$ for each k. By Proposition 9.44, each $\varphi^{(ks)}(\mathcal{A})$ is either an annulus, an open disc, or all of $\mathbb{P}^1_{\text{Berk}}$. If it is an annulus, it is contained in $\mathcal{B}_{z_0}(\vec{v}_\beta)^-$ and has modulus $M^k r$. If it is a disc, it is equal to $\mathcal{B}_{z_0}(\vec{v}_\beta)^-$. If it is $\mathbb{P}^1_{\text{Berk}}$, it clearly contains $\mathcal{B}_{z_0}(\vec{v}_\beta)^-$.

Since $\overline{W}_\ell$ is connected and contained in V and since $\varphi(z_0) = z_0$ but $z_0 \notin V$, it follows that $z_0 \notin (\varphi^{(m)})^{-1}(\overline{W}_\ell)$ and in turn that $z_0 \notin \overline{W}_{\ell+m}$. Since $z_{\ell+m} \in \mathcal{B}_{z_0}(\vec{v}_\beta)^-$, we must have $\overline{W}_{\ell+m} \subset \mathcal{B}_{z_0}(\vec{v}_\beta)^-$.

If some $\varphi^{(ns)}(\mathcal{A})$ is $\mathcal{B}_{z_0}(\vec{v}_\beta)^-$ or $\mathbb{P}^1_{\text{Berk}}$, then trivially $\overline{W}_{\ell+m} \subset \varphi^{(ns)}(\mathcal{A})$. On the other hand, if each $\varphi^{(ns)}(\mathcal{A})$ is an annulus, then since the moduli of those annuli grow to ∞, we must have

$$\bigcup_{k=1}^{\infty} \varphi^{(ks)}(\mathcal{A}) = \mathcal{B}_{z_0}(\vec{v}_\beta)^- \setminus \{x_\beta\}$$

for some $x_\beta \in \mathbb{P}^1(K)$. Here, x_β depends only on $\varphi(T)$, z_0, and β, but not on W_ℓ. Hence we can assume that W_ℓ was chosen so that $\varphi^{(m)}(x_\beta) \notin \overline{W}_\ell$, which means that $x_\beta \notin \overline{W}_{\ell+m}$. By compactness, again there is some $\varphi^{(ns)}(\mathcal{A})$ which contains $\overline{W}_\ell$.

Choose a point $z_{\ell+m+ns} \in \mathcal{A}$ for which $\varphi^{(ns)}(z_{\ell+m+ns}) = z_{\ell+m}$, and let $W_{\ell+m+ns}$ be the component of $(\varphi^{(ns)})^{-1}(W_{\ell+m}) = (\varphi^{(\ell+m+ns)})^{-1}(W_0)$ which contains $z_{\ell+m+ns}$. By the discussion above,

$$\overline{W}_{\ell+m+ns} \subset \mathcal{A} \subset W_0.$$

Thus if we take $q = \ell + m + ns$ and put $U = W_{\ell+m+ns}$, then U is a $\varphi^{(q)}$-saturated simple domain with $\overline{U} \subset \varphi^{(q)}(U)$, and we can conclude the proof by using Theorem 10.83 as before. $\square$

REMARK 10.89. It is an open problem to prove the "classical" version of Theorem 10.88 for $\mathbb{P}^1(K)$, that is, to prove that the repelling periodic points of φ in $\mathbb{P}^1(K)$ are dense in the classical Julia set $J_\varphi(K)$ of φ. By a theorem of J.-P. Bézivin [22, Theorem 3], if there exists at least one repelling periodic point in $J_\varphi(K)$, then the repelling periodic points of φ in $\mathbb{P}^1(K)$ are dense in $J_\varphi(K)$. Hence the problem is equivalent to the *a priori* weaker assertion that if $J_\varphi(K)$ is nonempty, then there exists at least one repelling periodic point for φ in $\mathbb{P}^1(K)$.

10.8. Dynamics of polynomial maps

In this section, we characterize the Berkovich Julia set and the canonical measure in potential-theoretic terms in the case where $\varphi(T) \in K[T]$ is a polynomial. Here the characteristic of K can be arbitrary.

Let $\varphi(T) \in K[T]$ be a polynomial of degree $d \geq 2$.

DEFINITION 10.90. The *Berkovich filled Julia set* K_φ of φ is

$$K_\varphi := \bigcup_{M > 0} \{x \in \mathbb{A}^1_{\text{Berk}} : [\varphi^{(n)}(T)]_x \leq M \text{ for all } n \geq 0\}.$$

In other words, K_φ is the set of all $x \in \mathbb{P}^1_{\text{Berk}}$ for which the sequence $\{[\varphi^{(n)}(T)]_x\}_{n \geq 0}$ stays bounded as n goes to infinity. It is clear that K_φ is a compact subset of $\mathbb{P}^1_{\text{Berk}}$ not containing ∞. It is nonempty, because every preperiodic point of φ in $\mathbb{A}^1(K)$ is contained in K_φ. It is not hard to see that K_φ is just the closure in $\mathbb{P}^1_{\text{Berk}}$ of the classical filled Julia set

$$\bigcup_{M > 0} \{x \in K : |\varphi^{(n)}(x)| \leq M \text{ for all } n \geq 0\}.$$

It is also easy to see that the complement of K_φ coincides with the attracting basin for the attracting fixed point ∞ of φ.

Both K_φ and ∂K_φ are completely invariant under φ. Since the classical exceptional locus $E_\varphi(K)$ consists of attracting periodic points, which are contained in either the complement of K_φ (in the case of ∞) or the interior of K_φ, we know by Corollary 10.57 that

$$(10.70) \qquad J_\varphi \subseteq \partial K_\varphi .$$

We will see below that in fact the Berkovich Julia set of φ equals ∂K_φ.

Recall that the Call-Silverman local height function $\hat{h}_{\varphi,\infty} = \hat{h}_{\varphi,v,(\infty)}$ relative to the point ∞ and the dehomogenization $f_1(T) = 1$, $f_2(T) = \varphi(T)$ is defined for $x \in \mathbb{A}^1(K)$ by the formula

$$\hat{h}_{\varphi,\infty}(x) = \lim_{n \to \infty} \frac{1}{d^n} \max(0, \log_v([\varphi^{(n)}]_x)) .$$

It is a continuous subharmonic function on $\mathbb{A}^1_{\text{Berk}}$ which belongs to $\text{BDV}(\mathbb{P}^1_{\text{Berk}})$ and satisfies

$$\Delta \hat{h}_{\varphi,\infty} = \delta_\infty - \mu_\varphi .$$

Moreover, the definition of $\hat{h}_{\varphi,\infty}$ shows that $\hat{h}_{\varphi,\infty}(x) = 0$ for all $x \in K_\varphi$.

By (10.8) and continuity, we have the transformation law

$$(10.71) \qquad \hat{h}_{\varphi,\infty}(\varphi(x)) = d \cdot \hat{h}_{\varphi,\infty}(x)$$

for all $x \in \mathbb{A}^1_{\text{Berk}}$. From this and the fact that $\hat{h}_{\varphi,\infty}(x) - \log^+[T]_x$ is a bounded function on $\mathbb{P}^1_{\text{Berk}}$, one deduces easily that $\hat{h}_{\varphi,\infty}(x) > 0$ for all $x \in \mathbb{P}^1_{\text{Berk}} \backslash K_\varphi$. Thus $K_\varphi = \{x \in \mathbb{P}^1_{\text{Berk}} : \hat{h}_{\varphi,\infty}(x) = 0\}$.

Next consider the Green's function $g_\varphi(x, \infty) = g_{\mu_\varphi}(x, \infty)$. By formula (10.21), for $x \ne \infty$ we have

$$(10.72) \qquad g_\varphi(x, \infty) = \hat{h}_{\varphi,\infty}(x) + C$$

where $C = \hat{h}_{\varphi,v,(0)}(\infty) + \log_v(R)$. For each $x \in \mathbb{P}^1_{\text{Berk}}$, Theorem 10.18 and the fact that $\varphi^*((\infty)) = d \cdot (\infty)$ give

$$(10.73) \qquad g_\varphi(\varphi(x), \infty) = g_\varphi(x, \varphi^*(\infty)) = d \cdot g_\varphi(x, \infty) .$$

If $x \in K_\varphi$, the iterates $\varphi^{(n)}(x)$ all belong to the compact set K_φ, so the values $g_\varphi(\varphi^{(n)}(x), \infty)$ are uniformly bounded, and iterating (10.73) shows that $g_\varphi(x, \infty) = 0$. Hence $C = 0$ and $g_\varphi(z, \infty) = \hat{h}_{\varphi,\infty}(z)$ for all z.

By formula (10.21) and what has just been shown,

$$(10.74) \qquad g_\varphi(x, y) = \begin{cases} -\log_v(\delta(x,y)_\infty) + \hat{h}_{\varphi,\infty}(x) \\ \qquad + \hat{h}_{\varphi,\infty}(y) + \log_v(R) & \text{if } x, y \ne \infty , \\ \hat{h}_{\varphi,\infty}(x) & \text{if } y = \infty , \\ \hat{h}_{\varphi,\infty}(y) & \text{if } x = \infty . \end{cases}$$

From (10.74) and the fact that $\hat{h}_{\varphi,\infty} \equiv 0$ on K_φ, it follows that for any probability measure ν supported on K_φ, we have

$$\iint_{\mathbb{P}^1_{\text{Berk}} \times \mathbb{P}^1_{\text{Berk}}} g_{\mu_\varphi}(x,y) \, d\nu(x) \, d\nu(y)$$

(10.75)
$$= \log_v(R) + \iint_{\mathbb{P}^1_{\text{Berk}} \times \mathbb{P}^1_{\text{Berk}}} -\log_v(\delta(x,y)_\infty) \, d\nu(x) \, d\nu(y) \, .$$

Note that μ_φ is supported on J_φ, which is contained in K_φ by (10.70). By Theorem 10.8, the left-hand side of (10.75) is always nonnegative and is zero if and only if $\nu = \mu_\varphi$. On the other hand, by the definition of the logarithmic capacity and the equilibrium measure, the right-hand side is minimized by the equilibrium measure $\mu_{K_\varphi,\infty}$ of K_φ relative to ∞. It follows that $\mu_\varphi = \mu_{K_\varphi,\infty}$ and the capacity $\gamma_\infty(K_\varphi)$ is equal to R. Since μ_φ has continuous potentials, the Green's function $G(z,\infty;K_\varphi)$ is continuous. Thus $G(z,\infty;K_\varphi)$ is equal to the Call-Silverman local height $\hat{h}_{\varphi,\infty}(x)$, since both functions are zero on K_φ, have Laplacian equal to $\delta_\infty - \mu_\varphi$, are continuous on $\mathbb{P}^1_{\text{Berk}} \backslash \{\infty\}$, and have bounded difference from $\log_v([T]_x)$ as $x \to \infty$.

Recall that $R = |\text{Res}(F_1, F_2)|^{-\frac{1}{d(d-1)}}$. If
$$\varphi(T) = a_d T^d + a_{d-1} T^{d-1} + \cdots + a_1 T + a_0 \, ,$$
then an easy calculation shows that $|\text{Res}(F_1, F_2)| = |a_d|^d$, so $\gamma_\infty(K_\varphi) = R = |a_d|^{-\frac{1}{d-1}}$.

We can now show that $J_\varphi = \partial K_\varphi$. For this, it suffices to note that by Corollary 7.39, the connected component of $\mathbb{P}^1_{\text{Berk}} \backslash K_\varphi$ containing ∞ is $U_\infty = \mathbb{P}^1_{\text{Berk}} \backslash K_\varphi = \{x \in \mathbb{P}^1_{\text{Berk}} : \hat{h}_{\varphi,\infty}(x) > 0\}$. This is also the attracting basin of ∞. By Corollary 7.39(C),
$$J_\varphi = \text{supp}(\mu_\varphi) = \partial U_\infty = \partial K_\varphi \, .$$
By Proposition 7.37(A)(6), it follows that $\gamma_\infty(J_\varphi) = \gamma_\infty(K_\varphi)$.

Combining these facts, we have proved the following result:

THEOREM 10.91. *Let $\varphi \in K[T]$ be a polynomial with degree $d \geq 2$ and leading coefficient a_d. Then:*

(A) *For all $x \in \mathbb{P}^1_{\text{Berk}}$,*
$$g_{\mu_\varphi}(x,\infty) = G(x,\infty;K_\varphi) = \hat{h}_{\varphi,v,(\infty)}(x)$$
$$= \lim_{n \to \infty} \frac{1}{d^n} \max(0, \log_v([\varphi^{(n)}]_x)) \, .$$

(B) *The canonical measure μ_φ coincides with $\mu_{K_\varphi,\infty}$, the equilibrium measure of K_φ with respect to ∞.*

(C) $J_\varphi = \partial K_\varphi$.

(D) $\gamma_\infty(J_\varphi) = \gamma_\infty(K_\varphi) = |a_d|^{-1/(d-1)}$.

(E) *The connected component of F_φ containing ∞ is equal to $\mathbb{P}^1_{\text{Berk}} \backslash K_\varphi$, the basin of attraction of ∞ for $\varphi(T)$.*

10.9. Rational dynamics over $\mathbb{C}_p$

The field $\mathbb{C}_p$ is (up to isomorphism) the smallest complete and algebraically closed non-Archimedean field of characteristic 0 and positive residue characteristic p. It can be constructed as the completion of an algebraic closure of the field $\mathbb{Q}_p$ of p-adic numbers and is often viewed as the p-adic analogue of the complex numbers. It is thus of great importance in arithmetic geometry. Thanks to the work of Rivera-Letelier, Benedetto, and others, the dynamics of rational functions $\varphi(T) \in \mathbb{C}_p(T)$ are much better understood than those of rational functions over an arbitrary complete and algebraically closed non-Archimedean field K.

In this section, we summarize the main results of Rivera-Letelier from [**81, 82, 83, 84**] concerning rational dynamics on $\mathbb{P}^1_{\text{Berk},\mathbb{C}_p}$, formulating his results in the terminology of this book. There are many interesting results in those papers we will not touch upon, and we omit the proofs, so we encourage the reader to consult the original papers. For the reader's convenience, we repeat some of the results and definitions from earlier sections.

Let p be a prime number. The field $\mathbb{C}_p$ is special in several ways:

- It has characteristic 0 and residue characteristic $p > 0$.
- It has a countable dense subfield isomorphic to $\overline{\mathbb{Q}}$ (a fixed algebraic closure of $\mathbb{Q}$).
- Its value group $|\mathbb{C}_p^\times|$ is a rank 1 divisible group isomorphic to $\mathbb{Q}$.
- Its residue field is isomorphic to $\overline{\mathbb{F}}_p$, an algebraic closure of the prime field $\mathbb{F}_p$.

These properties have several consequences. In particular, they imply that $\mathbb{P}^1_{\text{Berk},\mathbb{C}_p}$ has points of all four types, I, II, III, and IV, that $\mathbb{P}^1_{\text{Berk},\mathbb{C}_p}$ has countably many points of type II, with countable branching at each such point, and that $\mathbb{P}^1_{\text{Berk},\mathbb{C}_p}$ (equipped with the Berkovich topology) is metrizable. However, most important for rational dynamics are the strong finiteness properties which follow from the fact that $\overline{\mathbb{F}}_p$ is a union of finite fields. For example, if ζ is a type II fixed point of $\varphi(T) \in \mathbb{C}_p(T)$ and $\tilde{\varphi}(T) \in \overline{\mathbb{F}}_p(T)$ is reduction of $\varphi(T)$ in a corresponding local coordinate, then the forward orbits of $\tilde{\varphi}(T)$ in $\mathbb{P}^1(\overline{\mathbb{F}}_p)$ are automatically finite. Moreover, if $\tilde{\varphi}(T)$ has degree 1, then for some $n \geq 1$, $\tilde{\varphi}^{(n)}(T)$ is the identity map on $\mathbb{P}^1(\overline{\mathbb{F}}_p)$.

These finiteness properties are the ultimate source of Rivera-Letelier's description of the Berkovich Fatou set in $\mathbb{P}^1_{\text{Berk},\mathbb{C}_p}$ as a disjoint union of attracting basins, the "quasi-periodicity domain" and its preimages, and wandering domains. Such a decomposition does not hold over an arbitrary complete and algebraically closed non-Archimedean field K.

Rivera-Letelier starts with a rational function $\varphi(T) \in \mathbb{C}_p(T)$ and studies the action of φ on $\mathbb{P}^1(\mathbb{C}_p)$ and on the "p-adic hyperbolic space" $\mathbb{H}_p$, which turns out to be naturally isomorphic to $\mathbb{H}_{\text{Berk}} = \mathbb{P}^1_{\text{Berk}} \backslash \mathbb{P}^1(\mathbb{C}_p)$ (see [**82**, §9.3]), though at first glance Rivera-Letelier's definition of $\mathbb{H}_p$ looks rather different from the definition of $\mathbb{P}^1_{\text{Berk}} \backslash \mathbb{P}^1(\mathbb{C}_p)$. Rivera-Letelier equips $\mathbb{H}_p$ with the "strong topology" induced by the path distance $\rho(x,y)$ as defined in §2.7. This topology is strictly finer than the Berkovich subspace topology coming from the inclusion $\mathbb{H}_p \subset \mathbb{P}^1_{\text{Berk}}$.

In [**81, 82, 83, 84**], Rivera-Letelier takes the point of view that $\mathbb{P}^1(\mathbb{C}_p)$ is the fundamental object of study and that $\mathbb{H}_p$ is auxiliary. He thus states his main results in terms of $\mathbb{P}^1(\mathbb{C}_p)$. In his later works with Favre [**46, 47, 48**], he adopts a point of view closer to ours, that $\mathbb{P}^1_{\text{Berk}}$ is the appropriate domain for studying non-Archimedean dynamics.

The fundamental idea and principal novelty of Rivera-Letelier's work is that one can deduce useful facts about how $\varphi(T)$ acts on $\mathbb{P}^1(\mathbb{C}_p)$ by studying its action on $\mathbb{H}_p$. In particular, fixed point theorems for the action of $\varphi(T)$ on $\mathbb{H}_p$ play a prominent role in Rivera-Letelier's work.

To each $x \in \mathbb{H}_p$, Rivera-Letelier associates a family of *bouts* (literally: ends). In the context of our definition of $\mathbb{H}_{\text{Berk}}$, a bout $\mathcal{P}$ can be thought of as a tangent vector emanating from x, and the collection of bouts at x can therefore be identified with the "tangent space" T_x at x. To each bout $\mathcal{P}$ at x, one can associate a ball $\mathcal{B}_\mathcal{P} \subset \mathbb{P}^1(\mathbb{C}_p)$ in a natural way: $\mathcal{B}_\mathcal{P}$ is the set of $y \in \mathbb{P}^1(\mathbb{C}_p)$ such that the path (x,y) lies in the equivalence class $\mathcal{P}$. Thus, if $\mathcal{P}$ corresponds to $\vec{v} \in T_x$, then $\mathcal{B}_\mathcal{P} = \mathcal{B}_x(\vec{v})^-$ in our notation above.

However, in Rivera-Letelier's construction of $\mathbb{H}_p$, bouts are defined before points. Rivera-Letelier's definition of a bout [**82**, §2] is as follows. Let

$$B_0 \subset B_1 \subset B_2 \subset \cdots$$

be an increasing sequence of closed balls in $\mathbb{P}^1(\mathbb{C}_p)$ whose union B is either an open ball or all of $\mathbb{P}^1(\mathbb{C}_p)$. The collection of open annuli (or balls, if $B = \mathbb{P}^1(\mathbb{C}_p)$) $\{B \backslash B_i\}_{i \geq 0}$ is called a *vanishing chain*. Two vanishing chains are considered equivalent if each is cofinal in the other, under the relation of containment. A bout $\mathcal{P}$ is an equivalence class of vanishing chains. The ball $\mathcal{B}_\mathcal{P}$ associated to a bout $\mathcal{P}$ is the (generalized) ball $B = \bigcup B_i$ attached to any of the vanishing chains in $\mathcal{P}$; it is independent of the vanishing chain.

For Rivera-Letelier, a point $x \in \mathbb{H}_p$ is a collection of bouts $\{\mathcal{P}_\alpha\}$ whose associated balls $\mathcal{B}_{\mathcal{P}_\alpha}$ are pairwise disjoint and have $\mathbb{P}^1(\mathbb{C}_p)$ as their union. He calls a point x (and the bouts in it) rational, irrational, or singular, according as x consists of infinitely many bouts, two bouts, or one bout, respectively. Under Berkovich's classification, these correspond to points of type II, III, and IV, respectively. If x is of type II, then for each bout $\mathcal{P}$ associated to x, $\mathcal{B}_\mathcal{P}$ is an open ball with radius in $|\mathbb{C}_p^*|$; if x is of type III, then $\mathcal{B}_\mathcal{P}$ is an irrational ball; and if x is of type IV, then $\mathcal{B}_\mathcal{P} = \mathbb{P}^1(\mathbb{C}_p)$.

Fundamental lemmas. Rivera-Letelier bases his theory on two fundamental results concerning the action of a rational function on $\mathbb{P}^1(\mathbb{C}_p)$. The first ([**82**], Proposition 4.1) enables him to define the action of φ on $\mathbb{H}_p$, by first defining its action on bouts (compare with Lemma 9.45):

LEMMA 10.92. *Let $\varphi(T) \in \mathbb{C}_p(T)$ have degree $d = \deg(\varphi) \geq 1$. Let $\mathcal{P}$ be a bout of $\mathbb{P}^1(\mathbb{C}_p)$. Then there exists another bout $\mathcal{P}'$ of the same type (rational, irrational, or singular) and an integer $1 \leq m \leq d$ such that for each vanishing chain $\{C_i\}_{i \geq 0}$ defining $\mathcal{P}$, there is an $N \geq 1$ such that*
(1) *$\{\varphi(C_i)\}_{i \geq N}$ is a vanishing chain defining $\mathcal{P}'$,*
(2) *for each $i \geq N$, the map $\varphi : C_i \to \varphi(C_i)$ is of degree m.*

Rivera-Letelier calls $m = m(\mathcal{P})$ the multiplicity of the bout $\mathcal{P}$ and writes $\varphi(\mathcal{P})$ for $\mathcal{P}'$. In our notation, if $\mathcal{P}$ corresponds to $\vec{v} \in T_x$, then $m(\mathcal{P}) = m_\varphi(x, \vec{v})$ and $\varphi(\mathcal{P}) = \varphi_*(\vec{v})$. Rivera-Letelier defines the action of $\varphi(T)$ on a point $x = \{\mathcal{P}_\alpha\}$ by setting $\varphi(x) = \{\varphi(\mathcal{P}_\alpha)\}$. He then defines a local degree $\deg_\varphi(x)$ at each $x \in \mathbb{H}_p$ which coincides with our multiplicity $m_\varphi(x)$. For example, for a type II point he defines $\deg_\varphi(x) = \deg(\tilde{\varphi}(T))$, where $\tilde{\varphi}(T)$ is a reduction of $\varphi(T)$ relative to x and $\varphi(x)$. (See [**82**, Proposition 4.4]; for the definition of $\tilde{\varphi}(T)$, see the discussion preceding Corollary 9.25.) He shows that for any bout $\mathcal{Q}$ belonging to y,

$$\deg_\varphi(x) = \sum_{\substack{\text{bouts } \mathcal{P} \text{ in } x \\ \text{with } \varphi(\mathcal{P}) = \mathcal{Q}}} m(\mathcal{P}) .$$

(Compare this with our Theorem 9.22.) We will henceforth write $m_\varphi(x)$ instead of $\deg_\varphi(x)$.

The second fundamental result ([**82**, Lemma 4.2]) describes the action of φ on the ball $\mathcal{B}_\mathcal{P}$ associated to a bout (compare with our Proposition 9.41):

LEMMA 10.93. *Let $\varphi(T) \in \mathbb{C}_p(T)$ have degree $d = \deg(\varphi) \geq 1$. Let $\mathcal{P}$ be a bout with multiplicity $m = m(\mathcal{P})$, and let $\varphi(\mathcal{P})$ be its image under φ. Then there is an integer $m \leq N \leq d$ (depending on $\mathcal{P}$), with $N = m$ if and only if $\varphi(\mathcal{B}_\mathcal{P}) = \mathcal{B}_{\varphi(\mathcal{P})}$, such that*
(1) *for each $y \in \mathcal{B}_{\varphi(\mathcal{P})}$ there are exactly N points $x \in \mathcal{B}_\mathcal{P}$ with $\varphi(x) = y$ (counting multiplicities),*
(2) *for each $y \in \mathbb{P}^1(\mathbb{C}_p) \backslash \mathcal{B}_{\varphi(\mathcal{P})}$ there are exactly $N - m$ points $x \in \mathcal{B}_\mathcal{P}$ with $\varphi(x) = y$ (counting multiplicities).*

Rivera-Letelier shows that for each noncritical point $x \in \mathbb{P}^1(\mathbb{C}_p)$, there is a maximal open affinoid containing x on which $\varphi(T)$ is injective, the *component of injectivity* of x ([**81**, Proposition 2.9]):

PROPOSITION 10.94. *Let $\varphi(T) \in \mathbb{C}_p(T)$ be nonconstant. Suppose $x \in \mathbb{P}^1(\mathbb{C}_p)$ is not a critical point. Then there is a maximal open affinoid $V \subseteq \mathbb{P}^1_{\text{Berk}}$ containing x on which $\varphi(T)$ is injective: $\varphi(T)$ is injective on V, and if X an open affinoid containing x on which $\varphi(T)$ is injective, then $X \subseteq V$.*

In [**82**, Corollaries 4.7, 4.8], he shows that $\varphi(T)$, acting on $\mathbb{H}_p$, is continuous relative to the path distance topology. In Theorem 9.35, we have established this for arbitrary fields K.

LEMMA 10.95. *Let $\varphi(T) \in \mathbb{C}_p(T)$ be a nonconstant rational function of degree d. Then for any $a, c \in \mathbb{H}_p$,*

$$\rho(\varphi(a), \varphi(c)) \leq \int_{[a,c]} m_\varphi(x) \, dx \ .$$

In particular, $\rho(\varphi(a), \varphi(c)) \leq d \cdot \rho(a, c)$.

Given $x \in \mathbb{H}_p$, write $\widehat{\mathcal{B}}(x, r)^- = \{z \in \mathbb{H}_p : \rho(x, z) < r\}$ for the open ball of radius r about x, relative to the path distance metric. The following result from [84], which Rivera-Letelier calls the "Incompressibility Lemma", shows in particular that $\varphi(T)$ is an open map on $\mathbb{H}_p$ relative to the path distance topology (see our Corollary 9.39):

LEMMA 10.96. *Let $\varphi(T) \in \mathbb{C}_p(T)$ be nonconstant, and let $x \in \mathbb{H}_p$. Then for each $r > 0$,*

$$\widehat{\mathcal{B}}(\varphi(x), r)^- \subseteq \varphi(\widehat{\mathcal{B}}(x, r)^-) \ .$$

The theory of fixed points. If $x \in \mathbb{P}^1(\mathbb{C}_p)$ is fixed by $\varphi(T)$ and if its multiplier is λ, it is called

(1) an *attracting* fixed point, if $|\lambda| < 1$,

(2) an *indifferent* fixed point, if $|\lambda| = 1$; such a fixed point is called *parabolic* if λ is a root of unity and is *rationally indifferent* otherwise,

(3) a *repelling* fixed point, if $|\lambda| > 1$.

If a point $x \in \mathbb{P}^1(\mathbb{C}_p)$ is fixed by $\varphi^{(n)}(T)$ for some n and if n is the smallest integer with this property, then x is called periodic of order n, and $C = \{x, \varphi(x), \ldots, \varphi^{(n-1)}(x)\}$ is called the cycle associated to x. It is well known that the multiplier of $\varphi^{(n)}(t)$ is the same for all points $t \in C$ and the multiplier of C is defined to be that number. A cycle is called *attracting*, *indifferent*, or *repelling*, according as its multiplier λ satisfies $|\lambda| < 1$, $|\lambda| = 1$, or $|\lambda| > 1$. An attracting cycle $C \subset \mathbb{P}^1(\mathbb{C}_p)$ is called *superattracting* if its multiplier is 0. An indifferent cycle C is called *parabolic* if its multiplier is a root of unity.

Rivera-Letelier calls a fixed point of $\varphi(T)$ in $\mathbb{H}_p$ *indifferent* if $m_\varphi(x) = 1$ and *repelling* if $m_\varphi(x) > 1$. (We have explained the motivation for this terminology in §10.7.) He shows that fixed points of type III or IV in $\mathbb{H}_p$ are necessarily indifferent [84, Proposition 5.2]; see our Lemma 10.80.

Repelling fixed points in $\mathbb{H}_p$ come in two kinds, *separable* and *inseparable* [84]. If $x \in \mathbb{H}_p$ is a repelling fixed point (necessarily of type II) and $\tilde{\varphi}(T) \in \overline{\mathbb{F}}_p(T)$ is a reduction of $\varphi(T)$ relative to x, then x is called separable if the extension $\overline{\mathbb{F}}_p(T)/\overline{\mathbb{F}}_p(\tilde{\varphi}(T))$ is separable, and it is called inseparable if $\tilde{\varphi}(T) = \tilde{g}(T^p)$ for some $\tilde{g}(T) \in \overline{\mathbb{F}}_p(T)$. If x is separable, then there are a finite number of tangent directions $\vec{v}$ at x for which $m_\varphi(x, \vec{v}) > 1$, while if x is inseparable, then $m_\varphi(x, \vec{v}) \geq p$ for every tangent direction. In terms of the local action of $\varphi(T)$ described in §10.7, if x is separable, then it has a finite number of repelling "jets", while if x is inseparable, then *every* direction is a jet. Inseparable fixed points play an important role in the study of wandering components of the Fatou set.

Periodic points of $\varphi(T)$ in $\mathbb{H}_p$ become fixed points of $\varphi^{(n)}(T)$ for some n, and they are called indifferent, repelling, separable, or inseparable if they have those properties as fixed points of $\varphi^{(n)}(T)$. All the points of a given cycle have the same type.

In Theorem 10.81, we have seen that all repelling periodic points belong to the Berkovich Julia set J_φ. In Lemma 10.54, we have seen that all attracting periodic points belong to the Berkovich Fatou set F_φ. Over $\mathbb{C}_p$, in contrast to the situation over $\mathbb{C}$, all indifferent periodic points belong to F_φ:

PROPOSITION 10.97. *Let $\varphi(T) \in \mathbb{C}_p(T)$ have degree $\deg(\varphi) \geq 2$. Then each indifferent periodic point of $\varphi(T)$, in $\mathbb{P}^1(\mathbb{C}_p)$ or in $\mathbb{H}_p$, belongs to F_φ.*

PROOF. Let $x_0 \in \mathbb{P}^1_{\text{Berk}}$ be an indifferent periodic point of period n. We claim that there are an $N \geq 1$ and a simple domain V containing x such that V omits a disc in $\mathbb{P}^1_{\text{Berk}}$ and $\varphi^{(nN)}(V) = V$. If this is granted, then x belongs to the Berkovich Fatou set of $\varphi^{(nN)}(T)$, since its forward images omit all points of $\mathbb{P}^1(\mathbb{C}_p)\backslash V$. We finish by noting that $F_\varphi = F_{\varphi^{(nN)}}$ by Lemma 10.51.

We now establish the claim. After replacing $\varphi(T)$ by $\varphi^{(n)}(T)$, we can assume that x_0 is a fixed point of $\varphi(T)$.

If x_0 is of type I, choose coordinates so that $x_0 \neq \infty$ and expand $\varphi(T)$ as a power series as in (10.67). Then because the multiplier λ of x_0 satisfies $|\lambda| = 1$, the theory of Newton polygons shows there is a disc $D(x_0, r)^-$ such that $\varphi(D(x_0, r)^-) = D(x_0, r)^-$ (see Corollary A.17). Let $V = \mathcal{D}(x_0, r)^-$ be the corresponding Berkovich disc. By continuity, $\varphi(V) = V$.

Next suppose x_0 is of type II. Let $\tilde{\varphi}(T) \in \overline{\mathbb{F}}_p(T)$ be a reduction of $\varphi(T)$ at x_0. Since $m_\varphi(x_0) = 1$, Corollary 9.25 shows that $\deg(\tilde{\varphi}) = 1$. Since $\overline{\mathbb{F}}_p$ is a union of finite fields, there is some ℓ such that $\tilde{\varphi}(T) \in \text{PGL}_2(\mathbb{F}_{p^\ell})$. This is a finite group, so there is an $N \geq 1$ for which $\tilde{\varphi}^{(N)}(T) = \text{id}$. This means $(\varphi^{(N)})_*$ fixes each $\vec{v} \in T_{x_0}$.

For all but finitely many tangent directions, we have $\varphi^{(N)}(\mathcal{B}_{x_0}(\vec{v})^-) = \mathcal{B}_{x_0}(\vec{v})^-$. (For example, this holds for each $\mathcal{B}_{x_0}(\vec{v})^-$ not containing a pole of $\varphi^{(N)}(T)$.) Let S be an index set for the remaining tangent directions (we may assume that $S \neq \emptyset$). For each $\vec{v}_\alpha$ with $\alpha \in S$, Lemma 9.33 shows that there is a segment $[x_0, c_\alpha]$ with initial tangent vector $\vec{v}_\alpha$ such that $\varphi^{(N)}([x_0, c_\alpha]) = [x_0, \varphi(c_\alpha)]$ is another segment with the same initial tangent vector $\vec{v}_\alpha = (\varphi^{(N)})_*(\vec{v}_\alpha)$. Furthermore, since $m_{\varphi^{(N)}}(x_0, \vec{v}_\alpha) = m_{\varphi^{(N)}}(x_0) = m_\varphi(x_0)^N = 1$, for each $y \in [x_0, c_\alpha]$ we have $\rho(x_0, y) = \rho(x_0, \varphi^{(N)}(y))$. Since any two segments with the same initial tangent vector have a common initial segment, after moving c_α closer to x_0 if necessary, we can assume that $\varphi^{(N)}(c_\alpha) = c_\alpha$ and that $\varphi^{(N)}(T)$ is the identity on $[x_0, c_\alpha]$. Let $\mathcal{A}_{x_0, c_\alpha}$ be the annulus associated to the segment (x_0, c_α). By Lemma 9.33, we have $\varphi^{(N)}(\mathcal{A}_{x_0, c_\alpha}) = \mathcal{A}_{x_0, c_\alpha}$. Thus, if V is the component of $\mathbb{P}^1_{\text{Berk}}\backslash\{c_\alpha\}_{\alpha \in S}$ containing x_0, it follows that $\varphi^{(N)}(V) = V$.

If x_0 is of type III, then T_{x_0} consists of two tangent vectors and by Lemma 10.80, φ_* fixes them both. If x_0 is of type IV, then T_{x_0} consists of a single tangent vector, and φ_* clearly fixes it. By an argument like the one above, in either case we can construct a simple domain V containing x_0 with $\varphi(V) = V$. □

The basic existence theorem for fixed points of $\varphi(T)$ in $\mathbb{P}^1(\mathbb{C}_p)$ is due to Benedetto [**13**, Proposition 1.2]:

PROPOSITION 10.98. *Let $\varphi(T) \in \mathbb{C}_p(T)$ be a nonconstant rational function. Then $\varphi(T)$ has at least two distinct fixed points in $\mathbb{P}^1(\mathbb{C}_p)$, and at least one of these is nonrepelling.*

PROOF. (Sketch) If $\varphi(T)$ has degree d, then φ has exactly $d+1$ fixed points $z_0, z_1, \ldots, z_d$ in $\mathbb{P}^1(\mathbb{C}_p)$, counting multiplicities. Since each z_i can have multiplicity at most d, at least two of the z_i must be distinct. If some z_i has multiplier $\lambda_i = 1$, it is clearly nonrepelling. Otherwise, transferring the Index Formula (see [**73**, Theorem 12.4]) from $\mathbb{C}$ to $\mathbb{C}_p$ using the fact that $\mathbb{C}_p \cong \mathbb{C}$ as fields, Benedetto concludes that

$$\sum_{i=0}^{d} \frac{1}{1-\lambda_i} = 1 .$$

By the ultrametric inequality, this cannot hold if each $|\lambda_i| > 1$. Hence $|\lambda_i| \leq 1$ for some i. □

Rivera-Letelier [**82**, Theorem B] proves an existence theorem for repelling fixed points in $\mathbb{P}^1_{\text{Berk}}$, complementing the one above:

PROPOSITION 10.99. *Let $\varphi(T) \in \mathbb{C}_p(T)$ have degree $\deg(\varphi) \geq 2$. Then $\varphi(T)$ has a repelling fixed point in $\mathbb{P}^1_{\text{Berk}}$ (which could lie in either $\mathbb{P}^1(\mathbb{C}_p)$ or $\mathbb{H}_p$).*

The existence of certain kinds of periodic points implies the existence of others. Concerning classical fixed points and cycles, Rivera-Letelier proves the following result [**81**, Corollary 3.17, p. 189]. While the statement is purely about $\mathbb{P}^1(\mathbb{C}_p)$, the proof makes crucial use of the action of φ on $\mathbb{H}_p$:

PROPOSITION 10.100. *Let $\varphi(T) \in \mathbb{C}_p(T)$ have degree $d \geq 2$. If $\varphi(T)$ has more than $3d-3$ attracting cycles in $\mathbb{P}^1(\mathbb{C}_p)$, then it has infinitely many.*

For example, $\varphi(T) = T^p$ has infinitely many attracting cycles, given by the roots of unity in $\mathbb{C}_p$ of order coprime to p.

In [**82**, Theorems A, A'], Rivera-Letelier shows:

PROPOSITION 10.101. *Let $\varphi(T) \in \mathbb{C}_p(T)$ have degree $\deg(\varphi) \geq 2$. If $\varphi(T)$ has at least two nonrepelling periodic points in $\mathbb{P}^1(\mathbb{C}_p)$ (counted with multiplicities), then it has a repelling periodic point in $\mathbb{H}_p$ and infinitely many nonrepelling periodic points in $\mathbb{P}^1(\mathbb{C}_p)$.*

Concerning indifferent periodic points, he shows ([**84**, Proposition 5.1], [**81**, Corollary 4.9]):

PROPOSITION 10.102. *Let $\varphi(T) \in \mathbb{C}_p(T)$ have degree $\deg(\varphi) \geq 2$. If φ has an indifferent periodic point in either $\mathbb{H}_p$ or $\mathbb{P}^1(\mathbb{C}_p)$, then it has infinitely many indifferent periodic points in both $\mathbb{H}_p$ and $\mathbb{P}^1(\mathbb{C}_p)$. It also has at least one repelling periodic point in $\mathbb{H}_p$. Each indifferent fixed point in $\mathbb{P}^1(\mathbb{C}_p)$ is isolated.*

Concerning inseparable periodic points in $\mathbb{H}_p$, he proves [**84**, Principal Lemma]:

PROPOSITION 10.103. *Let $\varphi(T) \in \mathbb{C}_p(T)$ have degree $d \geq 2$. Then $\varphi(T)$ has either 0, 1, or infinitely many inseparable periodic points in $\mathbb{H}_p$.*

He gives $\varphi(T) = T^p + pT^d$, with $d > p$, as an example of a polynomial with infinitely many inseparable periodic points.

Rivera-Letelier also proves theorems limiting the number of periodic cycles. A classical result of Fatou asserts that for a rational function $\varphi(T) \in \mathbb{C}(T)$ of degree $d \geq 2$, each attracting or parabolic cycle attracts at least one critical point of $\varphi(T)$, so there are at most $2d - 2$ such cycles. Any isomorphism between $\mathbb{C}_p$ and $\mathbb{C}$ takes 0 to 0 and roots of unity to roots of unity. Using this, he obtains [**81**, Theorem 1, p. 194]:

PROPOSITION 10.104. *Let $\varphi(T) \in \mathbb{C}_p(T)$ have degree $d \geq 2$. Then the number of superattracting and parabolic cycles of $\varphi(T)$ in $\mathbb{P}^1(\mathbb{C}_p)$ is at most $2d - 2$.*

A well-known result of Shishikura improves Fatou's theorem to say that over $\mathbb{C}$, there are at most $2d - 2$ nonrepelling cycles; however the example $\varphi(T) = T^d$, with d coprime to p, shows that this can fail over $\mathbb{C}_p$: the roots of unity contain infinitely many indifferent cycles.

Recall that a point $x \in \mathbb{P}^1_{\text{Berk}}$ is called *exceptional* if the union of its forward and backwards orbits is finite. The exceptional locus in $\mathbb{P}^1(\mathbb{C}_p)$ consists of at most two points. The following result (see [**84**, Theorem 4], or our Proposition 10.45) describes the exceptional locus for $\mathbb{H}_p$:

THEOREM 10.105. *Let $\varphi(T) \in \mathbb{C}_p(T)$ have degree $d \geq 2$. Then the exceptional set of $\varphi(T)$ in $\mathbb{H}_p$ consists of at most one point. It is nonempty if and only if the unique exceptional point is a repelling fixed point and after a change of coordinates, $\varphi(T)$ has good reduction at that point. In that case, the exceptional point is the only repelling periodic point of $\varphi(T)$ in $\mathbb{P}^1_{\text{Berk},\mathbb{C}_p}$.*

Finally, Rivera-Letelier classifies the cases with extreme behavior. The following result combines Theorems 1, 2, and 3 from [**84**].

THEOREM 10.106. *Let $\varphi(T) \in \mathbb{C}_p(T)$ have degree $d \geq 2$. Then the number of periodic points of $\varphi(T)$ in $\mathbb{H}_p$ is 0, 1, or ∞. Moreover:*

(A) $\varphi(T)$ has no periodic points in $\mathbb{H}_p$ if and only if $\varphi(T)$ has finitely many nonrepelling periodic points in $\mathbb{P}^1(\mathbb{C}_p)$. In that case, $\varphi(T)$ has a unique attracting fixed point $z_0 \in \mathbb{P}^1(\mathbb{C}_p)$, and all other periodic points of $\varphi(T)$ in $\mathbb{P}^1(\mathbb{C}_p)$ are repelling.

(B) $\varphi(T)$ has exactly one periodic point in $\mathbb{H}_p$ if and only if, after a change of coordinates, $\varphi(T)$ has inseparable good reduction. In that case, the unique periodic point is an inseparable exceptional fixed point, and all periodic points of $\varphi(T)$ in $\mathbb{P}^1(\mathbb{C}_p)$ are attracting.

An example of a function satisfying (A) is $\varphi(T) = (T^p - T)/p$ (see Example 10.120 below). A function satisfying (B) is $\varphi(T) = T^p$.

The proofs of the preceding fixed point theorems make use of two general results concerning subsets of $\mathbb{H}_p$ which are contracted (resp. expanded) by $\varphi(T)$. In [83, §8], Rivera-Letelier establishes the following fixed point property for subsets of $\mathbb{H}_p$ contracted by $\varphi(T)$ (see our Theorem 10.85):

PROPOSITION 10.107. *Let $\varphi(T) \in \mathbb{C}_p(T)$ be nonconstant, and let $X \subset \mathbb{H}_p$ be a connected set containing at least two points, with $X \supseteq \varphi(X)$. Then X contains either a type II fixed point of $\varphi(T)$ in $\mathbb{H}_p$ or a geodesic ray emanating from an attracting fixed point of $\varphi(T)$ in $\mathbb{P}^1(\mathbb{C}_p)$.*

In [84, Proposition 9.3], Rivera-Letelier establishes a fixed point property for subsets of $\mathbb{H}_p$ expanded by $\varphi(T)$ (compare with our Theorem 10.83). Given $V \subset \mathbb{H}_p$, denote its closure in the metric space $(\mathbb{H}_p, \rho)$ by $\text{cl}_\mathbb{H}(V)$ and denote its boundary by $\partial_\mathbb{H} V$.

PROPOSITION 10.108. *Let $\varphi(T) \in \mathbb{C}_p(T)$ be nonconstant, and let $V \subset \mathbb{H}_p$ be a connected open set for which $\text{cl}_\mathbb{H}(V) \subset \varphi(V)$ and $\varphi(\partial_\mathbb{H} V) = \partial_\mathbb{H}(\varphi(V))$. Then V contains either a type II fixed point of $\varphi(T)$ in $\mathbb{H}_p$ or a geodesic ray emanating from a repelling fixed point of $\varphi(T)$ in $\mathbb{P}^1(\mathbb{C}_p)$.*

The Fatou set and its components. For a subset $X \subset \mathbb{P}^1_{\text{Berk},\mathbb{C}_p}$, we will write $X(\mathbb{C}_p)$ for $X \cap \mathbb{P}^1(\mathbb{C}_p)$. In particular, if $X = F_\varphi$ is the Berkovich Fatou set, we have seen in Theorem 10.67 that $F_\varphi(\mathbb{C}_p)$ coincides with the classical Fatou set, defined as the set of all points $x \in \mathbb{P}^1(\mathbb{C}_p)$ for which the iterates $\varphi^{(n)}(T)$ are equicontinuous on a neighborhood $U(\mathbb{C}_p)$ of x (relative to the chordal metric $\|x, y\|$).

Recall that Rivera-Letelier works primarily with $\mathbb{P}^1(\mathbb{C}_p)$, not with $\mathbb{P}^1_{\text{Berk}}$. In [81], following Benedetto [14], he partitions $F_\varphi(\mathbb{C}_p)$ into subsets called *analytic components*. Since we are interested in $\mathbb{P}^1_{\text{Berk}}$, we will define analytic components slightly differently than Rivera-Letelier and Benedetto. Given a set $F \subset \mathbb{P}^1(\mathbb{C}_p)$ and a point $x \in F$, Rivera-Letelier defines the analytic component of x in F to be the union of the sets $X(\mathbb{C}_p)$, as X ranges over all connected open affinoids $X \subset \mathbb{P}^1_{\text{Berk}}$ for which $X(\mathbb{C}_p) \subset F$. However, we define the analytic component of x to be the open set $V \subset \mathbb{P}^1_{\text{Berk}}$ which is the union of the affinoids X above. Thus Rivera-Letelier's analytic component is our $V(\mathbb{C}_p)$.

Analytic components usually have dynamical significance, but this is not always the case. For example, if $\varphi(T)$ is a rational function with good reduction, its Julia set is $\{\zeta_{\text{Gauss}}\}$ and its Fatou set is $F_\varphi = \mathbb{P}^1_{\text{Berk}} \backslash \{\zeta_{\text{Gauss}}\}$, so $F_\varphi(\mathbb{C}_p) = \mathbb{P}^1(\mathbb{C}_p)$ and the analytic component of $F_\varphi(\mathbb{C}_p)$ containing any given $x \in \mathbb{P}^1(\mathbb{C}_p)$ is $\mathbb{P}^1_{\text{Berk}}$, which contains ζ_{Gauss}.

In [**83**], Rivera-Letelier introduces a more subtle notion of components of $F_\varphi(\mathbb{C}_p)$, which he calls *Fatou components*. Given a point $x \in F_\varphi(\mathbb{C}_p)$, he defines the Fatou component of x to be the union of all sets $X(\mathbb{C}_p)$, where X is a connected open affinoid with $X(\mathbb{C}_p) \subset F_\varphi(\mathbb{C}_p)$, satisfying the condition that

$$(10.76) \qquad \bigcup_{n=0}^{\infty} \varphi^{(n)}(X(\mathbb{C}_p)) \text{ omits at least three points of } \mathbb{P}^1(\mathbb{C}_p) \ .$$

Since we are interested in $\mathbb{P}^1_{\text{Berk}}$, we will define the Fatou component of x to be the union $U^{RL} \subset \mathbb{P}^1_{\text{Berk}}$ of the corresponding Berkovich open affinoids X. Thus, Rivera-Letelier's Fatou component is our $U^{RL}(\mathbb{C}_p)$. The fact that U^{RL} is dynamically meaningful follows from [**83**, Proposition 7.1], which says:

PROPOSITION 10.109. *Let $\varphi(T) \in \mathbb{C}_p(T)$ be a rational function, and suppose X_1, X_2 are connected open affinoids satisfying* (10.76). *If $X_1 \cap X_2$ is nonempty, then $X_1 \cup X_2$ also satisfies* (10.76).

We will now show that the Rivera-Letelier Fatou components U^{RL} coincide with the topological connected components of the Berkovich Fatou set F_φ. To see this, fix $x \in F_\varphi$, let U^{RL} be the Fatou component of x, and let U be the connected component of x in F_φ. Since each point of U^{RL} has a connected open affinoid neighborhood X satisfying (10.76), it follows from Theorem 10.56 that $U^{RL} \subseteq F_\varphi$. Since U^{RL} is connected, we have $U^{RL} \subseteq U$. Conversely, let $X \subset U$ be a connected open affinoid containing x. By Corollary 10.87, X satisfies (10.76), so $X \subset U^{RL}$. Since such sets X exhaust U, it follows that $U \subseteq U^{RL}$.

Rivera-Letelier identifies three kinds of Fatou components of particular interest: immediate basins of attraction, components of the domain of quasi-periodicity, and wandering components.

If $C \subset \mathbb{P}^1(\mathbb{C}_p)$ is an attracting periodic cycle, its *basin of attraction* $\mathcal{A}_C(\varphi)$ is the collection of all points in $\mathbb{P}^1_{\text{Berk}}$ which are topologically attracted by C. It is easy to see that each basin of attraction is an open set (see Lemma 10.41). The *immediate basin of attraction* $\mathcal{A}^0_{z_0}(\varphi)$ of a periodic point $z_0 \in C$ is the analytic component of the basin of attraction which contains z_0. It follows from [**83**, Proposition 6.1] that each immediate basin of attraction is a Fatou component (and therefore a connected component of F_φ).

A point $x \in \mathbb{P}^1(\mathbb{C}_p)$ belongs to the *domain of quasi-periodicity for* φ if there are a neighborhood $U(\mathbb{C}_p)$ of x and a sequence $n_j \to \infty$ such that $\varphi^{(n_j)}(z)$ converges uniformly to the identity on $U(\mathbb{C}_p)$. Rivera-Letelier defines the domain of quasi-periodicity to be the collection of all such x. However, for us the domain of quasi-periodicity $\mathcal{E}(\varphi)$ will be the union of the analytic components associated to that set, so Rivera-Letelier's domain of quasi-periodicity is our $\mathcal{E}(\varphi)(\mathbb{C}_p)$. By [**83**, Proposition 6.1], each analytic component of $\mathcal{E}(\varphi)$ is a Fatou component.

Note that by definition, $\mathcal{E}(\varphi)$ is open and $\varphi(\mathcal{E}(\varphi)) = \mathcal{E}(\varphi)$. It is easy to see that $\mathcal{E}(\varphi^{(n)}) = \mathcal{E}(\varphi)$ for each $n \geq 1$ [**81**, Proposition 3.9].

A point $x \in \mathbb{P}^1_{\text{Berk}}$ is *recurrent* if it belongs to the closure of its forward orbit $\{\varphi^{(n)}(x)\}_{n \geq 0}$. By definition, each point of $\mathcal{E}(\varphi)(\mathbb{C}_p)$ is recurrent. Rivera-Letelier shows that $\mathcal{E}(\varphi)(\mathbb{C}_p)$ is the *interior* of the set of recurrent points in $\mathbb{P}^1(\mathbb{C}_p)$ [**81**, Corollary 4.27].

Let $\mathcal{R}$ be the set of all recurrent points in $\mathbb{P}^1_{\text{Berk}}$, and let $\mathcal{R}^0$ be its interior. Note that by Theorem 10.59, the Berkovich Julia set J_φ is contained in $\mathcal{R}$. It follows from Proposition 10.117 below that $\mathcal{E}(\varphi) \subset \mathcal{R}^0$.

We claim that $\mathcal{E}(\varphi) = \mathcal{R}^0$, that is, $\mathcal{E}(\varphi)$ is the interior of the set of recurrent points in $\mathbb{P}^1_{\text{Berk}}$. Let $x \in \mathcal{R}^0$. Then there is a connected open affinoid neighborhood X of x with $X \subset \mathcal{R}^0$. In particular, $X(\mathbb{C}_p) \subset \mathcal{R}^0(\mathbb{C}_p) = \mathcal{E}(\varphi)(\mathbb{C}_p)$. By definition, this means X is contained in some analytic component of $\mathcal{E}(\varphi)$, so $x \in \mathcal{E}(\varphi)$. Thus $\mathcal{R}^0 \subset \mathcal{E}(\varphi)$.

A domain $\mathcal{D} \subset \mathbb{P}^1_{\text{Berk}}$ is called *wandering* if
(1) its forward images under φ are pairwise disjoint and
(2) it is not contained in the basin of attraction of an attracting cycle.

The following result, which mirrors Fatou's famous classification theorem in complex dynamics, is Rivera-Letelier's description of the classical Fatou set over $\mathbb{C}_p$ [**81**, p. 205]:

THEOREM 10.110 (Classification theorem). *Let* $\varphi(T) \in \mathbb{C}_p(T)$ *be a rational function of degree* $d \geq 2$. *Then* $F_\varphi(\mathbb{C}_p)$ *is the disjoint union of the following three sets:*

(A) *immediate basins of attraction and their preimages,*
(B) *the domain of quasi-periodicity and its preimages,*
(C) *the union of all wandering discs.*

If $\varphi(T)$ has simple reduction, then φ has no wandering discs [**81**, Corollary 4.33]. Benedetto has shown [**12**, Theorem 1.2] that if φ is defined over a field L which is a finite extension of $\mathbb{Q}_p$ and if φ has no "wild recurrent Julia critical points" (recurrent critical points belonging to $J_\varphi(\mathbb{C}_p)$, whose multiplicity is divisible by p), then φ has no wandering discs. In particular, this holds if $\deg(\varphi) \leq p$. On the other hand, Benedetto [**15**] has also given an example of a polynomial in $\mathbb{C}_p[T]$ which has a wandering disc.

Benedetto has conjectured that if $L/\mathbb{Q}_p$ is a finite extension and if $\varphi(T) \in L(T)$ has degree at least 2, then the Fatou set of $\varphi(T)$ has no wandering components. If true, this would be a non-Archimedean analogue of Sullivan's famous "No Wandering Domains" theorem in complex dynamics.

Let $W(\mathbb{C}_p)$ be the union of the wandering discs in $F_\varphi(\mathbb{C}_p)$. Not much is known about the structure of $W(\mathbb{C}_p)$ or the Fatou components corresponding to it. However, in [**81**, Lemma 4.29] Rivera-Letelier shows that the lim inf of the chordal diameters of the forward images of a wandering disc is 0.

In [**83**, Theorem A], Rivera-Letelier shows that each Fatou component associated to $W(\mathbb{C}_p)$ is a wandering domain:

PROPOSITION 10.111 (Classification of Periodic Components). *Suppose $\varphi(T) \in \mathbb{C}_p(T)$ has degree at least 2. Then any periodic Fatou component is either an immediate basin of attraction of an attracting periodic point $x_0 \in \mathbb{P}^1(\mathbb{C}_p)$ or an analytic component of the domain of quasi-periodicity.*

Basins of attraction. Rivera-Letelier gives the following description of the immediate basin of attraction of an attracting periodic point [**81**, Theorem 2, p. 196].

THEOREM 10.112 (Description of Immediate Basins of Attraction). *Let $\varphi(T) \in \mathbb{C}_p(T)$ be a rational function of degree $d \geq 2$. Let $\mathcal{A}^0_{z_0}(\varphi)$ be the immediate basin of attraction of an attracting periodic point $z_0 \in \mathbb{P}^1(\mathbb{C}_p)$. Then either*

(A) *$\mathcal{A}^0_{z_0}(\varphi)$ is an open disc $\mathcal{D}$ or*
(B) *$\mathcal{A}^0_{z_0}(\varphi)$ is a domain of* Cantor type, *meaning that its boundary $\partial \mathcal{A}^0_{z_0}(\varphi)$ is a Cantor set.*

Furthermore, the number of attracting periodic cycles whose immediate basin of attraction is of Cantor type is bounded [**81**, Proposition 4.8, p. 197]:

PROPOSITION 10.113. *Let $\varphi(T) \in \mathbb{C}_p(T)$ be a rational function of degree $d \geq 2$. Then φ has at most $d-1$ attracting cycles such that the points belonging to them have an immediate basin of attraction of Cantor type.*

Rivera-Letelier gives an explicit description of the action of $\varphi(T)$ on each immediate basin of attraction [**81**, pp. 199–200]:

PROPOSITION 10.114. *Let $\varphi(T) \in \mathbb{C}_p(T)$ have degree $d \geq 2$, and let $X \subset \mathbb{P}^1_{\text{Berk}}$ be the immediate basin of attraction of an attracting fixed point x_0. Then there is a decreasing sequence of neighborhoods $\{X_n\}_{n \in \mathbb{Z}}$ of x_0, which is cofinal in the collection of all neighborhoods of x_0, such that X_0 is a disc, $\bigcup_n X_n = X$, $\bigcap_n X_n = \{x_0\}$, and $\varphi(X_n) = X_{n+1}$ for each n.*

The domain of quasi-periodicity. Rivera-Letelier also gives an explicit description of components of the domain of quasi-periodicity [**81**, Theorem 3, p. 211].

THEOREM 10.115 (Description of the Domain of Quasi-periodicity). *Let $\varphi(T) \in \mathbb{C}_p(T)$ be a rational function of degree $d \geq 2$. Then each analytic component U of $\mathcal{E}(\varphi)$ is a strict open affinoid domain, that is, in suitable coordinates,*

$$U = \mathcal{D}(a,r)^- \setminus \bigcup_{j=1}^{n} \mathcal{D}(a_j, r_j) , \tag{10.77}$$

where $a, a_1, \ldots, a_n \in \mathbb{C}_p$ and $r, r_1, \ldots, r_n \in |\mathbb{C}_p^\times|$, with pairwise disjoint discs $\mathcal{D}(a_j, r_j)$. Moreover, each boundary point of U is a repelling periodic point.

If $n = 0$ in (10.77), then U is called a *Siegel disc*. If $n \geq 1$, then U is called an *n-Hermann ring*. Rivera-Letelier gives examples of rational functions $\varphi(T)$ with an n-Hermann ring, for each $n \geq 1$ ([**81**, Proposition 6.4, p. 225 and Proposition 6.7, p. 227]).

Since any preimage under φ of a strict open affinoid is itself a strict open affinoid (see Lemma 9.12 and its proof), each connected component of $\bigcup_{n \geq 0} \varphi^{-n}(\mathcal{E}(\varphi))$ is a strict open affinoid.

By calculus, for any function $f : \mathbb{R} \to \mathbb{R}_{>0}$ and any $x \in \mathbb{R}$,

$$\lim_{\delta \to 0} \frac{f(x)^\delta - 1}{\delta} = \ln(f(x)) .$$

Rivera-Letelier shows that if an analytic function $f(T) \in \mathbb{C}_p[[T]]$ induces an automorphism of a disc, then its iterates satisfy an analogous limit [**81**, Lemma 3.11, Proposition 3.16]:

LEMMA 10.116. *Let $f(T)$ be an analytic function, defined by a power series with coefficients in $\mathbb{C}_p$, which induces an automorphism of a disc $D(a, R)^-$. Suppose there is a $\gamma < 1$ such that $|f(z) - z|_p < \gamma R$ for all $z \in D(a, R)^-$. Then:*

(A) *There is a bounded analytic function $f_*(T)$ on $D(a, R)^-$, called the* iterated logarithm *of $f(T)$, such that for any sequence of natural numbers $\{n_k\}_{k \geq 0}$ with $|n_k|_p \to 0$, the sequence of functions $(f^{(n_k)}(T) - T)/n_k$ converges uniformly to $f_*(T)$ on $D(a, R)^-$.*

More precisely, there are a constant C and a number $\rho_0 > 0$ depending on γ (but not on f or R), such that for each $n \in \mathbb{N}$ with $|n|_p < \rho_0$ and all $z \in D(a, R)^-$,

$$\left| \frac{f^{(n)}(z) - z}{n} - f_*(z) \right|_p \leq CR |n|_p . \tag{10.78}$$

(B) *For each $z_0 \in D(a, R)^-$, we have $f_*(z_0) = 0$ iff z_0 is an indifferent periodic point of $f(T)$, and $f_*(z_0) = f'_*(z_0) = 0$ iff z_0 is parabolic. If $f_*(z_0) \neq 0$, then there is an integer $k_0 = k_0(z_0)$ such that on a suitably small neighborhood of z_0, $f^{(k_0)}(T)$ is analytically conjugate to a translation. In particular, the periodic points of f in $D(a, R)^-$ are isolated.*

In [**81**, Lemma 3.11], Rivera-Letelier shows (under the hypotheses of Lemma 10.116) that for each $w \in \mathbb{C}_p$ with $|w| < \rho_0$, there is a canonical automorphism $f^{(w)}$ of $D(a, R)^-$, which coincides with the n-fold iterate $f^{(n)} = f \circ \cdots \circ f$ when $w = n \in \mathbb{N}$. These automorphisms have the property that if $|w_1|, |w_2| < \rho_0$, then $f^{(w_1)} \circ f^{(w_2)} = f^{(w_1+w_2)}$, and for each $z \in D(a, R)^-$

$$\lim_{w \to 0} \frac{f^{(w)}(z) - z}{w} = f_*(z) .$$

He interprets this as saying that the map $w \mapsto f^{(w)}$, which interpolates the map $n \mapsto f^{(n)}$ for $n \in \mathbb{N}$, is a 'flow' attached to the 'vector field' $f_*(z)$.

Let us now consider what Lemma 10.116 says about the induced action of $f(T)$ on the Berkovich disc $\mathcal{D}(a, R)^-$. By [**81**, Lemma 3.11], one has $\sup_{z \in D(a,R)^-} |f_*(z)| \leq DR$, where $D = \max_{k \geq 1}(k\gamma^k)$. Put $B = \max(C, D)$; note that B depends only on γ. It follows from (10.78) and standard properties of non-Archimedean power series that if $D(b, r) \subset D(a, R)^-$ is a subdisc with $r \geq BR|n|$ and if $|n| < \rho_0$, then $f^{(n)}(T)$ induces an automorphism of $D(b, r)$. Thus it fixes the point $x \in \mathcal{D}(a, R)^-$ corresponding to $D(b, r)$ under Berkovich's classification theorem. Taking limits over nested sequences of discs, we see that $f^{(n)}(T)$ fixes each $x \in \mathcal{D}(a, R)^-$ with $\mathrm{diam}_\infty(x) \geq BR|n|$. Thus, for any sequence of natural numbers with $\mathrm{ord}_p(n_k) \to \infty$, the iterates $f^{(n_k)}(T)$ "freeze" larger and larger parts of $\mathcal{D}(a, R)^-$, moving the remaining points within subdiscs of radius less than $BR|n_k|$.

We can now describe the action of $\varphi(T)$ on a component of the domain of quasi-periodicity. The following is [**81**, Proposition 4.14, p. 200]:

PROPOSITION 10.117. *Let $\varphi(T) \in \mathbb{C}_p(T)$ be a rational function of degree $d \geq 2$. Let X be a connected closed affinoid contained in a component of the domain of quasi-periodicity $\mathcal{E}(\varphi)$. Choose coordinates on $\mathbb{P}^1$ so that $\infty \notin X$. Then there are a $k \geq 1$ and constants $\gamma < 1$ and $B > 0$ such that:*

(A) *$\varphi^{(k)}(T)$ induces an automorphism of X, which fixes ∂X and the main dendrite D of X.*

(B) *If we write $X \backslash (\partial X \cup D)$ as a disjoint union of discs $\mathcal{D}(a, R_a)^-$, then $\varphi^{(k)}(T)$ induces an automorphism of each $\mathcal{D}(a, R_a)^-$, and the action of $f(T) = \varphi^{(k)}(T)$ on $\mathcal{D}(a, R_a)^-$ is the one described in the discussion above, with the indicated γ and B, uniformly for all such discs.*

In particular, this shows that each point of X is recurrent and that each point $x \in X \cap \mathbb{H}_{\mathrm{Berk}}$ is fixed by some iterate $\varphi^{(n)}(T)$. Proposition 10.75, used in the proof that for $\mathbb{P}^1_{\mathrm{Berk},\mathbb{C}_p}$ the Berkovich Fatou set coincides with the Berkovich equicontinuity locus, is an immediate consequence of Proposition 10.117.

We conclude this section by giving Rivera-Letelier's formula for the number of fixed points in $\mathbb{P}^1(\mathbb{C}_p)$ belonging to a component of $\mathcal{E}(\varphi)$.

Let U be a component of $\mathcal{E}(\varphi)$, and let $k \in \mathbb{N}$ be such that $\varphi^{(k)}(U) = U$. Let $\zeta_1, \ldots, \zeta_m$ be the boundary points of U which are fixed by $\varphi^{(k)}(T)$. By Theorem 10.115, each ζ_i is a repelling fixed point of type II. For each ζ_i, we can change coordinates so that it becomes the Gauss point. Let $\tilde{f}_i(T) \in \overline{\mathbb{F}}_p(T)$ be the corresponding reduction of $\varphi^{(k)}(T)$. The tangent direction at ζ_i pointing into U corresponds to a fixed point α_i of $\tilde{f}_i(T)$ in $\mathbb{P}^1(\overline{\mathbb{F}}_p)$; let $n_i(U)$ be the multiplicity of that fixed point (that is, the order of vanishing of $\tilde{f}_i(T) - T$ at α_i). Rivera-Letelier [**81**, Proposition 5.10] shows:

PROPOSITION 10.118. *Let $\varphi(T) \in \mathbb{C}_p(T)$ have degree $d \geq 2$. Let U be a component of the domain of quasi-periodicity $\mathcal{E}(\varphi)$. Suppose that $\varphi^{(k)}(U) = U$, and let $\zeta_1, \ldots, \zeta_m$ be the points of ∂U fixed by $\varphi^{(k)}(T)$. Then each fixed point of $\varphi^{(k)}(T)$ in $U(\mathbb{C}_p)$ is indifferent, and the number of such fixed points (counting multiplicities) is exactly*

$$(10.79) \qquad 2 + \sum_{i=1}^{m}(n_i(U) - 2) \ .$$

10.10. Examples

In this section, we provide some examples illustrating the theory developed above. In Examples 10.119 and 10.124–10.126, K is arbitrary; otherwise we take $K = \mathbb{C}_p$.

EXAMPLE 10.119. Consider the polynomial $\varphi(T) = T^2$. The filled Julia set is
$$K_\varphi \ = \ \mathcal{D}(0,1) \ = \ \{z \in \mathbb{A}^1_{\text{Berk}} \ : \ [T]_z \leq 1\} \ .$$
The Julia set of φ is the Gauss point $\zeta_{\text{Gauss}} \in \mathcal{D}(0,1) \subset \mathbb{P}^1_{\text{Berk}}$. The canonical measure μ_φ in this case is a point mass supported at ζ_{Gauss}.

EXAMPLE 10.120. If $\varphi(T) = \frac{T^p - T}{p} \in \mathbb{C}_p[T]$, one can show that $K_\varphi = J_\varphi = \mathbb{Z}_p$, which is also the classical Julia set for φ. By Theorem 10.91, the canonical measure μ_φ coincides with the equilibrium distribution of $\mathbb{Z}_p$ with respect to the point ∞, which is known ([**88**, Example 4.1.24]) to be the normalized Haar measure on $\mathbb{Z}_p$.

EXAMPLE 10.121. The following example is due to Benedetto ([**14**, Example 3]). Let p be an odd prime, and choose $a \in \mathbb{C}_p$ such that $|a| = p^\epsilon > 1$, where $0 < \epsilon \leq \frac{p}{2p+2}$. If $\varphi(T) = T^2(T-a)^p \in \mathbb{C}_p[T]$, then φ does not have simple reduction, but the classical Julia set $J_\varphi(\mathbb{C}_p)$ is nonetheless empty. From Lemma 10.53, Corollary 10.60, and Theorem 10.91, we conclude that J_φ is an uncountable subset of $\mathbb{H}_p$ with the structure of a Cantor set.

Note that for a polynomial $\varphi(T) \in \mathbb{C}_p[T]$ of degree at most $p + 1$, Benedetto has shown that $J_\varphi(\mathbb{C}_p)$ is empty if and only if φ has simple reduction (Proposition 4.9 of [**14**]).

EXAMPLE 10.122. This example is due to Rivera-Letelier ([**82**, Example 6.3]). Let p be a prime, and take $\varphi(T) = (T^p - T^{p^2})/p$, with $K = \mathbb{C}_p$.

It is not hard to see that if $|a| \leq 1$ and $p^{-1/(p-1)} < r \leq 1$, then the preimage under φ of the disc $D(a, r)$ is a disjoint union of p closed discs $D(a_i, r')$ where $r' = (r/p)^{1/p} > p^{-1/(p-1)}$. Put $D = D(0, 1)$, and let $D_1, \ldots, D_p$ be the preimages of D under φ; each D_i has radius $p^{-1/p}$. Inductively, for each $m \geq 2$, $(\varphi^{(m)})^{-1}(D)$ is a disjoint union of p^m closed discs $D_{i_1,\ldots,i_m}$ of radius $r_m = p^{-(1-p^{-m})/(p-1)}$. Here we are indexing the discs by the sequences $\{i_1, \ldots, i_m\}$ with $1 \leq i_j \leq p$, and i_j is determined by the property that

$$\varphi^{(j-1)}(D_{i_1,\ldots,i_m}) \subset D_{i_j} \quad \text{for } j = 1, \ldots, m.$$

Clearly $D_{i_1,\ldots,i_m} \subset D_{i_1,\ldots,i_{m-1}}$, and $\varphi(D_{i_1,\ldots,i_m}) = D_{i_2,\ldots,i_m}$, that is, the action of φ on iterated preimages of D is conjugate to a left shift on the indices.

One has $D \supset \varphi^{-1}(D) \supset (\varphi^{(2)})^{-1}(D) \supset \cdots$ and

$$J_\varphi = \bigcap_{m=1}^\infty (\varphi^{(m)})^{-1}(D).$$

Under Berkovich's classification theorem, each $x \in J_\varphi$ corresponds to a nested sequence of discs $\{D_{i_1,\ldots,i_m}\}$, so the points of J_φ are in one-to-one correspondence with the sequences $\{i_j\}_{j\geq 1}$ with $1 \leq i_j \leq p$ for each j. Furthermore

$$\text{diam}_\infty(x) = \lim_{m\to\infty} p^{-(1-p^{-m})/(p-1)} = p^{-1/(p-1)}.$$

If the sequence of discs corresponding to x has empty intersection, then x is of type IV, while if it is nonempty, then x is of type II. Furthermore, the action of φ on J_φ is conjugate to a left shift on the index sequences, and it can be seen that x is of type II if and only if the corresponding sequence $\{i_j\}_{j\geq 1}$ is periodic. Thus, J_φ is isomorphic to a Cantor set contained in the set of points in $\mathbb{H}_{\text{Berk}}$ with $\text{diam}_\infty(x) = p^{-1/(p-1)}$.

Since $K = \mathbb{C}_p$, there are only countably many points of type II, so J_φ consists of countably many points of type II and uncountably many points of type IV, but no points of type I or III.

Let $X_p = \{1, \ldots, p\}$ and let μ_p be the probability measure on X_p which gives each element mass $1/p$. Equip the space of sequences $X_p^\mathbb{N} = \prod_{j=1}^\infty X_p$ with the product measure $\mu_p^\mathbb{N}$. The canonical measure μ_φ is obtained by transporting $\mu_p^\mathbb{N}$ to J_φ.

EXAMPLE 10.123. Let p be an odd prime, take $\varphi(T) = pT^3 + (p-1)T^2$, and let $K = \mathbb{C}_p$.

One checks easily that 0, -1, and $1/p$ are the fixed points of $\varphi(T)$ in $\mathbb{C}_p$, with $|\varphi'(0)| = 0$, $|\varphi'(-1)| = 1$, and $|\varphi'(1/p)| = p$. Thus, 0 is a superattracting fixed point, -1 is an indifferent fixed point, and $1/p$ is a repelling fixed point. It follows that 0 and -1 belong to the Fatou set F_φ, while $1/p$ belongs to the Julia set J_φ.

The reduction of $\varphi(T)$ at ζ_{Gauss} is $\widetilde{\varphi}(T) = -T^2$, so ζ_{Gauss} is a repelling fixed point, and it belongs to J_φ. As J_φ contains at least two points, it is infinite, perfect, and has uncountably many connected components. By Proposition 10.45, no point in J_φ is exceptional. This means $1/p$ and ζ_{Gauss} each have infinitely many preimages under the iterates $\varphi^{(k)}(T)$, and those preimages belong to J_φ. Hence $J_\varphi(\mathbb{C}_p)$ and $J_\varphi \cap \mathbb{H}_{\text{Berk}}$ are both infinite. Furthermore, Theorem 10.59 shows that every Berkovich neighborhood of $1/p$ contains preimages of ζ_{Gauss}, and every Berkovich neighborhood of ζ_{Gauss} contains preimages of $1/p$. In particular, $1/p$ is a limit of points in $J_\varphi \cap \mathbb{H}_{\text{Berk}}$.

The next group of examples comes from Favre and Rivera-Letelier [48], who describe a class of rational functions whose Berkovich Julia sets and canonical measures can be determined.

Identify $\mathbb{P}^1_{\text{Berk}}$ with $\mathbb{A}^1_{\text{Berk}} \cup \{\infty\}$, and for each $t \in \mathbb{R}$ let $S(t) \in \mathbb{A}^1_{\text{Berk}}$ be the point corresponding to the disc $D(0, q_v^t)$ under Berkovich's classification. Given a collection of numbers $a_1, \ldots, a_n \in K^\times$ satisfying $|a_1| < \cdots < |a_n|$ and positive integers $d_0, d_1, \ldots, d_n$, consider the rational function

$$\varphi(T) = T^{d_0} \cdot \prod_{k=1}^n \left(1 + (T/a_k)^{d_{k-1}+d_k}\right)^{(-1)^k}.$$

Then it is easy to see that $\varphi(T)$ has degree $d = d_0 + d_1 + \cdots + d_n$, and if $z \in K^\times$ and $|z| \neq |a_1|, \ldots, |a_n|$, then

$$|\varphi(z)| = |z|^{d_0} \cdot \prod_{k=1}^n \max\left(1, |z/a_k|^{d_{k-1}+d_k}\right)^{(-1)^k}.$$

Set $a_{n+1} = \infty \in \mathbb{P}^1_{\text{Berk}}$, taking $|\infty| = \infty$, and put $c_k = \prod_{j=1}^k a_k^{(-1)^k(d_{k-1}+d_k)}$. It follows from Proposition 2.19 that $\varphi(S(t)) = S(L(t))$ where $L(t)$ is the continuous, piecewise affine map defined by

$$L(t) = \begin{cases} d_0 t & \text{if } t \leq \log_v(|a_1|), \\ \log_v(|c_k|) + (-1)^k d_k t & \text{if } \log_v(|a_k|) \leq t \leq \log_v(|a_{k+1}|). \end{cases}$$

If $a_1, \ldots, a_n$ and $d_0, d_1, \ldots, d_n$ are chosen appropriately, then there will be a closed interval $I = [a, b] \subset \mathbb{R}$ and closed subintervals $I_0, \ldots, I_n \subset I$ which are pairwise disjoint except possibly for their endpoints, such that $L(t)$ maps I_j homeomorphically onto I for each $j = 0, \ldots, n$. In this situation $L^{-1}(I) = I_0 \cup \cdots \cup I_n$. Put $J_0 = S(I)$, and for each $m \geq 1$ put $J_m = S((L^{(m)})^{-1}(I))$, so $J_0 \supseteq J_1 \supseteq J_2 \cdots$. Favre and Rivera-Letelier show that $J_m = (\varphi^{(k)})^{-1}(J_0)$ and that

$$J_\varphi = \bigcap_{m=1}^\infty J_m.$$

They also show that if $\sum_{k=0}^n 1/d_k = 1$, then $J_\varphi = J_0$; on the other hand, if $\sum_{k=0}^n 1/d_k < 1$, then J_φ is a Cantor set contained in J_0.

Furthermore, if $J_\varphi = J_0$, then μ_φ is the measure on J_0 obtained by transporting the Lebesgue measure on I, normalized to have total mass 1. If J_φ is a Cantor set, then μ_φ is the measure described as follows. For each $m \geq 1$, $\left(L^{(m)}\right)^{-1}(I)$ can be written as a union of closed subintervals $I_{i_1,\ldots,i_m}$, where each i_j satisfies $0 \leq i_j \leq n$ and is determined by the property that

$$L^{(j-1)}(I_{i_1,\ldots,i_m}) \subset I_{i_j} \ .$$

Then $\mu_\varphi(S(I_{i_1,\ldots,i_m})) = d_{i_1} \cdots d_{i_m}/d^m$.

EXAMPLE 10.124 (Favre, Rivera-Letelier). Let $n \geq 2$ and take $a \in K$ with $|a| > 1$. Put $a_k = a^{k/n}$ (fixing any choice of the root) for $k = 1, \ldots, n-1$, and let $d_0 = \cdots = d_{n-1} = n$. Consider the ration function of degree n^2

$$\varphi(T) = T^n \cdot \prod_{k=1}^{n-1} \left(1 + (T/a_k)^{2n}\right)^{(-1)^k} \ .$$

Let $I = [0, \log_v(|a|)]$ and let $I_k = [\frac{k}{n}\log_v(|a|), \frac{k+1}{n}\log_v(|a|)]$ for $0 \leq k \leq n-1$. Then $L(I_k) = I$ for each k, and $\sum_{k=0}^{n-1} 1/d_k = 1$.

By the discussion above, $J_\varphi = S(I)$, and μ_φ is the transport of the normalized Lebesgue measure on I.

EXAMPLE 10.125 (Favre, Rivera-Letelier). Take $n = 2$, and choose $a \in K$ with $|a| > 1$. Put $a_1 = a^{1/2}$, $a_2 = a^{3/4}$ for any choices of the roots, and set $d_0 = 2$, $d_1 = 4$, $d_2 = 4$. Consider

$$\varphi(T) = \frac{T^2(1 + (T/a_2)^8)}{1 + (T/a_1)^6} \ ,$$

which has degree 10. Let $I = [0, \log_v(|a|)]$ and take

$$I_0 = [0, \frac{1}{2}\log_v(a)]\ , \quad I_1 = [\frac{1}{2}\log_v(a), \frac{3}{4}\log_v(a)]\ , \quad I_2 = [\frac{3}{4}\log_v(a), \log_v(a)] \ .$$

Then $L(I_0) = L(I_1) = L(I_2) = I$, and $1/2 + 1/4 + 1/4 = 1$. Again $J_\varphi = S(I)$, and μ_φ is the transport of normalized Lebesgue measure on I.

EXAMPLE 10.126 (Favre, Rivera-Letelier). Take $n = 1$, and fix $a \in K$ with $|a| > 1$. Let $d \geq 5$ be an integer, take $a_1 = a$, and put $d_0 = d-2, d_1 = 2$. Consider the function

$$\varphi(T) = \frac{T^{d-2}}{1 + (T/a)^d} \ ,$$

which has degree d. Then

$$L(t) = \begin{cases} (d-2)t & \text{if } t \leq \log_v(|a|) \ , \\ d\log_v(|a|) - 2t & \text{if } t \geq \log_v(|a|) \ . \end{cases}$$

If $I = [0, \frac{d}{2}\log_v(|a|)]$, $I_0 = [0, \frac{d}{2d-4}\log_v(|a|)]$, and $I_1 = [\frac{d}{4}\log_v(|a|), \frac{d}{2}\log_v(|a|)]$, then I_0 and I_1 are disjoint since $d \geq 5$, and $L(I_0) = L(I_1) = I$. Clearly $I_0 \cup I_1 \subset I$ and $1/d_0 + 1/d_1 < 1$. By the discussion above, J_φ is a Cantor set contained in $S(I)$.

Our final example concerns a rational function originally studied by Benedetto [**11**, p. 14]. It gives an indication of how complicated J_φ can be, in general.

EXAMPLE 10.127. Let p be a prime, take $K = \mathbb{C}_p$, and put

$$\varphi(T) = \frac{T^3 + pT}{T + p^2}.$$

Note that $\varphi(0) = 0$ and $\varphi'(0) = 1/p$ (so that $|\varphi'(0)| > 1$). Thus 0 is a repelling fixed point and it belongs to J_φ. Let $a_0 = 0$. Benedetto shows that for each $m \geq 1$, there is a point $a_m \in \mathbb{C}_p$ with $|a_m| = p^{-1/2^m}$ for which $\varphi(a_m) = a_{m-1}$. Hence, $a_m \in J_\varphi$ as well. The sequence $\{a_m\}_{m \geq 1}$ has no Cauchy subsequences, since if $m > n$, then $|a_m - a_n| = |a_m|$ by the ultrametric inequality, and $\lim_{m \to \infty} |a_m| = 1$. In particular, $\{a_m\}_{m \geq 1}$ has no limit points in $\mathbb{C}_p$, so $J_\varphi(\mathbb{C}_p)$ is not compact.

However, the point ζ_{Gauss}, which corresponds to $D(0,1)$ under Berkovich's classification theorem, is a limit point of $\{a_m\}_{m \geq 1}$ in $\mathbb{P}^1_{\text{Berk}}$. Since J_φ is compact in the Berkovich topology, we must have $\zeta_{\text{Gauss}} \in J_\varphi$. We can see this directly, by considering the reduction of $\varphi(T)$ at ζ_{Gauss}, which is $\widetilde{\varphi}(T) = T^2$. Since $\deg(\widetilde{\varphi}) > 1$, ζ_{Gauss} is a repelling fixed point for $\varphi(T)$ in $\mathbb{H}_{\text{Berk}}$, and it belongs to J_φ by Theorem 10.81.

Since $m_\varphi(\zeta_{\text{Gauss}}) = \deg(\widetilde{\varphi}) = 2$, but $\deg(\varphi) = 3$, the point ζ_{Gauss} must have a preimage under φ distinct from itself. Using Proposition 2.18, one can show that if x_0 is the point corresponding to $D(-p^2, 1/p^3)$ under Berkovich's classification theorem, then $\varphi(x_0) = \zeta_{\text{Gauss}}$. Similarly, for each $k \geq 1$, if x_k is the point corresponding to $D(-p^{2+k}, 1/p^{3+k})$, then $\varphi(x_k) = x_{k-1}$. As $k \to \infty$, the points x_k converge to the point 0 in $\mathbb{P}^1_{\text{Berk}}$. This gives an explicit example of a point in $J_\varphi(\mathbb{C}_p)$ which is a limit of points in $J_\varphi \cap \mathbb{H}_{\text{Berk}}$.

Continuing on, taking preimages of the chain of points $\{x_k\}_{k \geq 0}$, we see that each a_m is a limit of points in $J_\varphi \cap \mathbb{H}_{\text{Berk}}$. However, this is far from a complete description of J_φ. By Corollary 10.60, J_φ is uncountable, and by Theorem 10.88 each point of J_φ is a limit of repelling periodic points. The points $\{a_m\}_{m \geq 1}$ and $\{x_n\}_{n \geq 0}$ are all preperiodic but not periodic, so they are not among the repelling periodic points described in Theorem 10.88, but each one is a limit of such points. According to Bézivin's theorem [**22**, Theorem 3], each a_m is in fact a limit of repelling periodic points in $\mathbb{P}^1(K)$.

However, there are only countably many repelling periodic points in $\mathbb{P}^1(K)$. Furthermore, there are only countably many type II points in $\mathbb{P}^1_{\text{Berk},\mathbb{C}_p}$, and each repelling periodic point in $\mathbb{H}_{\text{Berk}}$ is of type II. Thus, J_φ contains uncountably many other points which we have not yet described.

10.11. Notes and further references

Much of the material in §10.2 also appears in [**5**, Appendix A].

Chambert-Loir [**35**] and Favre and Rivera-Letelier [**46**, **47**, **48**] have given independent constructions of the canonical measure on $\mathbb{P}^1_{\text{Berk}}$ attached to φ. Also, Szpiro, Tucker, and Piñeiro [**78**] have constructed a sequence of blowups of a rational map on $\mathbb{P}^1/\operatorname{Spec}(\mathcal{O}_v)$, leading to a sequence of discrete measures supported on the special fibers. When suitably interpreted in terms of Berkovich space, the weak limit of these measures gives another way of defining the canonical measure.

The proof we have given of Theorem 10.36 borrows some key ingredients from the work of Favre and Rivera-Letelier, but it is also rather different in several respects. (We make extensive use of the Arakelov-Green's function $g_\varphi(x, y)$ attached to φ, while Favre and Rivera-Letelier's proof does not use this at all.)

As with Theorem 7.52 (see the remarks in §7.10), Arakelov geometry provides another approach to global equidistribution results such as Theorem 10.24 (cf. [**35**],[**94**]).

Our treatment of Fatou and Julia theory on $\mathbb{P}^1_{\text{Berk}}$ owes a great deal to the comments of Juan Rivera-Letelier. We thank him in particular for pointing out Example 10.70, explaining the proof of Theorem 10.72 for wandering domains, and for teaching us about the theory of uniform spaces.

The potential-theoretic approach to Fatou-Julia theory in the classical complex setting began with the work of Brolin [**31**], who proved equidistribution of preimages for polynomial maps and other important facts. It was subsequently used by Lyubich [**72**] and Freire, Lopes, and Mañé [**54**] to prove equidistribution of preimages for rational maps and by Tortrat [**95**] to prove the equidistribution of periodic points. The (pluri)potential approach to complex dynamics is fundamental in higher dimensions, where the most convenient way to define the Julia set is as the support of a suitable current generalizing the canonical measure attached to a rational map in dimension one. For an overview of the pluripotential theoretic approach to complex dynamics in higher dimensions, see the papers [**50, 51, 52, 53**] by Fornaess and Sibony.

We have assumed throughout §10.5–§10.8 that $\operatorname{char}(K) = 0$. As noted in the text (e.g., Remark 10.37), one could give analogues of the results in those sections when $\operatorname{char}(K) = p$, but several of the statements (e.g., that of Proposition 10.45) would have to be modified and the proofs would be correspondingly more involved. Our decision to assume that $\operatorname{char}(K) = 0$ arose partly from our desire to illustrate the close parallels between rational dynamics on $\mathbb{P}^1(\mathbb{C})$ and on $\mathbb{P}^1_{\text{Berk}}$; some of those parallels would be less clear if we tried to deal systematically with the "pathologies" which can arise in characteristic p. As noted in the text, a number of the proofs in §10.5–§10.8 (e.g., the proofs of Corollary 10.47 and Lemma 10.54) extend verbatim to characteristic p if one uses the general form of Theorem 10.36 proved in [**48**].

An earlier version of this chapter served as the basis for §5.10 of Joe Silverman's recent book [**93**]. The treatment given here is a thoroughly revised version of that earlier work, taking into account Example 10.70 and correcting some errors in the original manuscript.

APPENDIX A

Some results from analysis and topology

In this appendix, we recall some results from analysis and topology which are used in the main text.

A.1. Convex functions

Let $-\infty \leq a < b \leq \infty$. A function $\psi : (a,b) \to \mathbb{R}$ is called *convex* (cf. [79, §2.6]) if for every $t_1, t_2 \in (a,b)$,

$$\psi((1-\lambda)t_1 + \lambda t_2) \leq (1-\lambda)\psi(t_1) + \lambda\psi(t_2)$$

for all $0 \leq \lambda \leq 1$. A convex function on (a,b) is automatically continuous. Also, if $\psi \in \mathcal{C}^2(a,b)$, then ψ is convex if and only if $\psi'' \geq 0$ on (a,b).

In the proof of Proposition 3.14(C), we use the following result:

LEMMA A.1. *Let $g : (a,b) \to \mathbb{R}$ be a function such that for each $t \in (a,b)$, both one-sided derivatives $g'_-(t) = \lim_{h \to 0^-}(g(t+h) - g(t))/h$ and $g'_+(t) = \lim_{h \to 0^+}(g(t+h) - g(t))/h$ exist. Suppose that*

(A) $g'_-(x) \leq g'_+(x)$ *for each $x \in (a,b)$ and*
(B) $g'_+(x) \leq g'_-(y)$ *for each $x, y \in (a,b)$ with $x < y$.*

Then g is convex on (a,b).

PROOF. Note that g is continuous, since the left and right derivatives exist at each point. Fix $x, y \in (a,b)$ with $x < y$, and let

$$L_{x,y}(t) = \frac{g(y) - g(x)}{y - x} \cdot (t - x) + g(x)$$

be the line through $(x, g(x))$ and $(y, g(y))$. We claim that $g(t) \leq L_{x,y}(t)$ for all $x < t < y$. Suppose to the contrary that there is a point $z \in (x,y)$ where $g(z) > L_{x,y}(z)$. Put $\alpha = (g(z) - g(x))/(z - x)$, $\beta = (g(y) - g(z))/(y - z)$. Then $\alpha > \beta$.

We claim there are a point $u \in (x,z)$ where $g'_+(u) \geq \alpha$ and a point $v \in (z,y)$ where $g'_-(v) \leq \beta$. This contradicts $g'_+(u) \leq g'_-(v)$.

First consider the interval (x,z). Let $L_{x,z}(t) = \alpha \cdot (t - x) + g(x)$ be the line through $(x, g(x))$ and $(z, g(z))$, and put $f(t) = g(t) - L_{x,z}(t)$. Then $f(x) = f(z) = 0$. We will now apply the argument in Rolle's theorem. Since f is continuous, there is a point $u \in (x,y)$ where f achieves a maximum or a minimum. If u is a minimum, then $f'_-(u) \leq 0$ and $f'_+(u) \geq 0$, so $g'_-(u) \leq \alpha$ and $g'_+(u) \geq \alpha$. If u is a maximum, then $f'_-(u) \geq 0$ and $f'_+(u) \leq 0$, so $g'_-(u) \geq \alpha$ and $g'_+(u) \leq \alpha$. Since $g'_+(z) \geq g'_-(u)$, it must be that $g'_-(u) =$

$g'_+(u) = \alpha$. In either case we have a point u where $g'_-(u) \leq \alpha$ and $g'_+(u) \geq \alpha$. This second inequality is the one we want.

Next consider the interval (z, y). By the same reasoning as before, there is a point $v \in (z, y)$ where $g'_-(v) \leq \beta$ and $g'_+(v) \geq \beta$. This time it is the first inequality that we want. $\square$

A.2. Upper and lower semicontinuous functions

Let X be a topological space. We denote by $\mathcal{C}(X)$ the space of continuous functions $f : X \to \mathbb{R}$ and by $\mathcal{C}_c(X)$ the space of compactly supported continuous functions $f : X \to \mathbb{R}$.

Recall that a real-valued function $f : X \to [-\infty, \infty)$ is *upper semicontinuous* if for each $x_0 \in X$,

$$\text{(A.1)} \qquad \limsup_{x \to x_0} f(x) \leq f(x_0) .$$

The limit superior in this context is defined by

$$\limsup_{x \to x_0} f(x) = \lim_V \sup_{x \in V \setminus \{x_0\}} f(x) = \inf_V \sup_{x \in V \setminus \{x_0\}} f(x) ,$$

where the limit (or infimum) is taken over all neighborhoods V of x. Note that while this is the standard modern usage of $\limsup_{x \to x_0} f(x)$, it is different from the usage in [**29**].

The condition (A.1) is equivalent to requiring that $f^{-1}([-\infty, b))$ be open for each $b \in \mathbb{R}$. The function f is called *strongly upper semicontinuous* if for each $x_0 \in X$,

$$\limsup_{x \to x_0} f(x) = f(x_0) .$$

If X is *compact*, then according to [**79**, Theorem 2.1.2], an upper semicontinuous function f on X is bounded above and achieves its maximum value on X.

If $f : X \to [-\infty, \infty)$ is locally bounded above, we define the *upper semicontinuous regularization* f^* of f by

$$f^*(x) = \limsup_{y \to x} f(y) .$$

See [**79**, Definition 3.4.1]. The function f^* is upper semicontinuous and $f^* \geq f$ on X. Moreover, f^* is the *smallest* function with these properties, in the sense that if g is upper semicontinuous and $g \geq f$, then $g \geq f^*$. In particular, if f is upper semicontinuous, then $f^* = f$.

Similarly, $f : X \to (-\infty, \infty]$ is *lower semicontinuous* if for each $x_0 \in X$,

$$\liminf_{x \to x_0} f(x) \geq f(x_0) ,$$

or, equivalently, if $f^{-1}((b, -\infty])$ is open for each $b \in \mathbb{R}$. It is *strongly lower semicontinuous* if for each $x_0 \in X$,

$$\liminf_{x \to x_0} f(x) = f(x_0) .$$

According to [**49**, Proposition 7.11(c)], we have the following result:

LEMMA A.2. *If $\mathcal{G}$ is a family of lower semicontinuous functions on X and if $f(x) = \sup\{g(x) : g \in \mathcal{G}\}$, then f is lower semicontinuous.*

Similarly, if $\mathcal{G}$ is a family of upper semicontinuous functions on X and if $f(x) = \inf\{g(x) : g \in \mathcal{G}\}$, then f is upper semicontinuous.

Furthermore, if X is a locally compact Hausdorff space and f is a nonnegative lower semicontinuous function on X, then by [**49**, Proposition 7.11(e)], we have
$$f(x) = \sup\{g(x) : g \in \mathcal{C}_c(X), 0 \leq g \leq f\}.$$

From this and the monotone convergence theorem, one deduces (see [**49**, Corollary 7.13]):

PROPOSITION A.3. *Let X be a locally compact Hausdorff space. If μ is a positive Radon measure and f is lower semicontinuous and bounded below on X, then*

$$(A.2) \qquad \int f\,d\mu = \sup\left\{\int g\,d\mu : g \in \mathcal{C}_c(X),\, 0 \leq g \leq f\right\}.$$

For a discussion of Radon measures, see §A.5 below.

A.3. Nets

Our primary reference for this section is [**49**, §4.3]; see also [**75**, §29].

In the theory of metric spaces, convergent sequences play a major role. For example, a real-valued function f on a metric space is continuous if and only if $x_n \to x$ implies that $f(x_n) \to f(x)$. But for a general topological space, this is not true: even if X is compact Hausdorff, there can be discontinuous functions on X for which $f(x_n) \to f(x)$ whenever $x_n \to x$. The problem is that the indexing set $\mathbb{N}$ for a sequence is too small to capture the complexities of an arbitrary topological space. To rectify the situation, one introduces the notion of a net, in which the index set $\mathbb{N}$ for sequences is replaced by an arbitrary directed set.

A *directed set* is a set A equipped with a reflexive and transitive binary relation $\geq$ with the property that, for any $\alpha, \beta \in A$, there exists $\gamma \in A$ with $\gamma \geq \alpha$ and $\gamma \geq \beta$. For example, if X is a topological space, the set of all open neighborhoods of a point $x \in X$ is a directed set, with $V \geq U$ if and only if $V \subseteq U$.

A subset B of a directed set A is called *cofinal* in A if for each $\alpha \in A$, there exists $\beta \in B$ such that $\beta \geq \alpha$. More generally, a function $h : B \to A$ is called *cofinal* if for each $\alpha_0 \in A$, there exists $\beta_0 \in B$ such that $\beta \geq \beta_0$ implies $h(\beta) \geq \alpha_0$.

If X is a set, a *net* in X is a mapping $\alpha \mapsto x_\alpha$ from a directed set A to X. We write $\langle x_\alpha \rangle_{\alpha \in A}$, or just $\langle x_\alpha \rangle$, to denote a net indexed by A. It is important to note that a sequence is the same thing as a net indexed by $\mathbb{N}$.

Let X be a topological space, and let $x \in X$. A net $\langle x_\alpha \rangle$ in X *converges* to x (written $x_\alpha \to x$) if for every neighborhood U of x, there exists $\alpha_0 \in A$ such that $x_\alpha \in U$ for all $\alpha \geq \alpha_0$. If X is a metric space with metric $\rho(x, y)$, then a net $\langle x_\alpha \rangle$ converges if and only if it is *Cauchy*: that is, for each $\varepsilon > 0$, there is an $\alpha_\varepsilon \in A$ such that $\rho(x_\alpha, x_\beta) < \varepsilon$ for all $\alpha, \beta \geq \alpha_\varepsilon$.

The following lemma is part of [**49**, Proposition 4.18]:

LEMMA A.4. *If $E \subseteq X$, then $x \in X$ belongs to the closure of E if and only if there is a net in E which converges to x.*

The following result (see [**49**, Proposition 4.19]) was alluded to above:

PROPOSITION A.5. *If X, Y are topological spaces and $f : X \to Y$, then f is continuous at $x \in X$ if and only if $f(x_\alpha) \to f(x)$ for each net $\langle x_\alpha \rangle$ converging to x.*

A topological space X is Hausdorff if and only if every net in X converges to at most one point.

A *subnet* of a net $\langle x_\alpha \rangle_{\alpha \in A}$ is a net $\langle y_\beta \rangle_{\beta \in B}$ together with a cofinal map $\beta \mapsto \alpha_\beta$ from B to A such that $y_\beta = x_{\alpha_\beta}$. Note that the map $\beta \mapsto \alpha_\beta$ is not required to be injective. A point $x \in X$ is a *cluster point* of $\langle x_\alpha \rangle$ if for every neighborhood U of x and every $\alpha \in A$, there exists $\beta \geq \alpha$ such that $x_\beta \in U$.

The following result gives an analogue of the Bolzano-Weierstrass theorem which holds in a general topological space [**49**, Proposition 4.20, Theorem 4.29]:

PROPOSITION A.6. *Let X be a topological space. Then:*

(A) *If $\langle x_\alpha \rangle_{\alpha \in A}$ is a net in X, then $x \in X$ is a cluster point of $\langle x_\alpha \rangle$ if and only if $\langle x_\alpha \rangle$ has a subnet which converges to x.*
(B) *X is compact if and only if every net in X has a convergent subnet (which happens if and only if every net in X has a cluster point).*

There is also a version of Dini's lemma for nets of functions:

LEMMA A.7. *Let X be a compact topological space, and suppose $\langle f_\alpha \rangle_{\alpha \in A}$ is a monotonically increasing or decreasing net in $\mathcal{C}(X)$. If f_α converges pointwise to a continuous function $f \in \mathcal{C}(X)$, then $f_\alpha \to f$ uniformly on X.*

PROOF. We may suppose that $\langle f_\alpha \rangle$ is monotonically increasing. Let $\varepsilon > 0$. For each $\alpha \in A$, set

$$U_\alpha = \{x \in X : f_\alpha(x) > f(x) - \varepsilon\}.$$

Then each U_α is open, $U_\alpha \supseteq U_{\alpha'}$ if $\alpha \geq \alpha'$, and $\bigcup_\alpha U_\alpha = X$. By compactness, there is an $\alpha_0 \in A$ such that $U_{\alpha_0} = X$. Thus $f(x) - \varepsilon < f_\alpha(x) \leq f(x)$ for all $x \in X$ and all $\alpha \geq \alpha_0$, so that $f_\alpha \to f$ uniformly on X. □

A.4. Measure-theoretic terminology

The following definitions and facts are taken from [**49**, §3.1].

Let X be a topological space, and let Σ be a σ-algebra on X. A *signed measure* on (X, Σ) is a function $\nu : \Sigma \to [-\infty, \infty]$, assuming at most one of the values $\pm\infty$, such that (a) $\nu(\emptyset) = 0$ and (b) if E_j is a sequence of disjoint subsets in Σ, then

$$\nu\left(\bigcup_{j=1}^{\infty} E_j\right) = \sum_{j=1}^{\infty} \nu(E_j) ,$$

with $\sum_{j=1}^{\infty} \nu(E_j)$ converging absolutely when $\nu(\bigcup_{j=1}^{\infty} E_j)$ is finite.

A *positive measure* (or, more properly, a *nonnegative measure*) is one for which $\nu : \Sigma \to [0, \infty]$.

A positive measure is *finite* if $\nu(X)$ is finite, and it is a *probability measure* if $\nu(X) = 1$.

Every signed measure on (X, Σ) can be written as a difference of two positive measures, at least one of which is finite. More precisely, one has the *Jordan Decomposition Theorem*: If ν is a signed measure, there exist unique positive measures ν^+ and ν^- on (X, Σ) such that $\nu = \nu^+ - \nu^-$ and $\nu^+ \perp \nu^-$. The latter condition means that there exists a partition of X into disjoint subsets $E^+, E^- \in \Sigma$ such that $\nu^-(E^+) = \nu^+(E^-) = 0$. The *total variation* $|\nu|$ of ν is the positive measure defined as $|\nu| = \nu^+ + \nu^-$.

A signed measure is *finite* if $|\nu|$ is finite.

If X is a topological space, then unless otherwise specified, a *measure* on X will mean a finite, signed Borel measure on X.

A.5. Radon measures

The idea behind the concept of a Radon measure is to axiomatize the properties of measures corresponding to positive linear functionals on the space of continuous functions on a locally compact Hausdorff space. Our primary reference concerning Radon measures is [**49**, §7.1].

Let X be a locally compact Hausdorff space, and let μ be a positive Borel measure on X. If E is a Borel subset of X, then μ is called *outer regular* on E if

$$\mu(E) = \inf\{\mu(U) : U \supseteq E, U \text{ open}\}$$

and *inner regular* on E if

$$\mu(E) = \sup\{\mu(E') : E' \subseteq E, E' \text{ compact}\} .$$

We say that μ is *regular* if it is both inner and outer regular on all Borel sets. We say that μ is a *Radon measure* if it is finite on all compact sets, outer regular on all Borel sets, and inner regular on all open sets.

If X is *compact* (or more generally if X is a countable union of compact sets), then μ is Radon if and only if it is regular and finite on all compact subsets of X.

More generally, a signed Borel measure is said to be a *signed Radon measure* if and only if its positive and negative parts are both Radon measures.

For many of the topological spaces one encounters in practice, every Borel measure is Radon. This is the case, for example, for any locally compact Hausdorff space in which every open set is a countable union of compact sets [**49**, Theorem 7.8]. In §A.6 below, we will see that if X is compact and if the Borel and Baire σ-algebras coincide, then every Borel measure on X is Radon. However, there are compact Hausdorff spaces which admit finite, non-Radon Borel measures (see [**49**, Exercise 7.15]).

Let $\mathcal{RM}(X)$ denote the normed vector space consisting of all signed Radon measures on X, together with the norm given by $\|\mu\| = |\mu|(X)$, and let $\mathcal{C}_0(X)$ (resp. $\mathcal{C}(X)$) denote the space of compactly supported (resp. all) continuous functions $f : X \to \mathbb{R}$.

THEOREM A.8 (Riesz Representation Theorem). *The map $\mu \mapsto \int_X f\, d\mu$ is an isometric isomorphism from $\mathcal{RM}(X)$ to $\mathcal{C}_0(X)^*$.*

COROLLARY A.9. *If X is a compact Hausdorff space, then $\mathcal{C}(X)^*$ is isometrically isomorphic to $\mathcal{RM}(X)$.*

Let μ_α be a net in $\mathcal{RM}(X)$. We say that μ_α *converges weakly* to $\mu \in \mathcal{RM}(X)$ if and only if $\int f\, d\mu_\alpha \to \int f\, d\mu$ for all $f \in \mathcal{C}_0(X)$. The corresponding topology on $\mathcal{RM}(X)$ is called the *weak topology*. (This is actually the weak* topology on $\mathcal{RM}(X)$, so we are proliferating a common abuse of terminology here. Some authors use the phrase "vague convergence" rather than "weak convergence" and call the corresponding topology the *vague topology*.)

Together with Alaoglu's theorem ([**49**, Theorem 5.18]), the Riesz Representation Theorem implies:

THEOREM A.10. *If X is a compact Hausdorff space, the unit ball*
$$\{\mu \in \mathcal{RM}(X) \,:\, \|\mu\| \leq 1\}$$
is compact in the weak topology.

As a corollary, we have:

THEOREM A.11 (Prohorov's theorem for nets). *If X is a compact Hausdorff space, then every net μ_α of Radon probability measures on X has a weakly convergent subnet.*

A.6. Baire measures

A subset E of X is called a *Baire set* if it belongs to the σ-algebra generated by the sets
$$\{x \in X \,:\, f(x) \geq \alpha\}$$
for $\alpha \in \mathbb{R}$ and $f \in \mathcal{C}_c(X)$. (This σ-algebra is called the *Baire σ-algebra* on X.) On a locally compact Hausdorff space, every Baire set is also a Borel set, but there are compact Hausdorff spaces where the class of Borel sets is

strictly larger than the class of Baire sets [**86**, §13.1, Problem 6]. A *Baire measure* on X is a measure defined on the σ-algebra of all Baire sets which is finite for each compact Baire set.

It is clear that one can replace $\{x \in X \,:\, f(x) \geq \alpha\}$ in the definition of a Baire σ-algebra by any or all of $\{x \in X \,:\, f(x) \geq \alpha\}$, $\{x \in X \,:\, f(x) > \alpha\}$, $\{x \in X \,:\, f(x) \leq \alpha\}$, and $\{x \in X \,:\, f(x) < \alpha\}$.

A useful fact about Baire sets is that if X is compact (or more generally if X is a countable union of compact sets), then every Baire measure on X is regular [**86**, Corollary 12, p. 340]. In particular:

PROPOSITION A.12. *If X is a compact Hausdorff space for which the Baire and Borel σ-algebras coincide, then every finite Borel measure on X is a Radon measure.*

A.7. The Portmanteau theorem

We make use of the following variant of the well-known "Portmanteau theorem" (Theorem 2.1 of [**23**]). We include the proof, for lack of a suitable reference for the precise statement needed.

THEOREM A.13. *Let X be a locally compact Hausdorff space, and let $\langle \mu_\alpha \rangle_{\alpha \in I}$ be a net of positive Radon measures, each of total mass $M > 0$, converging weakly to a positive Radon measure μ on X. Then:*

(A) $\mu_\alpha(A) \to \mu(A)$ *for every Borel set $A \subseteq X$ with $\mu(\partial A) = 0$.*

(B) *More generally, for every Borel set $A \subseteq X$ and every $\varepsilon > 0$, there exists $\alpha_0 \in I$ such that*

(A.3) $$\mu(A^\circ) - \varepsilon \;\leq\; \mu_\alpha(A) \;\leq\; \mu(\overline{A}) + \varepsilon$$

for all $\alpha \geq \alpha_0$.

PROOF. It suffices to prove (B), since if (B) holds and $\mu(\partial A) = 0$, then for every $\varepsilon > 0$, there exists α_0 such that for each $\alpha \geq \alpha_0$ we have

$$\mu_\alpha(A) \;\leq\; \mu_\alpha(\overline{A}) \;\leq\; \mu(\overline{A}) + \varepsilon \;=\; \mu(A) + \varepsilon \,,$$
$$\mu_\alpha(A) \;\geq\; \mu_\alpha(A^\circ) \;\geq\; \mu(A^\circ) - \varepsilon \;=\; \mu(A) - \varepsilon \,.$$

Thus $|\mu_\alpha(A) - \mu(A)| \leq \varepsilon$ for $\alpha \geq \alpha_0$, i.e., $\mu_\alpha(A) \to \mu(A)$.

To prove (B), we recall from Proposition A.3 that if f is lower semicontinuous and bounded below on X, then for all α we have

$$\int f \, d\mu_\alpha \;=\; \sup\left\{\int g \, d\mu_\alpha \,:\, g \in \mathcal{C}_c(X), 0 \leq g \leq f\right\},$$
$$\int f \, d\mu \;=\; \sup\left\{\int g \, d\mu \,:\, g \in \mathcal{C}_c(X), 0 \leq g \leq f\right\}.$$

So for any $\varepsilon > 0$ there exists $g \in \mathcal{C}_c(X)$ with $0 \leq g \leq f$ such that

$$\int g \, d\mu \;\geq\; \int f \, d\mu - \varepsilon/2 \,.$$

By weak convergence, there exists $\alpha_0 \in I$ such that for $\alpha \geq \alpha_0$ we have

$$\left| \int g \, d\mu_\alpha - \int g \, d\mu \right| \leq \varepsilon/2 \ .$$

In particular, since $0 \leq g \leq f$, for $\alpha \geq \alpha_0$ we have

(A.4) $$\int f \, d\mu_\alpha \geq \int g \, d\mu_\alpha \geq \int f \, d\mu - \varepsilon \ .$$

If $U \subseteq X$ is open, then the characteristic function χ_U of U is lower semicontinuous, and applying (A.4) gives

$$\mu_\alpha(U) = \int \chi_U \, d\mu_\alpha \geq \int \chi_U \, d\mu - \varepsilon = \mu(U) - \varepsilon$$

for all $\alpha \geq \alpha_0$.

If $B \subseteq X$ is closed, then since $\mu(X) = \lim \mu_\alpha(X) = M$, we have

$$\mu_\alpha(B) = \mu_\alpha(X) - \mu_\alpha(B^c) \leq M - \mu(B^c) + \varepsilon = \mu(B) + \varepsilon$$

for all $\alpha \geq \alpha_0$. $\square$

A.8. The one-point compactification

A reference for this section is [75, §29].

A basic result in topology is that any locally compact Hausdorff space has a *minimal compactification*. More precisely, we have the following result.

THEOREM A.14. *Let X be a locally compact Hausdorff space. Then there exists a compact Hausdorff space Y containing X such that $Y \backslash X$ consists of a single point. Moreover, if Y' is another such space, then there is a homeomorphism between Y and Y' which is the identity on X.*

The space Y given by the theorem is called the *one-point compactification* of X. As a set, we have $Y = X \cup \{\infty\}$ (disjoint union), where the symbol ∞ denotes an object which does not belong to X. The open subsets of Y are defined to be the open subsets of X, together with all sets of the form $Y \backslash C$, where $C \subset X$ is compact.

The converse of Theorem A.14 is also true, since an open subspace of a compact Hausdorff space is locally compact.

If X is already compact, then ∞ is an isolated point of Y. If X is not compact, then X is dense in Y.

For example, the one-point compactification of a locally compact field F is homeomorphic to $\mathbb{P}^1(F)$, the projective line over F. (In particular, the one-point compactification of $\mathbb{R}$ is homeomorphic to S^1, and the one-point compactification of $\mathbb{C}$ is homeomorphic to S^2.)

A.9. Uniform spaces

Uniform spaces were first introduced by André Weil, and there are several equivalent definitions in use today. For example, Bourbaki [29] defines uniform spaces in terms of "entourages", while John Tukey [97] defines them in terms of "uniform covers". There is also a definition due to Weil in terms of "pseudo-metrics". We will mostly use the uniform cover definition, though we also include a brief description of the entourage point of view. A general reference for this material is Norman Howes' book [60].

A.9.1. Uniform covers.
The definition of uniform spaces in terms of *uniform covers* is as follows.

Let X be a set. A *covering* of X is a collection of subsets of X whose union is X. A covering C *refines* a covering D, written $C < D$, if every $U \in C$ is contained in some $V \in D$. A covering C *star-refines* a covering D if for every $U \in C$, the union of all elements of C meeting U is contained in some element of D. For example, if X is a metric space and $\epsilon > 0$, the collection C of all open balls of radius $\epsilon/3$ in X star-refines the collection D of all open balls of radius ϵ. The *intersection* $C \cap D$ of two coverings is the covering $\{U \cap V \mid U \in C, V \in D\}$.

A *uniform structure* on a set X is a collection $\mathcal{C}$ of coverings of X, called the *uniform coverings*, satisfying the following two axioms:

(C1) If $C \in \mathcal{C}$ and $C < D$, then $D \in \mathcal{C}$.

(C2) Given $D_1, D_2 \in \mathcal{C}$, there exists $C \in \mathcal{C}$ which star-refines both D_1 and D_2.

It is easy to verify that in the presence of axiom (C1), axiom (C2) is equivalent to the following two axioms:

(C2') If $C_1, C_2 \in \mathcal{C}$, then $C_1 \cap C_2 \in \mathcal{C}$.

(C2'') Every $C \in \mathcal{C}$ has a star-refinement in $\mathcal{C}$.

If a collection $\mathcal{C}'$ of coverings of X satisfies (C2) but not (C1), we call it a *base* for a uniform structure. If $\mathcal{C}'$ is a base for a uniform structure, the *uniform structure generated by* $\mathcal{C}'$ is the set of all coverings of X having a refinement in $\mathcal{C}'$.

A *uniform space* is a pair $(X, \mathcal{C})$, where $\mathcal{C}$ is a uniform structure on the set X. A uniform space is called *Hausdorff* if it satisfies the following additional axiom:

(C3) For any pair x, y of distinct points in X, there exists $C \in \mathcal{C}$ such that no element of C contains both x and y.

If X is a metric space, one can define a canonical Hausdorff uniform structure on X by taking as a base for the uniform covers of X the set $\mathcal{C}' = \{C_\epsilon \mid \epsilon > 0\}$, where C_ϵ consists of all open balls of radius ϵ in X. The fact that $\mathcal{C}'$ satisfies axiom (C2) follows from the triangle inequality, which guarantees that $C_{\epsilon/3}$ is a star-refinement of C_ϵ. Axiom (C2) can thus be thought of as a version of the triangle inequality which makes sense without the presence of a metric.

If $(X, \mathcal{C})$ is a uniform space, then for $x \in X$ and $C \in \mathcal{C}$, one defines $B(x, C)$, the "ball of size C around x", to be the union of all elements of C containing x. By definition, the *uniform topology* on $(X, \mathcal{C})$ is the topology for which $\{B(x, C) \mid C \in \mathcal{C}\}$ forms a fundamental system of neighborhoods of $x \in X$. In other words, a subset $U \subseteq X$ is open in the uniform topology if and only if for each $x \in U$, there exists a cover $C \in \mathcal{C}$ with $B(x, C) \subseteq U$. If X is a metric space, then $B(x, C_{\epsilon/2})$ contains each open ball of radius $\epsilon/2$ about x and is contained in the open ball of radius ϵ about x, and thus the metric topology on X coincides with the uniform topology on X. The uniform topology on $(X, \mathcal{C})$ is Hausdorff iff $(X, \mathcal{C})$ satisfies axiom (C3).

A basic result in the theory of uniform spaces is that if X is a compact Hausdorff topological space, then there is a *unique* uniform structure on X compatible with the given topology [**60**, Theorem 2.7]. A base for the uniform structure on X is given by the collection of all finite coverings of X by open subsets. It is not hard to see that if $\mathcal{B}$ is any base for the topology on X, a base for the uniform structure on X is also given by the collection $\mathcal{C}_\mathcal{B}$ of all finite coverings of X by elements of $\mathcal{B}$.

If X is a topological space and Y is a uniform space, a collection $\mathcal{F}$ of continuous functions $f : X \to Y$ is called *equicontinuous* if and only if for every point $x \in X$ and every uniform covering $\mathcal{C}$ of Y, there is an open neighborhood U of x such that $f(U)$ is contained in some element of $\mathcal{C}$ for every $f \in \mathcal{F}$ (using Axiom (C2″), this definition is easily seen to be equivalent to the one in [**60**, p. 366]). In this definition, it suffices to consider a base of uniform coverings for Y.

If X is a topological space and Y is a compact uniform space, then a collection $\mathcal{F}$ of continuous functions $f : X \to Y$ is equicontinuous if and only if for every point $x \in X$ and every finite covering $\mathcal{C}$ of Y by open sets belonging to some base for the topology on Y, there is an open neighborhood U of x such that $f(U)$ is contained in some element of $\mathcal{C}$ for every $f \in \mathcal{F}$.

A.9.2. The entourage point of view. The definition of a uniform structure in terms of entourages (cf. Bourbaki [**29**]) is as follows.

DEFINITION A.15. A *uniform structure* $\mathcal{U}$ on a set X is a collection of subsets V of $X \times X$, called *entourages*, which form a filter and satisfy the following additional axioms:

(E1) Each $V \in \mathcal{U}$ contains the diagonal in $X \times X$.
(E2) If $V \in \mathcal{U}$, then $V' := \{(x, y) \in X \times X \mid (y, x) \in V\} \in \mathcal{U}$.
(E3) If $V \in \mathcal{U}$, then there exists $W \in \mathcal{U}$ such that $W \circ W \subset V$, where
$$W \circ W := \{(x, y) \in X \times X \mid \exists z \in X \text{ with } (x, z) \in W \text{ and } (z, y) \in W\}.$$

A *uniform space* is a set X together with a uniform structure on X.

Every metric space has a canonical uniform structure generated by the entourages $V_\epsilon = \{x, y \in X \mid d(x, y) < \epsilon\}$ for $\epsilon > 0$, and every uniform space is a topological space by declaring that the open neighborhoods of a point $x \in X$ are the sets of the form $U[x] := \{y \in X \mid (x, y) \in U\}$ for $U \in \mathcal{U}$.

The definitions of a uniform structure in terms of uniform covers or entourages are equivalent. The connection between the two definitions is as follows. Given a uniform space $(X, \mathcal{U})$ in the entourage sense, define a covering C of X to be a uniform cover if there exists an entourage $U \in \mathcal{U}$ such that for each $x \in X$, there is a $V \in C$ with $U[x] \subseteq V$. Conversely, given a uniform space $(X, \mathcal{C})$ in the uniform cover sense, define the entourages to be the supersets of $\bigcup_{V \in C} V \times V$ as C ranges over the uniform covers of X. These two procedures are inverse to one another and furnish an equivalence between the two notions of uniform space.

A.10. Newton polygons

In this section we present selected facts from the theory of power series over a complete and algebraically closed non-Archimedean field K. See [1, 26, 85] for further discussion.

Fix a number $q_v > 1$, and write $\log_v(t)$ for the logarithm to the base q_v. Let $|z|$ be the absolute value on K, and let $\operatorname{ord}_v(z) = -\log_v(|z|)$ be the corresponding valuation.

We first recall the theory of Newton polygons. Given a nonzero power series $f(T) = \sum_{n=0}^\infty a_n T^n \in K[[T]]$, the Newton polygon of f is defined to be the lower convex hull of the collection of points $\{(k, \operatorname{ord}_v(a_k))\}_{k \geq 0}$ in $\mathbb{R}^2$. If a_m is the first nonzero coefficient, the Newton polygon is deemed to have an initial side above $(m, \operatorname{ord}_v(a_m))$ with slope $-\infty$ and projection on the x-axis of length m. If $f(T)$ is a polynomial of degree N, the Newton polygon is viewed as having a terminal side beginning at $(N, \operatorname{ord}_v(a_N))$, with slope ∞ and projection on the x-axis of infinite length.

The *radius of convergence* of $f(T)$ is the number
$$R = \sup\{0 \leq r \in \mathbb{R} : \lim_{n \to \infty}(|a_n| r^n) = 0\} .$$

Clearly $R \geq 0$; we will only be interested in power series for which $R > 0$. In that case, $f(T)$ has either $D(0, R)^-$ or $D(0, R)$ as its *disc of convergence*, the set of all $z \in K$ for which $f(z)$ converges. If $R \in |K^\times|$, it converges on $D(0, R)$ if and only if $\lim_{n \to \infty}(|a_n| R^n) = 0$. If $R = \infty$ (for instance, if $f(T)$ is a polynomial), we take $D(0, \infty) = K$. If $\varphi(T) \in K(T)$ is a rational function defined at $T = 0$ and if $f(T)$ is the Taylor expansion of $\varphi(T)$ about 0, then R is the minimum of the absolute values of the poles of $\varphi(T)$, and $f(T)$ has disc of convergence $D(0, R)^-$. If α is any point in the disc of convergence, then $f(T)$ can be rearranged as a power series in $K[[T - \alpha]]$, with the same radius and disc of convergence.

The Newton polygon of f may either have infinitely many sides, each of which has slope less than $\log_v(R)$, or it may have finitely many sides and a terminal ray of slope $\log_v(R)$. If f converges on $D(0, R)$ and $R \in |K^\times|$, then its Newton polygon has a terminal side with slope $\log_v(R)$, and finitely many points $(n, \operatorname{ord}_v(a_n))$ lie on it. In that case, if $(N, \operatorname{ord}_v(a_N))$ is the rightmost point lying on the terminal side, the segment ending at that point is considered to be the final side of finite length, and the terminal ray is deemed to begin at that point.

The Newton polygon determines the absolute values of the roots of f:

PROPOSITION A.16. *Let K be a complete and algebraically closed non-Archimedean field, and let $f(T) \in K[[T]]$ be a power series with radius of convergence $R > 0$.*

Then the absolute values of the roots of $f(T)$ in its disc of convergence are determined by the sides of the Newton polygon having finite projection on the x-axis, and if there is a side with slope m and finite projection length ℓ, then $f(T)$ has exactly ℓ roots β for which $\log_v(|\beta|) = m$.

PROOF. First suppose $f(T)$ is a monic polynomial of degree $N \geq 1$, and let $\beta_1, \ldots, \beta_N \in K$ be its roots, listed with multiplicities. We can assume they are ordered in such a way that

$$|\beta_1| = \cdots = |\beta_{\ell_1}| \quad < \quad |\beta_{\ell_1+1}| = \cdots = |\beta_{\ell_1+\ell_2}|$$
$$< \cdots < \quad |\beta_{\ell_1+\cdots+\ell_{n-1}+1}| = \cdots = |\beta_{\ell_1+\cdots+\ell_n}|.$$

Let $S_{N-k}(x_1, \ldots, x_N)$ be the $(N-k)^{\text{th}}$ symmetric polynomial. Then $a_k = \pm S_{N-k}(\beta_1, \ldots, \beta_N)$ for $k = 0, \ldots, N-1$, and $a_N = 1$. By the ultrametric inequality, we have $|a_k| \leq \prod_{i=k+1}^{N} |\beta_i|$, and equality holds if $k = 0$ or $k = \ell_1 + \cdots + \ell_j$ for some j, since in that situation there is a unique term of maximal absolute value in the sum defining S_{N-k}. It follows that the Newton polygon has vertices at the points $(a_k, \text{ord}_v(a_k))$ for which

$$k \in \{0, \ell_1, \ell_1 + \ell_2, \ldots, \ell_1 + \cdots + \ell_n = N\}$$

and the segment between the vertices corresponding to $k = \ell_1 + \cdots + \ell_{j-1}$ and $k = \ell_1 + \cdots + \ell_j$ has slope $m_j = \log_v(|\beta_{\ell_1+\cdots+\ell_j}|)$ (we take $\ell_0 = 0$). This segment has projection length ℓ_j, and $f(T)$ has exactly ℓ_j roots β for which $\log_v(\beta) = m_j$.

Multiplying a polynomial by a constant $C \neq 0$ shifts its Newton polygon vertically by $\log_v(|C|)$, so the proposition holds for arbitrary polynomials.

Finally, if $f(T)$ is a power series and $0 < r \leq R$ is a number belonging to $|K^\times|$ such that $f(T)$ converges in $D(0, r)$, then the Weierstrass Preparation Theorem [**26**, Theorem 1, p. 201] shows that $f(T)$ has a factorization

$$f(T) = F(T) \cdot u(T),$$

where $F(T) \in K[T]$ is a polynomial whose roots all belong to $D(0, r)$ and $u(T) = 1 + \sum_{k=1}^{\infty} c_k T^k$ is a power series having no roots in $D(0, r)$, such that $|c_k| r^k < 1$ for each $k \geq 1$. (Such a power series is called a *unit power series for $D(0, r)$*, because it is a unit in the ring of power series converging on $D(0, r)$.) If $\deg(F) = N$, an easy computation using the ultrametric inequality shows that the part of the Newton polygon of f in the half-plane $x \leq N$ coincides with the Newton polygon of F and that each vertex of the Newton polygon of f in the half-plane $x > N$ lies above the ray beginning at $(N, \text{ord}_v(a_N))$ with slope $\log_v(r)$. Hence the proposition holds in general. □

The next result describes the image of a disc under a mapping defined by a power series.

COROLLARY A.17. *Let $f(T) = \sum_{k=0}^{\infty} a_k T^k \in K[[T]]$ be a nonconstant power series with radius $R > 0$. Given $0 < r < R$, put $s = \sup_{k \geq 1}(|a_k|r^k)$. Then*
$$f(D(0,r)) = D(a_0, s) , \quad f(D(0,r)^-) = D(a_0, s)^- .$$
Furthermore, there is an r_0 with $0 < r_0 \leq R$, such that for each $0 < r \leq r_0$:

(A) *If $a_1 \neq 0$, then*
$$f(D(0,r)) = D(a_0, |a_1|r) , \quad f(D(0,r)^-) = D(a_0, |a_1|r)^- .$$

(B) *If $a_1 = 0$ and $m \geq 2$ is the least index such that $a_m \neq 0$, then*
$$f(D(0,r)) = D(a_0, |a_m|r^m) , \quad f(D(0,r)^-) = D(a_0, |a_m|r^m)^- .$$

PROOF. It suffices to deal with the case of a closed disc, since each open disc $D(0,r)^-$ is the union of the discs $D(0,t)$ with $t < r$. Without loss of generality, we can also assume that $a_0 = 0$.

Since $f(T)$ is nonconstant, $s := \sup_{k \geq 1}(|a_k|r^k) > 0$. As $r < R$, we have $\lim_{k \to \infty} |a_k|r^k = 0$, and there is an index k for which $s = |a_k|r^k$; let ℓ be the largest such index. Then the Newton polygon of $f(T)$ lies on or above the line through $(\ell, \text{ord}_v(a_\ell))$ with slope $\log_v(r)$, and $(\ell, \text{ord}_v(a_\ell))$ is the last vertex on that line.

By the ultrametric inequality, for each $z \in D(0,r)$ we have $f(z) \in D(0,s)$. Conversely, if $w \in D(0,R)$, set
$$f_w(T) = f(T) - w .$$
Since $|w| \leq s = |b_\ell|r^\ell$, the Newton polygon of $f_w(T)$ lies on or above the line through $(\ell, \text{ord}_v(a_\ell))$ with slope $\log_v(r)$, and for $x \geq \ell$ it coincides with the Newton polygon of $f(T)$. Hence the Newton polygon of $f_w(T)$ has a segment with slope at most $\log_v(r)$ of finite projection length on the x-axis. By Proposition A.16, $f_w(T)$ has a root $\beta \in D(0,r)$.

We next prove (A). If the Newton polygon of $f(T)$ has a side with finite length beginning at $(1, \text{ord}_v(a_1))$, choose r_0 so that $\log_v(r_0)$ is less than the slope of that side. If there is no such side, then the Newton polygon has a terminal ray beginning at $(1, \text{ord}_v(a_1))$ with slope $\log_v(R)$. In this case, let r_0 be any number with $0 < r_0 < R$. Suppose $0 < r \leq r_0$. By construction $\max_{k \geq 2}(|a_k|r^k) < |a_1|r$, so $s = |a_1|r$.

The proof of (B) is similar. □

If $\varphi(T) \in K(T)$ is a nonconstant rational function, then by expanding $\varphi(T)$ as a power series and applying Corollary A.17, we obtain the following description of the mapping properties of a rational function:

COROLLARY A.18. *Let $\varphi(T) \in K(T)$ have degree $N \geq 1$. Write $\varphi(T) = P(T)/Q(T)$, where $P(T)$ and $Q(T)$ are coprime polynomials. Assume φ has*

no poles in the disc $D(\alpha, r)$, and write
$$P(T) = b_0 + b_1(T-\alpha) + \cdots + b_N(T-\alpha)^N,$$
$$Q(T) = c_0 + c_1(T-\alpha) + \cdots + c_N(T-\alpha)^N.$$
Put $s = (\max_{1 \leq k \leq N} |b_k| r^k)/|c_0|$. Then

(A.5) $\quad \varphi(D(\alpha, r)) = D(\varphi(\alpha), s), \quad \varphi(D(\alpha, r)^-) = D(\varphi(\alpha), s)^-.$

PROOF. Without loss of generality we can assume that $c_0 = 1$ and $\alpha = 0$. In this situation $u(T) := 1/Q(T)$ can be expanded as a unit power series for the disc $D(0, r)$. Write
$$f(T) = P(T) \cdot u(T) = \sum_{k=0}^{\infty} a_k T^k \in K[[T]].$$
Put $S = \sup_{k \geq 1} |a_k| r^k$; an easy computation with the ultrametric inequality shows that $S = \max_{1 \leq k \leq N} |b_k| r^k = s$. The result now follows from Corollary A.17. □

Applying the theory of Newton polygons in reverse, we have

COROLLARY A.19. *Let $g(T) \in K[T]$ be a polynomial. Suppose $g(T)$ has no zeros in a disc $D(a, r)$. Take $z \in D(a, r)$. Then $|g(z) - g(a)| < |g(a)|$, and $|g(z)| = |g(a)|$.*

PROOF. Write $g(T) = b_0 + b_1(T-a) + \cdots + b_N(T-a)^N$, where $b_0 = g(a)$. Since $g(T)$ has no zeros in $D(a, r)$, each segment of its Newton polygon has slope greater than $\log_v(r)$. This means that for each $k \geq 1$, the point $(k, \mathrm{ord}_v(b_k))$ lies strictly above the line through $(0, \mathrm{ord}_v(b_0))$ with slope $\log_v(r)$. Hence, for each $z \in D(a, r)$ and each $k \geq 1$, we have $|b_k(z-a)^k| < |b_0|$. By the ultrametric inequality, $|g(z) - b_0| < |b_0|$ and $|g(z)| = |b_0|$. □

When $\mathrm{char}(K) = 0$, there is a non-Archimedean version of Rolle's theorem. Our exposition follows [85, p. 316], where the proof is given in the case $K = \mathbb{C}_p$.

Given $f(T) = \sum_{k=0}^{\infty} a_k T^k \in K[[T]]$, let $f'(T) = \sum_{k=0}^{\infty} k a_k T^{k-1}$ be its formal derivative. It is easy to see that if $f(T)$ converges on a disc $D(0, r)^-$, then $f'(T)$ converges on that disc as well.

PROPOSITION A.20 (Non-Archimedean Rolle's Theorem). *Let K be a complete and algebraically closed non-Archimedean field of characteristic 0. Let $\tilde{K}$ be the residue field of K. If $\mathrm{char}(\tilde{K}) = 0$, put $\rho_K = 1$; if $\mathrm{char}(\tilde{K}) = p > 0$, put $\rho_K = |p|^{1/(p-1)} < 1$.*

Suppose $f(T) = \sum_{k=0}^{\infty} a_k T^k \in K[[T]]$ converges on a disc $D(0, r)^-$. If $f(T)$ has two or more roots in $D(0, \rho_K r)^-$, then $f'(T)$ has at least one root in $D(0, r)^-$.

REMARK A.21. The hypothesis that $\text{char}(K) = 0$ is needed so that $\rho_K > 0$ when $\text{char}(\tilde{K}) = p > 0$.

PROOF. If $f(T)$ has a multiple root $\alpha \in D(0, \rho_K r)^-$, then trivially α is a root of $f'(T)$ in $D(0, r)^-$ and we are done.

Otherwise, let $\alpha \neq \beta$ be roots of $f(T)$ in $D(0, \rho_K r)^-$, chosen in such a way that $|\beta - \alpha|$ is minimal among all such pairs of roots. After expanding $f(T)$ about α, we can assume that $\alpha = 0$ and $a_0 = 0$. We also have $a_1 \neq 0$, since α is not a multiple root.

Since $\beta \neq 0$ is a root of $f(T)$ with least positive absolute value, the Newton polygon of $f(T)$ has a segment of slope $\log_v(|\beta|)$ beginning at $(1, \text{ord}_v(a_1))$ with finite projection length on the x-axis. Let $(n, \text{ord}_v(a_n))$ be the right endpoint of that segment.

If $\text{char}(\tilde{K}) = 0$, then $\text{ord}_v(k) = 0$ for each integer $k \geq 1$, and the Newton polygon of $f'(T)$ is obtained by translating the Newton polygon of $f(T)$ left by one unit. Trivially the Newton polygon of $f'(T)$ has a segment of slope $\log_v(|\beta|) < \log_v(r)$ with finite projection length on the x-axis, so by Proposition A.16, $f'(T)$ has a root in $D(0, \rho_K r)^- = D(0, r)^-$.

If $\text{char}(\tilde{K}) = p > 0$, note that $(\text{ord}_v(a_n) - \text{ord}_v(a_1))/(n-1) = \log_v(|\beta|)$, so
$$\left|\frac{a_1}{a_n}\right| = |\beta|^{n-1} < (\rho_K r)^{n-1}.$$
Write $n = p^\nu m$, where m is coprime to p. If $\nu = 0$, trivially $(n-1)/(p-1) \geq \nu$. If $\nu \geq 1$, then
$$\frac{n-1}{p-1} = \frac{p^\nu m - 1}{p-1} \geq \frac{p^\nu - 1}{p-1} = p^{\nu-1} + \cdots + p + 1 \geq \nu.$$
In either case $\rho_K^{n-1} = |p|^{(n-1)/(p-1)} \leq |p^\nu| = |n|$, so
$$\left|\frac{a_1}{na_n}\right| < r^{n-1}.$$

Hence the line segment between the points $(0, \text{ord}_v(a_1))$, $(n-1, \text{ord}_v(na_n))$ has slope $m < \log_v(r)$. This segment lies on or above the Newton polygon of $f'(T)$, and $(0, \text{ord}_v(a_1))$ is the leftmost vertex of that Newton polygon, so the leftmost side of the Newton polygon has slope $m_1 \leq m < \log_v(r)$. Since the radius of convergence of $f'(T)$ is at least r, the leftmost side must have finite projection length on the x-axis. By Proposition A.16, $f'(T)$ has a root in $D(0, r)^-$. □

APPENDIX B

$\mathbb{R}$-trees and Gromov hyperbolicity

In this appendix, we present a self-contained discussion of $\mathbb{R}$-trees, including the connection with Gromov 0-hyperbolicity. Our main purpose is to provide a rigorous foundation for some of the topological properties of $\mathbb{P}^1_{\text{Berk}}$ which we use throughout this book. A secondary goal is to illustrate the central role played in the theory of $\mathbb{R}$-trees by the Gromov product, which is closely related to the generalized Hsia kernel (the fundamental potential kernel on $\mathbb{P}^1_{\text{Berk}}$).

Although most of the results here are known, our presentation of this material is not completely standard. However, we have been influenced by [**77**, §2.2], [**30**, Chapter III.H], [**38**, Chapter 2], and [**45**, Chapter 3]. Another standard reference for the theory of $\mathbb{R}$-trees is [**57**].

B.1. Definitions

A *geodesic segment* in a metric space (X, d) is the image $|\alpha| = \alpha([a,b])$ of an isometric embedding $\alpha : [a,b] \to X$ of a (possibly degenerate) real interval $[a,b]$ into X. We say that α is a geodesic from x to y, or that α *joins* x and y, if $\alpha(a) = x$ and $\alpha(b) = y$. The points x and y are called the *endpoints* of α. By abuse of terminology, we sometimes identify a geodesic segment $|\alpha|$ with the parametrizing map α.

An *arc* (or *path*[1]) from x to y is a continuous injective map $f : [a,b] \to X$ with $f(a) = x$ and $f(b) = y$. Again, we will sometimes identify an arc f with its image in X. The metric space X is called *arcwise connected* if every two points can be joined by an arc and *uniquely arcwise connected* if there is a unique arc between every two points.

A *geodesic space* is a metric space X in which any two points x, y can be joined by a geodesic segment. In general, there could be many different geodesic segments joining two given points of X.

An $\mathbb{R}$-*tree* is a metric space (X, d) such that for any two points $x, y \in X$, there is a *unique* arc from x to y and this arc is a geodesic segment. We will denote this geodesic segment by $[x, y]$.

A *rooted* $\mathbb{R}$-*tree* is a triple (X, d, ζ) consisting of an $\mathbb{R}$-tree (X, d) together with the choice of a reference point $\zeta \in X$ called the *root*.

---

[1] We use the terms *arc* and *path* interchangeably in this book, although in standard usage the two concepts are not the same. Similarly, we will use the terms "arcwise connected" and "path-connected" interchangeably.

A point x in an $\mathbb{R}$-tree X is said to be *ordinary* if the complement $X\setminus\{x\}$ has two connected components. A point which is not ordinary will be called a *branch point* of X. If $X\setminus\{x\}$ has just one connected component, the branch point x is also called an *endpoint* of X. A *finite $\mathbb{R}$-tree* is an $\mathbb{R}$-tree which is compact and has only finitely many branch points.

B.2. An equivalent definition of $\mathbb{R}$-tree

Recall that we defined an $\mathbb{R}$-tree to be a metric space X such that for any two points $x, y \in X$, there is a unique arc from x to y and this arc is a geodesic segment. Our goal in this section is to prove:

PROPOSITION B.1. *A metric space X is an $\mathbb{R}$-tree if and only if:*
  (A) *For any points $x, y \in X$, there is a unique geodesic segment $[x, y]$ joining x and y.*
  (B) *For any points $x, y, z \in X$, if $[x, y] \cap [y, z] = \{y\}$, then $[x, z] = [x, y] \cup [y, z]$.*

Before beginning the proof, we note the following simple facts:

LEMMA B.2. *If X is a metric space satisfying conditions* (A) *and* (B) *of Proposition B.1, then for any $x, y, z \in X$, we have $[x, z] \subseteq [x, y] \cup [y, z]$.*

PROOF. Given $x, y, z \in X$, let v be the first point where $[x, y]$ and $[z, y]$ intersect; more formally, if $[x, y] = \alpha([a, b])$ and $[z, y] = \beta([a', b'])$, then let $v = \alpha(c) = \beta(c')$, where

$$c = \inf\{d \in [a, b] : \alpha(d) \in |\beta|\}, \quad c' = \inf\{d' \in [a', b'] : \beta(d') \in |\alpha|\}.$$

Then $[x, v] = \alpha([a, c]) \subseteq [x, y]$ and $[v, z] = \beta([a', c']) \subseteq [y, z]$, and we have $[x, v] \cap [v, z] = \{v\}$. Thus $[x, z] = [x, v] \cup [v, z] \subseteq [x, y] \cup [y, z]$. □

COROLLARY B.3. *If X is a metric space satisfying conditions* (A) *and* (B) *of Proposition B.1, then for any points $x_0, x_1, \ldots, x_n \in X$ we have*

$$[x_0, x_n] \subseteq \bigcup_{i=1}^{n} [x_{i-1}, x_i].$$

PROOF. This follows from the lemma, together with induction on n. □

We can now prove Proposition B.1:

PROOF OF PROPOSITION B.1 (cf. [**38**, Lemma 2.1, Proposition 2.3]).
It is easily verified that an $\mathbb{R}$-tree, as we have defined it, satisfies conditions (A) and (B) of the proposition. So it suffices to prove that a metric space satisfying (A) and (B) is an $\mathbb{R}$-tree. Let X be such a metric space, let $x, y \in X$, and let $f : [a, b] \to X$ be an arc from x to y. If $[x, y]$ denotes the unique geodesic segment from x to y and $|f|$ denotes the image of f, we claim that $[x, y] \subseteq |f|$. Given this claim, the result follows, since then $f^{-1}([x, y])$ is a connected subset of $[a, b]$ containing a and b, so it is equal to $[a, b]$, and thus $|f| = [x, y]$.

To prove the claim, note that $|f|$ is a compact (hence closed) subset of X, so it suffices to prove that every point of $[x,y]$ is within distance ε of $|f|$ for every $\varepsilon > 0$. Given $\varepsilon > 0$, the collection $\{f^{-1}(B(x,\varepsilon/2)) : x \in |f|\}$ is an open covering of the compact space $[a,b]$, so by Lebesgue's covering lemma, there exists $\delta > 0$ such that any subset of $[a,b]$ of diameter less than δ is contained in some element of this cover. Choose a partition $a = t_0 < \cdots < t_n = b$ of $[a,b]$ so that $t_i - t_{i-1} < \delta$ for all i. Then $f([t_{i-1}, t_i]) \subseteq B(x_i, \varepsilon)$ for some $x_i \in |f|$, and in particular $d(f(t_{i-1}), f(t_i)) < \varepsilon$ for $i = 1, \ldots, n$. The triangle inequality then implies that $[f(t_{i-1}), f(t_i)] \subseteq B(x_i, \varepsilon)$ for $i = 1, \ldots, n$. The claim now follows from Corollary B.3. □

B.3. Geodesic triangles

Let X be an $\mathbb{R}$-tree. For every $x, y, z \in X$, let $\Delta = \Delta_{xyz} \subseteq X$ be the union of the three geodesic segments joining these three points:

$$\Delta = [x,y] \cup [y,z] \cup [x,z].$$

The proof of Lemma B.2 shows that Δ is a *tripod*, i.e., a finite $\mathbb{R}$-tree consisting of three (possibly degenerate) edges joined at a common vertex v.

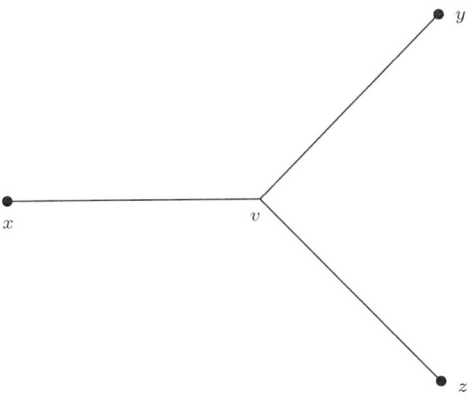

FIGURE B.1. A nondegenerate tripod with vertices x, y, z and center v.

Moreover, x, y, and z are the endpoints of the tripod. The point v of the tripod where the three edges meet is called the *center*; it is defined by the condition that

$$v = [x,y] \cap [y,z] \cap [x,z].$$

Note that $[x,y] \cap [y,z] = \{y\}$ if and only if $y = v$.

Following Gromov, we now establish a partial converse to the observation that every triangle in an $\mathbb{R}$-tree is a tripod, thereby obtaining a useful characterization of $\mathbb{R}$-trees.

Let X be a geodesic space. A *geodesic triangle* in X consists of a set $\{x, y, z\}$ of points of X (the "vertices" of the triangle), together with geodesic segments $|\alpha_{xy}|, |\alpha_{yz}|, |\alpha_{xz}|$ (the "edges" of the triangle) joining these

points pairwise. We usually identify a geodesic triangle with the underlying subspace $\Delta = \Delta_{xyz}$ of X given by
$$\Delta = |\alpha_{xy}| \cup |\alpha_{yz}| \cup |\alpha_{xz}|$$
and call x, y, z the *vertices* of Δ. For example, a tripod T is a geodesic triangle whose vertices are the endpoints of T.

If Δ_i is a geodesic triangle in the metric space X_i for $i = 1, 2$, a *simplicial map* $f : \Delta_1 \to \Delta_2$ is a continuous map taking vertices of Δ_1 to vertices of Δ_2 and edges of Δ_1 to edges of Δ_2.

DEFINITION B.4. A geodesic space X is called *strongly hyperbolic* if every geodesic triangle Δ in X is isometric to a tripod T_Δ via a bijective simplicial map $\chi_\Delta : \Delta \to T_\Delta$.

PROPOSITION B.5. *A geodesic space X is strongly hyperbolic if and only if it is an $\mathbb{R}$-tree.*

PROOF. We have already seen that an $\mathbb{R}$-tree is strongly hyperbolic. For the converse, we need to show that if X is strongly hyperbolic, then (A) there is a unique geodesic segment $[x, y]$ joining any two given points x and y and (B) if $[x, y] \cap [y, z] = \{y\}$, then $[x, z] = [x, y] \cup [y, z]$.

Suppose $\alpha, \beta : [a, b] \to X$ are geodesic segments joining x and y. Take any point $z \in |\alpha|$, let $c = \alpha^{-1}(z) \in [a, b]$, and consider the geodesic triangle Δ with vertices x, y, z formed by the geodesic segments $\alpha_1 = \alpha|_{[a,c]}$, $\alpha_2 = \alpha|_{[c,b]}$, and β. (So x, y, z correspond to a, b, c, respectively.) Choose an isometry $\chi : \Delta \to T_\Delta$ to a tripod T_Δ, and let d_T denote the metric on T_Δ. Let v be the point of X mapping to the center of the tripod T_Δ. If $\ell(\gamma)$ denotes the length of a geodesic segment γ, then we have
$$\begin{aligned}\ell(\alpha) &= \ell(\alpha_1) + \ell(\alpha_2) = d_T(\chi(x), \chi(z)) + d_T(\chi(z), \chi(y)) \\ &\geq d_T(\chi(x), \chi(y)) = \ell(\beta),\end{aligned}$$
with equality if and only if $\chi(z) = \chi(v)$, i.e., if and only if $z = v$. Since $\ell(\alpha) = \ell(\beta)$, it follows that $z = v$. But then $z \in |\beta|$, and since z was arbitrary, it follows that $|\alpha| \subseteq |\beta|$. By symmetry, we see that $|\alpha| = |\beta|$, which proves (A).

For (B), it suffices to note that if $[x, y] \cap [y, z] = \{y\}$, then the center of the (degenerate) tripod Δ_{xyz} must be y, and therefore $[x, z] = [x, y] \cup [y, z]$. □

It is sometimes useful to think of strong hyperbolicity in terms of quadrilaterals rather than triangles. If X is a geodesic space, $x_1, \ldots, x_4 \in X$, and α_{ij} is a geodesic segment joining x_i and x_j for each unordered pair (i, j) with $i \neq j$, we call the union of the segments α_{ij} a *geodesic quadrilateral* with vertices x_i. Thus a geodesic quadrilateral has four vertices and six sides ("four sides plus two diagonals").

There is also an obvious 4-endpoint analogue of a tripod, which we call a *quadripod*: it is a finite $\mathbb{R}$-tree obtained by gluing two tripods along a common edge:

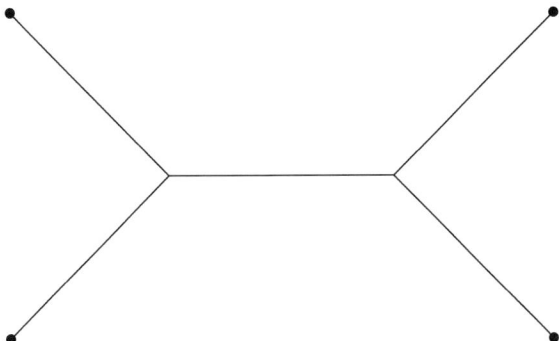

FIGURE B.2. A nondegenerate quadripod.

LEMMA B.6. *A geodesic space X is strongly hyperbolic if and only if every geodesic quadrilateral Q in X is isometric to a quadripod via a simplicial isometry.*

PROOF. The "if" direction is clear. For the reverse implication, suppose X is strongly hyperbolic. By Proposition B.5, there is a *unique* geodesic segment joining any two points of X. For any points $x, y, z, w \in X$, let Δ_{xyz} be the geodesic triangle corresponding to x, y, z, and choose an isometry $\chi_{xyz} : \Delta_{xyz} \to T_{xyz}$ to a tripod T_{xyz}. Let v be the point of X mapping to the center of T_{xyz}, and choose an isometry $\chi_{xvw} : \Delta_{xvw} \to T_{xvw}$. By Proposition B.5, there is a unique arc between x and v, and thus the point $v' \in X$ mapping to the center of T_{xvw} must belong to $[v, x]$. It follows easily that the isometries χ_{xyz} and χ_{xvw} can be "glued" to give an isometry from the geodesic quadrilateral with vertices x, y, z, w to a quadripod. $\square$

B.4. The Gromov product

Let (X, d) be a metric space, and for $x, y, z \in X$, define the *Gromov product* $(x|y)_z$ of x and y relative to z by the formula

$$(\text{B.1}) \qquad (x|y)_z \;=\; \frac{1}{2}\left(d(x,z) + d(y,z) - d(x,y)\right) .$$

To avoid any confusion, we emphasize that throughout this section and the next, $d(x, y)$ is an abstract metric on X, and when $X = \mathbb{P}^1_{\text{Berk}}$ it need not be the distance function on $\mathbb{P}^1_{\text{Berk}}$ introduced in §2.7.

As motivation for this definition, note that if (T, d_T) is a tripod with vertices x, y, z and center v, then $d_T(x, v) = (y|z)_x$, $d_T(y, v) = (x|z)_y$, and $d_T(z, v) = (x|y)_z$.

REMARK B.7. The Gromov product has the following "physical" interpretation. If we think of a tripod T as a (purely resistive) electrical network with resistances given by the lengths of the edges, then the Gromov product $(x|y)_z$ is the potential at y when one unit of current enters the circuit at x

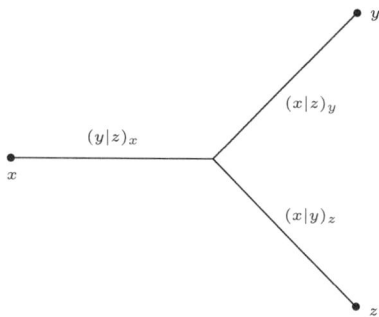

FIGURE B.3. The Gromov product on a tripod.

and exits at z (with reference voltage 0 at z). In the language of metrized graphs (§3.1), this means that $(x|y)_z = j_z(x,y)$ (cf. (3.4)).

DEFINITION B.8. A metric space X is called 0-*hyperbolic* (in the sense of Gromov) if for every $x, y, z, w \in X$, we have

(B.2) $$(x|y)_w \geq \min\{(x|z)_w, (y|z)_w\} \ .$$

In other words, X is 0-hyperbolic if and only if for any $w \in X$ and any real number $q > 1$, the function $d_w : X \times X \to \mathbb{R}_{\geq 0}$ given by

$$d_w(x,y) = q^{-(x|y)_w}$$

satisfies the *ultrametric inequality*

$$d_w(x,z) \leq \max\{d_w(x,y), d_w(y,z)\} \ .$$

REMARK B.9. More generally, if $\delta \geq 0$ is a real number, a metric space X is called δ-*hyperbolic* if for every $x, y, z, w \in X$, we have

(B.3) $$(x|y)_w \geq \min\{(x|z)_w, (y|z)_w\} - \delta \ ,$$

and X is called *Gromov hyperbolic* if it is δ-hyperbolic for some δ. See [**30**] for a detailed discussion of Gromov hyperbolicity, including a number of equivalent definitions.

Fix a base point ζ in the metric space X. Then X is called 0-*hyperbolic relative to* ζ if for every $x, y, z \in X$, we have

(B.4) $$(x|y)_\zeta \geq \min\{(x|z)_\zeta, (y|z)_\zeta\} \ .$$

By definition, X is 0-hyperbolic if and only if it is 0-hyperbolic relative to ζ for *all* $\zeta \in X$. We now prove the *a priori* stronger fact that X is 0-hyperbolic if and only if it is 0-hyperbolic relative to ζ for *some* $\zeta \in X$. First, we need the following lemma.

LEMMA B.10. *The metric space X is 0-hyperbolic with respect to the basepoint $\zeta \in X$ if and only if for any points $x, y, z, w \in X$, we have*

(B.5) $$(x|y)_\zeta + (z|w)_\zeta \geq \min\{(x|z)_\zeta + (y|w)_\zeta, (x|w)_\zeta + (y|z)_\zeta\} \ .$$

PROOF. If (B.5) holds, then since $(z|\zeta)_\zeta = 0$, taking $w = \zeta$ shows that X is 0-hyperbolic with respect to ζ.

Conversely, suppose X is 0-hyperbolic with respect to ζ. The desired inequality (B.5) is symmetric with respect to x and y and with respect to w and z. Let
$$M = \max\{(x|z)_\zeta, (x|w)_\zeta, (y|z)_\zeta, (y|w)_\zeta\}\,.$$
If $M = (x|w)_\zeta$, interchange the roles of w and z. If $M = (y|z)_\zeta$, interchange the roles of x and y. If $M = (y|w)_\zeta$, interchange the roles both of x and y and of w and z. In this way, we may assume without loss of generality that $M = (x|z)_\zeta$.

By 0-hyperbolicity, we have
$$(x|y)_\zeta \geq \min\{(x|z)_\zeta, (y|z)_\zeta\} = (y|z)_\zeta$$
and
$$(z|w)_\zeta \geq \min\{(x|z)_\zeta, (x|w)_\zeta\} = (x|w)_\zeta\,.$$
Thus
$$(x|y)_\zeta + (z|w)_\zeta \geq (x|w)_\zeta + (y|z)_\zeta\,,$$
which proves (B.5). □

For points $x, y, z, w \in X$, define their "cross ratio" relative to ζ to be
$$(x; y; z; w)_\zeta = (x|y)_\zeta + (z|w)_\zeta - (x|z)_\zeta - (y|w)_\zeta\,.$$

LEMMA B.11. *The cross ratio $(x;y;z;w)_\zeta$ is independent of ζ.*

PROOF. This is a direct calculation using the definition of the Gromov product; indeed
$$(x;y;z;w)_\zeta = \frac{1}{2}\Big(d(x,z) + d(y,w) - d(x,y) - d(z,w)\Big)\,.$$
□

REMARK B.12. There is an analogue of Lemma B.11 for metrized graphs: if Γ is a metrized graph, then the quantity
$$j_\zeta(x,y) + j_\zeta(z,w) - j_\zeta(x,z) - j_\zeta(y,w)$$
is independent of the choice of $\zeta \in \Gamma$; this follows from Proposition 3.3(C) and a straightforward calculation.

REMARK B.13. The cross ratio on the "Berkovich hyperbolic space" $(\mathbb{H}_{\text{Berk}}, \rho)$, and on its natural extension to $\mathbb{P}^1_{\text{Berk}}$, plays an important role in the work of Favre and Rivera-Letelier (see [**47**, §6.3]).

By Lemma B.11, we may write $(x;y;z;w)$ instead of $(x;y;z;w)_\zeta$. As a consequence of Lemmas B.11 and B.10, we obtain:

COROLLARY B.14. *If a metric space X is 0-hyperbolic with respect to some basepoint $\zeta \in X$, then it is 0-hyperbolic with respect to every basepoint $\zeta \in X$ (i.e., it is 0-hyperbolic).*

PROOF. By Lemma B.10, X is 0-hyperbolic with respect to ζ if and only if

(B.6) $$\min\{(x;y;z;w)_\zeta, (x;y;w;z)_\zeta\} \geq 0 \,.$$

But by Lemma B.11, condition (B.6) is independent of ζ. □

Consequently, X is 0-hyperbolic if and only if there are a point $\zeta \in X$ and a real number $q > 1$ for which the function $d_\zeta : X \times X \to \mathbb{R}_{\geq 0}$ given by

$$d_\zeta(x,y) = q^{-(x|y)_\zeta}$$

satisfies the ultrametric inequality.

We now come to the main result of this section, which says that a geodesic space is an $\mathbb{R}$-tree if and only if it is 0-hyperbolic. The proof we give is adapted from [**77**, §2.2] (see also [**30**, Chapter III.H]). Before giving the proof, we note the following easily verified facts. If Δ is a geodesic triangle in a metric space X, there is (up to isomorphism) a unique pair (T_Δ, χ_Δ) consisting of a tripod T_Δ and a surjective simplicial map $\chi_\Delta : \Delta \to T_\Delta$ whose restriction to each edge of Δ is an isometry. Moreover, χ_Δ is an *isometry* if and only if χ_Δ is *injective*.

PROPOSITION B.15. *If X is a geodesic space, then the following are equivalent*:

(A) *X is strongly hyperbolic.*
(B) *X is 0-hyperbolic.*
(C) *X is an $\mathbb{R}$-tree.*

PROOF. In view of Proposition B.5, it suffices to prove (A) $\Leftrightarrow$ (B). Suppose X is strongly hyperbolic, and let Q be any geodesic quadrilateral with vertices x, y, z, w. Since Q is isometric to a quadripod, it suffices to note upon inspection that a quadripod is 0-hyperbolic.

Conversely, suppose X is 0-hyperbolic. Let Δ be a geodesic triangle in X with vertices x, y, z, and let $\chi = \chi_\Delta : \Delta \to T_\Delta$ be the canonical piecewise isometry to the tripod T_Δ. Suppose $w, w' \in \Delta$ are points for which $\chi(w) = \chi(w')$. Without loss of generality, we can assume that $w \in \overline{xy}$ and $w' \in \overline{xz}$, where $\overline{xy}$ (resp. $\overline{xz}$) denotes the edge of Δ connecting x and y (resp. x and z). Since $\chi(w) = \chi(w')$, we have $d(x, w) = d(x, w') = r$ for some real number $r \geq 0$. Let v be the center of T_Δ; since $\chi(w) = \chi(w')$ is situated between $\chi(x)$ and v and since $d(\chi(x), v) = (y|z)_x$, we have $(y|z)_x \geq r$. Also, we have $(w|y)_x = (w'|z)_x = r$. As X is 0-hyperbolic, we see that

$$(w|w')_x \geq \min\{(w|y)_x, (y|w')_x\} \geq \min\{(w|y)_x, (y|z)_x, (z|w')_x\} \geq r \,.$$

On the other hand, by the definition of the Gromov product we have

$$(w|w')_x = r - \frac{1}{2}d(w, w') \,,$$

from which we obtain $d(w, w') \leq 0$. It follows that $w = w'$, and therefore χ is injective. As previously noted, this implies that χ is an isometry. □

B.5. ℝ-trees and partial orders

Let $(T, \leq)$ be a partially ordered set (poset) satisfying the following two axioms:

(P1) T has a unique maximal element ζ, called the *root* of T.
(P2) For each $x \in T$, the set $S_x = \{z \in T \ : \ z \geq x\}$ is totally ordered.

We say that T is a *parametrized rooted tree* if there is a function $\alpha : T \to \mathbb{R}_{\geq 0}$ with values in the nonnegative reals such that:

(P3) $\alpha(\zeta) = 0$.
(P4) α is *order-reversing*, in the sense that $x \leq y$ implies $\alpha(x) \geq \alpha(y)$.
(P5) The restriction of α to any full totally ordered subset of T gives a bijection onto a real interval.

(A totally ordered subset S of T is called *full* if $x, y \in S$, $z \in T$, and $x \leq z \leq y$ implies $z \in S$.)

Although inspired by it, this definition differs slightly from the one given in [**45**, Chapter 3].

Given two elements x and y of a parametrized rooted tree T, one can use the parametrization α to define the *least upper bound* $x \vee y$ of x and y. Indeed, since the set $S_x = \{z \in T \ : \ z \geq x\}$ is full and totally ordered, axiom (P5) provides a bijection from S_x to a real interval $[0, R_x]$. The intersection of S_x and $S_y = \{z \in T \ : \ z \geq y\}$ contains the root ζ, so it is a nonempty subset of S_x. Now define $x \vee y$ to be the unique point belonging to

$$S_x \cap \alpha^{-1}\left(\sup_{z \in S_x \cap S_y} \alpha(z)\right).$$

It is straightforward to show that $x \leq x \vee y$, $y \leq x \vee y$, and that if $z \in T$ is any point with $x \leq z$ and $y \leq z$, then $x \vee y \leq z$.

We now describe a one-to-one correspondence between rooted ℝ-trees and parametrized rooted trees.

Given a rooted ℝ-tree (T, d, ζ), one defines a partial order on T by declaring that $x \leq y$ iff y belongs to the unique geodesic segment $[x, \zeta]$ between x and ζ and a parametrization α on T by setting $\alpha(x) = d(x, \zeta)$. It is easy to check that the resulting structure satisfies axioms (P1)–(P5) for a parametrized rooted tree.

Conversely, given a parametrized rooted tree $(T, \leq)$ with root ζ and parametrization α, define a distance function d on T by

$$d(x, y) = \alpha(x) + \alpha(y) - 2\alpha(x \vee y).$$

It is easy to check that d defines a metric on T and that there exists a geodesic segment joining any two points $x, y \in T$, namely the concatenation of the segments $\sigma_x = \{z \ : \ x \leq z \leq x \vee y\}$ and $\sigma_y = \{z \ : \ y \leq z \leq x \vee y\}$.

PROPOSITION B.16. *(T, d) is an ℝ-tree.*

PROOF. First note that the Gromov product on (T, d) with respect to ζ satisfies the identity
$$(x|y)_\zeta = \alpha(x \vee y).$$
We claim that T is 0-hyperbolic with respect to ζ. This is equivalent to the assertion that for every $x, y, z \in T$, we have

(B.7) $\qquad\qquad \alpha(x \vee y) \geq \min\{\alpha(x \vee z), \alpha(y \vee z)\}.$

By axiom (P2), the set $\{x \vee y, x \vee z, y \vee z\}$ is totally ordered. Since (B.7) is symmetric in x and y, we may assume without loss of generality that $y \vee z \leq x \vee z$. But then
$$x \vee y \leq (x \vee y) \vee z = x \vee (y \vee z) \leq x \vee (x \vee z) = x \vee z.$$
Since α is order-reversing, it follows that $\alpha(x \vee y) \geq \alpha(x \vee z)$ and that $\alpha(y \vee z) \geq \alpha(x \vee z)$, proving (B.7), and thus that T is 0-hyperbolic with respect to ζ. By Corollary B.14, T is 0-hyperbolic. The result now follows from Proposition B.15. □

One verifies easily that the maps we have just defined
$$\{\text{rooted } \mathbb{R}\text{-trees}\} \rightleftarrows \{\text{parametrized rooted trees}\}$$
are inverse to one another, and therefore there is an equivalence between the notion of a rooted $\mathbb{R}$-tree and that of a parametrized rooted tree.

B.6. The weak and strong topologies

In this section, we mostly follow the exposition of [**45**, Chapter 5].

Let (X, d) be an $\mathbb{R}$-tree. The *strong topology* on X is defined to be the topology induced by the metric d. There is another natural and very useful topology on X, called the *weak topology*. In order to define it, we need some preliminary definitions.

Let $p \in X$. We define an equivalence relation on $X \setminus \{p\}$ by declaring that $x \sim y$ if and only if the unique geodesic segments from p to x and p to y share a common initial segment. An equivalence class is called a *tangent vector* (or *tangent direction*) at p, and the set of tangent vectors at p is called the (projectivized) *tangent space* of p, denoted $T_p(X)$. A point x in the equivalence class $\vec{v}$ is said to *represent* the tangent vector $\vec{v}$.

If $\vec{v} \in T_p(X)$ is a tangent vector at a point $p \in X$, define
$$\mathcal{B}_p(\vec{v})^- = \{x \in X \setminus \{p\} : x \text{ represents } \vec{v}\}.$$
By definition, the *weak topology* is the topology generated by all sets of the form $\mathcal{B}_p(\vec{v})^-$ (i.e., such sets form a subbase of open sets for the weak topology).

In [**40**], the weak topology is called the *observer's topology*, since one can think of $\mathcal{B}_p(\vec{v})^-$ as the set of all points which can be seen by an observer standing at p and looking in the direction $\vec{v}$. A fundamental system of open neighborhoods of a point $p \in X$ is given by sets of the form

(B.8) $\qquad\qquad \mathcal{B}_{q_1}(\vec{v}_1)^- \cap \cdots \cap \mathcal{B}_{q_k}(\vec{v}_k)^-$

with $p \in \mathcal{B}_{q_i}(\vec{v}_i)^-$ for all i. A set of the form (B.8) can be thought of as the set of all points which can simultaneously be seen by k observers situated at points of $X \setminus \{p\}$ and looking in the direction of p.

It is easy to see that the weak topology on a metric tree is always Hausdorff. As the names suggest, there is a relation between the weak and strong topologies. The following lemma and its proof are taken from [**45**, Proposition 5.5]. In the statement, we write X_{strong} for X endowed with the strong topology and X_{weak} for X endowed with the weak topology.

LEMMA B.17. *The strong topology on an $\mathbb{R}$-tree (X, d) is stronger (i.e, finer) than the weak topology. Equivalently, the natural map $\psi : X_{\text{strong}} \to X_{\text{weak}}$ which is the identity on the underlying sets is continuous.*

PROOF. Let x_n be a sequence in X which converges in the strong topology to x. Consider X as a rooted tree with x as the root, and let $\leq$ be the corresponding partial order. If $x_n \not\to x$ in the weak topology, then there must be a point $y < x$ such that $x_n \leq y$ for infinitely many n. But $d(x_n, x) \geq d(y, x)$ for such n, a contradiction. $\square$

There is also a relationship between connectedness in the strong and weak topologies. The following lemma and its proof are taken from [**40**, Proposition 1.7].

LEMMA B.18. *A subset of X is connected in the weak topology if and only if it is connected in the strong topology.*

PROOF. We need to show that X_{strong} and X_{weak} have the same connected subsets. A connected subset of X_{strong} is arcwise connected, and therefore it is both connected and arcwise connected in the weak topology by Lemma B.17.

Conversely, let C be a connected subset of X_{weak}, and assume that it is not connected in the strong topology. Then it is not convex, so that there exist points $x, y \in C$ and $z \in X \setminus C$ with $z \in [x, y]$. Let $\vec{v}$ be the tangent vector at z representing the path $[z, x]$. Then $V = \mathcal{B}_z(\vec{v})^-$ and $U = X \setminus (V \cup \{z\})$ are disjoint open sets covering C. Moreover, both V and U are nonempty, as $x \in V$ and $y \in U$. This contradicts the connectedness of C. $\square$

As a consequence, it is easy to see that elements of $T_p(X)$ are in one-to-one correspondence with the connected components of $X \setminus \{p\}$ in either the strong or weak topology.

We now show that arcwise connectedness in the weak topology is the same as arcwise connectedness in the strong topology. (This is not explicitly pointed out in [**45**] or [**40**].)

PROPOSITION B.19. *An injective mapping $i : [0, 1] \to X$ is continuous in the weak topology if and only if it is continuous in the strong topology.*

PROOF. Since the canonical map $\psi : X_{\text{strong}} \to X_{\text{weak}}$ is continuous, it is clear that if i is strongly continuous, then it is weakly continuous. To show the converse, suppose i is weakly continuous, and let Λ be the image of i. Let $x = i(0), y = i(1)$, and let σ be the unique geodesic segment from x to y. We claim that $\Lambda \subseteq \sigma$. Indeed, if not, then there exists $z \in \Lambda \backslash \sigma$. Since $z \notin \sigma$, it is easy to see that x and y are contained in the same connected component of $X \backslash \{z\}$. Let t be the unique point of $[0,1]$ for which $i(t) = z$. Since $i : [0,1] \to \Lambda$ is a continuous bijection between compact Hausdorff spaces (where Λ is endowed with the weak subspace topology), it must be a homeomorphism. Thus 0 and 1 lie in the same connected component of $i^{-1}(z) \subseteq [0,1] \backslash \{t\}$. This contradiction proves our claim that $\Lambda \subseteq \sigma$. It follows that $\psi^{-1}(\Lambda)$ is compact in the *strong* topology. The inverse of the map $\psi^{-1} \circ i : [0,1] \to \psi^{-1}(\Lambda)$ is $i^{-1} \circ \psi$, which is a continuous bijection between compact Hausdorff spaces, and therefore is a homeomorphism. Thus $\psi^{-1} \circ i$ is continuous, i.e., i is continuous in the strong topology. $\square$

COROLLARY B.20. *An $\mathbb{R}$-tree X is uniquely arcwise connected in its weak topology.*

COROLLARY B.21. *A subset of X is arcwise connected in the weak topology if and only if it is arcwise connected in the strong topology.*

APPENDIX C

A Brief overview of Berkovich's theory

In this appendix, we recall some definitions and results from Berkovich's general theory of analytic spaces. The primary references for this material are Berkovich's book [16] and his papers [17] and [19]; see also [39] and [42]. For the most part, we do not give proofs, although we do provide a self-contained proof of the compactness of Berkovich affinoids which is different from the one in [16].

All rings considered in this appendix will be commutative rings with identity.

C.1. Motivation

In rigid analytic geometry, one works with analogues of complex analytic spaces where the field $\mathbb{C}$ of complex numbers is replaced by a complete non-Archimedean field k. To deal with the fact that the canonical topology on an analytic manifold X over k is totally disconnected — which makes the naive notion of "analytic continuation" pathological — John Tate introduced rigid analytic spaces, whose building blocks are affinoid spaces of the form $\mathrm{Max}(\mathcal{A})$, where $\mathcal{A}$ is an affinoid algebra over K and $\mathrm{Max}(\mathcal{A})$ is the space of maximal ideals in $\mathcal{A}$. (Intuitively, an affinoid space is the common zero locus of a family of analytic functions on a closed polydisc.) Tate dealt with the problems which arise from the topological pathologies of X by introducing a certain *Grothendieck topology* on $\mathrm{Max}(\mathcal{A})$. A number of years later, Vladimir Berkovich had the insight that one could instead look at the larger topological space $\mathcal{M}(\mathcal{A})$ consisting of all bounded multiplicative seminorms on $\mathcal{A}$, together with a natural topology. There is a continuous embedding $\mathrm{Max}(\mathcal{A}) \hookrightarrow \mathcal{M}(\mathcal{A})$ having dense image, but the space $\mathcal{M}(\mathcal{A})$ has many extra points which allow one to recover the same coherent sheaf theory as in Tate's theory, but using actual topological spaces, as opposed to just a Grothendieck topology. One can glue the affinoid spaces $\mathcal{M}(\mathcal{A})$ together to obtain a global analytic space X^{an} functorially associated to any "reasonable" rigid analytic space X/k. In particular, every scheme X of finite type over k has a Berkovich analytification X^{an}. The analytic spaces in Berkovich's theory are very nice from a topological point of view. For example, Berkovich affinoids and the analytifications of projective varieties over k are compact Hausdorff spaces and are locally arcwise connected. Moreover, the analytification of a scheme X is arcwise connected if and only if X is connected in the Zariski topology, and the dimension of X^{an} is equal to the Krull dimension of X. Finally, the presheaf of analytic functions on a Berkovich analytic space is in fact a sheaf, and thus one has a good theory of "analytic continuation" on such spaces.

C.2. Seminorms and norms

A *seminorm* on a ring $\mathcal{A}$ is a function $[\cdot] : \mathcal{A} \to \mathbb{R}_{\geq 0}$ with values in the set of nonnegative reals such that for every $f, g \in \mathcal{A}$, the following axioms hold:[1]

(S1) $[0] = 0;\ [1] \leq 1$.
(S2) $[f + g] \leq [f] + [g]$.
(S3) $[f \cdot g] \leq [f] \cdot [g]$.

A seminorm $[\cdot]$ defines a topology on $\mathcal{A}$ in the usual way, and this topology is Hausdorff if and only if $[\cdot]$ is a *norm*, meaning that $[f] = 0$ if and only if $f = 0$. A seminorm $[\cdot]$ on a ring $\mathcal{A}$ is called *non-Archimedean* if

(S2') $[f + g] \leq \max([f], [g])$,

and it is called *multiplicative* if $[1] = 1$ and for all $f, g \in \mathcal{A}$, we have

(S3') $[f \cdot g] = [f] \cdot [g]$.

A *normed ring* is a pair $(\mathcal{A}, \|\cdot\|)$ consisting of a ring $\mathcal{A}$ and a norm $\|\cdot\|$. It is called a *Banach ring* if $\mathcal{A}$ is complete with respect to this norm. Any ring may be regarded as a Banach ring with respect to the *trivial* norm for which $\|0\| = 0$ and $\|f\| = 1$ for $f \neq 0$.

A seminorm $[\cdot]$ on a normed ring $(\mathcal{A}, \|\cdot\|)$ is called *bounded* if

(S4) there exists a constant $C > 0$ such that $[f] \leq C\|f\|$ for all $f \in \mathcal{A}$.

For a Banach ring, it is well known (see for example [**49**, Proposition 5.2]) that boundedness is equivalent to continuity. Moreover:

LEMMA C.1. *If $[\cdot]$ is multiplicative, then condition (S4) is equivalent to:*

(S4') $[f] \leq \|f\|$ *for all $f \in \mathcal{A}$*.

PROOF. Since $[f^n] \leq C\|f^n\|$, we have $[f] \leq \sqrt[n]{C}\|f\|$ for all $n \geq 1$. Passing to the limit as n tends to infinity yields the desired result. □

C.3. The spectrum of a normed ring

Let $(\mathcal{A}, \|\cdot\|)$ be a (commutative) Banach ring. We define a topological space $\mathcal{M}(\mathcal{A})$, called the *spectrum* of $\mathcal{A}$, as follows. As a set, $\mathcal{M}(\mathcal{A})$ consists of all bounded multiplicative seminorms $[\cdot]$ on $\mathcal{A}$. The topology on $\mathcal{M}(\mathcal{A})$ is defined to be the weakest one for which all functions of the form $[\cdot] \to [f]$ for $f \in \mathcal{A}$ are continuous.

It is useful from a notational standpoint to denote points of $X = \mathcal{M}(\mathcal{A})$ by a letter such as x and to denote the corresponding bounded multiplicative

---

[1] As Brian Conrad pointed out, it is desirable to allow the zero function on the zero ring to be a norm. This is why in axiom (S1) we allow $[1] = 0$. If $[1] = 0$, then (S1)–(S3) imply that $[f] = 0$ for all $f \in \mathcal{A}$; otherwise (S1)–(S3) imply that $[1] = 1$.

seminorm by $[\cdot]_x$. With this terminology, the topology on X is generated by the collection of open sets of the form

$$U(f,\alpha) = \{x \in X : [f]_x < \alpha\},$$
$$V(f,\alpha) = \{x \in X : [f]_x > \alpha\}$$

for $f \in A$ and $\alpha \in \mathbb{R}$.

Equivalently, one may define the topology on $\mathcal{M}(\mathcal{A})$ as the *topology of pointwise convergence*: a net $\langle x_\alpha \rangle$ in $\mathcal{M}(\mathcal{A})$ converges to $x \in \mathcal{M}(\mathcal{A})$ if and only if $[f]_{x_\alpha}$ converges to $[f]_x$ in $\mathbb{R}$ for all $f \in \mathcal{A}$. (See §A.3 for a discussion of nets.)

REMARK C.2. If $\mathcal{A}$ is a Banach algebra over the field $\mathbb{C}$ of complex numbers, then one can show using the Gelfand-Mazur theorem that $\mathcal{M}(\mathcal{A})$ is naturally homeomorphic to the Gelfand space of maximal ideals in $\mathcal{A}$.

THEOREM C.3. *If $\mathcal{A}$ is a nonzero Banach ring, then the spectrum $\mathcal{M}(\mathcal{A})$ is a nonempty compact Hausdorff space.*

PROOF. We will reproduce below Berkovich's proof (see Theorem 1.2.1 of [16]) that $\mathcal{M}(\mathcal{A})$ is nonempty and Hausdorff. First, however, we give a proof (different from the one in [16, Theorem 1.2.1]) of the compactness of $\mathcal{M}(\mathcal{A})$. It suffices to prove that every net in $X = \mathcal{M}(\mathcal{A})$ has a convergent subnet. Let T be the space $\prod_{f \in \mathcal{A}} [0, \|f\|]$ endowed with the product topology. By Tychonoff's theorem, T is compact. By Lemma C.1, there is a natural map $\iota : X \to T$ sending $x \in X$ to $([f]_x)_{f \in \mathcal{A}}$, and ι is clearly injective.

Let $\langle x_\alpha \rangle$ be a net in X. Since T is compact, $\langle \iota(x_\alpha) \rangle$ has a subnet $\langle \iota(y_\beta) \rangle$ converging to an element $(\alpha_f)_{f \in \mathcal{A}} \in T$. Define a function $[\cdot]_y : \mathcal{A} \to \mathbb{R}_{\geq 0}$ by $[f]_y = \alpha_f$. It is easily verified that $[\cdot]_y$ is a bounded multiplicative seminorm on $\mathcal{A}$ and thus defines a point $y \in X$. By construction, we have $\iota(y) = \lim_\beta \iota(y_\beta)$. This implies that $\lim_\beta [f]_{y_\beta} = [f]_y$ for all $f \in \mathcal{A}$, i.e., $y_\beta \to y$. Thus $\langle x_\alpha \rangle$ has a convergent subnet as desired.

We now prove that $\mathcal{M}(\mathcal{A})$ is nonempty (cf. [16, Theorem 1.2.1] and [76, Theorem 2.5]). It is well known that every maximal ideal M in a Banach ring $\mathcal{A}$ is closed, and therefore $\mathcal{A}/M$ is a Banach ring with respect to the residue norm. So without loss of generality, we may assume that $\mathcal{A}$ is a field. The set S of nonzero bounded seminorms on $\mathcal{A}$ is nonempty (since it contains the norm $\|\cdot\|$ on $\mathcal{A}$), and it is partially ordered by the relation $[\cdot] \leq [\cdot]'$ iff $[f] \leq [f]'$ for all $f \in \mathcal{A}$. Let $[\cdot]$ be a minimal element of S (which exists by Zorn's lemma). We will show that $[\cdot]$ is multiplicative.

Since S can only get smaller if we replace $\|\cdot\|$ by $[\cdot]$, we may assume (replacing $\mathcal{A}$ by its completion with respect to $[\cdot]$) that $[\cdot]$ coincides with the given Banach norm $\|\cdot\|$ on $\mathcal{A}$.

We claim that $[f]^n = [f^n]$ for all $f \in \mathcal{A}$ and all integers $n > 0$. If not, then $[f^n] < [f]^n$ for some $f \in \mathcal{A}$ and some $n > 0$; set $r = \sqrt[n]{[f^n]} > 0$. Let $\mathcal{A}\langle r^{-1}T \rangle$ be the set of all power series $f = \sum_{i=0}^{\infty} a_i T^i$ with $\sum_{i=0}^{\infty} \|a_i\| r^i <$

∞. Then $f - T$ is not invertible in $\mathcal{A}\langle r^{-1}T\rangle$: otherwise, we would have $\sum_{i\geq 0}[f^{-i}]r^i < \infty$, and thus

$$\sum_{i\geq 0}[f^{-in}][f^n]^i = \sum_{i\geq 0}[f^{-in}]r^{ni} < \infty,$$

whereas on the other hand $[f^{-in}] \geq [f^{in}]^{-1} \geq [f^n]^{-i}$, which implies that

$$\sum_{i\geq 0}[f^{-in}][f^n]^i \geq \sum_{i\geq 0} 1 = \infty,$$

a contradiction. Now consider the natural homomorphism

$$\varphi : \mathcal{A} \to \mathcal{A}\langle r^{-1}T\rangle/(f - T).$$

The composition of the Banach norm $\|\cdot\|_r$ on $\mathcal{A}\langle r^{-1}T\rangle/(f-T)$ with φ defines a bounded multiplicative seminorm $[\cdot]'$ on $\mathcal{A}$ with $[a]' = \|\varphi(a)\|_r \leq [a]$ for all $a \in \mathcal{A}$, and $[f]' = \|\varphi(f)\|_r = \|T\|_r = r < [f]$. This contradicts the minimality of $[\cdot]$ and proves the claim.

Next, we claim that $[f]^{-1} = [f^{-1}]$ for all nonzero elements $f \in \mathcal{A}$. If not, let f be a nonzero element of $\mathcal{A}$ with $[f]^{-1} < [f^{-1}]$, and set $r = [f^{-1}]^{-1}$. Then $f - T$ is not a unit in $\mathcal{A}\langle r^{-1}T\rangle$ by an argument similar to the one above. The natural map

$$\varphi : \mathcal{A} \to \mathcal{A}\langle r^{-1}T\rangle/(f - T)$$

again satisfies $\|\varphi(f)\|_r < [f]$, contradicting the minimality of $[\cdot]$ and proving the claim.

From this, it follows that $[\cdot]$ is multiplicative, since for nonzero $f, g \in \mathcal{A}$ we have

$$[fg]^{-1} = [f^{-1}g^{-1}] \leq [f^{-1}][g^{-1}] = [f]^{-1}[g]^{-1}.$$

This proves that $\mathcal{M}(\mathcal{A})$ is nonempty, as desired.

Finally, we prove that $\mathcal{M}(\mathcal{A})$ is Hausdorff. Let $x, y \in \mathcal{M}(\mathcal{A})$ with $x \neq y$. Then there exists $f \in \mathcal{A}$ with $[f]_x \neq [f]_y$. Without loss of generality, suppose $[f]_x < [f]_y$, and choose $r \in \mathbb{R}$ with $[f]_x < r < [f]_y$. Then

$$\{z \in \mathcal{M}(\mathcal{A}) : [f]_z < r\}, \quad \{z \in \mathcal{M}(\mathcal{A}) : [f]_z > r\}$$

are disjoint neighborhoods of x and y, respectively. □

REMARK C.4. In many cases of interest, such as when the Banach norm on $\mathcal{A}$ is already multiplicative, the fact that $\mathcal{M}(\mathcal{A})$ is nonempty is obvious.

There is an alternate way to view the points of $\mathcal{M}(\mathcal{A})$. A *complete valued field* is a field K together with an absolute value $|\cdot|$ on K with respect to which K is complete. A *character* χ on $\mathcal{A}$ with values in a complete valued field K is a bounded homomorphism $\chi : \mathcal{A} \to K$. Two characters $\chi' : \mathcal{A} \to K'$ and $\chi'' : \mathcal{A} \to K''$ are called *equivalent* if they factor through a common character $\chi : \mathcal{A} \to K$ (via some embeddings of K into K' and K'', respectively). A character $\chi : \mathcal{A} \to K$ defines a bounded multiplicative seminorm via $f \mapsto |\chi(f)|$, and equivalent characters define the same element of $\mathcal{M}(\mathcal{A})$. Conversely, a point x of $\mathcal{M}(\mathcal{A})$ defines a character via $A \mapsto \mathcal{H}(x)$, where $\mathcal{H}(x)$ is the completion of the quotient field of $\ker([\ ]_x)$. In this way, one gets a canonical bijection between points of $\mathcal{M}(\mathcal{A})$ and equivalence classes of characters on $\mathcal{A}$ with values in some complete valued field K.

C.4. Affinoid algebras and affinoid spaces

Let k be a field equipped with a nontrivial[2] non-Archimedean absolute value, and which is complete with respect to this absolute value. Following Berkovich, we want to define the category of k-affinoid spaces, which are local building blocks for general k-analytic spaces (much as affine schemes are the local building blocks for schemes).

In order to define k-affinoid spaces, we first need to define k-affinoid algebras. For a multi-index $\nu = (\nu_1, \ldots, \nu_n)$ with each ν_i a nonnegative integer, we let $|\nu| = \max_i(\nu_i)$, and write T^ν as shorthand for $T_1^{\nu_1} \cdots T_n^{\nu_n}$. Given positive real numbers $r_1, \ldots, r_n$, we define the corresponding *generalized Tate algebra*

$$k\langle r_1^{-1}T_1, \ldots, r_n^{-1}T_n\rangle := \{f = \sum_\nu a_\nu T^\nu \ : \ |a_\nu|r^\nu \to 0 \text{ as } |\nu| \to \infty\}.$$

The generalized Tate algebra $k\langle r_1^{-1}T_1, \ldots, r_n^{-1}T_n\rangle$ can be thought of as the ring of analytic functions with coefficients in k which converge on the polydisc $\{z \in K^n \ : \ |z_i| \leq r_i \text{ for all } i\}$ for every complete non-Archimedean field extension K/k.

A generalized Tate algebra becomes a Banach algebra over k when equipped with the *Gauss norm*

$$\|f\| = \max_\nu |a_\nu|r^\nu.$$

A surjective k-algebra homomorphism $\pi : k\langle r_1^{-1}T_1, \ldots, r_n^{-1}T_n\rangle \to \mathcal{A}$ from a generalized Tate algebra to a Banach algebra $\mathcal{A}$ over k is called *admissible* if the norm on $\mathcal{A}$ is equivalent to the quotient norm on $\mathcal{A}$ defined by

$$\|f\|_\pi = \inf_{g \in \pi^{-1}(f)} \|g\|.$$

A *k-affinoid algebra* is a Banach algebra $\mathcal{A}$ over k for which there exists an admissible surjective map $k\langle r_1^{-1}T_1, \ldots, r_n^{-1}T_n\rangle \to \mathcal{A}$ from some generalized Tate algebra to $\mathcal{A}$.

Generalizing results of Tate, one shows that k-affinoid algebras are Noetherian and that every ideal in such a ring is closed.

---

[2]Berkovich allows the trivial absolute value as well in his work, and somewhat surprisingly this "trivial" case has some interesting applications. But for simplicity, we will assume here that the absolute value on k is nontrivial.

REMARK C.5. In classical rigid analysis, one considers only Tate algebras in which $r_1, \ldots, r_n$ belong to the divisible group $\sqrt{|k^\times|}$ generated by the value group of k. In Berkovich's theory, these rings are called *strictly k-affinoid algebras*.

REMARK C.6. If K/k is any extension of complete non-Archimedean fields, there is a base change functor from k-affinoid algebras to K-affinoid algebras which takes $\mathcal{A}$ to $\mathcal{A}\widehat{\otimes}_k K$, where $\widehat{\otimes}$ is the *complete tensor product* [**16**, p. 12]. Berkovich defines an *affinoid k-algebra* (as opposed to a k-affinoid algebra!) to be a K-affinoid algebra for some K/k.

Let $\mathcal{A}$ be a k-affinoid algebra. The *Berkovich analytic space associated to $\mathcal{A}$*, considered as a topological space, is just the spectrum $\mathcal{M}(\mathcal{A})$ of $\mathcal{A}$ in the sense of §C.3. If $\mathcal{A} \neq 0$, $\mathcal{M}(\mathcal{A})$ is a nonempty compact Hausdorff space which is locally arcwise-connected, and it is arcwise-connected if $\mathcal{A}$ has no nontrivial idempotents. A *k-affinoid space* is a pair $(\mathcal{M}(\mathcal{A}), \mathcal{A})$, where $\mathcal{A}$ is a k-affinoid algebra.

A *morphism* $(\mathcal{M}(\mathcal{B}), \mathcal{B}) \to (\mathcal{M}(\mathcal{A}), \mathcal{A})$ between k-affinoid spaces is by definition a bounded k-algebra homomorphism $\psi : \mathcal{A} \to \mathcal{B}$, together with the induced map $\psi^\# : \mathcal{M}(\mathcal{B}) \to \mathcal{M}(\mathcal{A})$ on topological spaces defined by $|f|_{\psi^\#(x)} = |\psi(f)|_x$ for $x \in \mathcal{M}(\mathcal{B})$ and $f \in \mathcal{A}$. The category of k-affinoid spaces which we have just defined is anti-equivalent to the category of k-affinoid algebras (with bounded homomorphisms between them).

Just as in the theory of affine schemes, one can define a structure sheaf $\mathcal{O}_X$ on $X = (\mathcal{M}(\mathcal{A}), \mathcal{A})$ for which $H^0(X, \mathcal{O}_X) = \mathcal{A}$. To define the structure sheaf, we need to introduce the notion of an affinoid subdomain.

If $\mathcal{A}$ is a k-affinoid algebra, an *affinoid subdomain* of $\mathcal{M}(\mathcal{A})$ is a compact subset $V \subseteq \mathcal{M}(\mathcal{A})$, together with a k-affinoid algebra $\mathcal{A}_V$ and a bounded homomorphism $\varphi : \mathcal{A} \to \mathcal{A}_V$ with $\varphi^\#(\mathcal{M}(\mathcal{A}_V)) \subseteq V$, which satisfies the following universal property:

> If K/k is any extension of complete non-Archimedean fields and $\psi : \mathcal{A}\widehat{\otimes}_k K \to \mathcal{B}$ is any bounded homomorphism of K-affinoid algebras with $\psi^\#(\mathcal{M}(\mathcal{B})) \subseteq V$, there is a unique bounded homomorphism $\psi' : \mathcal{A}_V \to \mathcal{B}$ for which $\psi = \psi' \circ \varphi$.

One can show that if V is an affinoid subdomain of $\mathcal{M}(\mathcal{A})$, then $\mathcal{M}(\mathcal{A}_V)$ is homeomorphic to V, and therefore that the homomorphism $\varphi : \mathcal{A} \to \mathcal{A}_V$ is uniquely determined by V.

Examples of affinoid subdomains include the *Laurent subdomains*

$$V = \{x \in \mathcal{M}(\mathcal{A}) : [f_i]_x \leq p_i, [g_j]_x \geq q_j \text{ for } 1 \leq i \leq n, 1 \leq j \leq m\}$$

for $p_1, \ldots, p_n, q_1, \ldots, q_m > 0$ and $f_1, \ldots, f_n, g_1, \ldots, g_m \in \mathcal{A}$. The affinoid algebra $\mathcal{A}_V$ representing such a Laurent subdomain is

$$\mathcal{A}_V = \mathcal{A}\langle p_1^{-1}T_1, \ldots, p_n^{-1}T_n, q_1 S_1, \ldots, q_m S_m \rangle / (T_i - f_i, g_j S_j - 1) .$$

The Laurent subdomains containing $x \in \mathcal{M}(\mathcal{A})$ (and hence also the affinoid subdomains containing x) form a fundamental system of compact neighborhoods for x in $\mathcal{M}(\mathcal{A})$. The intersection of two affinoid (resp. Laurent) subdomains is again an affinoid (resp. Laurent) subdomain.

A closed subset $V \subseteq \mathcal{M}(\mathcal{A})$ is called *special* if it is a finite union of affinoid subdomains. If $V = \bigcup_{i=1}^{n} V_i$ is a finite covering of a special subset of $\mathcal{M}(\mathcal{A})$ by affinoid subdomains, one defines

$$\mathcal{A}_V = \ker\left(\prod_i \mathcal{A}_{V_i} \to \prod_{i,j} \mathcal{A}_{V_i \cap V_j} \right).$$

It can be shown that $\mathcal{A}_V$ is a Banach algebra over k which depends only on V and not on the choice of a finite covering. The association $V \mapsto \mathcal{A}_V$ is functorial.

It is a consequence of Tate's acyclicity theorem in rigid analysis that the correspondence $V \mapsto \mathcal{A}_V$ is a sheaf of Banach k-algebras in the "special G-topology" on X. Concretely, this means that the Cech complex

$$\text{(C.1)} \qquad 0 \to \mathcal{A}_V \to \prod_i \mathcal{A}_{V_i} \to \prod_{i,j} \mathcal{A}_{V_i \cap V_j}$$

is exact for any finite covering of a special subset V of X by special subsets V_i. (We refer the reader to [**26**] for a general discussion of G-topologies and sheaves on them.)

The structure sheaf $\mathcal{O}_X$ on $X = \mathcal{M}(\mathcal{A})$ is defined as follows. For an open subset $U \subseteq X$, define

$$\mathcal{O}_X(U) = \varprojlim \mathcal{A}_V ,$$

where the inverse limit is taken over all special subsets $V \subset U$. It is easy to see that $\mathcal{O}_X$ is a presheaf. The fact that $\mathcal{O}_X$ is in fact a *sheaf* is a consequence of the exactness of (C.1), together with the following lemma (cf. [**92**, Lemma 13]), whose proof is a straightforward application of the compactness of special subsets of X:

LEMMA C.7. *If $\{U_i\}_{i \in I}$ is an open covering of an open set $U \subseteq \mathcal{M}(\mathcal{A})$ and if $V \subseteq U$ is a special set, then there exists a finite covering of V by special subsets $V_1, \ldots, V_m$ with each V_i contained in some U_j.*

The fact that $\mathcal{O}_X$ is a sheaf means that the extra points in Berkovich's theory allow one to recover, in the non-Archimedean setting, a good theory of analytic continuation.

If $\mathcal{A}$ is a strict k-affinoid algebra, let $\mathcal{A}^\circ$ (resp. $\mathcal{A}^{\circ\circ}$) be the set of all $f \in \mathcal{A}$ such that $[f]_x \leq 1$ (resp. $[f]_x < 1$) for all $x \in \mathcal{M}(\mathcal{A})$. The quotient ring $\tilde{\mathcal{A}} = \mathcal{A}^\circ / \mathcal{A}^{\circ\circ}$ is a finitely generated algebra over the residue field $\tilde{k}$ of k. If $\mathcal{A}$ is nonzero and $x \in X = \mathcal{M}(\mathcal{A})$, the natural evaluation map $\mathcal{A} \to \mathcal{H}(x)$ induces a homomorphism $\tilde{\mathcal{A}} \to \widetilde{\mathcal{H}(x)}$ whose kernel belongs to $\tilde{X} = \text{Spec}(\tilde{\mathcal{A}})$;

this gives rise to a natural surjective *reduction map* $\pi : X \to \tilde{X}$. The map π is *anti-continuous*, in the sense that the inverse image of a Zariski open (resp. closed) set is closed (resp. open) in X. If ξ is the generic point of an irreducible component of $\tilde{X}$, then $\pi^{-1}(\xi)$ consists of a single point. The set $\{\pi^{-1}(\xi)\}$, as ξ runs through all such generic points, is precisely the *Shilov boundary* of X, i.e., the smallest closed subset of X on which $\sup_{x \in X}([f]_x)$ is achieved for every $f \in \mathcal{A}$.

EXAMPLE C.8. If $\mathcal{A} = k\langle T\rangle$, so that $X = \mathcal{M}(\mathcal{A})$ is the Berkovich unit disc $\mathcal{D}(0,1)$, then $\tilde{\mathcal{A}}$ is the polynomial ring $\tilde{k}[\tilde{T}]$ and $\tilde{X}$ is the affine line over $\tilde{k}$. The Gauss point of X (corresponding to the Gauss norm on $k\langle T\rangle$) is mapped onto the generic point of $\mathbb{A}^1_{\tilde{k}}$, and all other points of X are mapped onto closed points of $\mathbb{A}^1_{\tilde{k}}$. For a "type I" point $x \in D(0,1) \subset \mathcal{D}(0,1)$, $\pi(x)$ is just the image of x under the usual reduction map $k^\circ \to \tilde{k}$.

EXAMPLE C.9. If $\alpha \in k$ and $0 < |\alpha| < 1$, the spectrum of the strict k-affinoid algebra $\mathcal{A} = k\langle S, T\rangle/(ST - \alpha)$ is the Berkovich analytic space associated to the closed annulus $|\alpha| \leq |T| \leq 1$ (which is a Laurent domain in $\mathcal{D}(0,1)$). One can identify $\mathcal{A}$ with the ring of all generalized Laurent series $f = \sum_{i \in \mathbb{Z}} a_i T^i$ for which both $|a_i|$ and $|a_i| \cdot |\alpha|^i$ tend to zero as $|i| \to \infty$. We have $\tilde{\mathcal{A}} \cong \tilde{k}[\tilde{T}, \tilde{S}]/\tilde{T}\tilde{S}$, and $\mathrm{Spec}(\tilde{\mathcal{A}})$ has two generic points $\xi_{\tilde{S}}, \xi_{\tilde{T}}$. The unique inverse image of $\xi_{\tilde{S}}$ (resp. $\xi_{\tilde{T}}$) under π is the multiplicative seminorm sending $\sum a_i T^i$ to $\max |a_i|$ (resp. $\max |a_i| \cdot |\alpha|^i$).

C.5. Global k-analytic spaces

We now turn to the problem of gluing together k-affinoid spaces to make global k-analytic spaces. The category of k-analytic spaces is defined in Berkovich's book [16] in terms of locally ringed spaces (by analogy with the definition of schemes and complex analytic spaces). We will not give the precise definition here but instead refer the reader to Chapter 3 of [16]. The problem with the definition in [16] is that, unlike in the case of schemes or complex analytic spaces, the building blocks for global k-analytic spaces, namely the k-affinoid spaces, are *closed* rather than open. This makes the theory of sheaves and locally ringed spaces somewhat awkward, especially when one wants to understand the relationship between k-analytic spaces and rigid analytic spaces.

In [17], Berkovich gives a more satisfactory definition of a global k-analytic space.[3] The improved definition is based on the notion of a *net* (this usage of the term is different from that in §A.3). If X is a locally Hausdorff topological space, a *quasi-net* on X is a collection τ of compact Hausdorff subsets $V \subseteq X$ such that each point $x \in X$ has a (closed) neighborhood of

---

[3]Note that the categories of k-analytic spaces defined in [16] and [17] are not the same, the latter being more general. However, the analytification of an algebraic variety over k is always a k-analytic space in the sense of [16].

the form $V_1 \cup \cdots \cup V_n$, with $V_i \in \tau$ and $x \in V_i$ for all $i = 1, \ldots, n$. The quasi-net τ is a *net* if, for all $V, V' \in \tau$, the collection $\{W \in \tau \,:\, W \subseteq V \cap V'\}$ is a quasi-net on $V \cap V'$ [**17**, p. 14].

A *k-affinoid atlas* A on X with respect to the net τ is an assignment of a k-affinoid algebra $\mathcal{A}_V$ and a homeomorphism $V \xrightarrow{\sim} \mathcal{M}(\mathcal{A}_V)$ to each $V \in \tau$ such that for each $V, V' \in \tau$ with $V' \subseteq V$, there is a bounded homomorphism of k-affinoid algebras $\mathcal{A}_V \to \mathcal{A}_{V'}$ which identifies $(V', \mathcal{A}_{V'})$ with an affinoid subdomain of $(V, \mathcal{A}_V)$ [**17**, p. 16].

A *k-analytic space* is a triple (X, A, τ) with τ a net on X and A a k-affinoid atlas on X with respect to τ [**17**, p. 17]. If all $\mathcal{A}_V$ are strict k-affinoid algebras, then the triple is called a *strict k-analytic space*. When A and τ are understood, one often writes X as shorthand for the corresponding analytic space (X, A, τ).

EXAMPLE C.10. Let X be the set of multiplicative seminorms on $k[T]$ which extend the given absolute value on k, furnished with the weakest topology for which $x \mapsto [f]_x$ is continuous for each $f \in k[T]$. For each real number $r > 0$, let $X_r = \{x \in X \,:\, [T]_x \leq r\}$. Then restriction to $k[T]$ induces, for all $r > 0$, a homeomorphism $\iota_r \,:\, \mathcal{M}(k\langle r^{-1}T \rangle) \xrightarrow{\sim} X_r$, and in particular each X_r is a k-affinoid space. The collection $\tau = \{X_r\}_{r>0}$ forms a net on X, and the assignment $X_r \mapsto k\langle r^{-1}T \rangle$ furnishes a k-affinoid atlas A on X with respect to τ. The analytic space (X, A, τ) formed in this way is called the *Berkovich affine line* over k, or the *analytification of* $\mathbb{A}^1$ over k. (Compare with the discussion in Chapter 2.)

EXAMPLE C.11. A k-affinoid atlas for the Berkovich projective line $\mathbb{P}^1_{\text{Berk}}$ is given by the three sets consisting of the closed Berkovich unit disc $V_1 = \mathcal{D}(0,1)$, the complement $V_2 = \mathbb{P}^1_{\text{Berk}} \backslash \mathcal{D}(0,1)^-$ of the open Berkovich unit disc in $\mathbb{P}^1_{\text{Berk}}$, and their intersection $V_3 = \mathcal{D}(0,1) \backslash \mathcal{D}(0,1)^-$. Note that $V_1 \cup V_2$ is a compact neighborhood of the Gauss point ζ_{Gauss}, but ζ_{Gauss} does not have a single compact neighborhood of the form V_i. (This explains why one wants to allow finite unions in the definition of a quasi-net.)

EXAMPLE C.12. If $\mathcal{A}$ is a k-affinoid algebra and $X = \mathcal{M}(\mathcal{A})$, then the collection τ of all affinoid subdomains of X, together with the atlas given by the usual assignment $V \mapsto \mathcal{A}_V$ for $V \in \tau$, defines in a natural way a k-analytic space (X, A, τ) associated to the k-affinoid space $(X, \mathcal{A})$.

In Example C.12, we could also simply take $\tau' = \{X\}$ and $A' = \{\mathcal{A}\}$. One would like to consider the analytic spaces (X, A, τ) and (X, A', τ') as one and the same. To accomplish this, a good notion of a *morphism* of analytic space is needed for which the natural map from (X, A, τ) to (X, A', τ') becomes an isomorphism. One would also like the previously defined category of k-affinoid spaces to embed as a full subcategory of the category of k-analytic spaces.

From such considerations, Berkovich was led to the following definitions for the category of k-analytic spaces.

A *strong morphism* $\varphi : (X, A, \tau) \to (X', A', \tau')$ of k-analytic spaces is a continuous map $\varphi : X \to X'$ such that for each $V \in \tau$ there exists a $V' \in \tau'$ with $\varphi(V) \subseteq V'$, together with a compatible system of morphisms of k-affinoid spaces $\varphi_{V/V'} : (V, \mathcal{A}_V) \to (V', \mathcal{A}'_{V'})$ for all pairs $V \in \tau$ and $V' \in \tau'$ with $\varphi(V) \subseteq V'$. We define $\widetilde{k\text{-An}}$ to be the category whose objects are k-analytic spaces, with strong morphisms between them.

A strong morphism is said to be a *quasi-isomorphism* if φ is a homeomorphism between X and X' and if $\varphi_{V/V'}$ identifies V with an affinoid subdomain of V' whenever $V \in \tau$ and $V' \in \tau'$ with $\varphi(V) \subseteq V'$.

Finally, the category k-An of *k-analytic spaces* is the localization of the category $\widetilde{k\text{-An}}$ with respect to the quasi-isomorphisms, i.e., it is the category obtained from $\widetilde{k\text{-An}}$ by "formally inverting" the quasi-isomorphisms.[4]

Another way to describe the morphisms in k-An is to first show that every k-analytic space possesses a maximal k-affinoid atlas (whose elements are called *affinoid domains* in X); in terms of maximal atlases, one can describe morphisms in k-An similarly to the way that strong morphisms were defined above (see Section 1.2 of [17] for details).

REMARK C.13. Every k-analytic space X possesses a natural structure sheaf $\mathcal{O}_X$ generalizing the structure sheaf defined above on a k-affinoid space.

C.6. Properties of k-analytic spaces

Here are some of the properties of k-analytic spaces.

1. The functor $(X = \mathcal{M}(\mathcal{A}), \mathcal{A}) \mapsto (X, \{\mathcal{A}\}, \{X\})$ from the category of k-affinoid spaces to k-analytic spaces is fully faithful, so one does not obtain any new morphisms between k-affinoid spaces by thinking of them as objects of the larger category of k-analytic spaces.

---

[4]If $\mathcal{C}$ is a category and $\mathcal{S}$ is a class of maps in $\mathcal{C}$, the *localization of $\mathcal{C}$ with respect to $\mathcal{S}$* is a category $\mathcal{C}[\mathcal{S}^{-1}]$ together with a functor $Q : \mathcal{C} \to \mathcal{C}[\mathcal{S}^{-1}]$ such that (i) $Q(s)$ is an isomorphism for all $s \in \mathcal{S}$ and (ii) any functor $F : \mathcal{C} \to \mathcal{D}$ such that $F(s)$ is an isomorphism for all $s \in \mathcal{S}$ factors uniquely through Q. One can show that such a localization always exists. However, it is difficult in general to explicitly describe the morphisms in $\mathcal{C}[\mathcal{S}^{-1}]$. But if $\mathcal{S}$ admits a "calculus of right fractions", meaning that it satisfies a system of four axioms first described by Gabriel and Zisman in [56], then the localized category can be described rather concretely: its objects are the same as the objects in $\mathcal{C}$, and morphisms from A to B in $\mathcal{C}[\mathcal{S}^{-1}]$ are certain equivalence classes of diagrams

$$A \xleftarrow{s} C \xrightarrow{f} B$$

with $s \in \mathcal{S}$ and $f \in \text{Mor}(\mathcal{C})$. Every morphism in $\mathcal{C}[\mathcal{S}^{-1}]$ can thus be represented as a "right fraction" $f \circ s^{-1}$ with $s \in \mathcal{S}$ and $f \in \text{Mor}(\mathcal{C})$, and there is a rule similar to the usual one for equality of fractions. (In particular, if f, g are parallel morphisms in $\mathcal{C}$, then $Q(f) = Q(g)$ if and only if there is a $t \in \mathcal{S}$ such that $f \circ t = g \circ t$.) The family of quasi-isomorphisms in $\widetilde{k\text{-An}}$ admits a calculus of right fractions. See [56], [69, §1.3], and [17, §1.2] for further details.

2. The category k-An admits fiber products and, for each non-Archimedean extension field K/k, there is a *ground field extension functor* $X \mapsto X\hat{\otimes}K$ taking k-analytic spaces to K-analytic spaces. Given a point $x \in X$, there is an associated non-Archimedean field $\mathcal{H}(x)$ over k, called the *completed residue field of x*, and given a morphism $\varphi : Y \to X$, one can define in a natural way the *fiber Y_x* of φ at x, which is an $\mathcal{H}(x)$-analytic space. If $X = \mathcal{M}(\mathcal{A})$ is a k-affinoid space, then $\mathcal{H}(x)$ is the completion of the fraction field of the quotient of $\mathcal{A}$ by the kernel of the seminorm corresponding to x.

3. Each point of a k-analytic space has a fundamental system of open neighborhoods which are locally compact and arcwise connected. Every one-dimensional k-analytic space and every smooth k-analytic space (which we will not define) is locally contractible. (The latter result is quite difficult; see [20] and [21].)

4. There is an analytification functor which associates to each scheme X locally of finite type over k a k-analytic space X^{an}. The scheme X is separated iff X^{an} is Hausdorff, proper iff X^{an} is compact, and connected iff X^{an} is arcwise connected. If X is separated, its Krull dimension is equal to the topological dimension of X^{an}. If k is algebraically closed and X is a variety over k, then $X(k)$ (endowed with its totally discontinuous analytic topology) can be naturally identified with a dense subspace of X^{an}.

5. The analytification of $\mathbb{P}^n$ is contractible, as is the analytification of any smooth, proper, integral variety over k having good reduction. On the other hand, the analytification of an elliptic curve with multiplicative reduction is homotopy equivalent to a circle. A detailed description of the topological structure of one-dimensional Berkovich analytic spaces can be found in [16, Chapter 4]; see also [94]. For a discussion of the topological structure of A^{an} when A is an abelian variety, see [16, §6.5].

6. To any formal scheme $\mathcal{X}$ locally finitely presented over the valuation ring $k°$ of k, one can associate to it a paracompact strict k-analytic space $\mathcal{X}_\eta$ called the *generic fiber* of $\mathcal{X}$ and a reduction map $\pi : \mathcal{X}_\eta \to \mathcal{X}_s$, where $\mathcal{X}_s$ is the special fiber of $\mathcal{X}$.

Bibliography

[1] E. Artin. *Algebraic numbers and algebraic functions*. AMS Chelsea Publishing, Providence, RI, 2006. Reprint of the 1967 original.

[2] P. Autissier. Points entiers sur les surfaces arithmétiques. *J. Reine Angew. Math.*, 531:201–235, 2001.

[3] M. Baker. A lower bound for average values of dynamical Green's functions. *Math. Res. Lett.*, 13(2-3):245–257, 2006.

[4] M. Baker. An introduction to Berkovich analytic spaces and non-Archimedean potential theory on curves. In *p-adic Geometry (Lectures from the 2007 Arizona Winter School)*, volume 45 of *AMS University Lecture Series*. Amer. Math. Soc., Providence, RI, 2008.

[5] M. Baker. A finiteness theorem for canonical heights attached to rational maps over function fields. *J. Reine Angew. Math.*, 626:205–233, 2009.

[6] M. Baker and L. DeMarco. Preperiodic points and unlikely intersections. Preprint. Available at arXiv:0911.0918, 2009.

[7] M. Baker and X. Faber. Metrized graphs, Laplacian operators, and electrical networks. In *Quantum graphs and their applications*, volume 415 of *Contemp. Math.*, pages 15–33. Amer. Math. Soc., Providence, RI, 2006.

[8] M. Baker and L.-C. Hsia. Canonical heights, transfinite diameters, and polynomial dynamics. *J. Reine Angew. Math.*, 585:61–92, 2005.

[9] M. Baker and R. Rumely. Equidistribution of small points, rational dynamics, and potential theory. *Ann. Inst. Fourier (Grenoble)*, 56(3):625–688, 2006.

[10] M. Baker and R. Rumely. Harmonic analysis on metrized graphs. *Canad. J. Math.*, 59(2):225–275, 2007.

[11] R. L. Benedetto. *Fatou components in p-adic dynamics*. PhD thesis, Brown University, Providence, RI, 1998.

[12] R. L. Benedetto. p-adic dynamics and Sullivan's no wandering domains theorem. *Compositio Math.*, 122(3):281–298, 2000.

[13] R. L. Benedetto. Hyperbolic maps in p-adic dynamics. *Ergodic Theory Dynam. Systems*, 21(1):1–11, 2001.

[14] R. L. Benedetto. Reduction, dynamics, and Julia sets of rational functions. *J. Number Theory*, 86(2):175–195, 2001.

[15] R. L. Benedetto. Examples of wandering domains in p-adic polynomial dynamics. *C. R. Math. Acad. Sci. Paris*, 335(7):615–620, 2002.

[16] V. G. Berkovich. *Spectral theory and analytic geometry over non-Archimedean fields*, volume 33 of *Mathematical Surveys and Monographs*. Amer. Math. Soc., Providence, RI, 1990.

[17] V. G. Berkovich. Étale cohomology for non-Archimedean analytic spaces. *Inst. Hautes Études Sci. Publ. Math.*, 78:5–161, 1993.

[18] V. G. Berkovich. The automorphism group of the Drinfel'd half-plane. *C. R. Acad. Sci. Paris Sér. I Math.*, 321(9):1127–1132, 1995.

[19] V. G. Berkovich. p-adic analytic spaces. In *Proceedings of the International Congress of Mathematicians, Vol. II (Berlin, 1998)*, pages 141–151 (electronic), 1998.

[20] V. G. Berkovich. Smooth p-adic analytic spaces are locally contractible. *Invent. Math.*, 137(1):1–84, 1999.

[21] V. G. Berkovich. Smooth p-adic analytic spaces are locally contractible. II. In *Geometric aspects of Dwork theory*, pages 293–370. Walter de Gruyter and Co. KG, Berlin, 2004.

[22] J.-P. Bézivin. Sur les points périodiques des applications rationnelles en dynamique ultramétrique. *Acta Arith.*, 100(1):63–74, 2001.

[23] P. Billingsley. *Convergence of probability measures*. Wiley Series in Probability and Statistics: Probability and Statistics. John Wiley & Sons Inc., New York, second edition, 1999. A Wiley-Interscience Publication.

[24] Y. Bilu. Limit distribution of small points on algebraic tori. *Duke Math*, 89:465–476, 1997.

[25] E. Bombieri and W. Gubler. *Heights in Diophantine geometry*, volume 4 of *New Mathematical Monographs*. Cambridge University Press, Cambridge, 2006.

[26] S. Bosch, U. Güntzer, and R. Remmert. *Non-Archimedean analysis*, volume 261 of *Grundlehren der Mathematischen Wissenschaften [Fundamental Principles of Mathematical Sciences]*. Springer-Verlag, Berlin, 1984. A systematic approach to rigid analytic geometry.

[27] J.-B. Bost and A. Chambert-Loir. Analytic curves in algebraic varieties over number fields. Preprint. Available at arXiv:math.NT/0702593, 2008.

[28] S. Boucksom, C. Favre, and M. Jonsson. Valuations and plurisubharmonic singularities. *Publ. Res. Inst. Math. Sci.*, 44(2):449–494, 2008.

[29] N. Bourbaki. *General topology. Chapters 1–4*. Elements of Mathematics (Berlin). Springer-Verlag, Berlin, 1998. Translated from the French. Reprint of the 1989 English translation.

[30] M. R. Bridson and A. Haefliger. *Metric spaces of non-positive curvature*, volume 319 of *Grundlehren der Mathematischen Wissenschaften [Fundamental Principles of Mathematical Sciences]*. Springer-Verlag, Berlin, 1999.

[31] H. Brolin. Invariant sets under iteration of rational functions. *Ark. Mat.*, 6:103–144, 1965.

[32] G. S. Call and J. H. Silverman. Canonical heights on varieties with morphisms. *Compositio Math.*, 89(2):163–205, 1993.

[33] D. G. Cantor. On an extension of the definition of transfinite diameter and some applications. *J. Reine Angew. Math.*, 316:160–207, 1980.

[34] A. Chambert-Loir. Théorèmes d'équidistribution pour les systèmes dynamiques d'origine arithmétique. Panoramas et synthéses. Société Mathematique de France. Preprint.

[35] A. Chambert-Loir. Mesures et équidistribution sur les espaces de Berkovich. *J. Reine Angew. Math.*, 595:215–235, 2006.

[36] T. Chinburg, C. F. Lau, and R. Rumely. Capacity theory and arithmetic intersection theory. *Duke Math. J.*, 117(2):229–285, 2003.

[37] T. Chinburg and R. Rumely. The capacity pairing. *J. Reine Angew. Math.*, 434:1–44, 1993.

[38] I. Chiswell. *Introduction to Λ-trees*. World Scientific Publishing Co. Inc., River Edge, NJ, 2001.

[39] B. Conrad. Several approaches to non-archimedean geometry. In *p-adic Geometry (Lectures from the 2007 Arizona Winter School)*, volume 45 of *AMS University Lecture Series*. Amer. Math. Soc., Providence, RI, 2008.

[40] T. Coulbois, A. Hilion, and M. Lustig. Non-unique ergodicity, observers' topology and the dual algebraic lamination for $\mathbb{R}$-trees. *Illinois J. Math.*, 51(3):897–911, 2007.

[41] L. DeMarco. Dynamics of rational maps: Lyapunov exponents, bifurcations, and capacity. *Math. Ann.*, 326(1):43–73, 2003.

[42] A. Ducros. Espaces analytiques p-adiques au sens de Berkovich. *Astérisque*, (311):Exp. No. 958, viii, 137–176, 2007. Séminaire Bourbaki. Vol. 2005/2006.

[43] A. Escassut. *Analytic elements in p-adic analysis*. World Scientific Publishing Co. Inc., River Edge, NJ, 1995.

[44] B. Farkas and B. Nagy. Transfinite diameter, Chebyshev constant and energy on locally compact spaces. *Potential Anal.*, 28(3):241–260, 2008.

[45] C. Favre and M. Jonsson. *The valuative tree*, volume 1853 of *Lecture Notes in Mathematics*. Springer-Verlag, Berlin, 2004.

[46] C. Favre and J. Rivera-Letelier. Théorème d'équidistribution de Brolin en dynamique p-adique. *C. R. Math. Acad. Sci. Paris*, 339(4):271–276, 2004.

[47] C. Favre and J. Rivera-Letelier. Équidistribution quantitative des points de petite hauteur sur la droite projective. *Math. Ann.*, 335(2):311–361, 2006.

[48] C. Favre and J. Rivera-Letelier. Théorie ergodique des fractions rationnelles sur un corps ultramétrique. Preprint. Available at arXiv:0709.0092, 2007.

[49] G. B. Folland. *Real analysis*. Pure and Applied Mathematics (New York). John Wiley & Sons Inc., New York, second edition, 1999. Modern techniques and their applications, a Wiley-Interscience Publication.

[50] J. E. Fornaess and N. Sibony. Complex dynamics in higher dimension. I. *Astérisque*, (222):5, 201–231, 1994. Complex analytic methods in dynamical systems (Rio de Janeiro, 1992).

[51] J. E. Fornaess and N. Sibony. Complex dynamics in higher dimensions. In *Complex potential theory (Montreal, PQ, 1993)*, volume 439 of *NATO Adv. Sci. Inst. Ser. C Math. Phys. Sci.*, pages 131–186. Kluwer Acad. Publ., Dordrecht, 1994. Notes partially written by Estela A. Gavosto.

[52] J. E. Fornaess and N. Sibony. Complex dynamics in higher dimension. II. In *Modern methods in complex analysis (Princeton, NJ, 1992)*, volume 137 of *Ann. of Math. Stud.*, pages 135–182. Princeton Univ. Press, Princeton, NJ, 1995.

[53] J. E. Fornaess and N. Sibony. Complex dynamics in higher dimension. In *Several complex variables (Berkeley, CA, 1995–1996)*, volume 37 of *Math. Sci. Res. Inst. Publ.*, pages 273–296. Cambridge Univ. Press, Cambridge, 1999.

[54] A. Freire, A. Lopes, and R. Mañé. An invariant measure for rational maps. *Bol. Soc. Brasil. Mat.*, 14(1):45–62, 1983.

[55] J. Fresnel and M. van der Put. *Rigid analytic geometry and its applications*, volume 218 of *Progress in Mathematics*. Birkhäuser Boston Inc., Boston, MA, 2004.

[56] P. Gabriel and M. Zisman. *Calculus of fractions and homotopy theory*. Ergebnisse der Mathematik und ihrer Grenzgebiete, Band 35. Springer-Verlag New York, Inc., New York, 1967.

[57] É. Ghys and P. de la Harpe. Sur les groupes hyperboliques d'après Mikhael Gromov (Bern, 1988), volume 83 of *Progr. Math.*, Birkhäuser Boston, Boston, MA, 1990.

[58] D. Goss. The algebraist's upper half-plane. *Bull. Amer. Math. Soc. (N.S.)*, 2(3):391–415, 1980.

[59] E. Hille. *Analytic function theory. Vol. II*. Introductions to Higher Mathematics. Ginn and Co., Boston, Mass.-New York-Toronto, Ont., 1962.

[60] N. R. Howes. *Modern analysis and topology*. Universitext. Springer-Verlag, New York, 1995.

[61] L.-C. Hsia. Closure of periodic points over a non-Archimedean field. *J. London Math. Soc. (2)*, 62(3):685–700, 2000.

[62] L.-C. Hsia. p-adic equidistribution theorems. Manuscript, 2003.

[63] E. Kamke. *Theory of sets*. Dover, New York, 1950.

[64] E. Kani. Potential theory on curves. In *Théorie des nombres (Quebec, PQ, 1987)*, pages 475–543. de Gruyter, Berlin, 1989.

[65] J. L. Kelley. *General topology*. Springer-Verlag, New York, 1975. Reprint of the 1955 edition [Van Nostrand, Toronto, Ont.], Graduate Texts in Mathematics, No. 27.

[66] J. Kiwi. Puiseux series polynomial dynamics and iteration of complex cubic polynomials. *Ann. Inst. Fourier (Grenoble)*, 56(5):1337–1404, 2006.

[67] M. Klimek. *Pluripotential theory*, volume 6 of *London Mathematical Society Monographs. New Series*. The Clarendon Press, Oxford University Press, New York, 1991. Oxford Science Publications.

[68] M. Kontsevich and Y. Soibelman. Affine structures and non-Archimedean analytic spaces. In *The unity of mathematics*, volume 244 of *Progr. Math.*, pages 321–385. Birkhäuser Boston, Boston, MA, 2006.

[69] H. Krause. Derived categories, resolutions, and Brown representability. In *Interactions between homotopy theory and algebra*, volume 436 of *Contemp. Math.*, pages 101–139. Amer. Math. Soc., Providence, RI, 2007.

[70] S. Lang. *Fundamentals of Diophantine geometry*. Springer-Verlag, New York, 1983.

[71] S. Lang. *Algebra*, volume 211 of *Graduate Texts in Mathematics*. Springer-Verlag, New York, third edition, 2002.

[72] M. Lyubich. Entropy properties of rational endomorphisms of the Riemann sphere. *Ergodic Theory Dynam. Systems*, 3:351–385, 1983.

[73] J. Milnor. *Dynamics in one complex variable*. Friedr. Vieweg & Sohn, Braunschweig, 1999. Introductory lectures.

[74] P. Morton and J. H. Silverman. Periodic points, multiplicities, and dynamical units. *J. Reine Angew. Math.*, 461:81–122, 1995.

[75] J. R. Munkres. *Topology: a first course*. Prentice-Hall Inc., Englewood Cliffs, N.J., 1975.

[76] J. Nicaise. Introduction to Berkovich spaces. Preprint. Available at wis.kuleuven.be/algebra/artikels/intro_berkovich.pdf.

[77] K. Ohshika. *Discrete groups*, volume 207 of *Translations of Mathematical Monographs*. Amer. Math. Soc., Providence, RI, 2002. Translated from the 1998 Japanese original by the author, Iwanami Series in Modern Mathematics.

[78] J. Piñeiro, L. Szpiro, and T. J. Tucker. Mahler measure for dynamical systems on $\mathbf{P}^1$ and intersection theory on a singular arithmetic surface. In Fedor Bogomolov and Yuri Tschinkel, editors, *Geometric Methods in Algebra and Number Theory*, volume 235 of *Progress in Mathematics*, pages 219–250. Birkhäuser, 2005.

[79] T. Ransford. *Potential theory in the complex plane*, volume 28 of *London Mathematical Society Student Texts*. Cambridge University Press, Cambridge, 1995.

[80] J. Rivera-Letelier. Théorie de Fatou et Julia dans la droite projective de Berkovich. In preparation.

[81] J. Rivera-Letelier. Dynamique des fonctions rationnelles sur des corps locaux. *Astérisque*, (287):147–230, 2003. Geometric methods in dynamics. II.

[82] J. Rivera-Letelier. Espace hyperbolique p-adique et dynamique des fonctions rationnelles. *Compositio Math.*, 138(2):199–231, 2003.

[83] J. Rivera-Letelier. Sur la structure des ensembles de Fatou p-adiques. Preprint. Available at arXiv:math.DS/0412180, 35 pages, 2004.

[84] J. Rivera-Letelier. Points périodiques des fonctions rationnelles dans l'espace hyperbolique p-adique. *Comment. Math. Helv.*, 80(3):593–629, 2005.

[85] A. M. Robert. *A course in p-adic analysis*, volume 198 of *Graduate Texts in Mathematics*. Springer-Verlag, New York, 2000.

[86] H. L. Royden. *Real analysis*. Macmillan Publishing Company, New York, third edition, 1988.

[87] W. Rudin. *Principles of mathematical analysis*. McGraw-Hill Book Co., New York, third edition, 1976. International Series in Pure and Applied Mathematics.

[88] R. Rumely. *Capacity theory on algebraic curves*, volume 1378 of *Lecture Notes in Mathematics*. Springer-Verlag, Berlin, 1989.

[89] R. Rumely. On Bilu's equidistribution theorem. In *Spectral problems in geometry and arithmetic (Iowa City, IA, 1997)*, volume 237 of *Contemp. Math.*, pages 159–166. Amer. Math. Soc., Providence, RI, 1999.

[90] R. Rumely. The Fekete-Szegő theorem with splitting conditions. II. *Acta Arith.*, 103(4):347–410, 2002.

[91] R. Rumely and M. Baker. Analysis and dynamics on the Berkovich projective line. Preprint. Available at arXiv:math.NT/0407433, 150 pages, 2004.

[92] P. Schneider. Points of rigid analytic varieties. *J. Reine Angew. Math.*, 434:127–157, 1993.

[93] J. H. Silverman. *The arithmetic of dynamical systems*, volume 241 of *Graduate Texts in Mathematics*. Springer, New York, 2007.

[94] A. Thuillier. *Théorie du potentiel sur les courbes en géométrie analytique non archimédienne. Applications à la théorie d'Arakelov*. PhD thesis, University of Rennes, 2005. Preprint. Available at http://tel.ccsd.cnrs.fr/documents/archives0/00/01/09/90/index.html.

[95] P. Tortrat. Aspects potentialistes de l'itération des polynômes. In *Séminaire de Théorie du Potentiel, Paris, No. 8*, volume 1235 of *Lecture Notes in Math.*, pages 195–209. Springer, Berlin, 1987.

[96] M. Tsuji. *Potential theory in modern function theory*. Chelsea Publishing Co., New York, 1975. Reprinting of the 1959 original.

[97] J. W. Tukey. *Convergence and Uniformity in Topology*. Annals of Mathematics Studies, no. 2. Princeton University Press, Princeton, N. J., 1940.

[98] B. L. van der Waerden. *Algebra. Vol. I*. Springer-Verlag, New York, 1991. Based in part on lectures by E. Artin and E. Noether. Translated from the seventh German edition by Fred Blum and John R. Schulenberger.

[99] X. Yuan. Big line bundles over arithmetic varieties. *Invent. Math.*, 173(3):603–649, 2008.

[100] S. Zhang. Admissible pairing on a curve. *Invent. Math.*, 112(1):171–193, 1993.

Index

affinoid
  algebra, 409
  Berkovich, 9
  open, xx, 9, 29
  strict closed, 9, 29, 127, 140
  strict open, 9
  subdomain, 410
algebraically capacitable, ix, xxiii, 142, 181
analytic component, 364
annular segment, 274
annulus, 27
  Berkovich, 41, 265, 281
  open, 283
  rational, 283
  standard open, 284
Arakelov-Green's function, xxv, 225, 241, 246, 299, 305, 307, 375
arbre, 147
arc, 393
arcwise connected, *see also* path-connected
attracting basin, 320, 337, 365
Attracting Fixed Point Theorem, 350
attracting periodic point, 319
Autissier, Pascal, xi, 191

$\mathcal{B}_x(\vec{v})^-$, 41
Baire set, 382
ball
  closed, 76, 81, 84
  open, 75, 81, 84
BDV(Γ), xxi, 57, 204
BDV($\mathbb{P}^1_{\text{Berk}}$), xxi
BDV(U), 97, 217
Benedetto, Rob, xi, 292, 293, 303, 335, 357, 362, 364, 366, 367, 370, 374
Berkovich
  adelic neighborhood, 142, 178, 180
  adelic set, 141, 178, 184, 186

affine line, xv, 19, 413
classification theorem, xvii, 3, 21
curves, x, xi, 49, 117
equicontinuity locus, xxvi, 333
hyperbolic space, xvii, 38
projective line, ix, xv, 24
spaces, x, xv, xxvi, 410, 413
Berkovich, Vladimir, xxvi, 18, 47, 147, 405
Bézivin, J.-P., 354
Bilu's equidistribution theorem, xxiv
boundary derivative, 235
boundary point, 56
bounded potentials, 109, 246–248
bout, 358
branch point, 394
branching, 12
  countable, xiii
  of main dendrite, 147, 197
  uncountable, xiii

Call-Silverman local height, 295–297, 306
canonical distance, 83, 85
canonical height, 185, 306, 308
canonical measure, x, xii, xxvi, 240, 297
Cantor matrix, 156, 160
Cantor, David, 180
capacity, xxii
  global, 141, 178, 179, 182, 186
  logarithmic, xxii, 121
capped, 279, 281
Chambert-Loir, Antoine, x, 18, 191, 306, 374
Chebyshev constant, xxiii, 136
Christmas tree, inverted, 271
classical point, 5
cofinal, 379
coherent system of measures, 95, 98
completed residue field, 415

completely invariant, 329
component of injectivity, 359
continuous
  Hölder, 326–328
  Lipschitz, 271, 275, 326, 327
continuous potentials, 109, 240, 241, 298
convergence
  of Laplacians, 227
  pointwise, 219, 220, 225, 226
  weak, 123, 220, 223, 226, 382
converges
  moderately well, 63–66, 71
convex, 195, 377
convex hull, xx, 39
coordinate change, 32, 268, 283
CPA(Γ), 50

$D(U)$, 147
$\mathcal{D}(\Gamma)$, 56
DeMarco, Laura, ix
derivative
  boundary, 98
  directional, 50, 202
diameter, xviii, 11, 43, 73, 82
Dini's lemma, 229, 380
direct limit, 16, 21, 40
directed set, 379
Directional Multiplicity Formula, 265
Dirichlet pairing, 51, 69, 70, 105, 106, 246
Dirichlet problem, 155
disc
  Berkovich, 8, 20, 26, 279
  Berkovich closed, 74
  Berkovich open, 40, 74, 200
  closed, xvi
  image of a, 279, 389
  irrational, xvi, 5
  rational, 5
discrepancy
  v-adic, 310
domain, 40, 87
  associated to a finite subgraph, 200
domain of quasi-periodicity, 337, 366, 368–370
Domination Theorem, 213

E_φ, 318
electrical network, 52, 397
endpoint, 394
energy integral, xxii, 121

Energy Minimization Principle, xxv, 240, 242, 246, 299
equicontinuity, xxvi, 333, 336, 386
equidistribution theorem
  adelic, ix, x, xxvi, 184, 240, 306, 309
  Bilu's, ix, xiv, 145, 184
  for compact adelic sets, 187
  for preimages, xxvi, 318
equilibrium
  distribution, see also equilibrium measure
  measure, xxiii, 123, 160
    existence of, 124
    support of, 125, 169
    uniqueness of, 128, 153, 159
  potential, 132
equivalent
  multiplicative seminorms, 24
  paths, 12, 50
Evans function, 151
exceptional
  locus, 318, 324, 363
  point, 318

Fatou components, xxvi, 365
  classification of, 338, 366
Fatou set
  Berkovich, xxvi, 328
  classical, 333
Fatou-Julia theory, xxvi, 291, 375
Favre, Charles, xi, xii, 85, 93, 116, 219, 225, 226, 233, 234, 236, 240, 246, 306, 318, 325, 327, 358, 372, 374
Fekete-Szegö theorem, ix, xiv, xxiii, 141–143
  multi-center, 177, 180
finite subgraph, xx, 39, 76, 88
  domain associated to a, 200
  image of a, 275
finite-dendrite domain, 41, 155, 193, 199–204, 206, 210, 211, 219, 220
Fixed Point Property, 348
Frostman's theorem, xxiii, 132, 244
full subset, 11, 401
function
  CPA-smooth, 236
  piecewise affine, 50
  smooth, xxv, 118
function field, ix
functional equation
  for $g_\varphi(x,y)$, xxvi, 303
  for the canonical measure, xxvi, 297

$g_{\mu_\varphi}(x,y)$, 299, 305
$g_\varphi(x,y)$, 300, 301, 305
Gauss
  norm, 1, 37, 409
  point, 6, 18, 412
Gauss's lemma, xvi
Gelfand topology, *see also* topology, Berkovich
Gelfand-Mazur theorem, xvi, 407
generic value, 35, 36
geodesic
  quadrilateral, 396
  segment, 393
  space, 393
  triangle, 395
grand orbit, 318
graph of discs, 14
Green's function, xxiv, 164, 195
  lower, 181
  of a strict closed affinoid, 169
  pullback formula for, 174, 176, 289
Green's matrix, 179
Gromov product, 52, 116, 393, 397
  relation to generalized Hsia kernel, 82
Grothendieck topology, xv, 405

$\mathbb{H}_p$, 357
$\mathbb{H}_{\text{Berk}}$, xvii, 38
$\mathbb{H}_{\text{Berk}}^{\mathbb{Q}}$, xvii, 38
$\mathbb{H}_{\text{Berk}}^{\mathbb{R}}$, xvii, 38
Hadamard's inequality, 314, 316
harmonic, xxiv, 145, 197
  not strongly, 149
  preserved under pullbacks, 174
  strongly, xxiv, 145
harmonic measure, 158, 160
Harnack's
  inequality, 162
  principle, xxiv, 163
Hartogs's lemma, 230, 231
  for Arakelov-Green's functions, 233
height of an adelic set, 145, 186
Hermann ring, 368
Hölder continuous, 326–328
Hölder continuous potentials, 327
homoclinic orbit, 351
homogeneous dynamical height, 300, 308, 314
Hsia kernel, xix, xxii, 73, 85
  function-theoretic meaning, 76
  generalized, xix, 81, 194, 393
    geometric interpretation, 83
    relation to Gromov product, 82

Hsia's theorem, 334
Hsia, Liang-Chung, x, xii, 73, 85, 292, 293
hyperbolic
  0-, 398
  Gromov, 398
  strongly, 396

immediate basin of attraction, 337, 365, 367
Incompressibility Lemma, 278, 360
indifferent periodic point, 340, 342, 360, 361, 363
induction
  transfinite, 277, 347, 348
inverse limit, xi, xx, 16, 40
iterated logarithm, 368

$j_z(x,y)$, 52
  extension to $\mathbb{P}^1_{\text{Berk}}$, 78
  relation with Hsia kernel, 76, 398
$j_\nu(x,y)$, 67
Jonsson, Mattias, xii, 93, 116, 119, 219, 225, 226, 233
Jordan decomposition, 217, 227, 246, 381
Julia set
  Berkovich, xxvi, 328
  classical, 333
  filled, x, 354
  homogeneous filled, 314

Kani, Ernst, 191
Kiwi, Jan, xiii
Kontsevich, Maxim, xiii

λ-measure, 95
Laplacian, xv
  complete, 98
  generalized, 194
  is self-adjoint, 109
  on Berkovich curves, x
  on metrized graphs, xi, xvi, xx, 50, 54, 61
  on $\mathbb{P}^1_{\text{Berk}}$, xx
  on $\mathbb{R}$-trees, xii
  on subdomains of $\mathbb{P}^1_{\text{Berk}}$, 98
  pullback formula for, 288
Laurent domain, 36, 410
least upper bound, 401
limit superior, 378
line of discs, 13
local degree, 258, 359

local ring, 22, 27
localization of a category, 414

$m_\varphi(a)$, 249, 256
$m_\varphi(a, \vec{v})$, 264
main dendrite, 147, 196, 197, 209, 215, 218, 239
  and harmonic functions, 148
  and subharmonic functions, 195
  is finitely branched, 147
  majorized by harmonic functions, 199–202, 204, 206, 210, 211, 219, 220
Maria's theorem, xxiii, 131, 243
Mass Formula, 55, 103
Maximum Principle
  for harmonic functions, 150
  for subharmonic functions, xxv, 197
  strong, 151
measure, 381
  Baire, 93, 383
  discrete, 62, 118, 236, 254
  finite, 381
  has bounded potentials, 109, 246
  has continuous potentials, 109, 240, 241, 298
  positive, 381
  probability, 121
  Radon, 91–93, 119, 381
  regular, 381
  signed, 381
  total variation of a, 381
metric
  big, 13, 44
  path distance, xviii, 44
    is canonical, xix, 46
  small, 11, 43, 44
metrizability, xiii, 17, 19, 43
metrized graph, 49
  model for a, 49
mirror symmetry, xiii
Möbius transformation, xix, 31, 47, 284
modulus of an open annulus, 46, 281, 284
multiplicity
  algebraic, 249, 256
  analytic, xiv, xxv, 256, 261, 269
  in a tangent direction, 264
  of a bout, 359

nest, 127
net, xiii, 19, 379
Neumann problem, 155

Newton polygon, 320, 341, 361, 387–390
Northcott finiteness property, 187, 309

one-dimensional Hausdorff measure, 95, 107
one-point compactification, xvii, 23, 384
Open Mapping Theorem, 257, 278

φ-saturated, *see also* saturated domain
φ-small neighborhood, *see also* small neighborhood
parametrized rooted tree, 11, 43, 401
partial order, xvii, 9, 401
path, 50, 393
path-connected, 14, 393
  uniquely, xv, xvi, 12, 29, 404
Piñeiro, Jorge, 375
PL-domain, 181
Poincaré-Lelong formula, xxii, 99
Poisson formula, xxiv, 156, 158, 199
  classical, 155
Poisson-Jensen measure, 158, 160
  classical, 155
polar set, 215
Portmanteau theorem, 254, 383
potential function, xxiii, 100, 116, 128, 195, 214
  continuity of, 218
  generalized, 242
  need not be continuous, 130
potential kernel, 52
  extension to $\mathbb{P}^1_{\text{Berk}}$, 78
  relation with Hsia kernel, 76
potential theory, ix, xv
product formula, 141, 178, 186
  field, 185
Prohorov's theorem, 189, 311, 313, 382
Proj construction, xiii, 23
pseudo-equidistribution, x
pullback function, 285
pullback measure, xxv, 285
pushforward function, 258
pushforward measure, xxv, 285

quadripod, 396
quasar, 342

ρ, xviii, 13, 44
  is canonical, 46
$r_\varphi(a, \vec{v})$, 262
$\mathbb{R}$-tree, xv, xix, 9, 393
  finite, 394
  profinite, xiv, xix, 16
  rooted, 393

INDEX

ramification function, 270
rate of repulsion, 262
rational function
    action on $\mathbb{P}^1_{\mathrm{Berk}}$, 31, 37
reduction
    good, 269, 292, 298, 299, 303, 363
    nonconstant, 33–35, 268, 269
    simple, 298, 324, 325, 329, 332, 366, 370
    with respect to a coordinate change, 268
repelling fixed point, 344, 362
repelling periodic point, 340, 342, 343, 354, 360, 361
reproducing kernel, xxiv, 172
Repulsion Formula, 262
resultant, 30, 293, 294, 300, 302–304, 356
retraction map, xx, 16, 39, 89, 90
    is continuous, 90
Riemann Extension Theorem, 153
Riemann-Hurwitz formula, 319
Riesz Decomposition Theorem, xxv, 215
Riesz Representation Theorem, 87, 97, 382
Rivera-Letelier, Juan, xi, xxvi, 47, 85, 116, 234, 240, 246, 290, 292, 306, 318, 325, 327, 336, 337, 339, 357, 358, 364, 371, 372, 374, 375
RL-domain, 140, 142, 181
Robin constant, xxii, 121, 181
    global, 179, 182
    relative to μ, 244
Rolle's theorem, 322, 390

saturated domain, 259
saturation, 14
semicontinuous, 128, 378
    regularization, 207, 378
    strongly, 129, 195, 378
seminorm, 406
    bounded, 1, 406
    equivalent, 24
    evaluation, 3
    multiplicative, xvi, 1, 406
    normalized, 24
separability, 175, 360, 363, 364
separable space, 91, 93
sheaf of analytic functions, 19, 411
Shilov boundary, 412
Shishikura's theorem, 363
Silverman, Joe, 18, 292, 293, 375
simple domain, xx, 9, 41, 87

    image of a, 257
    inverse image of a, 257
    strict, 41, 148
simple subdomain, 159, 200
skeleton, 147
small neighborhood, 260
smoothing
    CPA, 234, 236
    Favre–Rivera-Letelier, 113, 114, 234, 235
Soibelman, Yan, xiii
special subset, 411
spectrum, 2, 406
spherical
    distance, 43, 79
    kernel, 79, 80
        geometric interpretation, 80
spherically complete, xvii, 5
strongly involutive, 344
subharmonic, xxiv, 119, 165, 193
    domination, xxv, 199, 206
    equals domination subharmonic, 200, 205
    strongly, xxiv, 193
subnet, 380
superattracting periodic point, 319
superharmonic, xxiv, 194
Szpiro, Lucien, 375

tangent
    direction, *see also* tangent vector
    space, 12, 41, 50, 402
    vector, 12, 41, 50, 261, 402
Tate algebra, 36, 409
Tate, John, xv, 405
test functions
    smooth, 118, 222
Thuillier, Amaury, x, xi, 116–119, 144, 147, 191, 219, 220, 248
topology
    Berkovich, xvi, 2, 7, 17, 406, 407
        neighborhood base for, xx, 42, 260, 411, 415
    direct limit, 40
    observer's, 402
    of pointwise convergence, 20, 219, 220, 225, 226
    of weak convergence, 220, 226
    on $\mathbb{A}^1_{\mathrm{Berk}}$, 19
    on $\mathbb{P}^1_{\mathrm{Berk}}$, 24, 28
    path distance, 42
    strong, 42, 45, 358, 402
    weak, 11, 16, 42, 382, 402

transfinite diameter, xxiii, 136
Transitivity Theorem, 332
tripod, 395
Tucker, Tom, 375
type, xvii, 5, 22
   preservation of, 32

$U(x;\vec{v})$, 12
Uniform Injectivity Criterion, 322
uniform space, 385, 386

vanishing chain, 358
Varley, Robert, 85
vertex set, 49

wandering
   component, 338
   domain, 366, 367
weak convergence, *see also*
   convergence,weak
Weierstrass Preparation Theorem, 2, 174
well-oriented subgraph, 200, 202, 204, 206
witch's broom, 12

Yuan, Xinyi, 191

ζ_{Gauss}, 6
$\zeta_{a,r}$, 5
$\text{Zh}(\Gamma)$, 54

Titles in This Series

159 **Matthew Baker and Robert Rumely,** Potential theory and dynamics on the Berkovich projective line, 2010

158 **D. R. Yafaev,** Mathematical scattering theory: Analytic theory, 2010

157 **Xia Chen,** Random walk intersections: Large deviations and related topics, 2010

156 **Jaime Angulo Pava,** Nonlinear dispersive equations: Existence and stability of solitary and periodic travelling wave solutions, 2009

155 **Yiannis N. Moschovakis,** Descriptive set theory, 2009

154 **Andreas Čap and Jan Slovák,** Parabolic geometries I: Background and general theory, 2009

153 **Habib Ammari, Hyeonbae Kang, and Hyundae Lee,** Layer potential techniques in spectral analysis, 2009

152 **János Pach and Micha Sharir,** Combinatorial geometry and its algorithmic applications: The Alcálá lectures, 2009

151 **Ernst Binz and Sonja Pods,** The geometry of Heisenberg groups: With applications in signal theory, optics, quantization, and field quantization, 2008

150 **Bangming Deng, Jie Du, Brian Parshall, and Jianpan Wang,** Finite dimensional algebras and quantum groups, 2008

149 **Gerald B. Folland,** Quantum field theory: A tourist guide for mathematicians, 2008

148 **Patrick Dehornoy with Ivan Dynnikov, Dale Rolfsen, and Bert Wiest,** Ordering braids, 2008

147 **David J. Benson and Stephen D. Smith,** Classifying spaces of sporadic groups, 2008

146 **Murray Marshall,** Positive polynomials and sums of squares, 2008

145 **Tuna Altinel, Alexandre V. Borovik, and Gregory Cherlin,** Simple groups of finite Morley rank, 2008

144 **Bennett Chow, Sun-Chin Chu, David Glickenstein, Christine Guenther, James Isenberg, Tom Ivey, Dan Knopf, Peng Lu, Feng Luo, and Lei Ni,** The Ricci flow: Techniques and applications, Part II: Analytic aspects, 2008

143 **Alexander Molev,** Yangians and classical Lie algebras, 2007

142 **Joseph A. Wolf,** Harmonic analysis on commutative spaces, 2007

141 **Vladimir Maz'ya and Gunther Schmidt,** Approximate approximations, 2007

140 **Elisabetta Barletta, Sorin Dragomir, and Krishan L. Duggal,** Foliations in Cauchy-Riemann geometry, 2007

139 **Michael Tsfasman, Serge Vlăduţ, and Dmitry Nogin,** Algebraic geometric codes: Basic notions, 2007

138 **Kehe Zhu,** Operator theory in function spaces, 2007

137 **Mikhail G. Katz,** Systolic geometry and topology, 2007

136 **Jean-Michel Coron,** Control and nonlinearity, 2007

135 **Bennett Chow, Sun-Chin Chu, David Glickenstein, Christine Guenther, James Isenberg, Tom Ivey, Dan Knopf, Peng Lu, Feng Luo, and Lei Ni,** The Ricci flow: Techniques and applications, Part I: Geometric aspects, 2007

134 **Dana P. Williams,** Crossed products of C^*-algebras, 2007

133 **Andrew Knightly and Charles Li,** Traces of Hecke operators, 2006

132 **J. P. May and J. Sigurdsson,** Parametrized homotopy theory, 2006

131 **Jin Feng and Thomas G. Kurtz,** Large deviations for stochastic processes, 2006

130 **Qing Han and Jia-Xing Hong,** Isometric embedding of Riemannian manifolds in Euclidean spaces, 2006

129 **William M. Singer,** Steenrod squares in spectral sequences, 2006

128 **Athanassios S. Fokas, Alexander R. Its, Andrei A. Kapaev, and Victor Yu. Novokshenov,** Painlevé transcendents, 2006

127 **Nikolai Chernov and Roberto Markarian,** Chaotic billiards, 2006

126 **Sen-Zhong Huang,** Gradient inequalities, 2006

TITLES IN THIS SERIES

125 **Joseph A. Cima, Alec L. Matheson, and William T. Ross,** The Cauchy Transform, 2006
124 **Ido Efrat, Editor,** Valuations, orderings, and Milnor K-Theory, 2006
123 **Barbara Fantechi, Lothar Göttsche, Luc Illusie, Steven L. Kleiman, Nitin Nitsure, and Angelo Vistoli,** Fundamental algebraic geometry: Grothendieck's FGA explained, 2005
122 **Antonio Giambruno and Mikhail Zaicev, Editors,** Polynomial identities and asymptotic methods, 2005
121 **Anton Zettl,** Sturm-Liouville theory, 2005
120 **Barry Simon,** Trace ideals and their applications, 2005
119 **Tian Ma and Shouhong Wang,** Geometric theory of incompressible flows with applications to fluid dynamics, 2005
118 **Alexandru Buium,** Arithmetic differential equations, 2005
117 **Volodymyr Nekrashevych,** Self-similar groups, 2005
116 **Alexander Koldobsky,** Fourier analysis in convex geometry, 2005
115 **Carlos Julio Moreno,** Advanced analytic number theory: L-functions, 2005
114 **Gregory F. Lawler,** Conformally invariant processes in the plane, 2005
113 **William G. Dwyer, Philip S. Hirschhorn, Daniel M. Kan, and Jeffrey H. Smith,** Homotopy limit functors on model categories and homotopical categories, 2004
112 **Michael Aschbacher and Stephen D. Smith,** The classification of quasithin groups II. Main theorems: The classification of simple QTKE-groups, 2004
111 **Michael Aschbacher and Stephen D. Smith,** The classification of quasithin groups I. Structure of strongly quasithin K-groups, 2004
110 **Bennett Chow and Dan Knopf,** The Ricci flow: An introduction, 2004
109 **Goro Shimura,** Arithmetic and analytic theories of quadratic forms and Clifford groups, 2004
108 **Michael Farber,** Topology of closed one-forms, 2004
107 **Jens Carsten Jantzen,** Representations of algebraic groups, 2003
106 **Hiroyuki Yoshida,** Absolute CM-periods, 2003
105 **Charalambos D. Aliprantis and Owen Burkinshaw,** Locally solid Riesz spaces with applications to economics, second edition, 2003
104 **Graham Everest, Alf van der Poorten, Igor Shparlinski, and Thomas Ward,** Recurrence sequences, 2003
103 **Octav Cornea, Gregory Lupton, John Oprea, and Daniel Tanré,** Lusternik-Schnirelmann category, 2003
102 **Linda Rass and John Radcliffe,** Spatial deterministic epidemics, 2003
101 **Eli Glasner,** Ergodic theory via joinings, 2003
100 **Peter Duren and Alexander Schuster,** Bergman spaces, 2004
99 **Philip S. Hirschhorn,** Model categories and their localizations, 2003
98 **Victor Guillemin, Viktor Ginzburg, and Yael Karshon,** Moment maps, cobordisms, and Hamiltonian group actions, 2002
97 **V. A. Vassiliev,** Applied Picard-Lefschetz theory, 2002
96 **Martin Markl, Steve Shnider, and Jim Stasheff,** Operads in algebra, topology and physics, 2002
95 **Seiichi Kamada,** Braid and knot theory in dimension four, 2002
94 **Mara D. Neusel and Larry Smith,** Invariant theory of finite groups, 2002
93 **Nikolai K. Nikolski,** Operators, functions, and systems: An easy reading. Volume 2: Model operators and systems, 2002

For a complete list of titles in this series, visit the
AMS Bookstore at **www.ams.org/bookstore/**.